THE COMPLETE ILLUSTRATED GUIDE TO
MINERALS, ROCKS & FOSSILS OF THE WORLD

THE COMPLETE ILLUSTRATED GUIDE TO
MINERALS, ROCKS &
FOSSILS OF THE WORLD

A comprehensive reference guide to over 700 minerals, rocks, and plant and animal fossils
from around the globe and how to identify them, with over 2000 photographs and illustrations

JOHN FARNDON & STEVE PARKER

southwater

This edition is published by Southwater, an imprint of Anness Publishing Ltd,
Blaby Road, Wigston, Leicestershire LE18 4SE; info@anness.com

www.southwaterbooks.com; www.annesspublishing.com

If you like the images in this book and would like to investigate using them for publishing, promotions or advertising,
please visit our website www.practicalpictures.com for more information.

Publisher: Joanna Lorenz
Editorial Director: Helen Sudell
Project Editors: Ann Kay and Catherine Stuart
Production Manager: Steve Lang
Editorial Readers: Alison Bolus and Jay Thundercliffe
Consultants: John Cooper (Booth Museum, Brighton), Dr Alec Livingstone (Former Keeper of Minerals, National Museum of Scotland, Edinburgh) and Dr John Schumacher (Department of Earth Sciences, University of Bristol, England)
Contributors: Vivian Allen, Julien Divay, Ross Elgin, Carlos Grau, Robert Randell, Jeni Saunders and Matt Vrazo
Book and Cover Design: Nigel Partridge
Artists: Andrey Atuchin, Peter Barrett, Stuart Carter, Anthony Duke, Samantha J. Elmhurst and Denys Ovenden (for these and other picture acknowledgements, see page 512)

A CIP catalogue record for this book is available from the British Library.

PUBLISHER'S NOTE

The authors and publishers have made every effort to ensure that all instructions contained within this book are accurate and safe. Persons handling and collecting rocks and mineral specimens do so at their own risk. This book offers guidance on the basic safety precautions to take when handling, identifying or collecting rock and mineral specimens; however, neither the authors nor the publisher can accept any legal responsibility or liability for any injury, loss or damage to persons and property that may arise as a result of these activities. It is especially important that collectors observe the protective laws applied to a particular geological site before entering the site or removing rock and mineral specimens, and note that these may vary from region to region, and from country to country. Useful information can, and should, be obtained from local government and environmental agencies before visiting any geological site.

Page 2: The Painted Hills in the John Day Fossil Beds National Monument, Oregon, USA.

CONTENTS

INTRODUCTION

The story of the Earth's rocks and minerals begins about four and a half billion years ago, when the newly formed planet – created from a spinning cloud of dust and gas – began to compact, cool and crystallize. As the Earth solidified, the components of the young planet settled into layers, with the densest materials falling towards the centre.

Today we have a planet with a massive central core of heavy iron and nickel, enveloped by a bulky outer rocky mantle made of lighter compounds. The outermost layer of the Earth is called the crust, and it consists mainly of rocks like basalt and granite. Our current technology only allows us to penetrate a relatively tiny distance beneath the surface of the crust into the mantle. Who knows what secrets may lie deep within the Earth's inner reaches, thus far beyond our skills to explore them?

Rocks and minerals

Minerals are naturally formed, crystalline, inorganic substances, and each has a consistent chemical composition. All rocks are composed of aggregations of minerals. The rocks and minerals of the Earth have played out a never-ending symphony of movement since they formed, and they have shaped our landscape endlessly

Below: Taken from near Mount Everest, this view of the Himalayas, Nepal, demonstrates the pure majesty of the mountain range.

Above: Fossils, like this Parkinsonia ammonite, tend to be found in sedimentary rock.

Above: The liquid resin, in which this insect was trapped millions of years ago, is now the mineral amber.

throughout geological time, and continue to do so. The landmasses, floating slowly on gigantic tectonic plates, have inexorably moved across the surface of the globe, sometimes crashing into each other and throwing up huge mountain ranges like the Himalayas, and at other times drifting away to isolate whole regions from the rest of the world. Volcanic eruptions, earthquakes, ice, wind, rain and the other forces of erosion have also played a part in forming, distributing and redistributing the rocks and minerals and shaping the Earth.

First life forms

About three and a half billion years ago – we cannot be exactly sure when – a dramatic event occurred: the first forms of life appeared on Earth. The ocean is probably the birthplace of life on Earth, and for three billion years, the major explosions of life all took place in the seas. From microscopic, primitive life forms slowly more complex creatures evolved. Some, like sharks, have changed little from when they first appeared, over 400 million years ago. Other creatures flourished for a time, and then disappeared from

Right: Dinosaur fossils, such as this one found in rock and sand, tell us much about the extinct life forms of ancient times.

the Earth forever. Animals and plants have only colonized the land for about 500 million years or so. The explosion of biodiversity is inextricably linked with the planet's rocks and minerals. Not only is our knowledge of the planet's structure and history written in the rocks, but so, too, is the record of past life on Earth. What we know about the creatures of the distant past comes from the evidence found in the form of fossils – the remains of long-dead creatures preserved in the rocks.

Fossil formation

However, for an animal or a plant to be preserved in the form of a fossil, conditions need to be exactly right. The organism must be quickly covered in sediment that prevents erosion, scavenging by animals or rapid decomposition by bacteria. If all of these criteria are met, then over time the hard parts such as bones, shells and teeth may become replaced by minerals, leaving a durable, stony cast or impression in the form of a fossil. Even when a fossil is formed, it may never be discovered, for the best conditions for fossilization occur in the water. Many of the fossils that have been found are in areas that were once seas, but over aeons of time have dried up, or been thrown up in the form of mountains when violent Earth movements have occurred. Although our knowledge of life in the past seems extensive, it is probably hugely incomplete; many creatures simply never formed fossils, or their fossils have not been found. It has been calculated that even among dinosaurs – some of the biggest of the animals to have been studied – we probably know of fewer than a quarter of all the species that existed in the late Cretaceous, the period when they became extinct.

The study of geology

The great advances in our knowledge of the Earth's structure – the science of geology – really began in the 19th

Above: This young geology enthusiast is using a brush to reveal any clues to formation that are contained within this piece of rock.

Above: Beaches and coastal areas supply many interesting samples of rocks, minerals and fossils.

century. Then, geologists such as Charles Lyell (1797–1875) began to study the composition and formation of the rocks and establish accurate dates for their ages – they proved the Earth to be far, far more ancient than had been previously thought.

Today, modern scientific techniques, such as carbon dating, have helped us to tell the age of the Earth's rocks a lot more accurately. Fossils present in such rocks can therefore be dated, too, providing us with when the creatures died. Conversely, the presence of the fossilized remains of creatures whose time on Earth has been dated can indicate the age of the rocks in which they are found.

Our fascination with rocks, minerals and fossils is enduring, and there are many reasons for this. Rocks and minerals are all about us, and the story they tell about the Earth's history is often easy to see. Many rocks and minerals are attractive, and simple to find and collect. Fossils also capture the imagination; to hold a fossil in your hand is to hold a tiny piece of your own long-distant natural heritage.

This superb book takes you on a journey through the history and structure of the Earth's rocks, minerals and fossils, and introduces you to many of the different types, inspiring you to go out and look for some of these natural treasures yourself.

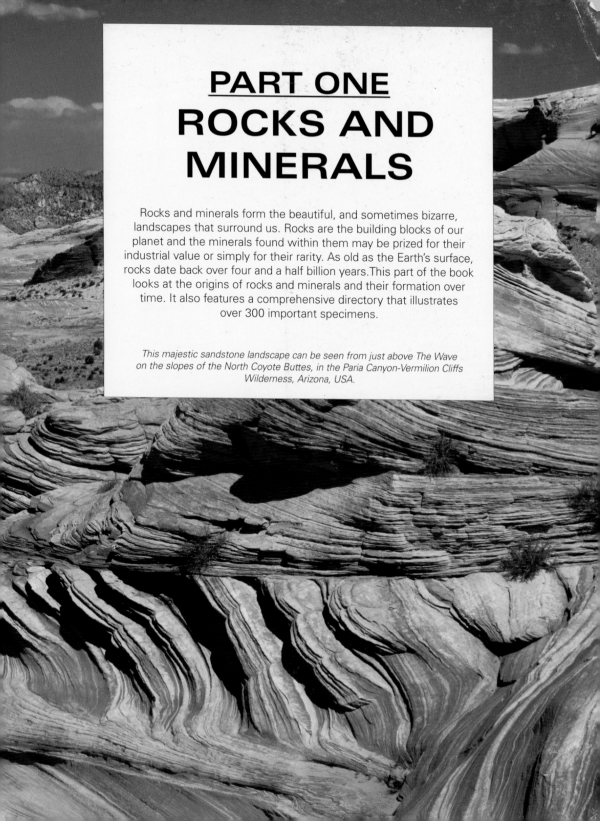

PART ONE
ROCKS AND MINERALS

Rocks and minerals form the beautiful, and sometimes bizarre, landscapes that surround us. Rocks are the building blocks of our planet and the minerals found within them may be prized for their industrial value or simply for their rarity. As old as the Earth's surface, rocks date back over four and a half billion years. This part of the book looks at the origins of rocks and minerals and their formation over time. It also features a comprehensive directory that illustrates over 300 important specimens.

This majestic sandstone landscape can be seen from just above The Wave on the slopes of the North Coyote Buttes, in the Paria Canyon-Vermilion Cliffs Wilderness, Arizona, USA.

INTRODUCING ROCKS AND MINERALS

Back in the early 19th century, when the whole idea of rock and mineral collecting was just getting under way, the famous Scottish poet Sir Walter Scott (1771–1832) described geologists thus, "Some rin uphill and down dale, knappin' the chucky stones to pieces like sa' many roadmakers run daft. They say it is to see how the world was made!" Even today many people consider hunting rocks and minerals a rather strange activity.

Yet go to a beach or any stream where the water has exposed sand and gravel banks. You will see stone upon

Below: Beautiful mineral specimens like these are rare treasures much sought after by rock hunters. Here are 'daisy' gypsum (top), 'dogtooth spar' calcite (left centre), 'blue john' fluorite (right centre) and pyrite (bottom), also known for obvious reasons as 'fool's gold'.

stone lying there. To start with, they probably all look dull and grey. But look closer and you begin to see subtle differences in colour. One might be pale cream. Another mottled brown. A third slightly stripy. To the untrained eye, they're still just stones, but to an experienced rock hunter, each has its own fascinating story to tell.

The pale cream stone could be limestone. Look at it through a magnifying glass, and you begin to see it is made from tiny grains with shiny surfaces, each a crystal of calcite precipitated out of tropical oceans hundreds of millions of years ago. Here and there in the stone you may actually see fossils of the sea creatures that swam in these ancient oceans.

Your rock hunter might go on to tell you that the mottled brown stone is granite – then show how through the magnifying glass you can see three different minerals. There are tiny black flecks of mica, glassy grains of quartz, and yellow feldspar – all forged in the fiery heat of the earth's deep interior millions of years ago. The stripy stone could be a schist, a rock formed when other rocks came under such intense pressure from earth movements that the crystals in them broke down and were made anew in different forms – squeezed into stripes by the pressure. A magnifying glass might reveal tiny red spots embedded in the rock. A rock hunter might identify these as garnets or even rubies, tiny versions of the beautiful gems that kings have fought and died for.

Above: Some specimens only reveal their true beauty when ground and polished like these specimens. At the top is the rock orbicular, a special variety of granite. Below are stones of the mineral onyx.

With all this and much more in just three stones picked up on a beach, it is not surprising that many people have become hooked on rock and mineral collecting. Many stones are collectable for their sheer beauty and rarity – not just the well-known gems such as rubies and diamonds, but also non-gem minerals such as crocoite and rose quartz. Many rocks and minerals are sources of the ores that provide us with metals, or of building materials. Yet even when neither beautiful nor valuable, stones have a fascination because of the story they have to tell.

Rocks through the ages

Rocks have long played a part in human history. Long ago, our ancestors were chipping the edges off hand-sized pebbles, perhaps to use as weapons. At least two million years ago, hominids began to use flints to make two-sided hand-axes, which is why the first age of humanity is known as the Stone Age. Finding good flints required a

Above: Insects from tens of thousands of years ago can be perfectly preserved in amber, a stone formed from ancient tree sap.

considerable practical knowledge of geology. Few people today would have a clue where to look for flints – yet these Stone Age people knew, and even dug mines to get at them underground.

Copper and gold were first used at least 10,000 years ago. They occur naturally as metals, and their distinctive colour makes them easy to see. Again, though, it required a real knowledge of geology to know where to look. Copper and gold were both too soft to make tools from, but the discovery that tin could be added to copper to make

Below: None of these stones is especially beautiful but each has a story to tell. Top left is a dripstone formed in caverns from minerals dissolved in rainwater. Top right is a limestone rich in fossils of bryozoans, sea creatures that lived hundreds of millions years ago. At the bottom is a limestone imprinted by ancient corals.

the tough alloy bronze about 5000 years ago initiated the first great age of metal use, the Bronze Age – and also the first great civilizations, such as that of Ancient Egypt. Tin only occurs in any quantity in the ore cassiterite, which has to be melted to extract the tin. Some cassiterite was found in river gravels where it often collects – but only in areas near granite. Some cassiterite was mined from veins deep underground, such as those in the Austrian Tyrol and in Cornwall. Again, locating these sources required a good knowledge of rocks and minerals.

Over the next 4000 years, miners acquired a wealth of hands-on practical geological skill as they hunted and exploited a range of metal ores and other materials. It is no accident that the word 'minerals' comes from miners. In the 1500s, the German mining engineer Georgius Agricola published the first great geology book, *De Re Metallica* (On Things Metallic).

The rock hunters

All the same, it was not until the late 18th century that geology emerged as a science, pioneered by the great Scottish geologist James Hutton (1726–97). In Hutton's day, most people still believed the Earth was just a few thousand years old. Hutton realized it is much, much older – it is now thought to be 4.5 billion years old – and that the slow processes we see acting on the landscape today were quite enough to shape it without invoking great catastrophes, as others did. Hutton demonstrated that landscapes are worn away by rivers and that sediment washed into the sea forms new sedimentary rocks. He also saw how the Earth's heat could transform rocks, lifting and twisting them to

Above and left: Very occasionally, the rock hunter is rewarded by the discovery of a gemstone, such as ruby (left) or even a diamond (above). Even if they are not especially valuable forms of the gems, the thrill of discovery is quite enough.

create new mountains. So the world is shaped by countless cycles of erosion, sedimentation and uplift – each new beginning often clearly marked by breaks in the rock sequence called unconformities (see Reading the Landscape, Understanding Rocks and Minerals).

Inspired by Hutton's ideas, more and more geologists ventured out into the field to explore rocks and the story they tell. Geology became a popular pastime for many Victorian gentlemen, who ventured out in stout boots with just a hammer and a strong bag for specimens. They included Charles Darwin, who brought his knowledge of geological history to bear in formulating his theory of evolution. A huge proportion of the rock and mineral species we know today were first identified and named by Victorian specimen hunters.

As the following pages show, our knowledge of geological processes has developed tremendously since those early days, and professional geologists are aided by a range of sophisticated equipment. Yet the amateur armed with a few basic tools and a sharp eye can find fantastic specimens. It is the aim of this book to help in this search.

UNDERSTANDING ROCKS AND MINERALS

Climb any mountain away from the city and gaze out on the landscape, and you see rivers winding down to the sea, hills and valleys, forests and fields. It seems a timeless landscape, but in terms of the Earth's history it is very young. The fields may be no more than a few centuries old, the forests a few thousand years old, and even the hills and valleys no more than tens of thousands of years old. Yet the history of the Earth's surface and the rocks and minerals which make it dates back over four and a half billion years. In geological terms, then, the landscape we see today is just a fleeting moment.

Geologists first began to understand the ever-changing nature of the Earth's surface about 200 years ago, as they started to realize just how old it was, and how it was shaped and reshaped continuously by the power of such forces as water, earthquakes and volcanoes, which have worn mountains down and risen new ones up. Yet it is only in the last half century that they have learned just how dynamic the Earth's geology really is, with the discovery that the entire surface of the planet is on the move – broken into 20 or so giant, ever-shifting slabs called tectonic plates. The movement of these plates is slow in human terms – barely faster than a fingernail growing – yet over the vastness of geological time, it is momentous: able to shift continents and oceans right around the globe.

Tectonic plate theory has revolutionized geologists' understanding of how rocks are made and remade over again, how mountains are built and knocked down, why volcanoes erupt, why earthquakes happen and much more besides.

Left: Sulphurous vapour rises from the crater of the active Sierra Negra volcano on Isabela Island, one of the Galapagos Islands. Volcanoes are among the world's most important mineral factories, forever bringing materials to the surface to form new minerals.

INSIDE THE EARTH

The Earth under your feet may seem solid, but recent research has shown that its interior is far more dynamic and complex than anyone ever thought. Beneath the thin rocky shell called the crust, the Earth is churning and bubbling like thick soup.

Half a century ago, scientists' picture of the Earth's interior was simple. In some ways, they thought, it seems like an egg. The outside is just a thin shell of rock called the crust. Immediately beneath, no more than a few dozen kilometres down, is the deep 'mantle' were the rock is hot and soft. Then beneath that, some 2,900km/1,800 miles down is the yolk or 'core' of metal, mainly iron and nickel. The outer core is so ferociously hot that it is always molten, reaching temperatures as high as the Sun's surface. The inner core, at the centre of the Earth, is solid because pressures there are gigantic.

The key to this structure is density. The theory is that when the Earth was young, it was hot and semi-molten. Dense elements such as iron sank towards the centre to form the core. Lighter elements such as oxygen and silicon drifted to the surface like scum on water and eventually chilled enough to harden into a crust.

Above: Meteorites big enough to cause a crater like this one, the famous Meteor Crater in Arizona, strike the Earth so hard they vaporize on impact. But some small meteorites survive and provide vital clues to the composition of the Earth's interior.

Below: Blacksmiths know that iron only melts at very high temperatures, and scientists have deduced that the temperatures of the molten iron in the Earth's outer core, where it is under intense pressure, may rise to 4,500K/7,600°F. The solid core could even reach 7,500K/13,000°F!

Some heavy elements such as uranium ended up in the crust despite their density because they unite readily with oxygen to make oxides and with oxygen and silicon to make silicates. Substances like these are called 'lithophiles' and include potassium.

Blobs of 'chalcophile' substances – substances like zinc and lead that join readily with sulphur to form sulphides – spread up and out to add to the mantle. Dense globs of 'siderophile' substances – substances such as nickel and gold that combine readily with iron – sank towards the core.

The only real complications to this picture seemed to be on the surface, where the crust is divided into continental and oceanic portions. The crust in continents can be very ancient – some rocks are nearly four billion years old – and quite thick. Although

it is just 20km/12 miles beneath California's Central Valley, it is 90km/54 miles thick under the Himalayas. The oceanic crust, on the other hand, is entirely made of young rocks – none older than 200 million years and some brand new – and is rarely more than 10km/6 miles thick.

Listening in

Discoveries in the last few decades have forced scientists to re-evaluate this fairly simple picture. The problem has been to see inside the Earth. A Japanese ship launched in 2005 is now beginning to drill the deepest hole ever, through the oceanic crust, hoping to reach the mantle, but this is barely scratching the surface. Yet there are other ways of telling. Astronomical calculations based on gravity tell us Earth's mass and show that the interior must be denser than the crust. Meteorites tell us a little about the mineral make-up of the interior, with the two kinds of meteorite, stony and iron, reflecting Earth's stony mantle and iron core (see Space Rocks, Directory of Rocks). Similarly, volcanoes throw up materials like olivine and eclogite from deep in the mantle. Yet the main clues come from earthquake (seismic) waves.

Long after an earthquake, its reverberations shudder through the Earth. Sensitive seismographs can pick them up on the far side of the world. Just as you can hear the difference between wood and metal when tapped with a spoon, so scientists can 'hear' from seismic waves what the Earth's interior is made of. Seismic waves are refracted (bent) as they pass through different materials. They also travel at varying speeds, shimmering faster through the cold hard rocks of the crust, for instance, than the warmer, soft rocks of the mantle.

Density and speed

One thing seismology has revealed is that there is another way of looking at the crust and upper mantle. Although they may be chemically different, their 'rheology' is not – that is, they distort and flow in much the same way. Fast seismic waves show the top 100km/60miles of the mantle is as stiff as the crust, and together upper mantle and crust form a rigid layer called the lithosphere. Below the lithosphere, slower waves show that heat softens the mantle to form a layer called the asthenosphere. Tectonic plates are huge chunks of lithosphere that float on the asthenosphere like ice floes on a pond.

About 220km/140 miles down pressure stiffens the mantle again to form the mesosphere. Farther down, pressure forces minerals with the same chemical composition through a phase change (like ice melting) into a denser structure. So below 420km/260 miles olivine and pyroxene are replaced by spinel and garnet. Deeper down still, beyond 670km/420 miles, even higher pressure changes mineral structure again or maybe the composition, this time to give perovskite minerals, from which the bulk of the mantle is made.

The core boundary

Down through the mantle, seismic waves move ever faster. Yet at the Gutenberg discontinuity 2,900km/1,800 miles down, there is a drop in speed, marking the transition to the core along the Core-Mantle Boundary (CMB). The change is dramatic. In just a few hundred kilometres temperatures soar 1,500°C/2,732°F, and the contrast in density between mantle and core is even more marked than that between air and rock.

The transition zone in the mantle down to the CMB is called the D" (pronounced D double prime) layer, and has attracted a lot of attention. The outer surface of this layer is marked by valleys and ridges, and lab tests have shown that it may be made of a unique form of perovskite dubbed post-perovskite. In 2005, scientists detected an increase in speed below the D", suggesting that the outer rim of the core may actually be solid.

Research into this whole CMB zone may have key implications for our understanding of how continents move and volcanoes erupt, because these could be tied in with deep circulation of material in the mantle.

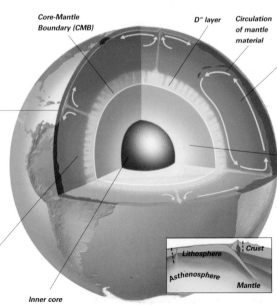

The Earth's interior
This globe reveals scientists' idea of the Earth's interior layers (not to scale), including the crust, mantle and core.

Core-Mantle Boundary (CMB)

D" layer

Circulation of mantle material

The crust *0–40 km/0–25 miles down is Earth's thin topmost layer, made mostly of silicate-rich rocks like basalt and granite. It is thinnest under the oceans and thickest under the continents. It is attached to the mantle's rigid upper layers, which floats in slabs on the soft mantle below.*

The lower mantle *is 670–2,900km/420–1,800 miles down. Here, huge pressures turn the lighter silicate minerals of the upper mantle into dense perovskite and pyroxene. Perovskite is the most abun-dant mineral in the mantle, and so in the Earth, since the man-tle makes up four-fifths of the Earth's volume.*

Inner core

Lithosphere
Asthenosphere
Crust
Mantle

The upper mantle *16–670km/10–420 miles down is so warm it is soft enough to flow. In the asthenosphere layer, below the lithosphere, pockets often melt to form the magma that bubbles up through the crust to erupt in volcanoes. The upper mantle is made mainly of the dense rock called peridotite.*

The Earth's core *2,900–6,370km/1,800–3,960 miles down is a dense ball of iron and nickel. The outer core is so hot, reaching temperatures of over 4,500K/7,600°F, that the metal is molten. The inner core is even hotter, up to 7,500K/13,000°F, but the pressures here are so great that the iron simply cannot melt.*

Lithosphere *(left) is the rigid outer layer of the Earth, broken into the tectonic plates that make up the surface. It consists of the crust and the stiff, cool upper portion of the mantle.*

CONTINENTS AND PLATES

The division of the world into land and sea seems so natural and so timeless that it is difficult to imagine how it could be any other way. Yet the very existence of continents is remarkable and Earth is the only planet we know to have them. Their foundations are even more astounding.

Like Earth, Venus and many other planets and moons in the solar system have rocky crusts. But the crusts on these other worlds are made almost entirely of basalt, and are very stable and ancient, barely changed since they formed billions of years ago. Earth has a basalt crust, too. The crust under the oceans is basalt, for instance. But Earth's crust is mostly neither stable nor ancient. In fact, nowhere is the ocean crust older than 200 million years, and much is forming even now.

Even more unusually, the Earth has giant chunks of crust made from granite-like rocks such as granodiorite. It is these chunks of granitic crust that form our continents, light enough to ride high above the surface and form landmasses on partially molten matter.

Making continents

No one knows for sure exactly how or why these chunks formed, but their formation is clearly a very slow process. The oldest pieces are almost four billion years old (the Earth is about 4.6 billion years old). Yet even now, granite crust still forms barely a quarter of the world's surface.

Basalt crust forms when magma (molten rock) from Earth's warm mantle cools and solidifies at the surface. Granite cannot form directly from mantle melts like this – only when basalt remelts, changing its chemistry and mixing in other substances met at the surface.

Geologists think this happens in two ways. The first is when hot, molten basalt magma wells up under the crust from the mantle, melting the crust to form granite magma. Less dense than basalt crust, the granite magma rises to the top before solidifying. The second process is when movement of the Earth's crust draws basalt crust back towards the mantle, remelting it and forming a new granite magma, which rises to form new continental crust.

This process initiated the evolution of the great continents that would support so much of Earth's living matter.

So just how does the Earth's crust move? The answer lies in the most powerful geological concept of the last century: plate tectonics. This is the idea that the Earth's rigid surface – all the lithosphere including the crust – is broken into 20 or so giant pieces of tectonic plates. These plates move slowly around the planet, carrying the continents and oceans with them.

Continental drift

The seeds for plate tectonic theory were planted in the early 20th century by a German meteorologist called Alfred Wegener. He had noticed the extraordinary way the west coast of Africa mirrored the east coast of South America. He also noticed amazing matches between widely separated continents of things such as geological strata and ancient animal and plant

The moving continents
Over the last 500 million years, Earth's continents have merged together then split and drifted apart. About 225 mya (million years ago) at the end of the Permian, all land was joined in one supercontinent geologists named Pangaea. Pangaea began to rift apart about the time dinosaurs appeared on Earth, so different kinds of dinosaur evolved in newly separated parts of the world. The maps here show the current continental shapes to help identify them, but in fact their shapes varied as low areas were flooded and mountain ranges rose up.

Permian 225 mya *During the Permian period all the world's landmasses moved together to form the giant continent Pangaea.*

Late Triassic 205 mya *During the Triassic, a wedge of ocean called the Tethys Seaway grew wider, elbowing into the east of Pangaea.*

Jurassic 150 mya *In the Jurassic, Pangaea began to split in various places, including the Tethys Seaway. South-western North America was flooded by the Sundance Sea.*

Cretaceous 80 mya *By the Cretaceous even southern Pangaea, called Gondwana, had split up to form today's southern continents. India began drifting northwards towards Asia.*

Present day *Over the last 50 million years, the North Atlantic has opened up to separate Europe and America, and India has crashed into Asia, throwing up the Himalayas.*

Above: In recent years, dramatic evidence of tectonic activity has been found deep under the sea. Fissures like this one in the East Pacific seabed, some 2,600m (8,500ft) below the surface, indicate the spreading of the oceans. Volcanic activity along these fissures warms the water and produces a rich mixture of chemicals, which creates a unique habitat for these white crabs, as well as other marine life.

fossils – especially from the Permian Period some 230 million years ago. Signs of ancient tropical species as far north as the Arctic circle only added to the impression.

Wegener guessed this was not mere coincidence. He realized that these matches might occur because today's separate continents were once actually joined together. He suggested that in Permian times they all formed one giant supercontinent which he called Pangaea, surrounded by a single giant ocean, which he called Panthalassa. At some time, perhaps some 200 million years ago Wegener thought, Pangaea split into several fragments and these have since drifted apart to form our present continental lands.

To many geologists of Wegener's time, the idea that continents drift around the world was ridiculous. The crust seems far too solid for this to happen. But over the next half century, the weight of evidence piled up. A key element was the discovery of grains of

Right: Looking at this bleak tundra on the Arctic island of Spitsbergen, it is hard to believe that lush tropical vegetation ever grew here. Yet fossils show that it did – not because the whole world had a tropical climate, but because Spitsbergen was once in the tropics, before continental drift took it way out into the chilly polar regions.

magnetite – a magnetic mineral – in ancient rock. These grains behaved like tiny compasses, lining up with the North Pole at the time when the rock formed. To geologists' surprise, these grains do not all point in the same direction. At first, geologists thought this must be because the magnetic North Pole had moved over time. Then they realized it was not the Pole that had moved, but the continents in which these grains were embedded, twisting this way and that. Geologists realized that with the aid of these ancient compasses or 'palaeomagnets' they could trace the entire path of a continent's movement through time – and this is how the maps below left were worked out.

Spreading oceans

A second key element was the realization that it was not just the continents moving but the entire surface of the Earth, including the oceans. Indeed, the continents are just passengers aboard the great, slowly sliding tectonic plates that make up Earth's surface. The breakthrough came in 1960 when American geologist Harry Hess suggested that the ocean floors are not permanent. Instead, he suggested, they are spreading rapidly out from a ridge down the middle of

the sea bed, as hot material wells up through a central rift, pushing the halves of the ocean bed apart. This does not make the crust bigger, because as quickly as new crust is created along the ridge, old crust is dragged down into the mantle and destroyed along deep trenches at the ocean's edge, in a process known by geologists as subduction.

Many were initially sceptical of Hess's idea, but evidence soon came when bands of magnetite were found in rocks on the sea floor in exactly matching patterns either side of the mid-ocean ridge. These bore witness to the spread of the ocean floor as truly as rings in a tree. Before long, the ideas of sea-floor spreading and continental drift were combined in the all-encompassing theory of plate tectonics.

It soon became clear that plate tectonics explained much more than coincidences of rocks and fossils. Earthquakes happen where plates shudder past each other. Volcanoes erupt where plates split asunder or dive into the mantle. Mountain ranges are thrown up where plates collide and crumple the edges of continents. In fact, the full implications of this revolutionary theory are only just beginning to be explored.

THE MOVING EARTH

Like a broken eggshell, Earth's surface is cracked into giant slabs of rock called tectonic plates. There are seven huge plates and a dozen or so smaller ones. The plates are not fixed, but ever shifting, breaking up and being made anew – almost imperceptibly slow in human terms, but dramatic geologically.

Tectonic plates are fragments of the lithosphere – the cool, rigid outer layer of the Earth, topped by the crust. The scale of some of these plates is staggering. They may be no more than 100km/60 miles thick, but some plates encompass entire oceans or continents.

The biggest is the Pacific plate, which underlies most of the Pacific Ocean. Interestingly, this is the only major plate that is entirely oceanic. All the others – in size order, the African, Eurasian, Indo-Australian, North American, Antarctic and South American – carry continents, like cargo on a raft. The remaining dozen or so plates are much smaller. But even the smallest of them, like North America's Juan de Fuca, is bigger than a country such as Spain.

It seems almost unimaginable that slabs of rock as gigantic as these could move. Yet they are moving, all the

The world's major plates
On this map showing the world's major plates, red arrows indicate the direction of movement.

time. Typically, they are shifting at about the pace of a fingernail growing – about a 1cm/0.4in a year, although the Nazca plate in the Pacific is moving 20 times as fast. Yet even a fingernail's pace has been fast enough to carry Europe and North America apart and create the entire North Atlantic in just 40 million years, which is a short time in geological terms. Accurate laser measurements can even detect the movements in a single month.

Why plates move
Scientists are not yet certain what it is that drives the plates, but most theories focus on the idea of convection in the Earth's mantle.

Inside the mantle, material is continuously churning around, driven up by the ferocious heat of the core, then cooling and sinking back. It was once thought that all these 'convection' currents moved in vast cells as big as the plates, and the plates simply rode on top of them like packages on

conveyor belts. Now more scientists are focusing on hot currents called mantle plumes that bubble up from the deep mantle, and in places lift the crust like a pie crust in an oven.

This may tie in better with other theories that suggest the driving force is the weight of the plates alone. Mid-ocean ridges are 2–3km/1–2 miles higher than ocean rims, so plates could be sliding downhill away from them.

Another idea is that plates are like a cloth sliding off a table. Hot new rock formed at the mid-ocean ridge cools as it slides away to the ocean rim. As it cools, it gets denser and heavier and sinks into the mantle again. It then pulls the rest of the plate down with it, just as the weight of a table cloth hanging over the edge can be enough to pull the rest off. It may be that all these mechanisms play a part.

Plate boundaries
Working out where one plate ends and another begins is far from easy. Seismic

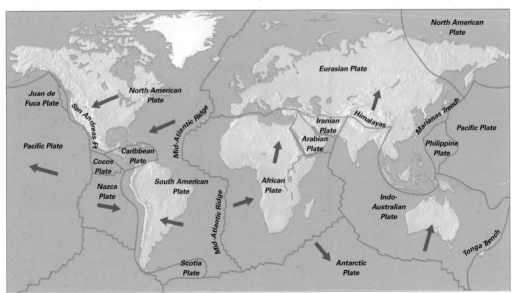

North American Plate

Eurasian Plate

Juan de Fuca Plate

San Andreas Ft

North American Plate

Mid-Atlantic Ridge

Iranian Plate

Himalayas

Marianas Trench

Pacific Plate

Pacific Plate

Cocos Plate

Caribbean Plate

Arabian Plate

Philippine Plate

Nazca Plate

South American Plate

Mid-Atlantic Ridge

African Plate

Indo-Australian Plate

Tonga Trench

Scotia Plate

Antarctic Plate

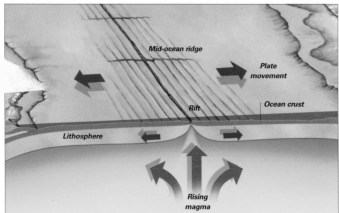

Mid-ocean ridge

Plate movement

Rift

Ocean crust

Lithosphere

Rising magma

Divergent boundaries *typically occur in mid-ocean. Right down the middle of the Atlantic Ocean, for instance, there is a ridge about 2km/1.2 miles down on the sea bed, forming a jagged line where plates meet. In the middle of this ridge is a trough some 500m/1,640ft deep and no more than 10km/6 miles wide. This trough is where the plates are moving apart, spreading the ocean wider and wider. Beneath this trough, magma is welling up from the asthenosphere. Some solidifies on the underside of the crust, creating rocks like gabbro. Some oozes up into vertical cracks created by the pressure of the magma to form wall-like sheets of basalt rock called dykes. Some magma spills out on the sea bed and freezes in the cold water into blobs called pillow lavas. As soon as it forms, new oceanic crust moves away from the ridge, more magma wells up and new crust forms.*

profiling using earthquake waves helps build a picture but it is often vague.

One of the best indicators of a plate boundary is where earthquakes start. Plates generate earthquakes as they judder past each other so nearly all major quakes occur in belts that follow plate boundaries. In May 2005, a Japanese geologist discovered a previously unknown plate under Japan's Kanto just by tracing the origin of over 150,000 small earthquakes.

Other features that identify a plate boundary include long chains of volcanoes, ranges of fold mountains, curving lines of islands and deep ocean trenches. Some plates meet along the coasts of continents, like the west coast

of South America. Coasts like these are called active margins. Other plates meet in mid-ocean. The continental crust is actually attached to the oceanic crust, and so its margin is passive.

Moving boundaries

All plates are on the move, some slow, some fast, and geologists identify three kinds of boundary, each with its own features. Divergent boundaries (see above) are where plates are pulling apart. Convergent boundaries (see below) are where they push together. Transforms are where they slide sideways past each other. Most transforms are short, linking segments

of mid-ocean ridge. But some, like the Alpine Fault in New Zealand, link ocean trenches. The famous San Andreas fault in California lies along the boundary where the Pacific Plate is rotating slowly past the North American plate.

Plate boundaries can be quite short-lived geologically. Some ancient continental plates, for instance, were thrust together so hard they welded into one solid piece. There was once a plate boundary in Asia, north of Tibet, for example. Geologists detected evidence of this via a seismological image of the edge of one of the plates involved thrust far down into the mantle beneath the boundary.

Convergent boundaries *typically occur along the edges of oceans. Here, as plates crunch together, the lighter continental plate rides up over the denser ocean plate, forcing it down into the mantle in a process called subduction. Deep ocean trenches open up where the descending plate plunges into the asthenosphere. As it goes down, the plate starts to melt, releasing hot, volatile materials, water and even molten rock, creating magma. Plumes of magma rise through the often shattered, faulted edge of the overlying plate, and may erupt to create an arc of volcanoes along the edge, or maybe even volcanic islands. The over-riding plate often acts like a giant mudscraper, scraping material off the sea floor on the subducted plate and piling it high in a wedge called an accretionary prism. As the subducted plate shudders down, the vibration can set off earthquakes, creating what is called a Wadati-Benioff zone, after seismologists Kiyoo Wadati and Hugo Benioff.*

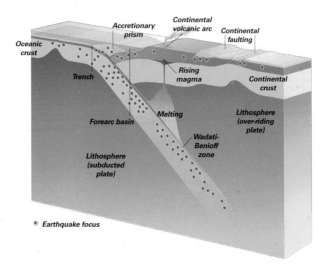

Accretionary prism

Continental volcanic arc

Continental faulting

Oceanic crust

Trench

Rising magma

Continental crust

Melting

Lithosphere (over-riding plate)

Forearc basin

Wadati-Benioff zone

Lithosphere (subducted plate)

• Earthquake focus

MOUNTAIN BUILDING

There are no more dramatic demonstrations of the dynamic power of the Earth's surface than mountains. Lifting the vast bulk of rock tens of thousands of metres upwards into towering ranges such as the Himalayas involves huge forces. Yet such forces have been deployed repeatedly in the Earth's history.

The remarkable thing about the world's biggest mountain ranges is just how young they are in geological terms. The Himalayas, the Andes, the Rockies and the Alps have all appeared within the last 50 million years. In other words, the dinosaurs were long dead before these so timeless-seeming mountain ranges began to rise up from the plain.

Until the coming of plate tectonics theory, geologists had no real idea of how these mountains came to be. They realized that major mountain belts or 'orogens' are created in mountain-building events or 'orogenies' that last tens of millions of years then stop. They also realized that when orogeny ceases, erosion can wear mountains back to sea level in only a little longer.

Below: The world's great mountain ranges, with lofty, snow-capped peaks like these in the Rockies, Colorado, were thrown up where tectonic plates crunch together, crumpling and fracturing rock in the collision zone. Mountains like these are called fold mountains. But isolated high mountain peaks, like Kenya's Mount Kilimanjaro – some 5895m (19,340 ft) high – are created by volcanoes.

In 1899, the famous geologist William Morris Davis developed a beautifully simple life history for mountains, called the Cycle of Erosion. This envisaged the creation of mountains in a brief and violent spasm of uplift in the landscape then a gradual decline through 'youth', 'maturity' and 'old age' as forces of erosion such as rivers and the weather did their slow, steady work. Once the mountains were worn flat, another spasm of uplift started the cycle again. It did not explain how uplift took place, but Davis's model seemed so elegant that it was widely accepted – until the development of plate tectonics in the 1960s began to reveal an entirely new picture.

Tectonic mountain-building

Mountains can be built in a number of ways (see below-right), but it now seems clear that most great ranges form along plate boundaries.

The world's longest range is actually the mid-ocean ridge that forms where plates are moving apart under the sea.

It winds all through the Atlantic and up into the Indian Ocean. All the high ranges on land, however, occur where plates are moving together. Like a rug rumpled against a wall, converging plates crumple the rocks in between, forcing them upwards and creating long folds all along the boundary.

When an oceanic plate slides under a continental plate, offshore volcanic arcs and other debris are swept against the continent. Too buoyant to subduct, they become welded to the edge of the continent as an 'accreted terrane'. As the oceanic plate goes on pushing under the continent, these terranes pile up higher and higher in fractured and folded mountain belts. The North American Cordillera formed like this.

Eventually, all the oceanic plate may slide into the mantle, leaving the two continents to collide head-on. The enormous force of their collision crumples up their edges to build the greatest mountain ranges of all.

In the distant past this happened when Africa and North America collided to throw up the Appalachians,

Right: Mountains are also created where plates are pulling apart or 'rifting'. As the plates separate, magma wells up under the rift, stretching the crust and cracking it. Blocks of rock can then drop away down these 'faults' to open up a rift valley, flanked by mountains formed from blocks that didn't drop. The Basin and Range Province of the south-west USA (right) formed in this way.

and also when what are now called Europe and North America collided to throw up the Caledonian mountains of Scotland and Norway. Both ranges are now worn down to a fraction of their former height. Now the same process is happening as India ploughs into Asia to build the Himalayas (see below).

A more complex picture

In recent years, however, geologists have begun to realize that this basic scenario is only half the story. One complication comes from the discovery that crustal rocks are not simply rigid and brittle, but actually flow, albeit slowly. So the Himalayas, for instance, are more like a ship's bow wave in front of India than a rumpled carpet.

Moreover, other factors are involved in the process besides plate movement. When British scientist George Airy was surveying India in the 19th century, deviations of his plumbline revealed that the mass of the Himalayas extends far below the surface. In fact, we now know that all mountains have deep 'roots' that protrude far down into the mantle. This is because mountains, like

all the crust, float on the mantle. But because they are so big and heavy, mountains sink farther.

As mountain ranges are worn away by erosion, however, they get lighter and so actually float up. This phenomenon is called isostasy. When recent precise surveys showed the Appalachian Mountains are gaining a few centimetres in height each century, geologists were at first convinced the figures must be wrong, since the Appalachians are far from any plate boundary. In fact, it seems the Appalachians are rising isostatically – because erosion of rock in the valleys

so lightens them that the entire range is floating upward.

Erosion is slow, but it can wear mountain ranges flat in much the same time it takes to build them tectonically. So the process of uplift can actually be assisted by erosion.

Moreover, erosion can be sped up or slowed by changes in climate which can be altered by the way continents move, and even by mountains themselves. So geologists now realize that mountain building involves a complex interaction between erosion, climate, tectonic movements and isostatic adjustment.

The Himalayas, the world's highest mountain range, began to form when the Indian plate crashed into southern Asia 55 million years ago. At the time Asia was made of softer, younger rock than India, and was powerless to resist the Indian advance. Long after the continents first crunched together, India has ploughed north at about 5cm/2in a year, pushing almost 2,000km/1,200 miles into the Asian plate. The Himalayas were thrown up as the crust doubled in thickness in the crumple zone. Because the Asian plate is warmer and lighter than the Indian plate, it is starting to ride over it, creating complex faults. But the creation of such high mountains interrupted the air flow over Asia, initiating India's famous monsoons. This intensifies erosion and is actually accelerating the ongoing uplift of the Himalayas as they are buoyed up isostatically by the weight of rock removed.

Himalayan Mountains

Kunlun Mountains

Continental crust

Tibetan Plateau

Continental crust

Complex thrust faults and folds

Strike-slip faults

Indian plate

Thin lithospheric bridge

Asian plate

Lithospheric mantle

Lithospheric mantle

Indian plate subducting beneath Asian plate

EARTHQUAKES AND FAULTS

The relentless movement of the Earth's tectonic plates can put the brittle rocks of the crust under such stress that every now and then they crack altogether, and great blocks slide past each other along fractures called faults, sending out shock waves that make the ground quake far around.

There is an earthquake somewhere in the world almost every day. Most are so tiny they are only detectable on the most sensitive of equipment. But a few are big enough to cause devastation, especially when they occur near major cities, or trigger giant waves called tsunami. Every year there are 20 quakes of the size that caused so much damage in the Turkish town of Izmit in 1999. Giant quakes such as that which triggered the tsunami that struck the coastal regions of Southern Asia on 26 December 2004 occur once every decade or so.

All kinds of things can trigger an earthquake, from a landslide or a volcanic eruption to the passing of a heavy vehicle, and they can occur almost anywhere. But most earthquakes – and nearly all major quakes – occur only in 'earthquake zones' which coincide with the edges of tectonic plates. In fact, an estimated 80 per cent of big quakes happen on the edges of the plates around the Pacific, and nearly all the rest happen on the boundary between the Eurasian and the Indo-Australian or Arabian plates. The mid-ocean rifts often send out tremors, but these are usually mild.

What causes earthquakes

Most quakes occur because of the immense forces generated as two plates grind past each other – either in subduction zones where one plate dives beneath another or along transforms where two plates slide sideways past each other. When one plate passes another, the rock either side of the crack may bend and stretch a little, but sooner or later the stress builds up to such a level that the rock suddenly snaps. The sudden rupture sends shock waves (seismic waves) shuddering out through the ground in all directions from the focus or hypocentre – the point where the rock snaps.

Above: Slicing right down through the state of California, the San Andreas is one of the world's most famous faults. Tremors set off by bursts of movement along the fault shake the cities of San Francisco and Los Angeles again and again. Seismologists feel it is only a matter of time before one of these cities is struck again with a quake as big or bigger than the one that devastated San Francisco in 1906. The San Andreas is not actually a single crack but a series of strike-slip faults (see Types of Fault, above right) that run along the transform boundary between two tectonic plates. To the west is the huge Pacific plate which runs right under the Pacific Ocean. To the east is the North American plate which makes up most of the continent of North America. Over the last 20 million years the Pacific plate has moved 560km/350 miles north – about 1cm/0.4in a year – but, disturbingly, the pace seems to have accelerated fivefold in the last century.

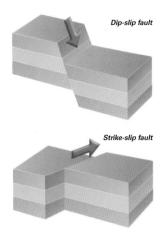

Dip-slip fault

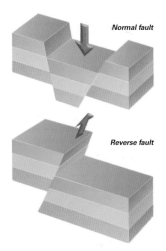

Normal fault

Strike-slip fault

Reverse fault

Types of fault

Geologists classify faults by how the rock moves. A 'dip-slip' fault is one in which the rock slips up or down. A 'strike-slip' fault is one in which the rock moves sideways. Big strike-slip faults are called transcurrent faults, and typically lie along transform plate boundaries, where tectonic plates are sliding sideways. Dip-slip faults, on the other hand, occur where the crust is being squeezed or stretched horizontally. 'Normal' faults occur where tension pulls rocks apart, so that one block slips down. Rift valleys form where parallel sets of normal faults open up as the crust is stretched by upwelling magma. Where rock is squeezed – maybe by converging plates – a block may slide over another to create a 'reverse' fault. A thrust is a reverse fault that slides up at a shallow angle.

The rupture spreads along the plate boundary like a crack spreading through glass. The longer the crack, the bigger the quake.

In the southern Asia quake of 2004, the rupture ripped 1,000km/620 miles along the Indo-Australian plate boundary. The massive Alaskan earthquake of 1964 moved entire mountains up 12m/40ft.

Most earthquakes only move the ground a few centimetres or so. Yet the cumulative effect of successive earthquakes has a bigger impact on the landscape. If the rocks either side of the crack or fault move just 10cm/4in a century, over a million years they can move up or down 1km/0.6 miles.

Earthquake waves

There is no chance of running from an earthquake. There are four kinds of shock wave – P or Primary and S or Secondary waves underground, and Love and Rayleigh waves on the surface. All move very fast. P-waves, which shake the ground up and down, are the fastest, roaring along at 5km/3 miles per second. S-waves, which snake from side to side, travel only a little slower. Surface waves are slower but these are what do the damage when an earthquake strikes cities. In solid rock, the waves move too fast for the eye to see. But they can turn loose sediments – in, for example, vulnerable areas such as landfill sites – into fluids so that waves can be seen rippling

across them like waves in the sea. These waves can capsize buildings. In both the Kobe (Japan) quake of 1995, and the San Francisco tremor of 1989, the worst damage was to buildings built on landfill sites.

Quake danger

Many of the world's cities – Los Angeles, Mexico City, Tokyo – are sitting on timebombs, because they are right in the middle of earthquake zones. People in these cities have learned to live with minor tremors. But sooner or later, one may be hit by a devastating quake that does terrible damage. For these people, learning to predict a quake is a race against time.

Below: The quake which devastated Marmara and Izmit in Turkey in 1999 seems to be one of a series moving west along the North Anatolian Fault and may reach Istanbul, just 50 miles (80km) away.

One approach is to look at the historical record. If there has not been a quake in an earthquake zone for some time, the chances are there will be one soon. The longer it has been quiet, the bigger the quake will be, as strain has been building in the rocks.

Most seismologists believe the key is to watch for strain building in rocks. In many earthquake zones, high-precision surveys monitor the ground for signs of deformation. Laser ranging from satellites such as the Japanese Keystone system can make acutely fine measurements. There is now some evidence that earthquakes occur in clusters, as one quake sets up stress further along a fault that will be in turn released as a quake. So a moving succession of quakes may be triggered off, as seems to be happening westwards along the North Anatolian Fault in Turkey, towards Istanbul.

VOLCANOES

Few sights are more awesome than a volcano erupting in a huge explosion of gas, ash and lava, and their effect on the landscape is immediate and dramatic. Volcanic activity is also going on beneath the Earth's surface. The combination of all this vulcanicity has a profound effect on the Earth's geology.

Volcanoes are places where red-hot magma (molten rock) wells up through the Earth's crust and erupts on to the surface. They are not randomly located around the world but clustered in certain places – where there is a ready supply of magma. Despite the inferno of heat in the Earth's interior, immense pressure keeps most rock in the mantle beneath the surface solid. But along the margins between the great tectonic plates that make up the Earth's surface, mantle rock melts into magma in huge volumes, and buoyed by its relatively low density wells up to the surface to erupt as volcanoes.

All but a few of the world's active volcanoes lie close to plate margins – mostly where plates are converging – and especially in a ring around the Pacific Ocean known as the 'Ring of Fire'. The exceptions are so-called 'hot-spot' volcanoes, such as Mauna Loa in Hawaii, which well up over fountain-like concentrations of magma called mantle plumes.

Above: Lava is the name for magma that has erupted on to the surface. It is so hot – over 1,100°C/2,000°F – the rock flows like a river.

Below: Most explosive eruptions begin with the blasting of a massive cloud of gas, ash and steam high into the atmosphere.

Volcanic eruptions vary widely in character. Along the ocean bed ridges where plates are moving apart – and over hotspots – runny, silica-poor 'basaltic' magma wells up through cracks that ooze red-hot lava almost continuously in gentle spouts. Some volcanoes belch ash and steam. Others eject showers of pulverized rock. Some unleash devastating mudflows as the heat melts ice, and some spew glowing avalanches of cinders and hot gas.

Explosive eruptions

The most terrifying and least predictable volcanoes of all tend to be those along the margins where plates are crunching together during subduction. Here, magma melting its way up through the thick plate margins becomes so contaminated with silica that it becomes 'dacitic' – so

thick and viscous that it frequently clogs up the volcano vent. It seems as if the volcano is sleeping or even dead – until, almost without warning, the pent-up pressure bursts through the plug in a cataclysmic explosion that hurls out shattered fragments of the plug (called pyroclasts), huge jets of steam and clouds of ash and cinder, as well as streams of lava.

The driving force in explosive eruptions is the boiling off of carbon dioxide gas and steam trapped within the reservoir of magma beneath the volcano called the magma chamber. The more gas and water present in the magma, the more explosive a volcano becomes. Magma near subduction zones often contains ten times as much gas as elsewhere. Gas in magma can expand to hundreds of times the volume of molten rock in a matter of seconds.

Types of eruption

No two volcanos are quite the same, but vulcanologists identify a number of distinctive styles of eruption. 'Effusive' eruptions occur when fluid basaltic lavas ooze from fissures and vents and flood far out over the landscape to form a plateau or a shallow dome. In some, lakes of lava pool up around the vent, while others shoot sprays of lava into the air. These sprays or fire fountains are driven by bubbles of gas in the lava, just like the droplets sprayed from a fizzy drink.

The explosive eruptions that occur with more viscous magmas are much more variable in character. Among the mildest are 'Strombolian' eruptions named after the island of Stromboli off the west coast of Italy. Gas escapes sporadically and the volcano repeatedly spits out sizzling clots of lava, but there is rarely a really violent explosion. 'Vulcanian' eruptions, named after Vulcano in the Italian Lipari islands, are much more ferocious. Here the magma is so viscous that the vent frequently clogs, in between roaring, cannon-like blasts that eject ash-clouds and fragments of magma followed by thick lava flows.

'Pelean' eruptions blast out glowing clouds of gas and ash called *nuées ardentes*, such as Mount Pelée, Martinique, in 1902. 'Plinian' eruptions are the most explosive of all, named after the Roman author Pliny, who witnessed the eruption of Vesuvius that buried Pompeii. Boiling gases blast clouds of ash and volcanic fragments high into the stratosphere.

Right: Hawaiian volcanoes like Kilauea – one of the most active on Earth – are famous for their spectacular jet-like sprays of liquid lava called fire fountains. They can occur in short spurts or last for hours on end. Occasionally they can shoot hundreds of yards into the air, and in 1958 one shot up to almost 610m/2,000ft. However, this was dwarfed by a fire fountain on the Island of Oshima, Japan, in 1986, which reached 1,524m/5,000ft!

Types of volcano

Volcanoes can also be classified by the kind of cone they create. The instantly recognizable cone-shaped volcanoes like Japan's Mt Fuji are composite or 'stratovolcanoes' created where sticky magma erupts explosively from a single vent. Successive eruptions build the cone from alternate layers of lava and the ash that rains down on it. Some cones, though, are built entirely of cinders and ash, such as Mexico's Paricutin. Shield volcanoes such as Hawaii's Mauna Loa are shallow, dome-shaped volcanoes formed where fluid lava spreads far out from a single vent. Fissure volcanoes are ridges created when runny lava oozes from a long crack. Large scale fissures occur along the mid-ocean ridges. Smaller ones burst through on the flanks of larger volcanoes.

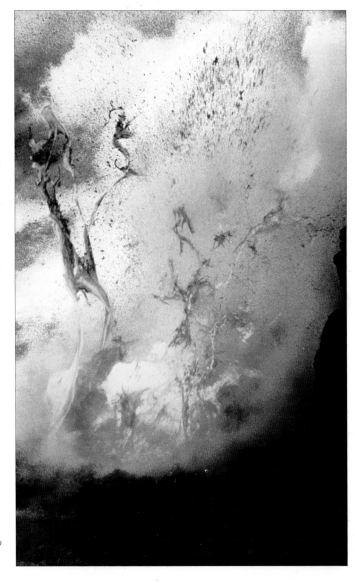

IGNEOUS FEATURES

Although eruptions are brief, volcanic activity leaves a lasting legacy as magma, lava and volcanic ash harden to form rock. Volcanic peaks and ash deposits are clearly visible on the surface, but much of the molten magma that wells up from the Earth's interior remains trapped underground, solidifying in situ.

Rock features formed by volcanoes above ground and magma underground are together termed igneous features. Features formed from magma trapped underground are termed intrusive igneous features. Those formed above ground are called extrusive features.

Intrusions underlie all the world's major continents, and in many places they have been exposed on the surface after erosion of the surrounding 'country rock'. Intrusive igneous rock, which includes granite and gabbro, is almost invariably very tough and crystalline, and endures the weather to stand proud long after the surrounding softer rocks have been worn away.

Massive intrusions

As seen in Types of Intrusion, opposite, the large intrusions forming deep underground are called plutons, and typically made of granite. Sometimes scores of plutons can coalesce over time to form monsters called batholiths (from the Greek for 'deep stone'), which lie like giant whale carcasses under most of the world's great mountain ranges. North America's Coast Range batholith extends 1,500km/932 miles under British Columbia and Washington. Batholiths mostly form as the convergence of two of the world's great tectonic plates generates huge quantities of magma underground. Sometimes, batholiths are topped by smaller protuberances called stocks and bosses. Erosion may expose several bosses on the surface, as at Dartmoor and Bodmin Moor in south-west England, which are linked underground to a single batholith.

Minor intrusions

Nearer the surface, smaller intrusions often occur as sheetlike formations called dykes and sills. Dykes are anything from a few centimetres to

hundreds of metres thick, and form as magma is injected into cracks in the rock. They are typically near vertical, but any sheet intrusion that cuts right across layers of country rock may be described as a dyke. Because they cut across existing structures, dykes are described as 'discordant'. Often, dykes form as the pressure of upwelling magma fractures overlying rock, opening cracks that fill with magma. A single intrusion may breed dozens of dykes like this in a 'dyke swarm'. Occasionally, these form in concentric downward-pointing cones or cone sheets crowning the intrusion, as famously on Mull in Scotland.

Ring dykes form like the sides of an upturned pan or cauldron as a round block of country rock drops away,

Above: In California's Yosemite Park, spectacular cliffs of grey rock rear up where erosion has stripped naked the ironhard granite batholith that underlies much of the Sierra Nevada mountains. The Half Dome, seen above, is the most impressive example.

leaving a gap to fill with magma. Classic ring dykes like this are seen in Glencoe in Scotland and Mount Holmes in Yellowstone Park.

Sills are typically horizontal or gently sloping sheets, forming as magma seeps between existing bedding planes. Because they follow existing structures, they are described as 'concordant' structures. Sometimes the magma arches up between rock beds to form dome-shaped laccoliths or warps the rock downward to form dish-shaped lopoliths.

Massive extrusions

Geologists have long debated over how vast intrusions form (see The Granite Problem under Granite, Directory of Rocks), but extrusions did not seem to pose the same problems, because volcanoes on land produce only small volumes of lava. Yet not only is the entire ocean floor made of extrusions of basalt erupted from mid-ocean fissures, but in places around the world there are solidified remains of gigantic lava flows.

Flood basalts are the solidified remains of huge floods of basalt lava that erupted in the past. The most spectacular is the Deccan Traps of India. Over 2km/1.2 miles thick and covering 500,000sq km/200,000sq miles, the Deccan Traps include half a million cubic km/120,000 cubic miles of lava – half a million times as much as erupted at Washington's Mount St Helens, USA, in 1980! The Traps' basalts erupted about 65 million years ago in a gigantic outpouring that has been blamed by some palaeontologists for the death of the dinosaurs.

Another huge flood basalt is the Columbia River plateau of north-western USA, which erupted 175,000 cubic km/42,000 cubic miles of lava about 16 million years ago. Others include southern Africa's Karoo and the Faroe Islands, North Atlantic.

Undersea floods

These flood basalts were originally thought to be rare exceptions. Then in the late 1980s, seismic surveys of the sea floor began to reveal even more spectacular examples of what came to be called Large Igneous Provinces or LIPs under the oceans.

The largest of these, called the Ontong Java Plateau, lies under the Pacific to the east of Borneo and covers nearly 5 million sq km/2 million sq miles, an area bigger than the whole of the USA. This enormous Igneous Province erupted less than three million years ago.

Above: India's Deccan Traps were the site of one of the greatest eruptions of lava in the Earth's history, about 65 million years ago. Traps is Dutch for staircase, referring to the steplike shape of the eroded layers of lava.

Geologists believe 'mantle plumes' cause these massive eruptions. These are fountains of hot matter that rise through the Earth's mantle, perhaps all the way from the D" layer on the core boundary (see Inside the Earth, this section). The theory is that as they rise, the resulting heat melts mantle rock to create huge quantities of molten magma which burns through the crust and floods on to the surface.

Types of intrusion

Intrusions vary widely in size and shape. Deep underground, diapirs (rising blobs of magma) open up huge spaces in the country rock to form lumps of igneous rock called plutons. These include drum-shaped stocks a few kilometres across and gigantic batholiths made from scores of plutons and stretching thousands of kilometres. Nearer the surface, magma intrudes into cracks in the rock to create igneous rocks in thin sheets (dykes and sills) or lens shapes (lopoliths and laccoliths). Where these minor intrusions cut across existing structures, as dykes often do, they are said to be discordant; where they follow existing structures, like sills, they are said to be concordant. Minor near-surface intrusions are often exposed on the surface by subsequent erosion of the surrounding country rock, but even deep-forming plutons may be exposed this way over a considerable period of time.

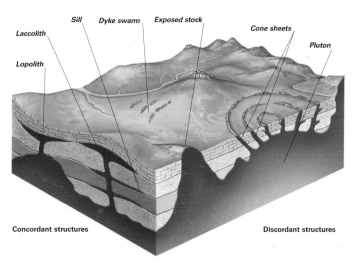

Laccolith / Sill / Dyke swarm / Exposed stock / Cone sheets / Pluton / Lopolith / Concordant structures / Discordant structures

THE ROCK CYCLE

Mountains and hills look so solid, it's hard to imagine they could ever change. Yet all the Earth's landscapes are being continually remodelled as rocks are attacked by the weather and worn away by running water, moving ice, waves, wind, and other 'agents of erosion'.

Occasionally, the landscape is reshaped suddenly and dramatically – by an avalanche or landslide. Yet most of the time it is remoulded slowly but relentlessly. One cold night does little damage. A single shower of rain seems to run straight off. Yet night after freezing night, shower after shower, repeated over millions of years, takes its toll. Research into rates of erosion has shown it to average at least 0.5mm/0.04 inches a year on land. At that rate, even a mountain range as high as the Himalayas can be worn entirely flat in just over 20 millions of years.

Some new landforms are created from old as rock is denuded or stripped away by weathering and erosion. Some are created by steady accumulation or deposition of the rock debris. Most hills and valleys are formed by a combination of both denudation and deposition.

Below. On coasts, shorelines are battered by waves packed with energy by wind blowing far over the ocean. Waves hurl tons of water filled with shingle against coastal rocks and ram air into cracks so that rocks are forced apart.

Weathering

As soon as rock is exposed to the weather, it gradually starts to break down under the assault of wind and rain, frost and sun. Sometimes the rock is corroded by chemical reactions caused by moisture in the air, or water trickling over it. Sometimes they are attacked by micro-organisms and lichens or chemicals released by plants. Sometimes they are broken down physically by, for instance, the effects of heat and cold. Water in cracks can expand so forcefully as it freezes that it can shatter rock. At -22°C/-7.6°F, ice can exert a pressure of 3,000kg/6,600lb on an area the size of a coin. This is called frost-shattering.

Geologists argue about the relative importance of chemical and physical weathering. Some limestone regions – known as 'karst' – show all the signs of chemical weathering which opens up potholes and spectacular caverns (see Rock Landscapes in this section). Other regions show strong signs of physical weathering. Frost-shattering

creates the jagged peaks in high mountain regions – as well as the piles of debris called scree below. In most places both chemical and physical processes are at work.

Erosion and deposition

All the debris created by weathering is gradually carried away by agents of erosion – the most important of which is water. Without running water to mould it, the landscape would be as jagged as the surface of the Moon. Rivers and streams slowly soften contours – wearing away material here, depositing it there. Over millions of years, a river can carve a deep canyon or spread out a vast plain of sediment as it flows towards the sea.

In deserts, however, running water is scarce. Intermittent streams carve out valleys, but the landscape is angular, and some landforms are carved into weird shapes by the blast of windblown sand. On coasts, it is the waves which are the dominant agents of erosion, and their continuous

Right: In many parts of the world, the landscape is shaped by water running over the land. Rivers can carve out deep valleys as they run down to the sea. A river's steady torrent washes small grains loose and grinds away solid rock with the stones it carries. Arizona's Grand Canyon (pictured here) was carved out as the Colorado river cut its way through a rising landscape over many millions of years.

pounding on the shore creates a distinctive range of coastal landforms, including steep cliffs where waves cut into hillsides, platforms of rock sliced out as the waves wear back into the cliff and stacks of rock left behind as the waves erode the cliff.

It is far from certain how much of today's landscape has been created by the processes we see operating today – and how much by more dramatic events in the past. Some geologists argue that some river-eroded features in deserts were carved out during especially wet periods in the past called pluvials. There is also no doubt that cold periods in the past known as ice ages were crucial in shaping the land in Europe and North America. Giant U-shaped valleys in the mountains of north-west USA and Scotland, fjords in Norway and Canada and vast deposits of 'till' (rock debris) covering the USA's Midwest can only have been created by moving ice.

Rocks remade

Just as erosion relentlessly destroys rock wherever it is exposed on the surface, so new rocks are created all the time. These new rocks are often forged from remnants of the old, so that rock is continually recycled in a process called the rock cycle (below).

Not all material moves through the cycle at the same rate. Some material in rocks in continental interiors may sit virtually unchanged for billions of years, while material in rocks near the active margins of continents – those

near coasts and near to subduction zones – has been through the mill again and again.

Some material eroded from rocks is taken back down into the mantle on subducted sea-beds. So too is rock forming oceanic plates. Some may melt as it descends and rise with magmas to form new igneous rocks. Some will be carried right down into mantle. Even there it is not necessarily lost forever. Convection makes material circulate right through the mantle, so it may re-emerge, eventually, as molten magma – even if the process takes hundreds of millions of years. Studies have shown that most of the atoms in rocks in the continental crust came up from the mantle over 2.5 billion years ago. Yet the rocks they are in are usually much, much younger, since these atoms have been recycled many times.

Rock breaking

New rock material is sometimes brought up from the mantle in magmas, but most surface rocks are made from ingredients that are continually recycled as rocks are made and remade. The ingredients can be large outcrops of rock, or small chunks, grains or even atoms, and geologists call the recycling process the rock cycle. There are many paths through the cycle. Igneous rock formed by the freezing of molten magma, for instance, might be broken up by the weather into fragments that are washed in rivers into the sea. There fragments pile up on the sea bed and eventually turn to sedimentary stone. This sedimentary rock may, in turn, be buried and squeezed or heated to form metamorphic rock. This too can be broken down and made into sedimentary rock.

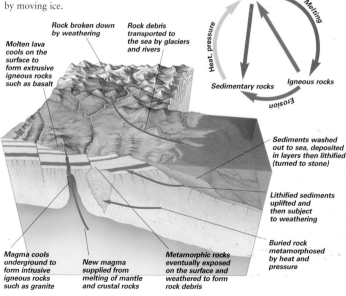

Metamorphic rocks

Melting

Heat, pressure

Sedimentary rocks

Igneous rocks

Erosion

Rock broken down by weathering

Rock debris transported to the sea by glaciers and rivers

Molten lava cools on the surface to form extrusive igneous rocks such as basalt

Sediments washed out to sea, deposited in layers then lithified (turned to stone)

Lithified sediments uplifted and then subject to weathering

Buried rock metamorphosed by heat and pressure

Magma cools underground to form intrusive igneous rocks such as granite

New magma supplied from melting of mantle and crustal rocks

Metamorphic rocks eventually exposed on the surface and weathered to form rock debris

HOW IGNEOUS ROCKS FORM

No rocks undergo quite such a change as they form as igneous rocks, the rocks that form almost all the ocean crust and a great deal of the continental crust. Before they solidified into their current form, these extraordinarily tough, crystalline rocks were glowing, searing hot liquid magma.

Igneous rocks are quite literally frozen magma, hot molten rock from the Earth's interior. Magma may be quite thick and sticky, but it is quite genuinely a fluid and flows like one. The process that turns it to solid rock is exactly the same that turns water to ice when the temperature drops to 0°C/32°F. The difference is that magma freezes at much higher temperatures than water – anywhere between 650°C/1,200°F and 1,100°C/2,000°F. Magma also contains a complicated mix of substances, each with its own freezing point, so magma does not freeze in one go like water but bit by bit.

In magma, elements such as silicon, iron, sodium, potassium, magnesium and so on occur in pure form, or as simple compounds. Yet as the magma cools, these elements and compounds join to form crystals of various minerals. The commonest minerals to form are quartz, feldspars, micas, amphiboles, pyroxenes and olivine. Magma also contains gases such as water vapour, sulphur dioxide and carbon dioxide, but these are driven off during cooling.

As magma solidifies, molecules which were vibrating wildly when it was liquid begin to calm down enough to form clusters. These clusters soon begin to grow and form crystals here and there in the melt. They grow especially quickly near the surfaces, which cool fastest. This is why a solid crust can form on the surface of lava which is hot and fluid on the inside.

Coarse and fine

The slower magma cools, the larger crystals grow. In intrusions below ground, cooling can take thousands, if not millions, of years, and crystals grow big enough to see with the naked eye. Rocks with such crystals are said to be phaneritic, or coarse-grained.

Cooling is usually much quicker in lavas – that is, magmas which erupt on the surface to form extrusive rocks. Cooling takes a matter of weeks or even days. Crystals have little time to grow and so are often only visible under a microscope. Rocks with such crystals are said to be aphanitic, or fine-grained. Sometimes a fine-grained rock may contain larger crystals that formed earlier, while the magma was still underground. These large crystals are called phenocrysts and the rock is said to be porphyritic.

Where lava is ejected in small blobs, it can cool in a matter of hours – so quickly that crystals cannot form at all. The result is a glass, like obsidian, in which there are no crystals at all.

How igneous rocks evolve

Each mineral crystallizes at a different temperature. Olivine and pyroxene set at over 1,000°C/1,832°F. Silicate minerals such as quartz don't freeze until temperatures are as low as 650°C/1,200°F. So magmas crystallize progressively, with some minerals forming earlier than others.

In the 1920s, laboratory tests conducted by Norman Bowen showed how minerals crystallize in a sequence (Bowen's Reaction Series), from high to low temperature: olivine, pyroxene, amphibole, biotite mica, quartz,

Left: There can be no more dramatic demonstration of the toughness of granite than these peaks in Torres del Paine, Chile. The softer country rock into which the granite magma was intruded has been stripped away by millions of years of erosion.

Above: Granite magmas cool slowly underground and crystals grow large enough to be seen with the naked eye, as in this specimen of Cornish granite. The white crystals are quartz, the black ones biotite mica and the pink ones potassium feldspar.

muscovite mica, potassium-feldspar and plagioclase feldspar.

This sequence shows how igneous rocks form and how each kind develops its own mineral make-up, according to the circumstances in which it forms. It also provides a mechanism for igneous rocks to evolve. Without such a mechanism, the granitic magmas that form our continents could never have developed.

The idea is that when it formed, the Earth was much like the Moon, made mostly of just a simple parent rock – a 'mafic' or 'ultramafic' rock low in

silica like basalt and peridotite, but unlike granite (which is silica-rich). From this basic start all the many other kinds of igneous rocks evolved by a process called fractionation. This is the way the composition changes during either melting or freezing of the rock, as different minerals melt or freeze first.

When a mafic rock melts, for instance, it splits into two fractions as low temperature minerals such as quartz and feldspar melt first and flow away, leaving high temperature minerals behind. Low temperature minerals are all silicates. So the melt is much richer in silicate minerals than the original rock. Successive refreezing and melting further boosts the silica-content to create silica-rich granite. Which of the many kinds of igneous rock that ultimately forms depends on a variety of factors, including the depth at which the magma is generated and its history of fractionation.

Where igneous rocks are found

Igneous rocks form only in certain places. Fractionation occurs mostly where tectonic plates are either moving apart at mid-ocean ridges, or pushing together at subduction zones. At mid-ocean ridges, fractionation of the parent magma from the mantle creates

Above: Clearly visible under a microscope in this fine-grained igneous rock are larger crystals or phenocrysts that crystallized slowly underground before the lava erupted to the surface.

basalt lava at the surface and gabbro deeper down. At subduction zones, partial melting of the subducted plate fractionates to create intermediate rocks such as diorite in island arcs. Further melting and remelting, especially beneath continents, creates granite. Granite only forms underground, but it can also melt to create the silica-rich lava rhyolite.

Below: All igneous rocks form when red-hot molten magma like this freezes, either on the surface or underground. The temperature at which the magma freezes, and the kind of rock that forms, depends on the balance of the various chemical elements in the magma.

HOW SEDIMENTARY ROCKS FORM

Over 90 per cent of the Earth's crust is made from igneous rock, but on land, 75 per cent of it is hidden beneath a veneer of sedimentary rock, rock formed from sediments laid down in places such as the sea-bed, buried and turned into stone over millions of years.

Sedimentary rocks start to form wherever sediments settle on the beds of oceans, lakes and rivers, or are piled up by moving sheets of ice or the wind in the desert. As sediments build up, layers are buried ever deeper, drying and hardening as water is squeezed out. Over millions of years, the pressure of overlying layers and the heat of the Earth's interior turns layers of sediment to solid rock in a process called lithifaction.

When sediments are powdery and soft and contain few hard sand grains, compaction alone is enough to turn them to stone. Very sandy sediments, however, are too hard to be compacted

so easily. To turn to stone, sandy sediments must be glued together by cements made from materials dissolved in the water from which the debris settled. The most common cements are silicate minerals, calcite and iron compounds which give rocks a rusty red look.

Beds and joints

Because sediments settle layer upon layer, outcrops of sedimentary rocks are usually marked by a distinctive layered or 'stratified' look. Where the rock has been undisturbed since turning into stone, these layers appear horizontal. You are not very likely to

Above: This ammonite fossil, preserved in chalk, once inhabited shallow waters.

see such uniform layering, however, as the movements of the Earth's crust twists them into contorted shapes.

The thinnest layers or beds are marked out by lines called 'bedding planes'. Beds may be the sediments laid down in just a single season. Yet the thickest layers or strata may have taken millions of years to build up. Sedimentary rocks may also be marked by 'joints', cracks across layers formed as the rock dried out and shrank.

Sedimentary rocks usually contain another distinctive feature – fossils, the remains of living things turned to

Enduring sand

Of the three major ingredients of clastic rocks, quartz is by far the most durable, surviving the destruction of its parent rock again and again. Once freed from igneous rocks by weathering, quartz accumulates in layers. Grains are then cemented together into sandstones (right) like the de Chelly beds of the buttes in Utah's famous Monument Valley (below). Tough though these rocks are, they too are broken down in time. Yet it is not the sand that is destroyed but just the cement. The liberated sand now scattered on the desert floor will, in time, form new sandstones.

Right: Sedimentation is very rarely continuous, but occurs in fits and starts. The result is that many sedimentary rocks, like these sandstones in Utah's famous Zion Canyon, are marked by countless bedding planes marking a brief pause in sedimentation.

stone. Fossils help geologists to determine the conditions in which certain rocks formed. They can tell chalks formed in shallow tropical seas, for instance, because they are studded with fossils of sea creatures that live only in such conditions. Moreover, different creatures and plants lived at various times in Earth's history. So geologists can also work out the relative ages of sedimentary rocks from the range of fossils they contain, a process known as biostratigraphy.

Kinds of sedimentary rock

Indeed, some sedimentary rocks, like limestone, are made almost entirely from the remains of living organisms, or else by chemicals created by them. Such rocks are described as organic. Chemical sediments such as evaporites are made from minerals precipitated directly from water. Most sedimentary rocks, however, are 'clastic' or 'detrital' which means they are made from fragments or clasts of rock broken down by the weather.

Rocks from debris

When rocks such as granite are weathered, they eventually crumble to form clasts. The different minerals crumble in different ways. Quartz crystals, for instance, are so hard they are left as distinct sand grains, while orthoclase feldspar breaks down into clay and plagioclase forms calcite. Although this debris all starts off together, the different kinds of clast are gradually separated as they are washed down by rivers into seas and lakes.

Moving water is like a natural sieve, sorting the rock fragments into small and large by carrying smaller grains farther and faster. The farther from the source they are carried, the more they

Right: Few rocks are so distinctive as white chalk, seen here in cliffs at Normandy, France. It is made of calcite-rich remains of countless marine plankton that floated in ancient seas.

become sorted. So typically only rocks that form fairly near the source, such as some conglomerates, contain a full range of particle or grain sizes. Most rocks contain particles predominantly of a particular size of grain.

The result is that clastic sedimentary rocks can be divided into three groups according to grain size: large-grained 'rudites' such as conglomerates and breccias; medium-grained 'arenites' such as sandstones; and fine-grained 'lutites' such as shale and clay.

As a river washes into the sea, or into a lake, the heaviest quartz grains are dropped nearest the shore, forming sandstones. Finer clay particles are

washed farther out, forming shales. Dissolved calcite is washed farther out still, and only finally settles to form limestone rocks. This only happens once the particles are taken from the water by living things which use them for building shells and bones. When they die, they take the calcite down to the sea floor in their remains.

In some rocks, such as wackes, there is a mix of grain sizes. However, even wackes may be banded into layers, with a gradation of grain size from fine at the top to coarse at the bottom. This 'graded bedding' develops because the largest, heaviest grains settle out of the water first.

HOW METAMORPHIC ROCKS FORM

When rocks are seared by the heat of molten magma or crushed by the enormous forces involved in the movement of tectonic plates, they can be altered beyond recognition. The crystals they are made from re-form so completely that they become new rocks, called metamorphic rocks.

Early geologists soon appreciated that igneous rocks formed from melts and sedimentary rocks from sediments. Yet there was a third very common kind which they could not quite pin down.

Slate used for roofs, for instance, has a colour and texture like shale. Yet it is much harder than shale and splits into sheets that are completely unrelated to bedding. An even harder rock called gneiss found in the Alps has strange swirls and bands. Beautiful white marble seems to be made of calcite like limestone, yet contains no fossils and has a dense crystalline structure not unlike granite.

Marble seemed to represent a cross between sedimentary and igneous rock and geologists began to realize that

this could provide the clue to its origin. Marble was indeed once limestone but it has been altered by heat. The calcite it is made of has been cooked and its crystals reformed in a way that looks more like an igneous rock. Marble, like many other rocks, is meta-morphosed or re-formed by heat, or pressure, or both. The word metamorphic comes from the Greek for 'change form'.

Cooking and crushing

It is now clear that just about any rock – igneous, sedimentary and metamorphic – can be metamorphosed into new rock. To create a metamorphic rock, the heat and pressure were extreme enough to

completely alter the original rock or 'protolith', but not so extreme as to melt it or break it down altogether. Heat intense enough to melt it would have created an igneous rock.

Rocks are subjected to heat and pressure either by being buried deep in the crust, or being crushed beneath converging tectonic plates. They are subjected to heat alone by proximity to hot magma.

Heat and pressure metamorphoses rocks in two ways. First, it changes the mineral content, by making minerals react together to form new ones. Some minerals are unique to metamorphic rock. When shale is metamorphosed to slate, for instance, its clay is changed to chlorite, a mineral only found in metamorphic rock.

Second, it changes the size, shape and alignment of crystals, breaking down old crystals and forming new ones in a process called recrystallization. The original rock might be made of just a single mineral. When this is metamorphosed the mineral recrystallizes in a different form. Pure quartz sandstone becomes quartzite. Pure calcite limestone becomes marble.

Metamorphic environments

Each kind of metamorphic rock has its own particular protolith or set of protoliths. Marble is only formed from pure calcite limestone. However, mylonite can form from practically any rock. Each kind of rock also forms only in specific conditions. Mild metamorphism turns shale to slate. Yet if the heat and pressure become more intense, it turns first to phyllite, then to schist and finally to gneiss.

Left: Like so much of the Canadian shield, the mountains of Churchill are made from ancient, very tough metamorphic rock. This quartzite landscape formed when sandstones were metamorphosed by intense heat and pressure, some two billion years ago.

It has become clear that particular combinations of conditions are likely to create particular metamorphic rocks. So geologists talk about environments of metamorphism. They also focus on sets of conditions that form particular combinations or facies of metamorphic minerals (see Metamorphic facies under Schists, Directory of Rocks). One of the most important environments is next to hot igneous intrusions. An intrusion may have a temperature of 900°C/1652°F, and literally cooks the rock it comes into contact with. This is called 'contact metamorphism' and involves heat alone without significant pressure.

In fault zones, rock is ripped apart by tectonic movements. Near the surface, this shatters rock to form a breccia or crushes it to a powder. Deeper down, rock is so warm that it smears out rather than breaks. As it does it is subjected to 'dynamic metamorphism' which creates the rock mylonite.

Regional metamorphism
The huge forces involved when continents collide and throw up mountain ranges can crush and cook rocks over large areas. Near the fringes, this regional metamorphism can be mild, and mudrocks are altered to low-grade metamorphic rocks such as slate and phyllite. Towards the heart of the mountain belt heat and pressure gradually increase. Under moderate heat and pressure, slate and phyllite are metamorphosed to schist. Intense heat and pressure (high-grade metamorphism) creates gneiss. Even more extreme conditions can partially melt the rock to create migmatite.

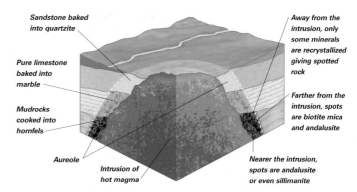

Sandstone baked into quartzite

Pure limestone baked into marble

Mudrocks cooked into hornfels

Aureole

Intrusion of hot magma

Away from the intrusion, only some minerals are recrystallized giving spotted rock

Farther from the intrusion, spots are biotite mica and andalusite

Nearer the intrusion, spots are andalusite or even sillimanite

Metamorphic grading
The most widespread metamorphic environment is where tectonic plates converge with the force to throw up mountains. Beneath the mountains, rocks are crushed and sheared, and heated by rising magma and their proximity to the Earth's mantle. The scale of this is enormous and is called 'regional metamorphism'. The intensity of regional metamorphism is described in terms of grades. Low-grade metamorphism means low temperature (below 320°C/608°F) and pressure. High-grade metamorphism means high temperature (above 500°C/932°F) and pressure. These different grades of metamorphism create different minerals and different structures.

Metamorphic rocks that form where plates converge have a very distinctive characteristic. Squeezed between the plates, new crystals are forced to grow flat, at right angles to the pressure.

Contact metamorphism
When rocks are cooked by the heat of an intrusion, the result is contact metamorphism. Around the intrusion is a ring or 'aureole' of affected rock. The way in which particular rocks are affected depends on how close they are to the intrusion and the intrusion's size.

The crystals are so intensely aligned that the rocks have layered structure, like the leaves of a book and so called foliation. Foliation means slate breaks easily into flat sheets, it gives schist a stripey look, called schistosity, and gneiss even more dramatic swirling bands, as the minerals are separated out into layers. Foliation is such a distinctive characteristic of regionally metamorphosed rocks that all metamorphic rocks – not just regionally metamorphosed rocks – are divided into foliated rocks (such as slate, phyllite, schist and gneiss) and non-foliated rocks (such as hornfels, quartzite and amphibolite).

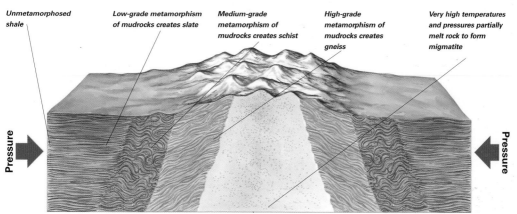

Unmetamorphosed shale

Low-grade metamorphism of mudrocks creates slate

Medium-grade metamorphism of mudrocks creates schist

High-grade metamorphism of mudrocks creates gneiss

Very high temperatures and pressures partially melt rock to form migmatite

Pressure

Pressure

ROCK LANDSCAPES

Every kind of rock and geological formation produces its own distinctive kind of landscape, from the jagged ice-capped peaks of young fold mountains and the spectacular gorges and caverns of some limestone regions to the gently rolling landscapes of chalk downland.

The relationship between rocks and landscapes is a complex one, and geologists have been trying to unravel it for centuries. Yet an experienced geologist can tell a great deal about the nature of the rocks and the rock formations simply by studying the shape of the land.

Some landforms give an instant clue to the kind of rock involved. Gorges with pale rock faces, spectacular caverns and deep potholes cannot be anything but limestone (see opposite page). Granite tors are also easy to spot. Similarly, long mountain ranges with soaring, jagged peaks are clearly fold mountains, created by the convergence of two tectonic plates.

Hills and vales

Most of the time, however, the relationship is more subtle. On the whole, the hardest rocks, such as granites, gneisses, sandstones and limestones tend to resist erosion and form hills. Softer rocks like clays and

Above: Spitzkoppe in Namibia is one of many steep-sided granite monoliths in the tropics called inselbergs. It was once thought only deep weathering in ancient climatic conditions could have produced such outcrops. Now it is believed that the shape simply reflects the way ordinary weathering attacks the structure of the original intrusion.

Below: Salt Cellar Tor in England's Peak District is one of several tors in this region. They are made of the tough sandstone and shale rock series millstone grit. These tors probably formed when rock was weathered underground by water seeping into joints – perhaps during the ice ages. Weathered rock was then stripped away to reveal the tor.

mudstones are worn away into valleys. There can also be valleys in hard rock, and tectonic movements can uplift soft rock to create hills or even mountains. Clay very rarely forms hills, however, because it is so quickly worn away.

In south-east England, and places in the Appalachians, gentle folding has tilted layers of sedimentary rock to create a distinctive 'belted' landscape of parallel ridges and valleys. Layers of softer rock, such as clay, are eroded faster to create valleys, while the harder rocks such as sandstones and chalk resist erosion to form ridges.

One side of the ridge is a steep 'scarp' slope where erosion has cut right across the strata. The other side, however, slopes gently down the top surface of the rock layer and is called the 'dip' slope. A ridge shaped like this, with one steep scarp and a gentle

dip slope, is called a cuesta. The angle of the dip slope gives an instant clue to the angle of the strata in the region.

Underground water

Often the landscape that forms from a particular kind of rock depends on the way water moves through it or over it. Sandstone, for instance, is a permeable rock. This means that it allows water to seep through it easily. This is not the same as being porous, though the terms are often confused. Porosity is the capacity of a rock to hold water in spaces in the rock – in other words, how full of holes it is. A rock like slate is barely 1 per cent porous; gravel is over 30 per cent porous. A rock that is porous is likely to be permeable – though not always – but a rock that is permeable is not necessarily porous.

Since sandstone is permeable, most water in sandstone landscapes soaks into the ground rather than flowing overland. This often makes sandstone landscapes quite angular because they are not rounded by water flowing over the land, except in times of flood. In wet regions, there may be too much water for the rock to soak up, so water erosion can do its work anyway.

Clays and shales, on the other hand, tend to be impermeable. They are often even more porous than gravel, but the pores are so small that water gets trapped. This means that the rocks easily become waterlogged and water remains on the surface to wear the rocks away, and round off contours.

Chalk and limestone

Like clay, chalk is not at all porous, and often contains clays which make it even harder for water to seep through. The same is true of limestone. Yet both these rocks are permeable. In particular, limestone usually has joints (cracks) so big that water can filter into the formation even if it can't penetrate the rock directly. It is a particular feature of limestone that the infiltrating water, called groundwater, corrodes away the rock to open up potholes and caverns.

Chalk has far fewer joints and is much less permeable than limestone and is not so susceptible to corrosion. So it rarely develops anything more dramatic than the occasional cavern. All the same, the chalk downlands of southern England do have their own special features including dry valleys called bournes. Bournes look like river valleys but contain no river. They may have formed in times when the climate

was wetter. Large hollows called combes may have been formed either by springs in these wetter times, or by frozen surface debris melting after the ice ages.

Granite

Like limestone, granite is tough enough to form mountains. In cold regions, granite intrusions are often left standing proud as dramatic monoliths when exposed on the surface. Yet in warmer regions, granite's feldspar content makes it prone to chemical weathering. Feldspar corrodes quickly in warm water. As with limestone, joints in granite allow water to penetrate deep into the rock, so tropical granite is corroded under-ground. When the weathered rock is stripped away, it leaves outcrops called kopjes. Similar features called tors form in cooler regions, such as in England's Dartmoor. It's thought that these are the result of climate change.

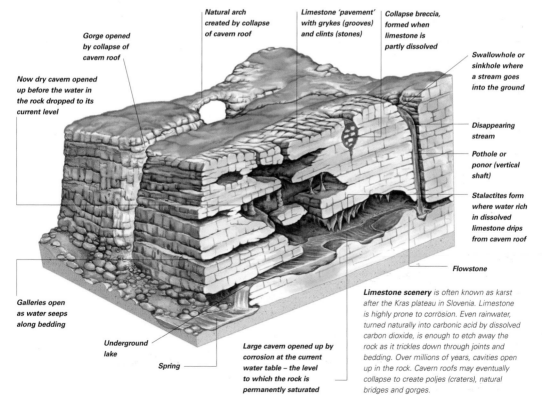

Gorge opened by collapse of cavern roof

Natural arch created by collapse of cavern roof

Limestone 'pavement' with grykes (grooves) and clints (stones)

Collapse breccia, formed when limestone is partly dissolved

Now dry cavern opened up before the water in the rock dropped to its current level

Swallowhole or sinkhole where a stream goes into the ground

Disappearing stream

Pothole or ponor (vertical shaft)

Stalactites form where water rich in dissolved limestone drips from cavern roof

Flowstone

Galleries open as water seeps along bedding

Underground lake

Spring

Large cavern opened up by corrosion at the current water table – the level to which the rock is permanently saturated

Limestone scenery is often known as karst after the Kras plateau in Slovenia. Limestone is highly prone to corrosion. Even rainwater, turned naturally into carbonic acid by dissolved carbon dioxide, is enough to etch away the rock as it trickles down through joints and bedding. Over millions of years, cavities open up in the rock. Cavern roofs may eventually collapse to create poljes (craters), natural bridges and gorges.

THE AGES OF THE EARTH

Written in the rocks of the crust is the entire history of the Earth's surface. The record is blurred in some places and lost in others. Nonetheless, by studying rocks in detail geologists have been able to piece together the remarkable story of how the Earth has changed through the ages.

Barely two centuries ago, most people thought the Earth was just a few thousand years old and little changed since it was created. But in the 19th century, geologists began to realize its age is immense. It is now thought to be nearly 4,600 million years old, and has undergone huge changes during its life. Its remarkable history is recorded in the rocks, if you know how to read it.

Pioneers of deep time

The great pioneers in the study of the Earth's distant past were James Hutton (1726–97) and William Smith (1769–1839). It was Scottish geologist Hutton who first suggested that the Earth is very, very old, and that the landscape has been shaped gradually by countless cycles of erosion and uplift.

Soon after, English surveyor William Smith was surveying routes for canals, and noticed that each layer of sedimentary rock contains its own range of fossils. He realized that all layers containing the same range of fossils must be of the same age. What's more, the 'principle of superposition' made it

possible to work out which layers are old and which young. In the 1600s, Danish geologist priest Nicolas Steno had realized that all sedimentary rocks were originally laid in flat beds, even if they've been tilted and broken since. Steno also realized beds were laid one on top of the other, so the oldest beds are always at the bottom and the

Above: Earth's rocks have been traced back over 3,800 million years. Rocks of this age are found exposed in western Greenland, like Sondre Strom fjord on the south-west coast.

youngest at the top. This is the principle of superposition.

Using this principle in conjunction with fossils, 'biostratigraphers' have built a detailed history of the Earth

Geologic time

This illustration gives a simplified picture of the major periods of Earth's history since the Cambrian Period 545 million years ago (mya) – the time when complex life forms first flourished.

4560–545 mya Precambrian *The oceans formed and the first single-celled life forms appeared including algae which gave oxygen to the air. Later on, multi-celled animals like sponges and jellyfish appeared.*

Blue-green algae (early life)

Orthonybyoceras (marine animals)

Trilobite (segmented creature)

545–495 mya Cambrian Period *An explosion of life in the sea including small invertebrates and the first animals with hard shells easily preserved as fossils.*

Cooksonia (land plant)

495–443 mya Ordovician Period *Melting of the polar ice caps flooded much of the land. Crustaceans (like crabs), early marine animals and coral reefs appear.*

Ichthyostega (amphibian)

Giant tree fern (first forest)

354–290 mya Carboniferous Period *The sea level was high and large areas were covered in tree-filled swamps. Limestones laid down. Amphibians and insects spread, and possibly the first reptiles.*

443–417 mya Silurian Period *The Caledonian Orogeny threw up mountains in what became N America and NW Europe. Fish with jaws and river fish appeared, and the first land plants.*

417–354 mya Devonian Period *Continents and mountains grew. Old Red Sandstones laid down. Forests of club mosses and tree ferns dominant. The first sharks. Animals began to live on land.*

stretching back half a billion years. If rock sediments remained undisturbed forever, it would be possible in theory to slice through them to reveal most of the sequence of Earth's history. If you could take a column through the sequence, you could read Earth's history like a book.

The geologic time scale

Although such a column exists nowhere on Earth, a detailed geologic time scale based on it is now widely used. This timescale is constantly updated as geologists make new discoveries. It is detailed only back to the start of the Cambrian Period, 545 million years ago. Only since this time have shelly and bony life forms been common enough to leave a good fossil record. We once knew very little about the four billion years of Earth history before that, known as Precambrian time. In recent years, discoveries have begun to fill in the picture.

Just as day is divided into hours, minutes and seconds, so geological time is divided into units. The longest are Eons, lasting at least half a billion years. Eons are divided into Eras, Eras into Periods, Periods into Epochs, Epochs into Ages and Ages into Chrons. Each unit in the timescale is given a name,

usually derived from the area where rocks of the period were first studied. The Devonian Period, for instance, was named after Devon in England where rocks of the age were first studied.

Biostratigraphy now provides a detailed system for matching rock sequences around the world. Yet it has its limitations. It can show one rock is older than another, but not exactly how old. In other words, it gives a relative, not absolute, date. By working out how fast sediments might have been deposited and other clues, geologists established rough dates for the geologic timescale. But it is only with the development of radioactive dating in the last 50 years we can be confident dates are fairly accurate.

Radiometric dating

Atoms of elements can occur in alternative varieties or isotopes, each with a different number of particles in its nucleus. The number of particles is given in the isotope's name, such as uranium-235. Radioactive or radiometric dating uses the way certain isotopes 'decay' naturally through time – that is, break down to become isotopes of different elements. This

decay begins from the moment a rock is formed. It happens at such a steady rate that it is possible to work out how long it has been going on by counting the relevant 'daughter' isotopes in the rock compared to those of the original 'parent' isotope. The steady rate is known as the half-life, the time it takes for half the parent isotopes to break down. The widely used rubidium-87 isotope (which breaks down to strontium-87) has a half-life of 47.5 billion years. Rubidium is a rare element, but is often found with potassium in minerals such as feldspars and micas, and rubidium-strontium dating is used to date granites and gneisses.

Different isotopes are used to date rocks of different ages. Potassium-40, which decays to argon-40, is used to date rocks under a million years. Uranium-235 is used for the oldest rocks. Traces of uranium-235 decay products are found in zircon, one of the few minerals that survives for billions of years unchanged. So geologists studying ancient rocks look for zircon. Zircon found in the Jack Hills of Western Australia dates back 4300 million years, almost to the Earth's birth.

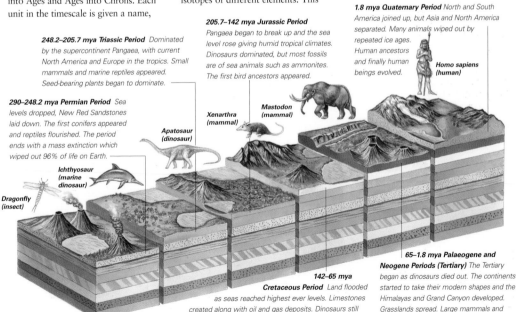

248.2–205.7 mya Triassic Period Dominated by the supercontinent Pangaea, with current North America and Europe in the tropics. Small mammals and marine reptiles appeared. Seed-bearing plants began to dominate.

290–248.2 mya Permian Period Sea levels dropped, New Red Sandstones laid down. The first conifers appeared and reptiles flourished. The period ends with a mass extinction which wiped out 96% of life on Earth.

205.7–142 mya Jurassic Period Pangaea began to break up and the sea level rose giving humid tropical climates. Dinosaurs dominated, but most fossils are of sea animals such as ammonites. The first bird ancestors appeared.

1.8 mya Quaternary Period North and South America joined up, but Asia and North America separated. Many animals wiped out by repeated ice ages. Human ancestors and finally human beings evolved.

Homo sapiens (human)

Xenarthra (mammal)

Mastodon (mammal)

Apatosaur (dinosaur)

Ichthyosaur (marine dinosaur)

Dragonfly (insect)

142–65 mya **Cretaceous Period** Land flooded as seas reached highest ever levels. Limestones created along with oil and gas deposits. Dinosaurs still dominated, but the first predatory mammals appeared.

65–1.8 mya Palaeogene and Neogene Periods (Tertiary) The Tertiary began as dinosaurs died out. The continents started to take their modern shapes and the Himalayas and Grand Canyon developed. Grasslands spread. Large mammals and primates appeared. Birds flourished.

MESOZOIC   *CENOZOIC*

GEOLOGICAL MAPS

A good geological map is probably the geologist's single most valuable tool. It displays what rocks appear where in the landscape and the major geological features. This not only helps the geologist identify rocks on the ground but provides a good indication of where to find particular minerals.

For the experienced geologist, a geological map does not simply show where rocks appear. With skilful interpretation, it is possible to work out the three-dimensional structure of rock formations, the way they relate to each other, and even some of their history and the way that the landscape has developed.

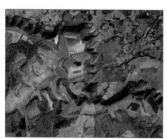

Most conventional geological maps are known as 'solid' maps, because they show the solid rocks under the surface. They help geologists to find the rock structures most likely to yield mineral ores and oil and gas deposits.

Geological maps can also show loose surface deposits, such as the sediment deposited by rivers in flood

Mapping the landscape
The same area can be mapped by geologists in four different ways. A satellite image (left) shows the landscape very clearly, allowing the geologist to interpret ground features relating to the underlying geology before surveying in the field. A step on from the satellite image is the digital terrain model (below), which combines satellite and ground survey data to build a 3D computer model. This gives an even clearer picture of the landscape and clues to the underlying geology.

or by glaciers. Maps like these are called 'drift' maps and are valuable for the construction industry since they reveal just how solid the ground on which they plan to build is likely to be. They are invaluable in the initial plotting of the course of a railway tunnel, for instance, before detailed survey work on the ground begins.

Colours and symbols
Geological maps usually use areas of different colours to represent the different kinds of rock or drift. Igneous rocks, for instance, are usually shown in various shades of purple and magenta, depending on whether they are intrusive or extrusive. Metamorphic rocks are typically shown in shades of pink and grey-green. Sedimentary rocks are usually shown, appropriately, in sandy colours – shades of brown and yellow, plus green – except for limestones, which are typically blue-grey.

Besides colours, maps often use types of shading or pattern called ornament. Each rock type is also given a set of letters to symbolize it on the map. This usually begins with a capital letter to show the age of the rock – for example J for Jurassic, K for Cretaceous, T for Tertiary and Q for Quaternary. The smaller letters indicate either the name of the particular formation or rock type.

Geology underground
Geological maps only actually show surface geology. Even solid maps only show outcrops – that is, where a solid rock of a particular kind reaches the surface, even if it is actually hidden beneath sheets of deposits. Outcrops that are not covered with deposits are called exposures. All the same, many maps have cross-sections showing a vertical slice through the rock formations, revealing how they are

arranged beneath the surface. Often these incorporate results from boreholes and seismic surveys. Yet an experienced geologist can construct a cross-section directly from the map.

Where there are a series of roughly parallel bands of sedimentary rocks, for instance, it is highly likely that they are simply where gently dipping (tilted) sedimentary strata reach the surface. Contours on the map reveal the shape of the land surface. If these reveal an escarpment with a gentle dip slope (see Rock Landscapes in this section), the angle of the dip (tilt) is easy to guess.

Geology in 3D

Cross-sections only show the structure of rocks in the landscape in two dimensions – as a thin slice – but

A third map (below) shows the rock underlying the same terrain, with different rocks indicated by coloured bands. A professional geologist can use this map to construct a cross-section (right) of the landscape from, say, A to B, plotting the height given by contours, then interpreting how the rock beds lie from their surface configuration.

geologists often want to know the whole structure of the area in three dimensions. In the past, they often constructed models called fence diagrams. These were made by constructing a series of cross-sections at right angles, interlocking like the fences of a square field. Sometimes these were created with card, and sometimes they were simply drawn.

Computer modelling

With the spread of computers, however, geologists are able to construct complete 3D models of the ground. These can be manipulated and

projected to reveal the subsurface structure in any way the geologist wants. Many geologists hunting for minerals now use these computer models instead of conventional maps wherever they exist.

Until recently, it was solid geology, not drift geology that received the attention from computer modellers. Now, however, organizations such as the British Geological Survey (BGS) use GSID (Geological Surveying and Investigation in 3D) software to create 3D models that show not only outcrops in their entirety but the surface deposits as well.

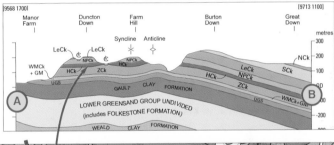

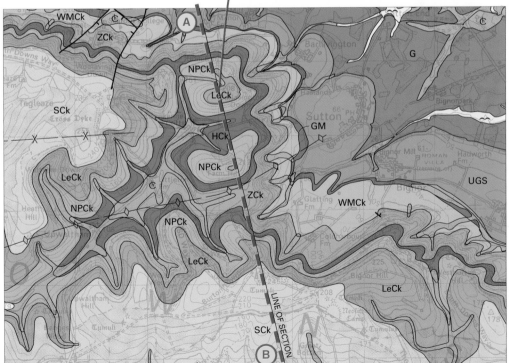

READING THE LANDSCAPE

Every bit of rock has a story to tell. Look at a craggy cliff face or pick up a stone on the shore, and if you know what to look for, you can read in it a great deal of its past, from scratches revealing episodes of glaciation to rock boundaries marking cataclysmic earthquakes.

In some ways, the geologist is like a detective. The idea is to look for clues, study the evidence and work out just what has happened. Sometimes the clues are so small they are visible only under a powerful microscope. Sometimes they are so big they are visible only from space. Fortunately most can be identified with no more specialist equipment than basic common sense.

If you find a stone, see what you can work out about its history. At first it might seem baffling and dull, but with a little thought you can often piece together some of its long story. Obviously it helps to identify the stone, and the Directory of Rocks later in the book should help you here. Yet even without a strong positive identification, you can often work out quite a bit about it.

First of all, you can probably guess whether the stone arrived naturally where you found it – or whether there was a human hand involved. If you see other similar stones nearby in a natural setting, the chances are it got there naturally. Exotic stones that look very different can arrive in their resting

The Great Unconformity

One of the world's most famous unconformities is North America's Great Unconformity. This dramatic gap in the geological sequence stretches all the way from Arizona to Alberta in Canada. Perhaps the best place to see this is in the Grand Canyon National Park (the Canyon's west rim is shown above), where the Colorado River has cut down through the strata over five million years to reveal the Unconformity at the canyon's foot. Here young Tapeats sandstone sits so directly on top of ancient, two-billion-year-old Vishnu schist that it is possible to touch them both with the span of a hand. The schists began life some two billion years ago as sediments. These sediments were metamorphosed to schist

about 1.7 billion years ago as they were penetrated by a magma intrusion and squeezed by tectonic movement. Slowly, over more than a billion years, this schist was worn flat, then sank beneath the sea. About 500 million years ago, the sediments that formed the Tapeats sandstone began to pile up on the sea floor. The sea floor began to subside under the weight of sediments, and sedimentation continued over hundreds of millions of years with little or no tilting, Eventually some 1,220 m/4,000 ft of level sedimentary beds were laid down. Since then the whole area has been uplifted, and the path of the Colorado river has exposed the whole sequence to reveal the Tapeats sandstone and the Great Unconformity at its base.

Below: Cliffs and chasms, such as this feature of the Blue Mountains in NSW, Australia, are one of the best places to see rock strata.

place naturally. Glaciers, for instance, carry entire giant boulders, called erratics, into areas of different rock. Flash floods and avalanches can carry quite large stones. Yet exotics like these are rarities. Most loose stones in a natural setting are not only similar to each other; they are often similar to the rocks nearby. If they resemble the solid rock around, you can be sure the stones were eroded from it.

Often, you can see where the stones came from immediately. Reasonably large stones are rarely carried far from their origin. Often, stones simply fall down a mountain slope to gather in a scree at the foot. Stones and pebbles can fall from a sea cliff on to the

beach below. There is often a marked difference between stones in a scree and stones on a beach. Scree stones are almost invariably sharp-edged and chunky, reflecting the way they were shattered from the mountain above by frost, often within the past few months. Beach stones, however, are often rounded pebbles. They are rounded because they have been rolled over and over in water laden with abrasive sand again and again. To get so rounded, they must have been in the water many thousands of years. Waves are powerful enough to carry stones some distance, so the pebbles on a beach have not necessarily come from the cliffs above.

How an angular unconformity forms

Sediments are usually laid down in a continual sequence but sometimes a break or unconformity develops. Shown below are the stages in the creation of an angular unconformity. It begins with the laying down of a sequence of sediments (1) which are then uplifted and folded into mountains (2). The mountains are worn down to a plain over a long period (3). The sea level rises, submerging the plain (4). New layers are laid down horizontally in the sea, over the contorted layers of the original sediments (5).

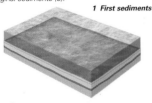

1 First sediments

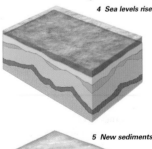

2 Mountains form

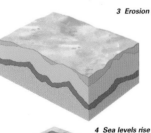

3 Erosion

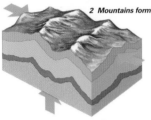

4 Sea levels rise

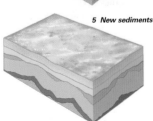

5 New sediments

Using rock layers

On a larger scale, you can tell a great deal about the history of rocks, especially sedimentary rocks, by stratigraphy – the study of rock layers and their contents. Earlier pages have shown how if rocks contain the same assemblage of key fossils they must be the same age. The law of superposition also shows how the uppermost layers are likely to have formed later in Earth's history. Another valuable rule of thumb is that the more contorted strata is, the older it is likely to be. Similarly, the law of cross-cutting relationships shows that whenever a geological feature like an igneous intrusion cuts across a sequence of rock layers, it is certain to be more recent than the rock layers.

Breaks in the sequence

One feature geologists look for in particular is an unconformity. On the whole, sediments are laid down one on top of the other in sequence, but every now and then you find a gap or even a dramatic break in the sequence. A young rock may overlay an old rock, with no intervening 'middle-aged' rocks, or a sedimentary sequence may overlay metamorphic rock. This gap or break is the unconformity, and it can be very revealing.

One kind of unconformity, easily seen in the field, is an angular unconformity. In this case, younger sediments sit on top of older sediments that have been tilted and deformed and then eroded to form a flat plain.

There are other kinds too. If the older sequence remains both level and undeformed, then the only sign of the unconformity may be the great age difference between rock layers, indicated by a significant gap in the fossil record. This is known as a parallel unconformity.

A disconformity is formed where the eroded surface of the older sequence is not a flat plain but has hills and valleys. In this instance the break in the sequence is marked by a corresponding wavy line. A non-conformity occurs where the sedimentary rock sequence has been interrupted by either igneous or metamorphic rock.

Stratification

When a geologist sees a sequence of rocks exposed in a cliff, he can examine the layers and build up a picture of their history, using familiar clues. The nature of the rock gives a clue to the environment in which they formed. Fossils in the rock layers may give a clue to their relative age. So too can their position within the sequence, while unconformities bear witness to dramatic breaks in the sequence, and past geological events.

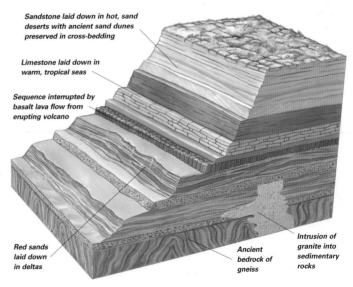

Sandstone laid down in hot, sand deserts with ancient sand dunes preserved in cross-bedding

Limestone laid down in warm, tropical seas

Sequence interrupted by basalt lava flow from erupting volcano

Red sands laid down in deltas

Ancient bedrock of gneiss

Intrusion of granite into sedimentary rocks

HOW MINERALS FORM

Minerals are the natural substances the world's rocks are made of. All of them are solid crystals with a particular chemical composition – some so tiny they can only be seen under a powerful microscope, others as big as tree trunks. Each kind forms under particular conditions in particular places.

There are 4,000 to 5,000 different minerals in the Earth's crust. Yet only 30 or so are very widespread. Most of the rest are present in rocks only in minute traces, and are only easy to see when they become concentrated in certain places by geological processes. It is concentrations like these that give us the ores from which many metals are extracted.

Furthermore, large crystals of minerals like those that illustrate this book, even common minerals, are so rare that a mineral hunter feels understandably excited to find one. Big, spectacular crystals need both time and space to grow – and a steady supply of exactly the right ingredients. Such a perfect combination is extraordinarily rare.

Mineral crystals form in four main ways. Some form as hot, molten magma cools and crystallizes. Some form from chemicals dissolved in watery liquids. Some form as existing minerals are altered chemically, and some form as existing minerals are squeezed or heated as rocks are subjected to metamorphism.

Below: The best crystals need space to grow, so they are often found in cavities, such as geodes. Geodes can look like dull round stones on the outside, but a tap with a hammer gives a tell-tale hollow sound. When cracked open, they reveal a glittering interior.

Minerals from magma

As magmas cool, groups of atoms begin to come together in the chaotic mix and form crystals. The crystals grow as more atoms attach themselves to the initial structure – just as icicles grow as more water freezes on to them. Minerals with the highest melting points form first, and as they crystallize out the composition of the remaining melt changes (see How Igneous Rocks Form, this section).

Chemicals that slot easily into crystal structures are removed from the melt first, and it is bigger, more unusual atoms that are left behind. It is these 'late-stage' magmas, the last portion of the melt to crystallize, that give the most varied and interesting minerals.

Above: This woman is inside the world's biggest geode – most are no bigger than a fist. Geodes are cavities that probably form from gas bubbles in lava (or limestone). The bubbles later fill with hydrothermal fluids in which large crystals such as amethyst grow.

Just what these minerals are depends on the original ingredients in the magma, and the way it cools. Large crystals tend to form in magmas that have cooled slowly. The biggest and most interesting often form in what are called pegmatites, which form from the fraction of melt left over after the rest has crystallized. Pegmatites typically collect in cracks in an intrusion or ooze into joints in the country rock, forming sheets of rock called dykes. The residual fluids in these late-stage magmas are rich in exotic elements such as fluorine, boron, lithium, beryllium, niobium and tantalum. These can combine to form giant crystals of tourmaline, topaz, beryl, and other rarer minerals. When the fluids are rich in boron and lithium, tourmaline is formed. When fluids are rich in fluorine, topaz is formed. When fluids are beryllium-rich, beryl forms.

Minerals from water

Water can only hold so much dissolved chemicals. When the water becomes 'saturated' (fully loaded), the chemicals precipitate out – they come out of the water as solids. Typically this happens when water evaporates, or cools down.

When sodium, chlorine, borax and calcium are dissolved from rocks, they may be carried by rivers to inland seas and lakes, which then evaporate, leaving mineral deposits of minerals such as salt, gypsum and borax.

Many other minerals form from the cooling of hydrothermal solutions – hot water rich in dissolved chemicals. Sometimes the water is rainwater that seeps down through the ground (meteoric water) and is then heated by proximity to either the mantle or a hot igneous intrusion.

Hydrothermal solutions also come from late-stage magmas, and so are rich in unusual chemicals. Such solutions ooze up through cracks in the intrusion and cool to form thin, branching veins.

Alteration minerals

Although some minerals such as diamond or gold seem to last forever, most have a limited life span. As soon as they are formed, they begin to react with their environment – some very slowly, some quite quickly. As they react, they form different minerals.

Metal minerals are often oxidized when exposed to the air, or oxygen-rich water. Iron minerals rust, like iron nails, turning to red and brown iron oxide. When water containing dissolved oxygen seeps down through the ground into rocks and veins containing metals, it creates an oxidation zone in the upper layers as the metals are altered. Cuprite, goethite, anglesite, chalcanthite, azurite and many other minerals form this way. Some sulphide minerals are oxidized to sulphates that dissolve in water. These sulphates may be washed down through the rock to be deposited lower down as different minerals which become often valuable ores such as chalcocite.

Minerals remade

Many minerals become unstable when exposed to heat and pressure and respond by altering their chemistry to become different minerals. This is known as recrystallization and is typically linked to metamorphism. When a rock is remade by metamorphism, its mineral ingredients are recrystallized.

While it takes hot magma or tectonic movements to metamorphose rocks, simple burial is often enough to alter minerals, since heat and pressure rise as depth increases. Minerals can also be recrystallized by contact with hydrothermal fluids.

In the simplest recrystallizations, the resulting minerals depend merely on the combination of heat and pressure, and the minerals in the original rock. However, new ingredients may seep in to alter the picture. Where magma intrudes into a limestone, for instance, the magma 'cooks' the limestone but also introduces new chemicals to create a complex mix. The limestone supplies calcium, magnesium and carbon dioxide, while the magma brings silicon, aluminium, iron, sodium, potassium and various other ingredients. The result is a 'skarn' rich in a huge variety of interesting silicate minerals.

Above: Topaz is one of many rare minerals that form when the last rare-mineral-rich residue of a magma finally crystallizes, particularly in dykes called granitic pegmatites. Topaz forms from residues rich in fluorine. Crystals formed in voids in the pegmatite, called miarolitic cavities, can occasionally grow very large.

Above: Fluorite is one of many minerals that form as hot hydrothermal solutions cool and precipitate some of their dissolved chemicals. Natural pipes carrying hydrothermal veins eventually completely crystallize to form veins.

Above: Cuprite is one of many minerals that form by oxidation reactions, caused by exposure to air, or oxygen-rich water. Cuprite forms a bright green crust on oxidized copper minerals.

Left: Rubies are among a number of rare and precious gem minerals that form when certain oxides are crystallized by the heat and pressure of metamorphism.

MINERAL CRYSTALS

Although mineral crystals found in the ground are often chunky, they rarely have the beautiful, regular geometric shapes you see in drawings. All the same, crystals grow in certain ways, and have particular symmetrical forms.

Mineral crystals all grow from countless tiny building blocks called unit cells. Every unit cell of a mineral is an identical arrangement of atoms, and it is the way these identical cells combine that gives the crystal its shape. If you could break up cubic salt crystals, for instance, you'd find they break into cubic grains, and the grains into tiny cubic unit cells.

The unit cells in crystals stack together to form an atomic 'lattice', a regular internal framework. This is what gives a crystal its essentially geometric shape or 'symmetry'.

All crystals are symmetrical in some way or other and crystallographers group crystals according to the manner of their symmetry. There are six basic crystal systems, each with a characteristic shape, further divided into classes. The simple, most symmetrical system is the isometric or cubic system. The other five

in order of decreasing symmetry are hexagonal (including trigonal), tetragonal, orthorhombic, monoclinic and triclinic. Crystals in each system appear in many different 'forms' (see below and right).

The symmetry of a crystal can be described in various ways. One way is in terms of axial symmetry. An 'axis of symmetry' is an imaginary line drawn through the crystal's centre from the

Crystal systems (with some possible forms)

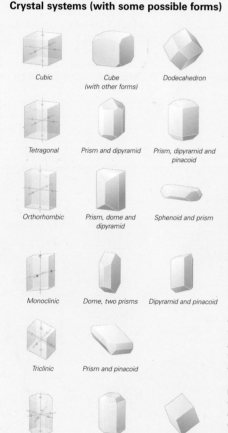

Cubic

Cube (with other forms)

Dodecahedron

Tetragonal

Prism and dipyramid

Prism, dipyramid and pinacoid

Orthorhombic

Prism, dome and dipyramid

Sphenoid and prism

Monoclinic

Dome, two prisms

Dipyramid and pinacoid

Triclinic

Prism and pinacoid

Hexagonal

Prism, pinacoid and dipyramid

Rhombohedron

Cubic or isometric
This is the most symmetrical possible, with three equal axes at right-angles; four-fold symmetry all round. The main forms are cubes, octahedron and rhombic dodecahedrons. E.g. Galena, halite, silver, gold, fluorite, pyrite (shown), garnet, spinel, magnetite, copper.

Tetragonal
Least common system, with three axes at right angles, two equal and one longer or shorter. One main four-fold axis of symmetry. E.g. vesuvianite (shown), chalcopyrite, zircon, cassiterite, rutile, wulfenite, scheelite.

Orthorhombic
Typically stubby matchbox or prisms maybe with pinacoid (see Crystal habit and form, opposite). Three unequal axes at right angles. Three axes of two-fold symmetry. E.g. barite (shown), olivine, topaz, sulphur, marcasite, aragonite, celestine, cerussite.

Monoclinic
Typically tabular. Three unequal axes, only two crossing at right angles. One two-fold axis of symmetry. E.g. selenite gypsum (shown), mica, orthoclase, manganite, hornblende, borax, azurite, orpiment, augite, diopside.

Triclinic
This is the least symmetrical system, with three unequal axes, none crossing at right angles. E.g. microcline feldspar (shown), plagioclases such as albite and anorthite, turquoise, kaolinite, serpentine, amblygonite.

Hexagonal (and trigonal)
Crystals in this system have three equal axes crossing at 60°, plus an unequal axis crossing at 90°. Hexagonal crystals have up to six-fold symmetry; trigonal crystals (sometimes considered a separate system) have up to three-fold symmetry. E.g. beryl (shown), quartz.

middle of opposite faces. Turn a crystal around one of its axes of symmetry and it always looks the same shape. Cubes are the most symmetrical of all shapes; you can draw an axis of symmetry through a cube in 13 ways.

Another way to think of symmetry is in planes or mirror images. Slice a crystal along a certain plane and each half forms a mirror image of the other. A perfect cube has nine planes of symmetry, which means that there are nine ways to slice it in half to create matching pairs. As you turn a crystal, matching images appear twice, three, four or six times, depending on the system. A slab-shaped crystal, for instance, may have to be turned 180 degrees to give a match. So the crystal is said to have two-fold symmetry. A cube needs only be turned 90 degrees for a match, so has four-fold symmetry. A hexagonal shape has to turn 60 degrees, so has six-fold symmetry.

One crucial feature of all crystals is the angles between the faces. These are always the same for each mineral, a feature called the Law of Constancy of Angle. It is so reliable that you can use the angle between faces to identify a mineral. Mineralogists measure it with an instrument called a goniometer, but you can make a home-made one with a protractor and a transparent ruler.

Crystal habit

Of course, crystals never achieve their perfect form in nature. Growing in cracks in rocks, hemmed in on all sides, their natural shape is distorted. Even crystals grown in the laboratory are distorted by gravity. Only in the zero gravity conditions of the International Space Station have scientists grown near-perfect crystals.

Though never perfect, crystals of each mineral tend to grow, or grow together (as an aggregate), in distinctive ways, or habits. Hematite, for instance, tends to grow in kidney-shaped masses, a habit called 'reniform' after its Latin name.

On the whole, each kind of mineral tends to form in particular conditions, and its habit reflects the conditions in which it forms. Some minerals, such as quartz, form in a variety of conditions so have several different habits.

Crystal form

WHOLE CRYSTAL (CLOSED FORM)

Isometric forms have various numbers of matching faces; e.g. tetrahedron (4 faces); cube (6 faces); octahedron (8 faces), dodecahedron (12 faces).

Non-isometric forms have non-matching faces; e.g. rhombohedron, dipyramid and scalenohedron. A rhombohedron is like a box squashed on one side e.g. rhodochrosite. A dipyramid is like two pyramids stuck base-to-base, with a variable number of faces. A scalenohedron has triangular faces, with unequal sides, e.g. calcite.

Crystal habit

Acicular Needlelike clusters of crystals, e.g. ulexite (left).

Arborescent Clusters of crystals in treelike branches, like native copper and silver. Similar to dendritic but chunkier.

Bladed Thin, flat crystals with a curved edge, like a butter knife blade. Barite crystals are often bladed.

Botryoidal Comes from the Latin for grapes, and means small, rounded masses, like a bunch of grapes, e.g. azurite, hemimorphite (above). Like reniform and mamilliform habits, but smaller.

Capillary Thin rods like hairs, like acicular but finer.

Columnar Clusters of columns or rods, e.g. aragonite (right). Usually parallel, but can be radiating.

Crusty Like a bread crust, e.g. limonite.

Cryptocrystalline Crystals too small to see without a microscope, e.g. chalcedony.

Dendritic Clusters of crystals in fernlike branches, like arborescent but finer, e.g. pyrolusite (left), psilomelane, copper and gold.

Drusy Thin layer of upright crystals coating a rocky surface like the pile of a carpet.

Fibrous Thin, fibrelike crystals, e.g. asbestos, sillimanite and strontianite. The crystals may point in all directions, but can be parallel bundles or radiating.

PART OF CRYSTAL (OPEN FORM)

Dome and sphenoid With two angled faces, like a ridge-tent.

Pedion Single flat face

Pinacoid A pair of parallel faces, like the top and bottom ends of a column.

Prism Multiple faces running parallel to the axis, like a faceted column e.g. beryl (left) and orthoclase feldspar.

Pyramid Like a pyramid, with 3 to 16 faces.

Foliated In thin leaves, e.g. mica (right).

Globular In ball shapes, e.g. prehnite.

Hopper Faces indented like the hopper trucks in mines, e.g. halite.

Lamellar Thin plates, e.g. chlorite.

Mamilliform Clusters of large rounded blobs, e.g. malachite. Bigger than botryoidal but smaller than reniform.

Massive Solid, without any obvious structure.

Nodular Growing in round lumps or nodules, e.g. chert and chalcedony (left).

Oolitic Like tiny fish-eggs, e.g. chamosite. Smaller than pisolitic.

Radial Spreading out like the spokes of a wheel or a fan, e.g. marcasite, stibnite and stilbite.

Reniform Shaped like kidneys, such as hematite (right). Bigger than botryoidal or mamilliform.

Reticulated Criss-crossing net or lattice of crystals, e.g. cerussite.

Rutilated Needlelike inclusions (tiny internal crystals) of rutile, e.g. rutilated quartz.

Stellated Cluster of long narrow crystals radiating in a star-shape, e.g. pectolite (right) and snowflake crystals.

Tabular Crystals form flat slabs, similar in shape to lamellar, but much thicker. E.g. barite, linarite.

MINERAL PROPERTIES: PHYSICAL

Every mineral has a particular range of characteristics that distinguish it from other minerals and help mineralogists to identify it. Many of these properties are physical, including its hardness, its density, and the way it bends or breaks.

Like all substances, minerals have a wide range of physical properties, from their melting points to their ability to conduct heat or electricity and reflect or absorb light. Mineralogists, however, concentrate mostly on those properties that help them distinguish various minerals in a practical way, particularly those that help them identify minerals in the field.

Perhaps the two most useful

properties for a mineralogist are hardness and density. In the Identifying Minerals chart, which precedes the Directory of Minerals, minerals are grouped according to their hardness and density to assist with quick identification.

Hardness can be established very easily in the field and gives an instant clue to a mineral's identity. Density can only be estimated in the field, but it is

very easy to measure at home (in terms of specific gravity), and narrows down the possibilities even further.

The ways a mineral bends and breaks are also useful distinguishing characteristics – particularly cleavage, fracture and tenacity, explained in the panel on the right.

One other property, useful for identifying a few minerals, is reaction to acids, given in the acid test.

Hardness and the Mohs Scale

When mineralogists talk of hardness, they usually mean scratch hardness – the mineral's resistance to being scratched. The simplest way to measure this is on the Mohs scale, devised by German mineralogist Friedrich Mohs in 1812. He selected ten standard minerals and arranged them on a scale of 1 to 10, with each one slightly harder than the preceding one. The softest on the scale (1) is talc; the hardest is diamond (10), the world's hardest mineral. All other minerals can be rated on this scale by working which they will scratch, and which not. Each will scratch a softer mineral but be scratched by a harder one.

What you need:
You can buy test kits incorporating samples of the ten standard minerals, but you can easily make up your own Mohs test kit using everyday items:

- Fingernail
- Brass coins
- Iron nail
- Glass
- Penknife
- Steel rasp
- Sandpaper
- Knife sharpener

To find the hardness of a material, scratch with each of the testers in turn, starting with the softest, a fingernail.

If it can't be scratched with a fingernail, try a brass coin. If the coin scratches it, it must be 2 or below on the scale.

If it can't be scratched with a coin, work through the harder testers in turn to find out exactly where it lies on the scale.

1 Talc	2 Gypsum	3 Calcite	4 Fluorite	5 Apatite	6 Orthoclase	7 Quartz	8 Topaz	9 Corundum	10 Diamond
Fingernail	Bronze coin	Iron nail	Glass	Penknife	Steel rasp	Emery sandpaper	Knife sharpener		

Bending and breaking: Cleavage

Cleavage is a mineral's tendency to break along lines of weakness. Because of the way crystals are built up from lattices of atoms, these lines of weakness tend to be flat planes. The majority of minerals have these characteristic planes of weakness, called cleavage planes. Most minerals tend to cleave in a certain number of directions. Mica, for instance, tends to break in just one direction, forming flakes. Fluorite, however, tends to break in four directions, forming octahedral (eight-sided) chunks.

Not all minerals break equally easily and cleanly. Mineralogists describe the cleavage as: good, distinct, poor and absent; very perfect, perfect, imperfect and none; or perfect, good, poor and indistinct. Fluorite has very good cleavage; quartz has none. Some minerals, like scheelite, cleave well in one direction and poorly in two others.

Cleavage in one plane only, like flat sheets: example, muscovite mica.

Cleavage in two planes, like a squarish rod: example, orthoclase feldspar.

Cleavage in three planes, like a block: example, halite.

In former times, miners used to describe minerals that split into rhombic shapes like calcite as 'spars'.

Cleavage in three planes at an angle to each other, like a three-dimensional diamond shape: example, fluorite.

Bending and breaking: Fracture

When struck with a hammer, minerals sometimes break roughly and in no particular direction, rather than along flat cleavage planes. This is called fracture, and can sometimes help identify the mineral. The fragments or fractured surface may be conchoidal (shell-like), hackly (jagged), splintery, fibrous or earthy.

Quartz varieties such as amethyst do not cleave but fracture into shell-like conchoidal fragments. So too do olivine, flint and glass.

Bending and breaking: Tenacity

When minerals are crushed, cut, bent or hit, they react in different ways. The way a mineral reacts is known as its 'tenacity' and is described by several terms. Most minerals are 'brittle', which means they crumble, break or powder when struck hard enough. Some are 'flexible' like molybdenite and talc (above) which means they bend. Some are 'elastic' like mica (above right), which means they bend and spring back.

'Malleable' minerals like native copper, gold and acanthite (below) can be hammered into sheets. 'Sectile' minerals like chlorargyrite can be sliced. 'Ductile' minerals can be drawn out or stretched into a wire.

Simple tests: Specific gravity

Although values vary slightly because of impurities, minerals can often be very precisely identified by their density. Yet it is hard to measure the density of an irregular chunk, so geologists use the mineral's specific gravity (SG), its density relative to water. You can measure this using a sensitive spring balance.

1 (left): To measure a sample's specific gravity, tie a thread around it and hang it from a spring balance to weigh it. Note the weight and call it A.

2 (right): Fully immerse the sample in a mug of water, and note the weight shown on the spring balance. Call this B. This should be less than A, as the water provides buoyancy. Subtract B from A, then divide A by the result. This gives the SG.

Simple tests: Acid test

One way of identifying minerals like calcite is with the 'fizz test'. If a little acid is dropped on a chip of carbonate mineral, for instance, it fizzes as it reacts with the acid. The fizzing is the carbon dioxide released as the carbonate dissolves. Sulphides such as galena and greenockite give the bad-egg smell of hydrogen sulphide gas. Powdered copper minerals such as covellite and bornite turn the acid greenish or bluish. Geologists use dilute hydrochloric acid, but strong household vinegar is safer. Vinegar (acetic acid) is weaker, so to get a reaction you need to either grind a sample of the mineral to powder, or warm the vinegar gently. If you wish to use hydrochloric acid, always wear rubber gloves and goggles to protect your eyes against splashes.

Use a dropper to drop a little acid on to the solid or powdered mineral.

MINERAL PROPERTIES: OPTICAL

Although it is not always possible to identify a mineral simply by the way it looks, every mineral has its own distinctive appearance – from the rich blue of azurite to the shimmering rainbow colours of opal – and these optical properties are often what gives a mineral its distinctive appeal.

The first thing most people notice about a mineral is its colour. A mineral gets its colour from the colours it reflects or transmits. These colours depend in turn on the bonds between atoms in the mineral.

Despite the rainbow array of colours you see in minerals, most are actually white or colourless in their pure state. Indeed, very few chemical elements are naturally strongly coloured. The colour in minerals tends to come almost entirely from what are called transition metals, such as cobalt, copper and manganese. Copper tends to give blue or green colours and iron red or yellow, but this is not always so.

The few minerals which are highly coloured in their pure state are called idiochromatic minerals. These minerals usually have one of the transition metals as a major constituent. Cobalt in erythrite turns it violet-red, for instance. Chromium makes crocoite orange. Copper makes azurite blue. Manganese makes rhodochrosite pink. Nickel makes annabergite green.

Most other minerals, however, tend to get their colour from traces of impurities. These minerals are called

Optical Effects

Adularescence
The bluish-white schiller effect on the surface of adularia moonstone.

Asterism
From the Greek word for star, asterism is a star-shaped sparkle in gems such as star sapphires, star rubies and star rose quartz, usually caused by minute needlelike rutile crystals included in the gem.

Aventurism
Glistening reflections of leaflike inclusions in minerals such as aventurine feldspar.

Birefringence
The way light is split as it passes through minerals like calcite, giving a double image as you look through it.

Chatoyancy
From the French for cat's eye, this is the cat's eye effect in stones such as andalusite, a shimmering band across the middle caused by the inclusion of minute crystals in the stone.

Dispersion
The degree to which minerals refract some colours more than others, dispersing the colours and creating the flashes of colour called fire in cut gems such as diamond, demantoid garnet and sphene.

Fire See Dispersion.

Iridescence
Shimmering rainbow colours produced by interference between waves of light in the surface layers of a mineral.

Labradorescence
Play of colours from blues to violets to greens and oranges in labradorite caused by twinning of sheets within the crystal.

Moonstone
Alkali feldspar gemstone with silvery or bluish iridescence, or opalescent variety of adularia.

Opalescence
Milky, lustrous shimmer on the surface of potch opal.

Phosphorescence
The way some minerals store light and glow some time after the activating ultraviolet light source has gone.

Play of colour
Shimmering rainbow colours that appear as a mineral is turned in the light. Effects include iridescence, schiller, labradorescence and adularescence. The play of colour on the surface of gem opal (left) is sometimes called opalescence. Strictly speaking, though, opalescence should refer only to the milky lustre in white, 'potch' opal.

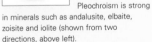

Pleochroism
The way some minerals show different colours, depending on the angle you look at them, because of how different colours of light are absorbed in different directions in the crystal. A three colour change is trichoic, a two colour change is dichoic. Pleochroism is strong in minerals such as andalusite, elbaite, zoisite and iolite (shown from two directions, above left).

Refraction
The way light passing through a transparent substance is bent.

Schiller
Bronze metallic lustre in orthopyroxenes such as enstatite caused by interference between light rays reflecting off internal mineral plates.

Sunstone
A kind of aventurine feldspar that gives a golden shimmer, caused by reflection off tiny platelike inclusions of hematite.

Thermoluminescence
The way some minerals glow when heated, e.g. apatite, calcite, fluorite, lepidolite and some feldspars.

Above: The variety of colours of quartz alone show just how unreliable a guide to identity colour can be.

allochromatic. Nearly all the famous coloured gems are allochromatic and get their colours from impurities. Beryl is turned into green emerald, for instance, by traces of chromium and vanadium. Corundum is turned to blue sapphire by titanium oxide and red ruby by chromium. Impurities can turn quartz almost any colour in the rainbow. Fluorite, too, can be many colours, but this is due to its special chemical structure, not impurities.

Many minerals are distinctively coloured. Some, like green malachite and blue azurite, are recognizable from their colour alone. Yet the fact that most get their colour from impurities means minerals are highly variable in colour, and colour alone is rarely a reliable guide to identity.

Above: The streak given by rubbing a mineral on unglazed porcelain gives a consistent colour identity. Clockwise from top: cuprite, pyrite, hematite, azurite and malachite.

The only way to be sure of mineral's colour is through its streak. This is a powdered version of the mineral. It is called a streak because it involves rubbing the mineral across the unglazed side of a white porcelain tile so that it leaves a distinctive streak. Hematite is often grey like galena, and often various other colours. Its streak though is always blood-red, while galena's is a lead-grey. So the streak test is very clear. Unfortunately, only

about a fifth of minerals have distinctive streak. Most translucent minerals have a white streak; most opaque minerals have a black streak.

Whatever their colour, minerals differ in the way they transmit light, a quality called diaphaneity. Some minerals are clear almost as glass in their pure state and let light shine clean through them. Minerals like these are said to be transparent. Tiny impurities make them less clear. Some minerals, like moonstone, are semi-transparent so that things seen through them look blurred. When a mineral lets light glow through but only obscurely, it is said to be translucent, like chrysoprase. Minerals that block off light completely like azurite, pyrite and galena are said to be opaque.

Above, left to right: The mineral barite can be transparent, translucent or opaque.

When some minerals are exposed to ultraviolet light, they glow in totally different colours than their normal daylight colour. This glow is called fluorescence, from the mineral fluorite, which varies tremendously in colour in daylight, but always glows blue or green in UV light. Other fluorescent minerals include willemite, benitoite, scheelite, adamite, sodalite and scapolite.

The same minerals under ultraviolet light (above) and in daylight (right). Clockwise from top: fluorite, hackmanite, willemite and scapolite.

Lustre

Lustre is the way the surface of crystal looks – shiny, dull, metallic, pearly and so on. The terms used to described lustre are not scientific, but useful, subjective guides and are mostly self-explanatory.

Adamantine
Brilliant and shiny like diamond and other gem crystals, e.g. cassiterite (right).

Dull
Any non-reflective surface, e.g. glauconite.

Earthy
Like dried mud, e.g. kaolinite.

Greasy
Like grease. The surface would be adamantine but for slight irregularities on the mineral's surface, e.g. nepheline (left).

Metallic
Gleaming like metal. Native metals have a metallic lustre. So to do most sulphides, e.g. stibnite (right).

Pearly
Milky shimmer like pearl, e.g. talc (left).

Resinous
Like sap or glue. Most minerals with a resinous lustre are yellow or brown, e.g. sulphur (right).

Silky
Subtle shimmer like silk because of small parallel fibres as in asbestos, e.g. gypsum variety satin spar (left).

Vitreous
The most common lustre, rather like glass, e.g. quartz (right).

Waxy
Like wax, e.g. serpentine.

MINERAL GEMS

Most mineral crystals are dull and quite small, but a few are so richly coloured and sparkling that they almost take the breath away. When such beautiful stones are hard enough to cut and fashion into jewellery, they are called gemstones and are the most sought-after mineral specimens of all.

Gemstones are incredibly rare. There are over 4,000 different minerals, yet only 130 are considered gemstones, and of these less than 50 are frequently used as gems. The rarest and most valuable of all are diamond, emerald (green beryl), ruby (red corundum) and sapphire (blue corundum). Slightly less

Below: No gem comes in such a huge range of colours as tourmaline, which the Ancient Egyptians called rainbow rock. Traces of different chemicals can give it over a hundred different colours of which just a handful are shown here.

rare stones, known as semi-precious stones, include aquamarine (blue beryl), chrysoberyl, sunstone, moonstone, garnet, topaz, tourmaline, peridot (gem olivine), opal (chalcedony), pearl (aragonite), jadeite, turquoise and lapis lazuli (lazurite).

Unusual geological conditions are required to create gemstones such as these, which is why they are so rare. They might form in volcanic pipes – or simply be found in them like diamonds in kimberlite and lamproite pipes. Gems are also found in pegmatites.

Above: Like pearl, opal and jet, amber is one of several gemstones that is not crystalline. Indeed, these gems are not strictly minerals since they form by organic processes. Amber is hardened droplets of sap from ancient pine trees that lived millions of years ago. When ground and polished, it makes one of the most beautiful of gemstones.

In the last stages of magma intrusion, pegmatites concentrate rare minerals to form gems such as beryls, rubies, sapphires, tourmalines, topazes and many others. Intense metamorphism may create garnets, emeralds, jades and lapis lazuli.

What defines a gem?

Gemstones are prized for their beauty, their durability and their rarity. Dealers often assess them in terms of what are called the four Cs: clarity, colour, cut and carat.

Clarity is the quality most prized in gems. A perfect gem is a flawless transparent crystal that sparkles brilliantly as it reflects light internally. Diamond, the most precious of all stones, is at its best when clear and colourless. Yet the best diamonds hide within them a rainbow of colours that flash with what is called 'fire' as light rays are 'dispersed' or split into colours inside the crystal. This fire cannot be seen in rough diamonds, and is only revealed once the stone is properly cut and faceted.

Opal gains a beautiful rainbow shimmer as light is reflected and split into colours by the tiny spheres of

Above: Most of the world's diamonds come from mines like these, which plumb the pipes of volcanic rock, such as the kimberlites of South Africa and the lamproites of Australia.

Above: Diamonds look unremarkable when rough from the ground. It takes careful cutting with 57 or 58 triangular faces to bring out their hidden brilliance and sparkle.

silica from which it is made. In some stones, particular flaws can enhance their beauty, such as the inclusions that create asterism in star sapphire and chatoyancy in chrysoberyl (see Mineral properties: Optical, in this section).

Gem colours

A vivid colour is also prized. Jade, turquoise and lapis lazuli are completely opaque, but their rich greens and blues make them as sought after as many clear gems. With most clear gems, except for diamond, colour is highly valued. Colourless beryl is only moderately valued, but emerald (green beryl) is one of the world's most valued stones.

Many gems are given a wide range of colours by trace elements. Peridot (gem olivine) is commonly green, but can vary from pale lemon to dark olive. Often the different colour varieties of a gem mineral have their own name, as with sapphire and ruby which are simply different-coloured varieties of corundum.

Cut and carat

For the collector, colour and sparkle are quite enough to make a stone attractive. Yet for a crystal to be a gemstone, it must also be tough enough to use in jewellery. All the major gemstones are at least as hard as quartz – over 7 on the Mohs scale – and diamond is the world's hardest mineral. A gemstone must be tough

enough to be cut to bring out all its sparkle and colour (see Olivine and Garnets: Gemstone cuts, Directory of Minerals).

On the whole, it is the largest stones that are the most highly prized. In the ancient world, gems were weighed with the seeds of the carob tree, which are remarkably constant in weight. Later, the carob seed became the basis of a standard weight called the carat, which is about a fifth of a gram.

A 632-carat Emerald known as the Patricia Emerald was found in the 1920s in Colombia. An 875 carat topaz was found in Ouro Preto in Brazil. Often the largest stones, though, contain flaws, so they are cut down to make smaller, flawless gems.

The world's largest cut diamond, called the Golden Jubilee, weighs 545.67 carats – a little over a hundred grams or about four ounces. It just beats the Great Star of Africa set in the British Royal Sceptre, which was one of several large diamonds cut from the gigantic 3,106 carat Cullinan Diamond found by Thomas Cullinan in 1905 at the Premier Mine, South Africa.

Famous gems

All the biggest and most famous gems have their own names and their own stories. St Edward's Sapphire in the British Imperial State Crown was reputedly worn by Edward the Confessor almost 1,000 years ago. Another stone in this crown is the

Black Prince's Ruby given to Edward the Black Prince in 1366 by Pedro the Cruel of Spain. It is actually thought to be a red spinel, not a ruby. The 109-carat Koh-i-Noor diamond was found in India seven centuries ago.

The Hope Diamond

Perhaps the most infamous of gems is the Hope Diamond, now in the Smithsonian Museum in Washington, DC. Found in India, this magnificent diamond weighed 112 carats when given to King Louis XIV in 1668. A few years later it was cut to a 67 carat heart-shape. Stolen during the French revolution in 1792, it resurfaced as a 45.5 carat gem in 1812, then mysteriously vanished again only to reappear in 1820 when it was bought by British King George IV. When George IV died it was bought by banker Henry Hope, after whom it was named. Over the next century it gained a reputation for bringing bad luck. One owner, an actress, was shot on stage while wearing it and another, a Russian prince, was stabbed to death by revolutionaries.

Below: Diamonds are tough enough to survive weathering to end up in river gravels, where they can occasionally be found by laboriously sorting the river gravels by panning.

MINERAL ORES

Our modern society could not function without metals extracted from rocks, yet even aluminium and iron, the most abundant metals, make up just a few per cent of rocks by weight. Fortunately, these and many other useful metals are concentrated by geologic processes in a few places as 'ore' minerals.

The first metals people used were native metals – copper, silver and gold – which occur as chunks of metal in the ground and could be fashioned into jewellery and knives. Then some unsung genius realized that there are minerals that while not actually metallic contain metals that can be extracted by heating until the metal melts out. This smelting process has provided us with metals ever since.

Minerals which contain enough metal for it to be easily extracted are called ore minerals. Galena contains a very high percentage of lead, so is a major lead ore. Iron comes mainly from hematite and magnetite. Pyrite is also rich in iron, but because it is bound so tightly to sulphur in the mineral it cannot be extracted, so pyrite is rarely an ore of iron.

To form an ore, the metal must be sufficiently concentrated in the mineral, and the mineral sufficiently concentrated in the ground for it to be worth extracting. Fortunately, geological processes ensure that such concentrations, known as ore deposits, do occur in certain places.

Below: Valuable minerals such as gold may sometimes be concentrated in iron-bearing ore deposits called gossans. They get their name from the Cornish for 'blood' because of their rusty red colour from the oxidized iron.

Above: Most ores are extracted from near-surface deposits in vast opencast mines like this copper mine in Bisbee, Arizona. Huge quantities of rock have to be dug out, much of which is later discarded.

Hot deposits

Many important ore deposits are linked to magma chambers, where melted rock collects. As the magma cools and begins to solidify, heavier minerals begin to sink to the bottom of the chamber. So when the magma finally freezes solid, heavy ore minerals may be concentrated at the base, especially sulphides ores, creating what are called magmatic sulphide ore deposits. Deposits like these tend to form only in relatively runny magmas such as komatiite basalt and gabbro. Famous sulphide deposits of this kind include South Africa's Bushveld Complex, Noril'sk in Russia, Jinchuan in China and Sudbury in Canada. Although their origins are not entirely understood, massive sulphide deposits can also occur in rock beds metamorphosed in the granulite facies (see Schists: Metamorphic facies in the Directory of Rocks). Australia's Broken Hill has superb lead, silver and zinc ores like this.

Rich ore deposits are also created by hydrothermal (hot-water) solutions circulating through magma or the rocks surrounding an intrusion. These fluids are typically rich in dissolved metals, which they deposit in fractures and pores, creating hydrothermal deposits.

If a hydrothermal deposit is dispersed throughout the intrusion, it forms a 'disseminated' deposit. If it follows cracks in the rock, filling them in with deposited minerals, it forms a hydrothermal vein deposit. Vein minerals include sulphide, oxide and silicate ores, as well as native metals such as gold, silver and platinum. Gold often appears as flakes in white quartz veins. Copper ores are often concentrated in porphyritic intrusions, forming porphyry copper deposits.

Some of the most valuable deposits are created when hydrothermal solutions percolate into limestone. When solutions seep into limestones and marbles around an intrusion, they create skarns, such as the tungsten skarns of Sangdong, Korea, King Island, Tasmania and Pine Creek in California.

The same process creates Mississippi Valley Type (MVT) deposits, which occur around the edge of sedimentary basins, deep down at the base of limestone beds. MVTs formed as hot solutions seeping through underlying rocks reacted with the limestone. North America's Tristate zinc district is the best known MVT, but there are also well known MVTs in Cumbria, England and Trepca, Serbia.

The hot waters flushed out of submarine volcanoes along the mid-ocean ridge are often rich in dissolved metals and sulphur. When the hot water meets the cold ocean, these dissolved chemicals often combine to form deposits of metal sulphides, which may be recovered in places where tectonic movements have lifted up ancient sea floor to much more accessible places.

Cool deposits

It is not just hot water that concentrates minerals. Even cool groundwater can dissolve metal ores as it seeps through rocks. The metal ores may have been widely scattered through the rock, but when the water re-deposits them, it may deposit them together in a concentrated deposit – a

Right: Lode gold is often found associated with quartz in veins like this one in siltstone in the Tanami desert of Australia's Northern Territory. This region is known for its gold and has recently become the target of seismic profiling, designed to reveal its geological structures and reveal likely sites for gold from the pattern of powerful vibrations sent through the ground by machines.

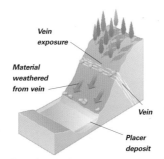

Placer deposits As veins are exposed at the surface, valuable minerals such as gold may be tumbled or washed down slope to collect in stream beds, forming placer deposits. Prospectors may sometimes recover these placer golds by panning the stream gravel.

process called secondary enrichment. This often happens when water seeping down through the ground reaches the water table, the level to which the ground is saturated.

In the opposite way, rainwater sinking into the ground may leach (dissolve) away particular chemicals, leaving behind concentrations of metals such as iron and aluminium. In the tropics where rainfall is heavy, the concentrations in these 'residual deposits' can become so intense that the soil itself becomes an ore, like bauxite aluminium ore.

Water can even help concentrate ore deposits without dissolving anything. After weathering has broken rocks up, rivers and streams carry away mineral grains. Heavier, more durable grains are dropped first by the water, and can accumulate in riverbed deposits. Gold, tin, diamonds and emeralds are among the minerals found in 'placer' deposits.

Some of the world's most important cold water ores formed in seawater. These include deep-sea manganese ores (see Aluminium and Manganese, Directory of Minerals) and the remarkable Banded Iron Formations (BIFs – see Iron Ores, Directory of Minerals) formed by bacteria almost two billion years ago, and now one of the world's major sources of iron.

Finding ores

In the old days, prospectors would scour the landscape hoping to stumble on 'shows' – exposures of ores on the surface, with little to guide them other than a knowledge of what geologic features they are often found in association with. Nowadays, geologists still read the landscape but have an array of sophisticated technology to help them, including aerial and satellite photographs. Kimberlite pipes that might yield diamonds often show as pale discs on the surface, for instance.

Once a potential site has been identified, the extent and richness of a deposit can be assessed by testing the ground's electrical conductivity and its magnetism. An instrument called a magnetometer may be used to locate deposits of magnetic ores like magnetite and ilmenite. Since ore minerals tend to be denser than average, measuring the local pull of gravity can also be revealing. Radioactive minerals and elements such as uranium and thorium might be detected with a Geiger counter. Since plants absorb traces of metals through their roots, it can even be worth analysing the plants in the area.

COLLECTING ROCKS AND MINERALS

You need very little special equipment to start building a rock and mineral collection – just sharp eyes for loose specimens when you're out walking. All the same, it helps to know where to look, and to acquire a few basic tools and storage systems for extracting good specimens and looking after them.

You can see rocks and minerals in many places. Office blocks often have polished granite faces. Houses might be built in sandstone and roofed with slate. Statues may be carved from marble. People wear precious gems made from mineral crystals. Many enthusiastic geologists get satisfaction from spotting such occurrences and identifying the rock.

Some enthusiasts build up a collection by looking for samples at rock shops and on the internet. But there is nothing to beat building up your own collection, by going out into the 'field', as geologists call it, to find your own samples.

Good, collectable specimens are not evenly spread around the Earth, but are concentrated in particular sites.

Some sites yield one or two kinds of mineral; others several hundred. The more you get to know local geology, the more likely you are to find good specimens. Good crystals are found in cavities and fissures, gemstones in pegmatites, gold in milky quartz veins and so on. Often minerals on the surface may be signposts to the real treasure beneath, like green copper.

What you need

Hammer
The key item in the geologist's toolbag is a good hammer. You can manage with an ordinary bricklayer's hammer, but it is really worth investing in a proper geological hammer. This typically has a square striking face and tapered tail (7). In the most common 'chisel' hammers, the tail is a flat edge at right angles to the handle, useful for levering samples out. In 'pick' hammers, the tail is a long, curved cutting edge, and is good for splitting rocks (6).

Chisel
There are three kinds of chisel. Cold chisels have a long, narrow handle and wedge, ideal for extracting crystals from cavities (8). Gad-point chisels are shorter and thicker, with a tapered point good for splitting, prising and wedging rocks apart. Broad-bladed chisels, bolsters and 'pitching tools' (10) are short and thick with a wide blade, good for splitting and trimming rock samples. If you only have one chisel, get a gad-point.

Safety equipment
Rocks can splinter when hit, so it is vital to wear good goggles when hammering. Goggles should have clear, hard plastic windows and flexible plastic sides (5). Avoid goggles without sides. Strong leather gloves are also useful protection and if you are visiting cliffs or quarries a safety helmet to protect the head is essential equipment (1).

Magnifying lens
A good magnifying glass or hand lens (3) not only helps you spot small crystals in the field but also helps you identify minerals and rocks you have found. A 3-times magnification lens is too weak to be of much help, but don't make the mistake of getting too powerful a lens. A 20-times gives too restricted a view to be useful in the field. The best compromise is a 5- or 10-times lens. Lenses are easy to lose, so tie your lens on a cord and attach it to your belt or collecting bag, or hang it round your neck.

Note-taking and recording
A notebook (2) and pen (9) are still the best way of noting down your finds. Sticky labels and markers (4) are good for labelling them on the spot. A small camera is useful for recording the site, and sometimes saves digging out a sample unnecessarily.

Way-finding
You need good local maps and perhaps geological maps to help you find your collecting site. A handheld Global Positioning System (GPS) is also useful for finding your way if you're straying far from the beaten track, and for recording precise locations and details of finds. A compass is a useful alternative – and enables you to test minerals for magnetism.

Collecting equipment
You need a strong bag for your samples. A small rucksack is ideal, leaving your hands free to pick up samples. Take bubble wrap and freezer bags to wrap samples.

Extra tools
In terms of additional portable items, a multi-purpose penknife and a small shovel are useful but not essential items. Shovels should have a shallow, pointed blade.

Hunting for rocks
Beaches and sea cliffs are good places to look for samples. The sea not only exposes the rock layers but pounds specimens free for you. But safety precautions are paramount. Always work at the cliff foot – never climb up – and wear a hard hat to protect your head from falling rocks. When hunting samples anywhere but a beach, always remember that most land belongs to someone and you may need permission to collect samples. Indeed, any samples found on someone else's land belong to the owner of the land. It's especially important to get permission if hunting in old quarries. Check also if the site is a nature reserve or protected in some way. Collection of samples may be forbidden. Even where you are allowed to collect samples, always be considerate. Don't ruin the site for others, clean out all samples or damage the rock.

Finding and cleaning specimens

There are two kinds of site to look for specimens: rock outcrops and deposits. Rock outcrops include cliffs, crags, quarries and cuttings. Deposits are where loose rock fragments have accumulated, including river beds, beaches, fields and even backyards. They include placer deposits where gold and gems such as diamonds have come to rest in stream sand and gravel (see Gold: Striking gold, in the Directory of Minerals).

Specimens taken from the ground are often filthy and should be cleaned before you stow them away. Make sure you have identified the mineral first, however, as some minerals, like halite, are soluble in plain water! For soluble specimens, scrub with a toothbrush, or dab with pure alcohol. If the sample is very soft, just use a blower brush like photographers use to clean lenses.

Below: The geologist's rule is to hammer as little as possible. Hammering promotes erosion and leaves a scar in the rock. You should never use a hammer for knocking out samples – only for breaking specimens up. Always wear goggles to protect your eyes from splinters.

With most minerals, fortunately, you can brush off loose dirt with a soft toothbrush, then rinse in warm (not hot) water. For greasy marks and stains, add a drop of household detergent to the water. If the specimen is encrusted with mud and grit, don't try to chip it off. Leave it to soak overnight to soften. It is fine to attack hard specimens like quartz with a nailbrush, but delicate specimens like calcite are all too easily damaged.

You can use vinegar to dissolve away unwanted calcite and limey deposits on most insoluble minerals. Iron stains can be removed with oxalic acid. You can get this from chemists, but it is poisonous so should be handled with care. It dissolves some minerals so test it on a fragment first.

Once your specimen is clean, you need to put it away in a cool, dry, dark place to keep it at its best. Don't store different minerals touching each other. Rocks and minerals 'breathe' and absorb and emit gases over time, and may alter accordingly. Keep your collection away from windows, room heaters, humid places such as bathrooms, and car exhaust, and try to keep conditions as stable as possible.

Some minerals, such as native copper and silver, oxidize and tarnish, especially in polluted city air. Some minerals, like borax, dry out, so store them in air-tight containers. Halite absorbs water from the air and gradually dissolves unless you keep it in an air-tight container with a little silica-gel to absorb any moisture.

Proper sample display drawers are expensive. Shallow drawers, wooden trays in cupboards, or glass and plastic boxes will do as alternatives. Identify every specimen with a number either on a sticky label or on a dab of white paint on an inconspicuous part of it. Enter this number, with the specimen profile, in your catalogue, which can be a computer file, cardfile or a notebook. You can group specimens by location or colour, but most geologists prefer to group by type: rocks into igneous, sedimentary and metamorphic types; minerals into chemical groups such as silicates and carbonates.

Cataloguing a collection

The minimum data for a catalogue is your own catalogue number, the mineral or rock name and for minerals a Dana number (see Classifying minerals in this section). But it is better to put as much data as you can. These are the details you should have:

1. Personal catalogue number
2. Dana number for minerals
3. Mineral or rock name
4. Chemical or mineral composition
5. Mineral or rock class
6. Exactly where you found it
7. Name of rock formation or kind of site where you found it
8. Date you found it
9. The collector's name (you)
10. Any other details, such as its history if you bought it, unusual characteristics of the specimen, and so on

DIRECTORY OF ROCKS AND MINERALS

Rocks and minerals are the raw materials of the landscape. Every valley, hill and mountain peak is made entirely of rocks and minerals. Furthermore every rock and mineral is also a clue to the Earth's history, for the characteristics of each depend on how and where it was formed – whether forged in the heat of the Earth's interior, transformed by volcanic activity and the crush of moving continents, or laid down in gently settling layers on the sea bed.

With this directory, geology enthusiasts will be able to identify rocks and minerals, and unravel the mysteries of our planet's history for themselves. In this section over 300 examples of rocks and minerals from all over the world are illustrated, each with a detailed colour photograph. Although it is not possible to include every single rock and mineral classified to date, offered here is a representative selection of the most important ones in each group, forming a stunning visual guide to their origins and identities.

The first part of the directory covers rocks, followed by the minerals section. Within each section, specimens are logically grouped according to their chemical composition and characteristics. Each entry summarizes the main features of the rock and its formation as well as topical information on how rocks, minerals and gems are used in industry and domestically. An authoritative data panel for each rock and mineral is included, forming a quick reference guide to all the standard characteristics. All entries also feature quick-reference identification aids, making this book an essential field guide for geology enthusiasts.

Left: The Giant's Causeway, County Antrim, Northern Ireland, UK, is formed of about 40,000 hexagonal interlocking basalt columns. It was created as a result of volcanic action that took place approximately 65–23 million years ago.

CLASSIFYING ROCKS

Most geologists agree on the basic grouping of rocks into igneous, sedimentary and metamorphic rocks, but when it comes to classifying rock species within those three broad groups, there is a great deal of controversy and there is no definitive system.

Classifying rocks is not simply a matter of identifying rocks and sorting them. Each classification system depends on a theory of how rocks are made. As ideas of how rocks are made change with new discoveries, so too do rock classification systems.

Over the last half century, for instance, there have been over 50 different classification schemes suggested for sandstone alone – and sandstone is by no means the most contentious and complex of rocks. The debate over igneous rocks has been, if anything, more heated. So the classifications presented here are essentially just a snapshot.

Igneous rocks

Rocks of this type are perhaps the most complex of all rocks to classify. Yet there is a surprisingly simple and easy-to-use basic classification that works well in the field. This is one based on colour and texture.

The texture or average grain size of an igneous rock depends largely on how long the melt from which it formed took to cool. So rocks that developed deep in the earth such as granite are coarse-textured or phaneritic, while rocks that formed on the surface such as basalt are fine-grained or aphanitic. Rocks which are basically fine-grained but contain large

crystals (phenocrysts) are termed porphyries. Using texture, it is, on one level, easy to group igneous rocks according to their origin, into coarse-grained 'plutonic' rocks formed at great depth, mixed and porphyritic 'hypabyssal' rocks formed at shallow depths, and fine-grained 'volcanic' rocks formed from lava on the surface.

Colour provides the other means of basic classification. For reasons that are not entirely clear, minerals at the top of Bowen's Reaction Series (see How Igneous Rocks Form in this section) like pyroxenes and amphiboles tend to be dark in colour while those at the bottom like quartz and plagioclase

Classifying igneous rocks

Colour and texture provide a good basic classification of igneous rocks, but geologists often need a more detailed system based on composition since colour and texture are not always enough to distinguish rocks. The Streckheisen system groups rocks according to the proportion of four minerals they contain: quartz, alkali feldspar, plagioclase feldspar and feldspathoids (foids). The percentages of each can be plotted on the diamond shaped diagrams. The corners represent 100% of the appropriate mineral.

Plutonic (intrusive) rocks
In the top half of the diamond for plutonic rocks are rocks rich in quartz – granitelike rocks such as granite, granodiorite and tonalite. Across the middle with a balance of quartz and foids are rocks like monzonite and anorthosite.

In the bottom half of the diamond for plutonic rocks are rocks rich in foids and poor in quartz – including ijolite, syenites and diorites. Diorites rich in plagioclase feldspar are gabbros.

Q (Quartz)
100% quartz
80% quartz
Alk Fp Granite
Granite
Granodiorite
Tonalite
20% quartz
Diorite (gabbro)
Anorthosite
Alk Fp syenite
QUARTZ
35% P Fp
Monzonite
Monzodiorite
65% P Fp
90% P Fp
Alk Fp
FELDSPAR
(Alk. feldspar)
F'OID
P (Plagio-
clase feldspar)
10% foid
Monzosyenite
Monzodiorite
Syenite
Diorite gabbro
80% foid
Ijolite
100% foid
F (Feldspathoid)

Volcanic (extrusive) rocks
In the top half of the diamond for volcanic rocks are rocks rich in quartz – rhyolites and dacites. Across the middle, with a balance of quartz and foids, are trachytes, latites and basalts. Quartz-rich basalt is andesite.

In the bottom half of the diamond for volcanic rocks are rocks rich in foids and poor in quartz – phonolites, tephrites and foidites.

Q (Quartz)
100% quartz
80% quartz
Alk Fp Rhyolite
Rhyolite
Dacite
20% quartz
Qz Alk Fp Trachyte
Alk Fp Trachyte
Quartz Trachyte
Quartz Latite
5% quartz
Andesite/ Basalt
P
10% foid
Trachyte
Latite
Basalt
A Fp
Fpd Trachyte
Fpd Latite
Fpd Basalt
Phonolite
Tephrite
80% foid
Foidite
100% foid
F (Feldspathoid)

Lutite (fine-grained): shale

Arenite (medium-grained): sandstone

Rudite (coarse-grained): breccia

QFL composition triangle for sandstone rocks

Classifying clastic sediments: texture
Above: The simplest way of classifying clastic sediments is by texture into three groups: lutites, arenites and rudites. Lutites are mainly silt and clay particles; arenites mainly sand; and rudites gravel, pebbles, cobbles and boulders.

Classifying clastic sediments: composition
Right: They can also be divided according to composition in terms of quartz, feldspar and lithics, QFL. Various percentages of QFL can be plotted on a triangular graph as for sandstones. The main division is between feldspar-rich arkoses and lithic-rich wackes.

feldspar tend to be light. The dark minerals separate out into what are called mafic magmas, because they are rich in MAgnesium and FerrIC (iron) compounds. The light minerals are concentrated in felsic magmas which are rich in FELdspar and SIliCa minerals. So mafic igneous rocks are dark in colour; felsic rocks are light. Because they are rich in silica, light-coloured rocks are also said to be silicic or acidic, while mafic rocks, which are low in silica, like basalt, are said to be basic.

Sedimentary rocks
This type of rock can form either from fragments of weathered rock or from minerals dissolved in water. Rocks made from rock fragments or 'clasts' are called clastic rocks. Rocks made from minerals dissolved in water are called chemical rocks, or biochemical rocks if they are made from chemicals derived from living things, such as the shells of shellfish.

Clastic rocks include sandstones and shales. They are made from fragments of rock that don't dissolve in water; those that do dissolve go on to form chemical rocks. Insoluble fragments are mostly silica-based minerals so clastic rocks are sometimes called siliclastic rocks.

Siliclastic rocks are typically classified according to the size of particle they are mostly made of. They mostly fall into three broad groups:

fine-grained lutites such as shale and clay; medium-grained arenites including many sandstones; and coarse-grained rudites such as breccia and conglomerate. But there are rocks that don't fit so neatly into these groups, such as wackes, sandstones made from a mix of grains. So some geologists prefer to classify according to the proportion of the three main grain types they contain – quartz sand, feldspar and lithics (rock fragments), sometimes known as QFL, plus the fine 'matrix' material. Using QFL, rocks such as sandstone are split into feldspar-rich arkose sandstones and lithic sandstones.

Chemical and biochemical rocks form into several groups. The biggest is the carbonates, including limestones and dolomite, made from calcium and magnesium carbonate. Others are chert, chalk, tufas and coal.

Metamorphic rocks
On a broad level, metamorphic rocks are divided into granular or non-foliated rocks and foliated rocks. With the exception of hornfels, granular rocks are made mostly from a single mineral, such as marble from calcite and quartzite from quartz. They are formed mostly by the heat of close contact with hot magma. Foliated rocks are more complex. They are characterized by their layered texture, the result of the intense compressional pressure brought about by regional scale

metamorphism. They are divided into low-grade metamorphic rocks such as slate; medium-grade rocks such as schist and high-grade rocks such as gneiss and granulite.

This works as a basic classification, but the composition of regionally metamorphosed rocks is especially complex and varied, so some geologists look at metamorphic rocks in terms of facies, conditions in which particular assemblages of minerals are formed (see Schists: Metamorphic facies in the Directory of Rocks). Or they might work in terms of the zones, such as the Barrovian and Buchan zones, in which particular rocks and minerals are formed (see Gneiss and Granulite: Mineral identifiers, also in the Directory of Rocks).

Below: Gneiss's dark and light zebra stripes of different minerals mark it out clearly as a foliated metamorphic rock. Non-foliated rocks show no such layered markings.

CLASSIFYING MINERALS

There are over 4,000 different kinds of mineral, and dozens more are being discovered and verified every year. Since each mineral has its own unique chemical identity, this huge array of minerals is usually divided into groups on the basis of chemical composition and internal structure.

Unlike rocks, minerals cannot be classified by any obvious visual characteristic. In fact, you usually have to identify a mineral first before you can slot it into any class. This is because although each mineral has its own chemical identity, it often appears in many guises, such as gypsum, or looks similar to another mineral that is totally different chemically. Indeed some minerals can be 'pseudomorphs', taking the same crystal shape as another mineral.

The Dana system

The only sure way to classify a mineral is by its chemical composition and its internal structure. J J Berzelius made the first chemical classification in 1824, but the system used today has its origins in the system devised by Yale University mineralogy professor James Dwight Dana in 1854. Chemically, all minerals are either elements, such as gold, or compounds (combinations of elements such as lead and sulphur in lead sulphide). So after making natural or 'native' elements one group and organic minerals another, Dana arranged the remaining minerals into seven groups of chemical compounds (see Main mineral groups, below).

Anion groups

In chemical compounds, elements typically occur not as atoms, but as ions. Ions are atoms that have either gained or lost electrons, which are tiny particles with a negative electrical charge. Certain elements, such as metals, tend to lose electrons to become positively charged ions or 'cations'. Other elements, such as oxygen, lose them to become negatively charged ions or 'anions'.

Just as opposite magnet poles attract each other, so do ions with an opposite electrical charge. So cations stick to anions. Most minerals are built up from unions of groups of cations with one or more anions. Chemically, halite, like table salt, is a compound of sodium and chloride. Sodium is the cation and chlorine is the anion.

All Dana's groups are anions or anion groups (unknown to Dana at the time). There is a good reason for this. Many minerals have a particular metal as a cation, such as the silver chloride (chlorargyrite). So you might think you could group minerals by the kind of metal they have as a cation. Yet silver chloride has very little in common with silver sulphide (acanthite) apart from its silver content. However, it has a lot in common with other minerals with the same anion, chlorine, such as sodium chloride (halite) and potassium chloride (sylvite).

All chlorides tend to form under similar geological conditions, for instance. So do many other anions. Sulphides tend to occur in vein or replacement deposits, for example, while silicates are major rock-formers. So it is logical to classify the major mineral groups according to their common anion.

Main mineral groups

The following groups are based on the old Dana classification that split minerals into nine groups. A newer Dana system divides minerals in 78 classes, all of which fall within the original nine groups. These new classes are indicated in brackets. In this system, every mineral is given a number which consists of the class number, followed by a subclass number and finally a species number. Thus pyrite is numbered 2.12.1.1 (2 for sulphides, 12 for its subclass and 1.1 for its species).

I Native elements (1)

Most minerals are compounds of chemical elements, but 20 or so occur naturally in pure or at least uncombined form. These native elements are divided into three groups: metals, semi-metals and non-metals. The main native metals are gold, silver and copper, plus platinum and iridosmine, and very rarely iron and nickel. The semi-metals include antimony, arsenic and bismuth. Non-metals are sulphur, and carbon as diamond and graphite.

II Sulphides and sulphosalts (2, 3)

Sulphides and sulphosalts are made from sulphur combined with a metal or metal-like substance and include some of the most important metal ores such as galena (lead ore), chalcopyrite (copper ore) and cinnabar (mercury ore). They are generally heavy and brittle. They form as primary minerals – that is, directly from melts or solutions rather than by alteration of other minerals. As soon as they are exposed to the weather, many are quickly altered to oxides.

III Oxides and hydroxides (4–8)

Oxides are a combination of metal with oxygen. They have the most varied physical character of any of the mineral groups, ranging from dull earths such as bauxite to rare gems such as rubies and sapphires. Hard primary oxides typically form deep in the Earth's crust. Softer earths tend to form near the surface as sulphides and silicates are broken down by exposure to the air.

IV Halides (9–12)

Halides are minerals that form when metals combine with halogen elements – chlorine, bromine, fluorine and iodine. They are very soft and dissolve easily in water. Yet they are so abundant that halide minerals such as halite (e.g. table salt) and fluorite are common world-wide.

Adding structure

Dana's system still provides the core of most mineral classifications, but it has been greatly revised and extended. It is now clear that chemical composition alone is not enough to determine a mineral's identity and classify it; its internal crystal structure is also crucial. A key breakthrough was the discovery in 1913 by British physicist Lawrence Bragg that X-rays could be used to look inside crystals and identify the arrangement of atoms. Bragg and Norwegian mineralogist Victor Goldschmidt went on to divide up the giant silicates group according to their internal structure (see Silicate shapes). X-ray crystallography remains a key way of confirming a mineral's identity.

Scientific testing

Although many minerals can be identified in the field or by some of the simple home tests described earlier, the only way to be absolutely sure of a mineral's identity is to subject it to a series of sophisticated laboratory tests and in particular X-ray crystallography. In this, X-rays are beamed through the specimen. Because every crystal has its own unique chemical structure, its identity is revealed by the way it diffracts (breaks up) the rays. Other tests that a lab can do include spectroscopy (testing which colours of light are absorbed) and the way the crystal polarizes and refracts (bends) light.

The Strunz system

The addition of crystallography to chemical composition was summed up by Hugo Strunz in 1941, and many mineral classifications today are based on Strunz, rather than Dana.

Strunz's classification system arranges minerals into ten groups: Elements; Sulphides and Sulphosalts; Halides; Oxides; Carbonates; Borates; Sulphates; Phosphates, Arsenates and Vanadates; Silicates; and Organic compounds. Within each group, minerals are split first into families according to their internal structure. Families are then subdivided into groups of minerals that are 'isostructural' – that is, have a similar internal structure.

Silicate shapes

The vast silicate group is divided into six groups by their internal atomic structure. All silicates are built from basic silicate units – a silicon atom ringed by four oxygen atoms in a tetrahedron (a three-sided pyramid; see below). The way these units join determines the class of silicate.

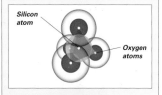

Silicon atom

Oxygen atoms

- **Nesosilicate (island)** – the simplest structure, made of just separate tetrahedrons, e.g. olivine and garnet.
- **Sorosilicate (group)** – the smallest class, made with pairs of tetrahedrons linked by an oxygen ion in an hour-glass-shape, e.g. epidote.
- **Inosilicate (chain)** – either single chains like pyroxene, or double chains linked by oxygen ions like amphiboles.
- **Cyclosilicate (ring)** – three, four or six tetrahedrons joined to form a ring, e.g. tourmaline.
- **Phyllosilicates (sheet)** – rings of tetrahedrons joined together in sheets, e.g. clay and mica.
- **Tectosilicate (framework)** – tetrahedrons interconnected like the framework of a building, e.g. feldspars, feldspathoids, quartz and zeolites.

V Carbonates, Nitrates, Borates (13–27)

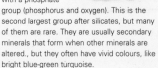

Carbonates are minerals that form when metals combine with a carbonate group (carbon and oxygen). The most abundant, calcite, is the major ingredient of limestone rocks. They are all soft, pale-coloured or even transparent. Most are secondary minerals, formed by the alteration of other minerals, though some form in carbonatite magmas, and in hydrothermal solutions or on the ocean bed.

VI Sulphates, Chromates, Molybdates (28–36)

Sulphates are minerals that form when metals combine with a sulphate group (sulphur and oxygen). They are all soft, translucent or transparent, and pale in colour. Sulphates such as gypsum, barite and anhydrite are very common.

VII Phosphates, Arsenates, Vanadates (37–49)

Phosphates are metals combined with a phosphate group (phosphorus and oxygen). This is the second largest group after silicates, but many of them are rare. They are usually secondary minerals that form when other minerals are altered., but they often have vivid colours, like bright blue-green turquoise.

VIII Silicates (51–78)

Silicates are metals combined with a silicate group (silicon and oxygen) and are the most common of all minerals. There are more silicates than all other minerals put together, both in mass and number. Almost a third of all minerals are silicates, and they make up 90 per cent of the Earth's crust.

Quartz and feldspar alone make up a huge proportion of most rocks. They are divided into subgroups according to their internal structure (see Silicate shapes, above).

IX Organic Minerals (50)

Organic minerals are naturally occurring solids formed either directly or indirectly by living organisms. They are not always accepted as minerals, simply because they are organic. These include amber, opal and jet.

ROCKS

The directory of rocks is sub-divided into three sections covering the major categories: igneous rocks, sedimentary rocks and metamorphic rocks. Within these sections, the rocks are then classified more specifically by composition: the igneous rocks include volcanic rocks, extrusive and intrusive rocks; the sedimentary rocks include clastic, biogenic and chemical rocks; and the metamorphic rocks include non-foliated, foliated and rocks altered by fluids and other means. There is a great deal of controversy surrounding the classification of rock species, so the categorization presented here refelects the widely accepted views at the time of writing. Included are ancient rocks such as greenstone, extremely rare rocks such as eclogite, and specimens such as pegmatites, long prized for their beauty. All entries include quick-reference identification checklists to aid classification, and practical information on how rocks are extracted and adapted for use by humankind.

Above from left: Monzonite, oolitic limestone, chiastolite slate.

Right: Strata in rock formation along Verzasca River valley in the Lepontine Alps, Switzerland.

HOW TO USE THE DIRECTORY OF ROCKS

The Directory of Rocks gives detailed profiles of over 100 species of rock from around the world. Using the identification tips in the following pages, you can begin to gauge the type of rock you have recovered from the field, then follow up these clues in the Directory to draw a more acccurate conclusion.

Rocks are made of countless grains packed together – that is, they are aggregates. Sometimes the grains are as least as big as sugar crystals and clearly visible to the naked eye. Others are too tiny to see except under a microscope. A few, like obsidian, have no grains at all.

A few rocks (such as the conglomerates and breccias) are made from large fragments of other rocks. More often, though, the grains in rocks are mineral crystals. Some rocks are made from just a single mineral. Marbles, for instance, can be almost pure calcite. Most, however, are made of at least two or three different minerals such as eclogite which is garnet and augite, and granite which is mica, quartz and feldspar. Most rocks

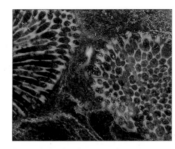

Above: A high fossil content often points to the sedimentary rock limestone, which forms from the remains of ancient marine creatures.

also contain a large number of 'accessory' minerals which are present in the rock but not in sufficient quantities to make a significant difference to its nature.

Rocks are typically divided into three major groups according to how they formed: igneous (from molten rock), sedimentary (from layers of sediment) and metamorphic (altered by extreme heat and pressure). This is how the specimens have been organized in the Directory.

Igneous rocks are further divided into extrusive (those that solidify from magma on the surface) and intrusive (those that solidify underground). Very loosely, the Directory section on Igneous rocks moves first through extrusive rocks from acidic, silica-rich or 'felsic' rocks like rhyolite, through intermediate rocks such as andesite to basic, silica-poor ('mafic' and 'ultramafic') rocks such as picrite. It then moves through intrusive rocks

IGNEOUS ROCKS

Rock name: may be plural if several varieties are profiled.

Identification: Notes the typical techniques used to identify specimens recovered from the field.

Data panel: Quick reference tool summarizing standard rock characteristics.

Norite

Norite is a similar rock to gabbro, based on a mix of plagioclase, pyroxene and olivine, and the two often form in the same large, layered intrusions as the mix separates during crystallization. Norite contains very slightly less plagioclase than gabbro, but the real difference is that gabbro's pyroxene is a clinopyroxene such as augite, while norite's is an orthopyroxene such as hypersthene. Unfortunately, the two can look so alike that they are impossible to distinguish without a microscope. Norite typically occurs in small, separate intrusions, or as layers along with other mafic igneous rocks such as gabbro.

Norite also formed in association with ancient basalt intrusions, beneath huge basalt dyke swarms. One famous norite intrusion is at Sudbury in Ontario. Here a cavity 30m/98ft deep has been excavated from solid norite to house the Neutrino Observatory to detect neutrinos, minute particles streaming from the stars. Norite is unusually low in natural radioactivity and acts as a shield to allow scientists to block out unwanted background radiation.

Identification: Norite is a dark grey rock with a slightly matted look dominated by quite long, prismatic black hypersthene or enstatite crystals. It looks very like gabbro, but the plagioclase feldspar tends to be sandy coloured, while in gabbro it is whiter.

Grain size: Phaneritic (coarse-grained), occasionally pegmatitic
Texture: Even-grained or porphyritic
Structure: Layering and xenoliths are common
Colour: Dark grey, bronze
Composition: Silica (58%), Alumina (17%), Calcium and sodium oxides (10.5%), Iron and magnesium oxides (11%), Potassium oxides (2%)
Minerals: Plagioclase feldspar (labradorite or bytownite); Pyroxene (hypersthene); Olivine; A little hornblende, biotite mica, quartz and alkali feldspar
Accessories: Magnetite, apatite, ilmenite, picotite
Phenocrysts: Plagioclase feldspar, hornblende
Formation: Intrusive: dykes, stocks, bosses, often with gabbro
Notable occurrences: Aberdeen, Banff, Scotland; Norway; Great Dyke, Zimbabwe; Bushveld complex, S Africa; Sudbury, Ontario

Grain size; Texture: Grain size is the first step to identifying igneous rocks. Are grains visible to the naked eye? Texture is variations in grain size and shape.

Structure: Large-scale structures such as layers.

Colour: Overall colour impression.

Composition: This is the overall chemical content. Silica-rich rocks are generally lighter in colour.

Minerals: The major minerals that define the character of the rock.

Accessories: Any mineral not essential to the rock's character.

Phenocrysts: Unusually large crystal.

Formation: Where the magma solidifies to form the rock.

Notable occurrences: Entries are country by country except for Canada and the USA which are listed last, state by state. Each site in a country is separated by a comma. Additional location information is given in brackets.

Profile: The main features of the rock, its formation and characteristics.

Specimen photograph: Some important features may be annotated.

that form large intrusions such as granite, to those that form small intrusions such as pegmatites.

Sedimentary rocks are divided into clastic (formed from rock fragments), biogenic (formed by living things) and chemical (formed from once-dissolved chemicals). Clastic rocks are ordered from fine-grained lutites such as shale

and clay through medium-grained arenites (sandstones) to coarse-grained rudites (conglomerates and breccias).

Metamorphic rocks are divided according to whether they display any pressure flaking, banding or striping into non-foliated and foliated, but there is considerable overlap, partly because some rocks, such as

amphibolite, can be either foliated or non-foliated. Foliated rocks are ordered, very loosely, from those that are formed under least pressure (low-grade rocks such as slate and phyllite) to the most (high-grade rocks such as schist and gneiss). Rocks formed from meteorites such as tektites and suevites make an additional category.

SEDIMENTARY ROCKS

Breccias

Sometimes called sharpstone, breccia is basically rubble turned into stone. The stones in breccia are jagged, caught up before there was time to round off any rough edges. Unlike conglomerates, breccias can form from almost any rock, soft and hard alike. But they almost always form near to their source and are said to be 'intraformational'. If the stones are washed any further away, they tend to get sorted and so do not form breccias. In mountain areas, breccias often form when screes are cemented together by finer sediment accumulating between the stones. Many breccias are formed rapidly by dramatic events – as when landslides and avalanches come to rest, or when flash floods or storm waves sweep masses of sediment into a beach or bar. Breccias also form when the roofs of limestone caves collapse, burying the floor in rubble. Coral reefs often contain extensive limestone breccias made of fragments broken off the reef. A few breccias are 'extraformational,' swept far from their source before consolidating and so have a very mixed composition.

Identification: Breccias are easily recognized by the large, angular stones they contain. It is not so easy to identify what the stones are or where they came from. The best place to start is a comparison with nearby rocks.

Grain size: Over 2mm/ 0.079in
Texture: Large angular stones in a finer matrix
Structure: Breccias are generally small, poorly stratified deposits
Colour: The colours are as varied as their source rocks
Composition: The stones are usually rock fragments – of almost any rock, including softer rocks such as marble
Formation: Some form in fast-flowing rivers, or on storm beaches. Others form from landslides and avalanches, both on land and under the sea.
Notable occurrences: Thessaly, Greece; Mexico; Vancouver Island, Midway, British Columbia; Platte Co, Wyoming; San Bernardino Co, California; Makinac Island, Michigan; Zopilote, Texas

Grain size and texture: Grain size is the first step to identifying sedimentary rocks. Are they clay-, sand- or pebble-sized? Texture is variations in grain size and shape.

Structure: Large-scale structures such as bedding planes.

Colour: Overall colour impression.

Composition: The balance of minerals making up the grains.

Formation: Where and how the sediments were laid down.

Notable occurrences: Entries are country by country except for Canada and the USA which are listed last, state by state. US states may be given in standard abbreviations where space is limited.

METAMORPHIC ROCKS

Inset detail: Illustrations of related species and special features.

Mica schists

The most common schists are mica schists. Flakes of mica give them the most marked schistosity, and a real shine. The flakes are typically about 0.5mm/0.02in thick and can often be prised off with a knife. There is plenty of quartz in mica schist too, often concentrated in mica-poor layers, and a fair amount of albite feldspar. Sometimes red garnet or green chlorite crystals are visible. The mica in mica schists can be muscovite, sericite or biotite. Biotite is usually brown; muscovite and sericite are pale coloured and called white mica. If a white mica is fine-grained, it is called sericite; if it is coarser it is called muscovite. Most mica schists contain all three, but one usually predominates. Muscovite and sericite schists develop where metamorphism is of moderate intensity, and are often associated with greenschists and phyllites. Biotite develops partly at the expense of muscovite (and chlorite) when metamorphism becomes more intense. Even more intense metamorphism edges the schist towards garnet schists.

Biotite schist: Biotite schist is brown and dark, but still has the mica gleam.

Muscovite schist: Muscovite schists, with sericite schists, are the lightest-coloured schists, coloured by white mica. Unlike sericite schist, the grains of mica in muscovite schist are clearly visible to the naked eye.

Mica (muscovite, sericite or biotite) schist
Rock type: Foliated, regional metamorphic
Texture: Medium to coarse-grained, sometimes with porphyroblasts. Thin schistose layering of mica flakes always marked.
Structure: Typically folded, on a small or large scale
Colour: Light grey, greenish (biotite schist is browner)
Composition: Quartz and mica (usually muscovite), plus kyanite, sillimanite, chlorite, graphite, garnet, staurolite
Protolith: Mainly mudrocks such as shale
Temperature: Moderate
Pressure: Moderate
Notable occurrences: Connemara, Ireland; Scotland; Scandinavia; Alps, Switzerland; Black Forest, Germany; Quebec; Duchess Co, NY; New Hampshire

Rock type: The degree of metamorphic banding and the mode of formation.

Texture: Overall grain size and variation.

Structure: Large-scale structures such as layers.

Colour: Overall colour impression.

Composition: The balance of minerals making up the grains.

Protolith: The original rock from which it was metamorphosed.

Temperature; Pressure: The conditions under which the rock was metamorphosed: low, medium or high-grade.

Notable occurrences: Entries are country by country except for Canada and the USA which are listed last, state by state.

IDENTIFYING ROCKS

Often dull greys, browns and blacks, rocks all look much the same at first glance. Yet they are as individual and distinctive as people, and once you know what to look for you will find it is quite easy to identify most of them and put them into their family groups.

Often the biggest clue to a rock's identity is where you found it. If you found it in an area of sandstone, it is likely to be sandstone. Before you begin to examine the specimen closely, see if you can spot formations of similar rock nearby that may yield useful clues. In cliffs and rock faces, the clear layering and beds of sedimentary rocks are usually unmistakable. Other rocks also create distinctive landscapes and landforms (see Rock Landscapes in Understanding Rocks and Minerals).

When you begin to examine your specimen in detail, the first task is to decide whether it is igneous, sedimentary or metamorphic.

Above: With a good eyeglass, you can often identify individual minerals within the rock, especially in medium- and coarse-grained igneous rocks such as granites.

Sedimentary rocks are pale in colour and tend to have similar grains, often held together by a cement. They may crumble as you rub them. Look for bedding planes and fossils.

Igneous rocks are identified by a harder, often shiny more compact look with a tightly packed interlocking mix of crystals. There should never be any layers or bands, except occasionally in layered granites and gabbros.

Some metamorphic rocks can look quite similar to igneous rocks, but foliation (layering and banding) is never found in igneous rocks. Granular metamorphic rocks tend to have a hard, shiny, sugary look and are more evenly dark or light, unlike igneous rocks which are quite often mottled.

The guides to identity given here are intended as a starting point, and should be used in conjunction with the clues given in the Directory.

Igneous rocks
COLOUR AND TEXTURE

	Light-coloured (silica-rich, acidic, felsic)	Medium-coloured (intermediate)	Medium-coloured (intermediate, feldspathic)	Dark-coloured (silica-poor, basic, mafic)
Fine-grained (aphanitic, volcanic, extrusive)	Rhyolite: White, grey, pink	Andesite: Salt and pepper (black and white)	Trachyte: Brownish-grey	Basalt: Dark grey to black
Medium-grained and porphyritic (hypabyssal, dyke, sill)	Quartz porphyry: White, grey, pink with light spots	Andesite porphyry: Dark grey, black with white spots	Monzonite: Dark grey with pale spots	Dolerite: Dark grey to black
Coarse-grained (phaneritic, plutonic, intrusive)	Granite: White, grey, pink; pinkish or whitish (tonalite')	Diorite: Salt and pepper (black and white)	Syenite: White, grey, pink; pinkish	Gabbro: Dark grey to black

OTHER TEXTURES AND COMPOSITIONS

Foam-like (vesicular) and glassy	Pumice: Whitish with fibrous look; very light	Scoria: Black to brown; very light	Vesicular basalt: Black to brown; heavy	Obsidian: Black, red, brown; glassy
Medium-to-coarse-grained carbonate and ultramafic rocks	Carbonatite: White with small grey spots	Dunite: Pale khaki to brown; ultramafic	Lamprophyre: Dark grey, with dark, shiny phenocrysts	Peridotite: Light to dark green; ultramafic

Metamorphic rocks

GRANULAR (NON-FOLIATED) ROCKS with no obvious layers, bands or stripes		FOLIATED ROCKS with obvious layers, bands or stripes	
Won't scratch glass	Will scratch glass	Grains often too small to see	Grains visible to the naked eye
Marble: Smooth feel; fizzes in dilute hydrochloric acid	Hornfels: Dark grey and black, dull, massive, conchoidal fracture	Slate: Dull grey, black, green, rings when struck; splits in thin sheets	Schist: Stripy bands or schistosity, platy cleavage
Dolomite marble: Smooth; powder fizzes in dilute hydrochloric acid	Metaquartzite: Pale translucent colours; fused quartz grains	Phyllite: Shiny grey, black, green; splits in thin sheets; may be striped	Biotite mica schist: dark; pale schist is muscovite mica
Greenstone: Greenish; harder than a fingernail unlike soapstone	Eclogite: Pale green pyroxene with red garnet	Mylonite: Streaky, smeared out texture	Gneiss: Tough, minerals separated into light and dark bands
Serpentinite: Greasy feel; green, yellow, brown or black	Amphibolite: Black, shiny crystals of amphiboles; also foliated	Glaucophane/Blue schist: Bluish colour, slender fibrous crystals	Granulite: No black mica, lenses of pale quartz and feldspar

Sedimentary rocks

No visible grains	Sand-size, visible grains	At least gravel-sized grains	Biochemical rocks that react with vinegar if powdered
Siltstone: Gritty feel; hard enough to scratch glass	Sandstone: Even, sand-sized grains; may be yellow, brown or red	Breccia: Large angular fragments set in mudlike mix	Chalk: White, powdery, leaves white mark, feels gritty
Claystone: Smooth feel; too soft to scratch glass	Ironstone: Even, sand-sized grains; very dark brown, red	Conglomerate: Large round pebbles set in mudlike mix	Oolitic limestone: Buff-coloured, tiny spheres like fish roe
Shale: Smooth feel; too soft to scratch glass	Greensand: Even, sand-sized grains; greenish colour	Boulder clay: Huge mix of stones set in muddy clay	Pisolitic limestone: Sand-coloured, tiny spheres like small peas
Marl: Earthy, slimy feel; too soft to scratch glass	Arkose: Even, sand-sized grains; won't crumble; pinkish, red		Fossil limestone: Pale grey, packed with fossils
Chert: Smooth, hard and glassy appearance	Greywacke: Mix of sand and fragments of rock		Dolomite: Dull grey; weathers pink or brown

VOLCANIC: SILICA-RICH ROCKS

Volcanic rocks are formed mainly when lava erupting from volcanoes cools and solidifies, but any material ejected from a volcano can form volcanic rock if it turns to stone, including ash, blobs of molten rock and froth. Because exposure to the air cools lava quickly, before crystals have time to grow, volcanic rocks are usually aphanatic (fine-grained) or even glassy. Volcanic rocks that are rich in silica (at least 55 per cent) are light in colour and include rhyolite, quartz porphyry and dacite.

Rhyolite

One of the most widespread volcanic rocks, rhyolite forms from the same silica-rich (70–78 per cent) magma that forms granite when it solidifies underground. This is the magma that melts its way up through the continental crust, so rhyolite is a continental rock. Rhyolite has been found on islands far from land, but such oceanic occurrences are rare. Rhyolite is the fine-grained, extrusive equivalent of granite, but there are subtle differences in chemistry. The mica in rhyolite is black biotite, not the brown muscovite seen in granite, and its potassium feldspar is sanidine while that of granite is orthoclase.

The high quartz content of rhyolite magmas means they are relatively cool and very viscous, and this sticky magma tends to clog up the volcanic vent. Sometimes, a plug is left behind long after the volcano has died, leaving a spire of rhyolite as it is gradually exposed by weathering. More often, the plug is blasted away in a mighty explosive eruption, which is why rhyolite is linked to some of the world's most explosive volcanoes, especially caldera complexes such as Tambora in Indonesia. Explosive eruptions are fuelled by the sudden expansion of steam and carbon dioxide in the magma. The explosion blasts away fragments of the plug in huge clouds of ash and deadly avalanches of pyroclasts. Gas bubbles turn parts of the lava to a froth that later solidifies as pumice. So rhyolite rock often forms from ash and pyroclasts rather than lava. Only once there is less gas in the magma can the rhyolite flow on to the surface as lava. Rhyolite lava piles up thickly in domes or coulées (tongues) around the vent – too sluggish to flow far. Rhyolite flows have broken, blocky surfaces, because the rind shatters as the inner mass creeps forward. Ancient rhyolites may have flowed farther, though, if, as some geologists argue, they were superheated and made less viscous by hotspots.

Because they erupt on the surface and cool rapidly, rhyolites are basically fine-grained. Indeed, where they have been quenched (cooled ultra-quickly), they are mostly glassy. Glassy rhyolites include obsidian, pitchstone and perlite. Yet because rhyolites are so viscous, they usually contain phenocrysts – large crystals that formed while the magma lingered in the volcano's magma chamber. Sometimes phenocrysts dominate so much that rhyolite can look like granite, with the microcrystalline groundmass visible only under a microscope. Rocks like these are called nevadites.

Banded rhyolite (below): When rhyolitic lava erupts on the surface, smaller crystals often respond to the flow by aligning themselves in bands, an effect called flow-banding. Banded rhyolite is sometimes called wonderstone, and valued by collectors – especially when it contains cavities filled with silica precipitates such as agate.

Spherulitic rhyolite (above): Sticky, rhyolitic lava often traps pockets of volatile vapours. In quickly cooled glassy rhyolites, some gas pockets develop into spherulites – balls of radiating needle-like crystals of quartz and feldspar – forming spherulitic rhyolite. Spherulites are typically a few millimetres across, but can be up to a metre.

Grain size: Aphanitic (fine-grained).
Texture: Phenocrysts common; alternating layers of grains common; flow-banding common (Banded rhyolite).
Structure: Vesicles and other remnant bubbles common; may contain spherules (Spherulitic rhyolite).
Colour: Usually light-coloured – pinkish or reddish brown, but also white, greenish, grey.
Composition: As for granite: Silica (74% average), Alumina (13.5%), Calcium and sodium oxides (less than 5%), Iron and magnesium oxides (less than 3.5%).
Minerals: Quartz; Potassium feldspar (sanidine) and plagioclase feldspar (oligoclase); Biotite mica.
Accessories: Aegerine, zircon, apatite, magnetite, amphibole or pyroxene.
Phenocrysts: Quartz, orthoclase and oligoclase feldspar, hornblende, biotite mica, augite.
Formation: Lava flows, dykes, volcanic plugs in continents.
Notable occurrences: Lake District, Shropshire, England; Snowdonia, Wales; Vosges, France; Black Forest, Saxony, Germany; Carpathian Mountains, Austria; Siebenbürgen, Romania; Tuscany, Italy; Iceland; Caucasus, Georgia; Rocky Mountains including Yellowstone; Arizona.

Rhyolite volcanoes include: Tambora, Indonesia; Mount Kilimanjaro, Kenya/Tanzania; Yellowstone, Wyoming; Crater Lake, Oregon.

Quartz porphyry

Quartz porphyry is a loose term for a rock with a similar chemical composition to granite and rhyolite that contains phenocrysts of white quartz (or, less often, orthoclase feldspar) that spot the rock like chunks of fat in a burger. The basic matrix of crystals around the phenocrysts is usually fine-grained, like rhyolite, because the melt containing the phenocrysts was fed into narrow dykes where it cooled and solidified quickly. So the average grain size is akin to granite. More recent quartz porphyries are dyke rocks, but ancient formations that formed in the Palaeozoic age of Earth's history – over 550 million years ago – were lava flows. So some geologists prefer to call ancient quartz porphyries palaeorhyolite. Many of these ancient quartz porphyries have been crushed and sheared by earth movements in their long history, giving them a striped look like schists. When the phenocrysts have been preserved, these rocks are called porphyry schists, and American geologists sometimes call them aporhyolite. They are well known from the Swiss Alps and from England's Charnwood Forest. The supposedly metamorphic halleflintas of Scandinavia may also have been formed like this.

Identification: Quartz porphyry is easily identified by the large white or grey blobs of quartz and feldspar in the reddish brown matrix of a rhyolitic mix of fine-grained crystals or even glass.

Rhyolite

Quartz phenocrysts

Grain size: Mixed.
Texture: Phenocrysts in fine-grained, microcrystalline or glassy matrix.
Structure: Vesicles rare
Colour: Usually light-coloured – red, brown, greenish.
Composition: As for granite: Silica (74% average), Alumina (13.5%), Calcium and sodium oxides (< 5%), Iron and magnesium oxides (< 4%).
Minerals: Quartz; Potassium feldspar (sanidine) and plagioclase feldspar (oligoclase); Biotite mica.
Accessories: Hornblende, augite, bronzite, garnet, cordierite, muscovite.
Phenocrysts: Quartz, orthoclase feldspar.
Formation: Dykes, or ancient lavas in continents.
Notable occurrences: Devon, Cornwall, Charnwood Forest (Leics), England; Westphalia, Germany; Alps, Switzerland; San Bernardino Co, CA; Lake and St. Louis Co, MN; Green Lake Co, Wisconsin; Pennsylvania.

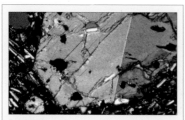

Porphyries

Porphyries are igneous rocks that contain large, conspicuous crystals in a groundmass of finer crystals. The phenocrysts, as the larger crystals are called, formed early on in the middle of the molten magma; the finer crystals formed later, typically after the magma containing them erupted or was injected into a dyke. In the micrograph of trachybasalt shown above, for example, large phenocrysts of olivine, clino-pyroxene (the large twinned crystal) and plagioclase feldspar are set in a fine-grained groundmass of the same crystals. The word porphyry now applies to any igneous rock with phenocrysts, but originally it referred to the beautiful red porphyries used by the Ancient Egyptians and the Romans in the time of the emperor Claudius. This rock, which the Romans called *porfido rosso antico*, was taken from a dyke 30m/98ft 5in thick on the Red Sea at Jebel Dhokan, and contained white or rose red pheno-crysts of plagioclase, dark black hornblende and plates of iron oxide, all in a dark red groundmass.

Dacite

Named after the Roman province of Dacia in modern Romania, dacite can be a beautiful rock when polished, but is commonly used for road-chippings. It forms from fairly viscous lava (55–65 per cent silica) in lava flows and dykes. It can also form massive intrusions in the heart of old volcanoes, creating lava domes such as Mount St Helens in Washington. It contains less quartz than rhyolite and so is intermediate in composition between rhyolite and the basic lava andesite. The quartz is often in the form of rounded phenocrysts in the groundmass, a little like quartz porphyry. Dacite also has andesine and labradorite as its feldspars rather than sanidine as rhyolite.

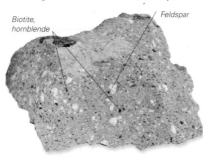

Biotite, hornblende

Feldspar

Grain size: Aphanitic (fine-grained) or even glassy.
Texture: Phenocrysts, alternating layers of grains and flow-banding common.
Structure: Vesicles and other remnant bubbles common.
Colour: Usually light-coloured – reddish or greenish.
Composition: Silica (65% av), Alumina (16%), Calcium & sodium oxides (8%), Iron & magnesium oxides (6.5%).
Minerals: Quartz; Potassium feldspar (andesine and labradorine) and plagioclase feldspar; Biotite; Hornblende.
Accessories: Pyroxene (augite and enstatite), hornblende, biotite, zircon, apatite, magnetite.
Phenocrysts: Quartz, feldspar, hornblende, biotite.
Formation: Lava flows, dykes.
Notable occurrences: Argyll, Scotland; Massif C, France; Saar-Nahe, Germany; Hungary; Siebenbürgen, Romania; Almeria, Spain; New Zealand; Martinique; Andes, Peru; Rocky Mountains; Nevada.

Identification: Dacite's rich hornblende and biotite are grey or yellowish, with white specks of feldspar like this. Augite and enstatite-rich dacites are darker.

ANDESITES

Midway between rhyolite and basalt in silica content, andesites are the most common volcanic rocks after basalt and are found all around the world near subduction zones. They get their name from the Andes in South America, and are associated with all the classic cone-shaped volcanoes, such as Mount Fuji in Japan and Mount Edgecumbe in New Zealand.

Andesite

Identification: Most andesite has a classic 'salt and pepper' look with white tablet-shaped grains of plagioclase feldspar visible to the naked eye set in a dark, often almost black, groundmass of fine-grained, occasionally glassy, minerals – mainly biotite mica, hornblende and pyroxene. Typically the dark minerals make up about 40 per cent of the rock by volume, much less than in basalt.

Andesites are found pretty much anywhere that an oceanic plate is subducted beneath a continent. Here they create a string of volcanoes along the continental margin, or an arc of volcanic islands along the edge of the continental shelf. Andesites are especially common in areas of recent mountain building. Not only the Andes, but the entire cordillera of mountains running from the Andes to the Rockies is predominantly andesite. In fact, wherever there are volcanoes in the 'Ring of Fire' around the Pacific, there is likely to be andesite.

Continental and island arc volcanoes spew out mainly andesite, dacite or rhyolite, depending on their silica content, while oceanic volcanoes emit basic lavas such as olivine, basalt and trachyte. Geologists call the line separating the two the andesite line. It runs roughly down the west coast of the Americas, and down from Japan to New Zealand via the Marianas, the Palaus, the Bismarcks, Fiji and Tonga. Although not as silica-rich as rhyolite, is still fairly viscous and often gets clogged up in the vent of the volcano. Consequently, andesite generates some of the world's most dramatic eruptions, as pressure builds up high enough to blast through the plug.

Andesitic volcanoes frequently produce devastating pyroclastic volcanoes, and most andesite volcanoes are stratovolcanoes – the classic cone-shaped peaks in which layers of lava alternate with layers of ash and pyroclasts, as lava pours out after the ash in each eruption.

Andesite lava flows better than rhyolite, but still sluggishly. When erupted, it first forms a mound around the vent. This creeps down the flanks of the volcano, advancing barely a few metres a day. The lava moves so slowly that the outside of the flow cools and solidifies. So, as it moves, the surface of the flow breaks up into a jumble of angular blocks that look like rubble. Even the hottest, most fluid andesites rarely flow farther than 10km/6 miles from the vent.

Porphyritic andesite: Many andesites are porphyritic – that is, they are spattered with large grains, clearly visible to the naked eye, that formed before the lava erupted. When these phenocrysts are especially large, the rock is called porphyritic andesite. The phenocrysts are typically plagioclase feldspar (white), but can be pyroxenes and amphiboles (usually greenish black).

Grain size: Aphanitic (fine-grained), occasionally glassy.
Texture: Often porphyritic.
Structure: Often displays flow structure; occasional vesicles.
Colour: Grey, purplish, brown, green, almost black.
Composition: Silica (59% average), Alumina (17%), Calcium and sodium oxides (10%), Iron and magnesium oxides (11%).
Minerals: Quartz; Feldspar: plagioclase feldspar (andesine) plus small amounts of potassium feldspar (oligoclase, sanidine); Biotite mica; Amphibole (hornblende); Pyroxene (augite).
Accessories: Magnetite, apatite, zircon, olivine.
Phenocrysts: Plagioclase feldspar, pyroxenes such as augite, amphibole, hornblende.
Formation: Extrusive lavas, ashes and tuffs in subduction zones and areas of mountain building.
Notable occurrences: Glencoe, Scotland; Lake District, England; Snowdonia, Wales; Vosges, Auvergnes, France; Rhineland, Germany; Siebenbürgen, Romania; Caucasus, Georgia; Andes; Rocky Mountains.
Andesite volcanoes include: Mount Fuji, Bandai-san, Japan; Krakatoa, Indonesia; Pinatubo, Philippines; Ngauruhoe, Mount Edgecumbe, Ruapehe, New Zealand; Citlaltépetl, Popocatépetl, Mexico; Mount Pelée, Martinique; Soufriere, St Vincent; Mount Shasta, California; Mount Hood, Oregon; Mount Adams, Washington.

Island arcs

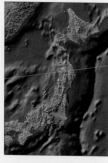

Some of the world's most dramatic and beautiful chains of islands occur in sweeping curves along the subduction zone at the outer edge of an oceanic plate. Scientists believe the curved edge of the plate may reflect the curvature of the Earth. There are many of these island arcs in the Pacific, including the Aleutians, the islands of Japan, the Marianas, Tonga and the Solomon Islands. The Antilles in the Caribbean also form an island arc. All these islands are volcanic and form when the subducted plate plunges down into the mantle, creating deep ocean trenches all along the margins of the plate. This satellite image of Japan (above) shows the Japan Trench, the dark area to the right of the islands, which forms part of the boundary between the Pacific and Eurasian plates. As the subducted plate sinks into the mantle, it melts and, in a complex process, forms magmas, including andesite, basalt and boninite. These hot magmas punch through the edge of the overriding plate like a needle stitching a hem, and as they penetrate, they erupt on the surface as volcanoes.

Boninite

This rarity takes its name from the Izu-Bonin-Mariana chain of islands south of Japan in the Pacific. Most of the boninites formed here between about 30 and 50 million years ago, but boninites are still forming today. They are associated almost exclusively with island arcs, and seem to require particular conditions for their formation. They are formed when an ocean tectonic plate is subducted beneath another oceanic plate. The subducted plate carries sea water down into the mantle with it as it plunges into the Earth, and the water alters the chemistry of the magma formed as the plate melts in the heat of the mantle. Geologists suggest that boninites form only if high temperatures are reached fairly quickly as the plate is subducted – otherwise, andesites will form. They may be linked to the early stage of subduction.

Grain size: Glassy.
Texture: Often porphyritic.
Structure: Often displays flow structure; occasional vesicles.
Colour: Dark grey, often black.
Composition: Silica (59% average), Alumina (17%), Calcium and sodium oxides (10%), Iron and magnesium oxides (11%).
Minerals: Quartz; Feldspar: plagioclase feldspar; Pyroxene (augite, bronzite); Amphibole (hornblende).
Accessories: Pentlandite, spinel.
Phenocrysts (small): Pyroxenes: augite, bronzite.
Formation: Extrusive lavas, dykes and sills in island arcs.
Notable occurrences: Crimea, Ukraine; Izu-Bonin-Marianas island chain, Pacific; North Tonga Ridge, New Hebrides; Setouchi, Japan. Possible continental margin occurrences: Isua, Greenland; Yukon, Canada; Glenelg, South Australia; Antilles, Caribbean.

Identification: Boninite is dark in colour, with small black phenocrysts set against a dark glassy groundmass.

Pyroxene andesite

There are actually several different kinds of andesite: the quartz-containing andesites normally thought of as dacite; the hornblende- or biotite-rich andesites; and the pyroxene andesites. Hornblende and biotite andesites are rich in feldspar and coloured pale pink, yellow or grey. The pyroxene andesites are by far the most common of the andesites and occur almost as widely as basalt, and often come from the same magma source. The pyroxene in pyroxene andesite is usually augite – giving augite andesite – but can be olivine. The augite gives these andesites a sparkle when the rock is broken, but they are frequently altered to hornblende. Sometimes, good-sized augite crystals can be found in pyroxene andesite tuffs – that is, deposits of ash and pyroclasts.

Biotite andesite: Biotite andesite is typically yellow, pinkish or grey, often with black phenocrysts of amphibole or pyroxene.

Identification: Pyroxene andesite can look quite like basalt, and its high pyroxene content means that it is chemically closer too. But pyroxene andesite usually contains phenocrysts and is usually slightly lighter in colour, with traces of white feldspar.

Grain size: Aphanitic (fine-grained), occasionally glassy.
Texture: Often porphyritic.
Structure: Often displays flow structure; occasional vesicles.
Colour: Grey, green, almost black.
Composition: Silica (59% average), Alumina (17%), Calcium and sodium oxides (10%), Iron and magnesium oxides (11%).
Minerals: Quartz; Feldspar: plagioclase feldspar (andesine) plus small amounts of potassium feldspar (oligoclase, sanidine); Biotite mica; Amphibole (hornblende) or Pyroxene.
Accessories: Magnetite, apatite, zircon, olivine.
Phenocrysts: Pyroxenes such as augite and olivine, amphibole, hornblende.
Formation: Extrusive lavas, ashes and tuffs in subduction zones and areas of mountain building.
Notable occurrences: See andesite.

TRACHYTES AND SPILLITE

Trachytes and phonolites are medium-coloured, fine-grained volcanic rocks that flow easily enough to form lavas. They occur in many of the same places as basalts, including rifts, but contain more lighter-coloured minerals, and a modicum of quartz. They are all alkaline rocks, containing sodium and potassium feldspars. Trachyte is the mildest alkaline, phonolite is the strongest.

Trachyte

This volcanic rock is medium-coloured and very fine-grained. It is the volcanic equivalent of syenite, and erupts along rifts, in oceanic settings, above hot spots and in back arc basins between island arcs and the continental land mass. Trachytes are often associated with basalt, and are sometimes thought to be characteristic of a volcano past its prime. The parasitic cones of the Hawaiian shield volcanoes often ooze trachyte lava, for instance. Yet trachytes can also form quite extensive lava flows in their own right, as they do in Saudi Arabia.

Trachyte is fine-grained, but unlike andesite and rhyolite, very rarely glassy. It seems that crystals nearly always form, even though they may be microscopically small. Remarkably, rectangular phenocrysts of white sanidine are often already formed in the lava when it erupts. These tend to line up with the direction of the lava flow, and through a microscope you can often see that the smaller crystals form flow patterns around them. This texture is described as trachytic, and is sometimes found in other lavas, such as Hawaiite.

Although they are too small to see, trachyte is full of tiny cavities left by gas bubbles. It is these that give the rock the slightly rough feel to which it owes its name, from *trachys*, the Greek for 'rough'. These cavities sometimes fill with tiny crystals of the silica minerals tridymite, cristobalite, opal and chalcedony. The bulk of trachyte, though, is mostly alkali (sodium- and potassium-rich) feldspars, notably sanidine (in rodlike microcrystals as well as phenocrysts), along with dark-coloured minerals such as biotite, amphiboles (hornblende) or pyroxenes such as aegerine and diopside. An increase in the silica content takes trachyte towards rhyolite; a decrease, with a corresponding increase in feldspathoids such as leucite, nepheline and sodalite, takes it towards phonolite.

Identification: Trachyte is a brownish grey, medium-coloured rock, which is microcrystalline but almost never glassy. The dark groundmass is usually spotted with thin white phenocrysts of sanidine. A tell-tale clue to its identity is the rough feel created by tiny gas bubbles.

Sanidine

Porphyritic trachyte: Trachyte is basically a medium-coloured rock, with a groundmass of dark minerals such as biotite, hornblende and pyroxenes, and light-coloured sanidine feldspar. Interestingly, the sanidine forms in two stages, and the rock may be spotted with large long white sanidine phenocrysts that formed early in the magma. Under a microscope, trachytic flow patterns are visible in the small crystals around them.

Grain size: Aphanitic (fine-grained), can occasionally be glassy.
Texture: Even, but often porphyritic.
Structure: Often displays flow structure called trachytic visible only under a microscope in which crystals of groundmass appear to flow around phenocrysts. Some specimens also have minute steam cavities, making the surface of the rock feel rough.
Colour: Usually grey, but can be white, pink or yellowish.
Composition: Silica (62% average), Alumina (17%), Calcium and sodium oxides (8%), Iron and magnesium oxides (6%).
Minerals: Quartz; Feldspar: potassium feldspar (sanidine) and plagioclase feldspar (oligoclase); Biotite mica; Amphibole (hornblende, often altered to magnetite and augite); Pyroxene (aegerine, diopside).
Accessories: Apatite, zircon, magnetite, leucite, nepheline, sodalite, analcime. Plus in cavities: tridymite, cristobalite, opal, chalcedony.
Phenocrysts: Tablet-shaped sanidine, often aligned with the flow.
Formation: Extrusive lavas, dykes and sills often in association with basalt.
Notable occurrences: Skye, Midland Valley, Scotland; Lundy Island, Devon, England; Eifel, Thuringia, Saar, Berkum, Drachenfels (Rhineland), Germany; Auvergne, France; Naples, Ischia, Sardinia, Italy; Iceland; Azores; Saudi Arabia; Ethiopia; Madagascar; Cambewarra (New South Wales), Australia; Hawaii; Black Hills, South Dakota; Colorado.

Phonolite

Phonolites are mostly quite recent rocks, all forming within the Tertiary Age – that is, in the last 66 million years. They are medium-coloured, fine-grained volcanic rocks. They split into thin slabs and have such a compact structure that the slabs ring when struck with a hammer, which is why they were once called clinkstone. Phonolites are quite similar to trachyte, and the two occur in similar places. But phonolites are richer in alkaline minerals. Their lower silica content favours the formation of feldspathoids such as nepheline, leucite and sodalite, rather than the potassium feldspars of trachyte. So phonolites are the extrusive equivalent of nepheline syenite, rather than plain syenite. Like trachyte, they contain two generations of crystals. The first generation are the large, flattened tablet-shaped crystals of sanidine and nepheline, which form slowly in the magma. These become phenocrysts when the lava is erupted, and smaller crystals quickly form around them, often with a microscopic trachytic flow structure. Sometimes, leucite replaces nepheline to create leucite phonolite, as found near Naples, which is often studded with blue hauyne crystals, and sphene.

Identification: Phonolite is usually a mottled grey, but tiny needles of pyroxene can turn it greenish. The way phonolite breaks into flat slabs is often a real clue to identity, especially if the slabs give a metallic clink when hit with a hammer.

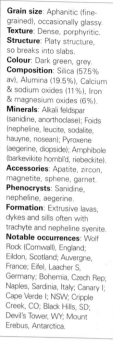

Grain size: Aphanitic (fine-grained), occasionally glassy.
Texture: Dense, porphyritic.
Structure: Platy structure, so breaks into slabs.
Colour: Dark green, grey.
Composition: Silica (57.5% av), Alumina (19.5%), Calcium & sodium oxides (11%), Iron & magnesium oxides (6%).
Minerals: Alkali feldspar (sanidine, anorthoclase); Foids (nepheline, leucite, sodalite, hauyne, nosean); Pyroxene (aegerine, diopside); Amphibole (barkevikite hornbl'd, riebeckite).
Accessories: Apatite, zircon, magnetite, sphene, garnet.
Phenocrysts: Sanidine, nepheline, aegerine.
Formation: Extrusive lavas, dykes and sills often with trachyte and nepheline syenite.
Notable occurrences: Wolf Rock (Cornwall), England; Eildon, Scotland; Auvergne, France; Eifel, Laacher S, Germany; Bohemia, Czech Rep; Naples, Sardinia, Italy; Canary I; Cape Verde I; NSW; Cripple Creek, CO; Black Hills, SD; Devil's Tower, WY; Mount Erebus, Antarctica.

Great East African Rift Valley
In few places is the power of tectonic plate movement more striking than in Africa's Great Rift Valley. The Valley is part of a huge set of fissures in Earth's crust called the East African Rift system, which threatens to split Africa in two. It started to open up about 100 million years ago, as plates on either side began to pull apart. As the crust stretched, volcanoes repeatedly burst through, and now dot the whole valley, including Erte Ale in Ethiopia and Ol Doinyo Lengai in Tanzania. Initially there were floods of basalt, then shield volcanoes emitting rhyolites and basanites, and finally volcanoes erupting trachyte and phonolite. The valley that formed is now over 6,000km/3,728 miles long and 50km/ 31 miles wide on average. The walls typically rise 900m/2,953ft above the valley floor, but at Mau in Kenya, cliffs soar to a height of 2,700m/8,858ft. Geologists believe that the Afar Triangle in Ethiopia, where the branches of the rift system meet, is the start of the world's next great ocean.

Spillite

Spillite is a medium-dark greenish black volcanic rock made from a groundmass of dark amphiboles such as actinolite and riebeckite, with occasionally bright cream phenocrysts of albite. It erupts mainly in oceanic locations, along with basalt, though it can often be found in ancient locations on land. It is one of the main kinds of magma that form pillow lavas. Pillow lavas are balls or tubes of lava that form where lava erupts slowly on the ocean bed. As the lava oozes up, contact with cold sea water quickly chills it to form a thin crust, and as the lava goes on pushing up it solidifies into a ball or tube, just like blobs of toothpaste squeezed from a tube.

Grain size: Aphanitic (very fine-grained).
Texture: Often porphyritic.
Structure: Platy structure, so breaks into slabs.
Colour: Dark green, black.
Composition: Silica (50% average), Alumina (16%), Calcium and sodium oxides (13%), Iron and magnesium oxides (18%).
Minerals: Plagioclase feldspar (albite); Amphibole (actinolite, riebeckite); Chlorite; Epidote.
Accessories: Apatite, zircon, magnetite.
Phenocrysts: Albite, actinolite.
Formation: Pillow lavas and tuffs.
Notable occurrences: Oceanic crust (worldwide); Cornwall, England; Alaska; California.

Identification: Spillite is dark, fine-grained volcanic rock quite similar to basalt. It can sometimes be identified by the way it breaks into slabs or by its formation as pillow lavas.

BASALTS

Black and fine-grained, basalt is the classic 'mafic' volcanic rock – rich in iron and magnesium minerals, and very low in silica. It is one of the earliest lavas to erupt from any volcano, coming straight up from the mantle in huge quantities, very hot and very fluid, and uncontaminated by the silicas that make other lavas much more viscous (less fluid).

Basalt

Alkali olivine basalt: The typical basalt rock is dark in colour with no visible grain structure. It tends to be black when freshly exposed, but turns reddish or greenish when weathered. Alkali or olivine basalt contains lots of olivine and augite in the groundmass (not as phenocrysts).

Ankaramite: This basalt gets its name from Ankara in Turkey where it has been found. It is an alkaline basalt, with lots of phenocrysts of both dark green olivine and black augite. Although basalt often has olivine phenocrysts, augite phenocrysts are rare because augite crystallizes late. All the same, augite forms up to half basalt's groundmass. Ankaramite is closely related to picrite.

Basalt is the most common rock on Earth. Much of it, though, is hidden away under the sea, for it forms the bulk of the ocean floor, which itself makes up 70 per cent of the Earth's surface. Basalt lava wells up through fissures in the ocean floor at the mid-ocean ridge as the two halves of the ocean floor pull apart. The lava freezes on to the receding edges of the two halves and ensures new rock is added as the ocean spreads away. It often forms pillow lavas here, as hot lava is suddenly chilled by the cold sea water to create countless cushion-like knobs of solid rock.

Basalt lava also wells up where hot spots penetrate the ocean floor to create island volcanoes such as Mauna Loa and Kilauea in Hawaii. Sometimes the hot basalt lava flows straight into the sea and is shattered as it suddenly freezes, and the fragments create black beach sands like those of Hawaii. Not all basalt is oceanic, though. Some basalt can erupt in continental fissures in a process no one quite understands. For example, huge floods of basalt have poured from fissures on to the surface to form gigantic plateaus, such as India's Deccan and North America's Columbia River Plateau. When it erupts, basalt lava is so hot and fluid that it can flow for tens of kilometres from the vent. When particularly hot it can even flow 500km/ 310 miles or so. This is why basalt habitually forms broad shield volcanoes, or flood basalt plateaux.

The shape the lava takes as it freezes depends on its temperature and its speed. When it is quite warm and fluid, surfaces wrinkle into rope-like ridges known by their Hawaiian name of *pahoehoe* (pronounced 'pa-hoy-hoy'). If they are a little more viscous or cooler, the surface tends to freeze and then break up into a jumble of rubble, a flow known as *aa* (pronounced 'ah-ah').

Grain size: Aphanitic (very fine-grained) or tachylytic (glassy).
Texture: Usually dense, with no visible mineral grains.
Structure: Often porphyritic, and tends to include xenoliths (large lumps of other minerals) of olivine and pyroxene. Frequently spongy with vesicles or amygdaloidal cavities. Large masses of basalt may be cracked into hexagonal columns, like the Giant's Causeway (N. Ireland).
Colour: Black or blackish grey when fresh, may weather to reddish or greenish crust.
Composition: Silica (50% average), Alumina (16%), Calcium and sodium oxides (13%), Iron and magnesium oxides (18%).
Minerals: Plagioclase feldspar (labradorite); Pyroxene; Olivine; Magnetite; Ilmenite. Tholeitic basalts (low in olivine): Plagioclase, pyroxene (hypersthene, pigeonite); Magnetite. Alkali basalts: (olivine-rich): Olivine, Pyroxene (augite); Magnetite.
Accessories: Countless.
Phenocrysts: Green glassy olivine or black shiny pyroxene, or occasionally white tabular plagioclase feldspar.
Amygdales: Zeolites, carbonates, and silica in the form of chalcedony and agate.
Formation: Extrusive lavas, dykes and sills. Most basalts occur as lava flows from volcanoes or as sheets building up lava plateaux in flood basalts. Surfaces are either smooth, ropey pahoehoe or clinkery aa. Under the ocean basalt is often in balloon-like masses of pillow lava.
Notable occurences: See opposite page.

Vesicular and amygdaloidal basalt

The Giant's Causeway
In the last stages of cooling, basalt lava flows often contract and fracture into extraordinary hexagonal columns. The most famous example of these is the Giant's Causeway in Antrim, Northern Ireland. Legend has it the stones were laid by the giant Finn MacCool to reach his lover in Scotland, but in fact they are a natural feature of basalt rock. About 65 million years ago, North America began to split apart from Europe, and basalt lavas welled up into the rift created. Here in Antrim, tholeitic lavas erupted over hundreds of thousands of years then stopped – only to start again abruptly. The lava poured into valleys and became so deeply ponded at the site of the Causeway that it formed a lava lake so deep that it cooled only slowly. As it slowly solidified and contracted, it developed regular hexagonal stress patterns. Soon six-sided columns 30–40cm/12–16in across permeated the whole cooling mass. Further eruptions followed, but this was the one that left its distinctive legacy.

The grains in basalt are usually so fine, they are invisible to the naked eye, and the impression of the rock is a black mass, created by a mix of dark minerals – essentially labradorite (plagioclase), pyroxene and olivine, with magnetite and ilmenite. But it can be porphyritic, or contain cavities called vesicles that form as gas bubbles expanded in the solidifying lava. Larger vesicles seem to form more commonly in basalt pahoehoe than any other lava. It may be because the lava is so hot and fluid enough for the bubbles to expand easily. Unlike vesicles in other, more viscous lavas, those in pahoehoe are pretty much round rather than long. Rock with lots of empty vesicles like these is called vesicular basalt. Once the lava is set, water percolating through the lava often begins to fill many of these cavities with mineral crystals. These infillings are called amygdales, from the Greek for almond for their typical shape. They can be anything from 1mm to 30cm/0.04 to 12in across and typically contain quartz, carbonates and zeolites. Basalt full of amygdales is called amygdaloidal basalt.

Vesicular basalt: Basalt is often filled with countless vesicles, which make it look as if some insects have been at it. Other igneous rocks do have vesicles, but they are especially numerous and rounded in basalt. A dark black groundmass indicates that a rock full of little holes like these is vesicular basalt.

Typical rounded vesicles

Alkali and tholeitic basalt

There is a wide spectrum of basalt rocks, but they can be divided into two broad groups according to their chemistry – the tholeitic basalts and alkali basalts. Tholeitic basalts are low in olivine and rich in calcium-poor pyroxenes such as hypersthene and pigeonite. Most basalts that ooze up from rifts and mid-ocean fissures are tholeitic. So ocean floors are tholeitic basalt. Flood basalts are tholeitic too. Alkali basalts contain more sodium and potassium and are rich in olivine, the feldspathoid nepheline and calcium-rich pyroxenes such as augite. Geologists believe they are more alkaline because minerals in the magma have been less divided by partial melting. In the Hawaiian hotspot volcanoes, lavas that build the initial undersea cone are alkali basalts, cooled quickly by seawater. But as the cone rises above the sea, streams of fluid tholeitic lava spew out to create a huge shield, cooling slower in the air and more affected by partial melting. As the volcano dies down again, spurts of alkali lavas resume, creating a cap of alkali basalt rock. In the million-year life cycle of Mauna Kea, Hawaii's biggest volcano, this last subdued alkaline stage has gone on for 40,000 years, and may yet last another 60,000 years.

Notable occurrences:
Tholeitic basalts: Deccan, India; Red Sea; Paraná Basin, South America; Palisades, New Jersey; Rio Grande Rift, Mexico; Columbia River Plateau, Washington-Oregon; Mauna Loa, Kilauea, Hawaii.

Alkali basalts: Ocean floors; Inner Hebrides, Scotland; Antrim, Northern Ireland; Iceland; Faroe Islands; Mauna Kea, Mauna Loa, Kilauea, Hawaii.

Leucite basalts: Italy; Germany; East Africa; Montana; Wyoming; Arizona.

Nepheline basalts: Libya; Turkey; New Mexico.

Amygdaloidal basalt: This kind of basalt with its white amygdale spots is easy to identify. Other igneous rocks do have amygdales, but they are rarely as large as in basalt. The black microscopically grained groundmass confirms its identity.

GLASSY ROCKS

When lava (or magma) cools quickly, there is no time for crystals to form within the mix, so the result is a glassy rock. Glassy rocks not only look like glass, though darker and cloudier, but they also shatter like it when struck with a hammer, giving sharp fragments. The main glassy rocks are obsidian, perlite and pitchstone, which are all inter-related and merge into each other, largely according to their water content.

Obsidian

Like rhyolite, obsidian forms from the same magma that solidifies as granite deep underground. When rhyolite magma approaches the surface, the reduction in pressure means that some of its water is lost as steam. This de-watered rhyolite magma becomes very thick and viscous. Indeed, it becomes so thick that crystals do not get a chance to grow before the erupted lava is chilled and frozen solid. The result is a rock that is just like solid glass except slightly harder, and often jet black. Obsidian lava is so thick that it advances at a snail's pace, and outcrops are usually quite small. It typically forms near the end of the volcanic cycle and creates just a small plug, a thin coating on rhyolite lava flows or a lining for rhyolite sills and dykes. Yet occasionally there are large flows of obsidian, as at Glass Buttes in Oregon and Valles Caldera in New Mexico, where there are layers of obsidian a few hundred metres thick. Such a flow occurred just 1,300 years ago at the Newberry volcano in Oregon.

Obsidian shatters like glass into sharp, conchoidal (curved) fragments. It quickly dulls with exposure, but when freshly broken obsidian gleams like polished glass. It is typically jet black, coloured by titanium oxide minerals, but streaks and swirls can make broken surfaces look like colour-flowed marbles, as in brown- and black-streaked 'mahogany' obsidian and 'midnight lace' obsidian, with its contorted streaks formed as the cooling lava rolled over and over. Iron oxides turn obsidian reddish or brownish, while gas bubbles and microcrystals make 'golden sheen' obsidians that shimmer iridescently in sunlight. Because it fractures with a beautifully curved glass-sharp edge, obsidian can easily be fashioned to make knife and axe blades, even better than flint. It can also take a high polish. As a result it was highly prized among early cultures. The Ancient Egyptians, the Aztecs and Mayans all used obsidian knives and arrowheads. So too did Native Americans. Because obsidian absorbs water once it is broken, some of these ancient obsidian artefacts can be accurately dated by measuring just how much water their surface layers have absorbed. Obsidian is rarely older than 20 million years, because it starts to alter as soon as it forms. In a process known as 'devitrification', the glass absorbs moisture and begins to form crystals and go cloudy.

Identification: Jet black obsidian, which looks like a lump of solid glass, is hard to mistake for any other rock, especially if it fractures conchoidally. But it can contain phenocrysts of quartz and microscopic crystals of feldspar.

Conchoidal fracture

Snowflake obsidian: Snowflake obsidian has white snowflake patches of the mineral christobalite. Sometimes, 'snowflakes' can form through devitrification, as moisture alters silica in the obsidian.

Grain size: None, obsidian is glassy.
Texture: Occasional small phenocrysts or microlites (tiny crystals).
Structure: Breaks conchoidally; occasional spherulites (tiny radiating clusters of needlelike crystals); flow banding with alternating glassy and devitrified layers. Contortion of flowbands in unset lava creates 'midnight lace' obsidian.
Colour: Usually jet black, but presence of iron oxides turns it reddish and brownish, and inclusion of tiny gas bubbles gives it a golden sheen. Dark banding and mottled grey, green and yellow. Microscopic feldspar crystals create 'rainbow obsidian'.
Composition: As for rhyolite: Silica (74% average), Alumina (13.5%), Calcium and sodium oxides (less than 5%), Iron and magnesium oxides (less than 3.5%). Obsidian always includes a certain amount of water, often in the form of minute bubbles of water vapour trapped in the glass. These bubbles are usually visible under a magnifying glass.
Minerals: Quartz; Potassium feldspar (sanidine) and plagioclase feldspar (oligoclase); Biotite mica.
Phenocrysts: Quartz; Christobalite.
Microlites: Feldspar.
Formation: Lava flows, dykes and sills.
Notable occurrences: Scotland; Eolie Island, Italy; Mount Hekla, Iceland; Mexico; Obsidian Cliff (Yellowstone), Wyoming; Arizona; Colorado; Valles Caldera, New Mexico; Big Obsidian Flow (Newberry), Glass Buttes, Oregon.

Perlite

Like obsidian, perlite is a natural glass that forms from rhyolite lava, but rather than being shiny black, perlite is grey like dirty snow. Also while obsidian contains very little water, perlite gets its name because it contains concentric cracks that make the rock break into tiny pearl-like balls. Unlike obsidian, perlite contains water (2–5 per cent) because it cools so quickly that water has no time to escape. Once formed, it goes on absorbing water from its surroundings. Each little pearl is like a balloon full of water, and this is what makes perlite a rather amazing material. When heated to 871°C/1,600°F, the water evaporates, and the steam turns each pearl into a bubble, inflating the perlite like popcorn up to 20 times its original volume. This creates an incredibly light, gas-filled material, which is used for all kinds of insulation, for both heat and sound. Many roofing tiles contain perlite as does pipe insulation. It is also used instead of sand to make lightweight concrete. Horticulturalists often grow plants in a perlite mix instead of soil because of its good aeration and water retention.

Identification: Perlite looks a bit like dirty ice, and lumps of it can look rather like old snowballs. It has a glassy texture, with no crystalline structure, but is often dotted with phenocrysts, much more so than obsidian.

Perlite pebble

Grain size: None, perlite is glassy.
Texture: Occasional small phenocrysts; when there are a lot of phenocrysts, the rock becomes 'vitrophyre'.
Structure: Concentric cracks, which mean the rock breaks into pearl-shaped balls.
Colour: Grey or greenish, but may be brown, blue or red.
Composition: As for rhyolite: Silica (74% average), Alumina (13.5%), Calcium and sodium oxides (<5%), Iron and magnesium oxides (<3.5%).
Minerals: Quartz; Potassium feldspar (sanidine) and plagioclase feldspar (oligoclase); Biotite mica.
Phenocrysts: Quartz, sanidine, oligoclase or, rarely, biotite or hornblende.
Formation: Lava flows, dykes.
Notable occurrences: Greece; Turkey; New Mexico; Sierra Nevada, California; Utah; Oregon.

Apache tears
Sometimes obsidian alters to perlite when it absorbs water during or after cooling. The water is gradually absorbed along cracks caused by the cooling process, turning more and more of the obsidian to balls of perlite. As more water is absorbed, these spread out in concentric circles through the obsidian. Eventually, all that is left is a small core of obsidian embedded in a mass of perlite. As the process goes on, the perlite is broken by weathering, leaving just a few isolated nodules of obsidian, rounded by wind and water into natural marbles, called 'Apache tears'. They got their name from the stones at Apache Leap Mountain near Superior in Arizona. Legend has it that in the 1800s Apache warriors were trapped at the top of a cliff on this mountain by pursuing US cavalry. Rather than surrendering to their enemies, the apaches leaped to their death. The tears of their wives and children are said to have fallen to the ground here and the Great Spirit, looking down, turned them into the Apache tears so that the courage of the warriors might be remembered forever.

Pitchstone

Pitchstone is a glassy volcanic rock occurring famously in the Hebridean islands of Scotland, where it was used in the Stone Age to make blades. Some pitchstone is high in silica and forms from the same granite-like magma as rhyolite. Other pitchstones are lower in silica and more akin to trachyte or even andesite. Unlike obsidian, pitchstone contains a lot of water (up to 10 per cent) and is dull in lustre – especially older pitchstones that have become almost completely devitrified (lost their glassiness) and look pretty much like rhyolite. Many pitchstones contain phenocrysts arranged in wavy tracks reflecting the flow of the magma. Pitchstones are often mixed in with crystalline volcanic rocks, and may have formed when water driven out of the crystallizing rock was taken up by the glassy pitchstone.

Grain size: Glassy, or cryptocrystalline.
Texture: Abundant phenocrysts; when phenocrysts dominate, the rock becomes 'vitrophyre' or pitchstone porphyry.
Structure: Wavy flow streaks. Breaks to poorly defined conchoidal fracture.
Colour: Streaked, mottled, or uniform black, brown, red, green.
Composition: Quartz (variable), Potassium feldspar and plagioclase feldspar, Biotite mica.
Phenocrysts: Quartz, potassium feldspar, plagioclase, or, rarely, pyroxene or hornblende.
Formation: Dykes and sills.
Notable occurrences: Arran, Eigg, Skye (Hebrides), Scotland; Chemnitz, Meissen, Germany; Lipari, Italy; Urals, Russia; Japan; New Zealand; Oregon; Colorado; Utah; California.

Identification: Pitchstone is glassy and dark – like solid tar – but much duller than obsidian and usually marked by wavy phenocrysts.

VOLCANIC FROTH AND ASH

Not all volcanic rocks are formed from molten lava. Pumices form from glassy lava so are filled with gas bubbles that it becomes a froth. Tuffs form from ash and pyroclasts – fragments of solid magma and rock shattered by an explosive eruption. Some falls to the ground and consolidates into rock only gradually. Some rushes out in flows so hot that material is literally welded together, creating solid ignimbrite.

Pumice

Floating rock: Pumice will float for several months before becoming waterlogged and sinking.

Pumice is the only rock that floats. It is solidified lumps of rhyolitic or dacitic lava froth so full of holes that it is less dense than water. When the lava erupted, the release of pressure made gases dissolved in it effervesce – like unscrewing a shaken fizzy drink bottle – and form bubbles that blew the lava up into a froth. Had it stayed under pressure it would have formed obsidian. Basalt and andesite lavas form froth rocks, too, called scoria, but because these lavas are so fluid, gases can escape. So scoria contains fewer holes than pumice and won't float. A basalt pumice does form in Hawaii, however – and it is even lighter than rhyolite pumice, and black! As volcanoes have erupted through time, tiny fragments of pumice have been scattered all over the world, and now coat the deep ocean floor. Some has come from undersea eruptions, but much came from fragments falling on the ocean after big eruptions, then floating for months before becoming waterlogged and sinking. Ground up, pumice is the abrasive used to 'stonewash' jeans among other things. Commercially the word pumice refers only to large stones; grains are called pumicite. Pozzolan is a pumicite mixed with lime to make cement.

Identification: Fresh pumice is very easy to identify since it is whitish, full of holes and so light that it actually floats, until it becomes waterlogged. However, pumice retains this lightness for only a short time geologically. Soon enough all the holes are infilled with secondary minerals and it is no longer buoyant, and the glassy solid material becomes devitrified.

Pumice is often full of air holes like this

Grain size: None, it is glassy.
Texture: Like a foam.
Structure: The solid glass forms threads and fibres surrounding rounded or elongated holes depending on the flow of the lava. Cavities may be infilled when percolating water deposits secondary minerals.
Colour: Usually white or light grey; scoria is black or brown.
Composition: The same as for rhyolite: Silica (74% average); Alumina (13.5%), Calcium and sodium oxides (<5%); Iron and magnesium oxides (2%).
Minerals: Quartz; Potassium feldspar (sanidine) and plagioclase feldspar (oligoclase); Biotite mica.
Formation: Lava flows and pyroclasts.
Notable occurrences: Pozzola, Italy; Greece; Spain; Turkey; Chile; Arizona; California; New Mexico; Oregon.

Banded tuff

Identification: Banded tuff is streaked with dark glass and welded ash.

Tuff is rock formed from consolidated volcanic ash. It typically shows layering as larger heavier particles land first in each eruption (unlike ignimbrite). But every now and then it can show a much more distinctive banded pattern. This is sometimes due to patterns created when hot ash is welded, and sometimes a result of last-minute mixing of magma from different sources. The two never mix perfectly, and the result is that when the pyroclasts finally settle, they accumulate in layers reflecting the different mixes of magma. Such banded tuffs were found after the cataclysmic 1912 eruption of Novarupta in the Valley of 10,000 Smokes in Katmai, Alaska.

Grain size: Welded into glassy or amorphous mass.
Texture: Said to be eutaxitic when it contains flammes (glassy pancakes of pumice).
Structure: Welding and mixing creates other bands.
Colour: Grey to black, may be turned pink by weathering.
Composition: Variable – usually from rhyolite or tachyte glass.
Formation: Ashfalls and pyroclastic flows.
Notable occurrences: Charnwood Forest (Leics), England; Mato Grosso, Brazil; Santa Cruz, California; Cripple Creek, Colorado; Katmai, Alaska.

Vesuvius and Pompeii
Mount Vesuvius is one of the world's most famous volcanoes. There have been eight major eruptions so far, the most recent in 1906 and 1944. Vesuvius is a composite volcano typical of subduction zones, supplied by trachyte and andesite magma. Its explosive, Plinian-type eruptions send up towering columns of ash, pumice and bombs, which smother the surrounding area, or collapse to push out devastating pyroclastic flows. In the terrible eruption of AD79, witnessed by the Roman writer Pliny, pyroclastic surges incinerated the mountainside town of Herculaneum (original settlement shown in the foreground, above), while ashfall completely buried Pompeii, killing the inhabitants, but preserving their homes perfectly. The eruption was catastrophic, but Vesuvius is a small cone sitting in the caldera of Monte Somma, a giant volcano that erupted 35,000 years ago with a force that would make the Pompeii eruption seem a mere puff. Over 30,000km²/11,583 sq miles of the land around the Bay of Naples is Campanian ignimbrite, created in a vast pyroclastic flow in this eruption.

Ignimbrite

No rock is created in such a rapid and dramatic way as ignimbrite. It is the rock that forms when pyroclastic flows and surges finally come to rest. Its name is Latin for 'fire cloud', and is very apt. Pyroclastic flows are clouds of glowing ash, cinders and hot gases that roar down from an eruption at jet plane speeds and temperatures of 450°C/842°F or more. Many flows are still so hot when they come to a halt and settle that volcanic fragments within them are instantly welded together. Towards the base of flows, the heat can squeeze pumice fragments flat to create pancake shapes called *flamme* (flames).

Grain size: Varied, mostly less than 2mm/0.08in.
Texture: Like a fruitcake. Said to be eutaxitic when it contains flammes (glassy pancakes of pumice).
Structure: Large pebbles of pumice in finer mass of glass fragments. Sometimes shows flowbanding and layering; welding creates other bands.
Colour: Grey to bluish grey, may be turned pink by weathering.
Composition: Variable – usually from rhyolite or tachyte glass.
Phenocrysts: Feldspar.
Formation: Pyroclastic flows.
Notable occurrences: Widespread; Mount Vesuvius, Italy; Hunter Valley (New South Wales), Australia; Coromandel, New Zealand; Ria Loa, Chile; Mount St Helens, Washington.

Identification: Ignimbrite has a distinctive dark fruitcake look, with its isolated phenocrysts and flammes of pumice, but can easily be mistaken for lava flow rock.

Lithic tuff

Ash thrown out by volcanoes settles on the ground like falling snow, building up in drifts. At first, ash-fall is just loose dust. But in time, it packs down and becomes consolidated into a soft, porous solid rock called tuff. Sometimes lithification (turning to stone) is helped by the way glass in the ash is turned by the weather into clay and zeolite cement. Tuffs vary widely in texture and composition, and older tuffs have lost most of their original texture through recrystallization. They can be classified into three kinds according to the predominant fragments in the ash: 'lithic' tuffs, made mostly of chunks of broken rock; 'vitric' tuff, made mostly of shards of volcanic glass; and 'crystal' tuff, made mostly of small crystals such as feldspar, augite and hornblende. Most ash grains in tuff are less than 2mm/0.08in across, but tuff can also contain pebble-sized fragments called lapilli. Wind can scatter ash over huge distances, but lapilli land close to the volcano. Those falling close enough may be still hot enough to weld ash together, forming welded tuff.

Identification: Tuff is much softer than any other volcanic rock, easy to scratch with a knife. Although there may be visible crystals in new tuffs, they are very unevenly distributed and rather shapeless.

Grain size: Varied, mostly less than 2mm/0.08in.
Texture: Like a dense sponge cake.
Structure: Tuffs are usually layered, rather like sedimentary rocks with the heaviest particles that drop first in each ashfall at the base of each layer.
Colour: Grey, black, very variable – older basaltic tuffs may be turned green as original minerals are altered to chlorite.
Composition: Variable.
Minerals: Variable.
Formation: Ashfall and pyroclasts.
Notable occurrences: Widespread, including Santorini, Greece.

VOLCANIC DEBRIS

When explosive volcanoes erupt, 90 per cent of the solid material ejected is not lava but pyroclastic material. The word 'pyroclastic' means 'fire-broken'. Pyroclasts are fragments of old magma, fresh magma and basement rock shattered into ash, lapilli (stones) and bombs (boulders) – collectively called tephra – by the explosive force of the eruption and scattered far and wide around the volcano.

Lapilli

Lapilli are pyroclasts usually thrown out by explosive volcanic eruptions. Lapilli is Italian for 'little stones' and geologists define them as pyroclasts 4–64mm/0.17–2.5in across – in other words, between the size of a pea and that of a walnut. Anything larger is classified as a bomb; anything smaller is ash. Some lapilli are globules of fresh, liquid magma. Some are fragments of exploded magma. Some are fragments chipped off basement rock by the eruption. Globules of magma can sometimes cool and solidify into a teardrop shape as they fly through the air. Lapilli like these are called 'Pelée's tears', after the Hawaiian goddess of volcanoes. Pelée's tears often trail a thread of liquid lava that chills in midair into a golden brown hair-like filament called Pelée's hair. Occasionally, nut-sized pellets called accretionary lapilli build up like hailstones in a thundercloud, as layers of ash cling to a drop of water. Froth in felsic lavas like rhyolite makes pumice lapilli that bob on water for months. Basaltic lava froth makes heavier scoria cinders, but occasionally forms reticulites. Reticulites are 98 per cent air bubbles – even lighter than pumice but so fragile they sink as the bubbles break and take in water.

Identification: Lapilli are little, light, cinder-like stones, often found in layers of ash or scattered around the foot of volcanoes. They may be glassy like these.

> **Size**: 4–64mm/0.17–2.5in.
> **Texture**: Usually glassy and vesicular – that is, containing gas bubbles. Silicic magmas typically produce pumiceous (pumice-like) lapilli. Basalt typically produces scoria lapilli or, occasionally, light air-filled reticulite.
> **Colour**: Black, grey or brown.
> **Composition**: Varies according to parent lava.
> **Formation**: Pyroclasts from fresh magma.
> **Notable occurrences**: Cumbria, England; Stromboli, Mount Etna, Mount Vesuvius, Mount Vulture, Italy; Santorini, Greece; Toba, Tambora, Krakatoa, Indonesia; Citlaltépetl, Popocatépetl, Mexico; Kilauea Volcano, Mauna Loa, Hawaii; Yellowstone, Wyoming; Alamo, Texas.

Tephra

This name was coined by Icelandic vulcanologist Sigurdur Thorarinsson in the 1950s. It comes from the Ancient Greek for 'ash', but is a general word used to describe all pyroclastic material thrown into the air by a volcano, including ash, lapilli and bombs of all kinds. The term is used only for material that falls from the air, but excludes pyroclastic flow material. Unlike tuff, tephra is loose material and turns into solid rock tuff only when cemented together. It varies in composition from scoria-fall (cinder) deposits to pumice-fall deposits. Scoria-falls are erupted in Strombolian-type eruptions and consist of basaltic to andesitic pyroclasts falling fairly close to the volcanic vent. These Strombolian scoria falls are typically dark in colour. Pumice-falls are blasted out in Plinian-type eruptions and consist of dacitic to rhyolitic pyroclasts often scattered over vast areas. Plinian-type pumice-falls are typically light in colour.

Identification: Tephra is a blanket term to describe any fragment thrown out by a volcano. It can be cinders like these, or ash and pumice, glass beads, and reticulite, to huge bombs and blocks.

> **Size**: Full range of sizes.
> **Texture**: Usually glassy and vesicular – that is, containing gas bubbles.
> **Structure**: Mixture of ash, lapilli and bombs, usually sorted into layers with largest particles at the base and finest at the top.
> **Colour**: Black, brown or grey.
> **Composition**: Varies according to parent lava.
> **Formation**: From falling pyroclasts.
> **Notable occurrences**: Many locations including: Surtsey, Hekla, Iceland; Stromboli, Mount Etna, Mount Vesuvius, Italy; Santorini, Greece; Toba, Tambora, Krakatoa, Indonesia; Citlaltépetl, Popocatépetl, Mexico; Yellowstone, Wyoming.

Blocks and bombs

Blocks and bombs are large fragments of pyroclastic material. Blocks are chunks of broken magma. Bombs are large blobs of molten magma. Big eruptions can hurl blocks heavier than a truck 1km/0.6 miles from the vent. They can fling smaller bombs 20km/12 miles or even 80km/50 miles – at speeds approaching 75–200m/s (250–650ft/s), faster than a bullet. Most land close to the volcano, though. Bombs and blocks are not as heavy as they look, because they are usually full of holes. Volcanic bombs are fluid as they fly through the air and take on various, quite diverse shapes, according to just how fluid they are. Some end up as long, flat ribbon bombs. Some are so fluid that they are streamlined by force of motion as they fly into spindle bombs. Viscous lava bombs solidify at the surface to create breadcrust bombs as gas bubbles in the liquid interior expand and crack the surface like (very) crusty bread. 'Cow-dung' bombs hit the ground still molten and spread out in pancakes. Bombs can form various kinds of rock when they land, including tuff breccia, made from 25–75 per cent bombs, pyroclastic breccia (over 75 per cent bombs and blocks), agglomerates (over 75 per cent bombs) and agglutinate (made from spatters of basaltic lava still molten when it hits the ground).

Breadcrust bomb: Breadcrust bombs are made from blobs of viscous lava that crack on the surface to look like crusty bread. They earn their bomb name by occasionally exploding in mid-air as internal gas bubbles expand.

Grain size: Greater than 64mm/2.5in.
Texture: Usually glassy and vesicular – that is, containing gas bubbles.
Structure: Varies according to the way it cools in the air or on the ground, from the gas expansion cracks on the surface of breadcrust bombs to the long hair-like trails of Pelée's hair.
Colour: Black or brown.
Composition: Varies according to parent lava.
Formation: Pyroclasts from fresh magma form into tuff breccia, pyroclastic breccia, agglomerate and agglutinate.
Notable occurrences: Stromboli, Mount Etna, Italy; Kluchevskoy, Tolbachinsky (Kamchatka), Russia; Mauna Loa, Hawaii; Vanuatu; Mount Lassen, California; Yellowstone, Wyoming; Craters of the Moon, Idaho; Red Bomb Crater, Oregon.

Perilous volcanic snow
Glowing streams of molten lava can be awe-inspiring, but tephra is also very dangerous. Ash quickly chokes people to death and buries vast areas under deep deposits. Rooftops covered in ash may collapse, crushing anyone beneath. Ash can also be a hazard to aircraft, as seen in 1982 when Java's Gallungang erupted, nearly bringing down two jet airliners as ash clogged their engines. Ash has also caused famine as it destroys vegetation. In 1815, Indonesia's Tambora sent ash up to 1,300km/808 miles away, and 80,000 people died of famine. The distance that ash travels depends on the height of the eruption column, the air temperature, and the wind direction and strength. The Krakatoa eruption of 1883 (Indonesia, above) spread 800,000km²/308,882 sq miles of ash, and burned the clothes of people 80km/50 miles away. The Toba (Indonesia) eruption of 75,000 years ago deposited 10cm/4in of ash 3,000km/1,864 miles away in India!

Block lava

Basalt lavas produce very fluid pahoehoe lava, which slurps along to freeze in rope-like coils, or more brittle, chunky *aa* lava. But more silicic lavas such as andesite and dacite produce a 'blocky' lava. Blocky lava is usually linked to stratovolcanoes with alternating layers of lava and pyroclasts. Lava flows from these volcanoes are very slow indeed, and the flow surface quickly congeals into a large rubble-like mass of blocks on top and at the front of the flow. The flow often follows narrow tongues called 'coulées' down the volcano, channelled by embankments or 'levées' of lava blocks either side. Rhyolite lavas are chunkier still and often develop dam-like ridges of blocks called ogives on the surface.

Pahoehoe lava: This fluid lava solidifies in rope-like coils from the hottest, most fluid basalt lava.

Texture: Chunky, angular blocks.
Structure: Biggest, most angular chunks occur on the edges of the flow.
Colour: Black, brown or grey.
Composition: When andesitic: Silica (59% average), Iron and magnesium oxides (7.5%).
Formation: Andesite, dacite and rhyolite lava flows.

Block lava: The surface of more viscous, silicic lava like andesite and dacite breaks into block-like chunks.

ULTRAMAFIC ROCKS

Ultramafic rocks are dark-coloured igneous rocks with even less silica than basalt, a lot of magnesium and iron, and made mostly of dark green or black olivines and pyroxenes. They form deep in the Earth's mantle and are usually brought to the surface in small quantities, often in masses no smaller than a fist or as large as a house, swept up as lumps in other magmas, or as the entire crust is uplifted by tectonic movements.

Picrite

Identification: Picrite is best identified by its setting, its dark green to black colour, its slightly shiny, sugary look and its even texture – not mottled like peridotite nor with plagioclase feldspar laths like dolerite (diabase), which are both often found with picrite.

Picrites are dark, heavy rocks rather similar to peridotite. Both form deep in the Earth's mantle, and both are rich in dark green olivine and brown augite. But while peridotite is often carried up into large intrusive masses, along with gabbro, norite and pyroxenite, picrite tends to be found in sills and intrusive sheets. Although picrite magma forms only under extreme pressures deep in the mantle, it is the one ultramafic rock that normally erupts on the surface as lava, as it did in the 1959 eruption of Kilauea in Hawaii. In this eruption, gigantic fire fountains shot out picrite lavas containing as much as 30 per cent olivine. Yet for picrite to erupt as lava like this, temperatures must be very high indeed – which is why it is often linked to hotspot volcanoes such as those in Hawaii.

Picrite is also often found in association with basalt as part of the ocean floor, which is why many of the best known occurrences of picrite are in ophiolites – chunks of the ocean floor brought to the surface by massive tectonic movements.

Occasionally, picrite can occur in substantial quantities in flood basalts, as it does in India's Deccan and South Africa's Karoo, but most flood basalts have a fairly low picrite content.

Picrite is very rich in magnesium and iron. Indeed, one definition of picrite is that it has at least 18 per cent magnesium oxide by weight. Komatiite is similar but has even less sodium and potassium oxide. Picrite's iron content can be so high that it is actually slightly magnetic. Some picrites, though, are especially rich in hornblende (see right). Others are especially rich in augite, like a few of those of Devon and Cornwall in England, which are sometimes called palaeopicrites because they formed well over half a billion years ago in the Palaeozoic era. Most picrites in this part of the world date more specifically to the Devonian period, 408–360 million years ago.

Grain size: Moderately fine-grained (salt-sized).
Texture: Granular
Structure: Evenly textured.
Colour: Dark green to black.
Composition: Silica (47% average), Alumina (10%), Calcium and sodium oxides (10%), Iron and magnesium oxides (31%).
Minerals: Olivine; Clinopyroxene (augite); Orthopyroxene (enstatite); Biotite mica; Hornblende; Plagioclase feldspar.
Accessories: Apatite, melilite, sphene, biotite, spinel, hornblende.
Phenocrysts: Green olivine or red brown augite.
Formation: In extrusive lavas at mid-ocean ridges and hotspots, dykes and sills. Appears in ophiolites and flood basalt lava plateaux.
Notable occurrences: Rhum (Hebrides), Inchcolm Island, Midland Valley, Scotland; Devon, Cornwall, Sark (Channel Islands), England; Wicklow, Ireland; Nassau, Fichtelberg, Germany; Troodos, Cyprus; Gran Canaria; Oman; Madagascar; Karoo, South Africa; Deccan, India; Dongwhazi, Kunlun, China; Tasmania; Hawaii; Yellowstone, Wyoming; Klamath Mountains, California; Oregon; Hudson River; Alabama; Montana.

Hornblende picrite: The minerals in picrite often decompose quite quickly. Olivine is replaced by green, yellow and red fibres of serpentine, while augite is replaced by chlorite or hornblende. Only hornblende remains unaltered, creating hornblende picrite. Many ancient picrites are like this, such as those found in Gwynedd and Anglesey in Wales and on Sark in the English Channel Islands.

Pyroxenite

Like picrite and peridotite, pyroxenite is an ultramafic rock formed from magmas that develop deep in the Earth's mantle. What makes pyroxenite different from these rocks – though similar to wehrlite and clinopyroxenite – is that it contains a high proportion of clinopyroxene (usually the mineral augite, clinopyroxenite) or orthopyroxene (enstatite or hypersthene, orthopyroxenite), at the expense of olivine. Sometimes, pyroxenites are inclusions in other magmas, and look rather like obsidian: shiny black and fracturing into the same sharp, conchoidal (curved) fragments. Pyroxenites rarely occur alone. Very often, they form layered complexes with other plutonic (deep-forming) igneous rocks such as gabbro and norite. In the Bushveld complex of South Africa, for instance, gabbro, norite and pyroxenite layers are interwoven, with more pyroxenite layers at the base and more gabbro layers at the top. Occasionally pyroxenite-like rocks may form when certain limestones are altered by contact with hot magma, but these are more properly called pyroxene hornfels.

Identification: Pyroxenites are very much plutonic, which means they are almost entirely coarse-grained, often containing individual crystals several centimetres long. They are hard to distinguish from similar dunites, hornblendites, and melilitites, except by laboratory tests.

Grain size: Medium to coarse.
Texture: Granular.
Structure: May be layered.
Colour: Green to black.
Composition: Silica (47% average), Alumina (10%), Calcium and sodium oxides (10%), Iron and magnesium oxides (31%).
Minerals: Clinopyroxene (augite); Orthopyroxene (enstatite, bronzite, hypersthene); Olivine; Biotite mica; Chromite; Hornblende.
Accessories: Plagioclase, chromite, spinel, garnet, iron oxides, rutile, scapolite.
Phenocrysts: Green olivine or red brown augite, orthopyroxene.
Formation: Small intrusions such as stocks and dykes and in bands in layered gabbro.
Notable occurrences: Shetland Islands, Scotland; Saxony, Germany; Bushveld complex, South Africa; New Zealand; Cortlandt (Hudson River), North Carolina.

Ophiolites

The problem with learning about the oceanic crust is that it is so inaccessible. But in places segments of the crust have been heaved up by tectonic movements and incorporated into mountains. These segments are called ophiolites, and although rarely complete they afford a rich opportunity to study the ocean crust and its material. They are quite literally a slice through the ocean crust, and always tend to have the same layers. At the top is a blanket of sediment less than 1km/0.6 mile thick, composed of clay grains and dead plankton. Beneath this is a layer of basalt pillow lavas, sitting on sheets of gabbro dykes. Further down are more gabbros along with norite and olivine-rich gabbro. Further down still (3–4.75km/2–3 miles) are layers of chromite-rich dunite, wehrlite, peridotite and pyroxenite. At the base is an unlayered agglomeration of serpentinized peridotite with dunite, harzburgite and olivine pyroxenite. There are famous ophiolites at Troodos in Cyprus. Others are at Josephine County in Oregon.

Pelagic sediments
Basalt pillow lavas
Gabbro sheeted dikes and sills
Serpentinite
Harzburgite
Lherzolite

Melilitite and nephelinite

These deep-forming ultramafic rocks are often linked to hot spots and rifts, where material from deep down is brought to the surface. They are typically 'coughed up' from the mantle as xenoliths, masses in other magmas, and probably form when mafic magmas mix with peridotite. Although they are rare, they are found all over the world. They are both highly alkaline, like carbonatite, lamprophyre and kimberlite. Nephelinite is essentially nepheline and clinopyroxene, plus olivine and iron and titanium oxides. Melilitite is basically the same and there is a gradation between the two, but melilitite has the mineral melilite instead of nepheline.

Grain size: Medium to coarse.
Texture: Granular.
Structure: May be layered.
Colour: Green, dark green to black.
Composition: Silica (less than 42% average), Alumina (15%), Calcium and sodium oxides (17.5%), Iron and magnesium oxides (18.5%).
Minerals: Feldspathoid (nepheline or melilite); Clinopyroxene (augite); Olivine; Perovskite.
Phenocrysts: Green olivine or red-brown augite.
Formation: Both intrusive and extrusive, forming xenoliths, lava or pyroclasts at hot spot and rift volcanoes.
Notable occurrences: Umbria, Italy; Rhine, Germany; East Greenland; Ol Doinyo Lengai, Tanzania; Western Cape, S Africa; Tasmania; Amazon; Sword Mount (Washington Co), Maryland.

Identification: Melilitite is almost as evenly dark as basalt, but the grains are coarse enough to be visible. It is often found in kimberlites or with carbonatites.

OLIVINE-RICH ROCKS AND CARBONATITE

Some igneous rocks formed deep in the mantle are especially rich in the green mineral olivine, which is one of the most common minerals on Earth, though most of it is in the mantle rather than in the crust. These rocks are brought up as xenoliths in other magmas, or form in lower parts of the sea floor or near the surface in small quantities as they separate from hot magmas, like those over hotspots and at rifts.

Peridotite

Peridotite is one of the main materials of the upper mantle, and both basaltic and gabbroic magma are thought to be produced by the melting of peridotite in the mantle. So ultimately, mantle peridotite is the source of most of the material that makes up the rocks of the Earth's crust. Laboratory tests have shown that if peridotite is heated to melting point, the melt is basalt. Partial melting seems to give picrites and komatiites.

Some peridotites appear low down in ophiolite complexes, layered along with pyroxenite. Others seem to accumulate at the bottom of gabbro intrusions when olivines crystallize and settle out. A few peridotites are squeezed upwards in deep volcanic pipes. Rare chunks of peridotite reach the surface as solid lumps, swept up as xenoliths in liquid magma – spinel peridotites in basalt, basanite, nephelinite and occasionally andesite, and garnet peridotites in kimberlites and lamproites.

Peridotite is the least siliceous of all the igneous rocks, with less than 46 per cent silica by weight, and contains almost no feldspar. So it is a very dark coloured rock, with small pale green crystals of olivine set in a mass of dark pyroxene and hornblende. There are actually several varieties of peridotite. Besides olivine, lherzolite contains both clinopyroxene (augite) and orthopyroxene (enstatite); wehrlite contains only clinopyroxene; harzburgite contains only orthopyroxene.

A special form of peridotite called kimberlite is, with related lamproite, the world's only primary source of diamonds (see Kimberlites and lamproites, right). But other peridotites are major sources of nickel and chromium minerals, platinum and talc, and chrysotile asbestos, too. In warm, humid places, peridotite has weathered to soils rich in iron, nickel, cobalt and chromium, which may one day be exploited as ores on a large scale.

Identification: Peridotite is a dark green to black, medium to coarse grained rock that looks rather like sugar stained with a dark green engine oil. Often it is studded with small pale green balls of olivine, or more rarely with red garnets.

Grain size: Medium to coarse (sugar-sized).
Texture: Granular. Frequently poikilitic, which means small round crystals (of pale green olivine) embedded in large irregular masses (of pyroxene and hornblende). Rarely porphyritic.
Structure: May be layered.
Colour: Dull green to black.
Composition: Silica (44–45.5%), Alumina (2–4%), Calcium and sodium oxides (4–6%), Iron and magnesium oxides (41–49%).
Minerals: *Wehrlite*: Olivine; Clinopyroxene (augite); *Lherzolite*: Olivine; Clinopyroxene (augite); Orthopyroxene (enstatite) *Harzburgite*: Olivine; Orthopyroxene (enstatite).
Minor and accessory minerals: Biotite mica, hornblende, chromite, garnet, picotite, corundum, platinum, awaruite.
Formation: Interlayered with other ultramafic rocks in ophiolite complexes; settled out of gabbroic intrusions; in volcanic pipes, dykes and other small intrusions; and as xenoliths in basalt.
Notable occurrences: Lizard (Cornwall), England; Anglesey, Wales; Skye, Ayrshire, Scotland; Harzburg, Odenwald, Silesia, Germany; Lherz (Pyrenees), France; Hungary; Norway; Finland; New Zealand; New York; Maryland.

Garnet

Garnet peridotite: Most peridotite has traces of garnet or spinel, but sometimes, especially in lherzolite, small crystals can form like cherries in a bun. Some are big enough to chip out as gems. Garnet and spinel peridotites, or rather lherzolites, are found as xenoliths in basaltic rocks and kimberlites, and are thought to form in the upper mantle.

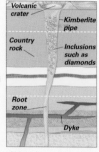

Volcanic crater
Kimberlite pipe
Country rock
Inclusions such as diamonds
Root zone
Dyke

Dunite

This rock is a kind of peridotite that is almost pure olivine. When freshly exposed it is olive green like olivine. But it gets its name from Mount Dun, the dun-coloured mountain in New Zealand, because dunite turns brown when weathered. Although some is formed in the mantle and appears in ophiolite complexes, much of it forms when olivine crystallizes out and sinks to the bottom of an intrusion of basic magma like gabbro. It is rich in a number of rare minerals, and is the world's major source of chromium ore.

Identification: Dunite is almost pure olivine, so when fresh it has a distinctive green colour as well as sugary texture. More usually though it is a tan brown, the colour it turns when exposed to the air.

Grain size: Medium to coarse (sugar-sized).
Texture: Granular.
Structure: May be layered.
Colour: Dull green, weathering to brown.
Composition: Silica (41%), Alumina (2%), Calcium and sodium oxides (1%), Iron and magnesium oxides (54.5%).
Minerals: Olivine; Clinopyroxene (augite); Orthopyroxene (enstatite).
Minor minerals: Chromite, magnetite, ilmenite, pyrrhotite, pyroxene.
Formation: At the base of gabbro intrusions or in the mantle.
Notable occurrences: Lizard (Cornwall), England; Anglesey, Wales; Skye, Ayrshire, Scotland; Harzburg, Odenwald, Silesia, Germany; Lherz (Pyrenees), France; Hungary; Norway; Finland; New Zealand; New York; Maryland.

Carbonatite

Almost all volcanic activity involves magmas based on silicates, so the existence of igneous rocks containing over 50 per cent carbonates is surprising. When first discovered in the early 20th century, these carbonate rocks, called carbonatites, were thought by geologists to be simply limestones moulded by the heat of silicate magmas, especially as many carbonatites occur interleaved with silicate rocks in volcanic structures called complexes. In fact, carbonatites originate as magmas in the mantle, just like other magmas. They are the coolest of all magmas, melting at just 540°C/1,004°F (compared to over 1,100°C/2,012°F for basalt), and carbonatite lavas look like liquid mud. Carbonatites are forcing geologists to rethink their understanding of mantle processes and magma production. There are 350 or so known carbonatite intrusions, over half of them in Africa. Most are quite small and occur beneath volcanoes, interweaving with formations of other magmas to create complexes. Even more surprisingly, there are a few places in the East African Rift Valley, and at Kaiserstuhl in the Rhine, where carbonatites have erupted on the surface. In 1960, it was realized that Oldoinyo Lengai in Tanzania is actually a carbonatite volcano.

Identification: Carbonatite often looks just like marble. It is the only white or even light grey igneous rock, and it is often possible to distinguish it from marble only by the place where it occurs and by complex laboratory tests.

Grain size: Medium to fine.
Texture: Granular.
Colour: White to grey (Sövite), cream/yellow (Beforsite), yellow-brown (Ferro-carbonatite).
Composition: Over 50% carbonates (<3% silicates).
Minerals: Calcite (Sövite carbonatite); Dolomite (Beforsite carbonatite) or Ankerite (Ferro-carbonatite). Plus apatite, phlogopite, aegerine, magnetite.
Accessory minerals: Many.
Trace elements: Includes barium, zircon, niobium, molybdenum, yttrium.
Formation: Intrusions in complexes with ijolites, syenites, fenites; eruptions of lava and ash with nephelinites.
Notable occurrences:
Intrusions: Fen, Norway; Kola, Russia; Loolekop, South Africa; Okurusu, Namibia; Ambar Dongar, India; Bayan Obo, Mongolia; Jacupiranga, Brazil; Oka, Quebec; Mountain P, CA
Volcanoes: Kaiserstuhl (Rhine), Germany; Kerimasi, Mosonik, Shombole, Oldoinyo Lengai, East African Rift Valley.

SYENITES

Unlike granites and their cousins, the syenites contain little or no quartz, but neither do they have large amounts of mafic minerals like peridotite and gabbro. Together, the syenites and their cousins make up a group called the alkaline igneous rocks, many of which are rich in feldspathoids, or 'foids', such as nepheline. The small amounts of mafic minerals they do contain are often in beautiful blues or greens.

Syenite

The name syenite was originally coined by the famous Roman scholar Pliny to describe the beautiful granite-like rocks quarried by the Ancient Egyptians at Syene (Aswan) on the Nile. In fact, these rocks are granite, and not syenite. But in the 18th century, the great German mineralogist A G Werner applied the term to some similar-looking rocks he found near Dresden, and the name now applies to igneous rocks like these.

The syenites are plutonic rocks that form massive intrusions rather like granite and, unlike the 'foid' syenites, are similarly rich in potassium feldspars. But syenites contain very much less quartz than granite. In fact, as the quartz content goes up they merge into granites, and since granites and syenites often form in the same places, quartz-rich syenites are hard to distinguish from quartz-poor granites.

Syenites are the underground equivalent of trachyte and are rich in alkali feldspars such as microcline and orthoclase. The orthoclase is white or pink and forms half the rock. So syenites, like the other alkaline igneous rocks, are all very light in colour – unlike mafic rocks that are low in quartz, such as basalt and peridotite, which are all dark.

Other minerals within syenites can make them among the most colourful and beautiful of igneous rocks, often displaying a shimmering iridescence when cut and polished. This is why they are such popular ornamental stones. Sodalite gives a lavender-blue tinge to soda syenite, while green aegerine can make aegerine syenite green. Syenites can sometimes be distinguished from granites because they contain dark blue needles of amphibole, while granite contains black prism-shaped amphiboles or micas.

Syenites are divided into three main kinds, according to which dark, ferromagnesian (iron and magnesium) mineral predominates – augite, hornblende or biotite mica. The Larvik area of southern Norway is famous for its beautiful rainbow-sheened red or grey augite syenites called larvikites or 'blue pearl granite', which are often used for pillars and facades. Similar rocks are found in the Sawtooth Mountains of Texas.

Sodalite syenite: Like granites, syenites can be divided into those in which the feldspars are mainly potash (potassium) or soda (sodium). Sodalite syenites such as pulaskite and nordmarkite are often perthitic, which means the sodium feldspar, mainly albite, intergrows with potash feldspar and gradually replaces it. Sodalite syenite can often be identified by lavender-blue specks of the feldspathoid sodalite.

Cancrinite syenite: This contains more of the feldspathoid mineral cancrinite than other syenites.

Grain size: Phaneritic (coarse- to medium-grained). Can be pegmatitic.
Texture: Usually even grains, but can often be porphyritic.
Structure: Often contains drusy cavities. Feldspars are can be perthitic – that is, they have intergrowths of potassium and sodium feldspar. May have veins of white albite.
Colour: Light red, pink, grey or white.
Composition: Silica (59.5% average), Alumina (17%), Calcium and sodium oxides (11%), Iron and magnesium oxides (8%), Potassium oxides (5%).
Minerals: Potassium feldspar (microcline, orthoclase); Plagioclase feldspar (albite, oligoclase, andesine); Biotite mica; Amphibole (hornblende); Pyroxene (augite). Can contain up to 10% quartz before it becomes granite.
Accessories: Sphene, apatite, zircon, magnetite, pyrites, plus feldspathoids nepheline, sodalite, cancrinite, feklichevite and leucite. Larger quantities of these grade the rocks towards 'foid' syenites.
Phenocrysts: Potassium feldspar, diopside, plagioclase.
Formation: Stocks, dykes and small intrusions, or interwoven with granites in large intrusions.
Notable occurrences: Larvik, Norway; Saxony, Germany; Alps, Switzerland; Piedmont, Italy; Azores; Kovdor Massif (Kola Peninsula), Russia; Ilmaussaq, Greenland; Pilanesburg (Bushveld complex), South Africa; Alaska; Dharwar, India; Montregian Hills, Quebec; White Mountains, Vermont; Sawtooth Mountains, Texas; Arkansas; Montana.

Nepheline or 'foid' syenites

Foyaite: Grey, green or red nepheline syenite with a lot of microcline.

Unlike ordinary syenites, the nepheline syenites contain no quartz whatsoever. Instead, they contain nepheline or another feldspathoid, or 'foid', such as sodalite or leucite – and the presence of quartz and these feldspathoids is mutually exclusive. Nepheline syenites are the intrusive equivalent of phonolite.

They are actually quite rare, and may form when soda-rich syenites and granites are altered in the melt, rather than forming distinct magmas. They often occur in ring dykes, and where granite magmas come into contact with limestones. There are many varieties of nepheline syenite, each with its own characteristics, including pink laurdalites from Laurdal in Norway, black mica-speckled miaskite from Miask in the Russian Urals, and nepheline-rich litchfieldite from Litchfield in Maine. But most come under the heading foyaites, after Foya in southern Portugal. The foyaites' odd chemistry means many rare minerals have been found in them, including eudialyte, eukolite, mosandrite, rinkite and lavenite.

Identification: Nepheline is often grey and can look like quartz, so it's easy to mistake nepheline syenite for granite or ordinary syenite. But nepheline syenite may have flow streaks and clots of dark minerals. And while quartz stays smooth on weathered surfaces, nepheline is pitted. The more nepheline there is, the greener it becomes. Blue sodalite and yellow cancrinite traces help to confirm the identity.

Grain size: Phaneritic (coarse). Can be pegmatitic.
Texture: Usually even grains, but often porphyritic.
Structure: Flow streaks and clots of dark minerals.
Colour: Usually grey, pink or yellow but can be greenish.
Composition: Silica (42% av), Alumina (15%), Calcium & sodium oxides (17%), Iron & magnesium oxides (18.5%), Potassium oxides (5%).
Minerals: Potassium feldspar (orthoclase, microcline); Plagioclase (albite, oligoclase); Nepheline; Biotite; Hornblende; Pyroxene (augite, aegerine).
Accessories: Sphene, apatite, zircon, magnetite, pyrites, plus sodalite, cancrinite & leucite.
Phenocrysts: Nepheline, albite.
Formation: Stocks, ring dykes and small intrusions, with granites and syenites.
Notable occurrences: Fen, Laurdal, Norway; Alnd I, Sweden; Foya, Portugal; Turkmenistan; Ilomba, Ulundi, Junguni, Malawi; Tasmania; Sierra de Tingua, Brazil; Quebec; White Mountains, VT; Beemerville, NJ; Magnet C, AR.

Ancient Egyptian masters of stone

No ancient culture used stone with such skill as the Ancient Egyptians. The land is blessed with many superb stones for construction and carving, including monzonite, black and red granite, quartzite, limestone, sandstone and graywacke (siltstone). Red granite quarried at Syeneh Aswan (pictured above) was cut in single giant blocks for obelisks. Quartzite quarried at Gebelein was used for the famous Colossi of Memnon. Limestone from Tura, Beni Hassan and Ma'asa was used for the first pyramid, Djoser's. The Egyptians had a knowledge of geology and geological strata well beyond any culture of the time – and unmatched until recently. They were also amazingly skilled at cutting and shaping the stones. No-one knows quite how they did, since, strangely, stone-working does not appear in hieroglyphics. Some think they cut stones with copper saws, or bronze wire held in a bow. Others think they used emery stone.

Monzonite

Monzonite gets its name from Monzoni in the Italian Tyrol. Like syenite, it is a plutonic rock that contains some, but not a lot of, quartz, and lies between the foids and the granites. In fact, it is actually what some geologists describe as the 'average' igneous rock, containing equal amounts of potassium and plagioclase feldspar, and lying halfway between the most acid and the most basic rocks in composition. Although by no means a rare rock, it only occurs in small masses, associated with and perhaps blending into gabbro and, at the margins of an intrusion, pyroxenite. There are three kinds of monzonite, depending on the presence of quartz, nepheline or olivine. Quartz monzonite is found in many mountain belts, and because it is such a tough rock often forms dramatic landforms.

Identification: Monzonite is light coloured like granite but contains less quartz.

Grain size: Medium.
Texture: Irregular plates of orthoclase embedded in plagioclase.
Structure: Crystals often have zones of different chemical composition.
Colour: Dark grey.
Composition: Silica (58.5% av), Alumina (17%), Calcium & sodium oxides (11%), Iron & magnesium oxides (9%), Potassium oxides (3%).
Minerals: Potassium feldspar (orthoclase); Plagioclase feldspar (labradorite, oligoclase); Pyroxene (augite, hypersthene, bronzite); Hornblende; Quartz, Nepheline or olivine; Biotite.
Accessories: Apatite, zircon, magnetite, pyrites.
Phenocrysts: Apatite, augite, orthoclase.
Formation: Stocks, dykes and intrusions.
Notable occurrences: Kentallen (Argyll), Scotland; Norway; Monzoni, Italy; Sakhalin Is, Russia; Yogo Peak (yogoite), Beaver Creek, Montana; Black Canyon, Colorado (quartz monzonite).

PLUTONIC: GRANITE

Granites are by far the most common of all the plutonic rocks, rocks that form deep underground as giant batholiths, tens or even thousands of kilometres across. Although granites form only deep underground, they are often seen on the surface because their high quartz content makes them so tough they may survive long after softer rock around them has been worn away by weathering.

Granite

Granite lasts. Long after other rocks have crumbled away, intrusions of granite stand proud, like islands above the sea. The pointing finger of Rio's Sugar Loaf mountain, the sheer cliffs of Yosemite's El Capitan and the wild wastes of England's Dartmoor all stand testament to granite's ability to endure. For all these eminences are batholiths – huge intrusions of magma that formed entirely underground, and have simply been exposed through erosion of overlying rocks. The moors of Cornwall and Devon are all just knobs on a single large batholith that will, in 100 million years' time, be thoroughly denuded just as the moors are now.

Granite batholith complexes can be huge. The Patagonian batholith underlying the southern Andes is 1,900km/935 miles long and up to 65km/40 miles wide. That underlying the Sierra Nevada is almost as vast. In fact, there are giant granite batholiths underlying most of the world's great mountain ranges, both ancient, such as the Appalachians of North America, and recent, such as the Himalayas. Indeed, granite is always closely linked to mountain building, and the margins of continents where subduction is going on.

Although the bulk of granite is in batholiths, it can also form dykes and sills, and veins and intrusions of one granite can cut across another. Granite intrusions frequently interweave with country rocks, changing them where they come into contact. In many places, granite blends imperceptibly into metamorphic granite gneisses. Granite intrusions engulf lumps of country rock, which become xenoliths, often partly altering them on the surface at least.

Granite is a light-coloured, speckled rock and is very 'acidic' – that is, it has a high silica content (at least 70 per cent), and a high proportion of quartz (at least 20 per cent). It is essentially a mix of white or pink feldspar, pale quartz and small specks of black muscovite mica. Because it forms by cooling slowly deep underground, crystals in granite are almost invariably large enough to be visible to the naked eye. Most are at least a few millimetres. The largest white feldspar crystals can be prisms up to 20cm/8in long.

White granite: Granites all have soft black flakes of mica, the first to crumble from weathered surfaces. They all have pale crystals of quartz, too, and this never crumbles. It is the large feldspar crystals that typically give the colour varieties.

Feldspar

Pink granite: Most granites are light grey or pinkish in colour, but can also be dark grey or red. They are light grey if they are dominated by a white alkali feldspar, but pink or red if their alkali feldspar is pink or red. If there are both white and pink feldspars, the pink one is likely to be alkali feldspar and the white one is likely to be plagioclase.

Grain size: Phaneritic (coarse-grained); often pegmatitic.
Texture: Normally granular, with large visible crystals and often porphyritic with large phenocrysts.
Structure: Typically uniform, but may be banded. Xenoliths are common. Near the crown of batholiths, may be cracked into massive rectangular blocks.
Colour: Usually light coloured – mottled white, grey, pink or red, with black specks.
Chemical composition: Silica (72% average), Alumina (14.5%), Calcium and sodium oxides (4.5%), Iron and magnesium oxides (less than 3.5%).
Minerals: Quartz; Potassium feldspar (microcline) and plagioclase feldspar (oligoclase); Mica; Hornblende.
Accessories: Aegerine, zircon, apatite, magnetite, amphibole or pyroxene.
Phenocrysts: Quartz, orthoclase and oligoclase feldspar, hornblende, biotite mica, augite.
Formation: Intrusive: batholiths, stocks, bosses, sills, dykes.
Notable occurrences: Syenogranite: Donegal, Ireland; Cairngorms, Scotland; northern Nigeria; Réunion Islands; Azores Islands; Canary Islands; Sugar Loaf Mountain (Rio de Janeiro), Brazil; Appalachian Mountains. Monzogranite: Cornwall, Devon, Lake District, England; Baltic shield, Finland and Sweden; Massif Central, France; Spain; Tatra Mts, Slovakia; Barrens, Newfoundland. Granodiorite: Andes; Rocky Mountains. Leucogranites: Himalayan Mountains.

The granite problem

Just how the world's giant granite batholiths formed has been the subject of fierce debate since the 18th century. The great Austrian geologist Eduard Suess could see they formed after the surrounding country rocks but, if so, how was such an enormous volume of granite accommodated? This 'room' problem has been at the heart of the controversy ever since.

In the mid-20th century, thinking developed into two opposing camps – the granitizers and the magmatists. The granitizers, like Doris Reynolds, believed that granite was formed when existing country rock is granitized (changed to granite) by 'metasomatic' processes. It all involved gases or fluids which came to be called 'ichor' after the lifeblood of the Greek gods. The idea was that country rock was metasomatized (altered) to granite by ichor emanating through it. That way, there was no 'room' problem. Magmatists, such as Norman Bowen, on the other hand, believed that granite formed from magma, not altered country rock. The magma was created at depth by the partial melting or 'anatexis' of the country rocks above.

Gradually, field observation of rock formations and laboratory experiments with melting rocks came down firmly on the side of the magmatists, and it is now widely accepted that most granites do indeed form by partial melting.

The theory is that it all takes place in the later stages of mountain building, as two tectonic plates crunch together. The collision not only buckles the plate edges to throw up mountain ranges, but also eventually creates such extreme pressures and temperatures that huge volumes of the mountain roots partially melt to create granite magma.

These hot granite melts are lighter than the rocks above, and so well up into them, like a blob of hot oil in a lava lamp, creating their own space. As they ascend they begin to cool and solidify, eventually turning to solid rock again.

Although the magmatists appear to have won this particular round of The Granite Problem, the controversy is by no means over. If granites do form by partial melting, for instance, then just which rocks have melted to form granite? Moreover, even among those who believed granite was a

magma, there were always those who believed that it formed not directly by melting of other rocks, but from basalt magmas changed gradually to granite as some chemicals crystallized and others didn't, a process known as fractional crystallization.

The evidence is that most continental granites – the great batholiths under mountains – do form by partial melting. But fractional crystallization may play a part in granites such as tonalites, which form in oceanic locations such as island arcs.

Granite varieties

Granite and granite-like rocks, called granitoids, come in so many subtly different varieties they have proved a nightmare to classify. One method is according to the balance of Quartz, Alkali Feldspar and Plagioclase feldspar (QAP) in their make-up. In the middle lies monzogranite in which alkali and plagioclase feldspar are fairly even. Syenogranite has more alkali feldspar, granodiorite has less. The extremes are alkali-rich alkali feldspar granite and plagioclase-rich tonalite. After the discovery that magmas making granites in eastern Australia came from both sedimentary and igneous rocks, and contained xenoliths of either type, geologists developed an 'alphabetic' way of dividing granites. I-type granites, rich in biotite mica and maybe hornblende, have a chemical make-up that implies they were formed from mafic igneous rocks. S-types, rich in both biotite and muscovite mica (plus garnet, cordierite and sillimanite), have a make-up that implies a sedimentary origin. M-types have a make-up implying they came from Earth's mantle. A-types have a make-up implying they are anorogenic – that is, they formed not in areas of mountain building but near hot spots and rifts.

Granophyre or porphyric microgranite: At the edges of intrusions, in thin intrusions and pegmatite, granite grains can be small. These microgranites often have phenocrysts of feldspar formed earlier and are related to graphic granites.

Graphic granite: Graphic granite is an extraordinary kind of granite that occurs in pegmatites. The entire rock is one large crystal of pale feldspar, typically microcline. Embedded in the feldspar are thin wedge-shaped growths of quartz, giving an effect that looks like the cuneiform writing of Ancient Sumeria. It is thought that both the feldspar and the quartz formed at the same time.

Quartz   *Feldspar*

GRANITE VARIETIES

Alkali feldspar granite: Granite with a very high proportion of alkali feldspar and almost no plagioclase.

Biotite granite: Granite containing up to 20% biotite mica.

Flaser granite: Granite-like gneiss, which has flattened feldspar crystals giving it a foliated look.

Augite granite: Granite rich in dark augite.

Tourmaline granite: Granite rich in black tourmaline.

Leucogranite: Light-coloured granite with less than 30% of the norm for mafic minerals.

Melanogranite: Dark-coloured granite with more than 30% of the norm for mafic minerals.

Granite gneiss: Gneiss formed from a sedimentary or metamorphic rock and with the same mineral composition as granite.

Peralkaline granite: Granite rich in alkaline feldspars and also alkaline amphiboles and pyroxenes such as aegerine and riebeckite.

GRANITOIDS

There are many varieties of granite and granitoids (granite-like rocks). Some form only small patches within other granite outcrops, such as rapakivi granite and orbicular granite, with their highly distinctive round markings. Others, such as granodiorite and tonalite, have their own particular chemical and mineral make-ups and are individual rock types in their own right.

Rapakivi granites

Rapakivi granites get their name from the Finnish for 'rotten rock' because they weather easily. They are granites – usually syenogranites – with an unusual grain pattern called rapakivi texture. In rapakivi texture, large, oval crystals of alkali feldspar such as sanidine are encased in a plagioclase feldspar such as albite. The alkali feldspar forms first, then is enveloped by a 'reaction rim' of plagioclase as it reacts with the surrounding magma. One theory is that alkali feldspars from rhyolite magma react with plagioclase from basaltic magma as the two magmas mix. Another is that the effect occurs as pressure drops in a rising magma. Some geologists believe that the conditions for rapakivis occurred along continental rifts that never quite developed. Rapakivi granites dating back 1,100 to 1,800 million years are found in a belt stretching right across Finland, Sweden and the Baltic to Labrador and the American south-west. They also occur in Brazil and Venezuela, and small pockets of more recent rapakivis are widely scattered.

Identification: Rapakivi granite is distinctive, with its pale rounds of feldspar embedded in a dark groundmass of mica, hornblende and quartz. When cut and polished, as here, rapakivi makes a popular decorative stone.

Grain size: Mixed.
Texture: Oval feldspar phenocrysts up to 2cm/0.8in across embedded in a groundmass of small crystals.
Structure: Typically uniform.
Colour: Pink or tan K feldspar crystals rimmed with white albite in a dark groundmass.
Chemical composition: Silica (72% average), Alumina (14.5%), Calcium and sodium oxides (4.5%), Iron and magnesium oxides (3.5%).
Minerals: Quartz; Potassium feldspar (sanidine) and plagioclase feldspar (albite); Mica.
Phenocrysts: Usually alkali feldspar, but can be quartz or plagioclase feldspar.
Formation: In syenogranites.
Notable occurrences: South Finland; South-east Sweden; St Petersburg, Russia; Estonia; Poland; Brazil; Venezuela; Labrador; Ontario; Maine; US mid-west and south-west.

Orbicular granite

Occasionally, granites may contain small patches with an unusual texture called orbicular granite. This looks a little like rapakivi, but the balls, or 'orbicules', are bigger and they are not phenocrysts, but formations that develop around cores of foreign material in the magma. Each core may be a grain of another igneous rock (a small xenolith), but could also be a grain of granite. Alternating layers of pale feldspar and dark biotite or hornblende grow around the core, a process called 'rhythmic crystallization' in which first one mineral crystallizes then another as conditions change in the magma.

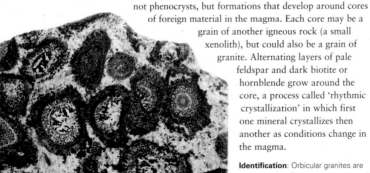

Identification: Orbicular granites are masses only a few metres across, and are easy to identify with their black and white rounds.

Grain size: Mixed.
Texture: Large round orbicules 2–15cm/0.8–6in across embedded in a groundmass of small crystals.
Structure: Typically uniform.
Colour: Black and white layered phenocrysts in a light grey groundmass.
Chemical composition: Silica (72% average), Alumina (14.5%), Calcium and sodium oxides (4.5%), Iron and magnesium oxides (3.5%).
Minerals: Quartz; Potassium feldspar (microcline) and plagioclase feldspar (oligoclase); Mica.
Notable occurrences: Finland; Sweden; Waldviertel, Austria; Riesengebirge, Poland; Japan; New Zealand; Peru; Vermont.

Granodiorite

Granodiorite is the intrusive equivalent of dacite and is the most abundant of all the granitoid rocks. It is very similar to granite but contains more plagioclase feldspar, and more mafic minerals (usually biotite and hornblende). In fact, granodiorite and granite often occur together, along with diorite, which contains even more plagioclase. In large batholiths, for instance, a single granodiorite magma may develop a granite heart and a skin of diorite or even tonalite, as minerals separate out in a particular way from the magma. This kind of process tends to happen only in large batholiths though. Where granite, granodiorite and diorite are all found in a smaller intrusion, the chances are they all came from separate magmas. Granitoid rocks such as granodiorite often occur in 'suites' – repeated associations of particular similar rocks. Granodiorite, for instance, is found in ancient, Archean formations along with tonalite and trondheimite. These TTG suites are among the world's oldest rocks, dating back more than two billion years, and are found all around the world, in places such as Lapland in Scandinavia, and the Big Horn Mountains of Wyoming.

Identification: Granodiorite is generally grey, and looks quite like granite, but it contains a higher proportion of dark minerals such as biotite mica and hornblende. Granite looks basically light grey with black specks. In granodiorite the grey and the black are more evenly balanced, giving a 'salt-and-pepper' look.

Grain size: Phaneritic (coarse).
Texture: Even texture; often porphyritic.
Structure: Typically uniform
Colour: Black and white with pink potassium feldspar.
Chemical composition: Silica (67%), Alumina (16%), Calcium & sodium oxides (7.5%), Iron & magnesium oxides (6%).
Minerals: Quartz; Plagioclase feldspar (oligoclase); Potassium feldspar (sanidine); Biotite mica; Hornblende.
Accessories: Zircon, apatite, magnetite, ilmenite, sphene.
Phenocrysts: Quartz or plagioclase feldspar.
Formation: Intrusive: stocks, bosses, batholiths, sills, dykes.
Notable occurrences: As for granite but also Aleutian Islands; Sonora, Mexico; Peninsular Mts, Baja California; Sierra Nevada, California.
TTG suites: Lapland, Finland; Barberton Mt Land, South Africa; Pilbara, Yilgarn, Australia; Big Horn Mts, Wyoming.

Tors

Granite-like rocks form underground, but are so tough they are often left standing proud after softer rocks are worn away. In some places, granite

gives huge bare-rock cliffs. In others it gives rounded hills topped by outcrops of bare rock the size of a house, usually called by the ancient Cornish name 'tors'. Tors are a distinctive feature of the moors of England's south-west, but occur in many other places such as Scotland's Cairngorms and in South Africa, where they are called 'castle koppies'. There are several theories on how Cornish tors formed, but all are connected with the pattern of cracks, or 'joints', that develops parallel to the surface of the rock. Tors are tough clumps of rock that have survived after weaker surrounding granite was stripped away. One theory is that the softer granite was weathered in an earlier tropically warm age by natural chemicals seeping into the joints deep below ground. Another is that it was weathered by frost in the Ice Ages. The debris was, in both cases, probably swept away at the end of the Ice Ages by a process called solifluction, in which the water in frozen ground melts, turning the soil into a liquid mush that flows easily away.

Tonalite

This rock takes its name from Tonale in the Italian Alps near Monte Adamello. It is the quartz-rich, granitoid equivalent of diorite. It has the least potassium feldspar and the most plagioclase of any of the granite-like rocks, and also the most of the dark mafic minerals such as hornblende and biotite. The hornblende is often greenish rather than brown, while the biotite is pleochroic – that is, shows different colours from different directions. Tonalite is actually quite like granodiorite, and the two frequently occur together in TTG suites (see Granodiorite above). They have the same black and light grey 'salt-and-pepper' look, and can be hard to tell apart – tonalite has more black pepper.

Identification: Tonalites look very like granite, but are slightly darker and browner.

Grain size: Phaneritic (coarse).
Texture: Even texture; often porphyritic.
Structure: Often threaded by veins of quartz and feldspar (aplites).
Colour: Pink or tan potassium feldspar crystals rimmed with white albite in a dark groundmass.
Chemical composition: Silica (58%), Alumina (17%), Calcium and sodium oxides (10%), Iron and magnesium oxides (11%).
Minerals: Quartz; Plagioclase feldspar (oligoclase); Biotite mica; Hornblende.
Accessories: Zircon, apatite, magnetite, orthite, sphene.
Phenocrysts: Quartz or plagioclase feldspar.
Formation: Intrusive: stocks, bosses, batholiths, sills, dykes.
Notable occurrences: Galloway, Cairngorms, Scotland; Ireland; Rieserferner and Traversella (Tyrol), Austria/Italy; Andes, Patagonia; Sierra Nevada, California; Alaska.
TTG suites: Lapland, Finland; Barberton Mt Land, South Africa; Pilbara, Yilgarn, Australia; Big Horn Mts, Wyoming.

GABBRO AND DIORITE

Gabbro and diorite occupy the middle ground of intrusive igneous rocks as far as the mineral balance goes. On the one hand, they contain only a little quartz and alkali feldspar – markedly less than the granitoid rocks. On the other, they contain only moderate amounts of olivine – much less than the ultramafics such as peridotite. Instead, they are made predominantly of plagioclase feldspar.

Diorite

Diorite is the coarse-grained, plutonic equivalent of andesite. It is darker than granitoids and contains heavier minerals, but it has a similar grain structure and forms in a similar way. Diorite is one of the igneous rocks that make their presence felt along continental margins where subduction of tectonic plates is throwing up mountain chains like the Andes. With granite, it forms the great long batholiths that underlie so many of these mountain chains. There is much less diorite than granite, and it often forms when rocks are caught up in a granite intrusion. Diorite contains much less quartz and alkali feldspar than granite, but more plagioclase feldspar than any other rock except anorthosite – over 75 per cent, even more than gabbro. The rest is mainly dark minerals such as hornblende and biotite. Diorite is similar to gabbro, but diorite's plagioclase tends to be oligoclase and andesine, whereas in gabbro much more of it is anorthite, bytownite and labradorite.

Dark diorite: Sometimes dark minerals can predominate over light in diorite.

Identification: Diorite looks very similar to gabbro and it can be quite hard to tell them apart. Yet even though it contains very little quartz, its high plagioclase feldspar content means that light minerals are usually more prominent in diorite than gabbro. Essentially, diorite is light grey with patches of black, while gabbro is black with patches of light grey, but this is by no means a hard and fast rule.

Grain size: Phaneritic (coarse-grained), occasionally pegmatitic.
Texture: Even-grained or porphyritic – often close together in the same rock.
Structure: Foliation and xenoliths common.
Colour: Speckled black and white, occasionally dark greenish or pinkish.
Composition: Diorite: Silica (58.5% average), Alumina (17%), Calcium and sodium oxides (10.5%), Iron and magnesium oxides (11%), Potassium oxides (2%). Monzodiorite: Silica (58% average), Alumina (17%), Calcium and sodium oxides (11%), Iron and magnesium oxides (10%), Potassium oxides (2%).
Minerals: Plagioclase feldspar (oligoclase or andesine); Biotite mica; Amphibole (hornblende); Small amounts of pyroxene (augite), quartz and alkali feldspar (sanidine). Feldspathoid diorite and feldspathoid monzonite: Foids (nepheline) instead of quartz.
Accessories: Magnetite, apatite, zircon, titanite, olivine.
Phenocrysts: Plagioclase feldspar, hornblende.
Formation: Intrusive: sills, dykes, stocks, bosses, plus xenoliths in granite.
Notable occurrences: Argyll, Scotland; Jersey (Channel Islands), England; Bavarian Forest, Black Forest, Harz, Odenwald, Germany; Finland; Washington; Massachusetts.

Monzodiorite

This rock is midway between monzonite and diorite in composition. That means most monzonite contains some quartz, and a little more plagioclase than alkali feldspar, and fewer dark minerals than diorite. There is another kind of monzonite – feldspathoid monzodiorite – which contains nepheline or other foids instead of quartz. Monzodiorite used to be called syenodiorite, but IUGS (International Union of Geological Surveys) recommended it should be called monzodiorite to avoid confusion with monzonite and monzosyenite, which also lie in between syenite and diorite in composition, but contain more alkali feldspar.

Identification: Monzodiorite has the same 'salt-and-pepper' look as diorite and monzonite, but has slightly more 'pepper' than monzonite and slightly less than diorite.

Gabbro

Polished gabbro: Rarely, gabbro is cut and polished and used as a decorative stone.

Named after a town in Tuscany, Italy, by the great German geologist Christian Leopold von Buch, gabbro is the coarse-grained, intrusive equivalent of basalt and dolerite. It is a dark rock, basically made up of plagioclase and pyroxene. More pyroxene and less plagioclase merges it into peridotite; less pyroxene and more plagioclase blends it into diorite. Gabbro is a very widespread rock, especially in the oceanic crust, where it forms part of the ophiolite sequence. This is the sequence of rocks down through the ocean bed that develops either side of a mid-ocean rift. As pillow lavas and sheeted dykes of erupted basalt form the top of the sequence, so gabbro continually freezes in clumps from molten peridotite on to the magma chamber walls beneath as the walls move apart. Over millions of years, this has created a layer of gabbro underneath all the world's oceans.

Gabbro can also flow into sills and dykes, and occasionally, huge sheet complexes called lopoliths, like that at the Bushveld complex in South Africa, Duluth in Minnesota and Rhum in Scotland. In many places, the sinking of heavier minerals as they crystallized has created distinct layers in the gabbro with dark minerals concentrated at the bottom and light minerals at the top of each layer.

The Sudbury structure
In places around the world, there are huge, mostly ancient layered intrusions in which mafic magmas have crystallized in layers of different chemical composition. The biggest by far is the Bushveld complex in South Africa, which covers 65,000km²/25,000 sq miles. One of the most fascinating is at Sudbury in Ontario, Canada. The area is one of the world's richest sources of nickel copper, found in association with gabbro, and was once thought to be entirely igneous in origin. Now geologists have realized that it is not actually igneous at all but the huge impact crater of a meteorite that struck the ground here 1,850 million years ago – the biggest ever. This is clear from shatter cones, rocks fractured in a cone shape by the impact (above). What makes this crater unusual, even for a meteor crater, is that it is oval – 200km/124 miles long and only about 100km/62 miles wide. Scientists believe, from mapping the geology of the area and creating various models, that it was created by a meteorite of some 10–19km (six to 12 miles) in width, which exploded on impact with the force of 10 billion Hiroshima bombs, fracturing the Earth's crust and bringing magma rich in mineral ores to the surface. This resulting heat melted the exposed granite and gneiss rocks into a glassy magma that deformed the crater. This mafic magma sits on top of the layered gabbro.

Identification: Gabbro can look similar to diorite, but tends to be darker as it contains the darker plagioclases (labradorite and bytownite) rather than the lighter plagioclases (oligoclase and andesine) of diorite. Gabbro is often ophitic (see texture, right), giving it a frosted look.

Gabbro is a very tough rock, which is why it is used widely for railway ballast and road metalling, but it is one of the least attractive of the intrusive igneous rocks, so is much less used as a decorative stone than the granites or syenites. However, it is virtually the only significant source of nickel, chromium and platinum minerals.

Grain size: Phaneritic (coarse-grained), occasionally pegmatitic.
Texture: Even-grained or porphyritic – often close together in the same rock mass. Frequently ophitic, which means long light plagioclase feldspar crystals are enveloped by dark pyroxene (augite) crystals.
Structure: Foliation and xenoliths common. Often forms alternating layers, with mostly light minerals at the tops and mostly dark at the bottoms of the layers.
Colour: Black and white or grey, occasionally dark greenish or bluish.
Composition: Silica (58% average), Alumina (17%), Calcium and sodium oxides (10.5%), Iron and magnesium oxides (11%), Potassium oxides (2%).
Minerals: Plagioclase feldspar (labradorite or bytownite); Pyroxene (augite); Olivine; Small amounts of amphibole (hornblende) and biotite mica; Very small amounts of quartz and alkali feldspar (sanidine).
Accessories: Magnetite, apatite, ilmenite, picotite, garnet.
Phenocrysts: Plagioclase feldspar, hornblende.
Formation: Intrusive: batholiths, lopoliths, sills, dykes, stocks, bosses, plus xenoliths in granite.
Notable occurrences: Shetland, Skye, Rhum, Aberdeen, Argyll, Scotland; Pembroke, Wales; Lake District, Lizard (Cornwall), England; Skaergaard, Greenland; Bergen, Norway; Odenwald, Harz, Black Forest, Germany; Wallis, Switzerland; Bushveld complex, South Africa; Jimberlana, Windimurra, Western Australia; Eastern Canada; Baltimore, Maryland; Peekskill, New York; Stillwater, Montana; Duluth, Minnesota.

Monzogabbro: Monzogabbro is gabbro with slightly less pyroxene and containing feldspathoids rather than quartz.

GABBROIC ROCKS

Gabbros are phaneritic (coarse-grained) rocks made of plagioclase feldspar, pyroxene and olivine. They are divided into various types according to how much of each they contain. Anorthosite is rich in plagioclase; troctolite is rich in plagioclase and olivine but not pyroxene; gabbro is rich in plagioclase and pyroxene but not olivine; essexite is rich in pyroxene. Norite is the midpoint.

Norite

Norite is a similar rock to gabbro, based on a mix of plagioclase, pyroxene and olivine, and the two often form in the same large, layered intrusions as the mix separates during crystallization. Norite contains very slightly less plagioclase than gabbro, but the real difference is that gabbro's pyroxene is a clinopyroxene such as augite, while norite's is an orthopyroxene such as hypersthene. Unfortunately, the two can look so alike that they are impossible to distinguish without a microscope. Norite typically occurs in small, separate intrusions, or as layers along with other mafic igneous rocks such as gabbro. Norite also formed in association with ancient basalt intrusions, beneath huge basalt dyke swarms. One famous norite intrusion is at Sudbury in Ontario. Here a cavity 30m/98ft deep has been excavated from solid norite to house the Neutrino Observatory to detect neutrinos, minute particles streaming from the stars. Norite is unusually low in natural radioactivity and acts as a shield to allow scientists to block out unwanted background radiation.

Identification: Norite is a dark grey rock with a slightly matted look dominated by quite long, prismatic black hypersthene or enstatite crystals. It looks very like gabbro, but the plagioclase feldspar tends to be sandy coloured, while in gabbro it is whiter.

Grain size: Phaneritic (coarse-grained), occasionally pegmatitic.
Texture: Even-grained or porphyritic.
Structure: Layering and xenoliths are common.
Colour: Dark grey, bronze.
Composition: Silica (58%), Alumina (17%), Calcium and sodium oxides (10.5%), Iron and magnesium oxides (11%), Potassium oxides (2%).
Minerals: Plagioclase feldspar (labradorite or bytownite); Pyroxene (hypersthene); Olivine; A little hornblende, biotite mica, quartz and alkali feldspar.
Accessories: Magnetite, apatite, ilmenite, picotite.
Phenocrysts: Plagioclase feldspar, hornblende.
Formation: Intrusive: dykes, stocks, bosses, often with gabbro.
Notable occurrences: Aberdeen, Banff, Scotland; Norway; Great Dyke, Zimbabwe; Bushveld complex, South Africa; Sudbury, Ontario.

Anorthosite

Identification: Anorthosite is the lightest coloured of all the gabbroic rocks. Dark and light minerals are often aligned in long crystals, giving anorthosite a streaky look.

Anorthosite is almost entirely plagioclase feldspar. Over 90 per cent is either bytownite or labradorite. Labradorite crystals may show an iridescence known as labradorescence. Although not as abundant as basalt and granite, anorthosite often occurs in huge formations such as in Labrador in Canada, and in giant complexes such as South Africa's Bushveld along with gabbro and norite. It is also one of the rocks that makes up the Moon's surface. While lunar seas are basalt, highlands are anorthosite. When the Moon was young its surface was melted, not only from heat within but by meteor impacts. Light plagioclase feldspar floated to the top and, when the lunar surface cooled, it solidified to form anorthosite. The Earth has much less anorthosite, but it is found in ancient rocks. On the early Earth, anorthosite may have been as abundant as on the Moon, but Earth's surface is so dynamic that most has long since vanished.

Grain size: Phaneritic (coarse-grained).
Texture: Long crystals often aligned.
Structure: Layering common.
Colour: Light grey.
Composition: Silica (51%), Alumina (26%), Calcium and sodium oxides (16%), Iron and magnesium oxides (5%).
Minerals: Plagioclase feldspar (labradorite or bytownite); Small amounts of pyroxene; olivine; magnetite and ilmenite.
Formation: Intrusive: dykes (rare), stocks, batholiths, often with gabbro.
Notable occurrences: Norway; Bushveld complex, South Africa; Sudbury, Ontario; Labrador; Stillwater, Montana; Adirondacks, New York; the Moon.

Bushveld Complex
South Africa's Bushveld complex in the former Transvaal is one of the world's great geological wonders. It is by far the largest layered intrusion, covering up to 65,000km²/25,097 sq miles and reaching up to 8km/5 miles thick. It is incredibly rich in minerals, containing most of the world's chromium, platinum and vanadium resources, as well as a great deal of iron, titanium, copper and nickel. The whole complex formed in a remarkably short time about 2,060 million years ago, as magmas were poured out on the surface and intruded into the ground to form large, complex layers of gabbroic and mafic rocks such as norite, anorthosite and pyroxenite. Some geologists thought it was created by the hotspot above a mantle plume; others, noting the coincidence of dates with the nearby Vredefort meteor crater, thought it was a meteorite impact feature. A recent theory uses both ideas, suggesting a meteorite impact triggered off a mantle plume to fountain huge amounts of magma up from the mantle – almost as if the meteorite had burst the Earth's crust.

Essexite

Named after Essex County in Massachusetts where it occurs, essexite is the rock used in Scotland in its porphyritic form to make curling stones. It is the gabbro that forms when there is less silica in the melt. It is less viscous than gabbro and flows into small intrusions near the surface, cooling quickly to form medium- and fine-grained rocks. The lack of silica means that nepheline forms in essexite instead of quartz, so it is actually a foid rock, like foyaite and nephelinite. Essexite is also richer than gabbro in pyroxene, and its pyroxene is titanaugite.

Grain size: Medium to fine.
Texture: Granular, sometimes porphyritic.
Structure: Layering common.
Colour: Light grey.
Composition: Silica (45%), Alumina (15%), Calcium and sodium oxides (17%), Iron and magnesium oxides (18.5%), Potassium oxides (5%).
Minerals: Plagioclase feldspar (labradorite or anorthite); Pyroxene (augite); Biotite; Hornblende; plus small amounts of nepheline and alkali feldspar.
Formation: Small intrusions, dykes, sills, often with gabbro.
Phenocrysts: Augite.
Notable occurrences: Lanarkshire, Ayrshire, Scotland; Kaiserstuhl, Baden, Germany; Oslo, Norway; Roztoky, Czech Republic; Tyrol, Italy; Essex Co, MA.

Identification: Essexite is a fine- to medium-grained grey rock often with slightly larger dark spots of augite. It is attractively evenly mottled.

Troctolite

This is a gabbroic rock which has almost no pyroxene. Instead it is made of plagioclase feldspar and olivine, midway between anorthosite and peridotite. It very often occurs in layered igneous complexes such as South Africa's Bushveld, and, most famously among geologists, the Isle of Rhum in Scotland's Hebrides. Layered complexes are intrusions that seem a layer-cake of related igneous rocks formed within a single magma chamber in the Earth's crust. The Rhum complex formed some 60 million years ago, during the birth of the North Atlantic Ocean, when ancient north-west Europe and North America began to drift apart, allowing magma to flood on to the surface. The original magma was probably an olivine-rich basalt, but as minerals crystallized in the magma chamber, heavier minerals probably sank to the bottom, creating layers of troctolite on top of layers of peridotite. Troctolites elsewhere probably formed in a similar way.

Identification: In German, troctolite is called *Forellenstein*, which means 'trout rock', and the name is apt, for it looks just like the skin of a trout. The medium-grained dark grey plagioclase looks like the trout's scales. Olivine forms black spots within it which, just like a trout's spots, can be red, green or brown when wholly or partly altered to serpentine by exposure to the weather.

Grain size: Medium to coarse.
Texture: Granular.
Structure: Layering common.
Colour: Grey studded with black, occasionally red or green.
Composition: Silica (51%), Alumina (26%), Calcium & sodium oxides (16%), Iron & magnesium oxides (5%).
Minerals: Plagioclase feldspar (labradorite or anorthite); Olivine; Small amounts of pyroxene, magnetite and ilmenite.
Formation: Intrusive: dykes, cone sheets, stocks, laccoliths, often with gabbro.
Notable occurrences: Rhum, Scotland; Cornwall, England; Oslo, Norway; Harz, Germany; Wolimierz (Silesia), Poland; Niger; Great Dyke, Zimbabwe; Bushveld complex, South Africa; Stillwater, Montana; Oklahoma.

DYKE, SILL AND VEIN ROCK

Fingers of magma ooze out into the country rock from every intrusion, either cutting across strata as dykes in which lamprophyres may form, or sliding between as sills to form rocks such as dolerites. As the intrusion begins to cool, cracks open in the solidifying rock. Residual fluids ooze into these cracks, altering the surrounding rock to greisens or solidifying to form veins of new rocks such as aplite.

Aplite

Aplites are unusually pale igneous rocks with fine, even grains that look just like unrefined sugar. They are closely related to pegmatites, and likewise form veins of crystalline igneous rock. Aplites, though, are fine-grained and tend to be much simpler in composition. In fact, they are basically quartz and potassium feldspar, with no mica, which is why they are so pale in colour. Moreover, while there are often complex zones of different composition in pegmatites, aplites are generally uniform throughout. Aplite veins form in almost every large granitic intrusion, striking like a pale scar across the host rock, and rarely more than a few centimetres across. As the intrusion cools and begins to crack, aplite veins develop when residual magma fills up the cracks. These aplites form at the lowest temperature of any igneous rock and water comes out of the melt as it loses pressure. So they crystallize very rapidly creating a texture that is remarkably fine-grained considering they form deep underground.

Identification: With its fine crystalline texture, aplite has a sugary look almost like pale sandstone. Unlike sandstone, though, the grains in aplite interlock, and there is none of the cement that glues sandstone grains together.

Grain size: Fine-grained.
Texture: Even-grained or, occasionally, porphyritic.
Structure: None.
Colour: Pale pink or whitish.
Composition: Silica (75% average), Alumina (14.5%), Calcium and sodium oxides (3.5%), Iron and magnesium oxides (5%).
Minerals: Quartz; Potassium feldspar (orthoclase or microperthite).
Accessories: Plagioclase feldspar, muscovite, apatite, tourmaline.
Phenocrysts: Quartz, orthoclase feldspar, tourmaline.
Formation: Aplite forms dykes in granite and granitic intrusions. It occasionally forms independent bosses, or on the rim of an intrusion.
Notable occurrences: Wherever there are large granitic intrusions.

Greisen

Identification: Greisen almost always occurs within granite, but is pale grey with almost no black mica.

Strictly speaking, greisen is a metamorphic, not igneous, rock, but it is always closely linked to granite, especially in tin-mining districts. In fact, it is granite that has been altered or 'metasomatized' by exposure to hydrothermal fluids or vapours rich in fluorine, lithium, boron and tungsten – just as basalt is metasomatized to spillite. As fluids flood through veins in granite, they alter the composition of the granite in the vein walls, destroying all the feldspar, and leaving just quartz and white mica. Gradually the vein becomes infilled with greisen. There is usually no definite boundary between the greisen and granite, and altered granite merges into unaltered granite imperceptibly. Greisens belong to the quartzolite family, the most quartz-rich of all rocks, with a quartz content of over 90 per cent.

Grain size: Medium-grained.
Texture: Even-grained or, occasionally, foliated.
Structure: None.
Colour: Grey or brown.
Composition: Silica (90% average).
Minerals: Quartz; White mica (muscovite, zinnwaldite, lepidolite, sericite).
Accessories: Topaz, fluorite, apatite, tourmaline, rutile, cassiterite, wolframite.
Formation: Short vein infillings no more than a few hundred metres long.
Notable occurrences: Skiddaw (Lake District), Cornwall, England; Galicia, Spain; Fichtelberg, Erzebirge, Germany; Portugal; Queensland, New South Wales, Tasmania, Australia.

Lamprophyres

Camptonite: Named after Campton in New Hampshire, this dark lamprophyre has a hornblende, labradorite feldspar and pyroxene groundmass with phenocrysts of the amphiboles kaersutite and ferrohornblende, along with titanaugite, olivine and biotite.

Lamprophyres get their name from the Greek for 'glistening mixture', and it is apt for these rocks are stuffed with large, gleaming phenocrysts of mica, amphiboles and olivine, giving the rock a very distinctive appearance. Unusually, they have no feldspar phenocrysts whatsoever; all their feldspar is in the fine groundmass. They are dark, or even ultramafic, rocks and probably form from cool melts of metasomatized (altered) mantle material. They are the classic dyke rocks. Most dyke rocks are simply versions of the same rocks that form larger intrusions. Lamprophyres alone form almost exclusively in dykes – though in recent years small lava flows and plutons have been found. They typically occur in dykes near tonalite and granodiorite plutons. The lamprophyres are a very varied group, and some geologists think they should be called a facies – a diverse group of rocks that simply crystallized under similar conditions. The most widespread form is minette.

Vosgesite: Named after the Vosges in Alsace, France, vosgesite has phenocrysts of hornblende along with augite and olivine. It is one of the calc-alkaline lamprophyres normally found along with rhyolites and basalts in island arcs and subduction zones.

Grain size: Mixed.
Texture: Porphyritic.
Colour: Dark grey with dark, even black phenocrysts.
Composition: Variable. Minette: Silica (47.5% average), Alumina (9.3%), Calcium and sodium oxides (11.5%), Iron and magnesium oxides (26%).
Groundmass: Plagioclase feldspar, feldspathoid, carbonates, monticellite, mellilite, mica, amphibole, pyroxene, olivine, perovskite.
Phenocrysts: Biotite/phlogopite, amphibole (hornblende, barkevikite, kaersutite), pyroxene (augite), olivine.
Formation: Mainly dykes.
Notable occurrences: Cairngorms, Cheviots, Scotland; Lake District; England; Ireland; Vosges, France; Black Forest, Harz, Germany; Wasatch, Utah.

Palisades Sill

New Jersey's Palisades are a dramatic line of brown cliffs that tower anything from 107m/350ft to 168m/550ft above the west bank of the Hudson River. The cliffs are the exposed margin of a vast sill of diabase (dolerite) rock 305m/1,000ft thick and 72km/45 miles long that dips away westwards. Radiation dating has shown the sill formed between 186 and 192 million years ago in the Early Jurassic, when a fat wedge of magma squeezed between layers of sandstone and shale. As the magma cooled and solidified, it cracked into the columns that characterize the cliff face – and earned the Palisades their name, given by explorers with Verrazano in 1524, who thought the cliffs resembled the forts of wooden stakes built by local Indians. In the 19th century, these rocks were ruthlessly exploited for building stone, and many a New York sidewalk is made of 'Belgian stone' from the Palisades. Eventually, in the 1930s, the area was designated an Interstate Park to halt the destruction.

Dolerite

A tough stone used for road metal, dolerite is a dark, mafic rock, the medium-grained equivalent of basalt and gabbro. It is known in the United Kingdom as diabase and is typically found in sills, such as New Jersey's vast Palisades sill. The famous bluestones of England's ancient stone circle Stonehenge were carved from dolerite cut from sills in the Prescelly Mountains of Wales. Usually when magmas cool and crystallize, dark minerals such as olivine crystallize first, followed by feldspar and mica, leaving quartz and any other silica to fill in the gaps. Lath-like crystals of feldspar form first, and the dark minerals are forced to fit in between them – often growing right around them in what is called ophitic texture.

Grain size: Medium.
Texture: Often ophitic, with large clinopyroxene (augite) crystals enclosing plagioclases.
Structure: Vesicles and amygdales common.
Colour: Dark grey, black, with green tinge when fresh. May be mottled white.
Composition: Silica (50%), Alumina (16%), Calcium and sodium oxides (13%), Iron and magnesium oxides (18%).
Minerals: Plagioclase feldspar (labradorite); Olivine; Pyroxene; Biotite; Magnetite; Ilmenite; Quartz; Hornblende.
Phenocrysts: Olivine and/or pyroxene or plagioclase.
Groundmass: Plagioclase and pyroxene with olivine or quartz.
Formation: Sills and dykes, often in large dyke swarms; occasionally lava flows.
Notable occurrences: Whin Sill, NE England; Lake Superior, Canada; Palisades, New Jersey.

Identification: Dolerite is best identified by its dark, greenish colour and its medium grain size.

PEGMATITES

Pegmatites are the cream of igneous rock, the places where all the biggest crystals and rarest minerals are concentrated when an igneous intrusion reaches its final stages of crystallization. Pegmatite formations are typically small pods and lenses no bigger than a house, but they are the source of some of the world's best gems and most valuable minerals.

Pegmatite features

Identification: Pegmatites are instantly recognizable from their gigantic crystals. What is harder to identify is the particular crystals within them, and so the particular kind. Large creamy or pink crystals are usually feldspar, white sugary crystals may be quartz, brown striped crystals are mica, black may be tourmaline. More colourful crystals are rarer minerals.

Quartz

Pink beryl

Amazonite feldspar

Lithium pegmatite: Many pegmatites are enriched with the mineral lithium, turning mica to lepidolite and creating the spodumene gems – lilac kunzite and green kunzite.

Lepidolite mica

Pink tourmaline

Pegmatites are perhaps the most fascinating bodies of rock in the world. No other rock formation contains such a wealth of large, spectacular crystals. All the world's largest natural crystals have been found in pegmatites. Even the average grains in pegmatites are not just clearly visible, as in coarse-grained granite, but substantial – at least the size of a grapefruit. Some pegmatite crystals are truly gigantic. Tourmaline and beryl crystals the size of a log are often found, while a spodumene crystal found in a pegmatite in South Dakota in the USA was a gigantic 13m/42ft long.

Pegmatites are also the source of an amazing range and variety of minerals. Over 550 different kinds of mineral have been found in pegmatites. Pegmatites are the source of many of the world's gems. Besides fabulous topaz and wonderful garnets, gems of all the beryl varieties (aquamarine, morganite, golden), all the tourmalines (pink, green, and multi-hued elbaite) and all the spodumenes (kunzite and hiddenite) are found in pegmatites. Pegmatites are also sources of rare elements such as beryllium, niobium, tantalum, rubidium, caesium and gallium, as well as tin and tungsten. Because they are so richly concentrated here, pegmatites are even major sources for more common minerals such as feldspar and quartz.

The term pegmatite was originally used in the early 19th century to describe graphic granites, which often occur in pegmatites. Now the term is used to describe any small body of igneous rock with crystals at least 1.3cm/0.5in across. They vary hugely in size and shape. Some are veins. Some are shaped like lenses. Some are shaped like knobbly turnips. The smallest are typically no bigger than a mattress but a few giant pegmatites are 3.2km/2 miles long and 0.5km/0.3 mile wide. Pegmatites are by no means isolated structures. In the famous Black Hills district of South Dakota in the United States, there are an estimated 24,000 pegmatite bodies in an area of 700km²/270 sq miles!

Grain size: Very coarse-grained. Crystals are at least 1–2cm/0.5in across, average 8–10cm/3–4in, and can be much bigger.
Texture: Hugely varied, with complex zoning and lots of vugs (open cavities).
Structure: None.
Colour: Pale pink or whitish.
Composition: Silica (75% average), Alumina (14.5%), Calcium and sodium oxides (3.5%), Iron and magnesium oxides (5%).
Minerals: Quartz; Potassium feldspar (albite and perthite).
Accessories: Plagioclase feldspar, muscovite, apatite, tourmaline, plus minerals containing elements such as tin, tungsten, niobium, tantalum, beryllium, gallium, rubidium and caesium.
Large crystals: Quartz, feldspar, tourmaline, beryl (aquamarine, morganite, golden), spodumene (kunzite and hiddenite), tourmaline.
Formation: Dykes, veins, pods, lens around the margins of intrusions, Pegmatites are typically either granite or syenite.
Notable occurrences: Pegmatites occur all around the world, wherever there are granite and syenite intrusions. Famous examples include those on the Island of Elba in Italy, Madagascar (especially Anjanbonoina), Pakistan and the Mesa Grande, California (notably the Himalaya Mine in Pala County). They are most abundant in mountain chains and stable continental shield areas, such as the Canadian Shield, Greenland and north Russia. Shield pegmatites are usually at least a billion years old. Mountain chain pegmatites such as those in the Himalayas are no more than 5 to 20 million years old.

Pegmatite formation

Pegmatites usually form at the margins of a large pluton, clustering like currents on its surface, or extending like fingers into the mass, or outwards into the surrounding country rock. Occasionally they are found completely separated from their parent intrusion in pockets in the country rock. It is thought that pegmatites form in the last stages of the crystallization of an intrusion. With the main body of rock formed, the bulk of the common minerals have already crystallized, leaving just a few small pockets to be filled in to create pegmatites. The melt left to create them is rich not just in rarer elements such as boron and fluorine, but also volatile liquids – and a great deal of water. It is this high water content that allows the crystals in pegmatites to grow so big. All this water dramatically increases the mobility of particles within the melt, and this means they can travel farther and faster as they become incorporated into crystals. Normally, large crystals form only when magmas cool very, very slowly from high temperatures, but the water in pegmatites means huge crystals grow rapidly at temperatures of no more than 100–200°C/212–392°F. Pegmatites can form from just about any kind of igneous intrusion including gabbro and diorite – and even in metamorphic gneisses and schists – but most form from granites and syenites and so have the same basic ingredients – quartz and potassium feldspar with a little muscovite mica. Pegmatites can be divided into simple and complex. Simple pegmatites are basically very coarse-grained equivalents of the parent rock, made of the same three basic ingredients, plus a little tourmaline. Complex pegmatites form later and have higher concentrations of rare minerals. Lithium, for instance, is typically found at concentrations of 30ppm (parts per million) but in complex pegmatites, lithium concentrations can reach over 700ppm. Complex pegmatites are typically divided into highly complex zones, with graphic granite in one place, tourmalines in another, and so on. They are often riddled with open cavities. It is the complex pegmatites that are the source of the most valuable and spectacular crystals.

Black tourmaline

Tourmaline pegmatite:
Tourmaline is a boron mineral and tourmaline pegmatites are created when the boron content is especially enriched in the last stages of the mix. An increase in fluorine often creates the gem topaz.

Varieties of pegmatites:
Pegmatites are very varied. Some are named after the main source rock. Others are named after a mineral or element that is particularly enriched in them.

Rock source pegmatites:
Granitic pegmatite, Syenitic pegmatite, Gabbroic pegmatite, Diorite pegmatite.

Mineral- or element-enriched pegmatites:
Tourmaline pegmatite, Lithium pegmatite, Beryl pegmatite, Emerald pegmatite, Spodumene pegmatite, Albite pegmatite, Quartz-albite pegmatite, LCT pegmatite.

Rock source and element-enriched pegmatites:
Phosphate granitic pegmatite, Boron granitic pegmatite, LCT granitic pegmatite.

LCT pegmatites: LCT (Lithium-Caesium-Tantalum) pegmatites are enriched not just with the elements lithium, caesium and tantalum, but also with rubidium, beryllium, gallium and tin. Granitic LCT pegmatites are host to many of the world's most precious gemstones, including emerald, chrysoberyl and topaz from Minas Gerais in Brazil, sapphire and ruby from Afghanistan and Pakistan, and gem tourmaline.

Isola d'Elba, Italy
The western end of the Island of Elba off the west coast of Italy is riddled with pegmatites that have been one of the world's richest sources of beryls and tourmalines, including the elbaite variety named after the island. In 1805, the first quarries were dug to extract the local granite to build houses and roads. Then in 1820, a local mineralogist called Captain Foresi noticed large, colourful crystals in the rock, later found to be tourmalines. In 1830, Foresi opened the first tourmaline mine at Grotta d'Oggi, or Cave of Today. He then traced the pegmatite zone and discovered many other sources of tourmalines and beryls such as the Masso Foresi (Foresi's Mass), the Fonte del Prete and la Speranza. Scores of quarries and mines were opened up and yielded many thousands of fine tourmalines and beryls. However, these localities existed within around 10 sq km (six sq miles) of each other and, unsurprisingly, by the end of the 19th century, the existing veins – sunk deep into granite rock – had been mined to exhaustion, and the best stones taken away. Many are now on display in mainland Italy, at the Florence Mineral Museum. Visitors to the island do occasionally still find fine elbaite crystals here, and many collectors believe there are countless new pegmatite-rich veins waiting to be discovered. However, with new laws to safeguard the landscape of the island against future mining, it is unlikely that the Isola d'Elba will regain its previous status as a mecca for those seeking pegmatite gems.

Yellow beryl

Beryl pegmatite:
This is rich in beryl, and may often contain gems like emeralds, such as in the pegmatites from Minas Gerais in Brazil.

Smoky quartz

Feldspar

CLASTIC: LUTITES

Sedimentary rocks are made from sediments of loose material that is gradually lithified (consolidated and turned to stone) over millions of years. Many sediments are fragments, or 'clasts', of older rocks broken down by weather. Clasts are mainly silicates (quartz, feldspar, mica). Rocks made from them are described as siliciclastic, and classified according to grain size. The finest grained are lutites or mudrocks, made mainly of clay- and silt-size grains. They include claystone, mudstone, siltstone and shale.

Claystone and mudstone

These siliciclastic rocks are the most abundant sediments on Earth. Half of all sedimentary rocks are clays and muds, and beds of clay stripe nearly every sedimentary formation. London and Paris both sit on vast dishes of clay that provide the bricks that built the cities and the rich soils of the farmlands around. Dull, flat and common claystones may be, but they are the most useful of all rocks. Impure clays are used for making bricks and tiles – and their organic content makes many self-firing. Pure clays such as kaolinite are so wonderfully mouldable they still provide the best materials for making pottery as well as fillers for papers.

No rocks are made from tinier grains than these. More than half the grains in claystone are clay-sized – less than 4μm/0.15mil across. Over two-thirds of the grains in mudstones are clay-sized. Tiny grains like these are the last remnants of rocks broken down by weathering. Being so light they are carried farthest from their source. When rivers flow into the sea and drop their sediment load, clays fall last. Many grains float right out into the deep ocean before sinking to become part of the ocean floor ooze.

Most claystones form from clay sediments that settle in shallow waters just off shore, in calm areas below the waves. Clays like these are dotted with fossils of sea creatures, from tiny shellfish to giant marine dinosaurs. Besides these marine sediments, claystones may also form on lake beds and where rivers flood. Some claystones are not sediments at all but 'residuals' developed as rocks are altered in situ to create soils such as laterites. Few claystones are old, however. Buried under later sediments, they are quickly consolidated first into shale and then to slate in a process called diagenesis. So clay tends to appear only in younger geological formations.

Clay minerals are divided into four broad groups: the kandites (such as kaolinite) formed by the breakdown of potassium feldspar; the illites formed from feldspars and mica; the smectites (including montmorillonite) formed from pyroxenes and amphiboles; and the chlorites. Each claystone contains its own mix of these four, with varying proportions of organic material as well. Illites and montmorillonites are most prevalent.

Claystone: The particles in clay are so tiny that, when pure, it feels smooth and slippery when wet, like plasticine. Claystones look like earthenware and range in colour from grey clays rich in plant material to red clays rich in iron oxide.

Black mudstone: Mudstones are defined as rocks made of one-third silt grains and two-thirds ultra-fine clay grains. They are basically hardened mud, and like most muds they are often rich with both plant and animal matter. It is this organic matter that often turns them black.

Grain size: Over 50% of grains are clay-sized, less than 4μm/0.15mil.
Texture: Even-grained, with fossils. No grittiness like silt. Can be plastic and often sticky when wet.
Structure: No fine layering like shale, but clay beds show larger scale stratification, including originally horizontal topset and bottomset beds formed on the top and beyond a delta, and originally sloping offset beds formed on the delta front. Sun cracks, rain prints etc are common. All clay particles are microscopically layered, which makes clays plastic and slippery when wet as layers slide over each other. Mudstones have a blocky, massive fabric.
Colour: Black, grey, white, brown, red, dark green or blue.
Composition: Mix of detrital quartz, feldspar and mica. Iron oxides turn clays red or brown. Organic matter turns them black.
Different groups of minerals: Kandites such as kaolinite; Illites; Smectites such as montmorillonite; Chlorites.
Formation: From clays and muds settling offshore, on lake beds and on river floodplains. Also as residuals as rock is altered in situ.
Notable occurrences: London Clay (London basin), Oxford Clay (Weymouth to Yorkshire), England; Paris basin, France; North German basin, Molasse basin, Upper Rhine, Germany; Kamchatka, Russia; Sydney basin (New South Wales), Australia; Trinity River, Texas; Appalachian Mountains; Newland, Montana; Muldraugh Hill, Kentucky.

Shale

Gray shale: Shale is often dark grey or brown with a thin, platy structure and no visible grains.

Like claystone and mudstone, shale is made from the fine particles that settled on the floor of shallow seas and lakes long ago. Yet unlike claystone and mudstone, shale has laminations like the pages of an ancient book, created as it was squeezed by the weight of overlying sediments. As a result, shale looks flaky like slate, and likewise splits easily into thin layers, a tendency called fissility. The layers vary from paper thin to card thick. Unlike slate, though, shale often contains the fossilized remains of sea life, buried in the mud and preserved forever, albeit somewhat flattened, as the mud turned to stone. Shale varies tremendously in colour according to the minerals it contains. Black shales are rich in carbon from organic remains (typically plankton and bacteria), which often turned to kerogen as the rock formed. Oil shales are black shales so rich in kerogen and bitumen (at least 20 per cent) that they yield oil if heated intensely. On average, 1 tonne/0.98 tons of rock can yield 750 litres/ 165 gallons of oil. Scientists have yet to find a way of extracting this oil economically.

Black shale: This shale is black because the remains of sea creatures it contains were never oxidized. Some formed in basins in which circulation was restricted. Others may have developed in times when global warming peaked, cutting down circulation of deep ocean currents. Black shales are famous for the extraordinary preservation of fossilized sea creatures. Even soft tissues often leave impressions.

Grain size: 50% of grains are clay-sized, less than 4μm/0.15mil.
Texture: Even-grained, with fossils. No grittiness like silt.
Structure: Splits easily into thin layers.
Colour: Black, grey, white, brown, red, green or blue.
Composition: Mix of detrital quartz, feldspar and mica. Iron oxides turn shales red or brown. Organic matter turns them black.
Formation: From clay sediments settling offshore, on lake beds and on river floodplains then undergoing diagenesis.
Notable occurrences: Many locations around the world: *Black shale*: Posidonia, Hünsruck, Germany; Chattanooga, TN; New Albany, PA; Great Plains of Kansas, Oklahoma, South Dakota; *Oil shale*: Torbane Hill, Scotland; Estonia; Lithuania; Israel; Tasmania; Green River, Colorado.

Jurassic mud
It was in the remarkable Oxford and London Clays of England that Victorian fossil hunters made many of the great finds that led to the discovery of dinosaurs and many other prehistoric creatures. Both contain a wealth of marine fossils, but the Oxford Clays are especially rich. The sediments there are made from developed some 140–195 million years ago in the Jurassic period. At this time, southern England was entirely covered by the waters of a tropical ocean, teeming with sea creatures. Oxford Clay is full of the fossils of fish and shellfish that swam there then, including the giant Leedsichthys fish, and countless ammonites and belemnites. But it is most famous for its marine dinosaurs, including the plesiosaurs *Cryptoclidus*, saucer-eyed *Opthalmosaurus* (pictured above), and the awesome *Liopleuridon*. At almost 25m/80ft long, *Liopleuridon* was the biggest carnivore that ever lived, with a mouth 3m/10ft long and teeth twice as long as those of the *Tyrannosaurus rex*.

Siltstone

Siltstones are much less common than clays and muds and rarely form thick beds. At least half the grains in siltstone specimens are coarser, silt-sized grains, 4–60μm/ 0.15–2.5mil across – large enough to be visible with a magnifying glass. They are mostly quartz, making the rock tougher than clay, and giving it a slightly gritty feel. Because the grains are heavier, siltstones form closer to the shore than clays, and often show ripple marks and crossbedding created by the interplay of the river currents and waves in shallow water. As the flow of water changed with the seasons, so sediment deposition varied. So pale siltstones are often found interlayered with darker mudstones.

Grain size: Over 50% of grains are silt-sized, in the range 4–60μm/0.15–2.5mil.
Texture: Even-grained, with fossils. Slightly gritty.
Structure: Often laminated. Shows crossbedding and ripple marks.
Colour: Pale grey to beige.
Composition: Mix of detrital quartz, feldspar and mica.
Formation: From clay sediments settling in river deltas, on lake beds and on river floodplains.
Notable occurrences: Many locations around the world including south-east England, The Great Plains of North America, China.

Identification:
Siltstone is easily identified by its pale colour, just-visible grains and slightly gritty feel. It is most often laminated with bands of dark mudstone.

MORE LUTITES (MUDROCKS)

Besides claystones, mudstones and siltstones, there is a wide range of other mudrocks, including marls, bentonites and boulder clays. Marls are earthy, crumbly mixtures of lime and silicate fragments. Bentonites are clays formed from volcanic ash that falls on the sea bed. Boulder clays are basically the debris left behind by moving sheets of ice.

Marl and marlstone

Green marl: Marls are often given a green tinge by the potassium mica mineral glauconite. These green marls are often very rich in fossils. There are extensive deposits in places such as England's Isle of Wight and North America's Atlantic seaboard.

Red marl: The lime and clay content of marl means that they are normally white, grey or brown, but some marls are very rich in iron, which turns them red. Strictly speaking, they are not marls, since they have a low lime content, but they have the same earthy texture.

Since many sediments form on sea and river beds, it is inevitable that they become enriched with a fair share of debris of shellfish and other marine life. As the rocks form, this debris turns to carbonates, first to calcite and aragonite, and eventually to calcite and dolomite in older rocks. Limestones and chalks are almost pure calcium carbonate or lime. Mudstones and claystones, however, contain only a little lime. Marlstone lies in between, rich in both lime and silicate fragments of weathered rocks.

Strictly speaking, marlstone is the rock, while marl is the soft, earthy material that forms as this and other rocks are weathered, but geologists often use the word marl as a general term for any hybrid of mudstone and fine-grained limestone. Extra lime turns marls into limestones; less turns them into clays and mudstones.

The mixed lime and clay content makes all marlstones soft and friable, even when they are not actually earths. Many disintegrate in water, and the lime content means they are easily dissolved in dilute hydrochloric acid or even vinegar.

In some marls, called shelly marls, the carbonate material is actual shell fragments. Shelly marls like these are much valued by farmers as a source of lime, because the lime is easy to extract. In others, the lime is a fine powder mixed in completely with the quartz and feldspar grains. Marls that form in freshwater are quite similar to those that form in the sea. They also often contain shell fragments, but most of the organic material usually comes from algae.

In England, there is a group of rocks called New Red marls that form beds up to 300m/1,000ft thick in places, as part of the Keuper system. These are iron-rich clays rather than pure marls, because they contain only a little calcium carbonate. They probably formed in salt lakes in desert conditions, and in places contain thick salt beds, such as those in Cheshire. Some hard slates in Germany are also described as marls, including the important copper-bearing marl-slates of the Mansfeld area.

Grain size: Over 50% of grains are clay-sized, less than 4μm/0.15mil.
Texture: Earthy, even-grained, with fossils. Marl has none of the grittiness of silt, feeling much softer. The lime is sometimes powdery, sometimes in the form of shell fragments.
Structure: No fine layering like shale, but clay beds show larger stratification. Sun cracks, rain prints etc are common. All clay particles are microscopically layered, which makes clays plastic and slippery when wet as layers slide over each other. The high lime content makes these rocks very friable and crumbly. The combination of lime and clay makes a very good basis for soil, which is why marl is often added to soils to improve fertility.
Colour: Various, including brown, white or grey; may also be red with iron content or green with glauconite.
Composition: Even mix of carbonates (mainly calcite) from organic sources and detrital quartz, feldspar and mica.
Formation: From clays and muds settling offshore, on lake beds and on river floodplains.
Notable occurrences: North Yorkshire, Leicester, Northamptonshire, Oxford, Exmouth, Vale of Eden, England; Valkenburg, Netherlands; Paris basin, France; Mansfeld, Bavaria, Germany; Sydney basin (New South Wales), Australia; Weka Pass, Canterbury Plains, New Zealand; Green River Formation, Wyoming; South Dakota; Atlantic coastal plain (New Jersey, Delaware, Maryland, Virginia).

Bentonite

Named after a kind of clay found near Fort Benton, Wyoming, in 1890, bentonite is a clay formed by the alteration of volcanic ash that has settled on the sea floor. Similar clays called tonsteins form when ash is weathered in the acidic waters of coal swamps. Bentonites and tonsteins consist of mostly smectite clays, but also contain unchanged volcanic fragments such as quartz grains and mica flakes. They may also contain beads of volcanic glass. There are two kinds of bentonite: sodium bentonite and calcium bentonite. Sodium bentonite is an incredibly useful material because it swells enormously when wet, creating a gelatinous mass that has been used for everything from sealing dams and drilling for oil to cat litter and detergents. Calcium bentonite makes the absorbent clay called fuller's earth. Bentonites are typically found interbedded with shallow marine limestones and shales, and represent a sudden and dramatic event when a volcano showered the sea floor with huge quantities of ash. Although beds up to 15m/45ft thick have been found, most are less than 0.3m/1ft. Although they are frequently altered beyond recognition, they can give a valuable insight into past volcanic events. In the Ordovician period, much of the eastern USA was covered by an immense ashfall 1m/3ft thick, leaving extensive bentonite deposits from Tennessee to Minnesota.

Bentonite: Bentonite looks like clay mud, but is generally a buff to olivegreen colour. If it absorbs water it will swell dramatically.

Grain size: Over 50% of grains are clay-sized, less than 4μm/0.15mil.
Texture: Earthy, even-grained. Slippery and plastic when wet. Greasy or waxy feel.
Colour: White to light olive green, cream, yellow, earthy red, brown and sky blue. Bentonite turns yellow on exposure to air.
Composition: Smectite clay minerals, quartz grains, mica flakes, volcanic glass beads, calcite and gypsum.
Formation: Bentonite formed by the alteration of volcanic ash falling on the sea bed. Tonstein formed from ash falling in coal swamps.
Notable occurrences: Redhill (Surrey), Woburn (Bedfordshire), Bath (Avon), England; Spain; Italy; Poland; Germany; Hungary; Romania; Greece; Cyprus; Turkey; India; Japan; Argentina; Brazil; Mexico; Saskatchewan; Wyoming; Montana; California; Arizona; Colorado; Black Hills, South Dakota.

Marl, the farmer's friend
Farmers have added marl to soil to improve its fertility for thousands of years. When added to acid soils, the lime in marl helps to neutralize the acidity. It also helps to glue sand grains together so that they retain heat and water better. When added to clay soils, wonderfully it has the opposite effect – helping to make the soil more crumbly and friable and allowing air, heat, water and roots to penetrate better. So marl promotes plant growth in a number of ways: it increases the food available for plants, and makes it easier for them to reach it. For centuries until artificial fertilizers began to take over, marl was dug from marlpits in large quantities. In the eastern USA, where marl is abundant, farmers often put 20–30 tonnes/19.7–29.5 tons of it on every 0.4 hectare/1 acre of land in the 19th century, giving magnificent potato, tomato, and berry crops. Clover grew especially lush on marled soils.

Boulder clay

Also known as tills and ground moraine, boulder clays are a legacy of the great ice ages that once covered much of northern Europe and North America in vast ice sheets. Mixed in like a fruit cake are large pebbles and boulders that were swept along beneath the glaciers and fine clay from rocks shattered by frost and stripped by moving ice. The materials in the boulder clay reflect where the ice travelled. Thus in Britain, boulder clays near Triassic and Old Red Sandstone areas are red, while near Silurian rocks they are buff or grey, and those near chalk can be white. Although the biggest boulder clay deposits were left by past ice ages, they are forming even today under glaciers and ice sheets in polar and mountain regions.

Grain size: Mixture of pebbles, boulders and clay-sized grains, less than 4μm/0.15mil. Boulders can weigh up to several tons.
Texture: Smooth clay embedded with angular stones.
Colour: Varies according to original rock – red, white, grey, brown, black.
Composition: Depends on the original rock.
Formation: Debris accumulated and swept along beneath glaciers and ice sheets.
Notable occurrences: All across northern Europe and northern North America, especially East Anglia in England and the North German plain.

Identification: With large stones and boulders set in a sticky clayey mass, boulder clay is unmistakable. The interesting task is to work out the origin of the material.

SANDSTONES

*Sandstones are second only to mudrocks in abundance, making up 10–15 per cent of all Earth's sediments,
and because they are so durable, they often form some of the most prominent hills and landmarks,
as well as providing valuable building stone. They are made mostly of sand-sized grains 60μm–2mm/
2.5–80 mil across. At least half the grains must be this size for it to be classed as a sandstone.*

Sandstone

Sandstone, as its name suggests, is made from grains of
sand – quartz, feldspar, or simply sand-sized
fragments of rock. Sometimes the sand was
piled up by desert winds, and the grains
were worn almost round as they were
buffeted along. Sometimes the sand
was laid down on river beds, beaches
or in shallow seas, and the grains
are a little more angular. The
sharpest sand of all came from glacial
debris or high up in rivers, where it
had not travelled far.

Beach sand is typically yellow, but each
sandstone is stained by the cement that binds the sand.
Limonite cement gives some sandstones a yellowish hue.
Calcite turns them white – perfect for glass. Bitumen turns
them black like the sandstones of Alberta, while iron oxides
stain them red and brown. It is these warm reds and browns
that are seen in New York's famous brownstone fronts, and
the rusty red mesas and buttes of Utah, Colorado and
Arizona. Occasionally, the cement is so weak that the rock
crumbles in your hands. Most of the time, it is much harder,
and sandstones resist erosion to create some of the world's
most dramatic landscapes – high ridges, steep bluffs and
towering tablelands. Sandstone's toughness also makes it the
perfect building stone, more widely
used than any other.

Sandstones can be like a 'book' displaying the history of
their formation. Cracks reveal where sand dried out in the
sun. Ripples show where waves rolled over it. Bedding
marks bear witness to the way sand deposition
changed continually season by season, year by year.
Desert sandstones, like those in Zion Canyon, Utah,
may even capture the shape of ancient wind-
blown sand-dunes. Most sandstones are also
rich in fossils of the creatures that burrowed in
the sand, or lived in the waters above it. In the
brownstone quarries of Portland, Connecticut,
huge footprints have been found that were made
by dinosaurs that walked over these sands long ago.

Identification: With their visible
grains of sand, sandstones are
easy to identify but it is much
harder to distinguish the kind of
sandstone. In a fresh surface, you
may be able to identify quartz and
feldspar. Quartz grains are milky
to clear, glassy with no cleavage
marks. Feldspars are usually
white or pinkish, with marked
cleavage planes. They may be
dissolved out leaving holes or
changed to clay.

Malachite sandstone: Sandstones are never pure quartz sand
or even quartz and feldspar. Most contain traces of minerals.
This sandstone from the Triassic period is specked with
green malachite.

Grain size: Over 95% sand-
sized grains, 60μm–2mm/
2.5–80mil.

Texture: Gritty texture like
solid sand. Grains well
sorted, often well
rounded. The amount
of cement between
grains varies widely.

Structure: Typically
occurs in blanket-shaped
deposits varying from a few
metres to several hundred
metres thick. Sandstones
are usually interbedded with
mudstones, limestones and
dolomites. They usually
display dramatic cross-
bedding and ripplemarks,
reflecting their formation in
high-energy environments.
Aeolian rocks often show
sand-dune shapes.

Colour: Variable – typically
red, brown, greenish, buff,
yellow, grey, white.

Composition: 40–95%
quartz, with feldspar and rock
fragments. Other
components include mica,
clay, organic fragments, plus
many heavy minerals. Quartz
and calcite cement.

Formation: Nearly all
sedimentary environments
from small alluvial fans to
vast deep-sea plains. Some
form in high-energy marine
environments such as
beaches. Some are formed
from aeolian (wind-blown)
sand-seas in deserts,
where there is a ready supply
of sand. A type called
ganisters forms when most
other kinds of grain are
leached away, leaving just
quartz sand.

Notable occurrences:
Western Highlands, Scotland;
Pennines, England;
Apennines, Italy; Carpathians,
Romania; Nile Valley, Egypt;
India; Appalachian Mountains;
Colorado and Allegheny
plateaux, Montana, US.

Old Red Sandstone

The Old Red Sandstones are among the most famous and studied of all rock formations. They are a gigantic sequence of rocks that formed from sediments piled up in a vast basin that stretched across what is now north-west Europe in the Devonian period from 408 to 360 million years ago. This was the time in the Earth's history when the first fish swam, the first land plants grew and the first insects crawled. As the massive Caledonian mountain range was slowly worn away, its remnants accumulated in this basin and in time turned to stone. So extensive were these sediments that geologists used to refer to them as the Old Red Sandstone Continent – and included the Catskill Mountains of North America in it, although these actually formed quite separately at roughly the same time. The Devonian sediments are by no means all sandstones, nor did they all form in the same way. Some were laid down by rivers, some in the sea and some in lakes. But the dominant beds are the massive layers of red sandstone. The basin lay in the baking tropics south of the Equator in Devonian times, and these sands piled up in desert sand-seas and alluvial fans. They acquired their distinctive red colour as moisture later rusted iron in them.

Identification: Old Red Sandstone formations are best identified by their location and the fossils they contain, such as the famous Devonian fish. These formations include shales and other mudrocks as well as sandstone. The sandstone can be recognized by its visible sand grains, often stained red by iron oxides.

Grain size: Over 95% sand-sized grains, 60μm–2mm/2.5–80mil.
Texture: Gritty texture like solid sand. Grains well sorted, often well rounded. The amount of cement between grains varies widely.
Structure: See sandstone.
Colour: Red, green, grey.
Composition: See sandstone.
Formation: Most of the red sandstones formed in vast alluvial fans spilling into desert basins, and in desert sand-seas.
   Notable occurrences: Shetland, Caithness, Midland Valley, Borders, Scotland; Fermanagh, Antrim, Northern Ireland; Mid-Wales; Shropshire, Devon, Somerset, England. North American red sandstones: western Canada; Catskill Mountains, NY.

Brownstone fronts
In the decades after the American Civil War in the 1860s, it was the fad to clad well-to-do houses in Boston and especially New York with brownstone. Entire districts became characterized by their brownstone fronts. Brownstone is a feldspar-rich sandstone that formed about 200 million years ago in the Triassic period. Iron-oxide cement coloured it a warm chocolate-brown. The most distinctive formation is near Portland in Connecticut, and in the late 1800s, the Portland quarries boomed. Yet the fad for brownstone did not last long – partly because of the coming of concrete, and partly because brownstone, which formed in horizontal layers, was set in buildings vertically face out or 'face-bedded.' Face-bedded like this, the brownstones quickly flaked off as water got in behind the layers and froze. Now these beautiful old buildings are cherished again, and the fronts are being restored with fresh stone from the reopened Portland Quarry and better mortar.

Greensand

This can be either a sandstone or mudrock turned green by tiny pellets of the clay mineral glauconite, which gets its name from the Greek for 'blue-green'. Some greensands are over 90 per cent glauconite with just a little quartz sand and clay. Glauconite is a potassium iron aluminium silicate and is a useful potash fertilizer and valuable water softener. It forms in shallow seas (50–200m/164–656ft deep) only at times when sediments are piling up slowly, allowing sealife to burrow widely. Pellets form as faeces or the insides of dead foraminifera shells are chemically altered. However, glauconite-rich rock is not always green because it is turned brown or yellow by weathering. Most greensands were laid down in the Jurassic and Cretaceous periods. In England, sandstone beds formed at this time are called greensand whether they contain glauconite or not.

Grain size: Clay to sand size.
Texture: Sometimes gritty and sandy, sometimes smooth like clay.
Structure: See sandstone.
Colour: Red, green, grey.
Composition: Mostly glauconite, with quartz sand and clay.
Formation: Greensands form on shallow, slowly sedimenting sea beds. They are often the last stage in a sedimentation sequence and typically appear just below an unconformity.
Notable occurrences: The Weald, Dorset, Berkshire, Oxfordshire, Bedfordshire, England; Boulonnais, France; New Jersey; Delaware.

Identification: Unweathered, greensand is coloured pale olive green by glauconite, but this turns brown or yellow when exposed to air, and so is less distinctive.

ARENITES AND WACKES

There are two main kinds of sandstone: arenites and wackes. Arenites are all sand-sized grains (60μm–2mm/2.5–80mil) with little cement. Wackes are less sorted, with sand chaotically embedded in silt and clay. Both arenites and wackes may be mostly quartz (like orthoquartzite), a mix of quartz and feldspar (like the arenite arkose), or 'lithic' – that is, made of various rock fragments (like the wacke greywacke).

Orthoquartzite (Quartz arenite)

Identification: With a bare minimum of cement materials, orthoquartzite looks like solidified sand, which it is. It looks even more like lump sugar because the predominance of quartz makes it very pale in colour.

Orthoquartzites, or quartz arenites, are among the most quartz-rich of all rocks, made almost entirely of sand-sized grains of quartz with a bare minimum of cement. Indeed, the definition of an orthoquartzite is a sandstone consisting of over 95 per cent quartz. There are also a few traces of 'heavy' minerals such as zircon, tourmaline and rutile, but these are fairly scanty.

Orthoquartzites form in high-energy environments. They are created where sand is dropped by waves crashing on beaches and by powerful rivers streaming into the sea. The often dramatic cross-bedding and ripple marks in orthoquartzites are a telling reminder of the way these strong currents tugged the sands to and fro. These sands accumulate above the level where the biggest storm waves start, piling up along the shoreline as beaches, dunes, tidal flats, spits and bars. Look closely at any orthoquartzite formation and you can often see the remnant form of these ancient coastal features.

Not all orthoquartzites form in water. Because they are made of essentially dry sand, they can form from sand-seas in deserts, where sands are piled high by desert winds. Water-formed orthoquartzites tend to be white or pale grey, because they are almost pure quartz. These wind-blown, or 'aeolian', orthoquartzites are often stained red or pink by fine powdered iron oxides, which coat the grains.

About a third of all sandstones are orthoquartzites, but their spread in space and time is patchy. Many formed in a surprisingly narrow time band in the Palaeozoic period (570–245 million years ago). It needed a plentiful supply of continental rock to be weathered, and an unusually long and stable period of weathering to provide all the sand to make them (as well as the removal of other impurities). This is why orthoquartzites are found on the stable margins of ancient continental cratons, such as central Australia, the Russian platform and the St Peters Sandstone of central North America. Some hugely thick orthoquartzites began as deposits where continents rifted slowly and moved apart, which were then folded up into mountain ranges. The Clinch sandstones of the Appalachians and the Tapeats of the Rockies are believed to have been formed by this process.

Grey orthoquartzite: Being almost pure quartz, washed by water or scrubbed by the wind over countless years, quartz arenite or orthoquartzite is often a remarkably clean, pale white or grey quartz colour.

Grain size: Over 95% sand-sized grains, 60μm–2mm/2.5–80mil.
Texture: Gritty texture like solid sand. Grains well sorted and well rounded.
Structure: Typically occurs in blanket-shaped deposits varying from a few metres to several hundred metres thick. They are usually interbedded with mudstones, limestones and dolomites. They usually display dramatic cross-bedding and ripplemarks reflecting their formation in high-energy environments. Aeolian rocks often show sand-dune shapes.
Colour: Water-formed orthoquartzites are typically white or pale grey. Aeolian orthoquartzites are often stained red, pink or brown by iron oxides.
Composition: Over 95% quartz, with a smearing of feldspar and carbonate cement. Also chert and metaquartzite, zircon, tourmaline and rutile.
Formation: Some form in high-energy marine environments such as beaches and spits around the edges of stable cratons. Some form on the continental shelf between rifting continents. Some are formed from aeolian (wind-blown) sand-seas in deserts, where there is a ready supply of sand.
Notable occurrences: Russian steppes; central Australia; St Peter sandstone, mid-west USA; Chilhowee, Tuscarora and Clinch formations in the Appalachian Mountains; Flathead and Tapeats formations in the Rocky Mountains.

Uluru

Australia's Uluru is the world's largest single block of freestanding rock. Towering to over 345m/1,100ft, Uluru looks like a giant boulder poking out of the sand of the Simpson Desert. In fact, the exposed rock is just the very tip of an ancient outcrop of arkose sandstone extending far under the desert. This outcrop formed when an ancient ocean floor called the Amadeus Basin was uplifted some 550 million years ago, initiating a dramatic period of erosion, and the deposition of arkoses. Later crustal movements have tilted these arkoses almost on end, at 80–85 degrees. Finally, Uluru's arkose was buried beneath the sediments of a shallow sea, and has only re-emerged as wind and water stripped these sediments away. Uluru was long known by its European name, Ayers Rock, but it is a sacred site for Aboriginal peoples, and in 1985 the Australian government gave its custodianship back to the Aboriginals and restored its Aboriginal name, Uluru.

Greywacke

Often called dirty sandstone, greywackes are tough, dark sandstones made from large, sharp grains of quartz, feldspar and rock fragments set in a mass of clay and silt. This unusually chaotic mix was often piled up by submarine avalanches or 'turbidity currents' that plunged from the continental shelf into the deep in a huge, churning mass of water and debris. Deposits are often thousands of metres thick and include the fossils of all kinds of deep-water creatures and plants caught up in the maelstrom. Greywackes were the main sandstones formed early in Earth's history, because land masses were so small at the time. Sandstones formed more recently are often better sorted.

Grain size: Mostly sand, mix of sizes from clay to gravel.
Texture: Chaotic mix of sand, gravel and silt. Poorly sorted but well graded.
Structure: Graded bedding. Beds folded and deformed. No cross-bedding. Forms sequences with laminated sandstones and shale.
Colour: Grey, green, brown.
Composition: Quartz (40–50%), feldspar (40–50%), mica, plus clay and rock.
Formation: Deposited by turbidity currents, and in other high-energy environments.
Notable occurrences: All fold mountain belts (except where there is lots of limestone), e.g. Wales; Scottish Uplands, Scotland; Cumbria, England; Schiefergebirge, Harz, Germany; Massif Central, France; Caples, Torlesse and Waipapa terrane, New Zealand; Coast Range, California; West Virginia.

Identification: The grey colour and chaotic mix of large fragments amid sand and clay make greywacke easy to identify.

Arkose (Feldspathic arenite)

Arkoses look so much like granite it can be hard to tell them apart. Often bedding marks are the only telltale signs that a rock is sedimentary arkose, not granite. This is because arkose is essentially reconstituted granite, with the same basic ingredients: quartz, feldspar and mica. It is the rock that forms when granite breaks down under particular conditions. What makes it different from other sandstones is that it contains feldspar. Under normal conditions, feldspar is weathered to clay, leaving just clay and quartz. Yet in arkose, feldspar is preserved. It was once thought this meant arkoses could form only in desert environments where there was too little moisture to destroy the feldspar. The Torridonian sandstones of north-west Scotland formed liked this. Now geologists know that feldspar may also be preserved if granite is being eroded and uplifted very rapidly. As a result, many arkoses formed as deltas and alluvial fans, where rivers spill out on to the grabens (depressions) created by the rifting of continents. Others occur along volcanic island arcs. So arkoses are linked with extremes in Earth's past – either extreme climates, or dramatic tectonic movements and high relief.

Identification: Arkose can look very like granite, with the same pinkish colour and the same assemblage of coarse-grained quartz, feldspar and mica minerals. The telltale signs are usually the shape of the formation, and evidence of bedding and layering.

Grain size: Mostly sand-sized grains, at least 1–2mm/ 40–80mil across.
Texture: Grains not as well sorted or rounded as ortho-quartzite, except desert arkoses.
Structure: In fan-shaped deposits a few metres deep. Less cross-bedding and ripple-marks than orthoquartzites. Aeolian rocks may show sand-dune shapes.
Colour: White, grey or pink reflecting feldspar content.
Composition: Quartz (40–50%), feldspar (40–50%), mica. In continental arkoses, orthoclase and microcline are the main feldspars; in island arc arkoses, plagioclase dominates.
Formation: As deltas and in river bars in areas of high relief and aeolian (wind-blown) deposits in deserts.
Notable occurrences: Torridon, Scotland; Pennines, England; France; Czech Rep; Uluru, Australia; Fountain Form, CO; California; eastern USA.

RUDITES

Sandstones, siltstones, mudstones and claystones are all made of small, fairly evenly sized grains.
However, some sedimentary rocks are made from a chaotic jumble of stones of many different sizes.
These jumbled, stone-filled rocks are called rudites, and are divided into two types: conglomerates and
breccias. In conglomerates, the stones are smooth and rounded. In breccias, they are sharp and angular.

Conglomerates

Sometimes called roundstone, conglomerates are basically round stones set in a matrix of finer sand and clay. The stones can be gravel (2–4mm/0.079–0.157in), pebbles (4–64mm/0.157–2.52in), cobbles (64–256mm/2.52–10.08in) and boulders (larger than 256mm/10.08in). Pebbles like these must have been tumbled along beaches or bowled down streams for countless years to round off all the sharp edges – and they had to be tough to survive this battering, so the stones in conglomerate are usually tough materials such as quartz, flint, chert and hard igneous rocks. In time, though, even the hardest stones are reduced to sand and clay. So conglomerates mark an interruption in the slow, steady process of deposition.

There are two kinds of conglomerate – orthoconglomerates and paraconglomerates. Orthoconglomerates are true sedimentary rocks and form where gravel and pebbles are dropped by flash floods in rivers or by storm waves on beaches. The stones in them are quite evenly sized and tightly packed. The spaces in between them are gradually filled up with finer sediment to cement them together, but the rock would be much the same shape with or without it.

Paraconglomerates are formed in one fell swoop and are a jumble of stones of all sizes scattered through a matrix. They are typically formed by landslides, by turbidity currents, and by glaciers – all dramatic events that move material wholesale without any sorting. Take away the matrix and all that is left is a pile of stones. Boulder clay is paraconglomerate.

Conglomerates are widespread, but deposits are usually small and localized. In some, dark pebbles stand out against the light cement like raisins in a pudding, earning them the name puddingstones. The brown puddingstones of Hertfordshire in England and Roxbury in Massachusetts, and the jasper puddingstones of St Joseph Island on Lake Huron in Canada, are all good examples of this.

Petromict conglomerate: The great majority of conglomerates are described as petromict or polymict. This means they contain a wide mix of different stones from a variety of sources, such as basalts, slates and limestones. They are mainly river deposits washed down from areas of high relief and dumped in alluvial fans.

Puddingstone: With their raisin-like pebbles, the puddingstones of Hertfordshire in England are very striking conglomerates. The white is a cement of quartz and feldspar; the pebbles are flints from the nearby chalk hills.

Grain size: Over 2mm/0.079in and can be granules, pebbles, cobbles or boulders.
Texture: Orthoconglomerates are mostly gravel-sized grains nearly touching and less than 15% sand and clay matrix. Paraconglomerates are at least 15% matrix and are really sand- or mudstones scattered with pebbles, cobbles and boulders.
Structure: Conglomerates are generally small, poorly stratified deposits with none of the bedding marks of finer sediments.
Colour: The colours are as varied as the rocks their stones came from. The stones are often markedly different in colour from the matrix. In jasper puddingstone, red stones are set in a pale matrix, like cherries in a cake.
Composition: The stones can be pure quartz or feldspar from sources such as pegmatites, but usually they are rock fragments – typically harder rocks such as rhyolite, slate and quartzite. The matrix can be silicates, calcites or iron oxides.
Formation: Orthoconglomerates form in fast-moving rivers and in shallow surf. Paraconglomerates are deposited by glaciers, landslides, avalanches and turbidity currents.
Notable occurrences: Hertfordshire, England; Kata Tjuta (Northern Territory), Australia; Huron, Ontario; Keeweenaw, Michigan; Ohio; Indiana; Illinois; Bahamas; Crestone (San Luis Valley), Colorado; Roxbury, Mass; Fairburn, S Dakota; Brooks Range, Alaska; Basin and Range, New Mexico; Van Horn, Texas; Death Valley, California.

Breccias

Sometimes called sharpstone, breccia is basically rubble turned into stone. The stones in breccia are jagged, caught up before there was time to round off any rough edges. Unlike conglomerates, breccias can form from almost any rock, soft and hard alike. But they almost always form near to their source and are said to be 'intraformational'. If the stones are washed any further away, they tend to get sorted and so do not form breccias. In mountain areas, breccias often form when screes are cemented together by finer sediment accumulating between the stones. Many breccias are formed rapidly by dramatic events – as when landslides and avalanches come to rest, or when flash floods or storm waves sweep masses of sediment into a beach or bar. Breccias also form when the roofs of limestone caves collapse, burying the floor in rubble. Coral reefs often contain extensive limestone breccias made of fragments broken off the reef. A few breccias are 'extraformational,' swept far from their source before consolidating and so have a very mixed composition.

Identification: Breccias are easily recognized by the large, angular stones they contain. It is not so easy to identify what the stones are or where they came from. The best place to start is a comparison with nearby rocks.

Grain size: Over 2mm/ 0.079in.
Texture: Large angular stones in a finer matrix.
Structure: Breccias are generally small, poorly stratified deposits.
Colour: The colours are as varied as their source rocks.
Composition: The stones are usually rock fragments – of almost any rock, including softer rocks such as marble.
Formation: Some form in fast-flowing rivers, or on storm beaches. Others form from landslides and avalanches, both on land and under the sea.
Notable occurrences: Thessaly, Greece; Mexico; Vancouver Island, Midway, British Columbia; Platte Co, Wyoming; San Bernardino Co, California; Makinac Island, Michigan; Zopilote, Texas.

Landslides
Every now and then a hill or cliff collapses suddenly in a landslide. Some landslides, like Black Ven in Dorset, are triggered as waves undercut the coast. Some are set off by a storm, like the thousands all over New Zealand after Cyclone Bola in 1986. Some are set off by volcanoes and earthquakes, like the 1989 Loma Prieta quake in California. Few events re-shape geology and remake rock material quite so quickly and dramatically. Soft rocks such as clays are very prone to landslides, but tougher rocks can also slide under certain conditions. They tend to fail along existing cracks such as joints. A key factor is often the presence of water, which pushes grains apart and reduces their cohesion. Local rains have caused landslides in the coastal town of Ventura, California (above). Very large rock falls can trap enough air to cushion the fragments, allowing them to travel far and fast. The 1970 Huascaran avalanche in Peru hurtled down the mountainside at a speed of over 320kph/200mph, killing 17,000 people in the towns in its path.

Volcanic, crush and impact breccias

Not all breccias are sedimentary. Volcanic breccias are tuffs that form from fragments blasted out by volcanoes. Crush breccias are formed when rocks are crushed underground by the sheer weight of formations above or by powerful tectonic movements. Some crush breccias are small scale, forming when veins and fissures are squeezed by crustal movements. Others occur on a much larger scale along faults, when the world's tectonic plates crunch past each other, or when layers of rock are folded in mountain building. Meteorite impacts create yet another kind of breccia when the huge force of an impact smashes crustal rocks to bits.

Grain size: Over 2mm/ 0.079in.
Texture: Large angular stones in a finer matrix.
Structure: Small, poorly stratified deposits.
Colour: The colours are as varied as their source rocks.
Composition: The stones are usually rock fragments.
Formation: Volcanic breccias form from pyroclasts. Crush breccias form underground when rocks are crushed by crustal movement. Impact breccias form from rocks smashed by meteorite impacts.
Notable occurrences: Volcanic breccias: Arizona; New Mexico. Crush breccias: Highlands, Scotland; Alps, Switzerland; Appalachian Mountains. Impact breccias: Haughton Impact Crater (Devon Island), Nunavut.

Identification: Volcanic breccia contains angular pyroclasts at least 2mm/0.079in across. The pyroclasts are often black glass.

BIOGENIC: BIOCHEMICAL ROCKS

Countless creatures are able to extract dissolved chemicals from seawater and use them to make shell and bone. Some use calcium and carbon to make carbonates. Others use dissolved silica to make silicates. When these creatures die, the solid material they created turns into sediments, which form 'biochemical' sedimentary rocks such as chert, flint, chalk and diatomaceous earth.

Bedded chert (biochemical chert)

Chert is made of quartz crystals so fine they can be seen only under a microscope. It is an incredibly hard rock, yet when hit with a hammer it cracks almost like glass into sharp conchoidal fragments – a quality that was much appreciated by prehistoric people for making cutting tools. Most beds of chert formed from the ooze that covers much of the deep ocean floor even today. The ooze is built up from the constant rain of plankton remains such as radiolarians, diatoms and microscopic sponges called spicules. Once the ooze is buried it slowly solidifies into chert. Relatively pure silica-rich oozes are known as radiolarian or diatomaceous oozes, depending on which microscopic organism is dominant. Slightly less pure oozes are known as sarls and smarls. Each forms a particular kind of chert. Ocean bed ooze chert is the top layer, above serpentines and basalts, in ophiolite sequences – segments of the sea floor thrown up on to dry land.

Identification: Chert is easy to recognize by its very fine-grained, almost glassy texture, and its tendency to break into sharp, conchoidal fragments when hit with a hammer.

Grain size: The crystals are cryptocrystalline (too small to be seen with the naked eye).
Texture: Almost glassy, with conchoidal fracture.
Structure: Biochemical cherts form in thin layers 1–10cm/0.4–3.9in thick. Typically massive or finely laminated (reflecting seasonal currents), but can show cross-bedding and scour marks from turbidity currents.
Colour: Black, white, red, brown, green, grey, depending on impurities.
Composition: Mostly pure quartz.
Formation: Forms when sea floor ooze solidifies.
Notable occurrences: Aberdeenshire, Scotland; Peaks, England; Bavaria, Harz, Schiefergebirge, Germany; Bohema, Czech Republic; La Salle County, Illinois; Marion Co, Arkansas; Ozarks, Missouri; Minnesota.

Flint (replacement chert)

Identification: Flint nodules look like white, knobbly pebbles on the outside, but once broken they look like black or treacle-toffee coloured glass – though they are much harder and break with very sharp edges.

Not all chert is biochemical in origin. Some is simply chemical. In other words, the silica is formed without any organisms, as calcite crystals in limestone are replaced. The best known of these replacement cherts are flints. Flints are nodules of black or toffee-coloured chert that form in limestones, especially the Cretaceous chalks of southern England and northern France. Any chert that is black may also be called flint. Both kinds of flint were widely used by prehistoric peoples for making tools, and also for striking sparks to make fire. Chert formed by replacement can also occur as a fine powder scattered throughout limestone, and it is also very occasionally found as the cement in sandstones.

Grain size: The crystals are cryptocrystalline, which means they are too small to be seen with the naked eye
Texture: Almost glassy, with conchoidal fracture
Structure: Nodules. Sometimes flint forms around a network of burrows like those of *Thalassinoides* (a branching burrow with Y- or T-shaped branches), so flint takes this shape.
Colour: Usually black
Composition: Nearly pure quartz
Formation: Flint forms from the solidification of sea floor ooze
Notable occurrences: North Yorkshire Moors, North and South Downs, England; Rugen, Germany; Mon, Denmark; Flint Ridge, Ohio

Chalk

Chalk is a white rock of almost pure calcite found in Europe and North America. About 100 million years ago in the Cretaceous period, large lowland areas of these continents were covered with tropical seas. Countless tiny floating algae left plate-like remains called coccoliths across the sea bed some 90–600m/300–2,000ft down, along with the shells of almost equally tiny organisms such as foraminifera. These algal plates and shell fragments turned quickly to almost pure white calcite. The sea bed remained undisturbed for a long time, and layer upon layer of these micro-organisms, along with the occasional larger shells such as ammonites, built up into thick layers of chalk, famously exposed in England's White Cliffs of Dover. Chalk is much softer than other limestones, and the vast beds that once covered most of north-west Europe have been stripped away, leaving bands of rounded hills. Chalks are porous rocks, though not very permeable, and these hills are marked both by dry valleys or bournes formed in wetter times, and also combes created by masses of crumbled rock flowing downhill during colder times.

Red chalk: Chalk may often be stained red by iron oxides.

Identification: Chalk's white colour is unmistakable. It looks like a fine powder, but the coccolith plates and foraminifera shells are clearly visible under a powerful microscope.

Grain size: Very fine-grained like mudstone.
Texture: Powdery grains.
Structure: Well stratified, with layers often shown up by beds of clays, shell layers and flint nodules. Often has burrow patterns. Occasional layers of crusted material called hardgrounds or Chalk Rock.
Colour: White, occasionally red.
Composition: Pure calcite.
Formation: Forms from the remains of marine algae and microscopic shells.
Notable occurrences: North Yorkshire Moors, Downs, Chiltern Hills, England; Champagne, France; Rugen, Germany; Mon, Denmark; South Dakota to Texas to Alabama.

Stone axes
Flints gave our human ancestors their first tools. Chipped to give a sharp edge, they made it possible to cut through tough hides to get at meat, or, later, to cut hide to make clothes and plants to make tools and shelters. The first stone toolmaker was *Homo habilis*, who appeared about 2.3 million years ago, but it was *Homo erectus* (1.8 million years ago) who made the first crafted stone hand axes. Named Acheulian axes after the French village where they were first found, these axes had two cutting edges and a round end for holding. Axes like these were widely used for over a million years, until half a million years ago a technique for creating a long narrow blade was devised. About 50,000 years ago, modern humans, *Homo sapiens*, made another key breakthrough in blade technology, creating stone knives. Getting a good edge from a flint stone, called knapping, was a tremendously skilled job, and there is evidence that factories were set up where the best knappers would work.

Diatomaceous earth

Diatoms are among the most abundant of all microscopic marine algae. When they sink to the bottom, their minute shells collect in the ooze and eventually turn to what is called diatomaceous earth. When this occurs in a more compact form as a soft, very light, porous, chalky rock it is called diatomite or kieselguhr. Miners sometimes call it white dirt because in bright sunlight it can look like fresh snow. Diatomaceous earth's remarkable purity and fine grain makes it a perfect filtration material, as well as a filler for paper, paint and ceramics. When sugars and syrups are clarified, diatomaceous earth is usually the filter. It is also used as a mild abrasive in toothpastes and polishes.

Grain size: Very fine-grained like mudstone.
Texture: Powdery grains. Diatom shells can be seen under a powerful microscope.
Structure: Well stratified, with layers often shown up by beds of clays.
Colour: White, yellow, greenish grey, sometimes almost black.
Composition: Silica shells of diatoms.
Formation: Diatomaceous earth forms from the remains of marine algae and microscopic shells.
Notable occurrences: Denmark; Lüneburger, Saxony, Halle, Germany; France; Central Italy; Russia; Algeria; Nevada; Oregon; Washington; Santa Barbara, California.

Identification: Diatomite looks a little like chalk but is so light it almost floats on water like pumice.

LIMESTONES (CARBONATE ROCKS)

Made up of at least half calcite (or the similar aragonite), limestones are distinctive whitish, grey or cream rocks. They are the third most abundant sedimentary rocks on Earth, after mudrocks and sandstones, and extend over vast areas of continents and continental shelves, dominating many mountain chains. Limestone 'karst' landscapes can often be very dramatic, with their caverns and gorges.

Limestone

Coral limestone: Few rocks are richer in fossils than limestones. Very often you can see perfectly preserved remains of sea creatures that swam and crawled in tropical seas long ago, or, like the coral polyps preserved in this rock, simply sat on the sea floor and waited for a meal to pass.

Limestones are a striking testament to the sheer profusion of life on Earth, especially in the sea. They are almost entirely the work of living things. Huge beds of limestone thousands of metres thick may be the accumulated remains of countless sea creatures piled up on the sea bed over millions of years, then slowly changed into rock as their chemistry alters. This accumulation is going on today, notably in places such as the Bahamas, and these remains too will in time turn to rock.

Living things contribute to the creation of limestone in two ways. Sometimes they contribute their 'skeletal' remains, their hard shells and bones, to the rock. Alternatively, like plankton and algae, they change the chemistry of the sea, and encourage the deposit of calcite. The key chemicals in limestones are carbonates – and in particular calcium carbonate in the form of calcite or aragonite. Carbonate sediments may be rich in either calcite or aragonite, but ancient limestones are almost always calcite-rich because aragonite alters over time to calcite.

Limestones form in many places – soils on old rocks, river flood plains, lakes – but most are the creation of shallow, clear tropical waters. Here there is not only an abundance of sea life, but the evaporation of such warm waters boosts the precipitation of calcium carbonates. This does not mean limestones are found only in the tropics, however. The continents have shifted so much through the ages that many places now nearer the Arctic were once in the tropics. During the Carboniferous period around 300 million years ago, much of what is now North America and Europe lay in the tropics, and was inundated by vast tropical seas. Huge beds of limestone, now visible in places such as Texas and the English Pennines, are the legacy of this time. In England, such limestones are called Carboniferous limestones.

Reef limestone: Reef limestones are the work of corals, those remarkable sea creatures still building up huge colonies like Australia's Great Barrier Reef. Reef limestones are made partly from their skeletons, built up over the ages, and partly from sediment trapped and bound by mats of microbes living on the reef. Unlike some other limestones, reef limestones contain no visible skeletal remains. They are also harder than other limestones, and are often left protruding as small hills after softer surrounding limestones has been weathered away.

Grain size: Varies from clay-sized to gravel-sized.

Texture: Highly variable, from very fine-grained, porcelain-like look to aggregate of large fossils.

Structure: Most limestones show the same range of structures as sandstones and mudrocks. Beds often include reef limestones, the fossils of coral reefs. Reef limestones show little bedding, although they preserve the growth pattern of corals and cavities filled by carbonate debris and cement. Patch reefs or 'bioherms' are oval lumps left by small round coral colonies. 'Biostromes' are large long limestone formations left by barrier reefs.

Colour: White, grey, cream plus red, brown, black.

Composition:
Skeletal remains: Algae and microbes (coccoliths and stromatolites); Foraminifera; Corals; Sponges; Byrozoans; Brachiopods; Molluscs; Echinoderms; Arthropods. Carbonate grains (overleaf): Ooids and pisoids; Peloids; Aggregates; Intraclasts. Lime mud: Bone, teeth and scale debris (phosphates); Wood, pollen and kerogen (carbonates); Cement (calcite, aragonite, dolomite).

Formation: Forms as carbonates form, mainly on the sea floor, either from the skeletal remains of sea creatures or by the precipitation of calcite.

Notable occurrences: Burren, Ireland; Pennines, Cotswolds, England; Slovenia; Italy; Swartberg, South Africa; Ratnapura, Sri Lanka; Laos; Thailand; Guilin, China; Victoria, Australia; Paparoa, New Zealand; New Mexico; Kentucky; Texas; South Dakota; Indiana; Onondaga, New York.

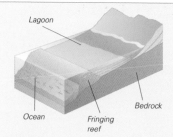

Lagoon

Ocean

Fringing reef

Bedrock

Coral reefs

Coral reefs are one of the wonders of tropical seas, teeming with an astonishing variety of sea creatures. The reefs themselves are made up from tiny sea anemone-like animals called polyps, which stay all their lives fixed in one place attached to a rock or to dead polyps. They take dissolved calcium carbonate from seawater and turn it into the mineral aragonite to build the cup-shaped skeleton or corallite in which they live. The skeleton becomes hard coral when they die. Coral reefs are made from millions of polyps and their skeletons, and can stretch for thousands of kilometres. Fringing coral reefs grow at a particular depth along the shoreline. Barrier reefs form a little way offshore. Coral atolls form around the edge of an island volcano. As the volcano sinks or the sea level rises, so the coral grows up and eventually leaves just a ring or atoll. Corals have been around since Cambrian times, and their reefs and fossils are abundant in limestone rocks of all ages since then.

Fossiliferous limestone: Bryozoan limestone

Like crinoids, bryozoans were sea creatures that lived in such vast numbers in the tropical oceans of the past that their remains have gone on to make a specific and abundant kind of limestone, bryozoan limestone. Over 15,000 species of bryozoan have been identified, of which 3,500 are alive today, living in many ocean shallows such as the western Pacific. They live in colonies of hundreds of animals or zooids, each secreting a short tube of lime to enclose its soft parts. A ring of about ten tentacles snakes out from the end of the tube to guide food into the animal's mouth. Bryozoan colonies look so like lace, they are also known as sea lace.

Grain size: Sand-sized grains with fossil remnants.
Texture: Highly variable, with sand-sized grains and partial and complete fossils.
Structure: Marked cross-bedding and ripple-marks. Layering from repeated cycles of sedimentation. Often broken into massive blocks divided by vertical joints and horizontal bedding planes.
Colour: White, cream.
Composition: Calcite.
Formation: From bryozoans in shallow tropical seas.
Notable occurrences: North Wales; Norfolk, England; Southern Sweden; Stevns Klint (Zealand), Denmark; Moravia, Czech Republic; Torquay and Geelong (Victoria), St Vincent (SA), off Tasmania, Australia; off Otago, Oamaru (South Island), New Zealand; Biscayne, Florida; Indiana.

Identification: Bryozoan limestone is identified from the lace-like colonies of bryozoans. Individual animals are tubes about 2mm/0.08in long.

Fossiliferous limestone: Crinoidal limestone

Many limestones consist largely of recognizable fossils of ancient sea creatures. Among the most widespread of these 'fossiliferous' limestones are crinoidal limestones. Sometimes called sea lilies, the crinoids of the past were animals that looked like long-stemmed flowers, with a central 'cup' containing the soft parts of the animal, numerous branching 'arms' and a stem up to 30m/98ft 5in long, which attached the animal to the ocean floor. In the Carboniferous period in particular, crinoid flowers grew in such extraordinary profusion that they created vast 'meadows' on the sea floor. When the animals died, ocean currents broke up most of their skeletal plates into sand-sized grains and rolled them together until they were cemented by calcite into thick deposits of limestone. Dramatic cross-bedding in these rocks bears witness to the shallowness of the seas in which the crinoids grew, and the power of the waves and currents that broke up their remains. Whole fossils are rare. The volume of crinoidal limestones around the world is staggering, and incorporates the remains of a huge number of crinoids. There are estimated to be at least 60,000km³/14,400 cubic miles of crinoid remains in the Mission Canyon-Livingstone formation in the Rockies alone.

Crinoid fossil

Identification: Crinoidal limestone is stuffed full of the fossils of crinoids. Although very few remain intact, there are usually enough of the cups, arms and stems surviving to be recognizable.

Grain size: Sand-sized grains with fossil remnants.
Texture: Highly variable, with sand-sized grains and partial and complete fossils.
Structure: Marked cross-bedding and ripple-marks. Layering from repeated cycles of sedimentation. Often broken into massive blocks divided by vertical joints and horizontal bedding planes.
Colour: White, grey.
Composition: Calcite.
Formation: From meadows of crinoids in shallow tropical seas.
Notable occurrences: North Wales; Derbyshire, Durham, Somerset, England; Austria; Nile Valley, Egypt; Timencaline Wells, Libya; Nepal; Namoi, Bingleburra, Australia; Mission Canyon-Livingston (Rocky Mountains), Canada–USA; Leadville, Colorado; Redwall, Arizona; Burlington, Iowa to Arkansas.

OOLITHS AND DOLOSTONES

There is a huge variety of carbonate rocks. While some limestones are largely fossiliferous or shelly – made largely of fragments of shell and bone – others consist of grains formed by the precipitation of calcite and aragonite from carbonate-rich sea-water. Like limestones, dolostones are carbonate rocks, but they are made of magnesium carbonate instead of calcium carbonate.

Grain limestone (Oolitic and Pisolitic limestone)

Limestones show the same range of grain sizes and textures as sandstones and mudrocks. Indeed, some geologists describe them using the same terms (lutites, arenites and rudites), adding 'calci-' or 'calca' to show they are limestones. So calcilutites are limestone muds, calcarenites lime sands and calcirudites gravels. Many calcarenites contain only a few shellfish remains. Instead, they are made mostly of calcite or aragonite grains precipitated out of water. Calcite grains can be washed into deposits just like sand and mud grains, but most form in situ. Ooliths or ooids are tiny balls made as layers of calcite build up on clay grains, kept round as they are rolled by underwater currents. Pisoids are gravel-sized balls that form in the same way. They look similar to grains called oncoids, but oncoids are actually created by microbes. Peloids are oval grains that usually started life as pellets of snail and shellfish faeces and were then altered to micrite (fine-grained calcite). Intraclasts are bits of broken calcite sediment. Limestones can be classified according to the dominant type of grain, as shown in the table below. They can also be classified by their texture (see table opposite).

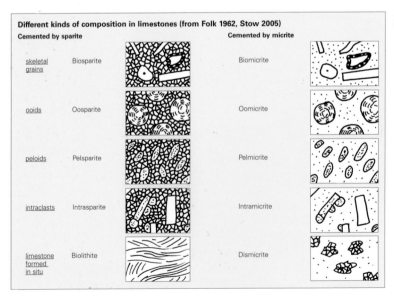

Oolitic limestone: Also known as roestone because it looks like fish roe, oolitic limestone is made of sand-sized grains called ooliths. Ooliths form in shallow, carbonate-rich tropical waters such as those around the Bahamas today, starting as aragonite and later changing to calcite. Wherever oolitic limestones appear, they are reminders that conditions were once like this.

Grain size: Ooids (0.2–0.5mm/7.87–19.69mil); pisoids and oncoids (over 2mm/78.8mil); peloids (over 1mm/39.4mil); intraclasts (1mm–20mm/0.04in–0.8in).
Texture: (see table opposite).
Structure: Grain limestones show the same range of structures as sandstones and mudrocks.
  Colour: White, grey, cream plus red, brown, and black.
Composition: Carbonate grains with lime mud cement (sparite and micrite).
Formation: Forms by the precipitation of aragonite in shallow, carbonate-rich tropical waters.
Notable occurrences: Dorset, Cotswolds, England; Luxembourg; Harz, Thuringia, Germany; Ukraine; Kertsch, Russia; Caucasus, Georgia; Newfoundland; Texas; Alabama.

Different kinds of composition in limestones (from Folk 1962, Stow 2005)				
Cemented by sparite			Cemented by micrite	
skeletal grains	Biosparite		Biomicrite	
ooids	Oosparite		Oomicrite	
peloids	Pelsparite		Pelmicrite	
intraclasts	Intrasparite		Intramicrite	
limestone formed in situ	Biolithite		Dismicrite	

Pisolitic limestone: Oolitic limestone is made from sand-sized grains 0.2–0.5mm/ 7.07–19.69mil in diameter; pisolitic limestone is made from larger, pea-sized grains at least 2mm/78.8mil across.

Dolostone (Dolomite limestone)

Ever since they were first identified in 1791 by Frenchman Deodat de Dolomieu in the Italian Dolomite mountains named after him, dolostone or dolomite limestone has intrigued geologists. While ordinary limestones are made of calcite or aragonite, dolostones are at least half made of the magnesium carbonate mineral dolomite. Until the 1960s, when it was found forming along the shore in the Arabian Gulf and in the Bahamas and Florida, no one had seen it actually forming directly from seawater. It seemed as if all dolostone formed by the chemical alteration of calcite in limestones by magnesium-rich solutions, a process called dolomitization. This is probably how most dolostones did form, but the process is now better understood. It seems to involve salty brines formed as evaporation concentrates seawater in tropical lagoons. These magnesium-rich briny waters sink seawards through limestones, slowly turning their calcite to dolomite. This process was more prevalent in the past, and most dolostones are Precambrian in origin (at least half a billion years old).

Identification: Dolomite is a much tougher rock than limestone and has a sugary white crystalline look. Recrystallization destroys fossils, so there are no visible organic remains.

Grain size: Varied – some forms are microcrystalline; others are sand-sized.
Texture: Dense sugary texture.
Structure: Stands out from ordinary limestone in rib-like beds because it is so tough. Coarse crystal dolostone shows the same structures as other limestones; fine crystalline dolostone does not.
Colour: White, grey, cream but weathers pink or brown.
Composition: At least half dolomite.
Formation: Thought to form when calcite in limestone is dolomitized (recrystallized and turned to dolomite).
Notable occurrences: Central England; Swabian and Franconian Jura, Rhineland, Germany; Dachstein, Austria; Dolomites, Italy; Niagara, Ontario; Arkansas; Iowa; Ohio; Kentucky.

Karst scenery
Limestone may have formed in water and yet the calcite it is made of is also quite easily dissolved by water that is slightly acidic. Rain and groundwater take up carbon dioxide from the air and soil, turning them into weak carbonic acid. Wherever limestone is near the surface, this acidic water seeps into cracks and begins to dissolve the rock. After thousands of years, huge cavities can be etched out often creating spectacular scenery, known as Karst after the Kras plateau in Slovenia, one of the many places where such scenery is found. Underground, huge potholes and caverns with stalactites and stalagmites are created. Above ground, cracks around blocks of rock on surfaces are etched out to create striking limestone pavements. Often cavern roofs collapse or potholes grow and merge to create deep gorges. Eventually, so much rock will be dissolved away that only distinctive, towerlike pillars are left, such as these in the famous Guilin Hills of China (above).

Different kinds of deposit texture in limestones

Original components not bound together

Mudstone (mud-supported, less than 10% grains)

Wackestone (mud-supported, more than 10% grains)

Packstone (grain-supported)

Grainstone (lacks mud and is grain-supported)

Original components bound together

Boundstone

No recognizable deposition texture

Crystalline

Original components not organically bound

Floatstone (matrix-supported, less than 10% sand-sized grains)

Rudstone (sand-supported, less than 10% sand-sized grains)

Original components organically bound

Bafflestone (organisms act as baffles)

Bindstone (organisms encrust and bind)

Framestone (organisms build a rigid framework)

This classification was devised by Dunham in 1962, then modified by Embry and Klovan in 1971 and Stow in 2005

CHEMICAL ROCKS

Chemical sedimentary rocks form neither from debris nor with the aid of living things, but entirely chemically as minerals precipitated out of water solutions. Many are left behind as solid 'evaporites' when the solution evaporates. Yet precipitation can occur whenever a solution becomes saturated and can no longer retain the minerals dissolved in it, forming rocks such as tufas, travertines and dripstones.

Tufa

Tufas are calcite deposits that build up around the rim of calcite-rich springs, rather like the limescale that builds up in baths and taps in areas of hard water. Tufa often builds towers underwater where springs bubble up into lakes or under the sea. If lake levels drop, these towers may be exposed, as in California's famous Mono Lake. Although tufas are chemical rocks, algae and other plant material does play a part in their formation. Tufa is always precipitated on to some surface or other, and quite often the surface is algae or plants. Indeed, algae actively spurs tufa to precipitate, forming algal mats or mounds called stromatolites made of tufa bound together by filaments of algae. Often the algae rots away leaving a sponge-like rock called a sinter. Tufa is sometimes called calcareous sinter to distinguish from siliceous sinter, a sinter that forms by the precipitation of opaline silica. Because it is so full of holes, tufa is light and easy to cut, which is why the Romans used it to line the Aqua Appia, the underground aqueduct they built in 312BC to supply the city of Rome with water.

Identification: Full of holes like a sponge, and quite light and soft, tufa is easy to recognize. It is usually white or a buff colour, but iron oxides can turn it red or yellow. Deposits are usually quite thin.

Grain size: Powdery.
Texture: Compact to earthy and friable.
Structure: Tufa is spongy and full of holes. Structures take the form of the places that they formed. Towers form around underwater springs. Algal colonies often form mounds.
Colour: White, buff, yellow, red.
Composition: Calcium carbonate in the form of calcite, or occasionally aragonite.
Formation: By precipitation from calcium-rich waters, typically in streams, around springs or algal mounds.
Notable occurrences: Glen Avon, Scotland; Ikka Fjord, Greenland; Great Rift Valley, Kenya; Kimberley, Western Australia; Mono Lake, Mojave Desert, California.

Travertine

Identification: Denser and more compact than tufa, with fewer holes, travertine looks a little like tofu, and is usually an attractive pale honey colour.

Tufa forms mainly around cool springs, typically when plants take carbon dioxide from the water and make less available to combine with calcium. Around hot springs, calcite is precipitated when hot water loses carbon dioxide as it cools. This leaves dense, hard crusts, such as those around Mammoth Hot Springs in Yellowstone Park, Wyoming. The terms tufa and travertine are sometimes used interchangeably, but geologists usually call the dense variety travertine, and the spongy variety tufa. Travertine is a pale honey colour, often with delicate banding. Many sculptors have used it as an easier-to-carve alternative to marble, and it is also cut into slabs and made into polished floors. The most famous travertine is Roman travertine, which gave the rock its name.

Grain size: Powdery.
Texture: Compact to earthy and friable.
Structure: Much denser than tufa, with only a few holes. Often banded.
Colour: Honey, red, brown.
Composition: Calcite, or occasionally aragonite.
Formation: By precipitation from calcium-rich waters around hot springs or in caves (see Dripstone).
Notable occurrences: Bohemia, Czech Republic; Aniene River, Italy; Pammukale, Turkey; Algeria; Thebes, Egypt; San Luis, Argentina; Baja, Vera Cruz, Mexico; Yavapai Co, Arizona; Yellowstone, Wyoming; Jemez, New Mexico; San Luis Obispo Co, California.

Evaporite

In arid conditions, salty water may evaporate to leave dissolved minerals as evaporite deposits. Some evaporites form when desert salt lakes dry up. They form on a larger scale when seawater evaporates in lagoons, coastal shallows and salt flats called sabkhas. The scale of some seawater evaporations is staggering. Evaporites are highly soluble, so they are rare on the surface, but there are ancient evaporites thousands of metres thick dating from the Cambrian, Permian, Triassic and Miocene periods. A depth of 1,000m/3,280ft of seawater needs to steam off to form each 15m/49ft of deposit. So, to build up these massive beds, coastal flats must have been flooded by the sea again and again over vast time spans. There are a large number of minerals dissolved in seawater, but only a few are abundant, and they always tend to be deposited in the same sequence, creating a bull's eye pattern of deposits. The sequence starts with the least soluble, dolomite, then goes through gypsum, anhydrite and halite (rock salt) to finish with the most soluble, potassium and magnesium salts called bitterns. The evaporites formed in salt lakes inland are typically dominated by halite, gypsum and anhydrite, but there are also many more minor salts.

Identification: Evaporites are usually crystalline, looking like solidified sugar or salt. Crystals can vary hugely in size. On lagoon and lake floors, selenite gypsum crystals can grow up to 1m/3ft 3in.

Grain size: Crystal size varies
Texture: Coarse/fine, earthy, friable, massive, sugary.
Structure: Lagoon deposits: cracked. Deep water deposits: laminated. Nodules of anhydrite form in sabhka gypsum, often leaving just a 'chicken wire' mesh of gypsum between them.
Colour: White, pink, red.
Composition: Dolomite, gypsum, anhydrite, halite or bittern.
Formation: Evaporation of salty waters in coastal salt flats, lagoons and salt lakes.
Notable occurrences: *Currently forming*: Caspian Sea, Georgia; Persian Gulf; Dead Sea; Great Salt Lake, Utah. *Ancient formations*: Green River, Wyoming. *Sabhka and shallow shelf*: northern Europe; Elk Pt, British Columbia; Salina, Michigan; Williston, Montana; Delaware, Texas. *Deep sea*: Mediterranean.

The great Mediterranean salt pan
In 1970, a team drilling in the Mediterranean Sea made an extraordinary discovery. There beneath the sea were the thickest evaporites ever found, many thousands of metres thick. It transpires that at the end of the Miocene epoch, about five million years ago, the movement of the continents greatly narrowed the Straits of Gibraltar. A brief ice age triggered a global drop in sea level, and suddenly the Atlantic stopped flowing into the Mediterranean to keep it topped up. Within a few thousand years, the entire Mediterranean – some 2.5 million km³ (599,782 cubic miles) of water – had evaporated to create one gigantic salt basin, like the Dead Sea but 4,000m/13,123ft deep! Buried in the sediments under the Nile in Africa is a great canyon 2,500m/8,202ft deep dating from this time – indicating that the Mediterranean had dried up entirely. This whole episode is known as the Messinian Event because the best known deposits from the time are under the port of Messina, Sicily (situated just south of the peninsula enclosing the Strait of Messina, shown in the satellite image above).

Dripstone and flowstone

Although most travertines form around hot springs, the most spectacular and beautiful are often those that form in limestone caverns. Here calcite-rich waters dripping from the ceiling create deposits called dripstones. Dripstones can build up in all kinds of fantastic formations, such as stalactites hanging from the ceiling and stalagmites projecting from the floor (see Calcites and Dolomite). Sliced across, these dripstones usually reveal how they were built up in layers, like the layers of an onion, in darker and lighter bands. Cavern walls and floors continually wet with running water may be coated in sheets of travertine called flowstone.

Grain size: Powdery.
Texture: Compact to earthy and friable.
Structure: Dense, compact. Stalactites and stalagmites show 'growth rings' caused by variations in precipitation.
Colour: Honey-coloured, red, brown.
Composition: Calcium carbonate in the form of calcite, or occasionally aragonite.
Formation: By precipitation from dripping and flowing calcium-rich groundwaters in limestone rock.
Notable occurrences: Kent's Cavern (Devon), England; Skocjan, Slovenia; Aggtelek, Hungary; Sorek, Israel; Reed Flute (Guilin), China; Philippines; Carlsbad, New Mexico; Mammoth Cave, Kentucky; Luray, Virginia.

Identification: Dripstones are often markedly layered, revealing variations in the seasonal flow of water down through the limestone.

ORGANIC ROCK

Coal is a very unusual sedimentary rock. Not only does it burn, which makes it a very useful fuel, but it is also almost entirely organic. It is made not from grains of minerals, like other sediments, but from the remains of plants that grew in tropical swamps hundreds of millions of years ago, transformed into solid black or brown carbon as they were buried.

Coal formation

Most of the coal resources in North America, Europe and northern Asia formed in the Carboniferous and early Permian periods, around 300 million years ago. At this time, these continents lay mostly in the tropics, and vast areas were covered in steamy swamps where giant club mosses and tree ferns grew in profusion. Waters moved through these swamps only sluggishly, so when plants died, their remains piled up on the swamp floor and rotted only slowly in the stagnant, poorly oxygenated water. Microbes began to turn the remains into peat. Peat is about half carbon and is a useful, if smoky, fuel when dried, but to transform peat into coal, it must be deeply buried, to a depth of at least 4km/2.5 miles.

Over millions of years, the peats from the Carboniferous swamp were buried under layers of accumulating sediment until they were not only squeezed completely dry, but began to cook in the heat of the Earth's interior. Cooking did not only destroy plant fibre – it drove out hydrogen, nitrogen and sulphur as gases, and gradually transformed the carbon compounds in the plants to pure carbon. The deeper and longer they were buried and the hotter they got, the more plants turned to carbon. Peat is quite soft and brown and only about 60 per cent carbon; anthracite, the deepest and oldest kind of coal, is hard, black and over 95 per cent carbon. In between come intermediate 'ranks' of coal – brown coal, or lignite (73 per cent carbon), and dull black bituminous coal (85 per cent carbon).

Most black coal dates from the Carboniferous and early Permian. The world's largest resources of these coals are in Russia and the Ukraine, which has almost half the world's entire coal reserves, but there are also huge black coal beds in the USA. Besides black coals, there are brown coals formed more recently, especially in the Tertiary, 1.6 to 64 million years ago. Although less rich in carbon, these coals are widespread, found in China, North America, especially Alaska, as well as southern France, central Europe, Japan and Indonesia.

Peat: All coal may have begun as peat. If so, the peat of old must have formed in tropical swamps. Peat today forms only in bogs in cool places. Here decomposition of plant material is so slow that thick layers can build up before the plants totally rot. Microbes do get to work, however, converting at least half of the plant material to carbon as it becomes compacted.

Lignite: Once peat is buried deeply, the process of 'coalification' begins. Microbial activity ceases, but pressure and heat begin to turn more of the plant remains to carbon. Lignite or brown coal is the first stage. It crumbles when exposed to the air and has a texture like woody peat. Most lignite is more recent than black coal, dating from the Tertiary. It is found nearer the surface than black coal, but is much less carbon-rich and burns with less heat and more smoke.

Grain size: Fine-grained, similar to mudrocks.

Texture: Varies with coal rank. The lower ranked coals contain many only partly altered plant remains; the highest ranked coals contain very few. Peat contains un-decomposed plants.

Structure: Coal occurs as beds or seams interlayered with other sedimentary rocks, often with a thin layer of carbon-affected material called seat-earth beneath. Seams are generally only a few metres thick, but can be several hundred metres. Coals form in stagnant environments so do not show any cross-bedding, but humic coals (see Formation below) contain bands from 1–10mm/0.039–0.39in thick. Each seam has its own banding profile, which can help to identify it almost like a fingerprint.

Colour: Brown, black.

Formation: Most coals are 'humic' and form when plant material piles up in situ in tropical coastal swamps and is then buried deep and converted by heat to carbon. Rarer 'sapropelic' coal forms when plant debris, spores, pollen and algae pile up far from their original source.

Notable occurrences: Some of the world's biggest coal reserves are in the heart of Siberia in Russia, in Kazakhstan and the Ukraine. There are also major reserves of black coal in northern Europe, the Damodar Valley in India, and Appalachians and Midwest of North America. Germany and China have huge resources of brown coal. Pennsylvania is famous for its anthracite deposits.

Coal types and components

Plants are made from a wide range of different components including massive, hard trunks, soft leaves and tiny seeds and spores. Once a plant dies and falls into a swamp, oxygen and microbes get to work on each of these components differently. These differences are most significant in the early stages of the coal formation process, when peat forms. Consequently, peats can differ widely in character according to the plant parts involved. Even once peat is buried and coalification proper begins, the variations in plant parts makes a difference to the nature of the coal. The plant components in coal are called 'macerals', and divided into three broad groups: vitrinite, liptinite and inertinite. Vitrinite comes from the woody parts of the plant – trunks, branches, roots. It's tough and shiny and is the major component in a type of coal called vitrain. Liptinite comes from the waxy and resinous parts of the plant – the seeds, spores and sap. It is softer and duller than vitrinite. Mixed with vitrinite, it makes a silky, laminated kind of coal called clarain. Inertinite comes from plant material much altered by oxidation during peat formation, or from parts affected by fungus. Mixed with liptinite, it makes a dull hard coal called durain. By itself it makes the soft, powdery charcoal-like kind of coal called fusain – easy to identify because it leaves your fingers smeared black.

Each coal seam contains varying amounts of these different kinds of coal, but the higher the rank of the coal – the closer to pure carbon anthracite it gets – the more they lose their distinctiveness as a result of the greater degree of 'coalification'.

Bituminous coal:
Bituminous or soft coal is second only in rank to anthracite, with a 75–85 per cent carbon content. It is dark brown to black and banded, and usually made of over 95 per cent vitrinite, which comes from plants' woody parts. This is the most widely used type of coal, but its high sulphur content can contribute to the creation of acid rain when it is burned.

Anthracite: Anthracite, or hard coal, is the highest ranked of all the coals – shiny black and over 95 per cent carbon. Temperatures in the ground have to reach over 200°C/392°F to turn bituminous coal to anthracite. It is the rarest, and usually most ancient, of coals but it has a very high energy content and burns almost without smoke.

How coal is mined
The way companies mine coal depends partly on the depth of the seam. With a seam less than 100m/328ft below the surface, the cheapest method is to simply strip off the overlying material with a giant shovel called a dragline. Brown coal tends to occur near the surface, and can often be mined economically by strip mining. The best bituminous and anthracite coal typically lies in narrow layers called seams, far below ground. To get at the coal, mining companies have to sink deep shafts to reach the seam. The Ashton pit in northern England plunged almost 1,000m/3,280ft. With the shaft dug, they then created a maze of horizontal or gently sloping tunnels to get into the seam and extract the coal. The surface of the exposed seam is called the coal face. Mining operations can be hazardous, with the constant danger of roof-falls or of explosions as methane gas forms from the coal. Miners can also suffer lung damage by inhaling coal-dust.

Composition:
Peat: Over 75% water by weight; Solid matter: over 50% carbon, under 50% dry mineral-free volatiles.
Lignite: 33–75% water by weight; Solid matter: 50–60% carbon, under 50% dry mineral-free volatiles.
Sub-bituminous: 10–32% water by weight; Solid matter: 60–75% carbon, 35–42% dry mineral-free volatiles.
Bituminous: Under 10% water by weight; Solid matter: 75–85% carbon, 18–37% dry mineral-free volatiles.
Anthracite: No water; Solid matter: 75–85% carbon, 18–37% dry mineral-free volatiles.

Plant components (macerals):
Vitrinite group (50–90%): Woody tissue – polymers, cellulose, lignin.
Liptinite group (5–15%): Waxy parts of plant – seeds, spores, resins.
Inertinite group (5–40%): Shiny black plant material highly altered during peat formation.

Coal types:
Vitrain: Glassy, brittle, bright bands, conchoidal fracture, dominated by vitrinites.
Clarain: Finely laminated, silky, bright and dull bands, smooth fracture, mix of vitrinite and liptinites.
Durain: Hard, dull, matlike, dull bands, mix of inertinites and liptinites.
Fusain: Soft, powdery, charcoal-like, dirties fingers, mostly inertinite.

Inorganic components:
Detrital quartz, heavy minerals, sulphates, phosphates, pyrite nodules, marcasite, siderite, dolomite, calcite.

NON-FOLIATED

Metamorphic rocks are formed from neither melts nor sediments but are created deep underground when other rocks are remade by heat and pressure, sometimes by direct contact with hot magma, sometimes by the tremendous forces present in the Earth's crust. The original rock's minerals are cooked and recrystallized in new forms or even as completely new minerals. Metamorphic rocks are divided into foliated (striped) rocks and non-foliated rocks, which include hornfels, metaquartzite and granofels.

Hornfels

Hornfels is a tough, splintery rock that gets its name from the German for 'horn rock' because broken edges are translucent like horn. Making it involves less stress than other kinds of metamorphism. The original rock or protolith is simply cooked by close contact with an intrusion. The heat is tremendous – typically about 750°C/1,350°F – but the rock is neither crushed, twisted nor pulled. So hornfels is free from foliation. Crystals are fine-grained and point in all directions. Indeed, hornfels can easily look like a volcanic rock. Small structures in the protolith are obliterated during metamorphism. The crystals reform in a tight, interlocking pattern like crazy paving called pfiaster structure, visible under a magnifying glass. Some hornfelses may also be distinctively 'spotted' with porphyroblasts (large crystals like phenocrysts in igneous rock), such as andalusite hornfels and cordierite hornfels.

Hornfels is not just a type of rock, though, but helps identify some of the various 'facies' of metamorphic rock – the particular combinations of minerals formed in different pressure and temperature regimes. Hornfels facies include the hornblende hornfels and the pyroxene hornfels facies. These facies are the array of minerals that form when pressure is low but temperatures are high. The exact composition depends on both the original rock and the temperature, often grading through different minerals the nearer to the intrusion the rock forms and the hotter it gets.

Hornfelses are often divided into three groups according to their protolith: those made from shales and clays; those made from impure limestone; and those made from igneous rocks such as dolerite, basalt and andesite. All these are fine-grained.

Shales and clays form biotite hornfelses specked with black biotite mica, though they also contain feldspar and quartz, and a little tourmaline, graphite and iron oxide. Geologists look in these rocks for the aluminium silicates andalusite, kyanite and sillimanite. Each forms at a particular temperature and pressure, so finding one reveals the conditions in which the rock formed.

Impure limestone hornfelses are tough rocks containing calcium-rich silicates such as diopside, epidote, garnet, sphene, vesuvianite and scapolite, as well as feldspars, pyrites, quartz and actinolite. The igneous hornfelses are, like their protoliths, rich in feldspar with brown hornblende and pale pyroxene, but they also contain streaks and patches of new minerals such as aluminium silicates.

Identification: Plain hornfels like this is very easily confused with basalt and other dark volcanic rocks. Sometimes, though, hornfels is unmistakably 'spotted' with porphyroblasts of minerals such as andalusite, cordierite, garnet or pyroxene.

Striped hornfels: Many hornfels rocks are streaked with crystals of aluminium silicates such as sillimanite and andalusite. These minerals are very characteristic of the hornfels facies, marked by high temperatures and low pressure.

Rock type: Non-foliated, contact metamorphic.

Texture: Even, fine-grained; sometimes contains porphyroblasts (large crystals), or poikiloblasts (large crystals enveloping smaller crystals).

Structure: Most small structures are obliterated by metamorphism, though bedding from protolith may be preserved.

Colour: Black, bluish, greyish, often speckled with dark porphyroblasts.

Composition: The matrix is too fine-grained for individual minerals to be easily distinguished, but tiny flakes of mica can sometimes be seen under a magnifying glass. Square black or red porphyroblasts of andalusite are visible in andalusite hornfels. If these crystals are cross-shaped, they are known as chiastolite, and the rock is called chiastolite hornfels. In cordierite hornfels, the rock is dotted with rice-grain-like porphyroblasts of dark cordierite. In pyroxene hornfels, there are porphyroblasts of pyroxene, andalusite or cordierite. Other common minerals may be garnets, hypersthene and sillimanite.

Protolith: Fine-grained rocks including shales, clays, impure limestones, dolerite, basalt, andesite.

Temperature: Very high.

Pressure: Low.

Notable occurrences: Comrie (Perthshire), Scotland; Cumbria, Dartmoor (Devon), Cornwall, England; Vosges, France; Harz Mountains, Germany; Elba, Italy; Nova Scotia; Sierra Nevada, California.

(Meta)quartzite

(Meta)quartzite is a tough, whitish, sugary-looking rock that looks rather like white marble, but is made from quartz, not calcite. Indeed, it is over 90 per cent quartz. It forms mainly from sandstone. Like the sandstone orthoquartzite, it is often simply called quartzite, and the two often grade into each other, depending on how much the original sandstone has been altered by metamorphism. During metamorphism, the quartz grains in sandstone recrystallize, creating new, larger grains. The cement and open pores in sandstone vanish, leaving only tightly interlocking grains. In fact, the quartz grains become effectively welded together so that when the rock breaks, it fractures right across the crystals, rather than breaking around the grains as sandstone does. Most metaquartzite is non-foliated. However, under extreme heat and pressure, it may be flattened or sheared in such a way that the grains are stretched out in a pancake shape, creating foliated metaquartzite.

Identification: (Meta)quartzite is a white rock that looks like marble but is much tougher. Unlike marble, it cannot easily be scratched with a coin or knife. White quartzite is also a little more brown than marble.

Rock type: Non-foliated, contact and regional metamorphic.
Texture: Even, medium-grained; sometimes granoblastic (grains are roughly shaped but even-sized).
Structure: Most small structures are obliterated by metamorphism, though bedding from the protolith may be preserved.
Colour: White, grey, reddish.
Composition: Tightly interlocking grains of quartz, with a little feldspar and mica.
Protolith: Sandstones and quartz-rich conglomerates.
Temperature: High.
Pressure: Low to high.
Notable occurrences: Islay, Grampians, Scotland; Anglesey, Wales; Norway; Sweden; Taunus, Harz, Germany; Wallis, Switzerland; Steiermark, Tyrol, Austria; North and South Carolina.

Granofels and charnockite

Granofels is one of the few non-foliated rocks to form under relatively high temperatures and pressures. This combination is generated only deep in the crust by tectonic forces that operate on a grand scale, so granofels is a product of regional, rather than contact, metamorphism. It is formed mostly from the granite family of rocks, or occasionally from thoroughly reconstituted clays and shales. Charnockite is a particularly widespread form of granofels. It was named by geologist T H Holland in 1900 after the tomb of Job Charnock, the founder of Calcutta, in St John's Church in Calcutta, India, which is made of this rock. Charnockite was once thought to be igneous, but it is now known to be metamorphic since despite the high temperatures and pressures, the original protolith never actually melted.

Stone aggregates
Virtually every construction project in the world, from the simplest house to the biggest suspension bridge, relies on 'aggregate' – the small chunks of rock that are cemented together to make bricks, concrete, asphalt and various other building materials. The average house contains over 50 tonnes/49.2 tons of aggregate. Some aggregates are readymade from sand and gravel deposits. Most are crushed rock – and the choice of rock is crucial. Soft rocks such as shale are really usable only for cement. The main hard rocks are basalt, gabbro and granite, limestone, gritstone and sandstone and the tough metamorphic rocks hornfels, amphibolite and gneiss. Road aggregates must not just be tough; they must be resistant to polishing by tyres, since this makes them slippery when wet, and must allow bitumen to stick to them. This rules out quartz-rich rocks such as granite. As a result road chippings tend to be limestone, basalt, hornfels or amphibolite. A furnace quarry at a stone aggregate mine is shown above.

Rock type: Non-foliated, regional metamorphic.
Texture: Coarse-grained.
Structure: Most small structures are obliterated by metamorphism.
Colour: Dark grey. Feldspar crystals may be dark green, brown or red; quartz may be bluish; hornblende may be brown to green.
Composition: Charnockite is made mostly of feldspar and quartz, but also contains the orthopyroxene hypersthene, plus hornblende and often pyrope garnet.
Protolith: Granitoids and altered shales and clays.
Temperature: High.
Pressure: High.
Notable occurrences: Scotland; Norway; Sweden; France; Madagascar; southern India; Sri Lanka; Brazil; Baffin Island; Labrador; Quebec; Adirondacks, New York.

Identification: Granofels is a dark grey, coarse-grained rock mottled with brownish feldspar and greenish hornblende crystals.

MAFIC METAMORPHIC ROCKS

When mafic igneous rocks such as basalt are subject to regional metamorphism, increasing heat and pressure progressively changes them from greenstone to greenschist, and then to amphibolite and finally to eclogite. This sequence can often be traced in the landscape, at right angles to the direction of the original pressure when the rock was metamorphosed deep underground.

Greenstone and greenschist

Greenstones are often very ancient indeed, and bands of greenstone rock called greenstone belts are found wrapped around granite in cratons, the billions-of-years-old cores of continents. Neither greenstone, nor the related greenschist, are single kinds of rock. Instead, greenstone encompasses any metamorphosed mafic igneous rock turned greenish by the presence of chlorite, epidote or actinolite. Greenschist is similar but foliated, marked by schist-like stripes. Under mild pressure, basalt simply recrystallizes to greenstone, leaving structures such as pillows and cavities intact. Further compression breaks these structures down to create the foliation of greenschist. Greenschist is also the name of one of the facies of metamorphic rock, and includes greenstone. The greenschist facies is the assemblage of minerals that is formed by low-grade regional metamorphism – low temperatures (300–500°C/572–932°F) and only moderately high pressure. In greenschist facies, minerals such as albite, epidote, chlorite, actinolite, titanite and pumpellyite totally or partially replace the major minerals in the original igneous rock such as pyroxene and plagioclase.

Identification: Greenschist's green colour is its most distinctive feature. Like many metamorphic rocks it has slightly sparkly crystalline appearance. Unlike greenstone, greenschist is slightly foliated, with signs of the banding called schistosity.

Rock type: Non-foliated, low-grade regional metamorphic.
Texture: Very fine-grained.
Structure: Phenocrysts, cavities and pillow structures from the original volcanic rock may be preserved.
Colour: Greenish.
Composition: Mainly actinolite, with other epidote group minerals such as chlorite.
Protolith: Mafic igneous rocks such as basalt (greenstone), or shale (greenschist).
Temperature: Low.
Pressure: Moderate.
Notable occurrences: Norway; Atlas, Morocco; Barberton, South Africa; Pilbara, Western Australia; Northwest Territories; Manitoba; Quebec; Ontario; Cascades; Rockies.

Amphibolite

Identification: The high pressures and temperatures that form amphibolites mean their texture is distinctively metamorphic. Crystals have a unique contorted form called crystalloblastic, which can be created only by high-grade metamorphism.

This is a coarse-grained rock composed mostly of plagioclase and hornblende. Strictly speaking, it is non-foliated, but geologists may call any plagioclase-hornblende rock amphibolite, whether foliated or not. Amphibolite is also one of the metamorphic facies, encompassing the assemblage of minerals that form in any rock under the huge pressures and moderate temperatures typical deep down during mountain building. Extreme conditions like this turn amphibolites into some of the toughest of all rocks, which is why they are often used for building roads. Some amphibolites are metamorphosed from dykes and sills cutting clean across softer sedimentary rocks. The tremendous pressure and heat that alters these intrusions to amphibolite transforms even the softest surrounding sediments into schists and gneisses. However, amphibolite is so resistant to shear stress that it remains unfoliated and survives intact as fragments within the other metamorphosed rocks.

Rock type: Non-foliated, regional metamorphic.
Texture: Medium- to coarse-grained. Sometimes contains porphyroblasts of garnet.
Structure: Hornblende crystals may be aligned, giving weak foliation.
Colour: Black, dark green, green, streaked white, or red.
Composition: Hornblende amphibole and plagioclase feldspar, plus mica almandine garnet and pyroxene.
Protolith: Mafic, intermediate igneous rocks: basalt, andesite, gabbro, diorite.
Temperature: Medium.
Pressure: High.
Notable occurrences: Donegal, Connemara, Ireland; Grampians, Scotland; Thuringia, Saxony, Germany; St Gothard Massif, Switzerland; Hohe Tauern, Austria; Quebec; Arizona; Adirondacks, New York.

Eclogite

Eclogites are among the rarest of metamorphic rocks, but they are also among the most interesting. They are very striking-looking rocks, typically made of red pyrope or almandine garnets embedded in a green pyroxene called omphacite. No other rock is so often full of interesting crystals and minerals.

There is little doubt that they formed under extreme conditions. The assemblage of minerals in eclogite, called the eclogite facies, could have formed only under high temperatures and pressures. They are closely linked with basalts, and a few geologists have argued that they are not metamorphic at all, but formed directly from basalt magmas deep underground in the Earth's mantle. Eclogites are never very large. Although there are instances of isolated blocks measuring 100m/328ft across in metamorphic rocks, most are xenoliths – chunks of foreign stone swept up from the lower depths in magmas. Xenoliths such as this often occur in diamond-bearing kimberlite and lamproites, and the diamonds are usually embedded in the eclogites themselves. It is now thought that there are actually three different types of eclogite, each forming in a different way. There are those that occur as xenoliths in kimberlite and basalt, as in Hawaii's Oahu crater. These formed at extremely high pressures and temperatures at least 100km/62 miles down in the mantle. Secondly, there are those that occur in bands and lenses in the midst of the most extremely metamorphosed gneiss, like those in west Norway and the Dabie mountains of China. The third type of eclogite occurs as blocks or bands in subduction zones along with blueschist, as in the Greek islands of the Cyclades. These crustal eclogites formed at lower temperatures and pressures. One theory is that these formed from massive gabbros in conditions where there was little water present; others suggest they formed from deeply subducted basalt crust.

Identification: Tough, dense and coarse-grained, eclogite is basically green and can look almost like solid gelatin. Often it is studded with large red porphyroblasts of garnet like this specimen.

Rock type: Non-foliated or foliated, regional metamorphic. May also be igneous, forming from basaltic magma.
Texture: Medium- to coarse-grained. Often contains porphyroblasts of garnet or pyroxene.
Structure: Very high density, massive, occasionally foliated.
Colour: Greenish, reddish, or green with red spots.
Composition: Dominantly omphacite pyroxene and almandine-pyrope garnet, with no plagioclase. Also includes quartz, kyanite, orthopyroxene, rutile, pyrite, white mica, zoisite and occasionally coesite. Xenoliths may contain diamonds.
Protolith: Basalt or gabbro, marl.
Temperature: Xenoliths in kimberlites, lamproites and orangeites above 900°C/1,652°F; eclogite lenses and xenoliths in ancient gneiss terranes 550–900°C/932–1,652°F; in blueschists near ocean trenches less than 550°C/932°F.
Pressure: High.
Notable occurrences: Glenelg, north-west Scotland; Greenland; West Norway; Saxony, Bavaria, Germany; western Alps, Switzerland; Carinthia, Austria; Apennines, Italy; Cyclades, Greece; DR Congo; South Africa; Botswana; Namibia; India; Borneo; Dabie Mountains, central China; Western Australia; north-west Canada; Oahu, Hawaii; California.

The ancient hearts of continents
Although some rocks forming the continents are quite young geologically, all of them have a very, very ancient core, or several cores. These cores are called cratons, and continents have grown around them over the ages to become the land masses they are today. The rocks in cratons are the oldest rocks on Earth, dating back at least 2.5 billion years. Gneisses metamorphosed from the volcanoes that created the first land masses are the oldest rocks. The Acasta gneiss of northern Canada is almost four billion years old. Almost as old are the distinctive greenstone belts, famous from Barberton in South Africa, from Pilbara in Australia and from northern Canada (above). Between 2.5 and 3.5 billion years old, these are rock islands of twisted greenstone wrapped around granite. Their origins are the subject of debate, but traces of pillow lavas are found in the greenstone, so many geologists think it may be pieces of ancient sea floor pushed up by a granite intrusion in ancient rift.

Retrograde eclogite: As eclogite xenoliths are brought nearer the surface, the minerals in them may sometimes be changed by retrograde metamorphism when they are affected by decreasing temperatures and pressures. Minerals such as amphibole may replace garnet and pyroxene.

MARBLE

Snowy white or cream with an extraordinary inner glow, marble is without doubt the most beautiful of all stones, cherished since the days of Ancient Egypt. For sculptors it is the finest of all stones, carved into shining statues such as Bernini's Ecstasy of St Theresa *and Michelangelo's* David. *Polished marble slabs have been used to face buildings from the Taj Mahal to the most modern skyscraper.*

Carrara marble

Identification: Pure marble is white but even pure marble can be stained grey by specks of graphite, or diopside like this. However, grey marble may be bleached snow white by contact with a hot intrusion.

Carrara marble: Those from the quarries near Carrara in Tuscany, Italy, are the most prized marbles of all. These snow-white rocks are almost pure calcite and were cherished not just by the great sculptor Michelangelo, but also in Roman times. The marbles occur in four main valleys in the Apennine mountains around Carrara.

For builders and sculptors, the word 'marble' covers a wide range of rocks. Limestones, serpentines and even quartzites may be called marble if they are pale in colour and can be carved or polished. For geologists, though, marble is a very specific kind of rock, made metamorphically under specific conditions. Even so, there is ambiguity. Some geologists describe as marble any rock that is metamorphosed from carbonate rocks. This includes metamorphosed dolomite, made from magnesium carbonate. A few geologists suggest that only rocks metamorphosed from pure calcium-rich limestones should be called marble.

This kind of true marble formed when beds of limestone were buried deep in the crust and altered by the heat of the Earth's interior and the pressure of overlying rocks. Marble is often brought up from deep mountain roots and continental collision zones interlayered with other ancient metamorphic rocks such as phyllites, quartzites and schists. Marble can be formed by contact as well as regional metamorphism, and small outcrops develop where hot granite intrusions have pushed their way into pure limestone beds.

Like metaquartzite, marble is made from reformed crystals of the same mineral as its protolith. During metamorphism, the calcite in limestone recrystallizes in larger, formless grains. Pore space between grains disappears, and grain and cement blur into one, leaving a tightly interlocking mass of calcite grains. The grains are odd shapes, and the texture can look sugary or even like the cracked pattern on ancient glazed porcelain. But this dense, uniform texture is just what makes marble so beautifully smooth for sculpture. The stone even seems to glow because it is slightly translucent and allows light to penetrate through the surface grains and reflect off internal grains.

Marble is soft enough to carve, but is tough enough to survive quite well in dry conditions. However, it is easily corroded by acid rain. Large masses of marble can be weathered into the same karst formations as limestone, and marble walls and statuary become pitted over time.

Rock type: Either foliated or non-foliated, low-grade regional metamorphic or contact metamorphic.
Texture: Medium- or coarse-grained, clearly visible to the naked eye. Even-grained, often sugary in appearance. Translucent in slabs up to 30cm/12in thick.
Structure: Old bedding structures and even fossils are occasionally preserved. More often, though, marble is evenly massive. Pure marble is rarely foliated, but because it flows under high pressure, coloured minerals may be stretched out to give highly contorted stripes.
Colour: Occasionally pure white, but often stained different colours by minerals in the protolith. Pyroxene turns it green; garnet and vesuvianite turn it brown; and sphene, epidote and chondrodite turn it yellow. Other minerals and the process of metamorphism usually add waves, flecks, grains and stripes of colour.
Composition: Mainly calcite or, in dolomitic marble, dolomite. Additional minerals include quartz, muscovite and phlogopite mica, graphite, iron oxides, pyrite, diopside and plagioclases such as albite, labradorite and anorthite. Other minerals include scapolite, vesuvianite, forsterite, wollastonite tremolite, talc, chondrodite, brucite, apatite, sphene, grossular garnet, zoisite, tourmaline, epidote, periclase, spinel, pyrrhotite, sphalerite and chalcopyrite.
Protolith: Limestone. Carbonate limestone gives pure marble; dolomitic limestone gives dolomitic marble.
Temperature: Low to high.
Pressure: Low to high.

Marble varieties

Pure marble is white, and made mostly of calcite with minor traces of other minerals. The commonest additional minerals are small rounded grains of quartz, scales of pale muscovite and phlogopite mica, dark, shiny plates of graphite, iron oxides and pyrite. Different metamorphic conditions and different minerals in the original limestone can give it all kinds of different colours and patterns. Often these impurities are smudged out into wonderful whirling streaks, like ripple ice cream, as the rock flowed slightly during metamorphism. Common additional minerals are green diopside, pale green actinolite, plagioclase feldspars and many more (see data panel). Sometimes the entire mass of marble may be stained by impurities. Pyroxene turns marbles green. Garnet and vesuvianite turn it brown. Sphene, epidote and chondrodite turn it yellow. Graphite can turn marble grey or even black.

Once formed, marble can often be altered by both chemical and physical stresses. As it is attacked chemically, minerals such as hematite may develop, staining it red, while limonite stains it brown and talc stains it green. A particularly attractive alteration is when the marble is coloured by patches of green or yellow serpentine altered from diopside and forsterite in the original marble. This variety is called ophicalcite or verd antique.

The rock called onyx marble is not actually marble at all. Neither is it onyx, which is banded chalcedony. It is rings of calcite deposited from cold mineral-rich solutions around springs in crevices and caves, often as stalagmites. Onyx marble is also called alabaster and was widely used for carving in the ancient world. Reddish Siena marble from Tuscany is onyx marble, as is the Algerian marble used in the buildings of Carthage and Ancient Rome.

Dolomite marble: Dolomite marble is not true marble since it was metamorphosed from doleritic limestone and is made mostly from dolomite (magnesium carbonate) rather than calcite (calcium carbonate). This usually makes it a little greyer.

Ophicalcite: This marble gets its name, like 'serpentine', from the classical word for snake, and is basically serpentinized marble. Exposure to chemical attack changes forsterite and diopside to serpentine, giving the rock an appearance of snakeskin.

The artist's stone
Because of its softness, glow and beautiful colours, marble has been prized as a stone for sculpture since Egyptian times. Pentelic marble from Attica was the luscious white stone that Ancient Greek sculptors such as Phidias and Praxiteles used to make their wonderful statues, the first lifelike carvings of people. The famous Elgin marbles that once adorned the Parthenon in Athens were made of Greek marble, too. In the Middle Ages, Michelangelo carefully chose a block of pure white marble from the Carrara quarries in Tuscany, Italy, to make his great statue of David. Antonio Canova chose the same white stone for his famous *Three Graces*, and many others. Marble has been used even more widely to face buildings, ever since the Romans discovered how to stick it to walls with cement. So many of the buildings of Ancient Rome were covered in marble that the city shone even at night. Today, buildings such as Washington's National Gallery are clad in marble.

Notable occurrences:
Devon, England; Connemara, Ireland; France; Spain; Fichtelberg, Germany; Tyrol, Austria; Tyrol, Tuscany, Italy; Wallis, Switzerland; Talladega County, Alabama; Harford County, Maryland; Vermont; Georgia.

Pure marble: Mount Pentelicus (Attica), Greece; Carrara, Massa and Serravezza (Tuscany), Italy; Bergen, Norway; Alabama; Georgia; Maryland; Vermont; Yule, Colorado.

Dolomite marble: Glen Tilt, Scotland; Norway; Sweden; Fichtelberg, Germany; Steiermark, Austria; Tyrol, Italy; Karelia, Russia; Utah.

Ophicalcite: Sutherland, Scotland; Connemara, Ireland; Mona (Anglesey), Wales; Fichtelberg, Germany; Wallis, Switzerland; Alps, France; Piedmont, Italy; Estramadura, Portugal.

Onyx and stalagmite marble: Siena (Tuscany), Italy; Oued-Abdallah, Algeria; Tecali, Mexico; El Marmol, California.

Black marble (non-metamorphic limestone): Kilkenny, Galway, Ireland; Ashford (Derbyshire), Frosterley (Yorkshire), England; Shoreham, Vermont; Glen Falls, New York.

FOLIATED METAMORPHIC ROCKS

High or moderately high pressures during metamorphism can create rocks with distinct layers called foliation, including slate, schist and gneiss. Foliation makes some rocks stripy, like mylonite, migmatite and glaucophane schists, and makes others, like phyllite, liable to split into thin sheets. Foliation means either that some minerals have been separated into bands, or that crystals have been aligned in parallel.

Mylonite

Sometimes, the huge forces involved in crustal movement literally tear rocks apart and drag the broken edges past each other. Near the surface, rock is shattered into angular fragments along these fault zones and ultimately crushed to powder. Deep down, however, the heat of the Earth's crust makes rock too soft and plastic to break. So when rocks are sheared, they smear out like toffee to form streaky rocks called mylonites. Softer materials recrystallize as minute grains, while a few more robust larger crystals may survive, crushed and reduced within this fine matrix. Because the larger crystals have not recrystallized, they are called porphyroclasts, not porphyroblasts. It was once thought that the fine grains in mylonite were simply pulverized; however, it is now known that they are actually new crystals that form under the strain, in a process called syntectonic recrystallization. The word mylonite was coined by Charles Lapworth in 1885 to describe the streaked rock he found in the Moine Thrust fault zone of the Scottish Highlands. Now it is used to describe any rock with a smeared-out streaky texture like this, and bands can be anything from 1–2cm/0.4–0.8in to 2–3km/1.2–1.9 miles thick.

Identification: With their streaky, smeared-out texture, mylonites are quite easy to recognize, but there are many kinds of mylonite, including protomylonite in which many of the grains are porphyroclasts – pulverized but not recrystallized – and ultramylonite in which there are no porphyroclasts left at all.

Rock type: Foliated, dynamic metamorphic.
Texture: Smeared-out, streaky texture with larger, but still tiny porphyroclasts in a very fine-grained matrix.
Structure: Mylonites sometimes but not always split along the direction of the streaks.
Colour: Varied.
Composition: Varies with the original rock, but the matrix is typically quartz and carbonate, with feldspar and garnet porphyroclasts.
Protolith: All kinds of rock.
Temperature: Low.
Pressure: High shear.
Notable occurrences: Moine Thrust, NW Scotland; Alps, Switzerland; Turkey; Deccan Traps, India; Ross Sea, Antarctica; Canadian Shield; Sierra Nevada, California; Blue Ridge, Virginia; Adirondacks, New York.

Migmatite

Migmatites are often the most extremely metamorphosed of all rocks, forming deep down in the continental crust under even greater pressure and heat than gneiss. Indeed, conditions are so hot that the rock partially melts. Minerals that melt at low temperatures liquidize, turning into igneous rock. Migmatites were first identified in 1907 by Finnish geologist J J Sederholm, who named them after the Greek word *migma* for mixture. The name is apt for migmatites are really a mixture of metamorphic and igneous rock. They usually consist of dark gneiss, schist or amphibolite striped by bands of leucocratic (pale-coloured) rock such as granite. At first migmatites were thought to be just pockets within gneiss. Now a few geologists think they may also be the last vestiges of rock that melted to form a granite magma. Some geologists argue that the pale bands have been intruded into the rock from an external source, rather than melting in situ. The gneiss portion is therefore older and called the 'palaeosome'.

Identification: With its distinctive humbug stripes of dark metamorphic rock and pale igneous rock, migmatite is usually fairly easy to identify.

Rock type: Foliated, regional metamorphic.
Texture: Medium-grained.
Structure: Alternating stripes of dark metamorphic and pale (leucocratic) igneous rock.
Colour: Varied.
Composition: Typically gneiss with granite. Can also be schist or amphibolite.
Protolith: May be all gneiss (or schist or amphibolite) or may be gneiss and granite.
Temperature: Very high.
Pressure: High.
Notable occurrences: Sutherland, Scotland; Scandinavia; Auvergne, France; Black Forest, Bavaria, Germany; Cyclades, Greece; Lake Huron, Ontario; Adirondacks, New York; New Jersey; Washington.

Phyllite

When mudstone and shale are subjected to mild metamorphism, their crystals line up perpendicular to the direction of pressure, and the rocks turn to slate. If the pressure and heat become a little more intense, they turn to phyllite. Even more heat and pressure turns phyllite to schist. Phyllite gets its name from the Latin for 'leaf-stone', and like slate it is characterized by laminations similar to the leaves of a book. Like slate, phyllite is made from very tiny grains of mica, chlorite, graphite and similar minerals, which grow flat at right angles to the pressure. Yet while slate looks dull, phyllite almost glitters because the extra heat and pressure creates thicker flakes of mica, especially the muscovite mica sericite. This silky sheen is called phyllitic lustre. In phyllite, the leaves are so compressed that it does not split into sheets nearly as well as slate, especially in its most highly metamorphosed, protophyllite, form. All the same, it is sometimes used, like slate, as roofing tiles.

Identification: Like slate, phyllite is distinctly layered. The layers, though, are not completely flat as in slate, but slightly wrinkled. These 'crenulations' make phyllite look like crepe. Phyllite also has a silky, silver lustre quite unlike slate's drab grey.

Rock type: Foliated, regional metamorphic.
Texture: Fine-grained with porphyroblasts.
Structure: Marked laminations at right angles to pressure. May have slaty cleavage and split into sheets as thin as 0.1mm/3.9mil. When metamorphism has gone further, the cleavage is only apparent and the rock won't split. May show minor folds and corrugations.
Colour: Silver grey, greenish.
Composition: Mostly sericite mica and quartz.
Protolith: Shale, mudstone.
Temperature: Moderate.
Pressure: Moderate.
Notable occurrences: Donegal, Ireland; Grampians, Scotland; Anglesey, Wales; Cornwall, England; Scandinavia; Vosges, France; Fichtelberg, Bavaria, Harz Mts, Germany; Alps, Switzerland; Connecticut; New York; Appalachians.

The Moine Thrust
The discovery of the Moine Thrust along the coast of Sutherland, north-west Scotland, in 1907 was a key moment in the history of geology. Geologists were already familiar with simple thrust faults. These are shallow reverse faults created when the crust is squeezed, forcing one block of rock up over another. But the Moine Thrust is not a single thrust. In fact, it is a complex belt of thrusts. We now know that such belts develop when tectonic plate movement repeatedly forces layers of crust up and over each other, then pulls them back, creating a complex, broken, multi-layered formation. At Moine, this all happened between 410 and 430 million years ago, when Scotland was crushed by opposing tectonic plates, creating a thrust belt stretching 180km/112 miles from the Moine Peninsula to the Isle of Skye. Similar thrust belts have now been discovered along the edges of fold mountain ranges all around the world. They are often characterized by complex bands of mylonite rock.

Glaucophane schist: blueschist

Glaucophane schist is a rock turned blue by the amphibole mineral glaucophane. It is also called blueschist, but it is one of a variety of rocks that form in similar conditions known as the blueschist facies. Not all blueschist facies rocks are blue, but laboratory experiments have shown they all form when pressures are very high but temperatures are low. This is a surprising combination, since high pressures usually go hand in hand with high temperatures, forming greenschists. Geologists now believe the answer is that blueschists form in subduction zones. The theory is that as a cold basaltic ocean slab is shoved deep into the mantle, a wedge of material, called an accretionary wedge, is scraped off and pushed back to the surface by slab material descending behind. This all happens so quickly in geological terms that the rock in the descending slab is squeezed hard, altered to blueschist then lifted back on the surface before it has time to heat up.

Identification: Schistose banding gives away glaucophane as a schistlike rock. A bluish tinge may establish its identity.

Rock type: Foliated, regional metamorphic.
Texture: Fine, medium-grain.
Structure: Weakly schistose. May show folding.
Colour: Bluish, light violet.
Composition: Mostly glaucophane or lawsonite amphibole or epidote, quartz, or jadeite, with garnet, albite, talc, zoisite, jadeite and chlorite.
Protolith: Usually basalt or dolerite, but may be mudrock.
Temperature: Low.
Pressure: High.
Notable occurrences: Anglesey, Wales; Channel Islands, England; Spitsbergen, Norway; Calabria, Tuscany, Val d'Aosta, Italy; Alps, Switzerland; California.

Schist

Mica

SLATE

The word slate is sometimes used to describe any stone that splits into flat slabs and is used as roofing tiles. True slate, however, is a very distinctive dark grey, brittle metamorphic rock that flakes into smooth, flat sheets. It is created when shales and clays are altered by low-grade regional metamorphism at low temperatures and moderate pressures.

Slate

Getting its name from the Old German word for break, slate is essentially metamorphosed mudrock that has been strongly compressed deep underground, but in a low-grade regional metamorphic environment, away from the most intense metamorphism. Conditions such as these are found deep at the root of fold mountains, where the convergence of tectonic plates slowly but surely crushed rocks deep down. Most slates occur in old mountain chains, like the Appalachian Mountains of the USA, or Snowdonia in Wales. They tend to be Precambrian or Silurian in age. Occasionally, though, they form in more recent fold mountain chains, like the Alps.

Slate varies enormously in colour, though it is normally dark grey, or purplish or greenish grey. But it is always easy to recognize because of the way it cleaves into the flat sheets that make it so useful for roofing. This distinctive 'slaty' cleavage develops when mudrock is metamorphosed. As the rock is compressed, water is squeezed out and the rock is compacted. All the tiny clay and silt grains are not simply recrystallized as mica and chlorite but are reoriented at right angles to the pressure. This alignment occurs partly because just as the layered nature of clay crystals means they can be moulded, they can also be pressed flat, and partly because new crystals grow in this direction.

Metamorphism usually destroys most of the original sedimentary structures, so the cleavage planes are entirely unrelated to bedding planes and are probably at an entirely different angle. Fossils are usually destroyed by metamorphism, too. Only when the pressure is pretty much at right angles to the bedding are fossils preserved – though often rather dramatically flattened. Because slate's cleavage is produced by the same forces that fold mountains, slate cleavage usually clearly marks out the pattern of folding in the formation and the direction of compression. This makes it a very useful rock when studying the tectonic history and structural geology of an area.

Identification: A dull dark, smooth grey, turning shiny black when wet, slate is a distinctive rock, even before it is broken. But its tendency to break into thin, flat sheets marks it out even more clearly. No other rock has such a distinctive cleavage.

Clayslate: This is a very mildly metamorphosed rock halfway between shale and phyllite. Some geologists regard it as a sedimentary rock, but it has been metamorphosed enough to prevent it absorbing water and swelling like shale. Clayslates never smell earthy when damp like shale, and rarely have any fossils. They also split like true slates.

Rock type: Foliated, low-grade regional metamorphic.
Texture: Very fine-grained. The grains are so small that it is impossible to identify individual minerals even under a magnifying glass.
Structure: Slates are characterized by single, perfect flat, slaty cleavage that allows it to be split easily into sheets. Traces of bedding planes and other original protolith structures are sometimes revealed like a picture on the flat cleavage surfaces. Sometimes fossils are preserved but they are usually squeezed flat.
Colour: Grey, black, shades of blue, green, brown and buff. Limonite and hematite colour it brown; chlorite colours it green.
Composition: Mainly mica and chlorite, with quartz, pyrite and rutile. Minor minerals include calcite, garnet, epidote, tourmaline, graphite and dark carbonate minerals.
Protolith: Mudrocks (mostly shale) and volcanic tuff.
Temperature: Low.
Pressure: Moderate.
Notable occurrences: Wicklow Mts, Ireland; Highland Boundary Fault, Scotland; Snowdonia, North Wales; Cornwall, England; Ardennes, France; Fichtelberg, Thuringia, Germany; South Australia; Brazil ('rusty' slate); New Brunswick; Nova Scotia; Ontario; Martinsburg, Pennsylvania; western Vermont; eastern New York; central Virginia; central Maine; northern Maryland.

Spotted and chiastolite slate

Spotted slate: Just as shale is metamorphosed to slate by pressure on a regional scale, so slate is metamorphosed locally to form spotted slate by contact with a hot intrusion.

Slate is the first stage in the sequence of rocks that develop when mudrock is progressively metamorphosed on a regional scale. Moderate heat and pressure on mudrock alters it to slate, but if the pressure increases it develops into phyllite. Even more intense pressure turns phyllite to schist. Just like sedimentary and igneous rocks, slate can also be cooked and altered by contact with a hot granite intrusion. Right next to the intrusion, slates lose their distinctive cleavage, and develop into splintery, tough hornfels. Further away, they retain their cleavage but develop dark or light round spots of altered minerals, typically white mica or chlorite. Spotted slates such as these usually contain minerals such as andalusite, garnet or, more rarely, cordierite. Andalusite in the form of chiastolite is especially characteristic of these spotted slates. Chiastolite slates contain distinctive large porphyroblasts of andalusite with dark crosses embedded in a light crystal. These crystals can often be 7–8cm/2.3–3.1in long. Commonly, in exposed slates, though, they are weathered to white mica or kaolin.

Chiastolite slate: Chiastolite slate is a very distinctive rock. Like ordinary slate, though, it is tough yet brittle and breaks easily into flat sheets. Unlike ordinary slate, it is marked with porphyroblasts, created by the heat of contact with an intrusion. The porphyroblasts are pale crystals of andalusite, aligned randomly throughout the rock.

Rock type: Foliated, low-grade regional metamorphic.
Texture: Very fine-grained, but studded with porphyroblasts of minerals such as andalusite and mica.
Structure: Like ordinary slates, they are characterized by single, perfect, flat, slaty cleavage that allows them to be split easily into sheets.
Colour: Grey, black, shades of blue, green, brown and buff. Limonite and hematite colour it brown; chlorite colours it green.
Composition: Mainly mica and chlorite, with quartz, pyrite and rutile. Minor minerals include calcite, garnet, epidote, tourmaline, graphite and dark carbonate minerals.
Protolith: Mudrocks (mostly shale) and volcanic tuff.
Temperature: Moderate.
Pressure: Moderate.
Notable occurrences: Snowdonia, Wales; Skiddaw, Devon, England; Betic Cordilleras, Spain; Halifax, Nova Scotia; California.

The slate industry

Slate may be very brittle and split easily into sheets, but it is actually a tough rock that is very resistant to the weather. This is why it is often seen in craggy outcrops in mountain regions, standing out darkly, almost black when it rains. The combination of slate's weather resistance, and the ease with which it can be broken into flat sheets makes it a superb light and durable roofing material.

Buildings have been roofed with slate for thousands of years. The famous slates of

North Wales have been used since at least Roman times. The Roman fort of Segontium (modern Caernarfon) had its tiles replaced with slate in the 4th century, while at nearby Caer Lugwy a fort had a slate roof two centuries earlier.

By the Middle Ages, Welsh slate was being shipped all around the British Isles, and maybe even abroad. When Chester castle was renovated in 1358 under the supervision of Edward the Black Prince, 21,000 slates were shipped from Wales to cover its roof. Many other castles and large houses used slate from the Welsh quarries at Cilgwyn and Penrhyn. Further north in Scotland, a thick, more durable kind of slate was used on the roofs of medieval houses in Edinburgh.

The use of slate remained fairly small-scale, though, until the 19th century and the Industrial Revolution. As cities in both North America and Britain grew rapidly, houses were built in millions, and each one of them had a slate roof. In Britain, the output of slates from the Welsh quarries expanded enormously. By 1832, they were digging out 100,000 tonnes/98,420 tons a year. Half a century later, almost half a million tonnes of slate were coming out of the Welsh

quarries alone. In North America, the situation was similar, with huge quantities of slate dug out of the quarries of Vermont and Pennsylvania to cover the roofs of millions of houses.

Demand for slate has declined dramatically since its 19th-century peak, as natural slate has been replaced by cheaper, more regular, mass-produced artificial slate. The Welsh quarries now produce less than 5 per cent of what they did at their height, and Vermont quarries even less.

Slates are still made today by hand by slaters in the same way they have been for centuries. Large blocks are dug out, then cut with a saw across the grain into sections slightly longer than the finished tile. Then the blocks are 'sculped' or split into slabs with a mallet and a broad-faced chisel. Finally, the slabs are split into tiles using a mallet and two chisels, and trimmed down to exactly the right size and shape. A last touch may be to punch two nail holes in one end. These holes allow the slate to be fixed quickly in place by the roofer. This last task requires some care if the tile is not to split, and relies on the slater's estimate of the particular qualities of the slate.

SCHISTS

The resilient rock that provides a firm base for Manhattan's towering skyscrapers, schist is the most extreme form of the regional metamorphism of mudrocks. When pressures on slate and phyllite climb, and temperatures rise above 400°C/752°F, the rocks completely recrystallize and reform as schist. All schists form this way, but there are many kinds, depending on the minerals they contain.

Schists

Folded schist: Sometimes the layers in schist can be intensely contorted even on a small scale.

Schist is one of the most striking of all rocks, with its distinctive layering or 'schistosity'. Schists mostly develop deep in the roots of mountain ranges as they are being folded. Here pressure and temperatures reach the point where the original clay minerals in mudrocks, slates and phyllites are completely broken down. The chemicals in them then reform as larger crystals of minerals such as mica and chlorite. Schists are therefore usually medium- to coarse-grained rocks, with crystals that, though not as well defined, can be as large as those in granite.

Schistosity develops only partly because the different minerals separate into layers. The main reason is that the continuous squeezing and shearing of the rock during metamorphism allows crystals only to grow in one plane, at right angles to the pressure. Because the crystals that develop are tabular flakes such as mica, or needle-like crystals such as amphiboles, this layering effect is even more marked. Whenever schist is split, it tends to break along layers of mica within the rock, giving the slightly misleading impression that it is entirely made of glistening mica. Quartz is also an important constituent, but the quartz layers tend only to be seen clearly when the rock is cut across the grain.

Although schist breaks into sheets like slate, and is sometimes used as roofing tiles where slate is scarce, it does not cleave quite as easily. The extreme heat and pressure mean the layers are more tightly knitted together. They are often also slightly wrinkled, or 'crenulated', similarly to phyllite but even more so. In some schists, these contortions can become quite dramatic.

Although they develop mainly from mudrocks, they can form from any rock that contains the right constituent minerals to make mica. Like gneiss, they are found primarily in areas of ancient rock. Most date back to at least Precambrian times (older than 570 million years). Schists do also occur in more recent formations, though, such as in the young fold mountain belts of the Alps and Himalayas.

Chlorite schist: In chlorite schist, weakly developed schistosity is created by flaky green chlorite crystals and fine green needles of actinolite which often form radiating clusters. This rock forms under relatively mild, 'greenschist' regional metamorphic conditions.

Garnet schist: Under fairly intense metamorphism, large porphyroblasts of garnet may develop in mica schist. Garnets 10mm/0.4in or more across can sometimes be seen, as in this garnet schist, and may be big enough to chip out and use as gems. The schistose layers tend to bend around the garnet like a stream flows around rocks. The garnets are typically iron-rich, pink almandine metamorphosed from pyroxenes and other minerals.

Chlorite schist
Rock type: Foliated, regional metamorphic.
Texture: Fine to medium-grained, sometimes with porphyroblasts of albite or chloritoid. Not as markedly schistose as mica schist. Unlike gneiss, schist's crystal fabric is dominated by long 'planar' crystals. This is a useful clue to identity.
Structure: Typically folded, on a small or large scale.
Colour: Greenish grey.
Composition: Chlorite, actinolite, epidote, talc, glaucophane and albite feldspar. Little or no mica and quartz.
Protolith: Mainly mudrocks such as shale.
Temperature: Moderate.
Pressure: Moderate.
Notable occurrences: Argyll, Scotland; Lake Tauern, Austria; Tyrol, Piedmont, Lombardy, Italy; Sierra Nevada, California.

Garnet schist
Rock type: Foliated, regional metamorphic.
Texture: Medium- to coarse-grained. Garnet porphyroblasts common. Well-developed schistosity.
Structure: Typically folded, on a small or large scale.
Colour: Black, brown, reddish.
Composition: Garnet, plus biotite and muscovite mica and quartz. Garnet schist also contains the same range of other minerals as mica schists.
Protolith: Mainly mudrocks such as shale.
Temperature: Moderate to high.
Pressure: Moderate to high.
Notable occurrences: Connemara, Ireland; Scotland; Scandinavia; Alps, Switzerland; Black Forest, Germany; Quebec; Duchess Co, NY; New Hampshire.

Mica schists

The most common schists are mica schists. Flakes of mica give them the most marked schistosity, and a real shine. The flakes are typically about 0.5mm/0.02in thick and can often be prised off with a knife. There is plenty of quartz in mica schist too, often concentrated in mica-poor layers, and a fair amount of albite feldspar. Sometimes red garnet or green chlorite crystals are visible. The mica in mica schists can be muscovite, sericite or biotite. Biotite is usually brown; muscovite and sericite are pale coloured and called white mica. If a white mica is fine-grained, it is called sericite; if it is coarser it is called muscovite. Most mica schists contain all three, but one usually predominates. Muscovite and sericite schists develop where metamorphism

is of moderate intensity, and are often associated with greenschists and phyllites. Biotite develops partly at the expense of muscovite (and chlorite) when metamorphism becomes more intense. Even more intense metamorphism edges the schist towards garnet schists.

Biotite schist: Biotite schist is brown and dark, but still has the mica gleam.

Muscovite schist: Muscovite schists, with sericite schists, are the lightest-coloured schists, coloured by white mica. Unlike sericite schist, the grains of mica in muscovite schist are clearly visible to the naked eye.

Mica (muscovite, sericite or biotite) schist
Rock type: Foliated, regional metamorphic.
Texture: Medium to coarse-grained, sometimes with porphyroblasts. Thin schistose layering of mica flakes always marked.
Structure: Typically folded, on a small or large scale.
Colour: Light grey, greenish (biotite schist is browner).
Composition: Quartz and mica (usually muscovite), plus kyanite, sillimanite, chlorite, graphite, garnet, staurolite.
Protolith: Mainly mudrocks such as shale.
Temperature: Moderate.
Pressure: Moderate.
Notable occurrences: Connemara, Ireland; Scotland; Scandinavia; Alps, Switzerland; Black Forest, Germany; Quebec; Duchess Co, NY; New Hampshire.

Metamorphic facies
It is impossible to tell from a single mineral at what pressure and temperature a metamorphic rock formed, but you can tell from the groupings of minerals known as 'facies' – assemblages that form in certain conditions. These facies include: zeolite, greenschist, amphibolite, blueschist, eclogite, granulite and various hornfelses. Although most facies are named after a rock containing one of these groups of minerals, different rocks can be in the same facies. Amphibolite and hornblende schist both form in amphibolite facies, for instance, and the above thin section shows a greenschist to amphibolite facies in meta-basalt, in which most of the green crystals are in fact hornfels amphibole. Although the link isn't always definite, hornfels, granulite and eclogite are high-grade; greenschist and amphibolite medium-grade; and zeolite and blueschist low-grade. Hornfels facies are linked to high-temperature contact zones; granulite to the high temperatures and moderate pressures typical of mountain roots; amphibolite to moderate temperatures and pressures in mountain roots and continental interiors; greenschists and zeolite to mild conditions under continental interiors; and blueschist to low temperatures and high pressures in accretionary wedges along subduction zones.

Amphibole schist

Amphibole schist is schistose like mica schist, but the layers are formed not by flakes of mica but by parallel bands of long, thin amphibole crystals. If the crystals are not parallel, the rock is not schistose and is called amphibolite. The amphibole in amphibole schist is usually hornblende, and the rock is called hornblende schist, but it can also be actinolite or tremolite. All amphibole schists are much richer in feldspar than amphibolite. Other minerals in hornblende schist include chlorite, epidote, pyroxene and garnet. Garnet often forms dark red porphyroblasts.

Dark amphibolite crystal

Amphibole (hornblende) schist
Rock type: Foliated, regional metamorphic.
Texture: Medium- to coarse-grained, sometimes with porphyroblasts. Thin schistose layering of amphibole needles always marked.
Structure: Typically folded, on a small or large scale.
Colour: Dark black or brown, often streaked or flecked with white or red.
Composition: Hornblende (or actinolite or tremolite), plagioclase feldspar, quartz and biotite mica, plus pyroxene, epidote, muscovite mica and garnet.
Protolith: Basalt and dolerite, as well as mudrocks.
Temperature: Moderate.
Pressure: Moderate.
Notable occurrences: Connemara, Ireland; Scotland; Tyrol, Austria/Italy; St Gotthard Massif, Switzerland; Quebec; Mitchell Co, North Carolina.

GNEISS AND GRANULITE

The word gneiss (pronounced 'nice') comes from an old Slavonic word for 'sparkling', and that's exactly what gneiss does. Under a microscope it can be seen to be made of tightly packed, iridescent crystals forged by the most intense metamorphism of all. Gneiss and granulite are incredibly tough rocks and are found together in terranes that are the most ancient rock formations on Earth.

Gneiss

Identification: Gneiss can often be identified by its humbug stripes of dark and light minerals and its crystal fabric (see Texture, right). It is also incredibly tough, found in the most ancient and time-worn landscapes, such as the vast Canadian Shield and Scotland's Hebridean islands.

Almandine garnet

Granular gneiss: Granular gneiss is less distinctly banded than other gneisses. Its composition is often close to granite, and is made from high proportions of quartz, white and pink feldspar and white and dark mica.

Unlike schist, gneiss is not dominated by long, planar crystals. But it can often be the most markedly striped of all rocks, composed of alternating bands of light and dark minerals. In most cases these bands are just 2mm/0.08in thick, but they can be as wide as 1m/39in. The bands in gneiss are not like the layers in schist, which are made from sheets of mica, nor does gneiss split easily in the same way. The high temperatures it takes to create gneiss tend to destroy mica, and the banding is formed in an entirely different way – as minerals separate out and form into distinct bands. The light-coloured bands are formed by light-coloured, typically felsic, minerals such as quartz, feldspar and white mica (usually muscovite). The dark bands are made of dark, typically mafic, minerals like amphibole, pyroxene and biotite mica. Compositional bands like these form only at very high temperatures when minerals almost melt and so can move freely before recrystallizing. Gneisses form deep in the Earth in subduction zones or under the roots of fold mountains, and are bought to the surface only by massive tectonic movements, or the slow erosion of the overlying mountains.

Compositional bands in gneiss can also form when variations in the original rock survive through all the stages of metamorphism. A rock originally made of alternate narrow beds of shale and sandstone may be transformed by metamorphism into gneiss made of alternate bands of quartzite and mica.

Some gneisses, however, derive their bands in an entirely different way. These gneisses get their bands when thin floods of granitoid magma ooze their way in between layers of the protolith, or when there is local melting. These gneisses blend into migmatites, and have led some geologists to conclude that many ancient gneisses formed from granodiorite and tonalite magmas in this way. Such gneisses are closely linked to ancient greenstone belts, and often merge into them.

Gneiss is an incredibly tough rock, perhaps the toughest of all, and vast quantities of it have survived since the very early part of the Earth's history. Large areas of Greenland are made from gneisses at least three billion years old. The world's oldest known rock is Acasta gneiss from northern Canada, which has been dated to 3,900 million years ago.

Rock type: Foliated, regional or dynamothermal metamorphic.

Texture: Medium- to coarse-grained. Marked by striking alternate light and dark bands. The light bands are often coarsely granular. The dark bands are finer-grained and, when they contain biotite mica, may be foliated like schist. Unlike schist and granulite, gneiss has a mix of long 'planar' crystals and 'equant' crystals, crystals that measure much the same in all directions. Schist has more planar crystals and granulite more equant crystals. This difference in crystal fabric is a useful clue to identity.

Structure: May be marked by large-scale as well as small-scale dark and light bands. Often folded. Very often criss-crossed with granite and pegmatite veins.

Colour: Greyish, pinkish, reddish, brownish, greenish with dark stripes.

Composition: Varies with protolith, but typically abundant in feldspar and quartz and white mica, which forms the light layers, and biotite and hornblende which forms the dark layers. Other constituents include cordierite, garnet and sillimanite.

Protolith: Almost any other rock. Gneiss formed from igneous rocks is called orthogneiss; gneiss made from sedimentary rock is called paragneiss.

Temperature: High.

Pressure: High.

Notable occurrences: Lewis, Orkneys, Scotland; Greenland; Scandinavia; Vosges, Massif Central, Brittany, France; Bavaria, Erzebirge, Germany; Alps, Switzerland; Southern India; Thailand; Canadian Shield; Appalachians; Idaho.

Mineral identifiers
About a century ago, geologist George Barrow was investigating the rocks of the Scottish Highlands around Aberdeen (pictured above). Here shale, sandstone, limestone and mafic lava were crushed and folded during a powerful phase of mountain building that created the ancient Caledonian mountains. As he studied these ancient metamorphosed rocks, Barrow noticed how particular minerals appeared in rocks in a sequence across the landscape, reflecting different intensities of the metamorphosis of pelites (metamorphosed mudrocks). These 'index' minerals were, in order: chlorite, biotite, garnet, kyanite and sillimanite. Zones in which these index minerals appear are now called Barrovian metamorphic zones, and the boundary of each is marked by a line on a map called an isograd. Similar zoning was found running through andalusite, cordierite, staurolite and sillimanite. Zones in which these minerals are found are called Buchan zones. The Buchan sequence is thought to be created by lower pressures than Barrovian zones.

Granulite

Granulite is a tough, coarse-grained rock that, like gneiss, forms at very high temperatures and pressures. It is also the facies of metamorphic minerals that form under these extreme conditions, which tend to destroy mica and replace it with minerals such as pyroxene. The high pressure drives out any water, so the minerals that form are said to be anhydrous. It is thought that granulite formed at the base of the continental crust. Indeed, most of the underside of the continental crust is probably made of granulite. Granulite has mostly reached the surface either as small xenolith chunks in magmas, or when mountain ranges are worn away so far that their very roots are exposed. Most granulites are very ancient, and are found with gneisses in the granulite-gneiss terranes that contain the oldest rocks on Earth, dating back to billions of years ago.

Rock type: Mostly non-foliated, regional or dynamothermal metamorphic.
Texture: Coarse-grained. Often banded like gneiss but granulite has mostly 'equant' crystals (see Gneiss; Texture).
Structure: Marked by dark and light bands.
Colour: Light, almost white.
Composition: Pyroxene (diopside or hypersthene), quartz and feldspar, plus garnet, biotite, cordierite and sillimanite.
Protolith: All kinds of rock.
Temperature: High (>700°C).
Pressure: High.
Notable occurrences: NW Scotland; Greenland; Finland; Aldan shield (Yakut), Siberia; Ukraine; Limpopo, South Africa; Hopeh, Liaoning, China; Yilgarn, Australia; Enderby Land, Antarctica; Canadian Shield; British Columbia; Adirondacks, NY; Beartooth Mountains, Montana.

Identification: Granulite is a hard, sparkling rock made of coarse, rounded interlocking grains of mostly pale minerals.

Augen gneiss and other gneisses

Gneisses can be distinguished by their protolith, such as granite gneiss and syenite gneiss, or by their characteristic mineral, such as biotite gneiss and garnet gneiss. Some gneisses may be distinguished by their texture, such as platy gneiss and augen gneiss. The word augen comes from the German for 'eye', and refers to large, oval or eye-shaped crystals in the rock. The crystals are typically alkali feldspar, in a matrix of quartz, feldspar and mica, but can be quartz, or garnet (in which case the rock is known as garnet augen gneiss). Each eye, or auge, can be up to 10cm/3.9in across. Feldspar augen typically contain inclusions of minerals such as biotite. The augen in a rock tend to be much the same size and shape, and it is thought that they are survivors from an earlier stage, with a core too big to be affected by the recrystallization that aligned other minerals in bands. The bands flow around them, like a stream around a rock. Sometimes, garnet augen rotate as they grow during metamorphism, creating 'snowball' garnets, containing spiral inclusions of other minerals.

Identification: Both schists and gneisses can contain augen, which are large, oval crystals of feldspar, quartz or garnet. Augen gneiss is more common, and the augen are usually surrounded by dark, banded gneiss layers, rather than silvery, flaky schist.

Rock type: Foliated, regional or dynamothermal metamorphic.
Texture: Medium- to coarse-grained. Marked by large pale crystals or augen.
Structure: May be marked by large-scale as well as small-scale dark and light bands.
Colour: Greyish, pinkish, reddish, brownish, greenish with dark stripes.
Composition: Augen made of alkali feldspar or garnet. Matrix of feldspar and quartz forming light layers, and biotite forming the dark layers.
Temperature: High.
Pressure: High.
Notable occurrences: Lewis, Orkneys, Scotland; Greenland; Scandinavia; Vosges, Massif Central, Brittany, France; Bavaria, Erzebirge, Germany; Alps, Switzerland; Canadian Shield; Appalachians; Idaho.

ROCKS ALTERED BY FLUIDS AND OTHER MEANS

Not all metamorphic rocks are formed directly by the heat of contact with an intrusion, or by heat and pressure deep within the crust. Skarns and serpentinites are formed by the interaction of fluids with the country rock – skarns by fluid heated by an intrusion, serpentinites by cold water. Halleflintas are altered volcanic tuffs and fulgurites are rocks formed by the intense heat of lightning strikes.

Skarn

Skarns are treasure troves of unusual and valuable minerals such as grossular garnet and ores of iron, copper, lead, tungsten and zinc. They are typically patches of metamorphosed rock around a granite intrusion, but the term 'skarn' covers a wide range of different rocks and mineral deposits that originated in a variety of different ways. Some geologists use the word skarn to describe any calcium- and silicate-rich metamorphic rock containing unusual minerals. Most prefer to describe skarns as only metamorphic rocks that form when limestones and dolomites are altered by contact with hot granite intrusions. Granite intrusions generate hot fluids carrying copious amounts of silicon, iron, aluminium and magnesium either emanating directly from the intrusion or cooked up as the intrusion heats groundwater in the limestone. Infiltrating the limestone, this rich brew alters minerals to calcium, iron and magnesium silicates. This process is really metasomatism, not metamorphism, because the minerals in the rock are replaced by others as they come into contact with the hot fluids.

Apatite

Orange calcite

Identification: Skarns are very piebald in appearance. They are characterized by large patches of minerals of different colours, formed as they were concentrated by the hot fluids that oozed through the limestone during metasomatism.

Rock type: Non-foliated, hydrothermal metasomatic.
Texture: Fine-, medium- or coarse-grained.
Structure: Occurs in small patches, with minerals concentrated in nodules, lenses and radiating masses.
   Colour: Brown, black or grey but very variable.
   Composition: Pyroxene, garnet, idocrase, wollastonite, actinolite, magnetite, epidote. Skarns host copper, lead, zinc, iron, gold, tungsten, molybdenum and tin ores.
Protolith: Mostly limestones and dolomites.
Notable occurrences: Dartmoor (Devon), England; Central Sweden; Elba, Italy; Trepca, Serbia; Banat, Romania; Arkansas; Crestmore, California.

Halleflinta

Identification: No metamorphic rock looks more like flint than halleflinta. It is very fine-grained, almost cryptocrystalline, and splinters like flint when hit with a hammer.

Halleflinta gets its name from the Swedish for 'rock-flint'. It is a very hard, flinty, metamorphic rock so fine-grained that it is hard to identify individual minerals even under a microscope. It is basically a very intimate mix of quartz, feldspar and other silicate minerals. It forms in similar conditions to gneiss and schist, under intense heat and pressure, and often occurs in association with them in Scandinavia, where it was first identified. But it contains none of the banding of gneiss, nor the schistosity of mica. Indeed, it is almost glassy in texture, and breaks into sharp splinters like flint. The reason for this difference is almost certainly due to its protolith. It is probably metamorphosed volcanic tuff, and halleflinta often retains signs of the original layers of volcanic debris as it settled after successive eruptions. Extra silica has usually got into the rock during metamorphism.

Rock type: Foliated, regional or dynamothermal metamorphic.
Texture: Very fine- and even-grained, almost glassy, so that the rock splinters. May contain larger porphyroblasts of quartz.
Structure: Layering related to original volcanic deposit, but no schistosity or banding.
Colour: Grey, buff, pink, green or brown.
Composition: Quartz, feldspar, mica, iron oxides, apatite, zircon, epidote, hornblende.
Protolith: Volcanic tuff.
Temperature: High.
Pressure: High.
Notable occurrences: Sweden; Finland; Tyrol, Austria; Bohemia, Czech Republic; Galicia, Poland; Ukraine.

Serpentinite

Serpentinization is a process that alters rocks, but it is not like other forms of metamorphism. It gets its name because it creates a rock flecked like snakeskin called serpentinite. This consists mostly of fibrous serpentine minerals such as chrysotile, antigorite and lizardite. In serpentinization, it is not heat and pressure that alter the minerals but heat and water. It affects mostly ultramafic rocks such as peridotite and dunite, although serpentinite can form from gabbro and dolomitic limestone, as well. What happens is that water infiltrates the rock and alters iron-rich minerals such as olivines and pyroxene to create serpentine minerals. The water is cool but the chemical reaction is exothermic and generates its own heat. It was once thought serpentinite was quite rare, and occurred only above subduction zones, or within small ultramafic intrusions. Now it is realized that serpentinites pretty much underlay the entire ocean floor, forming part of the ophiolite sequence, as olivine-rich magmas oozing up through the mid-ocean rift are serpentinized by sea water. Hydration (the uptake of water) makes serpentinites light so they well up in many places, creating undersea mountains. They also well up in the accretionary wedges above subduction zones.

Identification: Serpentinites are dark green to black and look very much like the snakeskin that earned them their name. They are often quite coarse-grained, and green serpentine crystals are usually easy to see.

Rock type: Foliated, hydrothermal metamorphic.
Texture: Medium- to coarse-grained. Compact, dull, waxy. Fractures in splinters.
Structure: Often banded. Usually criss-crossed with veins of chrysotile serpentine.
Colour: Grey-green to black.
Composition: Serpentine (chrysotile, lizardite, antigorite), olivine, pyroxene, hornblende, mica, garnet, iron oxides.
Protolith: Peridotite, dunite, pyroxenite, and occasionally gabbro and dolomite.
Notable occurrences: Lizard (Cornwall), England; Shetland Islands, Scotland; Pyrenees, Vosges, France; Liguria, Italy; Montana; Oregon; California; Maine; The Lost City, Mid-Atlantic seabed; Izu-Bonin-Marian seamounts, Pacific.

The Lost City
Geologists have long known about 'black smokers', or hydrothermal vents (above). These are remarkable chimneys on the sea floor in the mid-ocean ridge. They are built up from deposits left by smoky clouds of sulphide-rich water that bubble up through the sea floor, superheated by magma. In 2001, oceanographers discovered an entirely different kind of smoker towering up from the Atlantic ocean floor. Forming what was dubbed the Lost City, these white smokers are made of carbonate, and develop in an entirely different way to black smokers. For a start, they form well away from the central ocean rift. More significantly, they are heated not by magma but by the heat generated from serpentinization reactions in peridotite rocks under the ocean floor surface. As sea-water infiltrates the peridotites, it not only turns them to serpentinite but also generates warm, alkali-rich waters that bubble up in white smokers to form brucite and calcite towers.

Fulgurite

Fulgurites are the most unusual and rarest of all metamorphic rocks. They get their name from the Latin for 'thunderbolt', and they are natural tubes or crusts of glass that form when lightning strikes. To fuse sand instantly into glass needs temperatures of 1,800°C/3,272°F, and lightning regularly reaches a searing 2,500°C/4,532°F. There are two kinds of fulgurite: sand and rock. Sand fulgurites form in the loose sand on beaches and in deserts. They are branching tubes that look like roots. They average 2.5cm/1in in diameter and can be up to 1m/39in long. Rock fulgurites are crusts or coats of glass that form when lightning strikes solid rock. Typically, they form branching marks across the rock surface, or line pre-existing fractures in the rock. Rock fulgurites are typically found on mountain tops, most famously on Oregon's Mount Thielsen, known as the Cascade's Lightning Rod due to evidence of strikes found there.

Rock type: Non-foliated, contact metamorphic.
Texture: Glassy.
Structure: Sand fulgurites: branching tubes of glassy sand in loose sand. Rock fulgurites: glassy crusts on solid rock in veins or fractures.
Colour: Grey-green to black.
Composition: The silica mineral lechtalierite.
Protolith: Sand fulgurites form in loose sand; rock fulgurites can form on any rock.
Notable occurrences: Sand fulgurites: Sahara Desert, Africa; Namib Desert; Botswana; Lake Michigan, Atlantic coast of North America; Utah deserts. Rock fulgurites: Isle of Arran, Scotland; Mt Blanc (Alps), Pyrenees, France; Mount Ararat, Turkey; Toluca, Mexico; Sierra Nevada, CA; Wasatch Range, UT; Mount Thielsen (Cascade Range), OR; South Amboy, NJ.

Identification: Sand fulgurites are branching knobbly tubes of glassy sand.

MINERALS

Often emanating from the parent rock in brilliant colours and weird and wonderful textures, minerals are the pride of any geological collection. From the green, gold and blue iridescent sheen of labradorescence to the large, wheat-sheaf shaped crystals formed by stilbite, many minerals are a visual treat. Unlike rocks, it is not possible to classify minerals by obvious visual characteristics; they must be classified according to their chemical composition. The main classifications followed in this part of the directory are: Native Elements, Sulphides and Sulphosalts, Oxides, Salts, Carbonates, Nitrates and Borates, Sulphates, Chromates and Molybdates, Phosphates, Arsenates and Vanadates, Silicates, Mineraloids, and Ore-forming Minerals. In addition to the individual specimen notes, identification aids and data panels for each mineral, special feature boxes build a bigger picture of the key factors behind mineral formation and of their relationship to rocks.

Above from left: Pyrite, rhodochrosite, halite.

Right: Salt flats of Death Valley, California, USA.

HOW TO USE THE DIRECTORY OF MINERALS

*The Minerals' Directory gives detailed portraits of over 250 species of mineral from around the world.
Using the clues on the following page, you can make a rough identification of a mineral, then follow up
those clues in the Directory to make a closer identification. The diagrams here show how entries work.*

Minerals are natural, solid chemicals. Most, but not all, are crystalline, and their crystals fit into one of the six or seven major systems of crystal symmetry, although each has its own particular habit or way of growing.

Most minerals are found in rocks, and form part of their chemical make-up, and so are called rock-forming minerals. Minerals valued primarily for their ores are also covered, in the final section of the Directory. 'Rock-formers' are ordered according to the major chemical groups associated with minerals; ore-formers according to their major metal constituent (see Classifying Minerals; Mineral Ores; Understanding Rocks and Minerals).

Above: Spectacular minerals like these needles of millerite growing in a cavity are very rare indeed.

In fact, many of the rock-formers profiled in the Directory occur only in very tiny amounts in rocks. There are just a few dozen minerals that are actually major constituents of rocks.

Major minerals in igneous rocks include quartz, feldspars, feldspathoids, micas, augite, hornblende and olivine. Major minerals in sedimentary rocks include salts and clays, plus various carbonates, sulphates and phosphates that occur only in sedimentary rocks as well as quartz and feldspars. Major minerals in metamorphic include many of the same as those in igneous rocks such as quartz, feldspars, mica and hornblende, plus many that are primarily characteristic of metamorphic rocks including: actinolite, andalusite, axinite, chlorite, epidote, the garnet group, graphite, kyanite, prehnite, sillimanite, staurolite, talc and tremolite.

Mineral name: may be plural if several varieties are profiled

Chemistry: The chemical formula and name is given for each mineral.

Inset detail: Illustrations of related species and special features.

Identification: Notes the most useful clues for identifying specimens found in the field.

Data panel: Quick reference tool summarizing standard mineral characteristics.

Crystal system: The way in which crystals of the mineral are symmetrical. The line diagram shows the form in which crystals of the mineral commonly grow.

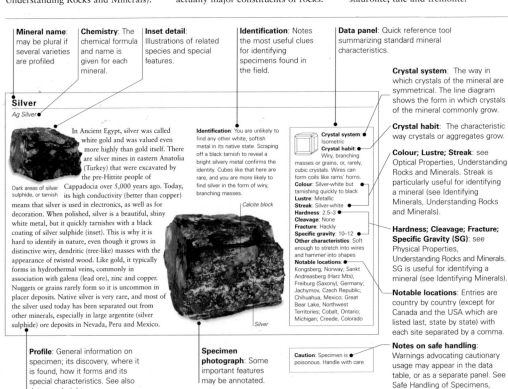

Silver
Ag Silver

In Ancient Egypt, silver was called white gold and was valued even more highly than gold itself. There are silver mines in eastern Anatolia (Turkey) that were excavated by the pre-Hittite people of Cappadocia over 5,000 years ago. Today, its high conductivity (better than copper) means that silver is used in electronics, as well as for decoration. When polished, silver is a beautiful, shiny white metal, but it quickly tarnishes with a black coating of silver sulphide (inset). This is why it is hard to identify in nature, even though it grows in distinctive wiry, dendritic (tree-like) masses with the appearance of twisted wood. Like gold, it typically forms in hydrothermal veins, commonly in association with galena (lead ore), zinc and copper. Nuggets or grains rarely form so it is uncommon in placer deposits. Native silver is very rare, and most of the silver used today has been separated out from other minerals, especially in large argentite (silver sulphide) ore deposits in Nevada, Peru and Mexico.

Dark areas of silver sulphide, or tarnish

Identification: You are unlikely to find any other white, softish metal in its native state. Scraping off a black tarnish to reveal a bright silvery metal confirms the identity. Cubes like that here are rare, and you are more likely to find silver in the form of wiry, branching masses.

Calcite block

Silver

Crystal system: Isometric
Crystal habit: Wiry, branching masses or grains or, rarely, cubic crystals. Wires can form coils like rams' horns.
Colour: Silver-white but tarnishing quickly to black
Lustre: Metallic
Streak: Silver-white
Hardness: 2.5–3
Cleavage: None
Fracture: Hackly
Specific gravity: 10–12
Other characteristics: Soft enough to stretch into wires and hammer into shapes
Notable locations: Kongsberg, Norway; Sankt Andreasberg (Harz Mts), Freiburg (Saxony), Germany; Jachymov, Czech Republic; Chihuahua, Mexico; Great Bear Lake, Northwest Territories; Cobalt, Ontario; Michigan; Creede, Colorado

Crystal habit: The characteristic way crystals or aggregates grow.

Colour; Lustre; Streak: see Optical Properties, Understanding Rocks and Minerals. Streak is particularly useful for identifying a mineral (see Identifying Minerals, Understanding Rocks and Minerals).

Hardness; Cleavage; Fracture; Specific Gravity (SG): see Physical Properties, Understanding Rocks and Minerals. SG is useful for identifying a mineral (see Identifying Minerals).

Notable locations: Entries are country by country (except for Canada and the USA which are listed last, state by state) with each site separated by a comma.

Profile: General information on specimen; its discovery, where it is found, how it forms and its special characteristics. See also data panel, right.

Specimen photograph: Some important features may be annotated.

Caution: Specimen is poisonous. Handle with care

Notes on safe handling: Warnings advocating cautionary usage may appear in the data table, or as a separate panel. See Safe Handling of Specimens, opposite

MINERAL LOCALITIES

Good mineral specimens are not found scattered evenly around the world but concentrated in a few locations where rare combinations of conditions create particular mineral formations.

Every mineral has a type locality. This is the place where the first formally recognized specimens of a mineral were discovered. Sometimes, a mineral takes its name from the type locality such as Franklinite, after the famous Franklin mines in New Jersey.

Very few type localities are host only to one kind of mineral. Some of the most famous sites are type localities for dozens of other minerals – as well as many other minerals found first elsewhere.

Besides type localities, there are sites famous either for the sheer number of different minerals found there, for the rarity of the species, or simply for the very special nature of the specimens found there. A few sites are famous for the large and spectacular crystals they yield, for instance, or their rare twins, or beautiful gems. Some, of course, are simply famous for the sheer quantity of accessible specimens they provide the collector. A list of the more famous locations appears to the right, some still producing, some now exhausted.

Right: The Isle of Skye, off the north-west coast of Scotland, offers many excellent opportunities for mineral collecting.

FAMOUS MINERAL LOCALITIES		
Europe	**Australia and Asia**	**USA and Canada**
Almaden, Ciudad Real, Spain	Bombay mines, India	Arizona copper mines
Binntal, Switzerland	Broken Hill, NSW, Australia	Bancroft Ontario
Black Forest, Germany	Coober Pedy, South Australia	Black Hills, South Dakota
Cornwall, England	Dundas, Tasmania	Boron mines of California
Cumbria, England	Kashmir	Cobalt, Ontario
Durham, England	Mogok, Myanmar (Burma)	Francon, Montreal, Quebec
Harz Mountains, Germany	NSW, Australia	Franklin & Sterling Hill, NJ
Ilmaussaq, Greenland	Ratnapura, Sri Lanka	Keeweenaw, Michigan
Kola Peninsula, Russia		Magnet Cove, Arkansas
Langban, Varmland, Sweden	**Africa and South America**	Mississippi Valley region
Langesundfjord, Norway	Katanga, Congo	New Hants phosphate mines
Leadhills, Lanarks, Scotland	Kimberley, South Africa	Rapid Crk & Big Fish, Yukon
Lavrion, Greece	Kivu, Congo	Mont Saint-Hilaire, Quebec
Pribram, Czech Republic	Transvaal, South Africa	San Benito Co, California
Skye, Scotland	Tsumeb, Namibia	San Diego Co, California
Strontian, Argyll, Scotland	Copiapo and Atacama, Chile	Sudbury, Ontario
Ural Mountains, Russia	Llallagua, Potosi, Bolivia	Tristate Mining District,
Mount Vesuvius, Italy	Minas Gerais, Brazil	Kansas-Missouri-Nebraska

SAFE HANDLING OF SPECIMENS

Handling mineral specimens safely is basically common sense. Only very few minerals are significantly poisonous or radioactive, and even these can be handled safely. You just need to make sure you handle them only in a well-ventilated room to avoid breathing in fumes, and wash your hands carefully afterwards to make sure contamination is not carried to your mouth. Minerals that are especially toxic or radioactive are identified in the directory. Of course, all mineral specimens should be kept out of children's reach, but especially these minerals.

Handling and storing radioactive minerals

Certain minerals are mildly radioactive, including carnotite, autunite and torbernite. They are mostly radioactive because of their uranium content. However, their radioactivity is very mild. You would have to hold a large chunk of high-grade uranium ore for a few hours to get the same dose of radiation as from a chest X-ray. You would have to hold it continuously for over four days to get the same dose as you get naturally from your surroundings over a year. All the same it is worth taking a few basic precautions when handling minerals that are classed as naturally radioactive:

1 Don't carry radioactive minerals in pockets, or bring them close to the reproductive organs.
2 Keep mineral specimens, and hands after handling them, away from the eyes.
3 Wash hands properly after handling specimens.
4 Never crush or grind specimens, since the dust can get in the air.
5 Store specimens out of children's and pets' reach.
6 Store specimens in screwtop glass containers in a cabinet in a well-ventilated room not used for everyday living.
7 Label specimens clearly so that everyone knows what they're touching.

IDENTIFYING MINERALS

Of over 4,000 different minerals in existence, a few of them, such as crocoite and malachite, are instantly recognizable. Many of the rest, though, can look very similar to one another. So how do you tell them apart and make a positive identification?

You might think that you could identify minerals in the same way as flowers by colour and shape alone. Unfortunately, you cannot in all but a few cases. Nearly every mineral is far too variable in colour and shape – and the differences between them either subtle or completely misleading. Fortunately, there are other more reliable clues to identity. There are also a number of simple tests you can conduct to confirm a mineral's identity.

Although a mineral's visual colour is variable, its streak is not, so you can narrow down the range of possibilities by first doing a streak test – scraping the mineral across the back of a white ceramic tile – then Mohs hardness test (see Mineral Properties: Physical in Understanding Rocks and Minerals). Find the streak colour in the table below, then find the mineral's hardness within the group. This should give you at most a dozen or so choices. If you can test its specific gravity (SG), you should be able to narrow the choice even farther.

Identifying in three steps

Mohs	Mineral	SG
NON–METALLIC LUSTRE		
White streak		
1	Talc	2.8
1–2	Scarbroite	2
1–2	Aluminite	1.7
1–2	Carnallite	1.6
1–2.5	Chlorite	2.5–2.9
1.5–2	Kaolinite	2.6
1.5–2	Nitratine	2.2–2.3
1.5–2.5	Chlorargyrite	5.5–5.6
2	Ulexite	1.5–2
2	Sylvite	2
2	Halite	2.1
2	Sulphur	2
2	Gypsum	2.3
2	Vivianite	2.6
2	Melanterite	1.9
2–2.5	Epsomite	1.7
2–2.5	Muscovite	2.8
2–2.5	Zinnwaldite	3.0
2–2.5	Hydrozincite	3.7
2–3	Gaylussite	1.9
2.5	Lepidolite	2.8
2.5	Jouravskite	1.95
2.5–3	Mendipite	7–7.5
2.5–3	Ferrimolybdite	4.–4.5
2.5–3	Cryolite	2.95
2.5–3	Phlogopite	2.95
2.5–3	Glauberite	2.7–3
2.5–3	Biotite	3
2.5–3	Thenardite	2.7
2.5–3	Gibbsite	2.4
3	Kainite	2.1
3	Calcite	2.7
3	Anglesite	6.3
3	Wulfenite	6.8
3	Vanadinite	7
3–3.5	Polyhalite	2.8
3–3.5	Cerussite	6.5
3–3.5	Celestine	4
3–3.5	Barite	4.5
3.5	Witherite	4.3
3.5	Adamite	4.3
3.5	Kieserite	2.6
3–4	Anhydrite	2.9
3–4	Serpentine	2.6
3.5–4	Pyromorphite	6.5–7
3.5–4	Mimetite	7.1
3.5–4	Strontianite	3.8
3.5–4	Scorodite	3.1–3.3
3.5–4	Magnesite	3
3.5–4	Wavellite	2.3
3.5–4	Dolomite	2.9
3.5–4	Ankerite	2.9–3.8
3.5–4	Aragonite	2.9
3.5–4	Alunite	2.5–2.8
3.5–4	Stilbite	2.2
3.5–4	Heulandite	2.2
3.5–4.5	Siderite	3.8–3.9
4	Fluorite	3.2
4	Rhodochrosite	3.5
4	Kyanite	3.6
4–5	Smithsonite	4.4
4–4.5	Jarlite	3.87
4–4.5	Magnesite	3
4.4.5	Phillipsite	2.2
4–5	Triphyllite	3.6
4.5	Colemanite	2.4
4.5	Chabazite	2.1
4.5–5	Scheelite	6
4.5–5	Stibiconite	3.5–3.9
4.5–5	Wollastonite	2.8
4.5–5	Apophyllite	2.3
5	Apatite	3.2
5	Hemimorphite	3.4
5–5.5	Monazite	4.9–5.3
5–5.5	Sphene	3.5
5–5.5	Analcime	2.2
5–5.5	Natrolite, Scolecite	2.3
5.5	Perovskite	4
5–6	Lazurite	3
5–6	Scapolite	2.6
5–6	Turquoise	2.6–2.8
5–6	Bronzite	3.3
5.5–6	Sodalite, Hauyne, Nosean	2.4
5.5–6	Anatase	3.8
5.5–6	Nephelite	2.6
5.5–6	Leucite	2.5
5.5–6	Rhodonite	3.5
5.5–6	Actinolite	3.1
5.5–6	Nephrite	3.1
5.5–6	Amblygonite	3–3.1
5.5–6.5	Opal	1.9–2.5
5.5–6.5	Diopside	3.3
6	Fassaite	3.3
6	Zoisite	3.3
6	Adularia	2.5
6	Orthoclase, Microcline, Sanidine	2.5
6	Albite	2.6
6–6.5	Anorthite	2.76
6–6.5	Plagioclase	2.6–2.8
6–6.5	Prehnite	2.9
6.5	Jadeite	3.2
6.5	Vesuvianite	3.4
6–7	Cassiterite	7
6–7	Kyanite	3.6
6–7	Epidote	3.3–3.5
6.5–7	Sillimanite	3.2
6.5–7	Olivine	3–2–4.3
6.5–7	Axinite	3.3
6.5–7	Spodumene	3.2
6.5–7	Diaspore	3.4
6.5–7.5	Garnet family	4
7	Quartz	2.65
7–7.5	Staurolite	3.7
7–7.5	Boracite	3
7–7.5	Cordierite	2.6
7–7.5	Tourmaline	3.1
7.5	Andalusite	3.1
7.5	Zircon	4.5
7.5–8	Beryl	2.7
8	Topaz	3.5
8	Spinel	3.7
8.5	Chrysoberyl	3.7
9	Corundum	4
10	Diamond	3.52
Yellow to brown streak		
1.5–2	Orpiment	3.4
2	Sulphur	
2	Autunite	3.1
2	Carnotite	4.5
2.5	Uranophane	3.8
2.5–3	Vanadinite	7
2.5–3	Crocoite	6
2.5–3.5	Jarosite	2.9–3.3
3	Wulfenite	6.8
3–4	Vanadinite	6.7–7.1
3.5–4	Pentlandite	4.6–5
3.5–4	Sphalerite	4
3.5–4	Copiapite	2.1
4	Zincite	5.4–5.7
4–4.5	Siderite	3.8
4.5	Xenotime	4.4–5
4–6	Pitchblende Uraninite	9–10.5
5	Goethite Limonite	4.3
5–5.5	Wolframite	7.3
5–5.5	Plattnerite	9.4+
5.5	Chromite	4–4.8
5.5–6	Hornblende	3.2
5.5–6	Brookite	4
5.5–6	Hypersthene	3.5
6	Columbite	5–8
6–6.5	Aegirine	3.5
6–6.5	Rutile	4.2

Yellow to brown streak

Minerals with distinctive colours

Some minerals exhibit particular colours that make them easier to recognize. When noting the outward characteristics of a specimen, be as descriptive as possible – for example, jotting down what the colour reminds you of. Getting into the habit of making acute observations may help you to recognize the mineral on sight the next time you encounter it.

Scarlet orange: crocoite

Yellows and golds
• gold: gold, pyrite, chalcopyrite, pyrrhotite, marcasite
• yellow: sulphur, carnotite
• lemon: adamite
• custard: jarosite
• yellow-green: autunite
• golden syrup: orpiment

Reds and oranges
• vermillion: cinnabar
• rose red: rhodonite
• scarlet orange: crocoite, vanadinite
• red wine: cuprite
• rosé wine: grossular garnet
• ruby red: ruby
• strawberry jam: proustite; pyrargyrite
• blood red: jasper
• jelly-red: rhodochrosite
• marmalade: citrine

Purples and blues
• purple: amethyst
• blue: diaboleite, chalcanthite, azurite, cyanotrichite, lazulite, chrysocolla, linarite, labradorite, sodalite
• sky blue: sapphire

Greens
• opaque green: malachite, varsicite, garnierite
• apple-green: atacamite

Opaque green: malachite

• jade green: jadeite, nephrite
• pale green or cyan: hemimorphite
• clear green: olivine, emerald, dioptase

6–7	Cassiterite	7

Orange streak

| 1.5–2 | Realgar | 3.5–3.6 |
| 2.5–3 | Crocoite | 6 |

Green streak

1–2.5	Chlorite	2.5–2.9
1.5–2	Vivianite	2.6
1.5–2.5	Annabergite	3
2–2.5	Garnierite	4.6
2–2.5	Torbernite	3.5
2–2.5	Autunite	3.1–3.2
2.5–3	Köttigite	3.3
3	Olivenite	3.9–4.4
3–3.5	Mottramite	5.7–6
3–3.5	Millerite	5.3–5.5
3–3.5	Atacamite	3.75
3.5	Antlerite	3.9
3.5–4	Malachite	4
4	Libethenite	3.6–3.9
4.5	Cuprotungstite	7
4.5	Conichalcite	4.3
5.5–6	Hornblende	3.2
5.5–6	Hedenbergite	3.5
5.5–6	Augite	3.4
6–6.5	Aegirine	3.5

Blue streak

1–3	Cyanotrichite	3.8
2	Vivianite	2.6
2–2.5	Proustite	5.6
2.5	Chalcanthite	2.2–2.3
2.5	Linarite	5.3
3–3.5	Boleite	5
3.5–4	Azurite	3.8

| 5–6 | Glaucophane | 3–3.2 |
| 5.5–6 | Lazulite | 3.1 |

Red or crimson streak

1.5	Realgar	3.5
2–2.5	Proustite	5.6
2–2.5	Cinnabar	8.1
2.5	Pyrargyrite	5.8
3–3.5	Polyhalite	2.8
3–3.5	Greenockite	4.5–5
3–3.5	Descloizite	5.9
3.5–4	Cuprite	6
4	Rhodochrosite	3.5
6	Columbite	5–8
6.5	Hematite	5.1
6.5	Piemontite	3.4

Grey to black streak

1.5	Covellite	4.7
2	Argentite	7.3
2.5–3	Chalcocite	5.6
3–3.5	Cerussite	6.5
3–4	Tetrahedrite	4.4–5.4
5–5.5	Wolframite	7.3
5–6	Psilomelane	4.5
5–6	Ilmenite	4.7
5.5–6	Ilvaite	4.1
5.5–6	Magnetite	5
5.5–6	Hornblende	3.2
5.5–6	Diallage	3.3
5.5–6	Hedenbergite	3.5
5.5–6	Augite	3.4
5.5–6	Hypersthene	3.5
5.5–6	Anthophyllite	2.8–3.4
6	Columbite	5–8
6.5	Thortveitite	3.5
6–7	Epidote	3.4

METALLIC LUSTRE

White streak

2–2.5	Muscovite	2.8
2.5–3	Biotite	3
5–6	Bronzite	3.3
5.5–6	Anatase	3.8
6–7	Cassiterite	7

Yellow to brown streak

2.5–3	Gold	15–19.5
3	Baumhauerite	5.3
3.5–4	Sphalerite	4
5–5.5	Niccolite	7.7
5–5.5	Wolframite	7.3
5.5	Chromite	4–4.8
5.5–6	Chloanthite	6.5
5.5–6	Hypersthene	3.5
5.5–6	Brookite	4
6–6.5	Rutile	4.2

Greenish black streak

| 3.5–4 | Chalcopyrite | 4.2 |

Red streak

2–2.5	Cinnabar	8.1
2–2.5	Pyrargyrite	5.8
2.5–3	Native copper	8.9
3–4	Tetrahedrite	4.4–5.4
3.5–4	Cuprite	6
5–6	Hematite	5.3
5–6	Ilmenite	4.5–5

Grey to black streak

1	Graphite	2.2
1.5	Molybdenite	4.8
1.5	Covellite	4.7
2	Stibnite	4.6
2	Argentite	7.3
2–3	Jamesonite	5.5–6
2–3	Pyrolusite	4.5
2.5	Cylindrite	5.5
2.5–3	Chalcocite	5.6
2.5–3	Digenite	5.6
2.5–3	Galena	7.4
2.5–3	Altaite	8.2
3	Bornite	5.1
3	Bournonite	5.8
3.5	Enargite	4.4
3–4	Tetrahedrite	4.4–5.4
3.5–4	Chalcopyrite	4.2
4	Stannite	4.3–4.5
4	Manganite	4.3
4	Pyrrhotite	4.6
4.5–5.5	Carrollite	4.6
4.5–5.5	Safflorite	7–7.3
5	Löllingite	7.3
5–5.5	Wolframite	7.3
5–5.5	Niccolite	7.7
5.5	Cobaltite	6.2
5–6	Ilmenite	4.7
5.5–6	Hypersthene	3.5
5.5–6	Arsenopyrite	6
5.5–6	Magnetite	5
5.5–6	Ilvaite	4.1
5.5–6	Skutterudite	6.6–7.2
5.5–6	Chloanthite, Rammels–bergite	6.6–7.2
5.5–6.5	Franklinite	5.1
6–6.5	Pyrite	5.1
6–7	Iridosmine	19–21
6–7	Sperrylite	10.6
6.5	Hematite	5.1

IDENTIFYING MINERALS BY SURROUNDINGS

Where you find a mineral specimen – its location and associations – can often provide many important clues to its identity. Not only are particular minerals characteristic of particular rocks and geological environments, but some minerals are nearly always found together.

Just as a certain range of animals live in a particular habitat, likewise certain minerals often occur together. Gold, for instance, is often found with milky quartz. Minerals that occur together are said to be associated. Some associations are the minerals that form a particular kind of rock, such as quartz, feldspars and micas in granite. Other associations might form in particular types of mineral-forming environment, such as a vein, or a cavity, or an encrustation.

Sometimes the association is built up in stages, with first one mineral added then another, through time. On a small scale this happens in balls of minerals found in sedimentary rocks called septarian nodules. These begin as mudballs surrounding decomposing

sealife. As they dry out they fill with minerals such as dolomite. These minerals begin to crack, and as they do, the cracks fill with veins of calcite.

On a larger scale, ore deposits in rocks are often altered by watery solutions seeping down through cracks. Chemicals weathered from pyrite ore, for instance, are washed down through the rock and react with minerals lower down. Ferric sulphate reacts with copper, lead and zinc minerals to create sulphates which can, in turn, be dissolved and carried on down by the water. Sulphuric acid dissolves carbonates and as these are washed down, they react with copper minerals to create a layer of minerals such as malachite and azurite. Even farther down in the rock – below the

water table (the level to which the rock is saturated) – the reaction of local minerals to the watery solutions can create a layer of copper minerals such as chalcocite, covellite, bornite and chalcopyrite. Because these minerals are very rich in copper, the process is called secondary enrichment.

Mineral assemblages

In other situations, mineral associations form at more or less the same time. This is called an assemblage and happens, for example, as igneous rocks form from melts, or when metamorphic or sedimentary rocks form, under particular conditions. In igneous rocks, minerals such as feldspars, quartz and micas form in granite, for example – partly because

---

Mineral associations

Distinctive associations
• Green adamite on orange, earthy limonite.
• Red almandine garnet crystals set in shiny black biotite.
• Green amazonite with smoky quartz.
• Purple amethyst with golden or clear calcite.
• White analcime with pinky serandite.
• Green apatite in orange calcite.
• Green apophyllite or turquoise cavansite with white stilbite.
• Fibrous artinite on masses of green serpentine.
• Vivid blue azurite with green malachite.
• Honey coloured barite on yellow calcite.
• Blue benitoite, brown neptunite with white natrolite.
• Blue boleite forming star-shaped crystals with cumengite.
• Blue celestine with yellow sulphur.
• Blue chrysocolla coated with quartz or patched with green malachite.
• Red copper with silver.

• Pink elbaite tourmaline with mauve lepidolite.
• Purple fluorite with black sphalerite.
• Black hübnerite with clear quartz.
• Deep black neptunite and egg white natrolite.
• Golden pyrite with milky quartz.
• Red ruby with green zoisite.
• Silver galena with yellow anglesite and sparkling cerussite.
• Silver sulphide clusters: polybasite, stephanite and acanthite.
• Grey willemite, white calcite and black franklinite.

Precious metals
• Gold: calaverite, krennerite, nagyagite, pyrite, quartz, sylvanite.
• Silver: acanthite, pyrargyrite, proustite, galena.

Typical associations
Sulphides and sulphosalts
• Arsenopyrite: gold, cassiterite, scorodite.
• Boulangerite: galena, pyrite, sphalerite, tetrahedrite, tennantite, proustite, quartz, carbonates.
• Bournonite: tetrahedrite, galena, silver, chalcopyrite, siderite, quartz, sphalerite, stibnite.
• Chalcocite: bornite, calcite, chalcopyrite, covellite, galena, quartz, sphalerite.
• Chalcopyrite: pyrrhotite, quartz, calcite, pyrite, sphalerite, galena.
• Cinnabar: pyrite, marcasite and stibnite.
• Enargite: quartz and sulphides such as galena, bornite, sphalerite, pyrite, chalcopyrite.
• Greenockite: prehnite, zeolites.
• Marcasite: lead and zinc minerals.
• Pentlandite: chalcopyrite, pyrrhotite.

Blue azurite and green malachite

• Pyrargyrite: calcite, galena, proustite, sphalerite, tetrahedrite.
• Pyrrhotite: pentlandite, pyrite, quartz.
• Realgar: orpiment; other arsenic minerals.
• Sphalerite; galena, dolomite, quartz, pyrite, fluorite, barite, calcites.
• Sylvanite: fluorite, other tellurides, sulphides, gold, tellurium, quartz.
• Sylvanite: gold tellurides calaverite and krennerite.

Oxides and hydroxides
• Brookite: rutile, anatase, albite.
• Cassiterite: wolframite, quartz, chalcopyrite, molybdenite, tourmaline, topaz.
• Cuprite: native copper, malachite, azurite, chalcocite, iron oxides.
• Diaspore: corundum, magnetite, spinel, bauxite, dolomite, spinel.
• Goethite: limonite, magnetite, pyrite, siderite.
• Manganite: barite, calcite, siderite, goethite.
• Uraninite: cassiterite, arsenopyrite, pitchblende.

the chemicals that they are composed of occur in the melt. The same melt also contains chemicals that under different conditions combine to make the range of rare minerals associated with pegmatites.

Similarly, shales are made from the assemblage muscovite-kaolin-quartz-dolomite-feldspar. But as shales are metamorphosed to hornfels, the heat alters them to create a new assemblage: garnet-sillimanite-biotite-feldspar. Particular conditions of metamorphism produce particular assemblages of minerals. Called facies, these include blueschists rich in glaucophane and albite, and greenschists rich in chlorite and actinolite.

Identity links

Because many minerals are likely to be found together, associations can be useful clues to identity. A purple-coated, coppery red mineral found in hydrothermal veins along with chalcopyrite, marcasite, pyrite and quartz is likely to be bornite, for instance. Alternatively, if you find a mineral in a pegmatite you think is vivianite, it might be worth a rethink if there is no triphylite around. Associations can also be useful signposts to particular minerals, such as ores. The lead and zinc ores galena and sphalerite, for instance, are often found in association with calcite and barite. So finding a vein of calcite and barite can often lead to the discovery of galena or sphalerite deposits.

Mineral environments

Minerals are often linked to particular geological environments and processes, and this can help you in narrowing down identities. You might find thin, pale blue crystals in sedimentary rocks, for example, and guess they are kyanite. If so, you'd probably be wrong because kyanite is forged by the heat of metamorphism. So you are unlikely to find it in sedimentary rocks but very likely to find it in schists and gneisses. Pale blue crystals in sediments are more likely to be celestine.

Where did you find it?

Igneous veins	Garnet	Limestone quarry
Sulphides such as pyrite	Tourmaline	Calcite
Coppers such as malachite and azurite	**Volcanic vents**	Gypsum
Gold and silver	Sulphur	Fluorite
	Sulphates	Galena
	Hematite	Sphalerite
Igneous intrusions	**Hot springs**	Marcasite
Quartz	Travertine	Hematite
Feldspar	Gypsum	
Mica	Selenite	**Metamorphic rocks**
Pyroxenes	Salts	Sulphides
Amphiboles		Garnet
	Volcanic debris	Mica
Pegmatites and cavities in lava	Pumice	Calcite
	Olivine	Chromite
Quartz	Augite	Quartz
Feldspars	Obsidian	Spinel
Mica		Chlorite
Sulphides	**River sands**	Andalusite
Siderite	Quartz	Sillimanite
Apatite	Gold	Kyanite
Beryl	Diamond	Antigorite
Sapphire	Emerald	Feldspars
Topaz	Cassiterite	Chrysotile
	Pyrite	Staurolite
	Magnetite	Talc

Halides

• Atacamite: malachite, azurite, quartz.

• Chlorargyrite: silver, silver sulphides. cerussite, limonite, malachite.

• Fluorite: silver and lead ores, quartz, calcite, dolomite, galena, pyrite, chalcopyrite, sphalerite, barite.

• Halite: anhydrite, gypsum, sylvite.

Carbonates

• Dolomite: lead, zinc, copper ores.

• Smithsonite: malachite, azurite, pyromorphite, cerussite, hemimorphite.

• Strontianite: galena, sphalerite, chalcopyrite, dolomite, calcite, quartz.

• Trona: halite, gypsum, borax, dolomite, glauberite, sylvite.

• Witherite: quartz, calcite, barite.

Phosphates, arsenates and vanadates

• Adamite: azurite, smithsonite, mimetite, hemimorphite, scorodite, olivenite, limonite.

• Carnotite: tyuyamunite.

• Mimetite: galena, pyromorphite, vanadinite, arsenopyrite, anglesite.

• Olivenite: malachite, goethite, calcite, dioptase, azurite.

• Pyromorphite: cerussite, smithsonite, vanadinite, galena, limonite.

• Vanadinite: galena, barite, wulfenite, limonite.

• Variscite: apatite, wavellite, chalcedony.

• Xenotime: zircon, anatase, rutile, sillimanite, columbite, monazite, ilmenite.

Sulphates and relatives

• Anhydrite: dolomite, gypsum, halite, sylvite, calcite.

• Barite: galena, sphalerite, fluorite, calcite.

• Brochantite: azurite, malachite, copper minerals.

• Crocoite: wulfenite, cerussite, pyromorphite, vanadinite.

• Ferberite: cassiterite, hematite,

arsenopyrite.

• Hübnerite: quartz, cassiterite, topaz, lepidolite.

• Scheelite: wolframite.

• Wulfenite: cerussite, limonite, vanadinite, galena, pyromorphite, malachite.

Silicates

• Andalusite: corundum, kyanite, cordierite, sillimanite.

• Anorthoclase: augite, apatite, ilmenite.

• Dioptase: limonite, chrysocolla, cerussite, wulfenite.

• Glaucophane: epidote, almandine, chlorite, jadeite.

• Haüyne: leucite, nepheline, nosean.

• Hemimorphite: smithsonite, galena, calcite, anglesite, sphalerite, cerussite, aurichalcite.

• Humite: cassiterite, hematite, mica, tourmaline, quartz, pyrite.

• Kaolinite: clays, quartz, mica, sillimanite, tourmaline, rutile.

• Lepidolite: tourmaline, amblygonite, spodumene.

Greeny yellow adamite on orange, earthy limonite

• Leucite: natrolite, analcime, alkali feldspar.

• Monticellite: forsterite, magnetite, apatite, biotite, vesuvianite, wollastonite.

• Natrolite: other zeolites, apophyllite, quartz, heulandite.

• Nepheline: augite, aegirine, amphiboles.

• Quartz: beryl, calcite, fluorite, gold, hematite, microcline, muscovite, pyrite, rutile, spodumene, topaz, tourmaline, wolframite, zeolite.

• Sillimanite: corundum, kyanite, cordierite.

• Spodumene: feldspar, micas, quartz, columbite-tantalite, beryl, tourmaline, topaz.

• Staurolite: garnet, tourmaline, kyanite or sillimanite.

• Talc: serpentine, tremolite, forsterite.

• Tourmaline: beryl, zircon, quartz, feldspar.

• Vesuvianite: grossular garnet, wollastonite, diopside, calcite.

• Willemite: hemimorphite, smithsonite, franklinite, zincite.

• Wollastonite: brucite, epidote.

NATIVE ELEMENTS: METALS

Most minerals occur in combinations of chemicals as compounds. However, 20 or so are 'native elements' that occur in small quantities by themselves in relatively pure form. Most native elements are metals such as gold, which does not readily combine with other materials. Many metals typically occur in conjunction with others in ores, but these less reactive metals are often found alone. Indeed, gold – unusually – is primarily found alone.

Silver

Ag Silver

In Ancient Egypt, silver was called white gold and was valued even more highly than gold itself. There are silver mines in eastern Anatolia (Turkey) that were excavated by the pre-Hittite people of Cappadocia over 5,000 years ago. Today, its high conductivity (better than copper) means that silver is used in electronics, as well as for decoration. When polished, silver is a beautiful, shiny white metal, but it quickly tarnishes with a black coating of silver sulphide (inset). This is why it is hard to identify in nature, even though it grows in distinctive wiry, dendritic (tree-like) masses with the appearance of twisted wood. Like gold, it typically forms in hydrothermal veins, commonly in association with galena (lead ore), zinc and copper. Nuggets or grains rarely form so it is uncommon in placer deposits. Native silver is very rare, and most of the silver used today has been separated out from other minerals, especially in large argentite (silver sulphide) ore deposits in Nevada, Peru and Mexico.

Dark areas of silver sulphide, or tarnish

Identification: You are unlikely to find any other white, softish metal in its native state. Scraping off a black tarnish to reveal a bright silvery metal confirms the identity. Cubes like that here are rare, and you are more likely to find silver in the form of wiry, branching masses.

Calcite block

Silver

Crystal system: Isometric.
Crystal habit: Wiry, branching masses or grains, or, rarely, cubic crystals. Wires can form coils like rams' horns.
Colour: Silver-white but tarnishing quickly to black.
Lustre: Metallic.
Streak: Silver-white.
Hardness: 2.5–3.
Cleavage: None.
Fracture: Hackly.
Specific gravity: 10–12.
Other characteristics: Soft enough to stretch into wires and hammer into shapes.
Notable locations: Kongsberg, Norway; Sankt Andreasberg (Harz Mts), Freiburg (Saxony), Germany; Jachymov, Czech Republic; Chihuahua, Mexico; Great Bear Lake, Northwest Territories; Cobalt, Ontario; Michigan; Creede, Colorado.

Copper

Cu Copper

A warm reddish gold in colour, copper is the most easily recognized of all metals. It is quite soft, and sometimes found pure in native form – which is why it was one of the first metals people learned to use. The oldest knives found, dating back 6,000 years, are made of copper. Pure copper is often found in sulphide-rich veins in warm desert areas, or in cavities in ancient lava flows. Like silver, it often grows in branching masses and tarnishes quickly. Yet the tarnish is bright green, not black, and a copper deposit is often revealed by bright green stains on rocks called copper bloom. As a native element, copper is quite rare, so most of the copper used today comes from ores such as chalcopyrite.

Identification: Copper is readily identified by its reddish gold colour. It appears darker when tarnished, as shown opposite.

Crystal system: Isometric.
Crystal habit: Typically wiry, branching masses or clusters.
Colour: Copper, green tarnish.
Lustre: Metallic.
Streak: Red copper colour.
Hardness: 2.5–3.
Cleavage: None.
Fracture: Hackly, ductile.
Specific gravity: 8.9+.
Other characteristics: Can be stretched and shaped.
Notable locations: Chessy, France; Siegerland, Germany; Turinsk, Russia; Namibia; Broken Hill (NSW), Australia; Bolivia; Chile; New Mexico; Keweenaw, Michigan; Arizona.

Gold

Au Gold

No metal is quite like gold. Because it rarely forms compounds with other elements, it occurs naturally in almost pure form, and remains shiny and untarnished for thousands of years. It was one of the first metals used by mankind, and some of the world's oldest metal artefacts are gold. So far, about 150,000 tonnes of gold has been taken from the ground; a further 2,500 tonnes or so are mined each year. About a quarter of this is locked up in vaults, primarily in the USA and Europe, to provide the reserves against which the world's currencies are valued according to the Gold Standard. In the past, most gold came from South Africa, but this required costly, deep mines to be sunk, such as that at Savuka in Witwatersrand which is almost 4km/2½ miles deep. In recent years, mining companies have begun to work deposits nearer the surface with opencast mines in places such as China, Russia and Australia. The Grasberg gold mine in Indonesia is now the world's largest. Most gold is still used for jewellery, but the electronics industry is consuming increasing amounts for use in computer and communications technologies.

Identification: Shiny and gold in colour, gold is instantly recognizable – but a few other minerals can be mistaken for gold, such as pyrites and other sulphides and some tellurides. Only gold forms gold grains and nuggets.

Crystal system: Isometric.
Crystal habit: Crystals are rare and tiny, typically cubic or octahedral. Usually forms grains and nuggets or wire, branchlike crystal clusters.
Colour: Golden yellow.
Lustre: Metallic.
Streak: Golden yellow.
Hardness: 2.5–3.
Cleavage: None.
Fracture: Hackly.
Specific gravity: 19.3+.
Other characteristics: Ductile, malleable and sectile.
Notable locations: Verespatak (Bihar Mountains), Romania; Siberia; Witwatersrand, South Africa; Grasberg, Indonesia; Bendigo & Ballarat (Victoria), Flinders Range (SA), Australia; Yukon; California; Colorado; South Dakota.

Gold

Small quartz crystals

Striking gold

Gold is found in two major kinds of deposit. It typically forms in volcanic veins or 'lodes' in igneous rocks, where it is often found in association with quartz or with sulphide minerals such as stibnite. All the world's greatest gold mines exploit lodes like these. Yet in the past, many people found gold by searching in rivers for placer deposits. These are small grains or nuggets of gold washed there after the rock in which they formed was broken up by the river. Because gold grains are so dense and resistant to corrosion, they accumulate in shoals in the stream bed. These grains can be recovered by the simple technique of panning. The process is painstaking and rarely produces any reward – yet panning still occurs in countries such as Chile, above. The idea is to scoop shingle from the river bed and gradually swill out all the sand with water. With skill, the lighter sand is discarded, leaving the heavier gold grains behind.

Platinum

Pt Elemental platinum

Platinum is one of the rarest and most precious of all native elements. It typically occurs in thin layers of sulphide metal ores in mafic igneous rocks (dark igneous rocks rich in magnesium and iron). Only very rarely are crystals of platinum found. More usually platinum is found as fine grains or flakes or, occasionally, as grains and nuggets, which, like gold, are found concentrated in placer deposits (see Striking gold, left). Unlike gold, it is rarely pure in nature and tends to be mixed in with other metals such as iridium and iron. It is typically associated with chromite, olivine, pyroxene and magnetite. Nearly all of the world's platinum comes from two very distinct parts of the world – the Urals in Russia and the Bushveld complex in South Africa.

Crystal system: Isometric.
Crystal habit: Crystals rare and tiny, typically cubic. Platinum is usually found as grains and flakes; occasionally nuggets.
Colour: Pale silver-grey.
Lustre: Metallic.
Streak: Steel grey.
Hardness: 4–4.5.
Cleavage: None.
Fracture: Hackly.
Specific gravity: 14–19+.
Other characteristics: Doesn't tarnish, may be weakly magnetic and is ductile, malleable and sectile.
Notable locations: Norlisk, Russia; Bushveld complex, S Africa; Colombia; Ontario; Alaska; Stillwater, Montana.

Identification: Platinum is best identified by its pale, silver-grey colour and its relative softness.

NATIVE ELEMENTS: METALS AND SEMI-METALS

It is not just metals that occur as native elements. So too do some semi-metals, such as antimony and bismuth. Semi-metals are metallic in appearance, but not always as shiny as true metals. They are also brittle, shattering when struck with a hammer, and are less conductive than proper metals.

Antimony

Sb Elemental antimony

Antimony is a silvery grey semi-metal that only rarely occurs as a native element, typically in hydrothermal veins. In fact, its name comes from the Greek words *anti* and *monos*, which mean 'not alone', because it combines so readily with other minerals, such as sulphur. The chemical symbol Sb comes from the Latin word for its most common ore, stibnite. Antimony is found in about 100 minerals, and traces are sometimes extracted from silver, copper and lead ores when they are smelted. Yet the only major ore is stibnite, three-quarters of which comes from China. Antimony was not identified as a separate element until the 17th century, even though the alchemists knew how to purify it from stibnite. Yet in the form of stibnite it has a long history. Powdered to make kohl, it has been used as black eye make-up since the time of the Ancient Egyptians. Today antimony is mainly used to impregnate plastics, rubbers and other materials to make them fireproof. It is also used to toughen up the lead in lead acid batteries.

Identification: Antimony is often found as small silvery plate-shaped crystals and masses on quartz. It can easily be confused with bismuth.

Quartz

Antimony

Crystal system: Trigonal
Crystal habit: Crystals look cubic but it is typically found in botryoidal, lamellar and radiating masses
Colour: Whitish grey but can be darker when tarnished
Lustre: Metallic but dull when tarnished
Streak: Whitish grey
Hardness: 3–3.5
Cleavage: Perfect in one direction, basal
Fracture: Uneven
Specific gravity: 6.6–6.7+
Other characteristics: Melts at the relatively low temperature of 630°C/1,166°F
Notable locations: Carinthia, Austria; Sala (Vastmanland), Sweden; Auvergne, Brittany, France; Malaga, Spain; Lombardy, Italy; Lavrion, Greece; Wolfe Co, Quebec; Kern Co, California; Arizona
Caution: Antimony is mildly poisonous; handle with care

Bismuth

Bi Elemental bismuth

Like antimony, bismuth is a silvery white semi-metal that rarely occurs as a native element. It is found typically in high-temperature veins along with quartz, and minerals formed by metals such as cobalt, silver, iron and lead, or along the edge of granite intrusions where they meet limestone. More commonly, bismuth occurs in its main ore minerals, bismuthinite and bismite. The name bismuth probably comes from the German for 'white mass' – although the famous chemist Paracelsus said it came from the German for 'white meadow', because it was found in the fields of Saxony. Like antimony, bismuth shares with water the rare quality of expanding rather than contracting as it freezes. This property makes it very useful in soldering, because it expands to fill gaps as it solidifies. Because it is non-toxic, bismuth has replaced lead in many uses such as plumbing. Its low melting point makes it the perfect plug for fire-sprinkler systems. It is also used as a medicine for stomach upsets.

Identification: Bismuth exhibits a rainbow-coloured iridescent tarnish and has a pink tinge when freshly broken.

Crystal system: Trigonal
Crystal habit: Crystals are rare. Typically occurs in foliated masses.
Colour: Silver-white with an iridescent tarnish
Lustre: Metallic
Streak: Greenish black
Hardness: 2–2.5
Cleavage: Perfect in one direction, basal
Fracture: Uneven, jagged
Specific gravity: 9.7–9.8
Other characteristics: Pinkish tint on broken surfaces
Notable locations: Devon, England; Saxony, Germany; Wolfram (Queensland), Australia; San Baldomero, Tasna, Bolivia; South Dakota; Colorado; California
Caution: Bismuth is mildly poisonous; handle with care

Arsenic

As Elemental arsenic

Arsenic is a silvery grey semi-metal. In nature it occurs as a brittle metallic mineral, but in the laboratory it can be made into a white powder – and this is the kind of arsenic that poisoners have used since the days of the Ancient Greeks. Arsenic rarely occurs as a native element and is more typically found in sulphides and sulphosalts such as arsenopyrite, orpiment, realgar and tennantite. When it is found, native arsenic is usually mixed in with silver and antimony. Arsenic is a common by-product when silver ore is mined. It usually occurs in masses with onion-like layers called 'shelly' arsenic or 'scherbencobalt', or in kidney-like crusts. Very occasionally arsenic crystals are found, typically trigonal in form. However, arsenic exhibits polymorphism (has more than one shape) and in Saxony, Germany, some orthorhombic crystals have been found. These orthorhombic arsenic crystals are called arsenolamprite.

Identification: Arsenic gives off a characteristic garlic smell when hit with a hammer. This specimen is a dull black mass of arsenic, containing proustite.

The mineral proustite is a compound of silver, arsenic and sulphur.

Crystal system: Trigonal.
Crystal habit: Typically in round banded or botryoidal masses.
Colour: Pale grey. Tarnishes to dark grey or black.
Lustre: Metallic but dull when tarnished.
Streak: Black.
Hardness: 3–4.
Cleavage: Perfect in one direction, basal.
Fracture: Uneven.
Specific gravity: 5.4–5.9+.
Other characteristics: Garlic smell when hit or crushed.
Notable locations: Saxony and Harz Mountains, Germany; Sainte-Marie-aux-Mines (Vosges Mountains), France; Kongsberg, Norway; Fakui (Honshu), Japan; Atlin, British Columbia; Washington Camp (Santa Cruz Co), Arizona.
Caution: Arsenic is very poisonous; handle with care. Wash your hands afterwards.

Metal crystals

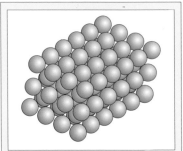

Metals are special because of the way that their crystals form. Unlike other minerals, metals tend not to form separate crystals. Instead, their atoms pack together in simple lattice-like structures, tied together by special chemical bonds called metallic bonds. In a structure like this, the bonds all work together, making metals strong but not brittle, which is why they can typically be hammered, bent and stretched into shape without breaking. The lattice arrangement also leaves electrons (tiny subatomic particles) only loosely attached in the spaces in between the atoms. This results in metals being good conductors of electricity (and heat), because there are plenty of 'free' electrons to transmit it through the lattice. In magnesium and zinc, the lattice is arranged in closely packed hexagons, whereas in aluminium, copper, silver and gold the atoms are arranged in cubes with an atom in the middle of each side of the cube.

Mercury

Hg Elemental mercury

Mercury is one of the few metals that is liquid at room temperature, along with caesium and gallium. In fact, it only freezes at less than -40°C/-40°F. Strictly speaking, it is not a mineral but a mineraloid because it does not normally form crystals. It is the only mineral besides water to occur naturally as a liquid. All the same, it doesn't often form pools but is found as tiny blobs on the mercury ores of cinnabar and calomel, typically lodged in crevices or attached by surface tension. Occasionally, pools of mercury can also be found filling rock cavities, usually in active volcanic regions. It has also been found as a precipitate from hot springs, along with cinnabar and other minerals. Because it is so rare, native mercury is not used as a source of mercury.

Mercury beads

Identification: Beadlike drops of mercury are instantly recognizable.

Crystal system: Hexagonal crystals below -40°C/-40°F.
Crystal habit: Droplets or pools of mercury liquid.
Colour: Bright silvery metallic.
Lustre: Metallic.
Streak: Liquid, has no streak.
Hardness: Liquid so does not have hardness.
Cleavage; Fracture: None.
Specific gravity: 13.5+.
Other characteristics: Mercury is a liquid metal.
Notable locations: Almadén (Ciudad Real), Spain; Idrija, former Yugoslavia; Almaden mine (Santa Clara Co), Socrates mine (Sonoma Co), California.
Caution: Mercury fumes are poisonous; handle only in a well-ventilated space.

NATIVE ELEMENTS: NON-METALS

Only a few non-metals occur as native elements, of which there are really only two: sulphur and carbon.
Carbon exists in various forms, including graphite, diamond and chaoite. Carbon and sulphur are among
the most interesting and important of all minerals, and carbon compounds play a vital role in the
chemistry of every living organism.

Sulphur

S Elemental sulphur

Sulphur is also known as 'brimstone' or burning stone because it burns easily, giving a blue flame when lit. (Don't do this though; it gives off a poison gas when it burns!) Pure, native sulphur is quickly identified from its bright yellow colour. It typically forms crust-like deposits around the margins of hot volcanic springs and smoky volcanic chimneys called fumaroles. Most of the world's sulphur is mined from beds of limestone and gypsum such as those under the Gulf of Mexico. Extracting the sulphur involves taking advantage of sulphur's low melting point, which is only a little above the boiling point of water. Wells are drilled and superheated water is injected into the sulphur formation, melting the sulphur. The resulting slurry is then pumped to the surface, where the water is evaporated to leave behind the sulphur residue. This is called the Frasch process. Unfortunately, the process destroys the good sulphur crystals that are often found in limestone beds. Around volcanic springs, sulphur typically forms crumbly-looking masses that are full of little bubbles and have a strong smell of rotten eggs.

Masses of small sulphur crystals

Identification: Sulphur's bright yellow colour, low melting point, rotten egg smell and softness make this an easy mineral to identify.

Crystal system: Orthorhombic.
Crystal habit: Typically massive or powdery forms but chunky crystals are also common. Sometimes found in an acicular (needle-like) form called rosickyite.
Colour: Bright yellow colour to yellow-brown.
Lustre: Vitreous as crystal to greasy or earthy as masses.
Streak: Whitish or yellow.
Hardness: 2.
Cleavage: Very poor.
Fracture: Conchoidal.
Specific gravity: 2.0–2.1.
Other characteristics: Can give off the smell of rotten eggs; very brittle. Has a low melting point and burns readily with a bluish flame. Often associated with the pungent (and poisonous) gas it produces when heated.
Notable locations: Sicily; Poland; France; Russia; Japan; Mexico; Yellowstone, Wyoming; Sulphurdale, Utah; Louisiana; Texas.

Graphite

C Elemental carbon

Graphite forms as carbon compounds interact in veins, and in metamorphic rocks as organic material in limestone is altered by heat and pressure. It is one of the softest of all minerals, with a value of less than 2 on the Mohs scale of hardness. In fact, its softness along with its black colour make it the perfect material for pencils – which earned its name from *graphein*, the Greek word for 'writing'. Yet despite its softness, graphite, like diamond, the hardest of all minerals, is a form of pure carbon. Graphite breaks into minute, flexible flakes that easily slide over one another. This 'basal cleavage' gives graphite a distinctive greasy feel that makes it a good dry lubricant. In nature, graphite is found in two distinct forms: flake graphite and lump graphite. Lump graphite is compact and lacks graphite's normal flakiness.

Identification: Graphite is dark grey, soft and greasy looking. It will leave black smudges on your fingers.

Crystal system: Hexagonal.
Crystal habit: Crystals are rare but typically form as flaky plates or masses in veins.
Colour: Black-silver.
Lustre: Metallic, dull.
Streak: Black to brown-grey.
Hardness: 1–2.
Cleavage: Perfect in one direction.
Fracture: Flaky.
Specific gravity: 2.2.
Other characteristics: Leaves black marks on fingers and paper. It also conducts electricity but only very weakly.
Notable locations: Borrowdale (Cumbria), England; Pargas, Finland; Mount Vesuvius, Italy; Austria; Galle, Sri Lanka; Korea; Ticonderoga, Bear Mountain, New York; Ogdensburg, New Jersey.

Diamond

C Elemental carbon

Diamond is virtually the hardest substance known to man. It gets its name from the Ancient Greek word *adamantos*, meaning 'invincible'. The only mineral that is harder is the very rare lonsdaleite formed in meteorite impacts. Like graphite and lonsdaleite, diamond is pure carbon. It glitters like glass because it has been transformed by enormous pressure underground. It is now possible to make diamonds artificially be squeezing graphite under extreme pressure. But pressures like these are rare in nature, occurring only deep in the Earth's crust and upper mantle. All diamonds found today are extremely old, being formed at least a billion years ago, with some more than three billion years old! They formed at least 145km/90 miles below the Earth's surface and were gradually carried up in pipes of hot magma. These volcanic pipes cooled to form blue rocks called kimberlites and lamproites, which are the source of most of the world's best diamonds. Diamonds can easily be weathered out of kimberlites to be washed away by streams into placer deposits. In their rough state diamonds look dull and it was only when jewellers began to cut them in the Middle Ages that their amazing brilliance was revealed.

Diamond

Kimberlite

Identification: Diamond looks non-descript when rough, but its hardness is unmistakable. It is the only clear crystal that will scratch glass deeply and easily.

Crystal system: Isometric (cubic).
Crystal habit: Typically forms cubes and octahedrons, but very varied.
Colour: Typically colourless, but also tinged yellow, brown, grey, blue or red.
Lustre: Adamantine, or greasy when rough.
Streak: White.
Hardness: 10.
Cleavage: Perfect in four ways.
Fracture: Conchoidal.
Specific gravity: 3.5.
Other characteristics: When cut, diamonds show 'fire' – rainbow colours that flash as light is reflected internally at different angles.
Notable locations: Mir, Yakhutsk, Russia; Kimberley, South Africa; Ellendale, Argyle (WA), Echunga (SA), Australia; Minas Gerais, Mato Grosso, Brazil; Murfreesboro (Pike Co), Arkansas.

The alter egos of carbon

Carbon is an extraordinary element. It has a remarkable capacity for forming bonds with other chemicals, which makes it the basis for all the chemicals of life, from simple sugars to complex proteins and enzymes. But even pure elemental carbon is found in a variety of forms, as its atoms have a rare capacity to form either single, double or triple bonds with each other. When the atoms of an element combine to form different-shaped molecules, the varieties are called allotropes. Until recently, carbon was thought to form just two very different allotropes: graphite and carbon. Now scientists know of a third, buckminsterfullerene, and possibly a fourth (carbon nanotubes). When crystals of the same mineral have a different structure, the varieties are called polymorphs. Carbon is found in a range of mineral polymorphs – not just diamond and graphite, but possibly also lonsdaleite, chaoite and fullerite.

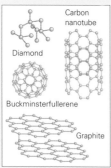

Carbon nanotube

Diamond

Buckminsterfullerene

Graphite

Chaoite

C Elemental carbon

When meteorites strike the ground, they create such extreme pressure and heat that minerals in the impact zone can be transformed into new ones such as suessite and martensite. This process is called shock metamorphism. Impacts are believed to change graphite to a rare kind of extra-hard diamond with a hexagonal crystal system called lonsdaleite. They are also thought to change graphite into carbon polymorphs called carbynes. In 1968, two geologists found a carbyne at the Ries impact crater in Germany. They called it chaoite and identified it as a new polymorph of carbon. Chaoite is recognized as a mineral by the International Mineral Association, but, as so few specimens have been positively identified, some geologists question whether chaoite should be given mineral status.

Crystal system: Hexagonal.
Crystal habit: Forms in flakes and small grains.
Colour: Dark grey.
Lustre: Submetallic.
Streak: Dark grey.
Hardness: 1–2.
Cleavage: Unknown.
Fracture: Unknown.
Specific gravity: 3.3.
Locations: Chaoite has been found in only three locations with any certainty: the Ries meteorite crater (Bavaria), Germany, and the Goalpur and Dyalpur meteorite craters, India.

Identification: Geologists have found it hard to identify chaoite positively, and some insist that elaborate X-ray analyses are necessary to be certain that a sample is chaoite.

SULPHIDES

Sulphides are compounds of sulphur with one or two metals, so they usually have a metallic lustre. They include some of the world's most important metal ores (see also Ore-forming minerals in this section for copper, silver, lead, zinc, cobalt and nickel). They are typically brittle and heavy and are formed from magmas or the hot fluids in hydrothermal veins. Acanthite, cobaltite, carrolite and digenite are all what are called simple sulphides because their molecules include just one sulphur atom.

Acanthite (and Argentite)

Ag₂S Silver sulphide

Acanthite, or 'silver glance', gets its name from the Greek *acantha*, for 'thorn', because of its spiky crystals. It is often listed as the same mineral as argentite, and argentite is often said to be the main ore of silver. In fact, both argentite and acanthite are 87 per cent silver and it is acanthite that is the main silver ore. Argentite and acanthite are simply polymorphs. Acanthite is the form that occurs at normal room temperatures in flat, monoclinic crystals. Argentite only forms at temperatures of over 173°C/343.4°F, forming cubic crystals. When forming from hot magmas, argentite crystallizes first. But as soon as the temperature drops below 173°C, the cubic argentite crystals will change to flatter, monoclinic acanthite. The change is not always so obvious, and some crystals retain argentite's cubic shape, even though their internal structure has changed. They look rather like argentite, but are actually acanthite. These are called 'pseudomorphs' (false shapes).

Identification: Shiny, dark grey crystals that tarnish to black on exposure to light. The tarnish on sterling silver is acanthite. It can also be cut with a knife.

Crystal system: Monoclinic (acanthite) and isometric (argentite).
Crystal habit: Distorted prisms and branching masses. Also cubelike pseudomorphs that look like argentite but are acanthite.
Colour: Dark grey to black.
Lustre: Metallic.
Streak: Shiny black.
Hardness: 2–2.5.
Cleavage: Absent.
Fracture: Conchoidal.
Specific gravity: 7.2–7.4.
Other characteristics: Sectile (can be cut with a knife) and malleable. Surfaces will darken on exposure to light.
Notable locations: Kongsberg, Norway; Black Forest, Freiberg, Schneeberg, Germany; Batopilas (Chihuahua), Guanajuato, Mexico; Comstock Lode, Nevada; Butte, Montana; Michigan.

Cobaltite

CoAsS Cobalt arsenic sulphide

Cubic cobaltite crystal

Cobaltite or cobalt glance is a very uncommon mineral that forms in sulphur-rich veins when the metals cobalt and arsenic join to form a sulphide. The very rarity of cobaltite crystals makes them much sought after by collectors. Cobaltite forms cubic crystals that look very similar to iron pyrite, although the actual structure is slightly different. Fortunately cobaltite is easily distinguished from pyrite by its colour. Pyrite is brassy yellow while cobaltite is silvery white. However, it is much harder to tell cobaltite from one of the other cobalt ores, skutterudite, which is also white and forms cubic crystals. Sometimes, when cobaltite has been exposed to the air, it will become covered with a bright pink or purple crust of other minerals such as erythrite. This colourful crust is called 'cobalt bloom' and is a strong sign of the presence of cobaltite or skutterudite. Cobaltite is one of the important ores of cobalt.

Identification: Cobaltite forms silvery cubes (here embedded in pyrrhotite). It may have a slightly sulphurous or garlic smell.

Crystal system: Appears isometric.
Crystal habit: Cubic, pyrite-like shapes. Grains and masses.
Colour: Silver-white with pink or purple tinge.
Lustre: Metallic.
Streak: Dark grey.
Hardness: 5.5.
Cleavage: Breaks into cubes.
Fracture: Uneven, subconchoidal.
Specific gravity: 6–6.3.
Other characteristics: Arsenic (garlic) smell.
Notable locations: Cornwall, England; Skutterud, Norway; Hankansbö, Tunaberg, Sweden; Siegerland, Rhineland, Germany; Bou Asser, Morocco; DR Congo; New South Wales, Australia; Sonora, Mexico; Cobalt, Ontario; Boulder, Colorado.

Carrollite

Cu₂S₄ Copper sulphide

Carrollite is a very rare copper sulphide mineral. It was first discovered in 1852 in the Papasco Mine, Finksburg, and the Mineral Hill Mine, Sykesville. Both of these mines are in Carroll County, Maryland, USA, earning the mineral its name carrollite. It typically forms very small crystals. Indeed, it is often cryptocrystalline – that is, the 'hidden' crystals are too small to be seen except under a microscope. Recently, some wonderful octahedral crystals up to 5cm/2in across were found in the Kambove Mine, Katanga, in the Congo. These were shiny silver and almost mirrorlike in lustre. Typical carrollite crystals are more likely to be dark grey and metallic in lustre, and are commonly intergrown with the minerals sphalerite, bornite and chalcopyrite.

Identification: Carrollite so rarely forms good big crystals that it is very hard to recognize, and only an expert can usually make a positive identification.

Carrollite

Crystal system: Isometric.
Crystal habit: Typically forms grains. Crystals are usually cubic but can be octahedral.
Colour: Grey, copper red.
Lustre: Metallic.
Streak: Dark grey to black.
Hardness: 4.5–5.5.
Cleavage: Poor.
Fracture: Conchoidal.
Specific gravity: 4.5–4.8.
Notable locations: Buskerud, Norway; Siegerland, Germany; Katanga, DR Congo; Hokkaido, Japan; Suan, North Korea; Yukon, Alaska; Carroll County, Maryland; Franklin, New Jersey.

Katanga, Africa's mineral treasure trove
Ancient metamorphic processes have endowed the gneiss, pelite and psammite rocks of the Katanga plateau in south-east DR Congo with some of the world's richest mineral deposits. There are huge deposits of copper, cobalt, manganese, platinum, silver, uranium and zinc ores here, plus occurrences of hundreds of rare minerals, including digenite, cuprosklodowskite, demesmaekerite, metatorbernite, swamboite, soddyite, stilleite, schoepite, cornetite, fourmarierite, thoreaulite and uranophane. The copper sulphide ores in Katanga's shale beds are among the world's largest copper reserves, and copper was dug here and traded through Africa centuries before European colonization in the 19th century. At its height in 1960, Katanga was producing 10 per cent of the world's copper, 60 per cent of its uranium and 80 per cent of its industrial diamonds. The uranium mines of Katanga were of huge importance to the early development of nuclear weapons in the United States, until alternative Canadian sources began to be exploited. All the 1500 tons of uranium oxide needed to get the Manhattan project up and running came from the Shinkolobwe mine in Katanga. Yet this mineral wealth has meant that Katanga has been fought over ever since the Belgians first began mining there in the 1880s, From 1971 until 1997, Katanga was called Shaba and was a focus of fighting in the civil wars. The political situation means it is now impossible to tell how much ore Katanga produces. In March 2004 the Central Bank of Congo stated that it produced just 783 tonnes of cobalt and yet the country recorded 13,365 tonnes of cobalt in exports. So maybe as much as 90 per cent of the mineral trade from Katanga is illicit.

Digenite

Cu₉S₅ Copper sulphide

Digenite is a rare ore of copper that is very similar to chalcocite. In fact, digenite is sometimes mistaken for chalcocite, and is usually found in association with it. However, the two minerals have distinct but similar chemical compositions. Chalcocite has ten copper atoms to every five sulphur atoms, whereas digenite has only nine copper atoms for every five sulphur atoms. The two are related by the fact that digenite is a high-temperature form of chalcocite that forms cubic crystals. Digenite was named after the Greek word *digenes*, meaning 'two kinds', because it was originally thought to be a mix of two different kinds of copper ore.

Crystal system: Isometric.
Crystal habit: Commonly forms masses or as grains in in sulphide-rich masses in sedimentary rocks. More rarely forms much-prized octahedral crystals.
Colour: Blue or black.
Lustre: Submetallic.
Transparency: Opaque.
Streak: Black.
Hardness: 2.5–3.
Cleavage: Good.
Fracture: Conchoidal fracture.
Specific gravity: 5.6.
Notable locations: Cornwall, England; Auvergne, Alpes-Maritimes, France; Sangerhausen, Thuringia, Germany; Carinthia, Salzburg, Austria; Katanga, DR Congo; Tsumeb, Namibia; Khingan Mts, China; Queensland, Australia; Kennecott, Alaska; Butte, Montana; Bisbee (Cochise Co), Arizona.

Pyrite

Rich mass of digenite

Identification: Digenite is usually identified by its black streak and the blue tinge to its black crystals – and the blue lustre it acquires when polished.

SIMPLE SULPHIDES

The simple sulphides are minerals made by a combination of a single sulphur atom with one or two other metal atoms. This group includes the economically important metal-bearing ores of galena (lead), cinnabar (mercury) and sphalerite (zinc) as well as greenockite, covellite, alabandite and millerite. They are typically brittle and heavy and form from magmas or in hydrothermal veins.

Greenockite

CdS Cadmium sulphide

Greenockite is a rare mineral made of cadmium sulphide, which occurs as small, brilliant honey-yellow crystals. It was first described in 1840 by Jameson and Connell, who found crystals during the cutting of the Bishopton tunnel on the Glasgow and Greenock railway in the United Kingdom. A specimen had been found over 25 years earlier but wrongly identified as sphalerite. The newly discovered mineral was named greenockite after Lord Greenock, on whose land it was discovered. Good crystals of greenockite have been found only around Glasgow, where they formed in amygdaloidal cavities in basaltic lava about 340 million years ago. Greenockite has been discovered elsewhere such as at Joplin in Missouri, but only as a powder or crust dusted over the top of sphalerite and other zinc minerals. Greenockite is actually the only proper ore of the metal cadmium, but it is so rare that all cadmium is obtained by extracting the trace amounts from lead and zinc ores. The only other cadmium-containing mineral is hawleyite, and the two can be hard to distinguish.

Identification: Greenockite has a honey colour, but looks very like sphalerite. In the field, the honey-coloured crusting, combined with its orange streak, is the best identifier. This enlarged view shows one of the rare complete crystals. Its absorption of certain wavelengths of white light creates the intense yellow colour.

 Crystal system: Isometric.
Crystal habit: Small tapering hexagonal crystals. Also seen as crusts or dustings over zinc and calcite ores.
Colour: Honey yellow, orange, red or light to dark brown.
Lustre: Adamantine to resinous.
Streak: Red, orange or light brown.
Hardness: 3–3.5.
Cleavage: Poor in one direction (basal), good in three others (prismatic).
Fracture: Conchoidal.
Specific gravity: 4.5–5.
Notable locations: Greenock, Scotland; Przibram (Bohemia), Czech Republic; Lavrion, Greece; Potosi, Bolivia; Silver Standard mine, British Columbia; Paterson, New Jersey; Joplin, Missouri; Kansas; Oklahoma; Illinois; Kentucky.

Covellite

CuS Copper sulphide

Covellite

Pyrite

Identification: Covellite is usually recognizable from its iridescent indigo-blue colour and the way it breaks into sheets.

Sometimes called covelline or indigo copper, covellite is a rare, deep indigo-blue mineral with shimmering iridescence that makes it much sought after by collectors. The blue colour turns to purple when wet, and crystals often have a black or purple tarnish. It was discovered in the early 19th century on Mount Vesuvius in Italy by Italian mineralogist Niccolo Covelli (1790–1829) and it is named after him. Covelli found his specimen in a fumarole where it probably formed directly from volcanic material. Yet it is more commonly found as a secondary mineral, formed by the alteration of other copper minerals in veins. Covellite typically forms in thin, hexagonal plates, and the best examples of crystals like these were found at Butte, Montana, USA, often called the richest hill on Earth. Even more strikingly coloured masses have been found at Kennicott in Alaska, USA.

 Crystal system: Hexagonal.
Crystal habit: Forms platelike, hexagonal crystals, and also occurs as grains and striplike masses.
Colour: Iridescent indigo blue, tarnishing black or purple.
Lustre: Metallic.
Streak: Grey to black.
Hardness: 1.5–2.
Cleavage: Breaks into sheets.
Fracture: Flaky.
Specific gravity: 4.6–4.8.
Other characteristics: Sheets of covellite bend like mica. Fuses when heated.
Notable locations: Dillenberg, Germany; Calabana, Sardinia; Bor, Serbia; Felsöbanya, Romania; Shikoku, Japan; Moonta, South Australia; Butte, Montana; Kennicott, Alaska.

Alabandite

MnS Manganese sulphide

Also known as manganblende and manganese glance, alabandite was first found in south-east Turkey and its name comes from Alabanda, the place where it was first identified. It typically forms in epithermal sulphide deposits (shallow, sulphide-rich vein deposits formed by hot fluids rising through the ground). Alabandite forms in the early stages of vein deposition along with arsenopyrite, pyrite and quartz. In the later stages, alabandite is replaced by rhodochrosite, the carbonate of manganese. Alabandite is also, less commonly, found in stony enstatite-type chrondrite meteorites. In fact, enstatite or E chrondrites are divided into EH (high) and EL (low) chondrites according to their sulphide content. The presence of alabandite identifies a meteorite as an EL chondrite. The presence of the associated mineral niningerite (magnesium manganese sulphide) identifies it as an EH chondrite.

Crystal system: Isometric.
Crystal habit: Small cubic and octahedral crystals in matrix, or in small masses or grains.
Colour: Iron black, or brown tarnish.
Lustre: Submetallic.
Streak: Dark green.
Hardness: 3.5–4.
Cleavage: Perfect.
Fracture: Uneven.
Specific gravity: 4.
Other characteristics: Brittle.
Notable locations: Alabanda, Turkey; Shakotan, Japan; Broken Hill (NSW), Nairne (SA), Australia; Allan Hills meteorite, Antarctica; Pierina, Peru; Mule Mts, Patagonia Mts, Tombstone, Arizona.

Identification: Alabandite is hard to identify except by its dark green streak.

Unearthly minerals
Meteorites fascinate mineralogists because they contain a unique mix of minerals that tells us both about other planets, and about the origins of minerals on Earth. The bulk of meteorites is made up from seven minerals that are common on Earth – olivine, pyroxene, plagioclase feldspar, magnetite, hematite, troilite and serpentine – and three minerals that are found only in meteorites – taenite (high-nickel iron), kamacite (low-nickel iron) and schreibersite (iron-nickel phosphide). There are also small amounts of millerite, alabandite and almost 300 other minerals found in meteorites. Of these, over 25 occur only in meteorites and nowhere else on Earth: barringerite; brezinaite; brianite; buchwaldite; carlsbergite; chaoite; daubreelite; farringtonite; gentnerite; haxonite; heidite; kosmochlor; krinovite; lawrencite; lonsdaleite; majorite; merrihueite; niningerite; oldhamite; osbornite; panethite; ringwoodite; roedderite; sinoite; stanfeldite; yagiite. A thin section of an iron meteorite, showing the rare minerals taenite and kamacite, appears above.

Millerite

NiS Nickel sulphide

Millerite was named after the British mineralogist W H Miller (1801–1880). It is sometimes used as an ore of nickel, but it is best known for its extraordinary acicular (needle-like) yellow crystals. Indeed, these are so distinctive that it is sometimes called 'capillary pyrites'. These typically grow like cobwebs inside cavities in limestone and dolomite, but millerite can form in veins carrying nickel and other sulphide minerals. Millerite needles are so thin that you would never know their crystals are trigonal. Millerite is also one of several minerals found in iron-nickel meteorites.

Crystal system: Trigonal.
Crystal habit: Typically grows in long, thin, acicular (needle-like) clusters, often in radiating sprays. Also forms fibrous coatings and granular masses.
Colour: Brassy yellow.
Lustre: Metallic.
Transparency: Opaque.
Streak: Greenish black.
Hardness: 3–3.5.
Cleavage: Perfect, but crystals are so thin this is rarely seen.
Fracture: Uneven.
Specific gravity: 5.3–5.5.
Other characteristics: Crystals inflexible and brittle.
Notable locations: Glamorgan, Wales; Wissen, Freiberg, Germany; Western Australia; Sherbrooke and Planet mines, Quebec; Ontario; Manitoba; South Central Indiana; Keokuk, Iowa; Sterling Hill, New Jersey; Antwerp, New York; Lancaster Co, Pennsylvania.

Siderite

Identification: Millerite is easily identified from its sprays of very thin brassy yellow needles.

IRON SULPHIDES

Iron sulphides are by far the most widespread of all the metal sulphide minerals, often found as beautiful, pale gold crystals. They form mostly in veins, where hot fluids bring minerals up through cracks in the ground, and in marine sediments where there is little oxygen present. Iron sulphide grains end up just about everywhere and are among the most common minerals in soil.

Pyrite

FeS₂ Iron sulphide

This shiny, yellow mineral can look so like gold it has fooled many prospectors into thinking they have struck it rich. No wonder, then, it is sometimes known as 'fool's gold'. It is among the most common of all minerals, found in almost every environment. Indeed, any rock that looks a little rusty probably contains pyrite. It comes in a vast number of forms and varieties, but the most common crystal shapes are cubic and octahedral. One sought-after form is flattened nodules found in chalk, siltstone and shale called 'pyrite suns' or 'pyrite dollars'. The nodules are usually made of thin pyrite crystals radiating from the centre. Pyrite gets its name from the Greek for 'fire' because it can give off sparks when struck – which is why it has been used to light fires since prehistoric times. Although it is rich in iron, it has never been used as an iron ore. In the past, though, it was used as a source of sulphur for making sulphuric acid.

Identification: Pyrite looks rather like gold, but it gives off sparks if struck hard with a metal hammer. Cubic crystals often have stripy marks or striations, often clearly visible in the larger cubes.

Crystal system: Isometric but huge variety of forms.
Crystal habit: Very varied. Crystals, cubes and pyritohedrons. Forms 'iron cross' interpenetrating twins. Also grains, radiating nodules.
Colour: Brassy yellow.
Lustre: Metallic.
Streak: Greenish black.
Hardness: 6–6.5.
Cleavage: Poor.
Fracture: Conchoidal.
Specific gravity: 5.1+
Other characteristics: Brittle, striations on cubic faces.
Notable locations: Elba, Italy; Rio Tinto, Spain; Germany; Berezovsk, Russia; South Africa; Cerro de Pasco, Peru; Chihuahua, Mexico; Bolivia; French Creek, Pennsylvania; Illinois; Missouri; Leadville, Colorado.

Marcasite

FeS₂ Iron sulphide

Marcasite

Marcasite can be very similar to pyrite. Indeed, it gets its name from the Arabic for 'pyrite'. It is actually a polymorph of pyrite, which means that, just like diamond and graphite, it has the same chemistry but a slightly different crystal structure. Just to cause further confusion, jewellers called pyrite 'marcasite'. It typically forms near the surface where acid solutions percolate down through shale, clay, chalk and limestone. Among the most distinctive marcasite crystals are spear-shaped twins, like those found in the chalk in Kent, England, and 'cockscomb' clusters of curved crystals. Collectors' specimens of marcasite often rust quickly, freeing sulphur to form an acid that speeds the crumbling of the specimen. There seems no way of preventing marcasite eventually turning to dust if it is left exposed to the air for any length of time.

Identification: Marcasite looks very, very similar to pyrite, but changes colour when exposed and soon begins to disintegrate.

Crystal system: Orthorhombic.
Crystal habit: Tabular, bladed or prismatic forms. Also massive, botryoidal, stalactitic and nodular.
Colour: Pale yellow, with greenish tint.
Lustre: Metallic.
Streak: Greenish brown.
Hardness: 6–6.5.
Cleavage: Poor in two directions.
Fracture: Uneven.
Specific gravity: 4.8+
Other characteristics: Sometimes has sulphur smell.
Notable locations: Pas de Calais, France; Russia; China; Guanajuato, Mexico; Peru; Joplin, Missouri; Wisconsin.

Arsenopyrite

FeAsS Iron arsenide sulphide

The name of this mineral comes from the Greek word *arsenikos*, which was the name the Greek philosopher Theophrastus gave to the mineral orpiment, another mineral containing arsenic. While arsenic is a poison, it can also be very useful in medicine (in the drug salvarsan) and in alloys. Arsenopyrite is a major ore of arsenic, yet it is rarely mined just for its arsenic. It is more commonly an unwelcome by-product in mining for other metals, such as in the nickel-silver mines of Freiberg, Germany, and in the tin mines of Cornwall, because it often forms in high temperature veins in association with other metals. The problem is how to dispose of the arsenic safely. In the copper and silver mines of Boliden in Sweden, special silos have been built to avoid spreading contamination.

Identification: Arsenopyrite looks very similar to marcasite, but fracturing it with a hammer releases arsenic's garlic smell.

Caution: Arsenic in arsenopyrite is poisonous; handle with extreme care and wash hands and surfaces.

Sphalerite

Arsenopyrite

Crystal system: Orthorhombic.
Crystal habit: Wedge-shaped or prismatic. Often twinned: sometimes cruciform, sometimes multiple.
Colour: Brassy white to grey, tarnishes to brown or pink.
Lustre: Metallic.
Streak: Dark grey to black.
Hardness: 5.5–6.
Cleavage: Distinct in two directions forming prisms.
Fracture: Uneven.
Specific gravity: 6.1+.
Other characteristics: Bitter garlic smell when powdered or broken.
Notable locations: Cornwall, England; Freiberg, Germany; Valais, Switzerland; Panasqueira, Portugal; Kyushu Island, Iname, Japan; Broken Hill (NSW), Australia; Bolivia; Wawa, Ontario; Edenville, New York; New Hampshire.

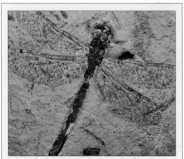

Life turned to pyrite
There are many ways living things can be preserved as fossils, but one of the most common is pyritization. Pyritization is a chemical process that involves the formation of iron sulphide minerals. What happens is that when an organism dies it is buried in sediments and the remains then begin to react chemically with fluids moving through the sediments. If the organism is buried rapidly – before it begins to disintegrate – and there are plenty of dissolved sulphates in the fluids, the organic tissue of the remains can be replaced molecule by molecule with iron sulphide minerals, essentially pyrite. In this way, over millions of years, the organic remains retain their shape but are transformed to pyrite. Pyritization can in this way preserve exquisite detail, even of soft tissues, which are not normally fossilized. Some of the best fossils of insects have been preserved by pyritization, including some wonderful dragonfly-like insects from the Cretaceous period.

Pyrrhotite

Fe₁₋ₓS Iron sulphide

Pyrrhotite gets its name from the Greek word *pyrrhos*, meaning 'reddish', yet it is not always red. In fact, it looks similar to other brassy-coloured sulphides like marcasite and chalcopyrite. It forms mostly in mafic igneous rocks and hydrothermal veins. Pyrrhotite is the only other common magnetic mineral besides magnetite. Not all specimens are strongly magnetic, but you can be sure you have pyrrhotite if it attracts a paper clip suspended from cotton, or moves a compass needle. Unusually, pyrrhotite's sulphur content can vary by up to 20 per cent. When the sulphur content is low, pyrrhotite crystals tend to be hexagonal in shape, whereas when the sulphur content is high, the crystals are usually flat plates.

Identification: Pyrrhotite's magnetism is usually enough to distinguish it from similar yellowish metallic minerals. But its slightly reddish tinge, its softness and its hexagonal or flat crystals can help to confirm the identification.

Crystal system: Hexagonal.
Crystal habit: Hexagonal or flat plates, but usually found in masses in rocks.
Colour: Bronze.
Lustre: Metallic.
Streak: Grey-black.
Hardness: 3.5–4.5.
Cleavage: None.
Fracture: Uneven.
Specific gravity: 4.6.
Other characteristics: Weakly magnetic, colour will darken with exposure to light.
Notable locations: Trentino, Italy; Andreasberg (Harz Mts), Germany; Kisbanya, Romania; Trepca, Serbia; Dalnegorsk, Russia; Japan; Kambalda, Western Australia; Morro Velho, Brazil; Chihuahua, Mexico; Sudbury, Ontario; Riondell, British Columbia; Standish, Maine; Ducktown, Tennessee; Pennsylvania; Franklin, New Jersey.

Intergrown hexagonal crystals

SULPHOSALTS

The sulphosalts are sulphide minerals in which a semi-metal – antimony, bismuth or arsenic – combines with a true metal such as lead or silver. Although nearly all these minerals are comparatively rare, they frequently form good crystals, because they crystallize slowly in cool pockets near the Earth's surface. Unfortunately, their very accessibility means that most good samples are hard to find.

Cylindrite

$FePb_3Sn_4Sb_2S_{14}$ Iron lead tin antimony sulphide

Crystal system: Trigonal, but this is disputed.
Crystal habit: Unique cylindrical crystals that look like tubes or rolls of metallic cloth, but occasionally found in masses.
Colour: Black to grey.
Lustre: Metallic.
Streak: Black.
Hardness: 2.5.
Cleavage: None.
Fracture: Conchoidal.
Specific gravity: 5.5.
Notable locations: Only found at Potosi and Mina Santa Cruz (Poopo) in Bolivia and in a few tin sulphide ores.

Identification: Cylindrite's tube-shaped crystals are unmistakable and impossible to mix up with any other mineral.

Cylindrite is one of the most unusual of all minerals. It owes its name to its almost unique crystal habit, which is cylindrical. Only chrysotile, a variety of serpentine, also forms tubular crystals, but these are microscopically thin with the appearance of tiny hairs. Crystals of cylindrite are actually coiled sheets that grow as if rolled into pipes. Sometimes the sheets can become uncoiled if put under pressure. Cylindrite is occasionally mined as an ore of lead, tin and the rare element indium, which is used by the microelectronics industry in the manufacture of transistors and silicon chips. However, cylindrite is collected primarily for its very unusual crystal habit. The majority of the best specimens of cylindrite crystals come from Bolivia – at Poopo in Oruro and at Potosi. Bolivian specimens are found in tin-rich mineralized veins and are typically associated with similar minerals in the sulphosalt group such as franckeite, incaite, potosiite and tellaite.

Jamesonite

$Pb_4FeSb_6S_{14}$ Lead iron antimony sulphide

Quartz

Jamesonite crystals

Identification: Jamesonite and similar sulphosalts can be identified by the dense, felt-like mats they form.

Jamesonite was named after the Scottish mineralogist Robert Jameson, who identified it in Cornwall, England. It forms in low-temperature lead-rich veins along with galena, sphalerite and pyrite, and in marbles. Jamesonite crystals are dense, felt-like, matted hairs sometimes called 'feather ores'. A number of other sulphosalts, including plagionite, zinkenite, boulangerite and jordanite, form in the same way and look almost identical. Boulangerite crystals can be bent while jamesonite crystals tend to snap, but otherwise almost the only way to tell these minerals apart is by chemical tests that reveal the iron in jamesonite. Jamesonite can also look quite similar to stibnite, but stibnite crystals tend to be better defined and break cleanly along the line of the crystals.

Crystal system: Monoclinic.
Crystal habit: Include loosely matted hairs and feathery masses.
Colour: Dark grey, tarnishing iridescent.
Lustre: Metallic and silky.
Streak: Grey-black.
Hardness: 2–3.
Cleavage: Perfect across the crystals at right angles.
Fracture: Uneven to conchoidal.
Specific gravity: 5.5–6.
Other characteristics: Crystals are brittle and snap easily.
Notable locations: Cornwall, England; Auvergne, France; Maramures, Romania; Kosovo; Mount Bischoff, Tasmania; Bolivia; Noche Buena mine (Zacatecas), Mexico; Colorado; South Dakota; Arkansas.

Pyrargyrite

Ag₃SbS₃ Silver antimony sulphide

Pyrargyrite is a deep red silvery colour giving both its nickname 'ruby silver' and its official name, from the Greek for 'fire' and 'silver'. Proustite is also called 'ruby silver' and the two minerals are very similar. Both are used as ores for silver and are often found together in low-temperature veins along with silver and other silver sulphides. In fact, pyrargyrite and proustite are 'isostructural', which means they have the same structure even though they are chemically slightly different. They both get darker when exposed to light, but pyrargyrite tends to be a deeper red than proustite. Pyrargyrite's tendency to darken on exposure to light means translucent crystals can quickly go opaque in the light, so good specimens are usually stored in a dark place. All the same, the dark coat can usually be cleaned off with a gentle washing in soap and water, or a quick dip in silver polish.

Identification: The combination of prismatic crystals and a dark red silvery colour is enough to identify a mineral as either pyrargyrite or proustite. Proustite is usually lighter in colour.

Crystal system: Trigonal.
Crystal habit: Typically prismatic crystals or massive forms.
Colour: Dark red when translucent but black when opaque.
Lustre: Adamantine.
Streak: Purplish red.
Hardness: 2.5.
Cleavage: Sometimes distinct in three directions forming rhombohedrons.
Fracture: Conchoidal, uneven.
Specific gravity: 5.8.
Other characteristics: Darkens upon exposure to light, crystals may be striated.
Notable locations: Harz Mts, Saxony, Germany; Alsace, Isère, France; Colquechaca, Bolivia; Castrovirrenya, Peru; Guanajato, Mexico; Silver City, Idaho; Comstock Lode, Nevada.

Boulangerite

Pb₅Sb₄S Lead antimony sulphide
Boulangerite forms pale bluish-grey hairlike fibres and is hard to distinguish from other 'feather ores' with similar crystals such as jamesonite. Unlike jamesonite, which snaps, boulangerite crystals bend. A variety of boulangerite with feathery plumes called plumosite was once thought to be a different mineral. Now it is simply classed as boulangerite.

Renierite

Cu(Zn)₁₁As(Ge)₂Fe₄S₁₆ Copper arsenic iron sulphide
Named after the Belgian geologist who discovered it in 1948, bronze-yellow renierite is found typically in Katanga in the DR Congo. Small grains are often found in granite and other igneous rocks.

Tetrahedrite

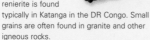

Cu₁₂Sb₄S₁₃ Copper antimony sulphide
Tetrahedrite is a minor ore of copper and silver. It gets its name from the tetrahedral (pyramid-shaped) crystals.

Baumhauerite

Pb₃As₄S₉ Lead arsenic sulphide

Baumhauerite is a very rare mineral that is seen embedded in dolomitic marble. It is found primarily in the Lengenbach quarry, Binnetal, in the Valais region of Switzerland. The mineral is named after Heinrich Baumhauer, who discovered it at Lengenbach in 1902. The Lengenbach quarry is famous among mineralogists for its array of rare minerals. Besides baumhauerite, many other rare arsenic sulphide and sulphosalt minerals have been found there, including marrite, bernardite, hatchite, novakite, smythite and many others. There is even a mineral that is named after the quarry, a sulphosalt discovered in 1904 called lengenbachite. Baumhauerite has also been found at Sterling Hill, New Jersey, USA, typically in association with molybdenite, and in massive aggregates at Hemlo, Thunder Bay, Ontario, Canada.

Identification: Baumhauerite is best identified by the striped prismatic shape of its crystals, which are typically about 1mm/¹⁄₂in long, and also by its density.

Crystal system: Triclinic.
Crystal habit: Striated prismatic, acicular crystals about 1mm/¹⁄₂in long with rounded faces. Also occurs as masses and grains.
Colour: Grey-black to blue-grey.
Lustre: Metallic to dull.
Streak: Dark brown.
Hardness: 3.
Cleavage: Indistinct.
Fracture: Conchoidal.
Specific gravity: 5.3.
Notable locations: Lengenbach quarry (Binnetal), Switzerland; Thunder Bay, Ontario; Franklin and Sterling Hill mines, New Jersey.

Realgar appears as red spots

Dolomitic marble

Baumhauerite

TELLURIDES AND ARSENIDES

The tellurides and arsenides are two small groups of minerals that are very similar to sulphur in chemical make-up – except that with tellurides, the element tellurium takes sulphur's place, and with arsenides, arsenic takes its place. Antimony and selenium can take sulphur's place in the same way, and so form two further small groups, the antimonides and selenides.

Tellurobismuthite

Bi_2Te_3 Bismuth telluride

Tellurium is a semi-metal, like antimony and bismuth. It was discovered in Transylvania in 1782 by Franz von Reichenstein. As an element it exists in two forms – as a silvery white, brittle, metallic-looking solid and as a dark grey powder. It is by no means common, and is usually found only in the tellurides of copper (kostovite), lead (altaite), silver (hessite and empressite), silver and gold (sylvanite, petzite and krennerite) and bismuth (tellurobismuthite). Tellurobismuthite is a rare silver-grey mineral typically found along with other tellurides in quartz-gold veins that form at high temperatures. It is very hard to distinguish from tetradymite, which is also a bismuth telluride but also contains an appreciable amount of sulphur. Like altaite, tellurobismuthite is a semiconducting material (a material like silicon that conducts electricity only under certain conditions). As a result, it is sometimes used in thermoelectric devices that produce electricity when heated.

Identification: Tellurobismuthite is typically a silvery white mineral that tarnishes to grey. It forms silvery flakes, as on this specimen, and usually occurs with tellurides.

Crystal system: Trigonal hexagonal.
Crystal habit: Foliated aggregates, irregular plates like mica, fine-grained fibres and masses.
Colour: Pale silvery grey.
Lustre: Metallic.
Streak: Pale lead grey.
Hardness: 1.5–2.
Cleavage: Perfect, flat leaves.
Fracture: Flexible.
Specific gravity: 7.8.
Notable locations: Clogau (Dolgellau Gold Belt), Mid Wales; Boliden (Västerbotten), Sweden; Tokke, Norway; Alpes-Maritimes, France; Larga (Metaliferi Mts), Romania; Kamchatka, Russia; Tohoku (Honshu), Kyushu, Japan; Dahlonega (Lumpkin Co), Georgia; Sylvanite (Hidalgo Co), New Mexico; Mule Mts, Arizona; Colorado.

Altaite

PbTe Lead telluride

Altaite gets its name from the Altai mountains in southern Siberia where it was discovered in 1854. Like other tellurides, it is frequently found in quartz-gold veins. It is a lead telluride related to another lead compound in the sulphide group, galena (lead sulphide). Like galena, altaite is an ore of lead, although it is much, much less abundant and widely extracted. Like galena it is unusually dense. Indeed, it is one of the few minerals that is actually denser than galena. This means it is often easy to identify by its heavy weight alone. It is distinguished from galena by the yellowish white colour. If the colour is not obvious, it is possible to identify it from the shape of the crystals. Although they are cubic as with galena, altaite crystals do not have the triangular pits in the faces that galena exhibits.

Altaite

Identification: Altaite and galena can be identified by their heavy weight. Altaite is yellow-white or silvery; galena is dark grey.

Crystal system: Isometric.
Crystal habit: Includes cubic and octahedral crystals, but more commonly found in masses and grains.
Colour: Tin white to yellowish white; tarnishes bronze yellow.
Lustre: Metallic.
Streak: Black.
Hardness: 2.5–3.
Cleavage: Perfect in three directions and forms cubes.
Fracture: Uneven.
Specific gravity: 8.2–8.3.
Notable locations: Transylvania, Romania; Przibram, Czech Republic; Altai Mountains, Zyrianovsk, Kazakhstan; Coquimbo, Chile; Moctezuma, Mexico; Greenwood, BC; Mattagami Lake, Quebec; Price County, Wisconsin; California; Arizona.

Rammelsbergite

NiAs₂ Nickel arsenide

Rammelsbergite is one of the group of rare minerals called arsenides. The two most common arsenides, niccolite and skutterudite, are both used as metal ores, niccolite for nickel and skutterudite for cobalt. So too is sperrylite, the rare platinum arsenide. Rammelsbergite is rare in itself, but is often found with other more common arsenides, and can be very hard to distinguish from them, especially the other nickel arsenides, niccolite, maucherite, dienerite and oregonite. All the arsenides tend to occur in similar conditions, but different minerals form according to the temperature and oxidization level in the vein. Arsenide minerals crystallize in the following order with increasingly oxidizing conditions: maucherite, niccolite, rammelsbergite, skutterudite, safflorite and löllingite. Rammelsbergite can also be hard to distinguish from nickel sulphides such as gersdorffite, millerite and pentlandite. It has a dimorphic cousin, pararammelsbergite, with the same chemistry but different crystal structure.

Identification:
Rammelsbergite looks very similar to most arsenides but can sometimes be identified by its red tinge, and its yellowy pink tarnish.

Crystal system: Orthorhombic.
Crystal habit: Masses, grains and radiating fibres. Rarer as flat tablets, stubby prisms or even cockscombs.
Colour: Silvery white with reddish tinge. Tarnishes yellow or pink.
Lustre: Metallic.
Streak: Grey.
Hardness: 5.5–6.
Cleavage: None.
Fracture: Uneven.
Specific gravity: 6.9–7.1.
Notable locations: Sainte-Marie-aux-Mines (Vosges de Alsace), France; Schneeburg (Harz Mountains), Germany; Lölling, Austria; Kongsberg, Norway; Binnetal, Switzerland; Bou Azzer, Morocco; Batopilas (Chihuahua), Mexico; Great Bear Lake, Northwest Territories; Cobalt, Ontario; Keweenaw, Michigan; New Jersey.

Telluride gold
Tellurium has a great affinity for gold, and gold tellurides are among the few significant ores of gold, which combines with few other elements but tellurium. The gold tellurides include petzite, coloradoite, melonite, krennerite, kostovite, nagyagite, stützite and montbrayite, but the most common are calaverite and sylvanite. Some gold tellurides are such good sources of gold that they have actually started gold rushes. Although it was not realized at the time, the gold in placer deposits (grains in river gravel) that started the famous Cripple Creek gold rush in the 1860s in Colorado came from telluride veins. When these telluride veins were discovered in the 1890s, it started a second gold rush (the above photograph of mining activity was taken in 1903). In Western Australia, Kalgoorlie's famous Golden Mile is mainly telluride gold. Cripple Creek is known for krennerite, sylvanite and calaverite. Kalgoorlie is known for calaverite. petzite and coloradoite. Chelopech in Bulgaria is known for kostovite.

Safflorite

CoAs₂ Cobalt arsenide or, more commonly, CoFeAs₂ Cobalt iron arsenide

Safflorite is closely related to rammelsbergite and both belong to the löllingite group of arsenide minerals. Safflorite is the cobalt-rich version of the group, rammelsbergite is the nickel-rich version and löllingite is the iron version, but each usually contains traces of the other two metals. In fact, safflorite can contain up to 50 per cent nickel before it becomes rammelsbergite, and contain 50 per cent iron content before it becomes löllingite. Safflorite gets its name from the German word *Safflor*, which means 'dyer's saffron', and may have got its name because trillings (groups of three intergrown crystals) of safflorite can look a little like tiny bunches of white crocuses (saffron flowers). It is often spotted close to erythrite, or 'cobalt bloom', the pinkish grey tinge that some cobalt minerals acquire when exposed to the weather.

Crystal system: Orthorhombic.
Crystal habit: Typically tiny flat tablet-shaped or prismatic crystals. Also in masses, grains, fibres, cockscombs or star-shaped trillings.
Colour: Bright white or grey but tarnishes to black.
Lustre: Metallic.
Streak: Black.
Hardness: 4.5–5.5.
Cleavage: Indistinct.
Fracture: Conchoidal.
Specific gravity: 7–7.3.
Other characteristics: Safflorite often forms twins that grow into groups of star-shaped crystals.
Notable locations: Nordmark (Varmland), Sweden; Schneeburg (Harz Mountains), Germany; Javornik, Czech Republic; Great Bear Lake, Northwest Territories; Oregon; Lafayette County, Wisconsin.

Identification: Safflorite can usually best be identified by its bright colour and black tarnish.

Safflorite *Calcite*

OXIDES

Oxygen is so common in the Earth's crust that 90 per cent of all minerals contain it. So all these are, in a way, oxides. To simplify things, geologists describe as oxides only those minerals that are a simple combination of a metal with oxygen, or a metal with oxygen and hydrogen (a hydroxide). Even so, oxides include everything from common ores such as bauxite to precious gems such as sapphires. Rutile, plattnerite, anatase and brookite are four simple oxides.

Rutile

TiO₂ Titanium oxide

Rutile was given its name by the famous 18th-century German mineralogist Abraham Gottlob Werner (1749–1814). It comes from the Latin *rutilus*, meaning 'reddish', from its coppery tinge. It is a

Rutilated quartz

minor ingredient in many plutonic rocks and crystalline slates, and of many basic pegmatites such as nelsonite. Famously, it is found alongside quartz, albite, chlorite, siderite, muscovite, ilmenite and apatite in vugs (pockets) in the Swiss Alps, as well as with its polymorphs brookite and anatase. Rutile is the most important ore of titanium, however, and these titanium ores come mainly from gravels and sands. Rutile can be often found as large crystals, but is best known for its presence as tiny needle inclusions in gemstones such as clear quartz, tourmaline, ruby, and sapphire, creating much-valued cat's eye and star effects. Clear quartz is turned into the beautiful ornamental stone rutilated quartz by golden needle rutile inclusions (inset).

Identification: Rutile can be identified by its copper tinge, bright sparkle and, often, by the long golden needles it forms.

Rutile

Calcite

Schist

Crystal system: Tetragonal.
Crystal habit: Typically stubby eight-sided prisms capped by low pyramids. Masses of long needles. Also inclusions.
Colour: Reddish brown in crystals. Yellow in needles and inclusions.
Lustre: Adamantine to metallic.
Streak: Brown.
Hardness: 6–6.5.
Cleavage: Good in two directions forming prisms.
Fracture: Conchoidal, uneven.
Specific gravity: 4.2+.
Other characteristics: Stripes on prism-shaped crystals. Very sparkly.
Notable locations: Alps, Switzerland; Urals, Russia; Minas Gerais, Brazil; Magnet Cove, Arkansas; Lincoln County, Georgia; Alexander County, North Carolina.

Plattnerite

PbO₂ Lead oxide

Identification: Plattnerite can be identified by its dark, black colour, and unexpected sparkle. It is also extremely heavy.

Plattnerite is a black, heavy mineral named after German mineralogist K F Plattner. It belongs to a small, important group of minerals known as rutiles, all of which have chain molecules that give distinctive prismatic crystals. Besides the mineral rutile itself, this includes a number of key ores such as cassiterite (tin ore) and pyrolusite (manganese ore), and also stishovite, which is a key indicator of meteorite impact. Plattnerite has a very high lead content, and this makes it one of the densest of all minerals. It is markedly denser than both galena and altaite. The lead in plattnerite also makes it sparkle. If this sounds odd, remember that lead is added to glass crystal to make it sparkle. Like many oxides, plattnerite is a 'secondary' mineral, and does not form directly in magma, but when other lead-bearing minerals are oxidized by exposure to the atmosphere.

Plattnerite may appear as small black needles

Crystal system: Tetragonal.
Crystal habit: Typically stubby prisms capped with pyramids. More commonly massive. Often forms a dry oxidation crust with tiny sparkling crystals on other lead minerals.
Colour: Black.
Lustre: Adamantine to metallic.
Streak: Chestnut brown.
Hardness: 5–5.5.
Cleavage: Good in two directions forming prisms. Poor basally.
Fracture: Conchoidal, uneven.
Specific gravity: 9.4+.
Notable locations: Leadhills (Lanarkshire), Dumfries and Galloway, Scotland; Alpes-Maritimes, France; Mapimi, Mexico; Shoshone Co, Idaho; Pima County, Arizona.

Anatase

TiO₂ Titanium oxide

Once known as octahedrite, anatase typically forms in low-temperature veins and in alpine fissures. The Binnetal district of the Alps is famous for its anatase. Anatase is the rarest of three titanium oxide minerals, anatase, brookite and rutile. These three are polymorphs, as they are chemically alike but have different crystal structures. Anatase is the version thought to form at the lowest temperatures, and is readily altered to rutile by exposure to higher temperatures. Where good crystals are found, in the quartz veins of the Diamantina district of Brazil, they are often only preserved as anatase where they are encased in quartz. Specimens of anatase like these are usually preserved as discovered, encased in quartz, and are much sought after. They are unmistakable, with blue-grey anatase forming long, double-pyramid-shaped crystals inside the clear or golden quartz.

Identification: The blue-grey double pyramid-shaped crystals of anatase are hard to mistake, especially when encased in quartz as shown below.

Quartz crystal

Anatase crystal

> **Crystal system**: Tetragonal.
> **Crystal habit**: Typically forms stretched double-pyramids.
> **Colour**: Blue-grey, also brown and yellow.
> **Lustre**: Adamantine to metallic.
> **Streak**: White.
> **Hardness**: 5.5–6.
> **Cleavage**: Perfect in four directions forming pyramids.
> **Fracture**: Subconchoidal, uneven.
> **Specific gravity**: 3.8–3.9.
> **Notable locations**: Tavistock (Devon), England; Binnatal, Switzerland; Alpes-Maritimes, France; Diamantina, Brazil; Somerville, Massachusetts; Gunnison County, Colorado.

Titanic titanium
In 1791, William Gregor, a vicar from Cornwall, England, found a magnetic black sand on the local beaches, and named it menachanite, after the nearby village of Menaccan. A few years later, the German chemist M H Klaproth separated a new metal element from the mineral, which he called titanium after the giants of Greek mythology. It wasn't until 1938, though, that German metallurgist W J Kroll developed a method of refining titanium from the ores ilmenite and rutile. Now titanium is becoming something of a miracle metal. It is three times as strong as steel and twice as light, making it ideal for the aerospace industry, where it is now widely used both by itself and in alloys. Titanium alloys are capable of operating at temperatures from sub zero to 600°C, and have been used in fuel tanks for space stations and in blades, shafts and casings in aircraft parts, including front fans and high pressure compressors. Because titanium is very resistant to corrosion, it is also the perfect metal for artificial hip joints, and architects are experimenting with it for building. A great deal of titanium, however, goes to making titanium dioxide, a white pigment for paints. Titanium dioxide paint is unrivalled for brightness and opacity and, unlike lead, which was once used for white paint, it is non-toxic.

Brookite

TiO₂ Titanium oxide

Brookite is one of the titanium oxide mineral trio, with anatase and rutile. It is much more common than anatase, and forms at the higher temperature of 750°C/1,382°F (650°C/1,202°F for anatase and 915°C/1,679°F for rutile). Brookite typically forms in quartz veins, but it can also be found as grains in sandy sediments, where large crystals can form, probably fed by cool solutions percolating through the rock. The best known specimens come from St Gotthard in Switzerland, where very thin, flat crystals are found. The quartz veins at Tremadoc in North Wales are also famous for brookite. In the USA, Arkansas's Magnet Cove quartzite yields very high quality brookite crystals.

Brookite

> **Crystal system**: Orthorhombic.
> **Crystal habit**: Flat, platelike crystals – except at Magnet Cove, where they tend to be fatter and more complex in shape.
> **Colour**: Dark reddish brown to greenish black.
> **Lustre**: Adamantine to submetallic.
> **Streak**: Yellow or white grey.
> **Hardness**: 5.5–6.
> **Cleavage**: Poorly prismatic.
> **Fracture**: Subconchoidal, uneven.
> **Specific gravity**: 3.9–4.1.
> **Notable locations**: Gwynedd, Wales; Carinthia, Salzburg, Tyrol, Austria; Liguria, Italy; Brittany, Savoie, France; St Gotthard, Switzerland; Urals, Russia; Nova Scotia; Magnet Cove, Arkansas; Somerville, Massachusetts; Ellenville, New York; Franklin, New Jersey; Pima Co, Arizona.

Identification: Brookite can usually be identified by its long, black, plate-like, roughly hexagonal crystals.

HYDROXIDES

Hydroxides are minerals formed from a combination of a metal with oxygen and hydrogen. They form at low temperatures typically in changing mineral-rich waters or in hydrothermal veins. The hydroxide minerals include the iron ore limonite, as well as brucite, gibbsite (an ingredient of the major aluminium ore bauxite), the manganese ore psilomelane and stibiconite.

Brucite

$Mg(OH)_2$ Magnesium hydroxide

Brucite is named after American mineralogist A Bruce, who first described it in 1814 in New Jersey, USA. It forms when ultramafic, magnesium-rich rocks and minerals, such as olivine and periclase, are altered during contact with hot, watery solutions – especially during serpentinization, when magnesium silicates are changed to the mineral serpentine. Serpentinization is common on the sea floor where ultramafic rocks come into contact with seawater, but brucite is common wherever there are serpentinized rocks, chlorite and talc schists, phyllites and marbles. Brucite is usually soft and breaks easily into layers because its molecules stack up in sheets of octahedrons of magnesium hydroxide. Brucite layers are often incorporated into crystals of serpentine, dolomite and talc. It was brucite's distinctive layer structure that gave the revolutionary architect Buckminster Fuller the idea for the ingenious octet truss structure now widely used in buildings.

Identification: Brucite typically forms soft white masses that flake off in plates when scraped with a fingernail.

Brucite plates

Crystal system: Trigonal hexagonal.
Crystal habit: Typically forms ill-defined plates. Also forms fibres and foliated masses.
Colour: White or colourless, sometimes with tinges of grey, blue or green.
Lustre: Vitreous or waxy; pearly lustre on cleavage surfaces.
Streak: White.
Hardness: 2–2.5.
Cleavage: Perfect in one direction, forming plates.
Fracture: Uneven, sectile (can be cut with a knife).
Specific gravity: 2.4.
Notable locations: Unst (Shetland Islands), Scotland; Filipstad, Nordmark, Jakobsberg, Sweden; Aosta, Italy; Urals, Russia; Asbestos, Quebec; Tilly Foster Mine (Brewster), New York; Lancaster County, Pennsylvania; Wood's Mine, Texas; Gabbs, Nevada.

Gibbsite

$Al(OH)_3$ Aluminium hydroxide

Identification: Gibbsite is soft and white and breaks easily into plates. It also has a very distinct clay smell.

Gibbsite is an important ore of aluminium, and one of the main ingredients of bauxite. Bauxite is often mistakenly thought of as consisting of a single mineral; however, it is actually a mixture of various aluminium minerals, one of which is gibbsite. Gibbsite is also one of the main minerals in tropical and subtropical soils. It usually forms when rocks rich in aluminium are weathered in the hot, wet conditions that are typical of the tropics, especially in densely forested regions. Occasionally, crystals of gibbsite form directly in aluminium-rich hydrothermal veins. Like brucite, gibbsite molecules are arranged in octahedral layers, making it soft and liable to break off in plates. Gibbsite also occurs as microscopic layers in the molecules of other minerals, such as illite and kaolinite.

Crystal system: Monoclinic.
Crystal habit: Typically massive but rare flat crystals are found. May form pisolites and other concretions.
Colour: White or colourless sometimes with tinges of grey, blue or green.
Lustre: Vitreous to dull; pearly lustre on cleavage.
Streak: White.
Hardness: 2.5–3.5.
Cleavage: Perfect in one direction, forming plates.
Fracture: Flexible, tough.
Specific gravity: 2.4.
Other characteristics: Has noticeable clay smell.
Notable locations: Vogelsberg, Germany; Gant, Hungary; Les Baux, France; Lavrion, Greece; Guyana; Dundas, Tasmania; Brazil; Surinam; Saline County, Arkansas; Alabama.

Psilomelane

Typically Ba(Mn+2)(Mn+4)$_8$O$_{16}$(OH)$_4$ Barium manganese oxide hydroxide

Psilomelane gets its name from the Greek words for 'smooth' and 'black' because of its tendency to form smooth black knobbly masses. Psilomelane is a general term used to describe various hard, massive mixtures of manganese hydroxides. Psilomelanes are mostly the mineral romanechite (from Romaneche-Thorins in France), which is essentially barium manganese hydroxide. Psilomelanes form when manganese-rich rocks are weathered – typically as concretions deposited by groundwater, or in swamp or lake bed deposits and clays. Psilomelanes also replace other minerals such as manganese carbonates and silicate minerals in limestones. Branchlike growths of psilomelane called manganese dendrites form along bedding planes between rocks. Psilomelane is often intergrown with iron oxides such as hematite and goethite. It also occasionally occurs in alternating black and grey bands with grey pyrolusite, making a very attractive stone when polished. Like pyrolusite, psilomelane is an important ore of manganese.

Identification: Psilomelanes are hard to tell from pyrolusite, but pyrolusite is generally shinier and softer, leaving marks on fingers. The key presence of barium is hard for an amateur to ascertain.

 Crystal system: Monoclinic.
Crystal habit: Forms in knobbly masses, nodules, concretions, dendrites, and tufts of hairlike fibres.
Colour: Metallic black to grey.
Lustre: Submetallic to dull.
Streak: Black or brownish black.
Hardness: 5–5.5.
Cleavage: None.
Fracture: Uneven.
Specific gravity: 4.4–4.5.
Other characteristics: Sometimes banded with the mineral pyrolusite.
Notable locations: Cornwall, England; Schneeburg (Harz Mountains), Germany; Romaneche-Thorins (Saône-et-Loire), France; Tekrasni, India; Ouro Prêto (Minas Gerais), Brazil; Wythe County, Virginia; Keweenaw, Michigan; Tucson, Arizona; Sodaville, Nevada.

Knobs and nodules
Most sedimentary rocks are not all one even mass. Instead, they often contain hard lumps of particular minerals that grew around a tiny nucleus – such as a shell fragment – as the rock was forming. Lumps have a host of common names such as crogs, knots, beetlestones, yolks and countless others. Some geologists call them concretions if they are lozenge-shaped with a smooth surface, and nodules if they are round and knobbly. The smallest concretions are oolites: sand-sized balls of calcite that form the basis of rocks such as oolitic ironstone (above). Gibbsite often forms oolites. Bigger concretions are called pisolites, from the Greek words for pea and stone. Pisolites are typically pea-sized, like the pisolites in bauxite, but can be the size of a pumpkin or bigger. The mineral composition of a concretion usually depends on the nature of the surrounding sediment and the conditions in which the sediment formed. Quite often, there are thin layers of concretions along bedding planes, where they formed when sedimentation was interrupted.

Stibiconite

Sb$_3$O$_6$(OH) Antimony oxide hydroxide

Stibiconite is a dirty white to yellowish mineral that forms when antimony-rich minerals such as stibnite are altered by exposure to warm air and oxidized. Stibiconite crystals are often pseudomorphs of stibnite – that is, minerals that adopt the same crystal structure as another as they form. In this case the stibnite crystals are replaced with stibiconite one by one in situ, as oxygen replaces sulphur, so that the shape of the stibnite crystal is retained. Because stibnite often forms spectacular swordlike crystals arranged in radiating clusters, so stibiconite forms matching spiky clusters. The only visible difference is that stibnite is a shiny steel grey whereas stibiconite is a dull dirty white in colour.

 Crystal system: Isometric.
Crystal habit: Typically earthy masses and crust but also as stibnite pseudomorphs in sword-like clusters.
Colour: White or grey tinged with brown or yellow.
Lustre: Earthy.
Streak: White.
Hardness: 4–5.5.
Cleavage: None.
Fracture: Earthy, except as pseudomorphs of stibnite, then brittle.
Specific gravity: 3.5–5.9.
Notable locations: Goldkronach, Germany; Oruro, Bolivia; Huaras, Peru; San Luis Potosi, Mexico; Wolfe County, Quebec; Nevada.

Identification: Stibiconite pseudomorph clusters look like stibnite, but they are a dull, dirty white colour.

Stibiconite clusters

OXIDE GEMS

Some oxides form deep down within the Earth's crust, in magma, or in very hot mineral veins. Oxides that form in this way include some of the toughest and most beautiful minerals, including corundum and its gem varieties ruby, sapphire and star sapphire. The rare gem taafeite and spinel in all its colour forms are also formed in this way as well as 'cat's eye' chrysoberyl.

Corundum

Al₂O₃ Aluminium oxide

Corundum is one of the world's hardest minerals – only diamond is harder. Powdered corundum is used to coat paper to make an abrasive paper finer than sandpaper, and blocks of corundum are used for knife sharpening. The main type of corundum used for abrasive paper and knife sharpeners is called emery, which forms masses underground. When exposed to the atmosphere it erodes and crumbles into a powder called black sand, which derives its colour from traces of iron. Not all industrial 'emery' is mined; some is made from ground-up crystals of pure corundum, or it can be manufactured synthetically. When it forms crystals, pure corundum is brown and translucent. The crystals are typically small, and shaped like two six-sided pyramids joined at the base.

Tapering corundum crystal set in syenite rock.

Identification: The simplest way to identify corundum is to use the Mohs test to confirm its extreme hardness (9).

Tablet

Spindle

Barrel

Crystal system: Hexagonal.
Crystal habit: Typically double-pyramid hexagons or flat hexagons, often elongated into barrels, or spindles. Also occurs as grains and masses.
Colour: Pure corundum is brown or brownish white; emery is black.
Lustre: Vitreous, adamantine.
Streak: White.
Hardness: 9.
Cleavage: None but splits basally and in two other ways.
Fracture: Conchoidal, uneven.
Specific gravity: 4+.
Notable locations: Myanmar (Burma); Thailand; Sri Lanka; numerous locations in Africa; North Carolina; Montana.

Ruby

variety of corundum, Al₂O₃ Aluminium oxide

Dubbed Rajnapurah, the king of gems, by the ancient Hindus, ruby is one of the most sought after of all precious gems. Large transparent rubies are even rarer and more valuable than diamonds. Rubies are a variety of corundum and get their rich red colour from traces of chromium. Traces of iron can make rubies slightly brown. The most sought-after rubies are deep blood red with a slightly purplish hue. These are known as 'Pigeon's blood ruby' or 'Burmese' ruby. For centuries, rubies like these have come from Mogok and Mong Hsu in Myanmar (Burma). Burmese rubies were originally embedded in marble and other metamorphic rock, but because ruby is so hard, the rubies have survived the breakdown of their parent rock and are typically found in river deposits. The majority of the good rubies today are brownish in colour and come from Thailand.

Rubies fluoresce red with long-wave UV

Identification: Rubies embedded in schists like this are unmistakable, but they are quite rare.

Tablet

Spindle

Barrel

Crystal system: Hexagonal.
Crystal habit: Found as prisms and double-pyramids.
Colour: Shades of red, often with a violet or purple tinge.
Lustre: Adamantine.
Streak: None.
Hardness: 9.
Cleavage: None.
Specific gravity: 4.
Notable locations: Mogok, Mong Hsu, Myanmar (Burma); Thailand; Sri Lanka; Cambodia; India; Pakistan; Tadjikistan; Madagascar; Umba River, Tanzania.

Sapphire

variety of corundum, Al₂O₃ Aluminium oxide

Corundum gems come in many colours besides the red of rubies. Geologists call all these non-red, coloured corundums sapphires. For jewellers, though, sapphires are only the most stunning blue corundum gems. These stones are coloured blue by traces of the titanium oxide mineral ilmenite. Different impurities give corundum a whole variety of colours including pink, yellow, orange and green. In the past, these other colours of sapphire were known as 'oriental' versions of other gems of the same colour, so green corundums were known as oriental emeralds, for instance. Nowadays, however, green corundums are simply called green sapphires. Only orange-pink sapphire has its own recognized name, padparadschah. The most famous sapphires are the cornflower-blue stones from Kashmir. Nowadays most sapphires come from Australia, although they also come from Myanmar (Burma), Thailand and Sri Lanka.

Identification: Like other corundum gems, sapphires are typically found in river deposits, where they stand out because of their blue colour and hardness.

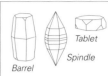

Barrel   Tablet   Spindle

Crystal system: Hexagonal.
Crystal habit: Found as prisms and double-pyramids.
Colour: Shades of blue.
Lustre: Adamantine.
Streak: None.
Hardness: 9.
Cleavage: None.
Specific gravity: 4.
Notable locations: Kashmir; Mogok, Myanmar (Burma); Thailand; Ratnapura, Sri Lanka; Australia; Judith Basin County, Montana.

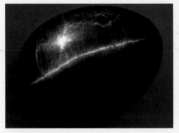

Stars and cat's eyes
Some gemstones display light effects created by light reflecting off the mineral's internal structure, or inclusions of other mineral crystals inside. These light effects show up well only when the stone is cut and polished in the right way. Asterism is the star effect caused by the inclusion of criss-crossing needle-like crystals. It is best known in star sapphires, but also occurs in rubies and diopsides. Another effect seen in some sapphires is chatoyancy, or 'cat's eye'. Here, mineral inclusions create a bright band across the stone that makes the stone look a little like a cat's eye. A number of other minerals exhibit light effects. Opalescence manifests as a milky shimmer, created as light is refracted and reflected by tiny silica spheres under the surface of opals and other minerals. Adularescence is the floating blue cloud effect, as seen in moonstone and some transparent opals. Labradorescence is the shimmering play of colour in labradorite as light is bounced around through lamellae (leaves) in the crystals. It is like the bronzy 'Schiller' effect created by internal plates in orthopyroxene.

Star sapphire

variety of corundum, Al₂O₃ Aluminium oxide

Although flawless, transparent sapphires are the most valuable, jewellers also prize sapphires that contain needle-like crystals of rutile. The rutile crystals often grow in three directions. White or silver light is reflected from these needles in such a way that it looks as if the stones contain a six-pointed (or occasionally twelve-pointed) star. Sapphires like these are known as star sapphires, and the effect is called asterism. In very rare black and gold star sapphires, asterism is caused not by rutile, but by hematite and ilmenite crystals. The sapphires themselves are black or dark brown, while the star is a deep golden colour.

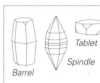

Barrel   Tablet   Spindle

Crystal system: Hexagonal.
Crystal habit: Found as prisms and double-pyramids.
Colour: Various shades of red.
Lustre: Adamantine.
Streak: None.
Hardness: 9.
Cleavage: None.
Specific gravity: 4.
Notable locations: Mogok, Myanmar (Burma); Tanzania; black and gold star sapphires are found only in Chanthaburi province, eastern Thailand.

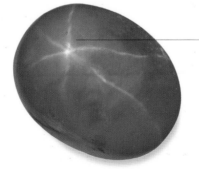

When reflection of white or silver light produces a six-pointed star like this, the effect is known as asterism

Identification: The star effect is immediately obvious when the stones are polished *en cabochon* – in other words, into a smooth, flat oval shape.

COMPLEX OXIDES

While some oxides are simply a combination of a metal and an oxide or hydroxide, a number of other oxides are more varied and complex, including at least two different metals in their chemical make-up. Aluminium, iron, manganese and chromium frequently form minerals containing two other metals in their chemical composition.

Perovskite

CaTiO₃ Calcium titanium oxide

Perovskite is one of the world's most abundant minerals, since it makes up the bulk of the Earth's mantle, which is 80 per cent of Earth's volume. Perovskite is less abundant in surface rocks, but still common. Here, perovskite is commonly found in mafic igneous rocks low in silica and aluminium, such as nepheline syenites, carbonatites, kimberlites and melilites, and in some schists. It was discovered in the Russian Urals in 1839 by Gustav Rose, who named it after Russian mineralogist Count Lev von Perovski. Mineral collectors look for its apparently cubic crystals. They are only apparently cubic, because although box-shaped, their internal structure is actually orthorhombic (longer in one direction), not isometric (equal length in all directions). Mining engineers seek out the massive variety of perovskite to use as an ore for titanium. It can also be a source of niobium and thorium, and rare earth metals like cerium, lanthanum and neodymium.

Identification: The simplest way to identify perovskite is from the nature of the local rocks. Dark-coloured, box-shaped crystals in syenites and carbonatites are likely to be perovskite.

Calcite

Schist rock

Perovskite

Crystal system: Orthorhombic (pseudocubic).
Crystal habit: Box-shaped crystals, plus blades, grains and masses.
Colour: Dark grey or brown to black. Occasionally with orange or yellow tinge.
Lustre: Submetallic to adamantine, greasy or waxy.
Streak: White to grey.
Hardness: 5.5.
Cleavage: Imperfect.
Fracture: Subconchoidal, uneven.
Specific gravity: 4.
Notable locations: Medelpad, Sweden; Zermatt, Switzerland; Lombardy, Italy; Eifel, Germany; Gardiner complex, Greenland; Zlatoust, Urals, Russia; São Paulo, Brazil; Riverside County, San Benito County, California; Bearpaw Mountains, Montana; Magnet Cove, Arkansas.

Spinel

MgAl₂O₄ Magnesium aluminium oxide

Identification: Spinel is usually identified by its twinned crystals or, failing that, its hardness in the Mohs test. It gets its name from the crystal's sharp points.

Spinel crystal

Spinels are a group of complex oxide minerals, including magnetite, franklinite and chromite, that have a similar structure. The best known is the gem spinel, which commonly forms in metamorphic rocks, especially marbles and calcium-rich gneisses, but can also form in pegmatites and as phenocrysts (spots) in lava. Gem spinel comes in many colours, such as green gahnite and black galaxite, but it is typically red and rivals the colour of ruby. In fact, many gems thought to be rubies have proved to be spinels. The most famous example is the great Black Prince's Ruby, set in the British imperial state crown. Spinel and ruby are chemically quite similar – the former is essentially magnesium aluminium oxide while ruby is aluminium oxide – and both get their red colour from chromium. Spinel typically forms in eight-sided crystals, but it often forms twins in a way no other mineral does, with two mirror-image planes.

Crystal system: Isometric.
Crystal habit: Typically octahedral, but can be found as dodecahedrons (12 sides) and other isometric forms. Also found as rounded grains in river deposits.
Colour: Typically red, but also green, blue, purple, brown or black.
Lustre: Vitreous.
Streak: White.
Hardness: 7.5–8.
Cleavage: None.
Fracture: Conchoidal to uneven.
Specific gravity: 3.6–4.
Notable locations: Sweden; Italy; Madagascar; Turkey; Myanmar (Burma); Sri Lanka; Afghanistan; Pakistan; Lake Baikal, Russia; Brazil; Amity, New York; Franklin mine, New Jersey; Galax, North Carolina.

Franklinite

(Zn,Fe,Mn)(Fe,Mn)₂O₄ Zinc iron manganese oxide

Franklinite is a mineral of the spinel group rather similar to magnetite, one of the chief ores of iron. Like magnetite it is magnetic, but franklinite's magnetism is quite weak, and this weak magnetism is a simple way of telling these two similar-looking minerals apart. Franklinite gets its name from the famous Franklin mine in New Jersey, USA, where it was discovered in 1819. It has since been found in neighbouring mines such as the Sterling mine in Ogdensburg, but nowhere else in the world in any quantity. At Sterling and Franklin, it is abundant, commonly occurring in either thick pure beds or mixed in with zinc minerals (zincite and willemite) in crystalline limestone when it is mined as an ore of zinc (with a 5 to 20 per cent zinc content). After the zinc is extracted, the residue is used to make *spiegeleisen*, or 'mirror iron', a manganese-iron alloy important in steelmaking.

Identification: The best indicator of franklinite is where it is found, at Franklin, New Jersey, but its dark colour and weak magnetism help to confirm the identity.

Franklinite

Crystal system: Isometric.
Crystal habit: Typically large octahedral crystals with rounded edges, but it has been found with 12 and even 16 sides. It is more commonly found as grains and masses.
Colour: Black.
Lustre: Metallic, submetallic.
Streak: Brownish black to reddish brown.
Hardness: 5.5–6.5.
Cleavage: None.
Fracture: Conchoidal to uneven.
Specific gravity: 5–5.2.
Other characteristics: Slightly magnetic.
Notable locations: Found only in the Franklin and Sterling Hill mines, Sussex County, New Jersey.

The jewel mine
No other single place in the world can boast such an extraordinary array of rare and fascinating minerals as the famous Franklin and Sterling Hill mines in Sussex County, New Jersey. Over 300 different minerals have been found there, including 60 new ones. Manganese and zinc are the key metals but they were transformed a billion years ago by the cataclysmic tectonic events that threw up the Appalachian Mountains and later by hydrothermal action, creating an astonishing range of minerals. Rare manganese and zinc oxides such as franklinite, and silicates such as willemite, are characteristic of the area. Franklin is especially renowned for its fluorescent minerals such as willemite, esporite, clinohedrite and hardystonite, which turn the mines into magic caves when lit by ultraviolet light. Visitors to the mine are given a dirt bucket and encouraged to pan for gems, particularly rubies, rhodolites, sapphires and garnets. They get to keep any they find. The mines were first opened in the 18th century, and reached their zenith in the late 19th century. They were mined partly for franklinite, which is an iron ore, but especially for zinc, and played a key part in America's Industrial Revolution. Bright red zincite was often found in large masses and lenses in black franklinite. One spectacular 8-tonne mass found in the 1840s was almost pure zincite, which is 80 per cent zinc metal. This great lump of ore was shipped at great expense to the Great Exhibition in London's Crystal Palace in 1852, where it attracted a great deal of attention and won a prize. The Franklin mine finally closed in 1954 and the Sterling Hill mine in 1986.

Chrysoberyl

BeAl₂O₄ Beryllium aluminium oxide

Chrysoberyl is an extremely hard gemstone typically found in pegmatite dikes, mica schists and where granites meet mica schists, and in river gravels. It looks a little like gem beryl and its name is derived from the Greek *chrysos* meaning 'gold' – that is golden beryl. There are three varieties of chrysoberyl: the less popular greenish clear chrysoberyl, cymophane and the highly sought-after alexandrite. Yellow-green or brown cymophane is the most distinctive of all gems displaying chatoyancy, the cat's eye effect created by inclusions of other minerals. Alexandrite was named in honour of the Russian Tsar Alexander II. It is coloured green under natural light as a result of traces of chromium, but, uniquely, it changes in colour in different lighting conditions, surprisingly changing to crimson in artificial light.

Crystal system: Orthorhombic.
Crystal habit: Often elongated prisms and tablets, or twinned either in v-shapes or more complex forms.
Colour: Yellow, green or brown. Alexandrite becomes violet-red in artificial light.
Lustre: Vitreous.
Streak: White.
Hardness: 8.5.
Cleavage: Fair in one direction, poor in another.
Fracture: Uneven to conchoidal.
Specific gravity: 3.7+.
Other characteristics: Pleochroic (displaying colours from different viewing angles).
Notable locations: Urals, Russia (largely worked out); Sri Lanka; Myanmar (Burma); Tanzania; Brazil; Colorado; Connecticut.

Chrysoberyl

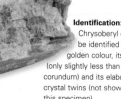

Identification: Chrysoberyl can often be identified by its golden colour, its hardness (only slightly less than corundum) and its elaborate crystal twins (not shown in this specimen).

SALTS

Halides are minerals that form when a metal combines with one of the five elements called halogens – fluorine, chlorine, bromine, iodine. (There is another halogen, astatine, but this is never found naturally.) The best known by far is halite or rock salt. Like rock salt, all the halides, including sylvite, chlorargyrite, and mendipite, are salts. Salts dissolve easily in water, which is why many halides occur only in special conditions. However, rock salt is so abundant it is found in huge deposits the world over.

Halite

NaCl Sodium chloride

Halite is common salt or rock salt and is the source of the salt we use on our tables. It is forming all the time as water evaporates from salty lakes. Most salt is mined from thick underground beds left behind long ago when ancient oceans evaporated. When halite crystallizes, it usually forms cube-shaped crystals, but it dissolves so easily in water that large crystals are rare. Where they do occur, they can be white, orange and pink. Some colour changes are created by bacteria and some are created by exposure to natural radiation. Gamma rays, for instance, turn halite first amber, then deep blue. The blue colour comes from specks of sodium metal, created when radiation knocks electrons towards sodium ions. When halite does form crystals, it often takes unusual habits, such as hopper crystals. Hopper crystals have a dent in each face that makes them look like the hoppers on a mine conveyor belt. The indentation occurs because the edges of the crystal grow faster than the centres of the faces.

Identification: Halite can be identified by its salty taste, but there is a risk of poisoning if you make a mistake, so it is better to identify it by its softness and the cube shape of its crystals.

Crystal system: Isometric.
Crystal habit: Mainly cubes or in massive sedimentary beds, but also grains and fibres. Also forms hopper crystals.
Colour: Clear or white, but can be orange, pink, purple, yellow or blue.
Lustre: Vitreous.
Streak: White.
Hardness: 2.
Cleavage: Perfect in three directions forming cubes.
Fracture: Conchoidal.
Specific gravity: 2.1+.
Other characteristics: Soluble in water.
Notable locations: Stassfurt, Germany; Salzburg, Austria; Galicia, Poland; Mulhouse, France; Uyuni, Bolivia; Bogota, Colombia; Great Salt Lake, Utah; Searles Lake, California; Gulf of Mexico; Retsof, New York.

Sylvite

KCl Potassium chloride

Sylvite is chemically a chloride very similar to halite, and like halite it formed in massive deposits on ancient sea beds. But while halite is sodium chloride, sylvite is potassium chloride. Sylvite's potassium content has long made these ancient sylvite beds a major source of 'potash' for fertilizers. A quarter of the world's sylvite is mined in Saskatchewan in Canada. Sylvite crystals do occur, but they are quite rare. Although basically white, they often have an attractive reddish tinge. Sylvite and halite can be so similar that it can be hard to tell them apart. The corners of sylvite's cubes are truncated (cut off) more often than with halite. With massive beds, another way to tell them apart is to slice them with a knife: halite powders but sylvite does not.

Identification: Sylvite forms pale reddish white cubic crystals that look like halite, but more often have truncated corners.

Crystal system: Isometric.
Crystal habit: Cubes or octahedral (or rather a cube with flattened corners). Commonly massive and granular.
Colour: Colourless or white, tinged red, blue or yellow.
Lustre: Vitreous.
Streak: White.
Hardness: 2.
Cleavage: Good in three directions forming cubes.
Fracture: Uneven.
Specific gravity: 2.
Other characteristics: Crystal corners often cut off.
Notable locations: Mount Vesuvius, Italy; Stassfurt, Germany; Spain; Kalush, Russia; Saskatchewan; New Mexico; Texas; Kern County, California.

Chlorargyrite

AgCl Silver chloride

Chlorargyrite, once called cerargyrite, is a silver mineral that forms when the surface of silver ores is exposed to the air and oxidized. Where there is plenty of chlorine present, such as in desert conditions, the silver in the ores combines to make chlorargyrite, which is silver chloride. Where there is more bromine around, the silver forms bromargyrite, silver bromide. In both cases, the effect is to concentrate the silver in a process called 'supergene enrichment', making the ore much more economically viable than it would otherwise be. As a result, chlorargyrite was once an important ore of silver in places like Mexico, Peru, Chile and Colorado, USA. These deposits have almost been completely worked out, and there are now very few unworked veins close enough to the surface to provide good specimens. Moreover, chlorargyrite rarely forms good crystals, so when they are found, they are much treasured.

Identification: Chlorargyrite can usually be identified by its tendency to turn dark when exposed to light and by the ease with which it can be sliced. It can form horn-like masses, which earned it its old name of horn silver.

Chlorargyrite may appear as brown-grey masses

Crystal system: Isometric.
Crystal habit: Rare cubes; more commonly massive crusts or columns.
Colour: Colourless when pure on fresh surfaces; turns pearly grey, brown or violet-brown on exposure to light.
Lustre: Resinous or adamantine.
Streak: White.
Hardness: 1.5–2.5.
Cleavage: Poor.
Fracture: Subconchoidal.
Specific gravity: 5.5–5.6.
Other characteristics: Crystals darken on exposure to light. Plastic and sectile.
Notable locations: Harz Mountains, Germany; New South Wales, Australia; Atacama, Chile; Peru; Mexico; Treasure Hill, Comstock Lode, Nevada; Colorado; San Bernardino County, California.

Underground salt city
Salt has been a precious commodity for health and a preservative since the earliest times, and salt mines are among the oldest of all mines. The famous Wieliczka mine in southern Poland near Krakow has been worked since the 13th century and has over 200km/124 miles of underground passages, and 2,000 chambers hundreds of metres/yards below the ground in the thick salt beds. Over the centuries, miners have carved churches, altars, bas-reliefs, giant statues – even chandeliers (above) – out of the glistening white salt. Wieliczka is now a World Heritage site, and one of the most popular tourist destinations in southern Poland. Yet salt is very soluble, and even the water vapour in the ventilation air is enough to dissolve the salt. So the amazing features are slowly being lost.

Mendipite

Pb₃Cl₂O₂ Lead chlorate

Discovered in 1839 in the Mendip Hills in south-west England, mendipite combines lead with chlorine and oxygen. It is a rare white, grey or pinkish white mineral that forms around volcanic vents and in hydrothermals. It is commonly found in association with calcite, cerussite, malachite, manganite, pyrolusite and pyromorphite, and masses of white mendipite are often found scattered in black masses of manganese oxide minerals miners call 'wad'. Deposits of mendipite are usually fibrous or column-like masses, often with radiating needles. Mendipite is both soft and dense.

Wad

Pink mendipite mass

Crystal system: Orthorhombic.
Crystal habit: Typically fibrous masses, in columns, or radiating like a star.
Colour: Bronze.
Lustre: Adamantine, pearly.
Streak: White.
Hardness: 2.5–3.
Cleavage: Distinct in two directions.
Fracture: Uneven fracture giving small, conchoidal fragments.
Specific gravity: 7–7.2.
Notable locations: Mendip Hills (Somerset), England; Värmland, Sweden; Lavrion, Greece; Ruhr, Sauerland, Germany; Hampshire County, Massachusetts; Chester County, Pennsylvania.

Identification: Mendipite can be identified by its pinkish white fibrelike crystals, typically set in manganese oxide. Separate masses are both soft and very heavy.

HYDROXIDES AND FLUORIDES

Among the other halides or salts are minerals that contain the halogen element fluorine. These fluorides, many of which were found originally in Greenland, include cryolite and jarlite. Other halides include hydroxides that are made with a halogen element such as chlorine. These include atacamite and boleite, both of which are often strikingly coloured.

Cryolite

Na_3AlF_6 Sodium aluminium fluoride

Cryolite was discovered by Danish geologists in Greenland in 1794, but was long known to the Eskimos, who saw it as a kind of ice because it melts so easily, even in a candle-flame. It was this ice-like quality that earned its name, from the Greek for ice and stone. It is called *Eisstein* (ice-stone) in Germany. It usually occurs in colourless or snow-white masses, often tinged brown or red by iron oxide, and occasionally black. It is normally translucent with a waxy lustre, but becomes virtually transparent when immersed in water. Cryolite is found in pegmatite veins at Ivigtut in south-west Greenland (now exhausted) and only a few other places, but it is vital to the aluminium industry. It is used as a flux when melting the aluminium in bauxite to lower the melting temperature and help draw out impurities. It is also used to make soda, and in the manufacture of very tough glass and enamelled ware. Much cryolite is now made artificially.

Identification: Cryolite can often be identified by its pseudocubic crystals.

Small brown siderite cleavages

Colourless cryolite

Crystal system: Monoclinic.
Crystal habit: Usually massive and, rarely, as pseudocubic crystals with deep striations.
Colour: Clear or white with red or brown tinges, but can also be black or purple.
Lustre: Waxy.
Streak: White.
Hardness: 2.5–3.
Cleavage: Good, can break as if into cubes.
Fracture: Uneven.
Specific gravity: 2.95.
Other characteristics: Melts very easily. Normally translucent or transparent. It seems to vanish in water because its refractive index is so close to that of water.
Notable locations: Ivigtut, Greenland; Spain; Miyask (Ilmen Mountains), Russia; Mont Saint-Hilaire, Francon Quarry (Montreal) Quebec; Yellowstone, Wyoming; Pikes Peak, Colorado.

Atacamite

$Cu_2Cl(OH)_3$ Copper chloride hydroxide

Atacamite

Atacamite is a bright green copper chloride mineral. It gets its name from the Atacama Desert in Chile, where some of the best specimens are found. The Atacama is one of the world's driest places, receiving on average less than 1cm/½in of rain in a year. Atacamite forms only in very dry places where copper sulphide minerals are exposed to the air. Typically, atacamite forms when these copper minerals are oxidized in very arid conditions. It is commonly associated with malachite, cuprite, limonite and azurite, as well as rarer minerals such as chrysocolla, connellite, pseudomalachite, libethenite, cornetite and brochantite. Atacamite has an unusual property in that it absorbs water very rapidly, and in the days before blotting paper was invented, atacamite was often used for drying ink blots.

Blue chrysocolla

Identification: Atacamite is typically a distinctive dark green colour that forms thin needle-like crystals or as a coating on other minerals. It forms only in very dry places.

Crystal system: Orthorhombic.
Crystal habit: Includes slender needles, prisms or tablets and also fibres.
Colour: Dark green or emerald green.
Lustre: Vitreous.
Streak: Light green.
Hardness: 3–3.5.
Cleavage: Perfect in one direction.
Fracture: Conchoidal, brittle.
Specific gravity: 3.75+.
Other characteristics: Crystals often have striations.
Notable locations: Atacama, Chile; Mount Vesuvius, Italy; Wallaroo, South Australia; El Boleo (Baja California), Mexico; Pinal County, Arizona; Tintic, Utah; Majuba Hill Mine, Nevada.

Jarlite

Na(Sr,Ca)₃Al₃F₁₆ Sodium aluminium fluoride

A glassy-looking mineral, jarlite is named after Carl Frederik Jarl (1872–1951), once President of the Danish Cryolite Company, who discovered the mineral. It is one of a number of fluoride minerals found at Ivigtut in Greenland besides cryolite, and it typically forms in vugs (pockets) in cryolite deposits along with other fluorides. Jarlite usually occurs in masses, but can also occur as simple, very small monoclinic crystals, typically no bigger than 1mm/¹⁄₂₄in long or so. Sometimes the crystals form in almost flat plates or sheaves. They can also form radiating layers inside druses, where they are intermixed with white barite, brick-red iron stain and, occasionally, the white powdery mineral gearksutite. Jarlite is typically white or pinkish white. A variety of the mineral called metajarlite is more greyish.

Identification: Jarlite is perhaps best identified by its glassy lustre and when it forms sheaves of platy white crystals.

Crystal system: Monoclinic.
Crystal habit: Jarlite typically forms flat plate- or sheaflike crystals. It is often also found as round aggregates or in masses.
Colour: White to greyish white to pinkish white.
Lustre: Vitreous or waxy.
Streak: White.
Hardness: 4–4.5.
Cleavage: Poor.
Fracture: Uneven, with flat surface breaking in an uneven way.
Specific gravity: 3.87.
Other characteristics: Forms in cavities in cryolite pegmatite.
Notable locations: Ivigtut (Arsuk Fjord), Kitaa, Greenland.

Minerals from the frozen north
The ancient rocks of Greenland have yielded a remarkable number of rare minerals. Although the most famous mines here are now all but exhausted, they have produced some fabulous specimens in the past and many collections contain some beauties from Greenland. The best known of Greenland's mineral sites is Ivigtut on the Arsuk Fjord in Kitaa in south-west Greenland. Hundreds of different minerals have been found here, including rare sulphides and sulphosalts such as eskimoite, matildite, vikingite and gudmundite. Ivigtut is particularly renowned for its halide minerals, found in pegmatites. The area was mined from 1854 until 1987 for its cryolite, which was used as a flux in aluminium smelting. Other halides from Ivigtut include acuminite, böggildite, bøgvadite, cryolithionite, gearksutite, jarlite, pachnolite, prosopite, ralstonite, stenonite and thomsenolite. For each of these, it is the type locality (the source in which the mineral was originally found).

Boleite

KPb₂₆Ag₉Cu₂₄Cl₆₂(OH)₄₈ Hydrated lead copper silver chloride hydroxide

Named after the place it was first discovered – Boleo in Baja California, Mexico – boleite is a rare halide mineral that has a very complex chemistry. Each molecule contains multiple atoms of lead, copper, silver and chlorine as well as numerous hydroxide groups and three molecules of water. Boleite is a very minor ore of copper, silver and lead, but it is valued by collectors because of its distinctive indigo-blue coloured crystals, which are sometimes cut to make gemstones. Boleite crystals are unusual because they look like cubes, but are actually always twinned rectangular crystals that pair in such a way that they look like cubes. Crystals like these are called pseudocubes.

Crystal system: Tetragonal.
Crystal habit: Rectangular crystals typically form pseudocubes.
Colour: Indigo blue to deep navy blue.
Lustre: Vitreous, pearly.
Streak: Greenish blue.
Hardness: 3–3.5.
Cleavage: Perfect in one direction.
Fracture: Uneven and brittle.
Specific gravity: 5.
Other characteristics: Notches or interpenetrating angles can be seen in some specimens, revealing their true twinned nature.
Notable locations: Broken Hill (New South Wales), Australia; El Boleo (Baja California), Mexico; Mammoth District, Arizona.

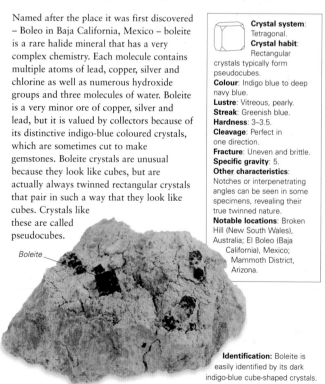

Boleite

Identification: Boleite is easily identified by its dark indigo-blue cube-shaped crystals.

FLUORITE

Fluorite displays a wider range of colours than any other single mineral – varying from the typical purple through to blue, green, yellow, orange, pink, brown and black, with all kinds of pastel shades in between. Yet remarkably, pure fluorite is actually colourless. All fluorite's rainbow of colours comes from traces of impurities of various metals taking the place of calcium in the molecule.

Fluorite

CaF₂ Calcium fluoride

Green fluorite (below): Green is one of the major fluorite colours. It comes in many hues of green, but they all tend to be an acid or mint green, rather than grass green.

Purple fluorite (below): The mauve cubic crystals displayed here are characteristic of the English fluorite mined in the North Pennines, but purple fluorites are found in many other locations, such as, historically, at Regensburg in Germany, and now in China. Purple fluorites tend to be more fluorescent than the green varieties.

Fluorite

Brown sphalerite masses

Fluorite is chemically calcium fluoride, a compound of the elements calcium and fluorine. But even though it contains fluorine, its name does not come from its chemical composition. Instead, it was originally named fluorspar by the famous German mineralogist Georg Agricola in 1546. Agricola named it from the Latin word *fluere*, which means 'to flow', because fluorite melts easily. A spar is the name given by mineralogists to any clear or pale crystal that breaks easily. It is this meltability that has made fluorite valued since Roman times in making steel, glass and enamel as a flux – a substance that lowers the melting point of a material and makes it easier to work. Most industrially extracted fluorite, however, goes to making hydrofluoric acid – the basis for all substances containing fluorine (including for dental care).

Fluorite is usually fairly pure, but can have up to a fifth of the calcium replaced with rare-earth metals such as yttrium and cerium. Yttrium-rich fluorite is called yttrofluorite; yttrium- and cerium-rich fluorite is called yttrocerian fluorite. Fluorite is found in many different environments. In southern Illinois, for instance, it is found in thick veins in limestone beds, where it formed at low temperatures and developed into simple, but many-hued, crystals. In other places it is found around hot-springs, in cavities and pegmatites.

But by far the most typical fluorite-forming environment is metal-rich veins – especially those with lead or silver. Here the higher temperatures encourage fluorite to crystallize in a whole variety of different forms, varying from octahedral to dodecahedral.

Cubic crystals (above): The barely yellowish tinge of this fluorite specimen indicates that it is relatively pure. In fact, clear fluorite can be so good optically that it is sometimes used for microscope lenses because it eliminates colour distortion.

Crystal system: Isometric.
Crystal habit: Typically forms cubes or octahedrons, or both together. Twins are common, and penetration twins often look like two cubes grown together.
Colour: Varies hugely, the widest colour range of any mineral. Colours include intense purple, blue, green, or yellow, reddish orange, pink, white and brown. A single crystal can be multicoloured. Pure, unflawed fluorite crystals are colourless.
Lustre: Vitreous.
Streak: White.
Hardness: 4.
Cleavage: Perfect in four directions forming octahedrons.
Fracture: Flat conchoidal.
Specific gravity: 3–3.3.
Other characteristics: Fluorite is translucent or transparent. It is typically fluorescent blue or, more rarely, green, white, red or violet. Fluorite is also thermoluminescent, phosphorescent and triboluminescent.
Notable locations: Alston Moor (Cumbria), Weardale (Durham), Castleton (Derbyshire), Cornwall, England; Harz Mountains, Wölsendorf (Bavaria), Germany; Tuscany, Italy; Göschenen, Switzerland; Nerchinsk (Urals), Russia; Maharashtra, India; Hunan, China; Naica, Chihuahua, Mexico; Hastings County, Ontario; Elmwood, Tennessee; Rosiclare and Cave-in-Rock (Hardin County), Illinois; Ottawa County, Ohio; Grant County, New Mexico.

Hardness and colour

Although beautifully coloured, fluorite is only rarely used as a gemstone because it is quite soft and fragile. Some collectors regard it as a worthwhile challenge to cut and polish it. Yet fluorite is so consistent in its relative hardness that Friedrich Mohs used it as the standard number 4 on his hardness scale.

Octohedral crystals (above): This green fluorite displays slightly frosted octahedral crystals. These fluorites are less common than those with cubic crystals.

Fluorite gets its extraordinary range of colours because of something called a 'colour centre'. A colour centre is a small region in the crystals where there is a slight defect in the network of atoms. These crystal defects interrupt light in a particular way, absorbing and reflecting only particular wavelengths of light. So fluorite's colour varies according to the pattern of its colour centres. Heat and radiation can both induce the defects that create colour centres – so heat and radiation can change fluorite's colour. The rare earths like yttrium that are found in many fluorites may also influence fluorite's colour – especially under ultraviolet light.

Botryoidal fluorite (above): Sometimes fluorite grows in extraordinary lemon-yellow botryoidal (grapelike) balls like this fluorite ball in a cavity from India.

The best fluorites

The most highly prized fluorites are perhaps the pink to red octahedral fluorites from alpine clefts in the Swiss and French Alps. Here they are often found with smoky quartz, and form from mineral-rich solutions circulating through the rock as it is metamorphosed. Some of the crystals found here are over 10cm/4in long.

The most famous locations for fluorite crystals, however, are in Germany and England. In Germany, small but beautiful fluorite crystals ranging from green to yellow are found in metal-rich veins. In England, the classic locations were the lead and iron mines in Cumbria and the tin mines in Cornwall. The majority of the highest quality English fluorites are purple, and they are among the best at displaying fluorescence. These sources are now largely exhausted. In the USA, the best Illinois fluorites have also been fully exploited, and America's most highly prized fluorites now originate from deposits found in Elmwood, Tennessee.

Fluorite geode (above): The best fluoride crystals often grow on the inside of geodes, and are revealed only when the geode is cracked open.

Fluorite's special glows

Fluorite is not only unique in the range of colours it displays under normal lighting – it also glows in the dark in various different ways. When some minerals are exposed to ultraviolet light, they glow purple, blue or green – no matter what colour they are in normal daylight. This glow is called fluorescence, after fluorite, because fluorite glows like this more readily than any other mineral. Fluorite usually fluoresces blue or green, though it can also glow white, red or violet. Minerals often fluoresce because of tiny impurities they contain. Fluorite is thought to fluoresce because of the traces of uranium and rare-earth metals it contains. Calcite fluoresces bright red when it contains traces of manganese. Fluorite also glows when gently heated – a property called thermoluminescence. And when some fluorites are taken out of direct sunlight and placed in a dark room, they glow – a quality called phosphorescence. Fluorite may even glow when crushed, scratched or rubbed – a quality called triboluminescence – as the pressure distorts colour centres.

Blue John (left): Most fluorites are a single colour, but a few have colour bands in line with the mineral's crystals. One of the best known banded fluorites is Blue John, found only in the Peak District, Derbyshire, England. Blue John gets its name from the French description of the rock, Bleu Jaune, which means 'blue yellow'. It was discovered in the 18th century when miners were exploring caves to look for sources of lead.

CARBONATES

Carbonates are typically light-coloured, often transparent, soft, brittle minerals that tend to dissolve in acids. They form when metals or semi-metals join with a carbonate – a combination of a carbon atom with three oxygen atoms. Some carbonates are brought to the surface from deep in the Earth by hot fluids. Many more form by the alteration of other minerals on the surface, as they are attacked by the mild acidity of the air. There are 80 kinds, including the aragonites and related minerals.

Aragonite

CaCO₃ Calcium carbonate

Aragonite is a common white mineral discovered in Aragon, Spain. Many sea creatures secrete it naturally to make their shells. Geologically, it crystallizes from low-temperature solutions in many sediments, metamorphic rocks such as schist and basic igneous rocks such as serpentinite. It is usually the last mineral to form in veins. It is also a product of the weathering of siderite and ultrabasic and magnesium-rich rocks.

Groups of tapering crystals tinged red by hematite inclusions.

Crystals of aragonite also form around hot springs, or in caves, where they may create stalactites and grow in coral-like shapes called 'flos ferri' (flowers of iron). Aragonite is a polymorph of calcite – chemically identical, but with a different crystal structure. Calcite crystals are trigonal, but those of aragonite are orthorhombic. Some calcite and aragonite crystals are too small to see, requiring complex scientific tests to tell them apart. Aragonite changes to calcite if heated above 400°C/725°F.

Identification: Aragonite and calcite both fizz in dilute acid. It can be hard to tell their crystal system apart, but each has different crystal habits. This crystal is a triplet of prismatic twins that join to make it look hexagonal. It is said to be pseudohexagonal.

Crystal system: Orthorhombic.
Crystal habit: Prismatic with wedge-shaped ends. Often forms triplets of twins that look hexagonal.
Colour: White or colourless. Tinges of red, yellow, orange, brown, green or blue.
Lustre: Vitreous to dull.
Streak: White.
Hardness: 3.5–4.
Cleavage: Distinct one way.
Fracture: Subconchoidal.
Specific gravity: 2.9–3.
Other characteristics: Fizzes in cold dilute hydrochloric acid. It is also fluorescent.
Notable locations: West Cumbria, England; Aragon, Spain; Mount Vesuvius, Italy; Agrigento, Sicily; Styria, Austria; France; Honshu, Japan; Tazoula, Morocco; Tsumeb, Namibia; Australia; Baja California, Mexico (Mexican onyx); Bisbee (Cochise Co), Arizona; Socorro County, New Mexico; White Pine County, Nevada.

Witherite

BaCO₃ Barium carbonate

Identification: Witherite is best identified by its triplets of twinned crystals and by its reaction to dilute acid.

Witherite was named after Dr Withering (1741–1799), the Birmingham physician who discovered it in 1784 at Alston Moor in Cumbria, England. It occurs here in low-temperature hydrothermal veins of lead ore or galena, and this is typical. It also forms in masses, deposited in limestone and other calcium-rich sediments. Unusually, witherite crystals are always twinned in groups of three, giving a double-pyramid shape. Although quite rare, witherite is the second most common barium mineral after barite. Because it dissolves easily in sulphuric acid, it is preferred to barite for some manufacturing uses such as making rat poison, in glass, porcelain and steelmaking, and, in the past, for the refining of sugar.

Crystal system: Orthorhombic.
Crystal habit: Triple twins form double-pyramid twin. Also occurs in botryoidal, massive and fibrous forms.
Colour: White, colourless, grey, yellowish or greenish.
Lustre: Vitreous to dull.
Streak: White.
Hardness: 3–3.5.
Cleavage: Distinct one way.
Fracture: Uneven.
Specific gravity: 4.3+.
Notable locations: Alston Moor (Cumbria), Hexham (Northumberland), England; Tsumeb, Namibia; Thunder Bay, Ontario; Rosiclare (Hardin Co), Illinois.
Caution: Witherite is mildly toxic. Wash hands after use.

Strontianite

SrCO₃ Strontium carbonate

Strontianite gets its name from the location of its first discovery, Strontian in Argyll and Bute, Scotland, where it was found in ores mined from the local lead mines in 1764. In 1790, Andrew Crawford separated the substance called strontium from the mineral, and in 1808 Sir Humphrey Davy showed that strontium is an element. Strontianite is almost the only mineral containing strontium; celestite is the only other significant source, which is used for the red in fireworks and signal flares, for refining sugar and as a painkiller. Strontianite typically occurs not as developed crystals but in masses of radiating fibres or tufts, although it can occasionally form twins rather like those of aragonite. These are usually white, but can be pale green, yellow or grey. The crystals are soft and brittle, and are commonly associated with galena and barite in low-temperature hydrothermal veins, or cavities in limestones.

Identification: Strontianite is best identified by its radiating needle-like crystals, and its reaction to acid when powdered.

Barite

Strontianite

Crystal system: Orthorhombic.
Crystal habit: Typically forms in radiating needle-like clusters and tufts, or in concretions. Occasionally forms triplets of twinned crystals like aragonite.
Colour: White, colourless, grey, yellowish or greenish.
Lustre: Vitreous to greasy.
Streak: White.
Hardness: 3.5–4.
Cleavage: Good in one way.
Fracture: Uneven, brittle.
Specific gravity: 3.8.
Other characteristics: Fizzes in warm dilute acid or, when powdered, in cold dilute acid.
Notable locations: Strontian, Scotland; Yorkshire, England; Drensteinfurt, Black Forest, Harz Mountains, Germany; Styria, Austria; Mifflin Co, Pennsylvania; San Bernadino Co, California; Schoharie, NY.

Mother of pearl
Aragonite is the main mineral in nacre, or 'mother-of-pearl' – the beautiful, iridescent substance that lines the shells of many shellfish. Many molluscs make mother-of-pearl but the main commercial sources are oyster-like species. In mother-of-pearl, aragonite is chemically mixed and bonded with water and an organic horn substance, known as conchiolin. Conchiolin binds the microcrystals of aragonite together to form the mother-of-pearl. The result is sometimes harder than inorganic aragonite and sometimes softer, depending on the ratio of the mixed chemicals. It can vary in colour depending on the species of mollusc and the character of the water. Colours range from soft white to pink, silver, cream, gold, green, blue or black. It always has the same pearly lustre and iridescent play of colours, which is caused by a film of conchiolin and the way overlapping platelets of aragonite interfere with light.

Cerussite

PbCO₃ Lead carbonate

The name cerussite comes from *cerussa*, the Latin for 'white lead', and cerussite is often called white lead ore. It was once used in white pigments. Queen Elizabeth I of England painted her face with a paste made from cerussite, in keeping with the fashion of the time for pale faces. Unfortunately, it was so poisonous it scarred her face and, without knowing the danger, she put more cerussite on her face to cover the damage. The lead content of cerussite means that in clear form (it is often white) it sparkles like lead crystal glass, and has one of the highest densities of any clear mineral crystal. Like other aragonite group minerals, cerussite forms twins, but those of cerussite are especially spectacular, including spoked star and snowflake shapes, and elbow or chevron-shapes. Cerussite is typically found where lead deposits are exposed to the air, and often forms crusts around galena (lead ore).

Crystal system: Orthorhombic.
Crystal habit: Typically forms needles, plates and spikes. Star-shaped and chevron-shaped twins are very distinctive. Also forms crusts on galena.
Colour: Usually colourless or white, also grey, yellow and even blue-green.
Lustre: Adamantine to greasy.
Streak: White or colourless.
Hardness: 3–3.5.
Cleavage: Good in one direction.
Fracture: Conchoidal, brittle.
Specific gravity: 6.5+.
Other characteristics: Very sparkly when clear.
Notable locations: Ems, Germany; Montevecchio, Sardinia; Murcia, Spain; Touissit, Morocco; Tsumeb, Namibia; Broken Hill (New South Wales), Dundas (Tasmania), Australia; Oruro, Bolivia; Socorro, New Mexico; Arizona.

Identification: Cerussite is usually striking for the weight, sparkle and twinning of its crystals. But it can often form white masses of needles as shown here.

CALCITE GROUP AND DOLOMITE

The calcite group is an important group of minerals that typically form in large masses, or in pearly,
hexagonal crystals. They are formed when a carbonate compound (carbon and oxygen) combines with
certain metals, such as calcium, cobalt, iron, magnesium, zinc, cadmium, manganese and nickel.
The calcite group includes magnesite, rhodochrosite, siderite and smithsonite, as well as calcite itself.

Calcite

CaCO₃ Calcium carbonate

Calcite is one of the world's most common minerals. It is the main component of limestone, marble, tufa, travertine, chalk and oolites. It may also be mixed in with clay to form marl. Calcite is also the fur deposited in kettles and boiler scale in hard-water districts, and the material from which bones and fossil shells are made. When dissolved in water, it can be deposited in crevices as grains and fibres, or precipitated from drips in limestone caves as stalagmites, stalactites and other speleothems (see Cave Formations, right). Most calcite forms in masses and aggregates, but good crystals do form in hydrothermal veins, alpine fissures, and pockets in basalt and other rocks.

There are over 300 kinds of calcite crystal, including Iceland spar, dogtooth spar and nailhead spar. Nailhead spar is a form of calcite beautifully described by its name. The flat-topped crystals look just like the heads of nails. Calcite is trimorphic (same chemistry, different crystal structures) with aragonite and the rare mineral vaterite, often formed by microbes.

Identification: Dogtooth spar calcite is very easy to recognize. Each crystal is as big and pointed as a canine tooth. Crystals like this form in clusters in standing pools in limestone caves. The crystal shape is called a scalenohedron, because the sides are scalene triangles – triangles in which each side is a different length.

Crystal system: Trigonal.
Crystal habit: Most calcite is found in masses, but crystals are found in a huge variety of forms, more than any other mineral.
Colour: Usually white or colourless.
Lustre: Vitreous to resinous to dull.
Streak: White.
Hardness: 3.
Cleavage: Perfect in three directions.
Fracture: Subconchoidal, brittle.
Specific gravity: 2.7.
Other characteristics: May be phosphorescent, thermoluminescent and triboluminescent.
Notable locations: Cumbria, Durham, England; Eskifjord, Iceland (Iceland spar or optical calcite); Harz Mountains, Germany; Bombay, India; Tristate, Kansas-Missouri-Oklahoma; Keeweenaw, Michigan; Franklin, New Jersey.

Magnesite

MgCO₃ Magnesium carbonate

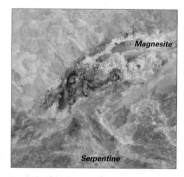

Magnesite

Serpentine

Identification: When a whitish porcelain-like mass appears in dolomite or magnesian limestone, it is likely to be magnesite.

Magnesite is thought to have got its name from the region of Magnesia in Turkey. It is the main ore of the metal magnesium, and is also used in making rubber and fertilizers. It typically forms when hot water alters limestone and dolomite rocks rich in magnesium minerals, such as serpentine. When this happens, solid white veins of magnesite are often created in the rock. It can also form when magnesium-rich solutions alter calcite. Magnesite rarely forms crystals, and the magnesite used industrially comes from very fine-grained massive deposits that look a little like porcelain and stick to the tongue when licked because they contain tiny holes. Crystals of magnesite can be hard to tell from calcite or dolomite.

Crystal system: Trigonal.
Crystal habit: Typically fine-grained, porcelain-like masses. Crystals are rare but typically rhombohedral or hexagonal.
Colour: White, grey, yellowish brown.
Lustre: Vitreous.
Streak: White.
Hardness: 3.5–4.
Cleavage: Perfect in three directions forming rhombs.
Fracture: Conchoidal, brittle.
Specific gravity: 3.
Other characteristics: Unlike calcite, does not fizz in cold hydrochloric acid.
Notable locations: Styria, Austria; Bahia, Brazil; Korea; China; Coast Ranges, California; Staten Island, NY.

Rhodochrosite

MnCO₃ Manganese carbonate

Rhodochrosite is one of the most easily recognized minerals. It is nearly always rose pink, although when exposed to air it often develops a coating of dark black manganese dioxide. Indeed, the name comes from the Greek *rhodon* for 'rose' and *chroma* for 'colour'. It typically forms in hydrothermal veins along with sulphide ores of copper, lead and silver, and occasionally in pegmatites. Massive deposits of rhodochrosite provide one of the major ores of the metal manganese. But rhodochrosite commonly forms striking crystals in cavities in veins. The ancient Inca silver mines of Catamarca province in Argentina are famous for stalactites made from rhodochrosite. These pink 'icicles' can be sliced like cucumbers to reveal beautiful rings of rhodochrosite in different shades of pink. They form when manganese minerals (or calcite) are dissolved by groundwater, combine with a carbonate material and then drip off the cave ceiling and from crevices.

Identification: Rhodochrosite is instantly recognizable by its rose-pink colour. This is stalactite rhodochrosite sliced to reveal the bands.

Crystal system: Trigonal.
Crystal habit: Can be massive, but crystals are typically rhombohedrons and scalenohedrons with rounded or curved faces. It also forms globules, fills veins and nodules or forms stalactites.
Colour: Rose pink.
Lustre: Vitreous to resinous.
Streak: White.
Hardness: 3.5–4.
Cleavage: Perfect in three directions.
Fracture: Uneven.
Specific gravity: 3.5.
Notable locations: Cornwall, England; Harz Mountains, Germany; Trepca, Serbia; Romania; Russia; Gabon; Hotazel, South Africa; Japan; Catamarca, Argentina; Huaron, Peru; Mont Saint-Hilaire, Quebec; John Reed mine (Lake Co), Sweet Home Mine (Park Co), Colorado; Butte, Montana.

Cave formations

Many caves in limestone are filled with spectacular rock formations called speleothems. They are created by the slow dripping or ponding of water rich in calcite dissolved from the limestone. The calcite comes out of the water and hardens in all kinds of shapes and forms, made mostly of calcite or aragonite. The best known speleothems are icicle-like stalactites that hang from the cave roof, and post-like stalagmites that grow up from the cave floor. But there are dozens of others, including draperies, which form like curtains from overhanging walls, cave pearls, which grow when calcite forms balls around a grain of sand, helictites (twisted stalactites), mushroom-shaped bell canopies, rimstones, which form around the rims of ponds, and soda straws (early stalactites that look just like drinking straws).

Dolomite

CaMg(CO₃)₂ Calcium magnesium carbonate

Dolomite is named after French mineralogist Deodat de Dolomieu, who discovered it in 1791. Massive beds of the mineral dolomite several hundred feet thick are found around the world. These are called dolomitic limestone, or simply dolostone – and their existence is a bit of a mystery (see Sedimentary Rocks: Dolomite, in the Directory of Rocks). Dolomite crystals form in marbles and hydrothermal veins. They also form in small deposits where water rich in magnesium filters through limestone. The magnesium replaces about half the calcium in calcite or aragonite to form dolomite, a process called dolomitization. Dolomite looks similar to calcite, but does not fizz in cold, dilute acid solutions like calcite.

Crystal system: Trigonal.
Crystal habit: Typically massive, but crystals can be prismatic or rhombohedral. It also forms distinctive saddle-shaped rhombohedral twins.
Colour: Colourless, white, pinkish, or light tints.
Lustre: Vitreous to dull.
Streak: White.
Hardness: 3.5–4.
Cleavage: Perfect in three directions.
Fracture: Subconchoidal.
Specific gravity: 2.8.
Notable locations: Cumbria, England; Binnetal, Switzerland; Styria, Austria; Saxony, Germany; Pamplona, Spain; Trepca, Serbia; Piedmont, Trentino, Italy; Algeria; Namibia; Brazil; Lake Erie, Ontario; Tristate, Kansas-Missouri-Oklahoma.

Identification: Dolomite typically forms pinkish or colourless rhombohedral crystals rather like calcite. The rhombs may have very slightly curved faces.

HYDRATED AND COMPOUND CARBONATES

*Hydrated and compound carbonates never form directly in rocks; they are always secondary minerals
that form when other minerals are altered. The brightly coloured compound carbonates malachite and
azurite form when copper minerals are altered. Hydrated carbonates such as scarbroite and gaylussite
form when other minerals are altered by water.*

Scarbroite

$Al_6(CO_3)(OH)_{13}.5H_2O$ Hydrated aluminium carbonate hydroxide

Scarbroite was discovered in 1829 by the English geologist
Doctor James H Vernon. He found the mineral on the coast
at Scarborough in England, and the mineral is named after
the town. Scarbroite is an aluminium carbonate
hydroxide mineral that has absorbed plenty of
water. It forms in vertical fissures in
sandstone, or as nodules. When
scarbroite forms crystals, they are
hexagonal, although they appear to
be rhombohedral. More typically it
forms very soft white, earthy
masses, only a little harder than
talc. In these, the crystals are
visible only with a microscope.
Scarbroite is also found in a
platy sheet form a bit like
micas, but this is rare.

Identification: Scarbroite is a
white knobbly mineral found in
clumps in sandstone. It looks a
little like cotton wool and is
very soft.

Crystal system:
Triclinic.
Crystal habit:
Scarbroite can be
either pseudorhombohedral
or hexagonal but is
typically found as
microcrystalline masses.
Colour: White.
Lustre: Earthy.
Streak: White.
Hardness: 1–2.
Cleavage: Poor.
Fracture: Uneven.
Specific gravity: 2.
Notable locations:
Scarborough (North
Yorkshire), Chipping
Sodbury (Gloucestershire),
Weston Favell
(Northampton); East
Harptree (Somerset),
England; Pilis Mountains (Pest
Co), Hungary; Soria, Spain.

Malachite

$Cu_2(CO_3)(OH)_2$ Copper carbonate hydroxide

Crystal system:
Monoclinic.
Crystal habit: Its
massive forms are
botryoidal,
stalactitic or
globular. Crystals are typically
tufts of needles.
Colour: Green.
Lustre: Vitreous to dull
in massive forms and silky
as crystals.
Streak: Pale green.
Hardness: 3.5–4.
Cleavage: Good in one
direction but rarely seen.
Fracture: Subconchoidal to
uneven, brittle.
Specific gravity: 4.
Notable locations: Chessy,
France; Sverdlovsk, Urals,
Russia; Katanga, DR Congo;
Morocco; Tsumeb, Namibia;
Burra Burra, South Australia;
Greenlee County, Pima
County, Arizona.

Unusually for a carbonate, malachite is bright green, and
gets its name from the Greek for 'mallow leaf'. It is a
secondary mineral of copper, which means it forms when
copper minerals are altered. It forms, for instance, when
carbonated water interacts with copper, or when a copper
solution interacts with limestone. Malachite is the tarnish on
copper and serves as a bright green signpost to the presence
of copper ore deposits. It also forms tufts of needle-like
crystals and various other masses. The classic malachite
specimens are rounded masses with concentric bands of light
and dark green, revealed when a specimen is polished or
cut open. Because of its beauty and relative softness,
polished, banded malachite has been carved into
ornaments and worn as jewellery for
thousands of years. Malachite is also
popular as an ornamental stone,
especially in Russia, where an
enormous deposit of rounded malachite
was found in the Ural Mountains
long ago. This malachite was used
to make the columns of St Isaac's
Cathedral in St Petersburg.

Identification: Malachite's bright
green colour is so distinctive it is
easy to identify, even when just
in velvety crusts like this
specimen here. But when a
specimen shows concentric
internal banding, the identity is
established beyond doubt.

Azurite

$Cu_3(CO_3)_2(OH)_2$ Copper carbonate hydroxide

Azurite, like malachite, is formed by the weathering of other copper minerals. The copper gives its basic colour, but the presence of water in the crystal helps turn it bright blue rather than green. It was this brilliant blue that made it so popular with painters in the Renaissance as a pigment. It probably gets its name from *lazhward*, the ancient Persian word for 'blue', and in the past it was often confused with blue lapis lazuli and lazurite. Azurite is actually 55 per cent copper, and in Arizona and South Australia, large masses were once worked as copper ores. Like green malachite, bright blue stains of azure act as a colourful sign of the presence of copper. Malachite and azurite often occur together, but azurite is the rarer of the two since it is altered to malachite by weathering. Azurite typically forms velvety masses and rosettes of needle-like crystals, or tablet-like masses.

Identification: Azurite is immediately recognizable by its bright blue colour, its softness and associated minerals such as malachite.

Azurite coating

Minor malachite

Barite

Crystal system: Monoclinic.
Crystal habit: Crystals are typically stubby prisms and tablets. These often have many faces, often over 45 and sometimes as many as 100. Also found as crusts, tufts and earthy masses.
Colour: Deep blue crystals; pale blue masses and crusts.
Lustre: Vitreous to dull.
Streak: Sky blue.
Hardness: 3.5–4.
Cleavage: Good in one direction.
Fracture: Conchoidal, brittle.
Specific gravity: 3.7+.
Notable locations: Chessy, France; Touissit, Morocco; Tsumeb, Namibia; Katanga, Congo; Sinai, Egypt; Guangdong, China; Burra Burra (SA), Alice Springs (NT), Australia; Lasal, Utah; Bisbee (Cochise Co), Arizona; New Mexico.

Ancient colours
People learned long ago to make coloured pigments by grinding minerals into a paste. It may be that the world's oldest mine is Lion Cavern in Swaziland in southern Africa. Here, over 40,000 years ago, African bushmen mined for specularite, which they then powdered and rubbed on their heads to make them shimmer. By 20,000 years ago, cave painters were using four mineral pigments – red ochre from hematite, yellow ochre from limonite, black from pyrolusite and white china clay from kaolin. These four pigments were popular in many tribal civilizations, like those of the San people of South Africa (including the exquisite paintings above in the Drakensberg Mountains), first discovered by European settlers some 350 years ago. As the first civilizations arose, people learned to exploit other minerals to create a rich range of colours – realgar for red, orpiment for yellow, malachite for green, azurite for blue and precious lapis lazuli for ultramarine. The best pigments were highly coveted.

Gaylussite

$Na_2Ca(CO_3)_2.5H_2O$ Hydrated sodium calcium carbonate

Gaylussite was first identified in soda deposits at Lagunillas, Venezuela, and is named after the famous 18th-century French chemist and physicist Joseph-Louis Gay-Lussac, who pioneered the study of gases and made many advances in applied chemistry. It is a carbonate mineral, made of hydrated sodium and calcium carbonate, and is one of several carbonate minerals that form as evaporites far away from the sea, typically in soda lakes. It is thought that gaylussite has been forming in California's famous Mono Lake since 1970, when the salinity rose above 80 per cent. Gaylussite looks very much like all the other inland evaporites minerals such as nahcolite, pirssonite, thermonatrite and trona, and often X-rays are the only way to tell them apart.

Crystal system: Monoclinic.
Crystal habit: Includes intricately faceted prism- and tablet-shaped crystals, but also forms masses and crusts.
Colour: Colourless or white.
Lustre: Vitreous.
Streak: White.
Hardness: 2–3.
Cleavage: Perfect in two directions.
Fracture: Conchoidal.
Specific gravity: 1.9–2.
Notable locations: Tuscany, Italy; Kola Peninsular, Russia; Guateng, South Africa; Lake Chad, Central Africa; Gobi, Mongolia; Lagunillas (Merida), Venezuela; Deep Spring, Owens Lake, Searles Lake, Borax Lake, Mono Lake, China Lake, California; Soda Lake, Nevada.

Identification: The best indicator of gaylussite is the environment where it formed – an inland soda lake, like the famous lakes of California.

NITRATES AND BORATES

Nitrates, iodates and borates are not actually carbonates, but are grouped with them because their chemical structure is very similar. All three groups form mostly in very dry places – nitrates and iodates such as nitratine are found in the deserts of Chile, and borates such as boracite, colemanite and ulexite in the Mojave Desert and Death Valley in the south-west USA.

Nitratine

NaNO₃ Sodium nitrate

Nitratine, or soda niter, is rare in most places because it is very soluble in water. In fact, nitratine is so easily dissolved that it can turn liquid simply by absorbing moisture from the air, a phenomenon called deliquescence. This is why specimens have to be kept in airtight containers containing a dessicant such as silica gel. Not surprisingly, nitratine forms mostly in desert regions. In the Atacama Desert of northern Chile, it forms under the soil in beds about 2–3m/6½–10ft thick called caliche. Caliche is a hard, cemented mix of nitrates, sulphates, halides and sand that builds up as water rich in dissolved minerals evaporates in the dry conditions. Nitratine also forms as growths on dry cave and mine walls. Nitratine was once an important source of nitrates for fertilizers and explosives, and a century ago, there were over 100 nitrate extraction plants in the Atacama Desert near Antofagasta in Chile. Nowadays, most nitrogen comes from the air.

Identification: Nitratine is best identified by its source location. It is found in very dry conditions, and is fairly soft. It also dissolves easily if there is any moisture in the air. A drop of water makes a small chunk seem to vanish.

Crystal system: Trigonal.
Crystal habit: Typically in beds and soil deposits in deserts. Rare rhombohedral crystals are found.
Colour: White or grey.
Lustre: Vitreous.
Streak: White.
Hardness: 1.5–2.
Cleavage: Perfect in three directions forming rhombohedrons.
Fracture: Conchoidal.
Specific gravity: 2.2–2.3.
Other characteristics: Deliquescent.
Notable locations: Kola Peninsula, Russia; Antofagasta, Atacama, Tarapaca, Chile; Bolivia; Bahamas; Niter Butte, Nevada; Pinal County, Maricopa County, Arizona; San Bernardino County, California; Dona Ana County, Luna County, New Mexico.

Boracite

Mg₃B₇O₁₃Cl Magnesium borate chloride

Identification: Boracite is best identified by its colour, its relative hardness and its association with other evaporite minerals.

First discovered on Luneberg Heath near Hanover, Germany, in 1789, boracite is rich in the element boron, and so is often used as a source for borax, which is important in everything from a healthy human diet to making fibreglass. Although boracite can form quite attractive crystals, it is rarely used as a gem because it loses its shine very easily in damp conditions. It is an evaporite mineral, which means it is left behind as a deposit when water evaporates. Boracite is therefore often found mixed in with other evaporites such as anhydrite, gypsum, hilgardite, magnesite and halite, and boracite crystals are commonly embedded in the crystals of these other evaporites. Boracite crystals show double refraction, like Iceland spar calcite.

Surface of small spheres covered with micro-crystals

Crystal system: Orthorhombic.
Crystal habit: Typically pseudocubic and octahedrons, but also forms masses, fibres and embedded grains.
Colour: White to colourless, tinges of blue green with increased iron.
Lustre: Vitreous.
Streak: White.
Hardness: 7–7.5.
Cleavage: None.
Fracture: Conchoidal, uneven.
Specific gravity: 2.9–3.
Other characteristics: Slightly soluble in water. Piezoelectric and pyroelectric.
Notable locations: Boulby (Yorkshire), England; Stassfurt, Germany; Lorraine, France; Inowroclaw, Poland; Kazakhstan; Khorat, Thailand; Tasmania; Cochabamba, Bolivia; Muzo, Colombia; Chactaw Salt Dome, Louisiana; Otis, California.

Colemanite

CaB₃O₄(OH)₃.5H₂O Hydrated calcium borate hydroxide

Colemanite is a borate that occurs either as brilliant colourless or white crystals much cherished by collectors, or in large masses. It was discovered in 1882 in Death Valley, Inyo County in California, one of the world's hottest, driest places. Like other borates, it is an evaporite and typically forms in desert lakes called playas, which fill up in the rainy season as water rich in boron runs off nearby mountains. When the rains stop, the lake waters

evaporate, leaving borate evaporites behind. Interestingly, the colemanite does not form directly. Instead, minerals such as ulexite form first in beds. Then groundwater trickling through the ulexite reacts with it to form colemanite, which is deposited in cavities. Colemanite also occurs in large masses in ancient clays and sandstone – from 1–1.6 million years old – and these provide industrial sources of borax.

Identification: Colemanite is hard to tell from many other colourless or white minerals, but a specimen found among borates in playa lake deposits is likely to be colemanite.

Crystal system: Monoclinic.
Crystal habit: Include stubby equant crystals and prisms with complex facets. Also occurs as masses, sheets and grains.
Colour: White to clear.
Lustre: Vitreous.
Streak: White.
Hardness: 4.5.
Cleavage: Perfect in one direction, distinct in another.
Fracture: Uneven to subconchoidal.
Specific gravity: 2.4.
Other characteristics: Transparent to translucent.
Notable locations: Balevats, Serbia; Atyrau, Kazakhstan; Panderma, Turkey; Salinas Grandes, Argentina; Boron, Death Valley (Inyo Co), Daggett (San Bernadino Co), California; Nevada.

Salt flats
Deserts are not always entirely dry. In fact, from time to time, they contain some of the largest (and shallowest) lakes in the world. These lakes form when heavy rain falls in surrounding mountains and fills them up. Soon, though, the rains stop and the dry conditions steam off all the water in the lake, leaving behind a hard crust of salty minerals. These salt pans, or playas, are probably the world's flattest pieces of land, which is why they are used for land speed record attempts, like Utah's salt lake. They typically slope no more than 20cm/8in over 1km/0.6 miles – and when the rain comes, vast areas may be covered with just a few centimetres/inches of water. The evaporite deposits on the lake bed vary from place to place. In California's Death Valley, they are typically borate minerals. Elsewhere they may be halides or sulphates such as gypsum and epsomite.

Ulexite

NaCaB₅O₆(OH)₆.5H₂O Hydrated sodium calcium borate hydroxide

Ulexite is a borate evaporite that forms in desert playa lakes as boron-rich water evaporates. It is used as a source of borax, and forms in masses as well as tufts of needle-like crystals called 'cotton balls'. Yet it is best known among collectors for the remarkable crystals found at Boron in California, dubbed 'TV rock'. Here huge chunks of ulexite are found in veins of tightly packed fibre-like crystals. When cut about 2.5cm/1in thick and polished, the crystals behave like optical fibres and transmit light, so you see on the crystal surface a stunningly clear picture of what is on the far side of the stone.

Identification: With its light-transmitting, straight, fibre-like crystals, 'TV rock' ulexite is unmistakable.

Crystal system: Triclinic.
Crystal habit: Tufts of acicular crystals and masses of straight fibre-like crystals. Also masses.
Colour: White to light grey.
Lustre: Silky.
Streak: White.
Hardness: 2.
Cleavage: Perfect in one direction.
Fracture: Uneven.
Specific gravity: 1.6–2.
Notable locations: Harz Mts, Germany; Balevats, Serbia; Atyrau, Kazakhstan; Chinghai, China; Salinas Grandes, Argentina; Tarapacá, Chile; Arequipa, Peru; Boron, California; Mojave, Nevada.

SULPHATES, CHROMATES AND MOLYBDATES

Sulphates are a group of over 200 minerals that are made from a combination of one or more metals with a sulphate compound (sulphur and oxygen). They form when sulphides are exposed to air, either in evaporites or in deposits formed by hot volcanic water. All are soft, light coloured and are commonly translucent or transparent. The barite group is a small but important group of sulphates – barite, anglesite, celestine and hashemite. All but hashemite are important ores for the metals they contain.

Barite

BaSO₄ Barium sulphate

A sulphate of the metal barium, barite (or, in industry, barytes) is unusually heavy for a non-metallic mineral. Indeed, it has long been known as 'Heavy spar', and its name comes from *baryos*, the Greek word for 'heavy'. It is this very density that makes it such a useful material. Barite was once used in making asbestos goods. Now millions of tonnes are used in heavy muds pumped into the apparatus for drilling oil wells. Ground barite mud is forced down the drill hole to carry back rock cuttings and prevent water, gas or oil entering the hole before drilling is completed. Powdered barite is also used to help bulk out white paper and give it a smooth finish. Barite is so dense and unreactive it absorbs gamma rays, so special concrete and bricks are made with barite for shielding radioactive sources in hospitals.

Barite occurs along with quartz and fluorite in hydrothermal ore veins – especially those containing lead (galena) and zinc (sphalerite). It is also found as nodules in clay left by the weathering of some limestones and forms the cement that holds many sandstones together. Barite even occurs in massive beds that are thought to have been formed as sediments. Beds like these in China, Morocco and Nevada, USA, are the main sources of the world's industrial barytes. Some of the finest large barite crystals, however, come from igneous rocks in Cumbria in England and Felsöbanya in Romania.

Barite forms a huge variety of crystal shapes. It often forms as large, tablet-shaped crystals, but it also grows as prisms, fans, tufts and many other forms. It sometimes grows in thin blade-like crystals that cluster in a cockscomb formation. These crested barite cockscombs are often mixed with sand and stained red with iron, earning them the name desert rose (see Cherokees' tears, right).

Identification: The largest barite crystals often appear in tabular form, as seen in the specimen shown to the left.

Crystal system: Orthorhombic.
Crystal habit: A huge variety of forms include tablet and prism shapes. Also occurs as grains, plates, cockscombs and rosettes. Masses and fibres are also common. Banded nodules are known as oakstone because of their striking resemblance to oak wood.
Colour: Colourless, white, yellow, but also reddish, bluish or multicoloured and banded.
Lustre: Vitreous.
Streak: White.
Hardness: 3–3.5.
Cleavage: Perfect in one direction.
Fracture: Uneven.
Specific gravity: 4.5.
Other characteristics: In 1604, a cobbler from Bologna, Italy, found that concretions of barite phosphoresced (glowed) when heated. These are called Bologna stones.
Notable locations: Many including: Alston Moor (Cumbria), North Pennines, England; Strontian, Scotland; Felsöbanya, Romania; Harz and Black Mountains, Germany; Bologna, Italy; Alps; Ugo, Japan (hokutolite barite); Peitou (Hokuto), Taiwan (hokutolite barite); Morocco; Australia; Norman, Oklahoma; Nevada; Stoneham, Colorado; Elk Creek, South Dakota.

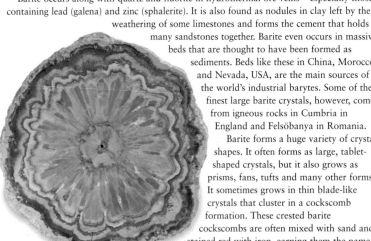

Rings of barite crystals

Identification: The specimen above is an example of botryoidal, or layered, barite. The crystals form rings like the growth layers seen in a piece of oak. The resemblance can be so striking that, when cut and polished, oakstone barite has even been mistaken for petrified wood.

Anglesite

PbSO₄ Lead sulphate

Named after the Welsh island of
Anglesey, Anglesite is an unusually pure
form of lead sulphate, and is a minor ore
of lead, sought after in Britain by the
Romans long ago. It usually forms when
galena is exposed to air, creating a ring-
banded mass around a galena core.
Cerussite (lead carbonate) often forms at
the same time or soon after. Although it
was first found in Wales, the best crystals
now come from Tsumeb in Namibia and
Touissit in Morocco. The brilliant lustre
associated with many lead minerals (as in lead crystal glass)
is apparent in many anglesite specimens and makes the
mineral especially appealing to collectors. Where it occurs
in its yellow form, it is particularly attractive. Frequently,
though, crystals are turned grey or black by galena
impurities. When colourless or white, it can look a little like
barite, but its lead content makes it much, much heavier.

Identification: The best way to
identify anglesite is by its
association with galena, its high
density and, commonly, the grey
colour of the crystals.

Crystal system:
Orthorhombic.
Crystal habit:
Crystals take a wide
variety of forms but
are typically tablet- or prism-
shaped, sometimes
elongated. It also forms
crusts, grains and masses.
Colour: Usually colourless,
white or yellow, but also pale
grey, blue or green.
Lustre: Adamantine.
Streak: White.
Hardness: 2.5–3.
Cleavage: Perfect in
one direction.
Fracture: Conchoidal, brittle.
Specific gravity: 6.4.
Other characteristics:
Fluoresces yellow.
Notable locations: Caldbeck
(Cumbria), England; Black
Forest, Germany; Sardinia;
Touissit, Morocco; Tsumeb,
Namibia; New South Wales,
Australia; Joplin, Missouri.

Cherokees' tears

Barite is just one of
several minerals that
form rosette shapes,
including
chalcedony, selenite
(gypsum), hematite
and aragonite. Yet
barite roses, created by
groundwater in sandstone,
can have the most perfect rose-like form and
may be coloured pink by traces of iron. They are
also linked to one of the saddest chapters in
American history, the Cherokees' Trail of Tears
(1838–9), or in Cherokee *Nunna daul Tsuny*.
When President Andrew Jackson ordered the
removal of the Cherokees from their Georgia
homeland, they were forced to trek far
westwards to Oklahoma through one of the
bitterest winters on record – harried at every step
by the relentless cavalry. On the way 4,000 men,
women and children, a fifth of the Cherokee
nation, perished in the snow. '*Long time we
travel on way to new land. People feel bad when
they leave Old Nation. Womens cry and make
sad wails, children cry and many men cry... but
they say nothing and just put heads down and
keep on go towards West. Many days pass and
people die very much.*' (A Cherokee survivor.)
Cherokee legend has it that God, in his pity,
decided wherever the blood of the braves and
the tears of the Cherokee maidens fell to the
ground they should turn to stone in the shape of
a rose. The rose gave the Cherokee women the
strength to care for their children alone. The
barite rose is now the state rock of Oklahoma.

Celestite

SrSO₄ Strontium sulphate

Discovered in Pennsylvania
in 1791, celestite, or
celestine, is popular with
collectors for its typical sky-
blue, 'celestial' colour,
unique in the mineral
world. It is often found in
colourful combinations
with minerals such as
yellow sulphur. Although celestite can look
similar to barite, it is actually strontium
sulphate, and has long been used as a source
of strontium, for fireworks, glazes and metal
alloys. Like calcite and dolomite, celestite
forms in sediments under the sea
– not as they are deposited
but in pockets and
fissures afterwards as
water trickles
through them.
Celestite may
also be found
in the cavities
of fossils. It is
known to
form, for
example, in
fossilized
ammonites.

Red celestite:
Despite its name,
celestite may
also be red.

Crystal system:
Orthorhombic.
Crystal habit:
Crystals take a wide variety
of forms but are typically
tablet-, prism- or plate-
shaped. It also forms crusts,
nodules, grains and masses.
Colour: Usually blue but may
be colourless, yellow, reddish,
greenish or brownish.
Lustre: Vitreous.
Streak: White.
Hardness: 3–3.5.
Cleavage: Perfect in
one direction.
Fracture: Uneven.
Specific gravity: 4.
Other characteristics: Burns
red in flame tests (see below).
Notable locations:
Gloucestershire, England;
Bohemia, Czech Republic;
Tarnowitz, Poland; Sicily;
Madagascar; San Luis Potosi,
Mexico; Ohio, Michigan; NY.

Identification: Celestite is very
easy to recognize when
coloured sky blue. Other-
wise, to distinguish it from
barite, try a flame test.
Soak a thin wooden splint
in water overnight, then dip
the soaked end in a little
powdered celestite. Wearing
safety goggles, hold it over a
gas flame. Celestite burns red;
barite burns lime green.

EVAPORITE SULPHATES

A range of sulphates form when salty waters evaporate from salt lakes and lagoons near the sea.
Typically these form large masses. Alternatively, fascinating crystals can be formed where salty waters
evaporate through rocks and soils. These evaporite sulphates include the many forms of gypsum, one of
the most common and useful minerals in the world, as well as anhydrite and glauberite.

Gypsum

CaSO₄.2H₂O Hydrated calcium sulphate

Gypsum is a very common mineral that occurs all over the world in a variety of forms. It is most commonly a soft white mineral that forms in thick beds where salty water evaporates. Most large deposits formed either on the beds of shallow seas, deposited along with anhydrite and halite, or in salt lakes. Gypsum also forms where anhydrite deposits are moistened by surface water. As the gypsum takes up water, it swells, so the beds are commonly contorted. In beds like these, gypsum is massive and fine-grained (see Alabaster, right), and this is the form that, when heated and dried, turns to the powder used as a base for most plasters, including plaster of Paris, and in most cements. It is also used in a wide range of other applications, such as in fertilizers and as a filler in paper.

Gypsum may also form clear or silky white crystals with fibre-like needles. These crystals are called satin spar or 'beef' and are treasured for carving into jewellery and ornaments. (Spar is a word geologists use to describe any white or light-coloured crystals that are easily broken.)

The more common crystalline form of gypsum, however, is selenite. Selenite crystals are transparent, and usually white or yellowish. They are typically tablet-shaped and form spearhead or swallowtail twins. They can also be prism-shaped, and these prisms can be curved or bent. Long thin crystals may twist in spirals known as ram's horn selenite.

In hot deserts, water often evaporates from shallow, salty basins. Under such conditions, gypsum can grow around grains of sand to form flower-like clusters of flat, bladed crystals. These clusters are called desert roses. Cockscomb barite forms similar roses, but the 'petals' in gypsum are usually better defined. Namibia in Africa is famous for its desert roses.

Crystal system: Monoclinic.
Crystal habit: Gypsum occurs in three main forms: crystals of selenite; fibrous satin spar; and fine-grained masses such as alabaster. It also forms elaborate 'daisies' and sand rosettes on various surfaces in dry places. Crystals include tablet, blade or prism shapes. Tabular crystals are often twinned, either as spearheads or swallowtails. Prisms may be bent.
Colour: Usually white, colourless or grey, but can also be shades of red, brown or yellow.
Lustre: Vitreous to pearly.
Streak: White.
Hardness: 2.
Cleavage: Good in one direction and distinct in two others.
Fracture: Splintery.
Specific gravity: 2.3+.
Other characteristics: Thin crystals can be bent slightly. A crystal of gypsum feels warmer than a crystal of quartz.
Notable locations: Nottinghamshire, England (satin spar); Thuringia, Bavaria Germany (swallowtail twins); Volterra, Bologna, Pavia, Italy; Montmartre, Paris; Sahara, Africa (desert roses); Whyalia, Tasmania; Pernatty Lagoon, South Australia; Naica (Chihuahua), Mexico (large bladed crystals in Cave of Swords); Nova Scotia; Alfalfa County, Oklahoma; Ellsworth (Mahoning Co), Ohio (isolated whole 'floating' crystals); Mammoth Caves, Kentucky (gypsum flowers); Lockport, New York.

Selenite (above): This form of gypsum gets its name from the Greek word *selene*, which means 'moon', and bladed crystals of selenite do indeed look like half moons.

Selenite swords (above): Selenite may form long prismatic 'swords', most famously in Mexico's spectacular Cave of Swords in Chihuahua, where there are crystals up to 2m/6ft long.

Well-defined, daisy-shaped clusters of evaporite gypsum crystals

Daisy gypsum (left): When gypsum forms from small pockets of moisture on the surface of rocks, it can often grow in radiating, overlapping patterns of crystals. These 'radiating aggregates' look so much like daisies that they are usually called daisy gypsum.

Anhydrite

CaSO₄ Calcium sulphate

Like gypsum, anhydrite is a white powdery mineral that typically forms in thick beds when water evaporates. In fact, anhydrite is commonly the mineral that forms when gypsum dries out – which is why it is harder – but it was recognized as a mineral in its own right in the 18th century, and can form independently. When concrete is made from gypsum, the gypsum is turned to anhydrite by heating. In nature, the drying-out process makes anhydrite shrink, so layers of anhydrite are often contorted, and sometimes riddled with caverns and smaller cavities. Although anhydrite forms mostly as masses in beds, it can form crystals in hydrothermal veins and alpine fissures or, more commonly, in druses with zeolites in basaltic rocks. Nevertheless, crystals of anhydrite are rare, because contact with water can easily turn them to gypsum. Beautiful lilac blue anhydrite is called angelite, because of its 'angelic' colour.

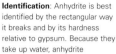

Identification: Anhydrite is best identified by the rectangular way it breaks and by its hardness relative to gypsum. Because they take up water, anhydrite specimens should be kept in an airtight container with silica gel.

Crystal system: Orthorhombic. **Crystal habit**: Usually fine-grained masses. Rare crystals include tablet and prism shapes, typically rectangular. **Colour**: White, grey or colourless but also bluish or purplish or even reddish. **Lustre**: Vitreous. **Streak**: White. **Hardness**: 3.5. **Cleavage**: Perfect in one direction, good in two others, forming rectangles. **Fracture**: Anhydrite is very, very brittle, breaking conchoidally. **Specific gravity**: 3. **Notable locations**: Lower Saxony, Germany; Pyrenees, France; Switzerland; Tuscany, Italy; Naica (Chihuahua), Mexico; Peru; Nova Scotia; New Mexico; Texas-Louisiana.

Alabaster
Massive gypsum commonly occurs as the hard white stone alabaster. This beautiful stone has been used since the time of Ancient Egypt for carving and engraving (and trade continues to thrive; see left), and there is a famous sphinx dating from 1700 BC which is made entirely from alabaster. Alabaster is a wonderful material for carving because it is so soft and easily shaped, although it is also easily damaged. The stone is beautifully translucent, shining through the colours and gilts that were often applied to it in the past. In the Middle Ages, it was widely used for carving in churches across Europe. Another English location, the city of Nottingham, was particularly famous for its alabaster carving in medieval times, exporting altarpieces as far afield as Iceland and Croatia. Italy was another famous medieval source for alabaster, while Roman alabaster originated from Egypt and Algeria. Italian alabaster was often known as Florentine marble while marble from the Middle East was known as Oriental alabaster. Today, alabaster from Mexico is known as Mexican onyx, and is widely used in the carving of ornaments and jewellery.

Glauberite

Na₂Ca(SO₄)₂ Sodium calcium sulphate

Glauberite gets its name from the sodium sulphate salt it contains. This salt, called Glauber's salt, was first made by Johann Rudolf Glauber from Hungarian spring waters to use as a mild laxative. Glauberite typically forms from evaporating salty water in the same places as halite, gypsum and calcite. Although glauberite crystals are rare, their 'ghosts', or pseudomorphs, are common. Because it is soluble in water, glauberite crystals often dissolve away to be replaced in exactly the same shape by other minerals, including opal. Glauberite has a distinctive crystal habit, so the pseudomorphs are easy to identify. Opal pseudomorphs of glauberite in Australia are known as pineapples.

Crystal system: Monoclinic. **Crystal habit**: Includes steeply pointed, inclined, flattened double-pyramid shaped crystals. **Colour**: White, yellow, grey or colourless. **Lustre**: Vitreous, greasy to dull. **Streak**: White. **Hardness**: 2.5–3. **Cleavage**: Perfect in one direction. **Fracture**: Conchoidal. **Specific gravity**: 2.7–2.8. **Other characteristics**: In water, it turns white and partially dissolves. **Notable locations**: Lorraine, France; Salzburg, Austria; Stassfurt, Germany; Villarrubia (Toledo), Spain; Kenya; Salt Range, Pakistan; India; Atacama, Chile; Gypsumville, Manitoba; Camp Verde, Arizona; Saline Valley (Inyo Co), Searles Lake (San Bernardino Co), California; Great Salt Lake, Utah. Pseudomorphs and casts found notably in Australia and in Paterson and Great Notch, New Jersey, USA.

Identification: Glauberite is best identified by the distinctive shape of its crystals – flattened prisms typically marked with grooves.

HYDROXIDE SULPHATES

When many sulphates (and other groups of minerals) crystallize, they include water in their structure and are said to be 'hydrated'. Some minerals lose this 'water of crystallization' if heated and become new anhydrous, or 'dry', sulphates, like barite and celestine. The hydroxide sulphates featured here – anterlite, linarite, alunite and jarosite – retain water in the form of OH.

Antlerite

$Cu_3(SO_4)(OH)_4$ Copper sulphate hydroxide

Antlerite is a copper mineral that was first found in the Antler mine in Mojave County, Arizona, USA. It typically occurs as bright emerald-green, gem-like, stripy crystals, but can also occur as a fine-grained pale green crust on other copper minerals. It is one of a number of minerals that form when copper minerals are oxidized (changed by exposure to oxygen in the air). Where there is plenty of carbon, the minerals formed tend to be copper carbonates such as malachite. If there is less carbon, then copper sulphates such as antlerite and brochantite are formed. These secondary minerals commonly occur in the same place and are quite hard to distinguish. Brochantite can only be definitively distinguished from antlerite by laboratory tests. Antlerite was once thought to be very rare until it was realized that the copper ore being mined in Chile's Chuquicamata is antlerite. Antlerite has since been confirmed at various other copper mines around the world.

Crystal system: Orthorhombic.
Crystal habit: Long slender prisms, or fibre-like crystals in tufts. Also found in veins and as masses, grains and crusts.
Colour: Emerald to very dark green.
Lustre: Vitreous.
Streak: Pale green.
Hardness: 3.5.
Cleavage: Perfect in one direction.
Fracture: Uneven.
Specific gravity: 3.9.
Other characteristics: Crystals are often striated, and it does not fizz in dilute hydrochloric acid.
Notable locations: Chuquicamata (Antofagasta), Chile; Mexico; Antler mine (Hualpai Mts, Mojave Co), Bisbee (Cochise Co), Arizona; Nevada; California; New Mexico; Utah.

Antlerite

Identification: Antlerite's bright green colour is a clear clue to its identity as a copper mineral. Striations and slightly elongated crystals help to narrow the identity down to antlerite.

Linarite

$PbCu(SO_4)(OH)_2$ Lead copper sulphate hydroxide

Identification: The bright blue colour immediately identifies a specimen as linarite or azurite. Linarite does not react to dilute hydrochloric acid but azurite does.

Linarite derives its name from the town of Linares in Spain. Like antlerite, linarite is a copper sulphate mineral, formed by the oxidation of copper minerals such as chalcopyrite – only with linarite there is lead (usually galena) involved as well. It is lead that gives linarite its bright blue colour, so cherished by collectors. In fact, it looks pretty much the same colour as the much better known azurite, and these two bright blue minerals are frequently confused – particularly since they are both found in the same kinds of locations. Sometimes the only way to tell them apart is to test them with dilute hydrochloric acid – azurite reacts, linarite does not. Linarite typically forms as a coating of tiny, bright blue crystals, but can form large masses in association with other copper minerals and is occasionally used as a copper ore.

Crystal system: Monoclinic.
Crystal habit: Tiny tablet-shaped crystals and long prisms. More typically in tiny needle-like layers and crusts. All crystals tend to have multiple facets.
Colour: Bright blue.
Lustre: Subadamantine to earthy.
Streak: Blue.
Hardness: 2.5.
Cleavage: Perfect in one direction, but only in larger crystals.
Fracture: Conchoidal.
Specific gravity: 5.3+.
Notable locations: Cumbria, Cornwall, England; Leadhills (Lanarks), Scotland; Black Forest, Germany; Linares, Spain; Tsumeb, Namibia; Argentina; Chile; Tiger (Mammoth, Pinal Co), Bisbee (Cochise Co), Arizona; Butte, Montana; Juab County, Utah.

Alunite

KAl₃(SO₄)₂(OH)₆ Potassium aluminium sulphate hydroxide

Known also as alum stone, alunite has been mined for making alum at Tolfa near Rome since the 15th century. Alum, a sulphate powder, was used in Roman times as a mordant (binder) for dyes and as a medical astringent. Nowadays, alum means just aluminium sulphate, but it is still used for a range of purposes, including water purification. In 1825, alum stone was found to contain a metal that came to be known as aluminium, and alunite is still a minor ore for the metal though much better sources have since been found. Alunite typically occurs as veins and replacement masses in potassium-rich volcanic rocks such as trachyte and rhyolite. In a process called 'alunitization', sulphuric acids in hydrothermal solutions react with metal sulphides in the rocks to create alunite. Large masses of alunite can be created in this way, and the white powdery masses can easily be mistaken for dolomite or limestone rock. Alunite may also form around fumaroles.

Identification: Masses of alunite look very like dolomite and limestone rock – but unlike these it does not fizz in dilute acid.

 Crystal system: Trigonal.
Crystal habit: Typically forms earthy masses and crusts. Crystals rare and pseudocubic (look like cubes).
Colour: White or grey to reddish.
Lustre: Vitreous to pearly.
Streak: White.
Hardness: 3.5–4.
Cleavage: Most crystals are too small.
Fracture: Conchoidal, uneven.
Specific gravity: 2.6–2.8.
Other characteristics: Some specimens fluoresce orange. Also piezoelectric.
Notable locations: Tuscany, Tolfa, Italy; Hungary; Bulladelah (New South Wales), Australia; Marysvale, Utah; Goldfield district, Nevada; Red Mountain (Custer Co), Colorado.

Mineral water on Mars?
In March 2004, NASA announced that the Mars rover *Opportunity* had found jarosite. In December of the same year, *Opportunity's* companion rover *Spirit* found the similar mineral

goethite on the far side of the planet in a rock dubbed Clovis. The discovery of both minerals may indicate there was once water on Mars's surface – jarosite because of the way it forms, and goethite because it is a hydrated mineral that can contain up to 10 per cent water. Another of *Opportunity's* mineral finds was tiny 'blueberries': balls that may be hematite concretions – further evidence of water. The rover took the microscopic image shown above, where the 'berries' appear as dark grey globules on rougher-textured, popcorn-like matter. Known as 'moqui marbles', similar balls are found on Earth in Utah and are thought to develop from groundwater. Earlier Mars missions had found evidence of significant quantities of carbonate minerals on the planet – signs too that there has been water there. However, in 2000 the *Mars Global Surveyor*, which was orbiting the planet, showed at least 2,589,988km²/1,000,000 square miles of the green mineral olivine. Since water reduces olivine quite quickly to clay and serpentine, this suggests that the Martian atmosphere has been dry for a very long time.

Jarosite

KFe₃(SO₄)₂(OH)₆ Potassium iron sulphate hydroxide

Jarosite is a distinctive custard-coloured mineral formed by weathering and as deposits around hot springs, and tends to occur mainly in arid regions. It gets its name from the Barranco del Jaroso (Jaroso Gorge) in the Sierra de la Almagrera region, Spain, where it was first found in 1852 by the German mineralogist August Breithaupt. Like natrojarosite, jarosite has the same trigonal crystal structure as alunite and these minerals form the alunite group. Jarosite occurs only as microscopic crystals, and is popular among collectors who specialize in 'micromounts'. In 2004, jarosite was discovered on Mars by the *Opportunity* rover and scientists think this is further evidence that the planet's surface once had large amounts of liquid water present.

 Crystal system: Trigonal.
Crystal habit: Normally occurs as earthy masses or crusts. Crystals are always tiny, and are typically hexagonal or triangular, or sometimes pseudocubic.
Colour: Yellow or brown.
Lustre: Vitreous to resinous.
Streak: Pale yellow.
Hardness: 2.5–3.5.
Cleavage: Good in one direction but only in very rare large crystals.
Fracture: Uneven.
Specific gravity: 2.9–3.3.
Notable locations: Barranco del Jaroso (Sierra Almagrera), Spain; Clara (Black Forest), Germany; Burra Burra, South Australia; Huanani, Bolivia; Sierra Gordo, Chile; Iron Arrow mine (Chaffee Co), Colorado; Maricopa County, Arizona; Custer County, Idaho; Mono County, California.

Identification: Jarosite is usually best identified by its striking custard colour, tiny crystals and its association with minerals such as hematite, limonite, variscite, pyrite, galena, barite and turquoise.

HYDRATED SULPHATES

Hydrated (wet) sulphates that incorporate molecules of water in their structure when they crystallize include colourful minerals such as cyanotrichite, jouravskite and copiapite, as well as the duller-coloured aluminite. Minerals like these tend only to form in damp environments underground. They may lose their water and disintegrate in dry air so need to be kept in airtight containers.

Aluminite

$Al_2(SO_4)(OH)_4 7H_2O$ Hydrated aluminium sulphate hydroxide

Aluminite is also known as websterite (after geologist Webster) and hallite (from Halle in Germany, where it was discovered in 1730). It is a white or grey mineral that typically forms in bauxite (aluminium ore) deposits and fissures or joints, commonly in association with gypsum, calcite, boehmite, limonite and quartz. Aluminite occurs in knobbly masses or nodules that look a little like cauliflower. These masses consist of fibrous crystals, but the individual crystals are generally too small to see. Consequently, aluminite is sometimes referred to as being a nodular mineral. It is a hydrated sulphate, which means its crystals contain water. In the caves of the Gaudalupe Mountains in New Mexico, aluminite is found as a bright to brilliant white and bluish white, paste-like to powdery, finely crystalline deposit on the cave walls. Milky deposits such as these are thought to form when aluminium-rich fluids crystallize then effloresce (dry out).

Identification: Aluminite can often be identified by the white cauliflower-like nodules it forms.

> **Crystal system:** Monoclinic.
> **Crystal habit:** Usually forms masses of tiny fibrous crystals, typically, rounded 'mamillary' form, like cauliflowers, botryoidal (grapelike). Also forms earthy clay-like mass.
> **Colour:** White, greyish white.
> **Lustre:** Earthy.
> **Streak:** White.
> **Hardness:** 1–2.
> **Cleavage:** None.
> **Fracture:** Irregular, uneven.
> **Specific gravity:** 1.7.
> **Other characteristics:** Fluorescent.
> **Notable locations:** Sussex, England; Halle, Germany; Ile-de-France; Mount Vesuvius, Italy; Zhambyl, Kazakhstan; Kinki (Honshu), Japan; Mount Morgan (Queensland), Australia; Miranda, Venezuela; Tristate, Missouri; Alum Cave, Tennessee; Utah.

Copiapite

$(Fe,Mg)Fe_4(SO_4)_6(OH)_2 - 20H_2O$ Hydrated iron magnesium sulphate hydroxide

Identification: Yellow encrustations tend to be iron sulphates, particularly on pyrite or pyrrhotite, but it can be hard to specifically identify copiapite.

Coquimbite

Copiapite

Copiapite gets its name from Copiapo in Chile, where it was first discovered. It is a secondary mineral, forming as an alteration product of other minerals – typically as pyrite, pyrrhotite and other iron sulphide minerals are exposed to the air. Many other hydrated iron sulphate minerals form in the same way – including melanterite, ferricopiapite, rozenite, szomolnokite, fibroferrite, halotrichites and bilinite. It can be very hard to distinguish copiapite from any of these other iron sulphates without X-ray studies. They can all be yellowish crusts, and yellow uranium minerals can even look quite similar but tend to be a darker yellow. Like some other hydrated minerals, copiapite loses water easily and disintegrates to a powder. Specimens should therefore be kept in an airtight container.

> **Crystal system:** Triclinic.
> **Crystal habit:** Typically grainy crusts and scaly masses. Individual crystals are rare.
> **Colour:** Yellows to olive green.
> **Lustre:** Pearly to dull.
> **Streak:** Pale yellow.
> **Hardness:** 2.5–3.
> **Cleavage:** Perfect in one direction.
> **Fracture:** Uneven.
> **Specific gravity:** 2.1.
> **Other characteristics:** Dissolves in water, relatively low density.
> **Notable locations:** France; Spain; Hunsrück, Black Forest, Germany; Elba, Italy; Copiapo (Atacama), Chile; Utah; California; Nevada.

Cyanotrichite

$Cu_4Al_2(SO_4)(OH)_{12}.2H_2O$ Hydrated copper aluminium sulphate hydroxide

Cyanotrichite is one of the most strikingly coloured of all minerals and gets its name from its deep 'cyan'-blue colour. The 'trich' part comes from the Greek for 'hair', because it typically occurs as crusts or radiating sprays or balls of fine crystals. The crystals in the sprays are tiny (1mm/^{1}/$_{24}$in long), but may cover an area as big as a postcard. Just as copiapite forms by the oxidation of iron sulphide minerals, so cyanotrichite forms by the oxidation of copper sulphide minerals. Like copiapite, too, it is similar to copper minerals that form in the same way, including other copper sulphates such as antlerite and brochantite, the carbonate mineral malachite and the halide mineral atacamite. Copper gives all these minerals a green colour; the water in cyanotrichite keeps it a vivid blue. It is therefore very important to keep it in an airtight container, to prevent it losing its water and beautiful colour.

Identification: Cyanotrichite's sky-blue colour, its velvety look and its association with other copper minerals such as malachite and smithsonite are clear signs of its identity.

Crystal system: Orthorhombic.
Crystal habit: Crusts, as well as radial sprays and balls of tiny needle-like crystals and small tabular crystals.
Colour: Sky blue.
Lustre: Vitreous to silky.
Streak: Blue.
Hardness: 1–3.
Cleavage: None.
Fracture: Uneven.
Specific gravity: 3.7–3.9+.
Other characteristics: Transparent to translucent.
Notable locations: Cornwall, England; Leadhills, Scotland; Black Forest, Germany; Auvergne, France; Romania; Lavrion, Greece; Russia; Broken Hill (New South Wales), Australia; Arizona; Nevada; Utah.

Jouravskite

$Ca_3Mn(SO_4,CO_3)_2(OH)_6.13H_2O$ Hydrated calcium manganese sulphate hydroxide

Something in the air
The oxygen in the air is one of the most reactive of all elements, and whenever metal minerals are exposed to the air, their surface begins to change as the oxygen begins to react with it. Iron rusts like this. This reaction is called oxidation, and it is always linked with a mirror image process called reduction. Oxidation and reduction often involve a swapping of oxygen between substances. As coal burns, for instance, its carbon joins with oxygen in the air to make carbon dioxide. Indeed, oxidation once referred to any chemical reaction in which a substance combines with oxygen. Burning, rusting and corrosion are all reactions like this. But now the definitions have been widened so that oxidation is when a substance loses electrons and reduction is when it gains them. Many new, secondary minerals form as primary minerals are exposed to the air and oxidized. Minerals formed by the oxidation of ores of metals like copper, vanadium, chromium, uranium and manganese are among the most colourful of all minerals.

Just as copiapite forms by the oxidation of iron minerals and cyanotrichite from copper minerals, jouravskite forms by the oxidation of manganese minerals. It forms distinctive lemon-yellow masses and crusts. It is a rare mineral and was first identified in the famous Tachgagalt manganese mine in Ouarzazate in the Atlas mountains of Morocco only as recently as 1965. It was named in memory of the French geologist Georges Jouravsky, head of the Moroccan geology division, who died the previous year. Jouravskite has been found in only a limited number of other places, also associated with manganese, including the Kalahari manganese fields in South Africa.

Crystal system: Hexagonal.
Crystal habit: Typically forms in grainy masses, but crystals are double-pyramids.
Colour: Greenish orange, greenish yellow, yellow.
Lustre: Vitreous.
Streak: Greenish white.
Hardness: 2.5.
Cleavage: Good.
Fracture: Uneven.
Specific gravity: 1.95.
Other characteristics: Translucent.
Notable locations: Tachgagalt mine (Ouarzazate), Morocco; N'Chwaning mine (Kalahari manganese fields, Northern Cape Province), South Africa.

Identification: Jouravskite is best identified by its lemon-yellow colour, its sugary texture – as is apparent in this enlarged view of the specimen – and its association with manganese ores.

WATER-SOLUBLE SULPHATES

A small number of sulphate minerals dissolve easily in water, including epsomite, chalcanthite, melanterite and thenardite. They are more common than you might think, even in relatively moist parts of the world, but specimens will quickly deteriorate unless kept in airtight containers with a dessicant (drying agent) such as silica gel.

Epsomite

MgSO₄.7H₂O Hydrated magnesium sulphate

Crystal system: Orthorhombic.
Crystal habit: Does not usually form crystals, but occurs as masses, grains, fibres, needles, crusts, grapelike clumps and stalactites, typically as cave efflorescences and in playa lakes.
Colour: Colourless, white, grey.
Lustre: Vitreous silky, earthy.
Streak: White.
Hardness: 2–2.5.
Cleavage: Perfect in one way.
Fracture: Conchoidal.
Specific gravity: 1.7.
Other characteristics: Very soluble in water.
Notable locations: Epsom (Surrey), England; Mount Vesuvius, Italy; Stassfurt, Germany; Herault, France; Sahara, Africa; Central Australia; Kruger Mountain, Washington; Carlsbad, New Mexico; Alameda County, California.

Epsomite is well known from the medicinal Epsom salts, first discovered in mineral waters at Epsom in England. Most Epsom salts are now made artificially, and epsomite dissolves so easily that it is rare in wetter regions. Epsomite does, however, occur as thick sedimentary beds and there is epsomite in sea salt deposits in South Africa. More usually it occurs as an efflorescence – a powdery deposit from mineral waters – in dry limestone caves out of the rain, or in desert regions around playa lakes. Epsomite can also be found growing on the walls of coal and metal mines and on abandoned equipment. Epsomite also occurs around hot springs, and fumaroles such as those on Mount Vesuvius in Italy. Large crystals are extremely rare and very fragile. Specimens are best cleaned with a little alcohol and kept in a sealed container. Yet although moisture is epsomite's main enemy, so too is dryness: losing even a single molecule of its water can change it to another mineral, called hexahydrite, with different monoclinic crystals.

Identification: Epsomite is best identified by its often fibrous habit, its location, colour, low density, solubility in water.

Chalcanthite

CuSO₄·5H₂O Hydrated copper sulphate

Chalcanthite comes from the Greek words for 'copper' and 'flower'. Chalcanthite is the basis of the classic copper sulphate solutions often seen in school chemical laboratories and is often used for demonstrations of how crystals grow. Indeed, chalcanthite crystals are so easy to grow that most specimens for sale are artificial. In nature, it forms through the oxidation of copper sulphide minerals such as chalcopyrite, covelite, bornite, chalcocite and enargite. It often forms crusts and stalactites on timbers and walls of copper mines. Chalcanthite is rare in wet regions because it is so soluble in water, but is common enough in arid regions such as the Chilean deserts to be used as a minor ore of copper.

Caution: Chalcanthite is poisonous; wash hands after contact.

Crystal system: Triclinic.
Crystal habit: Natural crystals are rare but typically stubby prisms and thick tablets. Typically occurs as grapelike and stalactite-like masses, and in veins and crusts.
Colour: Bright and deep blue.
Lustre: Vitreous, silky.
Streak: Pale blue to colourless.
Hardness: 2.5.
Cleavage: Poor (basal).
Fracture: Conchoidal.
Specific gravity: 2.2–2.3.
Other characteristics: Very soluble.
Notable locations: Minas de Rio Tinto, Spain; Chuquicamata, El Teniente, Chile; Bingham Canyon, Utah; Ducktown, Tennessee; Imlay (Pershing Co) Nevada; Bisbee (Cochise Co), Ajo, Arizona.

Sandstone rock

Identification: Chalcanthite is best identified by its bluish 'copper sulphate' colour, a dry environment, and the stalactites and crusts it forms.

Melanterite

FeSO₄-7H₂O Hydrated iron sulphate

Melanterite is a sulphate of iron that is typically formed by the oxidation of iron ore, particularly the iron sulphides pyrite, pyrrhotite, marcasite and the copper ore chalcopyrite. It typically forms white or green powdery deposits, encrustations, stalactites and, occasionally, small clusters of crystals. Miners in iron mines commonly see powdery crusts of melanterite along the walls of the mine's shafts where the ore has been altered. Melanterite is sometimes known as copperas, from the Greek for 'copper water'. It gets this name because when melanterite is dissolved in water, it creates exactly the same iron sulphate solution as when iron is dropped in dissolved chalcanthite (copper sulphate), precipitating metallic copper. Less commonly, melanterite gets a bluey colour from copper impurities – the more copper, the bluer it goes.

Identification: Melanterite is best identified by its greenish white colour, and its association with iron and copper sulphide minerals. Specimens lose water and crumble on exposure to air so they should be kept in a sealed container.

Crystal system: Monoclinic.
Crystal habit: Typically forms crusts, but crystals are shaped or prismatic, often twinned, sometimes forming crosses or stars.
Colour: White, green, or blue-green.
Lustre: Vitreous to silky.
Streak: White.
Hardness: 2.
Cleavage: Perfect in one direction.
Fracture: Conchoidal, brittle.
Specific gravity: 1.9.
Other characteristics: Soluble in water.
Notable locations: Minas de Rio Tinto, Spain; Rammelsberg, Harz Mts, Germany; Falun, Sweden; Ducktown, Tennessee; South Dakota; Colorado; Bingham Canyon, Utah; Comstock Lode, Nevada; Butte, Montana.

California's dry treasure lake

Searles Lake in San Bernardino County in California is one of the most famous locations in the world for evaporite minerals. It is a playa lake in the Mojave desert, one of the world's driest regions. It is now mostly dry or shallow, but it formed in the last Ice Age, when rain conditions made it the centre of a large drainage network. It was named after the Searles brothers, John and Dennis, who discovered borax here in 1863, and began working its huge deposits ten years later. So far Searles Lake has yielded a billion dollars' worth of chemicals. Besides borax, there are large deposits here of the evaporite trona or 'natron', naturally occurring sodium bicarbonate, after which the nearby town of Trona is named. It is said that nearly all the 90 or so naturally occurring elements can be found at Searles Lake, and there are deposits of rare minerals such as hanksite and pink halite. The area around Trona is home to over 500 tufa spires, which have been the setting for films such as *Star Trek V*, *Planet of the Apes*, and Disney's *Dinosaur*.

Thenardite

Na₂SO₄ Sodium sulphate

Thenardite is named after the French chemist Louis-Jacques Thenard. It is found in huge deposits near salt lakes and playas throughout the dry regions of south-west USA, Africa, Siberia and Canada, where it has been formed by evaporation. It is also found as a powdery deposit on desert soils and as a crust around fumaroles. Thenardite is natural sodium sulphate, a vital chemical for making soaps and detergents, paper and glass. Huge amounts of thenardite are mined to satisfy industrial demand. Searles Lake in California alone contains about 450 million tonnes of thenardite. The Great Salt Lake in Utah contains 400 million tonnes. But some of the biggest deposits are at Hongze in eastern China's Jiangsu province.

Crystal system: Orthorhombic.
Crystal habit: Typically forms crusts, grains and massive rockbeds. Also forms distorted intergrown clusters of crystals. Rare individual crystals are slender tablet- or prism-shaped.
Colour: White, yellowish, grey, brown.
Lustre: Vitreous.
Streak: White.
Hardness: 2.5–3.
Cleavage: Perfect in one direction.
Fracture: Splintery like hornblende.
Specific gravity: 2.7.
Other characteristics: Soluble in water, and fluoresces white.
Notable locations: Espartinas, Spain; Mount Etna, Sicily; Bilma Oasis, Niger; Khibiny and Lovozero massifs (Kola Peninsula), Russia; Kazakhstan; Hongze (Jiangsu), China; Pampa Rica, Chile; Searles Lake, California; Great Salt Lake, Utah; Camp Verde (Yavapai Co), Arizona.

Identification: Thenardite is best identified by its location and its white fluorescence.

CHROMATES, MOLYBDATES AND TUNGSTATES

These heavy, soft and brittle minerals have the same chemical structure as sulphates, except that chrome, tungsten and molybdenum replaces the sulphur in the molecule. Besides ores such as wolframite and scheelite, they include some of the most striking of all minerals – crocoite and wulfenite – and two rarities – ferrimolybdite and cuprotungstite.

Crocoite

$PbCrO_4$ Lead chromate

Blood red or orange, crocoite is one of the most beautifully coloured of all minerals. It is a chromate, in which chromium replaces sulphur, and it is the chromium that gives it its distinctive colour. It was first found at Berezovsk near Ekaterinburg in the Russian Urals in 1766, and was named crocoite in 1832 from the Greek for 'crocus' or 'saffron', because of its colour. In the Urals, crocoite is found in quartz veins running through granite and gneiss, and is associated with the similar minerals phoenicochroite and vauquelinite. Vauquelinite is named after the French chemist L N Vauquelin, who, in 1797, at the same time as H Klaproth, discovered the element chromium in crocoite. For a while, crocoite was the main ore of chromium. Now it is too rare. The most famous crocoite specimens come from Dundas in Tasmania, where crystals up to 20cm/8in long were once found. Most specimens, though, are made of small crystals like splinters, known as 'jackstraw' crocoite.

Identification: Crocoite's blood-red, often splinter-like, crystals are unmistakable. It can be confused with wulfenite but its crystals are prismatic and have a lower specific gravity.

Crystal system: Monoclinic.
Crystal habit: The long, slender, splinter-like crystals are distinctive. Crystal ends are sometimes hollow. Also forms grains and masses in granite and other igneous rocks.
Colour: Bright orange-red to yellow.
Lustre: Adamantine to greasy.
Streak: Orange-yellow.
Hardness: 2.5–3.
Cleavage: Distinct in two directions.
Fracture: Conchoidal to uneven, brittle.
Specific gravity: 6.0+.
Other characteristics: Very high index of refraction.
Notable locations: Urals, Russia; Umtali, Mashonaland, Zimbabwe; Luzon, Philippines; Dundas, Tasmania; Ouro Prêto, Brazil; Inyo Co, Riverside Co, California.

Wulfenite

$PbMoO_4$ Lead molybdate

Wulfenite is a very distinctive mineral and is named after the Jesuit mineralogist Xavier Wülfen (1728–1805) who discovered it at Carinthia, Austria, in 1785. Chemically it is a lead molybdate that forms where lead and molybdenum ores oxidize, but what makes it unique is the striking shape of one of its crystal forms. These crystals form in square, transparent interlocking slices that look almost like plastic counters. They are typically yellow, but can be white, orange and even red. The counter-shaped wulfenite crystals are the most memorable, but wulfenite forms a variety of crystal shapes, making it a very interesting mineral to collect. The most brilliant orange wulfenite crystals come from Red Cloud Mine, Yuma, Arizona, and from Chah-Kharbose in Iran.

Identification: When it forms square, flat yellow crystals like counters, wulfenite is readily identifiable.

Crystal system: Tetragonal.
Crystal habit: Very thin square crystals like plastic counters, and also grains and hollow masses.
Colour: Orange-yellow.
Lustre: Vitreous.
Streak: White.
Hardness: 3.
Cleavage: Perfect in one way.
Fracture: Subconchoidal to uneven.
Specific gravity: 6.8.
Other characteristics: Index of refraction: 2.28–2.40 (high but typical of lead minerals).
Notable locations: Carinthia, Austria; Slovenia; Czech Rep; Morocco; Tsumeb, Namibia; Zaïre; Australia; Sonora, Durango, Chihuahua, Mexico; Pinal Co, Yuma Co, Gila Co, Arizona; Stephenson Bennett mine, New Mexico.

Ferrimolybdite

Fe₂O₃.3MoO₃.8H₂O Hydrated iron and molybdenum oxide

Like molybdenite, powellite and wulfenite, ferrimolybdite is rich in molybdenum. It typically forms when molybdenite is weathered, especially in quartz veins, and is commonly found in layers in molybdenite. It can also form on spoil heaps. What marks ferrimolybdite out is its iron content, which is why it is commonly found in association with pyrite. Weathered ore deposits are often a mix of red hematite, yellow jarosite and ferrimolybdite, brown goethite and black oxides. Officially, ferrimolybdite was recognized by Pilipenko in 1914 in Khakassiya, eastern Siberia, but the process of discovery goes back much earlier. In 1800, the existence of 'molybdite' was noted in Europe, and German geologists described 'molybdenum ochre' in molybdenum layers. A few years before the Civil War, American geologist David Dale Owen found a specimen in California and noted its iron content. In 1904, Schaller insisted that although there was an iron-rich molybdenum oxide, it had yet to be proved that it occurred naturally, which is what Pilipenko did in 1914.

Identification: Ferrimolybdite is best identified by its association with other molybdenites, its yellow colour and its tiny, flat, needle-like or fibrous crystals.

Ferrimolybdite crust

Crystal system: Orthorhombic.
Crystal habit: Dull, clay-like masses of tiny, flat, needle-like or fibrous crystals.
Colour: Yellow.
Lustre: Silky.
Streak: Light yellow.
Hardness: 2.5–3.
Cleavage: Good, but natural crystals are too small.
Fracture: Uneven.
Specific gravity: 4–4.5.
Notable locations: Bohemia, Czech Republic; Saxony, Germany; Hohe Tauern Mountains, Austria; Bipsberg, Bastnas, Sweden; Lake Iktul (Khakassiya), Russia; Dundee (New South Wales), Australia; Climax, Colorado; Arizona; Texas; New Mexico.

Dundas mines, Tasmania
The mines of Dundas, near Zeehan in Tasmania, have long been famous for their magnificent crocoite specimens – like the Red lead mine shown here. Well over 90 per cent of the world's crocoite comes from here, and crocoite is Tasmania's state mineral emblem. Crocoite was first discovered here in 1896 by James Smith and W R Bell, when silver and lead miners dug deep into the ore bodies here and found them honeycombed with cavities full of cerussite and extraordinary blood-orange 'jackstraw' crystals of crocoite. The crocoite was so plentiful here that many beautiful crystals were sent off to be used as flux in smelters. Eventually, though, both the ores and the crocoite were worked out and the mines were largely abandoned. Then in the 1970s and in 2002, new deposits of wonderful crocoite and pyromorphite specimens were found at Dundas in the Platt Mine and the famous Adelaide Mine. Both these mines now harvest crystals for the collectors' market.

Cuprotungstite

Cu₂(WO₄)₂(OH)₂ Copper hydroxytungstate

Cuprotungstite was discovered in 1869 in Chile, and tends to be more common in drier environments such as Baja California, Broken Hill in Australia and Namibia, although good specimens have been found in places such as the Old Gunnislake mine in Cornwall, England. It is a combination of copper and tungsten often found as a tan- or green-coloured coating on copper minerals. It typically forms when scheelite deposits (tungsten ore deposits) contain copper sulphides. Cuprotungstite then forms when these two sets of minerals interact as they are weathered and oxidized.

Crystal system: Tetragonal.
Crystal habit: Typically fine, fibre-like crystals or masses.
Colour: Emerald green or brown.
Lustre: Vitreous, waxy.
Streak: Green.
Hardness: 4–5.
Fracture: Conchoidal, brittle.
Specific gravity: 7.
Notable locations: Cornwall, Cumbria, England; Black Forest, Germany; Namibia; South Africa; Broken Hill (New South Wales), Australia; Honshu, Japan; China; La Paz (Baja California), Sonora, Mexico; Cochise Co, Maricopa Co and Pima Co, Arizona; Deep Creek Mountains, Utah.

Cuprotungstite

Identification: Cuprotungstite is best identified by its tan or green colour and its association with copper and tungsten ores.

PHOSPHATES, ARSENATES AND VANADATES

The phosphates, arsenates and vanadates are a group of interesting and often very distinctive minerals. The phosphates include many minerals rich in rare-earth elements found in few other minerals such as yttrium, caesium, thorium and even less known samarium and gadolinium. The vanadates include some of the most amazing crystals such as those of descloizite and vanidinite. The phosphates featured here are 'primary' (minerals that form directly), and are among the few phosphates that are anhydrous (contain no water).

Xenotime

YPO₄ Yttrium phosphate

Xenotime is one of a handful of minerals with a name starting with 'x', and its name contains a few historical errors. It comes, supposedly, from the Greek for 'empty honour' – because it was once thought to have the honour of containing a brand new element yttrium. It does contain yttrium, but it was not a new element, and *xenos* is the Greek word for 'strange'; *kenos* is the Greek for 'empty' – so it should really have been called kenotime! However, it remains a very interesting mineral and one of the very few to contain yttrium, which can be replaced by erbium. Xenotime forms brown, glassy crystals in clusters and rosettes in pegmatites within granites and gneisses, just like the very similar mineral monazite. Once the rock it formed in has been weathered away, it sometimes turns up in river sands and beaches, just like monazite, because it is relatively dense. However, it is softer and lighter than monazite and is therefore much rarer in these deposits.

Identification: Xenotime is best identified by its hardness and crystal habits.

Crystal system: Tetragonal.
Crystal habit: Crystals are typically prisms ending in slanted double-pyramids. Also forms rosettes and radiating clusters.
Colour: Shades of brown; also greyish, greenish or reddish.
Lustre: Vitreous to resinous.
Streak: Pale brown.
Hardness: 4–5.
Cleavage: Perfect in two ways.
Fracture: Uneven to splintery.
Specific gravity: 4.4–5.1.
Other characteristics: Traces of uranium and other rare-earth elements may make crystals slightly radioactive.
Notable locations: Arendal, Hittero, Tvedestrand, Norway; Sweden; Madagascar; Brazil; Colorado; California; Georgia; North Carolina.
Caution: Xenotime is radioactive. Handle with care and store appropriately.

Monazite

(Ce, La, Th, Nd, Y)PO₄ Cerium lanthanum thorium neodymium yttrium phosphate

Monazite is actually a blanket term for a range of phosphates that include varying amounts of rare-earth elements such as cerium, thorium, lanthanum and neodymium, as well as smaller amounts of praseodymium, samarium, gadolinium and yttrium. It forms as small brown or golden crystals in pegmatites within granites and gneisses, like those in Norway and Maine, or in alpine cavities with quartz (a combination called turnerite). When the rock is broken down by weathering and washed seawards, monazite normally settles close to the shore because it is both relatively tough and heavy. Monazite grains are commonly concentrated in beach sands, known as monazite beaches, most famously along India's Malabar Coast. These beach deposits are excavated primarily for thorium, but also for all the other rare earths that monazite contains.

Yellow beryl crystal

Feldspar

Monazite

Smoky quartz

Identification: Monazite is best identified by its complex red-brown splintery crystals – though it is hard to identify as sand grains.

Caution: Monazite is highly radioactive, so should be handled with care and stored in an appropriate container.

Crystal system: Monoclinic.
Crystal habit: Typically masses or grains. Crystals flat, splinter-like wedges and tablets – complex twinning.
Colour: Crystal red-brown or golden; grains yellow-brown.
Lustre: Resinous, waxy.
Streak: White, yellow-brown.
Hardness: 5–5.5.
Cleavage: Perfect in one way.
Fracture: Conchoidal.
Specific gravity: 4.9–5.3.
Notable locations:
Pegmatites: Norway; Finland; Callipampa, Bolivia; Madagascar; Minas Gerais, Brazil; Maine; Connecticut; Amelia, Virginia; Climax, Colorado; New Mexico
Alpine cavities: Switzerland
Sands: Malabar, India; Sri Lanka; Malaysia; Nigeria; Australia; Brazil; Idaho; Florida; North Carolina.

Triphylite

Li(Fe,Mn)PO₄ Lithium iron manganese phosphate

Triphylite is a relatively rare phosphate mineral that forms bluish or glassy masses. It gets its name from the Greek for 'family of three' because of the three elements it contains besides phosphate – lithium, iron and manganese. It is very similar to another phosphate mineral, lithiophilite, which contains the same three elements. The only real difference is that lithiophilite contains more manganese, and so is pinker in colour and denser, while triphylite contains more iron, and is bluer in colour and denser. What makes both these minerals interesting is not the minerals themselves but their products of weathering. They both form in phosphate-rich pegmatites within granite. When weathered, they change to various striking minerals such as eosphorite, vivianite, strengite, purpurite, wolfeite and sicklerite. In hydrothermal solutions, they can change to siderite and rhodochrosite.

Identification: Triphylite is best identified by its blue-grey colour and its situation in phosphate-rich pegmatites in granite. This triphylite mass has green surface patches of vivianite.

Vivianite

Tryphylite mass

Crystal system: Orthorhombic.
Crystal habit: Mostly forms glassy masses.
Colour: Blue or blue-grey.
Lustre: Vitreous.
Streak: White to grey-white.
Hardness: 4–5.
Cleavage: Near perfect in one direction (basal).
Fracture: Uneven.
Specific gravity: 3.58.
Notable locations: Verutrask, Sweden; Bavaria, Germany; Mangualde, Portugal; Buckfield, Poland; Karidid District, Namibia; Buranga (pegmatite), Rwanda; Namaqualand, South Africa; Rajasthan, India; Pilbara, Western Australia; Rio Grande do Norte, Brazil; Yellowknife, Northwest Territories; San Diego County, California; Maine; North Groton, New Hamps; Custer, S Dakota; Branchville, CT.

Purpurite-Heterosite

MnPO₄ Manganese phosphate

Crystal system: Orthorhombic.
Crystal habit: Mostly forms earthy masses, grains or crusts.
Colour: Purple, brown or red.
Lustre: Vitreous.
Streak: Deep red to purple.
Hardness: 4–4.5.
Cleavage: Good in one direction (basal).
Fracture: Uneven.
Specific gravity: 3.3.
Notable locations: Sabugal, Portugal; France; Koralpe Mts, Austria; Namibia; Northern Cape Province, South Africa; Pilbara, Western Australia; Faires Tin mine (Kings Mt, Gaston Co), North Carolina; Portland, Connecticut; Yavapai County, Arizona.

Officially discovered in 1905 in Gaston County, purpurite gets its name from its often stunning purple colour. It may have been used as a pigment as long ago as the Renaissance, but it is an extremely rare mineral and even today is used only sparingly by painters. Purpurite forms a series with the phosphate mineral heterosite, with purpurite at the manganese-rich end and heterosite at the iron-rich end. Purpurite is usually no more than a dusty coating or crust on other minerals, because it forms when the primary phosphate minerals that crystallized in granite pegmatites are altered over time by exposure to the air. In fact, it usually forms by the alteration of another rare mineral, lithiophilite, which is why it is itself so rare.

Identification: The deep purple colour of purpurite and heterosite usually makes them fairly easy to find in granite pegmatites – but there is no easy way of telling them apart.

Radioactivity

The nuclei of some atoms, especially large atoms, are naturally unstable and tend to break down or 'decay' to other more stable atoms. This process results in surplus energy being emitted as radiation, typically as tiny alpha and beta particles, but also as gamma rays (high-energy electromagnetic waves similar to light). This is called radioactivity and is measured with a Geiger counter. Radioactivity was discovered in 1896, and is known for its association with nuclear power and weapons. Yet some rock formations such as granite plutons produce quite high levels of radiation and a number of minerals are also naturally radioactive, including xenotime, monazite and autunite. Radioactive minerals require careful handling and storage, because even low-level radiation is a health hazard. For advice on how to handle radioactive specimens safely, see Handling and storing radioactive minerals at the beginning of the Directory.

VIVIANITES

Vivianite gives its name to a small group of rare hydrated phosphate minerals with very similar crystal structures. Vivianite itself is renowned for its blue colour, but all the members of the group – erythrite, annabergite, köttigite and baricite – are all very colourful. Erythrite is a striking crimson and annabergite is fresh apple green, while köttigite is russet coloured and baricite is baby blue (with occasional yellow patches).

Vivianite

$Fe_3(PO_4)_2.8H_2O$ Hydrated iron phosphate

Blue radiating crystals of vivianite.

Vivianite was first discovered as a mineral at the famous Wheal Jane tin mine near Truro in Cornwall, England, by J G Vivian, after whom the species was later named. It typically forms in iron ore veins and phosphate-rich pegmatites – either very late in the crystallization process, as in Idaho, Utah and Colorado, or as the original minerals, such as triphylite and manganese oxides, are weathered. Some of the best crystals are found in cavities within tin ore veins in Bolivia. Vivianite also forms concretions in clay, and when minerals are altered in recent sediments, in lignite and peat, and in fossils, such as a mammoth skull in Mexico (see Mammoth turquoise, right). When it first forms, vivianite is almost colourless, but it turns blue as soon as it is exposed to light. Indeed, it gradually turns completely black and also becomes brittle. Specimens should therefore be kept in a dark place.

Identification: Vivianite is best identified by its blue colour – and the fact that it gets darker when exposed to light. Vivianite beads become magnetic when heated.

Vivianite

Ferruginous sandstone

Crystal system: Monoclinic.
Crystal habit: Radiating clusters of prism-, needle-, or fibre-like crystals. Also forms earthy masses and crusts. May also line fossil shells.
Colour: Colourless to green, blue and indigo, darkening on exposure to light.
Lustre: Vitreous.
Streak: White or bluish green.
Hardness: 1.5–2.
Cleavage: Perfect in one direction.
Fracture: Uneven.
Specific gravity: 2.6–2.7.
Other characteristics: Thin crystals are flexible.
Notable locations: Wheal Jane mine (Cornwall), England; Trepca, Serbia; Crimea, Ukraine; Japan; Anloua, Cameroon; Bolivia; Brazil; Mullica Hill, New Jersey; Leadville, Colorado; Maryland; Utah; Idaho; Maine.

Erythrite

$Co_3(AsO_4)_2.8H_2O$ Hydrated cobalt arsenate

Identification: A crust of erythrite is readily identified by its dark pink colour, with the appearance of raspberry jam smeared on toast.

Erythrite is a striking crimson-coloured mineral that is formed by the weathering of cobalt-rich minerals such as cobaltite. Its bright crimson colour is very noticeable and miners used it to help them locate veins of cobalt-, nickel- and silver-bearing ores. Erythrite may occur as radiating crystals and concretions but more commonly forms a crust on cobalt minerals known as 'cobalt bloom'. Just as substances completely intermixed in liquids form liquid solutions, so components of solids can intermix to form solid solutions. A solid solution series is a range of minerals in which components swap around. Erythrite is part of a solid solution series, with annabergite at the other extreme, in which nickel changes place with cobalt. As the nickel content increases, the colour lightens to white, grey or pale green annabergite.

Crystal system: Monoclinic.
Crystal habit: Typically earthy crusts or masses. Rare crystals in slender prisms or clusters of long flat needles.
Colour: Crimson to lighter pink in crusts.
Lustre: Vitreous.
Streak: Pale red.
Hardness: 1.5–2.5.
Cleavage: Perfect in one direction.
Fracture: Uneven, sectile.
Specific gravity: 3.
Notable locations: Cornwall, Cumbria, England; Schneeburg, Germany; Czech Republic; Bou Azzer, Morocco (with skutterudite); Queensland, Australia; Alamos, Mexico; Cobalt, Ontario.

Annabergite

Ni₃(AsO₄)₂.8H₂O Hydrated nickel arsenate

Long known as nickel ochre, or 'nickel bloom', annabergite was given its name in 1852 by H J Brooke and W H Miller, after Annaberg in Saxony in Germany, one of the best locations for the mineral. It is the nickel-rich equivalent of erythrite, and the nickel turns annabergite apple green or even white – in marked contrast to erythrite's deep crimson. However, there are so many variations in between that it is not always so easy to distinguish them, except with the aid of laboratory analysis. Annabergite is a rare mineral that forms near the surface of cobalt nickel-silver ore veins, typically forming a thin greenish film as the nickel ore is weathered. Unlike erythrite, annabergite has never been found as large crystals. Even small crystals are rare and treasured. One of the few places they are found is Lavrion in Greece, where the specimen shown here was found. Lavrion annabergite crystals are unofficially known as cabrerite after Sierra Cabrera in Spain, where similar crystals have been found.

Identification: Annabergite forms as an apple-green coating and is associated with erythrite and nickel minerals such as skutterudite, niccolite and gersdorffite.

Annabergite

Calcite

Crystal system: Monoclinic.
Crystal habit: Typically earthy crusts and films. Rare small crystals like tiny straws.
Colour: Pale apple green to pink.
Lustre: Silky, glassy, dull.
Streak: Pale green or grey.
Hardness: 1.5–2.5.
Cleavage: Perfect in one direction.
Fracture: Flaky.
Specific gravity: 3.
Notable locations: Teesdale, England; Lavrion, Greece; Sierra Cabrera (Almeria), Spain; Allemont (Isère), Pyrenees, France; Annaberg (Saxony), Black Forest, Hesse, Germany; Carinthia, Salzburg, Austria; Cobalt, Ontario; Humboldt, Nevada.

Mammoth turquoise
In the Middle Ages, French Cistercian monks created a turquoise-blue gemstone to use in church decorations. They called the gemstone 'toothstone' because they made it by heating the fossilized tusks of mastodons, which they thought were giant teeth. Mastodons were an elephant-like creature that lived in Europe 13 to 16 million years ago. These fossilized tusks came from ancient sediments near the Pyrenean mountains. The monks thought toothstone was really the semi-precious stone turquoise, because it looked so much like turquoise. In fact, toothstone was a substance now called odontolite, and is chemically quite different from turquoise. It is essentially fluorapatite, with traces of manganese, iron and other metals. It was once thought that odontolite turned blue as it was heated by changing to vivianite. Recently, though, scientists at the Louvre in France subjected odontolite to spectroscopic analysis. No vivianite was found and they now believe that the colour comes from the alteration of manganese particles.

Köttigite

Zn₃(AsO₄)₂-8H₂O Hydrated zinc arsenate

Köttigite is a rare mineral, first found in the Schneeburg region of Saxony, Germany. Like erythrite and annabergite, it is arsenate that forms by the weathering of metal ores. While erythrite is cobalt-rich and annabergite is nickel-rich, köttigite is zinc-rich. However, there is no real gradation among köttigite and either of the other two, as there is between erythrite and annabergite. This is because zinc particles do not exchange places so readily with nickel or cobalt particles as nickel and cobalt do with each other. Köttigite is found in many locations in small quantities, but the best sites are Schneeburg, Germany, and Mapimi in Durango, Mexico.

Köttigite crystals

Crystal system: Monoclinic.
Crystal habit: Typically forms crusts or powdery masses. Rare, small crystals are flattened blades or radiating needles.
Colour: Reddish or white and grey.
Lustre: Vitreous.
Streak: Pale green or grey, in grey specimens.
Hardness: 2.5–3.
Cleavage: Perfect in one direction.
Fracture: Uneven.
Specific gravity: 3.3.
Other characteristics: Thin crystals are flexible.
Notable locations: Black Forest, Harz Mountains, Schneeburg, Germany; Hohe Tauern, Austria; Bohemia, Czech Republic; Lavrion, Greece; Otjikoto, Namibia; Honshu, Japan; Flinders Range, South Australia; Waratah, Tasmania; Mapimi (Durango), Mexico; Franklin, New Jersey; Churchill, Nevada.

Identification: Köttigite is best identified by its needlelike crystals, its occurrence as a crust and its association with zinc minerals such as smithsonite.

ANHYDROUS PHOSPHATES WITH HYDROXIDE

The hydroxide phosphates include apatite – one of the most abundant minerals, but also the white mineral that forms bones and teeth in all animals. Amblygonite is also typically white, but copper and magnesium turn two other hydroxide phosphates, libethenite and lazulite, into rich shades of green and blue, which makes them highly prized by collectors.

Apatite

Ca₅(PO₄)₃(OH,F,CL) Calcium (fluoro, chloro, hydroxyl) phosphate

Discovered by the famous German geologist Abraham Werner in 1786, apatite gets its name from the Greek *apatan*, which means 'deceive', because it can be confused with beryl, quartz and other hexagonal crystals. It is not actually a mineral species itself, but three similar minerals – fluorapatite, chlorapatite or hydroxyapatite – each of which gets its name from how much fluorine, chlorine or hydroxyl it contains. Together, the apatite trio is incredibly abundant around the world, found mostly in igneous rocks, but in metamorphic and sedimentary rocks as well. It is the main source of phosphorus in soil, which is needed by plants, and apatite and phosphorite are the only natural sources of phosphate minerals used for fertilizer. Although abundant, most apatite is found as tiny grains and crystals. Good, large crystals are rarer, occurring mainly in pegmatites, ore veins and igneous masses. Crystals come in many colours. Asparagus stone is a clear green gem variety of apatite that forms in pegmatites. Beautiful violet apatite forms in the tin veins of Ehrenfriedersdorf in Germany. Big gemlike yellow apatite is found in iron deposits in Durango in Mexico.

Identification: With its hexagonal crystals, apatite can look a little like beryl, tourmaline and quartz, but is much softer and often has a 'sucked sweet' (candy) look.

Apatite crystal

Crystal system: Hexagonal.
Crystal habit: Crystals are typically hexagonal, but apatite can also form tablet-, column- and globe-shaped masses, or form needles, grains and earths. Most commonly occurs as massive beds.
Colour: Typically green but also yellow, blue, reddish brown and purple.
Lustre: Vitreous to greasy.
Streak: White.
Hardness: 5.
Cleavage: Indistinct.
Fracture: Conchoidal.
Specific gravity: 3.1–3.2.
Other characteristics: Some apatites fluoresce yellow.
Notable locations: Saxony, Germany; Tyrol, Austria; Panasquiera, Portugal; Kola Peninsula, Russia; Campo Formosa (Bahia), Brazil; Copiapó, Chile; Durango, Mexico; Wilberforce, Ontario; Mount Apatite, Maine.

Libethenite

Cu₂PO₄(OH) Copper phosphate hydroxide

Identification: Libethenite is olive green, but can be hard to tell from other green copper minerals without complex tests.

Discovered by the famous German mineralogist Friedrich Breithaupt in 1823, libethenite gets its name from Libethen in Romania (now Lubietova in Slovakia), where it was first found. It is a rare copper mineral that is formed by the alteration of other copper minerals. It typically occurs in deeply weathered, concentrated copper sulphide ore bodies. Like many copper minerals it is green, but libethenite is a particularly deep, rich, olive green. It typically occurs in grains, crusts and microscopic crystals, but in 1975 spectacular large crystals were found at the Rokana mine in Zambia's copperbelt. Libethenite is isostructural with olivenite and adamite. This means they both have the same crystal shapes but different properties.

Above: Close-up view of libethenite crystals

Libethenite

Crystal system: Orthorhombic.
Crystal habit: Crystals typically diamond-shaped. Also forms needles, grains, masses and globules and druses.
Colour: Dark olive green.
Lustre: Resinous to vitreous.
Streak: Olive green.
Hardness: 4.
Cleavage: Good in two directions.
Fracture: Brittle.
Specific gravity: 3.6–3.9.
Other characteristics: Does not fizz like malachite in dilute hydrochloric acid.
Notable locations: Cornwall, England; Black Forest, Germany; Lubietova (Libethen), Slovakia; Urals, Russia; DR Congo; Zambia; Tintic, Utah; Gila, Pima and Pinal Cos, Arizona; California.

Amblygonite

(Li, Na)Al(PO₄)(F, OH) Lithium sodium aluminium phosphate fluoride hydroxide

First discovered in Saxony, Germany, by Friedrich A Breithaupt in 1817, amblygonite has now been found in many places around the world – typically in lithium- and phosphate-rich pegmatites. It usually forms large masses embedded in other constituents of these rocks such as quartz and albite. Because amblygonite looks rather like these minerals, it can make up a rather larger percentage of the rock than is perhaps at first apparent. The difference is that amblygonite contains lithium – although it might need a flame test to prove it. When powdered, amblygonite can be set alight with a gas flame, and the lithium burns bright red. Amblygonite gets its odd-sounding name from the Greek *amblus* for 'blunt' and *gouia* for 'angle' because of the shallow angles of its crystal cleavage. However, because amblygonite is usually found embedded in other minerals, it rarely forms good quality crystals. The exceptions are the gem-quality crystals that come from Minas Gerais in Brazil and those found in Myanmar (Burma).

Identification: It is quite difficult to tell amblygonite from other white minerals such as albite without more exhaustive tests, such as the flame test for its lithium content.

Crystal system: Triclinic.
Crystal habit: Typically large masses with irregular outlines, or fine, white crystals including short prisms, tablets and laths.
Colour: Typically creamy white or colourless, but can be lilac, yellow or grey.
Lustre: Vitreous.
Streak: White.
Hardness: 5.5–6.
Cleavage: Perfect in one direction, interrupted in others.
Fracture: Brittle, subconchoidal.
Specific gravity: 3–3.1.
Other characteristics: Some specimens fluoresce orange.
Notable locations: Varutrask, Sweden; Montebras, France; Sakangyi, Myanmar (Burma); W. Australia; Minas Gerais, Brazil; Yellowknife, Northwest Territories; Pala, California; Newry, Maine; Yavapai Co, Arizona; Black Hills, S. Dakota.

Apatite for bones and teeth
Teeth and bones are white because they are made mostly from calcium phosphate in the form of the mineral apatite. Bones are not solid like rock, of course, but comprise a honeycomb of cavities and criss-crossing struts. This composite structure makes bones both light and strong. The struts are actually a mix of organic materials and minerals – and the main mineral is the form of apatite called hydroxyapatite. Bonemaking cells called osteoblasts are constantly at work inside bones creating molecules of the organic material collagen – to renew old bone destroyed by bone-dissolving cells called osteoclasts. By a process called heterogeneous nucleation or heteronucleation, particles of hydroxyapatite dissolved in the surrounding fluid stick to the collagen particles and build up the bone. For bones to stay healthy and strong, we must eat food containing enough calcium and phosphate to keep the osteoblast supplied with hydroxyapatite.

Lazulite

(Mg, Fe)Al₂(PO₄)₂ (OH)₂ Magnesium iron aluminium phosphate hydroxide

Named for its resemblance to the blue gem lapis lazuli, lazulite is a rare and sometimes beautiful azure-blue mineral prized as an ornamental stone. It typically forms at high temperatures in hydrothermal veins, but can also form in phosphate-rich pegmatites, some metamorphic rocks and quartz-rich veins. Lazulite is closely associated with the mineral scorzalite. In fact, lazulite and scorzalite are simply the opposite extremes of a solid solution series (a gradation of mixtures). Lazulite is at the magnesium-rich end; scorzalite is at the iron-rich end. Most lazulite crystals are smallish and quite dull to look at, although occasionally specimens can be quite spectacular and may be as large as a hazelnut.

Crystal system: Monoclinic.
Crystal habit: Typically small wedge shapes. Also granular and massive.
Colour: Pale to deep azure.
Lustre: Vitreous to dull.
Streak: Pale blue to white.
Hardness: 5.5–6.
Cleavage: Distinct in one direction.
Specific gravity: 3.1.
Other characteristics: Clear gemmy crystals show strong pleochroism (yellowish, clear, blue). Crystals are slightly soluble in warm hydrochloric acid.
Notable locations: Zermatt, Switzerland; Horrsjøberg, Sweden; Salzburg, Austria; Copiapó, Chile; Diamantina, Minas Gerais, Brazil; Yukon; Graves Mount, Georgia; Death Valley (Inyo Co), California.

Quartz

Lazulite crystals

Identification: Lazulite is best identified by its dark blue colour, but it is hard to tell from lazurite, sodalite and other blue minerals.

HYDRATED PHOSPHATES

The hydrated phosphates include some fascinating minerals – not only the beautiful blue-green stone turquoise, one of the most ancient and most widely cherished gems, but also wavellite, with its distinctive starburst clusters, and two uranium-rich, radioactive minerals – dark green, blocky torbernite and greenish yellow, fluorescent autunite.

Turquoise

$CuAl_6(PO_4)_4(OH)_8.5(H_2O)$ Hydrated copper aluminium phosphate

Identification: Although often imitated by fakes such as chrysocolla, turquoise is generally very distinctive with its vivid blue-green colour and smooth, waxy appearance.

One of the most beautifully coloured of all stones, turquoise is a phosphate that combines copper and aluminium. The copper causes the exquisite blue-green, but the colour varies from green to yellowish grey. Pale sky blue is the most cherished gem colour – especially when infused with fine veins of impurities that show it is natural, not artificial. Although opaque, solid turquoise is actually crystalline, or rather, cryptocrystalline, since the crystals are too small to be seen by the naked eye. It typically forms in waxy veinlets where groundwater washes over weathered, aluminium-rich rock in the presence of copper. Thus it is often associated with copper deposits as a secondary mineral. Turquoise contains water, but won't form if conditions are wet, so most major deposits are in dry regions. For thousands of years, the best turquoise came from the deserts of Persia and was known as Persian turquoise. In the late 1800s, deposits of turquoise were discovered in the American south-west, and this is now the world's prime source of high-quality 'Persian' turquoise.

Coating of turquoise

Quartz

Crystal system: Triclinic.
Crystal habit: Typically tiny, cryptocrystalline forms as nodules and veinlets. Also forms crusts.
Colour: Shades of blue-green.
Lustre: Dull to waxy.
Streak: White with green tint.
Hardness: 5–6.
Cleavage: Perfect in two directions, but not often seen.
Fracture: Conchoidal.
Specific gravity: 2.6–2.8.
Other characteristics: Colour can change with exposure to skin oils.
Notable locations: Neyshabur, Iran; Afghanistan; Sinai; Broken Hill (NSW), Victoria, Australia; Baja California, Mexico; Arizona; Nevada; Colorado; San Bernardino Co, Inyo Co, Imperial Co, California.

Wavellite

$Al_3(PO_4)_2(OH)_3.(H_2O)_5$ Hydrated aluminium phosphate hydroxide

Named after William Wavell, the English country doctor who discovered it in Harwood, Devon, in 1805, wavellite is an aluminium phosphate. It typically forms in hydrothermal veins and crevices and on surfaces in limestone, chert and aluminium-rich metamorphic rocks, and is often associated with limonite, quartz and micas. The classic wavellite specimens are radiating 'starburst' clusters of sparkling yellow-green needle-like crystals. Sometimes called cat's eyes, these clusters typically form as nodules in crevices in limestone and chert, and are revealed when the crevices are split open to show them growing on surfaces like thin coins. If there was room for them to grow freely, they might form half-globes. Although the starburst cluster is best shown when the samples are split, some collectors prefer just a few split, leaving the rest as solid half-globes.

Identification: Wavellite's radiating starburst clusters are unmistakable. These clusters often form in globules, or botryoidal masses.

Crystal system: Orthorhombic.
Crystal habit: Radiating needles forming globules or botryoidal masses.
Colour: Yellow green or white.
Lustre: Vitreous.
Streak: White.
Hardness: 3.5–4.
Cleavage: Perfect in two directions.
Fracture: Uneven.
Specific gravity: 2.3+.
Notable locations: Devon, Cornwall, England; Zbiroh, Czech Republic; Ronneburg, Germany; Pannece, France; Llallagua, Bolivia; Garland County, Arkansas; Pennsylvania.

Torbernite

Cu(UO₂)₂(PO₄)₂-10(H₂O) Hydrated copper uranyl phosphate

Torbernite was first found in 1772 at Johanngeorgenstadt in Saxony, Germany, but is named after 18th-century Swedish scientist Torbern Bergmann. It typically forms small, dark green plates and square crystals embedded in and coating other crystals in cracks in pegmatites and granites. Torbernite is a uranium-rich mineral like autunite and usually forms by alteration of pitchblende, a black, massive form of uraninite, the main ore of uranium. The uranium content makes torbernite radioactive. The radioactivity is unlikely to do any harm, but specimens should be handled with care, and kept well away from children. Most uranium ores produce radioactive dust when handled, so touch specimens as little as possible and wash hands thoroughly afterwards. Keep specimens in an airtight container as well to contain any radon gas emitted, and open only outside. The main problem with torbernite is that samples disintegrate and lose their water easily to become crumbly metatorbernite – another reason why they should be handled infrequently and kept in an airtight container.

Identification: Torbernite is best identified by its dark green square crystals and its association with autunite and uraninite.

Caution: Torbernite is radioactive, so should be handled with care and stored appropriately.

Crystal system: Tetragonal.
Crystal habit: Typically square, tablet-shaped crystals, often stacked like books. Also forms crusts, micaceous, foliated and scaly aggregates.
Colour: Dark to light green.
Lustre: Vitreous to pearly.
Streak: Pale green.
Hardness: 2–2.5.
Cleavage: Perfect in one direction.
Fracture: Uneven.
Specific gravity: 3.2+.
Other characteristics: Radioactive.
Notable locations: Gunnislake (Cornwall) England; Trancoso, Portugal; Erzebirge, Germany; Bois Noir, France; Katanga, Zaïre; Mount Painter, South Australia; Mitchell County, North Carolina; Utah.

Ancient turquoise
Turquoise is probably one of the oldest gems known. The Egyptians cherished it 6,000 years ago, and established the world's oldest mines in hard rock, in the Sinai peninsula. When the tomb of the Egyptian queen Zer was excavated in 1900, she was found to be buried with a gold and turquoise bracelet – one of oldest known surviving pieces of jewellery. In Ancient Persia, where it was mined on Mount Alimersai, now in Iran, it was used as a talisman for good fortune. It probably came to Europe from Persia from the time of the Crusades onwards. The Europeans thought it originated in Turkey and it may have got its name from the French for Turkish. In the Americas, turquoise has been valued since at least 200BC by native peoples in the American south-west such as the Navajo, and by Indian tribes in Mexico. The Navajo made turquoise into beads, carvings and mosaics. The Navajo are said to believe turquoise is a piece of sky that has fallen to Earth, while the Apache think it combines the spirits of sea and sky to help hunters. The Aztecs also cherished turquoise, and the treasure of Montezuma contains a turquoise mosaic serpent.

Autunite

Ca(UO₂)₂(PO₄)₂-10H₂O Hydrated calcium uranyl phosphate

Named after Autun in France, autunite is a uranium mineral. It is closely related to torbernite, but is much more common, and therefore is frequently used as a uranium ore. It typically forms by the alteration of the surface of other uranium ores found in pegmatites. By daylight, it can be very difficult to see on the rock faces, but when it is exposed to ultraviolet light at night, it reveals its presence by fluorescing dramatically. Like torbernite, it is radioactive and tends to crumble easily when exposed to air (turning to meta-autunite), so it should always be handled with care and stored in an airtight container well out of the reach of children.

Crystal system: Tetragonal.
Crystal habit: Typically square, tablet-shaped crystals, often stacked like books.
Colour: Greenish yellow.
Lustre: Vitreous to pearly.
Streak: Pale green.
Hardness: 2–2.5.
Cleavage: Perfect in one direction.
Fracture: Uneven.
Specific gravity: 3.1–3.2.
Other characteristics: Radioactive and fluorescent.
Notable locations: Dartmoor, England; Trancoso, Portugal; Saône-et-Loire, Margnac, France; Katanga, DR Congo; Mount Painter, South Australia; Mount Spokane, Washington; Mitchell County, North Carolina.
Caution: Autunite is radioactive, so should be handled with care and stored appropriately.

Identification: Autunite is best identified by its fluorescence, its yellow-green square, tablet-shaped crystals and its association with torbernite and uraninite.

ARSENATES

Arsenates are very similar in make-up to the phosphates and vanadates, with arsenic often simply replacing phosphorus and vanadium in the mix. They include the rare and attractive green and greenish yellow minerals adamite, conichalcite, olivenite, scorodite and bayldonite. When powdered and burned with charcoal, they all give off arsenic's distinctive garlic smell.

Adamite

$Zn_2(AsO_4)(OH)$ Zinc arsenate hydroxide

Adamite is named after the 19th-century French mineralogist Gilbert-Joseph Adam, who discovered it at Chanarcillo in Chile. Its vividly coloured, lustrous crystals make it popular with collectors. It is a secondary zinc mineral that typically forms where zinc-rich ores are oxidized. Cobalt impurities give it a pink tinge; copper impurities turn it greenish. Pure adamite, uncontaminated by copper, is a distinctive yellow or white, and is also brilliantly fluorescent, glowing lime green under ultraviolet light. Typically, adamite is found lining cavities in limonite, but the classic location for adamite crystals is Mapimi, in the state of Durango, in Mexico. Mapimi is a limestone replacement deposit rich in rare arsenic minerals. Here, adamite is found not with limonite but in yellow straws and sprays with hemimorphite, austinite and rare minerals such as legrandite (see inset) and paradamite. All the best specimens of fluorescent adamite come from this location.

Legrandite

Identification: The lime-green colour and high lustre of adamite are hard to miss – and the brilliant fluorescence of pure adamite is unmistakable. Crusts of adamite can look like smithsonite.

Adamite

Crystal system: Orthorhombic
Crystal habit: Crystals are typically blunt-end wedge-like prisms, mostly in druses and radiating clusters in wheels and sheaves
Colour: Yellow or white when pure, tinged green by copper and pink by cobalt
Lustre: Adamantine
Streak: White to pale green
Hardness: 3.5
Cleavage: Perfect in two slanting directions
Fracture: Absent
Specific gravity: 4.3–4.4
Other characteristics: Fluoresces bright green
Notable locations: Cumbria, England; Tyrol, Austria; Hyeres, France; Reichenbach, Germany; Tuscany, Italy; Lavrion, Greece; Chanarcillo, Chile; Mapimi, Mexico; Inyo County, California; Gold Hill, Utah; Nevada

Conichalcite

$CaCu(AsO_4)(OH)$ Calcium copper arsenate hydroxide

Calcite

Conichalcite gets its name from the Greek for 'powder' (*konis*) and 'lime' (*chalx*). It was first identified in 1849 near Córdoba in southern Spain by the famous German mineralogist Friedrich A Breithaupt. It has a very distinctive grass-green colour unlike any other mineral, making it easy to identify. It coats limonite rock with a crust that looks rather like moss, but is entirely mineral in origin, and the green colour comes from the copper. This copper content means that it is occasionally used as a copper ore.

It forms where copper is oxidized when oxygen-rich water reacts with copper sulphide or oxide minerals. Sometimes the limonite is red or yellow, providing a striking colour combination. It is typically found in association with copper and arsenic minerals such as adamite, azurite, bayldonite, linarite, malachite, olivenite and smithsonite.

Coating of conichalcite

Identification: Conichalcite's striking grass-green colour and crust-forming habit, seen here on geothite.

Crystal system: Orthorhombic
Crystal habit: Typically forms crusts
Colour: Grass green
Lustre: Vitreous
Streak: Green
Hardness: 4.5
Cleavage: Absent
Fracture: Uneven
Specific gravity: 4.3
Notable locations: Gunnislake (Cornwall), Cumbria, England; Black Forest, Germany; Poland; Lavrion, Greece; Córdoba, Spain; Guanaco, Chile; Mapimi, Mexico; Lun County, New Mexico; Tintic Mountains (Juab County), Utah; Esmeralda, Eureka and Mineral Counties, Nevada; Cochise, Pima, Pinal, and Yavapai Counties, Arizona

Olivenite

Cu₂AsO₄(OH) Copper arsenate hydroxide

Despite its name, olivenite bears no relation to the mineral olivine. Its name comes entirely from its typical olive-green colour. It is a now rare, secondary mineral that crystallizes in small veins and vugs (pockets), where it forms by the alteration of copper ores and mispickel (arsenopyrite). It was once found in considerable quantities in association with limonite and quartz in mine dumps, and in the upper workings of copper mines, in Cornwall, England, especially near St Day and Redruth. It was in Cornwall, at the Carharrack mine at Gwennap, that olivenite was first identified in 1820. Cornish olivenite occurs as crusts of whitish colour-banded vertical needles that look like wood splinters, earning it the local name 'wood copper'. Wood copper is also found at Tintic in Utah, while the best olivenite crystals come from Tsumeb in Namibia. The arsenic in olivenite is often replaced by a proportion of phosphorus, forming a libethenite that looks like olivenite, as at Lubietova in Slovakia.

Identification: Olivenite is best identified by its coating of tiny, needle-like olive-green crystals. Wood copper olivenite forms distinctive wood splinter mineral crusts.

Quartzite   *Olivenite*

Crystal system: Orthorhombic.
Crystal habit: Tubby, prism-shaped crystals or crusts of needle-like crystals.
Colour: Dark olive green, yellow, brown, whitish.
Lustre: Vitreous to greasy.
Streak: Olive green.
Hardness: 3.
Cleavage: Rarely noticed.
Fracture: Conchoidal.
Specific gravity: 3.9–4.4.
Other characteristics: Soluble in hydrochloric acid.
Notable locations: Cornwall, England; Grube Clara (Black Forest), Germany; Lavrion, Greece; Tsumeb, Namibia; Tintic, Utah; Majuba Hill, Nevada.

Cornish tin mines
When the South Crofty mine based at Camborne in Cornwall, England, closed in the late 1990s, it marked the end of a tradition of tin mining there dating back to the Bronze Age. At its peak in the 1870s, Cornwall was the world's leading tin mining area, with over 2,000 pits at work. In the end Cornish tin proved uneconomic because it is so hard to access. As Cornwall's granite masses cooled, fissures opened up and were filled with hot molten material, which crystallized to form lodes rich with minerals such as tin, copper, zinc, lead and iron. Each vertical fissure had to be mined with a separate shaft, plunging deep and straight into the ground, often well below the water table – which is why powerful water pumps were needed, especially in mines perched on cliffs right next to the sea, like the Wheal Coates tin mine pictured above. Although the mines are now closed, specimens of rare minerals can be found on the old mine dumps, especially the Penberthy Croft at Wheal Fancy, where natanite, jeanbandyite, segnitite and bayldonite were first found. This site is now protected by law.

Scorodite and Bayldonite

Fe₃AsO₄·2H₂O Hydrated zinc arsenate (Scorodite)

Scorodite is an attractive mineral that forms pale grey-green clusters of crystals, or pale greenish earthy masses. It typically forms as a secondary mineral where zinc minerals in arsenic-rich veins are altered by exposure to oxygen. But it has also been found in crusts around hot springs. Scorodite forms a solid-solution series with mansfieldite in which aluminium replaces iron in the scorodite structure. Bayldonite (above) is a much rarer, but even more attractive, mineral that forms where copper and lead minerals in arsenic-rich veins are altered by exposure to oxygen. It was first discovered at the Penberthy Croft mine (Wheal Fancy) in Cornwall, England.

A thin green coating of bayldonite on quartz.

⬦ Scorodite
⬦ Bayldonite

Crystal system: Orthorhombic.
Crystal habit: Double pyramids that look like octahedrons. Also forms crusts and earthy masses.
Colour: Pale green, grey green, blue, and brown.
Lustre: Vitreous to sub-adamantine or greasy.
Streak: White.
Hardness: 3.5–4.
Cleavage: Generally poor.
Fracture: Conchoidal.
Specific gravity: 3.1–3.3.
Notable locations: Cornwall, England; Lavrion, Greece; Tsumeb, Namibia; Mapimi, Zacatecas, Mexico; Ouro Prêto, Brazil; Ontario; California; Utah.

Identification: Scorodite is best identified by its grey-green colour. It can look like zircon, but is not fluorescent.

Scorodite   *Quartz*

VANADATES

The vanadates are a group of minerals formed when vanadium and oxygen combine with various metals to form complex molecules. The conditions in which they form are very particular, so these minerals are quite rare. However, vanadinite and carnotite are often mined as ores of vanadium, radium and uranium, and many vanadates are prized by collectors for their brilliant colours.

Mottramite

PbCu(VO₄)(OH) Lead copper vanadate hydroxide

Originally named psittacine, mottramite was named in 1876 after the village of Mottram St Andrews in Cheshire, England, in 1876 by Sir Henry Enfield Roscoe (1833–1915). Roscoe was the chemist who first isolated metallic vanadium, then discovered that it occurs naturally in mottramite. Vanadium salts are obtained by 'digesting' mottramite with concentrated hydrochloric acid. Vanadium is now known to occur in vanadinite, descloizite, roscoelite and pucherite as well. Mottramite is a green to black mineral that usually occurs as a drusy crust on other rocks and minerals such as wulfenite and vanadinite. It is usually found in places where copper and lead minerals have been oxidized by contact with air and water, or just water. Large crystals of mottramite up to 2.5cm/1in long have been found in the Otavi triangle, Namibia. Mottramite is typically found in 'green' associations with other copper minerals such as descloizite, malachite and pyromorphite.

Identification: Dark, dirty green velvety crusts of mottramite are usually easy to identify, especially if associated with vanadinite, descloizite, malachite and wulfenite.

Crystal system: Orthorhombic.
Crystal habit: Typically tiny drusy crusts, radiating and stalactitic masses.
Colour: Typically various shades of green, but also more rarely black.
Lustre: Resinous.
Streak: Green.
Hardness: 3–3.5.
Cleavage: None.
Fracture: Conchoidal to uneven.
Specific gravity: 5.7–6.
Other characteristics: Nearly always found in association with descloizite.
Notable locations: Mottram St Andrews, England; Grootfontein, Otavi, Namibia; Bolivia; Chile; Bisbee, Tombstone (Cochise Co), Pinal County, Arizona; Sierra County, New Mexico.

Descloizite

PbZn(VO₄)(OH) Lead zinc vanadate hydroxide

Identification: Descloizite is best identified by the khaki or black velvety crusts it forms, its high density and its associations with zinc and vanadate minerals.

Dark grey crystals of descloizite

Discovered by A Damour in 1854 near Córdoba in Argentina, descloizite is named after famous French scientist Alfred Louis Oliver Legrand des Cloizeaux. Descloizite and mottramite are at the two opposing ends of a solid solution series – a range of solids in which certain chemical elements are mixed in varying degrees. Descloizite is the zinc-rich end of the range while mottramite is the copper-rich end, but there is a gradation in composition between; most descloizite specimens contain some copper and most mottramite specimens contain some zinc. The more copper thre is in descloizite, the more orange or yellow it is. Descloizite is a secondary mineral and often occurs on the surface of vanadinite. It typically forms small platelike crystals or velvety crusts in the oxidation zone of hydrothermal veins where lead, zinc and copper ores are associated with vanadates and other lead and zinc minerals.

Crystal system: Orthorhombic.
Crystal habit: Typically forming tiny platelike crystals, arrow-shaped crystals, velvety crusts, 'trees', small drusy crusts and stalactitic masses.
Colour: Cherry red, brown, khaki, black or yellowish.
Lustre: Translucent.
Streak: Orange to brownish red.
Hardness: 3–3.5.
Cleavage: None.
Fracture: Conchoidal to uneven.
Specific gravity: 5.9.
Notable locations: Cornwall, England; Carinthia, Austria; Tsumeb, Otavi, Namibia; Córdoba, Argentina; Bisbee, Tombstone (Cochise Co), Pinal County, Arizona; Lake Valley (Sierra Co), Grant County, New Mexico.

Carnotite

K₂(UO₂)₂(VO₄)₂.3H₂O Hydrated potassium uranyl vanadate

Named after the French chemist M A Carnot (1839–1920), carnotite is one of the most important sources of the radioactive elements radium and uranium. One hundred years ago, Marie and Pierre Curie, the French husband and wife pioneers of research on radioactivity, used carnotite from Colorado as their experimental material for radium – which is why it must be handled very carefully, like all radioactive minerals. For many years, the carnotite deposits of the American south-west were the world's major source of radium – until even richer sources of radium were found in Katanga, DR Congo, and Radium Hill, South Australia. Today the Colorado and Utah carnotite is a major source of uranium. They are thought to form when uranium and vanadium ores in red-brown sandstones are altered – often replacing fossil wood. Carnotite often forms a brilliant yellow coating on sandstone. It is closely related to tyuyamunite, but carnotite contains potassium while tyuyamunite contains calcium.

Identification: Carnotite is bright yellow and can be distinguished from autunite by a lack of fluoresce.

Caution: Carnotite is very radioactive, so should be handled and stored with great care.

Crystal system: Monoclinic.
Crystal habit: Includes microscopic platelike crystals, crusts, earthy masses and flaky or grainy aggregates.
Colour: Bright yellow.
Lustre: Pearly to dull or earthy.
Streak: Yellow.
Hardness: 2.
Cleavage: Perfect in one direction.
Fracture: Uneven.
Specific gravity: 4–5.
Other characteristics: Strongly radioactive and does not fluoresce.
Notable locations: Kazakhstan; Kokand, Ferghana, Uzbekistan; Katanga, Zaïre; Morocco; Radium Hill, South Australia; Wyoming, Colorado; Utah; Arizona; Grants County, New Mexico; Mauch Chunk (Carbon County), Pennsylvania.

Vanadinite

Pb₅(VO₄)₃Cl Lead chloro-vanadate

Namibia's mineral mecca
Tsumeb in Namibia and the nearby locations of Grootfontein and Otavi are among the world's richest sources of minerals. More than 240 different minerals have been found at Tsumeb alone! Until quite recently, the region's mines were a major source of copper, lead, silver, zinc and cadmium. Even the bushmen found copper here, in a malachite hill, which they bartered for tobacco with the Ovambo people. Now it is more famous for its wonderful crystals of dioptase, mottramite, descloizite, cerussite, aragonite, tennantite, enargite and many more. The mineral deposits centre on a hydrothermal vein, rich with lead, zinc and copper sulphides and arsenides, mixed in with rarer elements such as germanium and gallium. But what makes Tsumeb special is the limestone rock of the region, which allows water to percolate into the vein and oxidize the minerals at many different levels – as far as 1,500m/5,000ft into the ground. Each different level creates cavities containing different secondary minerals.

When it was first identified in Mexico in 1801, vanadinite was believed to contain a brand new element that was called erythronium. Only later was this found to be vanadium. In fact, vanadinite is now used as an ore of vanadium, and also a minor ore for lead. It is a secondary mineral that forms mostly in dry and desert regions, such as Namibia, Morocco and the south-west USA. It typically occurs where lead ores are weathered, and is associated with wulfenite, descloizite and cerussite. Sometimes, arsenic replaces some of the vanadium in vanadinite to create the mineral endlichite. When arsenic completely replaces vanadium in the crystal structure, it becomes mimetite.

Crystal system: Hexagonal.
Crystal habit: Typically small, thin six-sided prism-shaped crystals, or globular masses.
Colour: Mahogany red, brown, brownish yellow or orange.
Lustre: Resinous, adamantine.
Streak: Brownish yellow.
Hardness: 3–4.
Cleavage: None.
Fractue: Conchoidal.
Specific gravity: 6.7–7.1
Other characteristics: Very brittle.
Notable locations: Warlockhead (Dumfries and Galloway), Scotland; Carpathia, Austria; Urals, Russia; Mibladen, Morocco; Otavi, Namibia; Marico, South Africa; Chihuahua, Mexico; Tucson, Arizona; Sierra County, New Mexico.

Identification: Vanadinite is most easily identified when it occurs in the brown-red form. Association with descloizite, wulfenite and cerussite is a key indicator.

QUARTZ

Quartz is the single most common mineral in the Earth's crust and is a major ingredient of many igneous and metamorphic rocks. Because quartz is very tough, it doesn't break down easily and so it is also found as a major constituent of most sedimentary rocks other than those that form biogenically or chemically. Although quartz is basically colourless, impurities cause a huge variety of colours to be seen, ranging from purple amethyst to yellow citrine.

Quartz (crystalline and massive)

SiO$_2$ Silicon dioxide

Rock crystal (below): Rock crystal is the clear, colourless variety of quartz. It is one of the least expensive and most popular of all gemstones, simply because it is quite common – although the giant crystals from which fortune tellers' crystal balls were cut are very rare. Some collectors prefer natural clusters of rock crystal with arrays of glittering pinnacles to separate stones.

Smoky quartz (below): This quartz is one of the few brown gemstones. It comes in several varieties, ranging from pale brown to black, and its abundance makes it worth much less than either amethyst or citrine. In Scotland, dark brown quartz is known as Cairngorm, from the Cairngorm mountains, and is a popular ornamental stone, often carved into fireplaces or worn in brooches with Highland costume. Other popular varieties include black morion, and banded black and grey coon tail quartz. The dark colour comes from exposure to natural radiation deep underground. In fact, pale quartz can be artificially turned into smoky quartz by exposure to radiation. But if these artificially coloured stones are heated, they will turn pale again.

Quartz was one of the first mineral crystals known. In fact, the very word crystal comes from *krystallos*, the Ancient Greek word for 'rock crystal', the colourless form of quartz. Rock crystal (clear quartz) looks so much like ice that for a long time it was thought to be a form of ice that wouldn't melt. The word quartz itself is German in origin, but just what it meant is lost in the mists of time. Many varieties of quartz have been known since ancient times, such as carnelian, agate and chalcedony (see following pages). In the Middle Ages, large crystals of clear quartz were carved into balls for fortune tellers and alchemists. In the 16th and 17th centuries, barrel-loads of rock crystal were shipped from Brazil and Madagascar to Europe to be carved into vases, decanters and chandeliers. Now quartz has many uses, from radios and watches (see Electric quartz, right) to machine bearings.

Quartz forms in a huge variety of places. Most originally form when magmas rich in silica and water crystallize to form rocks such as granite. Larger quartz crystals form in pegmatites, but these are typically whitish in colour because they contain minute fluid-filled cavities. When these igneous rocks are broken down by weathering, the toughness of quartz grains ensures they survive to form the basis of most sedimentary rocks. It is these rocks that form the pure quartz sand that is the main source used in glassmaking. Large, collectable crystals of quartz form mainly in crusts, hydrothermal veins and in geodes in sandstone. Indeed, quartz is the dominant mineral in most mineral veins. Some of the best crystals come from cavities in the Swiss Alps known as crystal caves. There are literally hundreds of different varieties and names for quartz. The table opposite covers just the better known ones.

Crystal system: Trigonal.
Crystal habit: Widely variable, but most common is hexagonal prisms ending in six-sided pyramids. Can be cryptocrystalline.
Colour: Widely variable but clear is most common, followed by cloudy or milky quartz, purple amethyst and pink agate and brown smoky quartz.
Lustre: Vitreous or resinous.
Streak: White.
Hardness: 7.
Cleavage: Poor.
Specific gravity: 2.65.
Other characteristics: Piezoelectric (see box).
Notable locations:
Rock crystal: Aar Massif, Switzerland; Rhine Westphalia, Germany; Salzburg, Austria; Madagascar; South Africa; Brazil; Ouachita Mountains, Arkansas; Lyndhurst, Ontario.

Smoky quartz: Scotland; Alps, Switzerland; Brazil; Pikes Peak, Colorado.

Amethyst: Kapnik, Hungary; Urals, Russia; Zambia; Namibia; South Africa; Guerrero, Mexico; Bahia, Rio Grande do Sul, Brazil; Uruguay; Thunder Bay, Canada; Maine; Pennsylvania.

Chrysoprase: Marlborough (Queensland), Australia.

Tiger's Eye: Doorn Mountains, South Africa.

Citrine: Salamanca, Spain; Urals, Russia; Dauphine, France; Madagascar.

Citrine (below): Citrine is a semi-precious yellow, orange or brown variety of quartz. The name comes from *citrus*, the Latin for 'lemon', and the yellow colour comes from tiny particles of iron oxide suspended within it. It is the most valued quartz gem, but is sometimes thought of as 'imitation' topaz, which it resembles, but is slightly softer than. It can actually be created artificially by heating amethyst, and many of the citrine gems on the market are really heat-treated amethyst from Minas Gerais in Brazil. Heat-treated citrine tends to be more orange than natural citrine, which is typically pale yellow. Natural citrine is found in protruding clusters of small, pyramidal crystals, often in geodes, in places like the Russian Urals, Dauphine in France and Madagascar. Some naturally occurring citrine was once amethyst, but has been turned yellow-orange by exposure to hot magma. In ancient times, citrine was carried as a talisman against snake bites and evil thoughts.

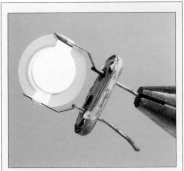

Citrine

Calcite

Basalt

Amethyst (right): Amethyst is one of the most attractive forms of quartz, a favourite with everyone from the pharaohs of Egypt to Catherine the Great of Russia. Tiny traces of iron in the quartz give it a colour varying from pale mauve to deep violet. It is found in most places where granite is exposed on the surface. The name amethyst is said to come from the Ancient Greek myth of the beautiful girl Amethyst. In a drunken rage after a party, so it goes, the god of revelling and drink Dionysus once swore that tigers would eat the next person that came along. When Amethyst was first to come by, the goddess Athene turned her to white stone to save her. Thoroughly remorseful, Dionysus cried into his drink and poured it over the stone Amethyst, turning it purple. It was later believed that amethyst would ward off drunkenness. Amethyst came to be a symbol of celibacy and piety and a key ornament for Catholics in the Middle Ages, worn by medieval bishops in their rings. Even today, the highest grade of amethyst is known as 'Bishop's Grade'. Leonardo da Vinci wrote that it 'dissipates evil thoughts and quickens the intelligence'. In Tibet, Buddhists make rosaries from it. The largest amethyst crystals come from South American countries such as Brazil and Uruguay, where they are found in gigantic geodes – some big enough to walk into (see How Minerals Form, Understanding Rocks and Minerals). The richest purple amethysts come from African countries such as Namibia and Zambia.

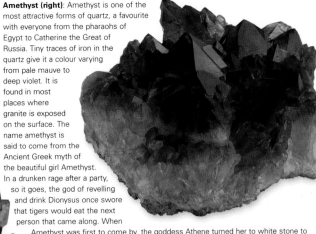

Electric quartz

Better than any other mineral, quartz demonstrates the phenomenon of piezoelectricity. This means that when pressure is applied to a quartz crystal, a positive electrical charge is created at one end of the crystal and a negative charge at the other. In the same way, if an electric current is applied to a quartz crystal, it bends or changes its shape slightly to and fro. In other words, it vibrates or oscillates, and as it does so it generates little pulses of electricity. This oscillation is so perfectly regular that it can be used as a highly accurate timer in watches, especially if shaped like a tiny tuning fork.

Crystalline and massive varieties of quartz

Name	Habit	Colour	Transparency	Cause of colouring, inclusions etc
Quartz, common vein quartz	massive crystals	white, grey, yellowish	dull, opaque	gases and liquids, cracks, etc.
Rock crystal	crystals	colourless	transparent	—
Smoky quartz	crystals	brown	transparent	colouring caused by radiation
Morion	crystals	brown to nearly black	translucent	colouring caused by radiation
Amethyst	crystals, also massive	purple	transparent	colouring caused by radiation
Citrine	crystals	yellow	transparent	traces of very finely divided FeOOH
Rose quartz	massive, crystals rare	pink	translucent	minute rutile needles
Blue quartz	granular, massive	blue	translucent to transparent	minute rutile needles
Prase	massive	leek green	translucent	nickel silicates, actinolite needles
Aventurine	massive	iridescent, various colours	opaque	flakes of chromian muscovite or hematite
Tiger's eye	fibrous	blue to golden	opaque	silicified crocidolite asbestos

CHALCEDONY

When quartz forms at low temperatures in volcanic cavities, the crystals can be so small that the mineral looks more like porcelain. The general name for this 'cryptocrystalline' quartz is chalcedony. It comes in an astonishing array of colours and patterns, including blood-red carnelian, wine-red jasper, brown-banded agate, green-moss agate, apple-green chrysoprase and black and white onyx.

Chalcedony (cryptocrystalline quartz)

SiO₂ Silicon dioxide

Chalcedony is a form of chert, and perhaps the most widely used of all gems through the ages. In fact, apart from sticks, bones and plain rocks, chalcedonies may have been the earliest hard materials used by mankind, shaped into arrowheads, knives, tools, cups and bowls. Its natural beauty meant it was also used ceremonially and for decoration long ago. It was a sacred stone for Native Americans, said to promote stability and harmony.

Chalcedony is typically fibrous and splintery and occurs in rounded crusts, rinds or stalactites in both volcanic and sedimentary rocks, precipitating from solutions moving through the ground. It can also form from organic matter. For example, over millions of years, chalcedony can replace the wood in dead trees, preserving the form of the trees in stone. Sometimes, this kind of chalcedony preserves the rings of the original tree, and so is mistaken for agate, the banded form of chalcedony.

Jasper (below): Jasper is the opaque, red, or occasionally green, variety of chalcedony. The red colour comes from traces of iron oxide, while the green comes from microscopic fibres of actinolite or chlorite. Sometimes other minerals can turn jasper brownish or yellow, but its streak is always white. It typically forms as part of agate nodules, but you can often find jasper pebbles on beaches in areas of igneous rock. Although dull when found, jasper takes polishing well, and pebbles glisten red when wet.

There are many varieties of chalcedony, some of which are shown here and on the opposite page. Apart from agate, another strikingly beautiful stone is chrysoprase. This translucent, apple-green stone gets its colour from traces of nickel. It typically occurs in cavities in serpentine. It was once found mainly in Silesia in central Europe. It was a particular favourite of Frederick the Great, and can be seen decorating many buildings in Prague, including the Chapel of St Wenceslas. Now most chrysoprase comes from Australia. Bloodstone or heliotrope is a dark-green stone containing the spots of red jasper that earned it its name. It was highly prized in the Middle Ages, when its 'blood' spots linked it to martyrdom and flagellation. Plasma is a semi-translucent, very fine-grained variety of chalcedony that owes its green colour to silicate particles such as amphibole and chlorite.

Carnelian (above): Carnelian is the orange-red, translucent variety of chalcedony. It gets its reddish colour from traces of hematite (iron oxide) or goethite and its colour can be enhanced by baking and dyeing with iron salts. It was valued long ago by the Greeks and Romans, who used it for signet rings. It is a close relative of russet-coloured sard, which got its name from the ancient Lydian city of Sardis. Together carnelian and sard were known as sardion in the Middle Ages. Sard with bands of white chalcedony is known as sardonyx and was once the most precious of all stones.

Crystal system: Trigonal.
Crystal habit: Cryptocrystalline with a fibrous structure.
Colour: Widely variable from green chrysoprase and moss agate, through red carnelian and jasper to brown sard and black and white onyx.
Lustre: Vitreous or resinous.
Streak: White.
Hardness: 7.
Cleavage: Poor.
Fracture: Brittle and conchoidal.
Specific gravity: 2.65.
Notable locations: Jasper: Urals, Russia; Oregon; Arizona; California.

Carnelian: Ratnapura, India; Warwick (Queensland), Australia; Campo de Maia, Brazil.

Sard: Ratnapura, India.

Bloodstone: Kathiawar Peninsula, India.

Onyx: India; Brazil

Plasma: Bavaria, Germany; Egypt; Madagascar; India; China; Australia; Brazil.

Agate: Idar-Oberstein, Germany; Salzburg, Austria; Botswana; Madagascar; China; India; Queensland, Australia; Minas Gerais, Rio Grande do Sul, Brazil; Uruguay; Chihuahua, Mexico; Fairburn, South Dakota; Nipomo, California; Oregon; Idaho; Montana; Washington.

Thunder egg agate: Jefferson County, Oregon.

Fire agate: Mexico; Deer Hill, Arizona.

Chrysoprase: Silesia, Poland; Urals, Russia; Australia; Brazil; California.

Agate (left): Agate forms when traces of iron, manganese and other chemicals create bands in chalcedony. Although the bands in agate form naturally, agates sold in shops are often stained artificially. Agates come mostly from cavities in basalt. They typically form in frothy basalt lavas that solidify so quickly as they flood on to the surface – especially where they meet water – that they trap gas bubbles. Silicate minerals dissolve from the lava and filter into the gas bubbles. As the lava cools, these dissolved silicate minerals coagulate as a gel inside the bubbles. Then iron and manganese compounds from the surrounding rock infiltrate the gel, creating layers of iron hydroxide. Eventually the whole bubble hardens and crystallizes, forming distinctive nodules of banded agate. The bands are revealed best when the nodule is cut into thin slices. There are scores of varieties of agate, including blue lace agate, with its wavy bands of mauve-blue and white, fire agate, with limonite inclusions, thunder eggs, with star-shaped brown and yellow bands, scenic agate, which looks like a woodland scene, and many more.

The distinctive brown and white bands of sardonyx make it a popular gem.

Onyx (above): Onyx is one of the most striking of the banded chalcedonies, known as the agates. Onyx is distinguished by its straight black and white bands but is sometimes confused with the similar-looking building material, striped travertine or 'marble onyx'. It gets its name from the Greek for 'fingernail' and has been popular since Roman times for cameos and other small carvings because a skilful stone cutter can carve out the different coloured bands to create a dramatic contrast between the picture and its background. It is one of the 12 stones mentioned in the famous passage in the Bible (Exodus, 28) where the high priest Aaron's breastplate is described. Onyx forms in the same way as other agates, but near the bottom of agate nodules in horizontal layers. The best natural onyx, as with jasper, is found in river gravels in India, where it was deposited after the breakdown of the basalt rocks of the Deccan. Most onyx specimens are not natural, however, but made by soaking natural agate in sugar water and heating gently for a few weeks – a technique known for 4,000 years. This makes brown onyx; black onyx is made by then treating the stone with sulphuric acid. Carnelian onyx is a form of carnelian with red and white bands; sardonyx (see left) has brown and white bands.

Idar-Oberstein

The little town of Idar-Oberstein in the Nahe river valley in Germany has been famous since the Middle Ages for its high quality agate work. It was especially famous for its carved agate bowls. Originally the agate came from local river pebbles, or from the nearby Setz quarry, where it was found in cavities in basalt. It was then ground and polished using wheels driven by the river. However, the Setz deposit was largely exhausted by the end of the 19th century, and so the Idar cutters turned to Brazil as a source for agate to cut, and as a result they exploited the other quartz stones there such as amethyst and citrine. Idar-Oberstein is now known throughout the world as a centre for the cutting of all kinds of gemstones.

Varieties of cryptocrystalline quartz

Name	Habit	Colour	Transparency	Cause of colouring, inclusions, etc.
Chalcedony	compact	pale blue, pale grey	translucent	even coloured or slightly banded
Carnelian	compact	yellowish to deep red	translucent	very finely divided hematite
Sardonyx	compact	brown	translucent	very finely divided iron hydroxide
Chrysoprase	compact	apple green	translucent	hydrated nickel silicates
Agate	compact	multi-coloured	translucent	finely banded, filling cavities
Onyx	compact	grey, black and white	opaque	bands thicker than in agate
Enhydros	compact	as chalcedony	translucent	cavities partly filled with water
Jasper	compact	multi-coloured	opaque	many impurities of clay, iron, etc.
Plasma	compact	leek green	opaque	chlorite, horn-blende and other green minerals
Heliotrope	compact	green with red flecks	opaque	red flecks due to hematite
Flint	compact	white, grey and other colours	opaque	like jasper, containing some opaline material

FELDSPARS

Together, the two kinds of feldspar, potassium (K) feldspar and plagioclase, make up almost two-thirds of the Earth's crust. K feldspars like orthoclase and sanidine are major ingredients in granite and other 'acidic' igneous rocks, as well as metamorphic rocks such as gneiss and sediments like arkose. Plagioclase feldspars, mostly albite and anorthite, are major ingredients in 'basic' igneous rocks such as gabbro.

Orthoclase

KAlSi₃O₈ Potassium aluminium silicate

Amazonite, a gem variety of microcline, is also known as amazon jade.

Like all feldspars, orthoclase is basically an aluminium silicate. Orthoclase, microcline and sanidine are all potassium- or K-rich, which is why they are called the potassium feldspars. These three are chemically identical, and the only real difference between them is their crystal structure. In particular, it can be almost impossible to tell microcline and orthoclase apart but by X-ray analysis – except when microcline is in its green form, known as amazonite (inset, top left). Orthoclase is typically white, but can be yellow and forms mainly in granites and syenites, as well as moderate to high grade metamorphic rocks such as gneiss and schist. Although it is incredibly common and a major raw material for the porcelain industry – especially orthoclase from aplite – almost all orthoclase is small grains and masses. Nevertheless, crystals do form in veins and porphyries in dikes. Good size yellow crystals ('noble orthoclase') are found in pegmatites in Madagascar. Colourless orthoclase crystals are named adularia after Adular in Switzerland, where they were first found in cavities in metamorphic rock. Moonstone is a rare and prized gem form of adularia and other feldspars that has a coloured surface sheen called adularescence.

Identification: Orthoclase looks like many light-coloured silicates, but it can be distinguished from spodumene by its blocky cleavage and from plagioclases by the lack of striations on the surface.

Crystal system: Monoclinic.
Crystal habit: Typically massive and small grained, but crystals are tablet-shaped; adularia forms flattened tablets. Often forms simple twins of the Carlsbad and Baveno type. Twinning common.
Colour: Off-white, sometimes yellow or reddish.
Lustre: Vitreous.
Streak: White.
Hardness: 6.
Cleavage: Good in two directions, forming prisms.
Fracture: Conchoidal or uneven.
Specific gravity: 2.53.
Notable locations: Carlsbad, Czech Republic; Adular, Disentis, Switzerland; Baveno, Italy; Lake Baikal, Siberia; Mount Kilimanjaro, Tanzania; Madagascar; Bernallio County, New Mexico; Robinson, Colorado; Clark County, Nevada.

Sanidine

KAlSi₃O₈ Potassium aluminium silicate

Identification: Sanidine looks very like orthoclase, but has a glassier, less grainy texture. It looks translucent or even transparent.

Orthoclase and microcline form in magmas that cool only moderately quickly, such as granite and syenite, crystallizing at temperatures between 400°C/750°F and 900°C/1,652°F. Sanidine forms in magmas that reach the surface and cool rapidly from temperatures of over 900°C, like rhyolite and trachyte, and also in rocks that metamorphose at high temperatures, such as gneiss. Rapid cooling allows crystals little time to grow, and consequently sanidine is almost glassy in texture. Chemically, it is also the potassium-rich end of a series of feldspars that form at high temperatures and are rich in either potassium or sodium. At the sodium-rich end is the high-temperature form of the plagioclase feldspar albite. Anorthoclase feldspar lies in between the two.

Crystal system: Monoclinic.
Crystal habit: Typically massive, but rare crystals are tablet- or prism-shaped.
Colour: Off-white or clear or grey; often transparent.
Lustre: Vitreous.
Streak: White.
Hardness: 6.
Cleavage: Good in two directions, forming prisms.
Fracture: Conchoidal, uneven.
Specific gravity: 2.53.
Notable locations: Elba, Italy; Caucasus Mountains, Russia; Ragged Mountain (Gunnison Co), Colorado; Grant Co, New Mexico; Bisbee, Arizona.

Albite

NaAlSi₃O₈ Sodium aluminium silicate

The green, gold and blue iridescent 'sheen' known as labradorescence is clearly visible in this labradorite specimen.

Albite commonly occurs in igneous and metamorphic rocks. At high temperatures it mixes with K feldspars and forms a series with sanidine. At low temperatures it separates out as layers inside K feldspar crystals to create a rock called perthite. When albite substitutes orthoclase in granite-like rocks, the rock becomes quartz monzonite. In pegmatites, albite can form bladelike aggregates called clevelandite. Albite is very common, but good crystals are rare, except in pegmatites and lavas called spilites. Albite and anorthite are the only plagioclase feldspars that ever form crystals; labradorite, andesine, bytownite and oligoclase occur only as aggregates. Labradorite exhibits an attractive array of colours called labradorescence (inset). Albite is one of the last feldspars to crystallize from a magma and so is often associated with some rare minerals. Twins are common in all feldspars, but albite crystals in igneous rocks are nearly always twinned.

Identification: Plagioclases can be told from K feldspars by the striations made by twinning, but it is hard to tell between the plagioclases except by accurate specific gravity measurements. Albite is the lightest with an SG of 2.63; oligoclase is 2.65; andesine 2.68; labradorite 2.71; bytownite 2.74 and anorthite 2.76.

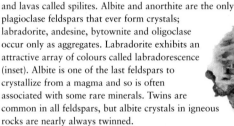

Clevelandite

Crystal system: Triclinic.
Crystal habit: Most albite is grains in rocks but when crystals form they are nearly always twinned. Crystals are typically square in cross-section. Bladelike aggregates (cleavelandite) form in pegmatites.
Colour: Off-white, sometimes yellow or reddish.
Lustre: Vitreous.
Streak: White.
Hardness: 6–6.5.
Cleavage: Good in two directions.
Fracture: Conchoidal.
Specific gravity: 2.63.
Notable locations: Alps, Switzerland; Tyrol, Austria; Minas Gerais, Brazil (cleavelandite); Francon, Mount-St-Hilaire, Quebec; Labrador; Amelia County, Virginia; San Diego County, California.

Perfect pottery
Porcelain is one of the most remarkable of all synthetic materials, light and strong yet clean and smooth, and wonderfully translucent. In fact, when it was first brought to Europe from China, many assumed it was natural. But the Chinese had discovered how to make it around 1,400 years ago by heating a mixture of a feldspar-rich rock called petuntse and kaolin (china clay). They kept their secret, and many Europeans spent a fortune trying to unravel it, for Chinese porcelain once fetched a higher price than gold. They knew china clay was involved, but they were missing the petuntse, the orthoclase feldspar-rich rock powder that made the clay hard. Some thought the missing ingredient might be ground bone, and around 1800 Josiah Spode added bone to china clay to make bone china. Another successful European equivalent was developed in Meissen in Germany by Johann Böttger and Ehrenfried von Tschirnhaus. Finally, in the 19th century, the role of orthoclase feldspar was revealed.

Anorthite

CaAlSi₃O₈ Calcium aluminium silicate

Each of the plagioclase feldspars has a different proportion of sodium or calcium. Albite has the most sodium; anorthite has the most calcium. The others – oligoclase, andesine, labradorite and bytownite – lie in between. While albite is typical of granite-like rocks that form at low temperatures, anorthite is common in basic rocks such as gabbros that form at high temperatures, and also meteorites and moon rocks. Anorthite crystals are always twinned but quite rare, occurring only in lavas in places such as Miyake-jima, Japan, and near Mount Vesuvius in Italy.

Crystal system: Triclinic.
Crystal habit: Usually massive, but crystals are short prism or tablet shapes and always twinned.
Colour: Colourless, whitish, grey, pinkish, greenish.
Lustre: Vitreous.
Streak: White.
Hardness: 6–6.5.
Cleavage: Good in two directions.
Fracture: Conchoidal.
Specific gravity: 2.76.
Notable locations: Mount Somma (Mount Vesuvius), Italy; Miyake-jima, Japan; Franklin, New Jersey; Aleutian Islands, Alaska; Lake Co, Nevada Co, California.

Anorthite

Identification: The best ways to identify anorthite are by its occurrence in lavas, its striated twinned crystals and its specific gravity of 2.76.

FELDSPATHOIDS

The feldspathoids is a group of minerals very similar to feldspars, including sodalite, haüyne, nepheline, lazurite, leucite, cancrinite and nosean. They are all minerals that would have become feldspars if there was more silica present when they formed. Feldspars tend to contain three times as much silica. So feldspathoids do not occur in silica-rich rocks such as granite but typically form in volcanic lavas.

Sodalite

$Na_8Al_6Si_6O_{24}Cl_2$ Sodium aluminium silicate with chlorine

Sodalite is a feldspathoid mineral that occurs with nepheline and cancrinite in basic igneous rocks such as nepheline syenites, and also in silica-poor dikes and lavas. It gets its name from its sodium content, and generally forms when aluminium silicate-rich rocks are invaded from below by waters rich in sodium chloride. Sodalite's colour ranges from royal blue to light blue and white and makes it very distinctive and likely to be confused only with lazurite and lazulite. Some sodalite is fluorescent and, under longwave UV light, nepheline syenite rock usually shows glowing patches of sodalite. A colourless variety called hackmanite is 'tenebrescent', turning red after exposure to UV then fading to pink. Although crystals are rare, rich blue masses of granular sodalite, big enough for ornamental stone, are found near Bancroft in Ontario, Canada. The Bancroft sodalite was famously excavated in 1906 to decorate Marlborough House in London. Even bigger, bluer masses are found in Bahia, Brazil, while there are thin veins of sodalite at Ice River in British Columbia and smaller masses in Maine. Colourless crystals are found in altered limestone blocks ejected by Mount Vesuvius, Italy.

Identification: Sodalite is usually easy to identify by its blue colour, similar only to lazurite and lazulite. Greyer and paler sodalites are typically fluorescent, gaining colour under longwave UV light, before fading again in daylight.

Crystal system: Cubic.
Crystal habit: Rare dodecahedral crystals but usually masses in rock.
Colour: Blue, white, grey.
Lustre: Vitreous or greasy.
Streak: White.
Hardness: 5.5–6.
Cleavage: Perfect in one direction.
Fracture: Uneven.
Specific gravity: 2.1–2.3.
Other features: Hackmanite is tenebrescent, briefly changing colour after exposure to longwave UV.
Notable locations: Ilimaussaq, Greenland; Mount Vesuvius, Italy; South Africa; Bahia, Brazil; Bancroft, Ontario; Ice River, British Columbia; Litchfield, Maine.

Haüyne

$Na_6Ca_2Al_6Si_6O_{24}(SO_4)_2$ Sodium calcium aluminium silicate sulphate

Crystal system: Cubic.
Crystal habit: Rare rhombo-decahedral crystals but usually masses in rock.
Colour: Blue, white, gray.
Lustre: Vitreous or greasy.
Streak: White.
Hardness: 5.5–6.
Cleavage: Perfect in three directions.
Fracture: Conchoidal.
Specific gravity: 2.4–2.5.
Notable locations: Mount Vulture, Mount Somma (Mount Vesuvius), Italy; Niedermendig Mine (Black Forest), Germany; Tasmania.

A rare but attractive mineral, haüyne is named after René Just Haüy (1743–1822), the pioneering French crystallographer, who discovered it on Monte Somma, Italy, among the Vesuvian lavas. It is a member of the sodalite group with sodalite and nosean. But unlike sodalite and nosean, haüyne contains calcium. It is usually a striking electric-blue colour, but can be green, red, yellow or even grey. Like all feldspathoids, haüyne usually occurs in igneous rocks low in silica and rich in alkalis such as sodium and calcium, which provide the basic raw materials for it to form. The typical environment for haüyne is silica-poor lavas such as phonolite and trachyte. Apart from Monte Somma, haüyne has been found more generally in the volcanoes of Lazio and notably in Tasmania in Australia. It has also been found at the Niedermendig Mine near Eifel in the Black Forest, Germany, where it occurs in volcanic bombs.

Identification: Startling electric blue haüyne crystals stand out clearly from the volcanic bombs or lava flow in which they are usually found.

Nepheline

(Na,K)AlSiO₄ Sodium aluminium silicate with chlorine

Nepheline is by far the most common of the feldspathoids and occurs in many silica-poor, alkali-rich igneous rocks. In fact, the presence of nepheline in a rock is generally taken as an indication that a rock is alkali-rich. In some rocks, nepheline is such a major component that they are named accordingly: nepheline syenite, nepheline monzonite and nephelinite. These rocks are also distinguished by how much of the various feldspars they contain. Although found in masses in rocks around the world, the most notable occurrences of crystals are in plutonic rocks at Bancroft, Ontario in Canada, and Karelskaya (Karelia) in Russia, and in cavities in limestone blocks thrown out by Mount Vesuvius. In recent years, massive nepheline has been used as a source material for soda, silica and alumina for the glass and ceramics industries. Because nepheline-bearing rocks do not contain quartz, they melt well. Nepheline can also be synthesized into a high-temperature mineral called carnegieite.

Identification: Distinctive six-, sided crystals like these formed in pegmatitic nepheline syenite dikes are rare. More usually, nepheline is seen as white grains in nepheline rock.

Hexagonal nepheline crystals

Crystal system: Hexagonal.
Crystal habit: Forms column-shaped hexagonal crystals but masses and grains typical.
Colour: Off-white to grey.
Lustre: Greasy to dull.
Streak: White.
Hardness: 5.5–6.
Cleavage: Poor.
Fracture: Conchoidal to uneven.
Specific gravity: 2.6.
Notable locations: Monte Somma (Mount Vesuvius), Italy; Karelskaya (Kola peninsular), Russia; Australia; Brazil; Bancroft, Ontario; Ouachita, Arkansas; Kennebec County, Maine; Colfax County, New Mexico.

Lazurite

(Na,Ca)₈Al₆Si₆O₂₄(S,SO₄)₂ Sodium calcium aluminium silicate sulphur sulphate

Its beautiful, rich blue colour makes lazurite one of the most distinctive and attractive of all minerals. Its chemical make-up is quite complex, but the blue colour comes from the sulphur that takes the place of some silicate atoms. It is soft and brittle and easily ground to make the rich blue pigment ultramarine, but it is most famous as the major ingredient of the gemstone lapis lazuli, famed since the days of Ancient Egypt. Little spots of golden pyrite look like stars in the deep blue lazurite of lapis lazuli. Lazurite is very rare and found in any quantity only in the Kokcha valley of Afghanistan and also high up in the nearby Pamir Mountains by Lake Baikal in Siberia.

Crystal system: Cubic.
Crystal habit: Forms column-shaped hexagonal crystals but masses and grains typical.
Colour: Off white to grey.
Lustre: Greasy to dull.
Streak: White.
Hardness: 5.5–6.
Cleavage: Poor.
Fracture: Uneven.
Specific gravity: 2.6.
Notable locations: Mount Vesuvius, Italy; Lake Baikal, Pamir Mountains, Russia; Kokcha River, Afghanistan; Ovalle, Chile; Sawatch Mountains, Colorado; Cascade Canyon and Ontario Peak, San Gabriel Mountains, San Bernardino County), California.

Identification: Lazurite is usually easy to identify by its vivid blue colour. It can be distinguished from sodalite by its association with pyrite, and from lazulite by its lower specific gravity.

Lapis lazuli
Lapis lazuli is one of the oldest, most treasured of all gemstones. Its name is a combination of the Latin *lapis* for 'stone' and the Arabic *azul* for 'sky' or the ancient Persian *lazhuward* for 'blue'. It usually occurs as lenses and veins in white marble. Consisting largely of lazurite with spots of pyrite, it has a mottled look. Crystals are sometimes found, but more usually it is massive, and is carved to make jewellery, cups and other decorative objects. It was first mined over 6,000 years ago at Sar-e-Sang in the Kokcha valley in Afghanistan, still the source of the world's finest lapis lazuli. The ancient royal tombs of the Sumerian city of Ur contained over 6,000 beautifully carved lapis lazuli statues, and it was a favourite stone of the Ancient Egyptians, much used in the tomb decorations of Tutankhamun. The Roman writer Pliny the Elder described it as 'a fragment of the starry firmament'. Today, lapis lazuli is mined near Lake Baikal in Siberia and at Ovalle in Chile as well as in Afghanistan. Chilean lapis contains flecks of calcite.

ZEOLITES

Zeolites are a group of 50 or so minerals including heulandite, stilbite, phillipsite, harmotome, natrolite, chabazite and analcime. Popular with collectors for their rarity and beauty, they are also of industrial importance for their ability to act as filters and chemical sponges. Similar to both clay and feldspars, they typically form when the minerals in cavities of volcanic rocks are altered by moderate heat and pressure.

Heulandite

$(Ca,Na)_{2-3}Al_3(Al,Si)_2Si_{13}O_{36}.12H_2O$ Hydrated calcium sodium aluminium silicate

Heulandite is one of the most common and best known zeolites, named after the English mineral dealer John Henry Heuland, who hunted for it in Iceland. It forms distinctive, cream-coloured, pearly lustred crystals shaped like the coffins in old horror films, with a wide waist. The crystals develop from a hydrothermal solution percolating through magnesium- and iron-rich volcanic rocks, especially basalts, and form in cavities left by gas bubbles trapped as the lava cools. Heulandite forms when the minerals in the bubble left by the solutions are altered by mild metamorphism. It can form in the same way in pegmatites, tuffs, metamorphic rocks and also in deep-sea sediments. It does form in other places, too, but the crystals are generally much smaller. The best specimens are from the basalts of Berufjördhur in Iceland and the Faroe Islands, and the Deccan traps of the Sahyadri mountains near Bombay. Red heulandite crystals have been found on Campsie Fells in Scotland, the Fassathal in the Austrian Tyrol, Gunnedah in New South Wales (Australia) and in eastern Russia.

Identification: Heulandite is best identified by its creamy coloured crystals, its pearly sheen and its coffin-shaped crystals.

Crystal system: Monoclinic.
Crystal habit: Crystals typically tablet-shaped, like an old-fashioned coffin, or pseudo-orthorhombic.
Colour: White, pink, red, brown.
Lustre: Vitreous or pearly.
Streak: White.
Hardness: 3.5–4.
Cleavage: Poor.
Fracture: Uneven.
Specific gravity: 2.1–2.3.
Notable locations: Campsie Fells, Scotland; Berufjördhur, Iceland; Faroe Islands; Fassatal, Austria; Urals, Russia; Sahyadri Mts, India; Gunnedah (NSW), Australia; Partridge Island, Nova Scotia, Canada; Paterson, New Jersey; Idaho; Oregon; Washington.

Stilbite

$NaCa_2Al_5Si_{13}O_{36}.14H_2O$ Hydrated sodium calcium aluminium silicate

Crystal system: Monoclinic.
Crystal habit: Crystals typically thin plates that can aggregate in wheat-sheaf shapes. Cruciform (cross-like) twins common.
Colour: White, pink, red, brown, orange.
Lustre: Vitreous or pearly.
Streak: White.
Hardness: 3.5–4.
Cleavage: Perfect in one direction.
Fracture: Uneven.
Specific gravity: 2.1–2.3.
Notable locations: Kilpatrick, Scotland; Berufjördhur, Iceland; Poona, India; Victoria, Australia; Rio Grande do Sul, Brazil; Bay of Fundy, Nova Scotia; Paterson, New Jersey.

Stilbite is another common zeolite very closely related to heulandite. It is very popular among collectors for the extraordinary stilbite crystals that grow in fantastic shapes resembling wheat sheaves or a stack of bow ties. Stilbite crystals are not always shaped like this, but when they are they are almost unique. Only the rare related zeolite stellerite has similar crystals. Crystals are usually whitish, but bright orange examples have been found.

Like heulandite, stilbite forms in cavities in basaltic lava, besides ore veins and pegmatites – and so tends to be found wherever there were large floods of basalt lava, such as in the Deccan traps of India. However, it is also the zeolite most likely to form in situations that are not typical for zeolites, such as mineral seams in granite.

Identification: Stilbite is best identified by its very distinctive bow-tie or wheat sheaf stacks of crystals.

Phillipsite

(K,Na,Ca)₁₋₂(Si,Al)₈O₁₆.6H₂O Hydrated potassium sodium calcium aluminium silicate

Named by Lévy in 1825 after English mineralogist W Phillips (1775–1829), Phillipsite is not one of the most common natural zeolites – though it is often made artificially. Yet it is popular with collectors because, besides crystals, it sometimes forms aggregates of little white balls with a silky surface. These balls form in bubble cavities in lava, like other zeolites, and can be seen in the phonolites near Rome in Italy. Phillipsite can form quite quickly; crystals have actually developed in cavities in the masonry of the hot mineral baths at Plombières in France. It can also form around hot springs. Phillipsite also forms in calcite-rich deep-sea sediments, and in 1951 specimens were dredged by the *Challenger* survey ship from the bottom of the Pacific, where they formed by the alteration of lava. Phillipsite and another rare zeolite, harmotome (inset), are often hard to distinguish.

Harmotome is a glassy mineral whose crystals may form attractive twins.

Identification: Phillipsite is most easily identified when it forms silky or rough little balls lining the insides of cavities, but can also form crystals that are not so easy to identify.

Crystal system: Monoclinic.
Crystal habit: Crystals typically twinned in groups of four or more. Often forms aggregates of small balls.
Colour: White, clear, yellowish, reddish.
Lustre: Vitreous, silky.
Streak: White.
Hardness: 4–4.5.
Cleavage: Imperfect in one direction.
Fracture: Uneven.
Specific gravity: 2.2.
Notable locations: Moyle (Antrim), Northern Ireland; Capo di Bove, Mount Vesuvius, Italy; Groschlattengruen, Germany; Cape Grim, Tasmania; Jefferson County, Colorado.

Chabazite

Ca₂(Al₄Si₈O₂₄).13H₂O Hydrated calcium potassium sodium aluminium silicate

Getting its name from *chalaza*, the Greek word for 'hailstone', chabazite can look very similar to other zeolites and often forms in similar places. Like phillipsite and harmotome, it forms when the bubbles trapped in lava are subjected to gentle metamorphism. The most dramatic chabazite crystals from Poona in India formed like this in the Deccan Traps, a large igneous province of flood basalt. In these small cavities, chabazite can occur with a huge range of other minerals, especially zeolites such as natrolite, scolecite, heulandite and stilbite. The chabazite crystals are actually rhombohedral, but look like little cubes. Chabazite can also form small crystals around hot springs.

Crystal system: Trigonal.
Crystal habit: Crystals rhombohedral but look like cubes.
Colour: Usually white or yellowish, can be pink or red.
Lustre: Vitreous.
Streak: White.
Hardness: 4.5.
Cleavage: Poor.
Fracture: Uneven.
Specific gravity: 2–2.1.
Notable locations: Kilmacolm, Scotland; Berufjördhur, Iceland; Oberstein, Germany; Switzerland; Poona, India; Richmond (Victoria), Table Cape (Tasmania), Australia; New Zealand; Nova Scotia; New Jersey; Yellowstone, Wyoming; Bowie, Arizona.

Zeolite marvels
Zeolites have been found to be such remarkable minerals that a whole industry has built up around making the most of them. They are aluminium silicates like clay, but unlike clay they have a rigid, honeycomblike crystal structure with networks of tunnels and cages. Water can move freely in and out of these pores without affecting the zeolite framework. The pores are of a uniform size, so the crystal can act like a molecular sieve, filtering out particles of the wrong size. The zeolite can also exchange ions well. The result is that zeolites can act as both microsponges able to retain water, just like the best soil, and as filters able to clean up the tiniest particles. They are so useful that artificial versions are made to take advantage of the properties of particular natural zeolites, especially chabazite, clinoptilolite and phillipsite. Zeolites are used for air filters, water treatment, cleaning oil spills, cleaning up radioactive waste and much more. They are also used as an artificial medium for growing plants.

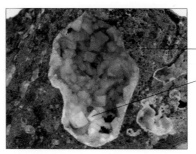

— Basalt rock

— Chabazite

Identification: When found in a bubble cavity, chabazite is clearly a zeolite. Its cubic-looking crystals mark it out as chabazite.

PYROXENES

Pyroxenes are present in most igneous and metamorphic rocks, and a major ingredient in darker mafic rocks such as gabbro and basalt. They were given the name 'pyroxene' by French mineralogist R J Haüy from the Greek for 'fire' and 'stranger', because he was surprised to see these dark green crystals ready-formed in lavas. In fact, they simply crystallize at high temperatures, before the lava erupts.

Diopside

$CaMgSi_2O_6$ Calcium magnesium silicate

Pyroxenes and amphiboles
Pyroxenes and amphiboles are closely related silicate minerals that look similar. Pyroxenes form when there is little water present; water turns the same minerals to amphiboles. Pyroxenes form shorter, stubbier, prism-shaped crystals that break almost at right angles, while amphiboles break in a wedge shape – hornblends is a classic example. The difference is usually clear when specimens are viewed through a petrological microscope or sometimes even a very good magnifying glass.

Identification: Diopside can usually be identified by its light green colour, and its right-angle cleavage.

Diopside is a silicate that often crystallizes direct from basic magmas, which is why it is a major ingredient in some ultrabasic igneous rocks. It also forms when silicates in dolomitic limestones are metamorphosed and in iron-rich skarns (metamorphosed ore deposits). Most of this diopside is grainy or locked into the rock. Distinct crystals are rarer, but can be attractive, especially green, chrome-rich ones. The name diopside comes from the Greek *dis* and *opsis* for 'double view', because crystals can be strongly birefringent (give a double image). Some display a chatoyancy (cat's eye) effect, and some are chatoyant in two directions, giving a cross. Some ancient civilizations believed spirits observed you through these star crosses.

Crystal system: Monoclinic.
Crystal habit: Usually short, prism-shaped crystals, but also grainy aggregates, columns and masses.
Colour: Green, khaki, colourless, light blue.
Lustre: Vitreous, dull.
Streak: White, light green.
Hardness: 6.
Cleavage: Perfect prism.
Fracture: Brittle, conchoidal.
Specific gravity: 3.3–3.6.
Notable locations: Fassa, Ala and Ossola Valleys, Mount Vesuvius, Italy; Binnetal, Switzerland; Zillertal, Austria; Outokumpu, Finland; Baikal, Russia; North Korea; Madagascar; De Kalb, New York; Riverside County, California.

Augite

(Ca,Na)(Mg,Fe,Al)(Al,Si)₂O₆ Calcium sodium magnesium iron aluminium silicate

Identification: Augite is usually recognizable by its dark colour and blocky, prismatic crystals.

Augite is closely related to diopside, but contains a good deal of sodium and aluminium, and much more magnesium. It gets its name from the Greek for 'lustre' because of the sheen it gives to the surface of augite-rich rocks. It is the most common pyroxene mineral, and an important ingredient in many dark-coloured, basic igneous rocks, notably basalts, dolerites, gabbros and peridotites. Some of the best augite crystals are phenocrysts (extra-large, preformed crystals) in basalt. The tuff deposits around some volcanoes are almost entirely augite crystals, and big crystals can be picked out of weathered lavas in the craters of some Italian volcanoes. Augite occurs in rocks metamorphosed at high temperatures, too, such as granulites and gneisses, and also in meteorites rich in basaltic material, which is why augite crystals are found in the Bushveld complex in South Africa (see Gabbroic rocks, Directory of Rocks). Although common, most big augite crystals are quite dull.

Crystal system: Monoclinic.
Crystal habit: Usually short, prism-shaped, rectangular crystals, but also grains, columns and masses.
Colour: Dark green, brown, black.
Lustre: Vitreous, dull.
Streak: Greenish white.
Hardness: 5–6.
Cleavage: Perfect prism.
Fracture: Uneven.
Specific gravity: 3.2–3.6.
Notable locations: Mounts Vesuvius, Etna, Stromboli, Lazio, Italy; Auvergne, France; Eifel, Germany; Bushveld complex, South Africa; St Lawrence, Ramapo Mountains, New York.

Spodumene

LiAlSi₂O₆ Lithium aluminium silicate

Spodumene gets its name from the Greek for 'burnt to ashes', because of its commonest grey colour. It occurs almost exclusively in lithium-rich granite pegmatites, where it is usually associated with lepidolite mica, elbaite tourmaline, caesium beryl, amblygonite, quartz and albite. Massive spodumene is the main ore of lithium, the lightest of all metals, used in ceramics and lubricants, and a key ingredient in anti-depressive drugs. Rough crystals of spodumene can be huge, and some have been found that are nearly 15m/49ft 3in long and weigh over 90 tonnes. Brilliant, glassy, small gem crystals are quite rare, however. They have never been that popular for jewellery because their colour fades when exposed to sunlight, but they are cherished by collectors and museums. Impurities taking the place of aluminium in the crystal structure give a wide range of colours – iron gives yellow to green spodumene, chromium gives deep green spodumene and manganese gives lilac spodumene. Violet spodumene is known as kunzite, and green spodumene is called hiddenite.

Identification: Spodumene is best identified by its long, striated, prism-shaped crystals, and its occurrence in lithium-rich pegmatites along with lepidolite mica and elbaite tourmalines.

Crystal system: Monoclinic.
Crystal habit: Usually massive; crystals can be large and long, usually prismatic with striations.
Colour: White or greyish white, also green, pink, yellow, lilac.
Lustre: Vitreous.
Streak: White.
Hardness: 6.5–7.
Cleavage: Perfect prismatic.
Fracture: Splintery.
Specific gravity: 3–3.2.
Notable locations: Varuträsk, Sweden; Afghanistan; Pakistan; Namibia; Madagascar; Brazil; Black Hills, South Dakota; Dixon, New Mexico; North Carolina.

Green jade
Jade has long been cherished in China, where it was a symbol of royalty and has been carved into jewellery, ornaments and small statues for thousands of years. It is so tough that it was even used to make knives and axes. It has an almost equally long history in South America. Indeed it gets its name from the Spanish *piedra de ijada*, which means 'stone of the side', because the Spaniards who came across it in Central America were told that it cured kidney problems. Europeans applied the name jade to the similar-looking ornamental green stone that came from China and from South America. It was only in 1863 that mineralogists realized there were two minerals involved – 'hard' jadeite and 'soft' nephrite (as in the Maori pendant, above), both of which are found in China and South America. The most highly prized jade is the emerald-green Imperial Jade, produced by traces of chromium. Nephrite can be pale green or white, while jadeite can be green, white, yellow or even violet in colour.

Jadeite

Na(Al,Fe)Si₂O₆ Sodium aluminium iron silicate

Jadeite is one of two minerals called jade. The other is nephrite. Jadeite is the more valued and is very rare. Most jadeite is opaque, and the best translucent stones are rarer still. Jadeite forms at high pressures in rocks subjected to intense metamorphism, which creates this beautiful green stone by altering minerals such as nepheline and albite in sodium-rich rocks. It is associated with the minerals glaucophane and aragonite, but jadeite is very rarely found in situ, as in Guatemala. Instead, it is more normally found as water-worn pebbles, freed from the parent rock by weathering, as is the case in China and Burma. In its pure form, jadeite is pure white in colour, but trace quantities of chrome, iron and titanium give it a green, blue or lavender hue.

Crystal system: Monoclinic.
Crystal habit: Usually massive or granular.
Colour: Green or white, blue and lavender.
Lustre: Vitreous, dull.
Streak: White.
Hardness: 6.5–7.
Cleavage: Good but rarely seen.
Fracture: Splintery to uneven.
Specific gravity: 3.25–3.35.
Notable locations: Urals, Russia; Tawmaw, Myanmar (Burma); China; Kotaki, Japan; Sulawesi; Motagua River, Guatemala; Mexico; San Benito County, California.

Identification: Jadeite is readily identified by its colour and beauty, and by its hardness. It can be distinguished by its specific gravity from nephrite, which is less dense at 2.9–3.3.

SILICATE GEMS

The hardness of silica means that when some silica minerals are combined with the right chemical 'colouring agents' they can turn into exquisite gems such as tourmaline, beryl and topaz. Under the headings tourmaline and beryl come a host of other gems, each transformed into unique gems such as elbaite, emerald, aquamarine and morganite by minute traces of particular chemicals.

Tourmaline

$Na(Li,Al)_3Al_6Si_6O_{18}(BO_3)_3.(OH)_4$ Sodium lithium aluminium boro-silicate hydroxide

Crystal system: Trigonal.
Crystal habit: Typically elongated triangular or six-sided prisms, with striations on the surface. The ends may be of different shapes.
Colour: Very variable, but commonly black or bluish black, blue, pink, green.
Lustre: Vitreous.
Streak: White.
Hardness: 7.5.
Cleavage: Very poor.
Fracture: Very poor.
Specific gravity: 3–3.2.
Notable locations: Elba, Italy; Urals, Russia; Chainpur, Nepal; Afghanistan; Yinniethara, Western Australia; Bahia, Brazil; Maine; San Diego County, California.

Tourmaline is the most variably coloured of all gems. The Ancient Egyptians called it rainbow rock, believing it gathered all the colours of the rainbow as it worked its way up through the Earth. Its name comes from the Sri Lankan *tur mali*, meaning 'multi-coloured stone'. Traces of different chemicals transform it into over 100 different colours. Indeed, single specimens can come in a mix of colours. Watermelon tourmaline has a bright pink centre and a vivid green rind. The 19th-century art critic John Ruskin said tourmaline's make-up was 'more like a medieval doctor's prescription than the making of a respectable mineral'. Tourmaline is not a single mineral species, but a group of species including minerals such as schorl (coloured black by iron) and dravite (made brown by magnesium), but most gem tourmaline is lithium-rich elbaite, named after the Italian island of Elba where many good tourmaline crystals were once found. Tourmaline is usually formed in granitic pegmatites, but it is also formed where limestones are metamorphosed by contact with granite magma. Because it is tough, tourmaline can be found in river gravel deposits after its parent rock is broken down by weathering.

Identification: Tourmaline is easy to identify when it is multi-coloured, like watermelon tourmaline. Otherwise the best clue to a tourmaline's crystal identity is its triangular cross-section.

Beryl

$Be_3Al_2Si_6O_{18}$ Beryllium aluminium silicate

Identification: Beryl's hardness, hexagonal crystals and association with pegmatites mean it can easily be confused only with topaz and quartz. Colour then helps to confirm the identity.

Yellow beryl

Quartz

Unlike tourmaline, beryl is a single mineral species, yet although pure beryl or goshenite is colourless, impurities give it a colour palette as rich as any other gem mineral. Chromium and vanadium turn it to brilliant emerald. Blue beryl is known as aquamarine, while yellow beryl is heliodor and pink beryl is morganite. The name beryl is reserved for the red and golden variety. Beryl is a tough crystal that forms deep down in the Earth. Most is found in granite pegmatites, typically 'frozen' in masses of quartz-feldspar, but sometimes occurring as free crystals in cavities. It also occurs in skarns, schists intruded by pegmatites and, as in Colombia, introduced by hydro-thermal waters into carbonate-sediments. Most crystals are small, but a few of the Colombian beryls are huge, measuring up to 5.5m/18ft in length. One beryl from Madagascar was recorded as weighing a staggering 36.6 tonnes.

Crystal system: Hexagonal.
Crystal habit: Crystals are long, six-sided prisms, typically with flat pyramid tops and striations on their surfaces.
Colour: A wide range including green, yellow, gold, red, pink and colourless.
Lustre: Vitreous.
Streak: White.
Hardness: 7.5–8.
Cleavage: Basal, poor.
Fracture: Conchoidal.
Specific gravity: 2.6–2.9.
Notable locations: Galicia, Spain; Russia; Namibia; Madagascar; Pakistan; Chivor, Muzo, Colombia; Minas Gerais, Brazil; Hiddenite, North Carolina; San Diego County, California.

Opal

SiO₂.nH₂O Silicon dioxide, with up to 21 per cent water

Unlike most other minerals, opal never really forms crystals. In fact, it is a hardened silica gel, containing about 5–10 per cent water. It forms when silica-rich fluids solidify in cavities to form nodules, crusts, veinlets and masses. Sometimes opal replaces bones or wood during fossilization. Sometimes it forms around hot springs and in volcanic rocks. Dull yellow, red or black 'potch' opal is incredibly common, and is widely mined for use as an abrasive and filler. Precious opal is much rarer and only forms where minute silica spheres settle completely undisturbed. Potch opal may contain spheres, but they are a varied hotch-potch. Precious opal is made of even-sized spheres, and light is diffracted through these spheres to create opal's distinctive shimmering colours or 'opalescence'. Until the 19th century, most precious opal came from Slovakia, where it was found in andesite. Now 90 per cent of the world's opals come from the sandstones and ironstones of the Australian outback, notably at Coober Pedy.

Identification: Opal's silvery sheen and play of colours usually makes this gem easy to identify. However, there are many varieties from dull browns to silvery white. Some are entirely opaque; some are completely transparent.

Crystal system: None.
Crystal habit: Massive, typically kidney-shaped, round, or sheetlike.
Colour:
Common, potch opal: Various colours, but lacks opalescence.
Precious opal: White or black and opalescent.
Fire opal: Red and yellow flamelike reflections.
Hyalite: Colourless.
Hydrophane: Transparent in water.
Lustre: Vitreous, pearly.
Streak: White.
Hardness: 5.5–6.
Cleavage: None.
Fracture: Conchoidal.
Specific gravity: 1.8–2.3.
Notable locations:
Cervenica, eastern Slovakia; Lightning Ridge (NSW), Coober Pedy (SA), Australia; Ceará, Brazil; Honduras; Mexico; Washington; Idaho; Virgin Valley, Nevada.

The greenest gem
Emeralds get their name from the Ancient Greek *smaragdos*, meaning 'green'. They are one of the oldest known gems, found with mummies in Egyptian tombs. The oldest Egyptian emerald mines, some 3,500 years old, were rediscovered by the French adventurer Caillaud in 1816. Then in 1900, Cleopatra's mines near the Red Sea were found. The Roman emperor Nero is said to have watched through an emerald as gladiators fought to the death. The best emeralds come from South America, especially the famous Chivor and Muzo mines in Colombia, first mined by the Chibcha Indians long before the Spanish arrived. The Spanish conquistadors pillaged many fine Chivor emeralds from the Aztecs and Incas, and then in 1537 discovered the mine for themselves. Colombia still provides half of the world's emeralds. It was here that the largest gem-quality emerald was found in 1961, weighing 7,025 carats (about 1.4kg/3lb).

Topaz

Al₂SiO₄(OH,F)₂ Aluminium fluohydroxisilicate

Topaz may have got its name from the Sanskrit word *tapaz* meaning 'fire', or from the legendary island of Topazios in the Red Sea, now known as Zebirget. It comes in many colours, including rare pink topazes, and prized yellow topazes from Brazil. Blue topaz can look like aquamarine. Topaz is formed from fluorine-rich solutions and vapours and is typically found in granite pegmatites, or seams in granite altered by fluorine-rich solutions, often accompanied by fluorite, tourmaline, apatite and beryl. It also forms in gas cavities in rhyolite. Because it is so hard, it may be found deposited in river gravels, washed there after its parent rock has been weathered away. Water-worn pebbles of colourless topaz can easily be mistaken for diamonds. In fact, the best known topaz, the 1,649-carat Braganza set in the old crown of Portugal, was once thought to have been a diamond. Most crystals are small and grainy, but gigantic specimens weighing up to 100kg/220lb exist.

Crystal system: Orthorhombic.
Crystal habit: Crystals usually prism-shaped, with two or more long prisms. Also masses and grains.
Colour: Colourless, pale yellow, blue, greenish, pink.
Lustre: Vitreous.
Streak: White.
Hardness: 8.
Cleavage: Perfect basal.
Fracture: Conchoidal.
Specific gravity: 3.5–3.6.
Notable locations:
Erzgebirge, Germany; Sanarka River, Urals, Russia; Pakistan; Minas Gerais, Brazil; San Luis Potosí, Mexico; Thomas Range, Utah; San Diego County, California.

Identification: Topaz is best identified by its hardness, its high density and its range of pale colours. Its association with fluorite in pegmatites is also a useful indicator.

GARNETS AND OLIVINES

Garnets and olivines are dark, dense minerals from deep within the Earth, forming only under conditions
of intense heat and pressure. Sometimes, they are brought to the surface when rock formed at these
depths is uplifted as a result of tectonic activity; sometimes they are carried to the surface in lava, where
garnet and olivine crystals can be seen glistening ready-formed in the newly erupted material.

Garnets: Andradite

$Ca_3Fe_2(SiO_4)_3$ Calcium iron silicate

Garnet got its name originally because red garnet looks just like seeds
of the pomegranate fruit. It is not actually a single mineral but a
group of over 20, including uvarovite, grossular and andradite
(calcium-rich 'ugrandite' garnets), and pyrope, almandine
and spessartine (aluminium-rich 'pyralspite' garnets).
All form under extreme conditions, in high-grade
metamorphic rocks such as schist and in deep-
forming igneous rocks such as peridotite.
Garnet can become so concentrated
in peridotite that it creates garnet
peridotite, a dark rock studded with
tiny red or brown garnets, with the
appearance of cherries in dark
chocolate. Andradite, named after
the 18th-century Portuguese mineralogist

Andradite

d'Andrada de Silva, can make great gems. It occurs in granite pegmatites, in carbonatites
(calcium-rich plugs of deep-formed magma), in metamorphic hornfels and skarns, and on
seams and in crusts on serpentinites. It gets its various colours from differing amounts of
calcium, iron and other metals in its make-up. Traces of chromium create the green gem
demantoid. Demantoid is the most valuable of all the garnets and gets its name from its
diamond-like brilliance – though when it was first found in the Russian Urals, it was dubbed
'Uralian emerald' for its green colour. Traces of titanium create black melanite which occurs
in igneous rocks such as phonolite and leucitophyre. Topazolite is yellow andradite.

Crystal system:
Isometric.
Crystal habit:
Crystals are
typically 12-faced rhombic
shape or 24-faced trapezoid
shape, or both.
Colour: Typically greenish
grey to green but also black,
yellow and rarely colourless.
Lustre: Vitreous.
Streak: White.
Hardness: 6.5–7.5.
Cleavage: None.
Fracture: Conchoidal, brittle.
Specific gravity: 3.8.
Notable locations: Northern
Italy; Valais, Switzerland;
Nizhni-Tagil (Urals), Russia;
Gtr Hinggan Mts, China;
Korea; Mt St Hilaire, Quebec;
San Benito County, California;
Arizona; Franklin, New Jersey;
Magnet Cove, Arkansas;
Chester Co, Pennsylvania.

Identification: Andradite is best
identified by its colour, hardness
and association with serpentine,
diopside, wollastonite, albite,
calcite, orthoclase and micas.

Garnets: Grossular

$Ca_3Al_2(SiO_4)_3$ Calcium aluminium silicate

Grossular is the calcium aluminium garnet. It gets its name from the Latin for 'gooseberry',
because one of its colour forms looks like cooked gooseberries. But it comes in a wider
range of colours than any other garnet. The most attractive, perhaps, is orange hessonite,
sometimes known as cinnamon stone. Tsavorite (from
Tsavo in Kenya) is turned green by chromium.
African 'jade' is massive veins of opaque
green grossular found in the Transvaal,
South Africa, that looks rather like jade.
Grossular is thought to form in limestones
that have been metamorphosed to marble,
or in skarns formed when ores in limestone
are metamorphosed by contact with magma.

*Hessonite
(cinnamon
stone)*

Identification: Grossular is best identified by
its colour and hardness. The association with
calcite, diopside, vesuvianite, and wollastonite
is a key indicator.

Crystal system:
Isometric.
Crystal habit:
Crystals are
typically 12-faced rhombic
shape or 24-faced trapezoid
shape, or both.
Colour: Colourless, yellow,
orange, green, red,
grey, black.
Lustre: Vitreous.
Streak: White.
Hardness: 6.5–7.
Cleavage: None.
Fracture: Conchoidal.
Specific gravity: 3.5
Notable locations: ,
Scotland; Ala, Italy;
Hesse, Germany; Tsavo,
Kenya; Transvaal, South
Africa; Sri Lanka; Chihuahua,
Mexico; Asbestos, Quebec,
Canada; Lowell, Vermont.

Garnets: Almandine

Fe₃Al₂(SiO₄)₃ Iron aluminium silicate

Almandine probably gets its name from Alabanda, an ancient city in Anatolia famous for its gem cutting. With pyrope and spessartine, it is one of the three aluminium-based garnets, but while pyrope is magnesium-rich and spessartine is manganese-rich, almandine is iron-rich, and the most common of the three. The distinction between them is not always that clear, though – rhodolite, for instance, is an almandine with a fair amount of pyrope. All three tend to be reddish or brownish in colour. Pyrope is often the deep ruby red that earned its name from the Greek for 'fire', while spessartine is more peach coloured. Almandine tends towards brown or even black. Pyrope typically occurs embedded in igneous rocks such as dunite and peridotite, whereas spessartine is more usually found in mineralized pockets in rhyolite, pegmatites and marbles. Almandine typically forms in medium- to high-grade regional metamorphic rocks such as mica schists and gneisses. In the garnet-rich rock garnet schist, the garnet is usually almandine.

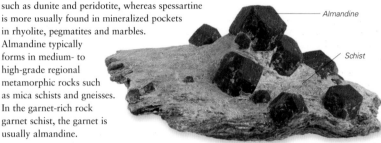

Almandine

Schist

Crystal system: Isometric.
Crystal habit: 12-faced rhombic or 24-faced trapezoid, or both.
Colour: Typically red to brown, or reddish black.
Lustre: Vitreous.
Streak: White.
Hardness: 6.5–7.5.
Cleavage: None.
Fracture: Subconchoidal.
Specific gravity: 4.3 (Pyrope 3.5; Spessartine 4.2).
Notable locations: Telemark, Norway; Zillertal (Tyrol), Austria; India; Sri Lanka; Broken Hill (NSW), Australia; Adirondacks Mts, New York; Wrangell Island, Alaska; Chaffee County, Colorado.

Identification: Almandine is best identified by its colour, specific gravity, occurrence in schists and with micas, staurolite, quartz, magnetite and andalusite.

Gemstone cuts

Rose cut | Marquise | Brilliant | Step cut | Baguette

Cabochon | Lentil cabochon | High cabochon | Pear | Oval

Until the late Middle Ages, most gemstones were simply polished along natural breaks. Jewellers then realized that they could produce brilliant effects by grinding or 'cutting' them into very specific shapes. Different styles of cutting are used to bring out a stone's best qualities. Opaque or translucent gemstones such as opals, jade and lapis lazuli are typically cut into a smooth oval, rounded on the top surface and flat underneath. This is called *en cabochon*. Clear precious stones, though, are typically cut with a series of mirrorlike facets to make them sparkle. Diamonds are 'brilliant-cut' with multiple triangular facets to make them glitter. Coloured stones, such as emeralds and rubies, are 'step-cut' to bring out the rich hues in the stone. All other cuts are variations of these basic two. Pear cuts are pear-shaped brilliants with a large flat top. Rose cuts or rosettes are brilliants without a 'pavilion' – a deep point at the back. Rose cuts were very popular for diamonds and garnets in Victorian times (1837–1901) but are less favoured today.

Olivine

(Mg,Fe)₂SiO₄ Magnesium iron silicate

Named for their olive-green colour, olivines are minerals rich in iron and magnesium that form at high temperatures. Olivine with a high magnesium content is forsterite; olivine with a high iron content is fayalite. Olivine is common in mafic rocks such as basalt and gabbro, while ultramafic rocks such as peridotite and dunite are almost pure olivine. Peridotites are what the Earth's mantle is made from, so olivines are among Earth's most common minerals. However, in the crust, they typically occur only as tiny grains. Rare large gems, called peridots, are much sought after. Basalts sometimes hold nodules containing olivine dredged from the mantle as they erupted that give geologists a window inside the Earth. Olivine does not survive weathering long; the higher the temperature a silicate mineral was formed at, the quicker it weathers. Olivine is prone to alteration to serpentine which extends snakelike veins into olivine as it forms.

Crystal system: Orthorhombic.
Crystal habit: Short prisms.
Colour: Olive green; redder when weathered.
Lustre: Vitreous.
Streak: White.
Hardness: 6.5–7.
Cleavage: Indistinct.
Fracture: Conchoidal.
Specific gravity: Forsterite 3.2; Fayalite 4.3.
Notable locations: Møre, Snarum, Norway; Eifel, Germany; Mount Vesuvius, Italy; Zebirget (St Johns) Island, Egypt; Myanmar (Burma); Holbrook, Gila County, Arizona

Fayalite: Mourne Mountains, Ireland; Fayal, Azores; Yellowstone, Wyoming.

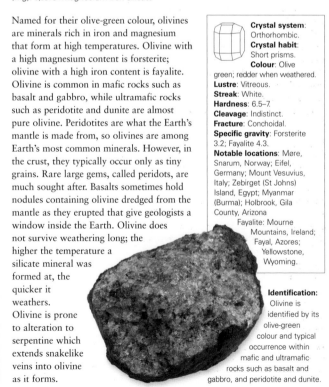

Identification: Olivine is identified by its olive-green colour and typical occurrence within mafic and ultramafic rocks such as basalt and gabbro, and peridotite and dunite.

NESO- AND SOROSILICATES

Silicates are all built up from the same basic SiO₄ atom groups, called silicate units, but can be classified by how these groups link up. Neso- and sorosilicates generally have the simplest arrangement – nesosilicates are built up from single SiO₄ units and sorosilicates from pairs (Si₂O₇). Nesosilicates include andalusite and sphene, as well as garnet and olivine. Sorosilicates include epidote and vesuvianite.

Andalusite, Sillimanite, Kyanite

Al₂SiO₅ Aluminium silicate

Sillimanite, named after American chemist B Silliman

Andalusite, sillimanite and kyanite are simple aluminium silicates found in metamorphic schists and gneisses, or loose as waterworn pebbles. They are chemically identical but have different crystal structures. Andalusite forms at low temperatures and pressures, kyanite at high ones. The presence of each of these minerals helps geologists tell under which conditions the rocks formed. All may be used as gems, but good crystals are rare, and it is in their massive forms they are useful, because they are so heat-resistant. Turned into mullite fibre, kyanite is used to make everything from insulators for car spark plugs to containers for molten steel. Andalusite, named after Andalusia in Spain, is the most common and widely used of the three. Under a microscope, its crystals may look orangey brown from one angle and yellowish green from another. Kyanite's French name is 'disthene', which means 'double hardness', because crystals have a hardness of 5 lengthways and 7 across.

Identification: The bluish blades of kyanite are very distinctive, while the square cross-section of andalusite crystals help to identify them clearly.

Andalusite

Kyanite

Crystal system: Andalusite (A), sillimanite (S): orthorhombic; kyanite (K): triclinic.
Crystal habit: A: square prisms; S: long prisms; K: flat blades. Also massive and granular.
Colour: A: russet, green, gold; S: yellow, white; K: blue to white.
Lustre: Vitreous.
Streak: White.
Hardness: 7.5.
Cleavage: A: Good, prismatic; S: Perfect, prismatic; K: Perfect, prismatic.
Fracture: Splintery.
Specific gravity: A: 3.1–3.2; S: 3.2–3.3; K: 3.5–3.7.
Notable locations: Tyrol, Austria; Bihar, Assam, India; Sri Lanka (K); Mogok, Myanmar; Xinjiang, China (A); Bim-bowrie, Australia (A); Minas Gerais, Brazil; California; Yancey, North Carolina.

Sphene

CaTiSiO₅ Calcium titanium silicate

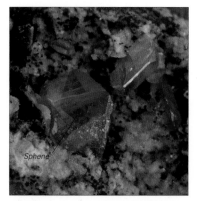

Identification: Sphene is best identified by its wedge-shaped crystals and its brown, green or yellow colour.

Sphene

Sphene gets its name from the Greek word for 'wedge', because of its typical wedge-shaped crystals. It is also called titanite, because of its titanium content. Sometimes, mineralogists call titanite 'common sphene', and reserve the name sphene for especially well-developed or twinned crystals. Common sphene is a minor ingredient of deep-forming igneous rocks such as granites, granodiorites and syenites, as well as gneisses, schists, pegmatites and crystalline limestones, where it commonly occurs in association with diopside, garnet and epidote. Beautiful yellow or pale green crystals of sphene with tremendous fire are found in fissures in schist, especially in the Swiss and Austrian Alps. These are cut as gemstones, although they are really too soft.

Crystal system: Monoclinic.
Crystal habit: Crystals typically flattened or wedge-shaped. Usually massive.
Colour: Brown, green or yellow, or white, or black.
Lustre: Adamantine.
Streak: White.
Hardness: 5–5.5.
Cleavage: Indistinct in two directions.
Fracture: Conchoidal.
Specific gravity: 3.3–3.6.
Other characteristics: Pleochroic if strongly coloured.
Notable locations: Alps, Switzerland; Tyrol, Austria; Madagascar; Minas Gerais, Brazil; Renfrew, Ontario.

Epidote

Ca₂(Al,Fe)₃(SiO₄)₃(OH) Calcium aluminium iron silicate hydroxide

Zircon is a nesosilicate but has a lot in common with epidote.

The epidote group includes a dozen or so silicate hydroxide minerals, but epidote is the only common one. Best known for its pistachio-green colour – though it can also occur in yellow-green and even black – epidote gets its name from the Greek word *epidosis* which means 'increase', because one side of the crystal is longer than the other. Epidote is a major ingredient of many metamorphic rocks, and typically forms when plagioclase feldspar, pyroxenes and amphiboles are altered by contact metamorphism – though usually only in the presence of hot solutions. It can also be found in association with vesuvianite, garnets and other minerals in contact with metamorphosed limestones and skarns (metamorphosed ores). Epidote sometimes occurs in cavities in basalt, too, or in the seams that form when granite cools and shrinks, allowing gases to escape.

Identification: Epidote is best identified by its pistachio green colour, its single cleavage direction, and its association with vesuvianite and garnets in metamorphosed rocks.

Epidote

Crystal system: Monoclinic.
Crystal habit: Crystals prism-shaped with striations; also forms crusts, masses, grains.
Colour: 'pistachio' green, yellow-green, brown, black.
Lustre: Vitreous.
Streak: White to grey.
Hardness: 6–7.
Cleavage: Good in only one direction.
Fracture: Uneven to conchoidal.
Specific gravity: 3.3–3.5.
Notable locations: Untersulzbachtal (Tyrol), Austria; Namibia; Afghanistan; Baja California, Mexico; Prince of Wales Island, Alaska.

Vesuvianite

Ca₁₀(Mg,Fe)₂Al₄(SiO₄)₅(Si₂O₇)₂(OH)₄ Calcium magnesium iron aluminium silicate hydroxide

Crystal system: Tetragonal.
Crystal habit: Prism-shaped crystals with square cross-section. Occasionally massive.
Colour: Normally green, but can also be brown, yellow, blue and/or purple.
Lustre: Vitreous or greasy to resinous.
Streak: White.
Hardness: 6.5.
Cleavage: Poor in one direction.
Fracture: Conchoidal to uneven.
Specific gravity: 3.3–3.5.
Notable locations: Vala Ala, Mount Vesuvius, Italy; Pitkäranta, Yakutia (Siberia), Russia; Asbestos, Quebec; Eden Mills, Vermont.

Identification: Vesuvianite is recognized by its greenish colour, and the square cross-section of its crystals, typically completely embedded in limestone calcite.

Vesuvianite was named in 1795 after Mount Vesuvius in Italy by the famous German mineralogist Abraham Werner, who found crystals of it in limestone blocks that had been ejected from the volcano. It was once also called idocrase, from the Greek for 'mixed form', because its chemistry includes elements of both neso- and sorosilicates. It typically forms when limestones are metamorphosed by contact with hot magma. Here it is typically associated with garnet (grossular and andradite), diopside and wollastonite. It can also occur in metasomatic seams and lenses of serpentinized ultramafic rocks, associated with the same minerals and with chlorite. Very occasionally it is also found in pegmatites. It is usually crystalline, but, when massive, it can look a little like jadeite.

Special twins

Twinning occurs during crystallization. Instead of a normal single crystal, twins double and seem to grow out of each other like Siamese twins. This is not random but follows rules called twin laws. There are two kinds of twins: contact and penetration. In contact twins, like those in sphene, there is a distinct boundary between the crystals so they look like mirror images. In penetration twins, it looks as if the two crystals grow right across each other. Staurolite shows one of the most amazing examples of twinning. In staurolite, two crystals interpenetrate so completely that it looks as if they grew out of each other. There are two kinds – one in which the crystals cross at 60 degrees to each other, and another, highly sought-after, version in which the crystals cross at right angles (above). The right-angle cross gave the mineral its name, from the Greek for 'cross'. It also gave the mineral its Christian associations with the Maltese cross, and its reputation as a good luck 'Fairy Cross'. Staurolite forms in metamorphic rock, and over a third of crystals found are twinned.

AMPHIBOLES

The amphiboles are a large and complex group of nearly 60 minerals that typically form wedge-shaped crystals quite similar to pyroxenes, but longer. Amphiboles occur in many igneous rocks, and many metamorphic rocks – especially those altered from dolomites and mafic igneous rocks. They vary hugely in chemical make-up and appearance, but can be hard to tell apart without the aid of laboratory tests.

Tremolite, Actinolite

$Ca_2Mg_5Si_8O_{22}(OH)_2$ Calcium magnesium silicate hydroxide (Tremolite)
$Ca_2(Mg,Fe)_5Si_8O_{22}(OH)_2$ Calcium magnesium iron silicate hydroxide (Actinolite)

White tremolite fibres in green chlorite

Tremolite and actinolite belong to a series of silicate hydroxide minerals in which iron and magnesium swap places. Tremolite is at the magnesium-rich end, and ferro-actinolite is at the iron-rich end, with actinolite in between. Tremolite is clear to white; as little as 2 per cent iron in place of some of its magnesium can change it to green actinolite. Although normally crystalline, they occur in a fibrous form, being the first minerals to be called asbestos, a term now applied to similar fibrous amphiboles. One peculiar fibrous variety of tremolite is called 'mountain leather' and looks just like felt. One kind of actinolite is the tough, smooth green nephrite, one of the two jade minerals. Both tremolite and actinolite form at moderately high pressures and temperatures in a watery environment. Tremolite often forms when dolomites are metamorphosed by contact with hot magma. Actinolites may form when basalt and diabase are metamorphosed to schists. Both may also form when pyroxenes are altered in igneous rocks.

Identification: In crystal form actinolite and tremolite can easily be confused with pyroxenes such as wollastonite, but when it forms green fibres actinolite is quite distinctive.

Long, finely fibrous actinolite crystals in calcite

Crystal system: Monoclinic.
Crystal habit: Usually long-bladed or prismatic crystals; may be fibrous or massive.
Colour: White or grey (tremolite), green (actinolite).
Lustre: Vitreous or silky.
Streak: White.
Hardness: 5–6.
Cleavage: Perfect in two directions.
Fracture: Uneven.
Specific gravity: 2.9–3.4.
Notable locations: Tremolite: Piemonte, Italy; Tremola V, Switzerland; Haliburton Co, Ontario; St Lawrence Co, New York. Actinolite: Tyrol, Austria; Baikal, Russia; Moonta, Wallaroo, Western Australia; Mont Saint Hilaire, Quebec; Windsor Co, Vermont.
Caution: Tremolite and actinolite often form asbestos fibres. Keep away from nose and mouth and wash hands carefully.

Glaucophane

$Na_2(Mg,Fe)_3Al_2Si_8O_{22}(OH)_2$ Sodium magnesium iron aluminium silicate hydroxide

Glaucophane gets its name from the Greek *glaukos*, which means 'blue', and *fanos*, which means 'appearing', because of its typical colour. It is found in rocks such as schist, marble and eclogite, and is generally formed in metamorphic zones known as blueschist facies. Blueschist facies typically occur along continental margins underthrust by shifting oceanic plates and in areas of high volcanic and seismic activity. Some of the best-known examples are in Japan, California, the Mediterranean and the Alps. Glaucophane gives these rocks a bluish hue.

Identification: Glaucophane is best identified by its colour, fibrous crystals, location near plate margins, and its association with minerals such chlorite, epidote, aragonite, jadeite, muscovite and garnet.

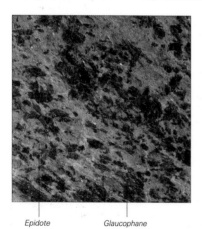

Epidote   Glaucophane

Crystal system: Monoclinic.
Crystal habit: Rare prism and needle-shaped crystals; usually fibres, grains or masses.
Colour: Blue to dull grey.
Lustre: Vitreous to pearly.
Streak: Pale grey to blue.
Hardness: 5–6.
Cleavage: Imperfect in two directions.
Fracture: Conchoidal to splintery.
Specific gravity: 3–3.2.
Notable locations: Anglesey, Wales; Valais, Switzerland; Val d'Aosta, Italy; Syra Is, Greece; Orange River, South Africa; Ishigaki, Japan; Flinders River, S Australia; Coast Range, California; Kodiak Is, Alaska.
Caution: Glaucophane may form asbestos fibres. Keep well away from nose and mouth; wash hands carefully.

The hazards of asbestos
Remarkable fibres called asbestos are made by the serpentine mineral chrysotile and many amphiboles, notably actinolite, tremolite, anthophyllite, amosite, crocidolite and riebeckite. *Asbestos* is Greek for 'indestructible', and the Greeks and Romans knew of its fire-resistant properties, using it in lamp wicks and in napkins that could be cleaned by throwing them into the fire. They also knew of its health hazards. But it was only in the 1880s that the modern use of asbestos began – when huge deposits of chrysotile were mined in Canada (an abandoned mine appears above) and Russia – and only after World War II that its use became widespread. By the 1970s, the USA was producing around 300 million tonnes of asbestos a year for producing everything from car brake pads to fireproof roofs. In the 1980s, however, the terrible long-term health effects of asbestos fibres – especially on the lungs – became clearer, and its use was banned in America and Europe. It is still widely used elsewhere, as in China.

Hornblende

Ca₂(Mg,Fe,Al)₅(Si,Al)₈O₂₂(OH)₂ Calcium magnesium iron aluminium silicate hydroxide

Hornblende gets its name from the German for 'horn', because of its dark colour, and *blenden*, which means 'to dazzle'. The term *blende* was often used to refer to a shiny non-metal. It is a common ingredient in igneous rocks, especially intrusive rocks such as granites and granodiorites, diorite and syenites. Medium-grade metamorphic gneisses and schists can also be rich in hornblende. Rocks with a predominance of amphiboles (essentially hornblende) are called amphibolites. In rocks such as diorite, the hornblende grows into the space between plagioclase crystals. In granites and granodiorites, hornblende forms tiny, well-formed crystals. Large crystals are rare but are usually stubby and prism-shaped. Hornblende has a large and varied chemical make-up, in which the mix of calcium and sodium, or magnesium and iron, varies. Hornblende with less than 5 per cent iron oxides is grey or white and called edenite, after Edenville, New York, where it was identified.

Identification: Hornblende can often be identified as an amphibole by the way it breaks into wedge-shapes, and as hornblende by its very dark colour.

 Crystal system: Monoclinic.
Crystal habit: Stubby, prism-shaped crystals; mostly masses, grains, fibres.
Colour: Black to dark green.
Lustre: Vitreous to dull.
Streak: Brown to grey.
Hardness: 5–6.
Cleavage: Prismatic, good.
Fracture: Uneven.
Specific gravity: 3–3.5.
Notable locations: Arendal, Norway; Bilina (Bohemia), Czech Republic; Falkenberg (Urals), Russia; Mt Vesuvius, Italy; Murcia, Spain; Bancroft, Ontario; Edenville, New York; Franklin, New Jersey.

Anthophyllite

(Mg,Fe)₇Si₈O₂₂(OH)₂ Magnesium iron silicate hydroxide

Anthophyllite gets its name from the Latin for 'clove' because of its distinctive clove-brown colour. It is usually confused only with similar amphiboles such as cummingtonite. Anthophyllite is formed by metamorphism and found in gneisses and schists derived by the alteration of magnesium-rich igneous or dolomitic sedimentary rocks. It also forms from metamorphic rocks, and by the alteration of minerals such as olivine by water. Under different conditions (if there were more water), serpentine would be the mineral produced from the alteration of olivine. Well-formed crystals of anthophyllite are rare, but some aggregates can be impressive.

Fibrous sheaves of anthophyllite on schist

Identification: Anthophyllite is best identified by its colour and fibrous masses.

Crystal system: Orthorhombic.
Crystal habit: Rare, prism-shaped crystals and fibrous, asbestos masses.
Colour: Brown, off-white grey.
Lustre: Vitreous or silky in fibrous forms.
Streak: Grey.
Hardness: 5.5–6.
Cleavage: Good in two directions.
Fracture: Easy, splintery.
Specific gravity: 2.8–3.4.
Notable locations: Nuuk, Greenland; Elba, Italy; Kongsberg, Norway; Butte, Montana; Franklin, North Carolina; Tallapoosa, Alabama; Cummington, Massachusetts (Cummingtonite).
Caution: Anthophyllite often contains asbestos fibres, which are a major health hazard. Keep well away from nose and mouth and wash hands carefully.

CLAYS

The term clay can describe any kind of fine particle, but also specifically a large group of minerals such as chlorite, kaolinite, talc and serpentine. Clay minerals are aluminium and magnesium silicates, and occur as fine particles that form when other minerals are broken down by weathering, water and heat. Their sheet-like molecular structure means that they absorb or lose water easily, making them very useful.

Chlorite

(Fe,Mg,Al)₆(Si,Al)₄O₁₀(OH)₈ Iron aluminium magnesium silicate hydroxide

The chlorites get their name from the Greek word for 'green', and they are a group of mainly pale green minerals, although they can be white, yellow or brown. They form when minerals such as pyroxene, amphibole, biotite mica and garnets are broken down by weathering or by exposure to hot solutions. Sometimes they literally take the place of these minerals within the rock, giving many igneous and metamorphic rocks a green look. Indeed, chlorite is a major ingredient in many schists and phyllites, especially the green chlorite schists. Sometimes they create earth-filled cavities and crevices. Eventually the whole rock might break down to create soil, and chlorite is a major ingredient of soils such as podzols. Occasionally, chlorite forms good, wedge-shaped crystals in cavities, but more typically it is massive and earthy. Chlorite can also form scaly flakes like mica, but they contain more water and no alkalis, making them much softer and lighter, and less well defined.

Identification: Chlorite is best identified by its green colour, its softness (almost as soft as talc) and in many specimens its scaly flakes.

Crystal system: Monoclinic.
Crystal habit: Rare crystals are tablet- or prism-shaped hexagons, but more usually earthy aggregates or scaly flakes.
Colour: Usually green, also white, yellow, brown.
Lustre: Vitreous, dull or pearly.
Streak: Pale green to greyish.
Hardness: 2–3.
Cleavage: Breaks into slightly flexible flakes.
Fracture: Lamellar (sheetlike).
Specific gravity: 2.6–3.4.
Notable locations: Carinthia, Austria; Zermatt, Switzerland; Piedmont, Italy; Guleman, Turkey; Renfrew County, Ontario; Brewster, New York; San Benito County, California.

Kaolinite

Al₂Si₂O₅(OH)₄ Aluminium silicate hydroxide

Crystal system: Triclinic; or monoclinic.
Crystal habit: Foliated and earthy masses. Crystals rare.
Colour: White, colourless, greenish or yellow.
Lustre: Earthy.
Streak: White.
Hardness: 1.5–2.
Cleavage: Perfect basal.
Fracture: Earthy.
Specific gravity: 2.6.
Other characteristics: Mouldable when wet.
Notable locations: Many including Cornwall, England; Dresden, Germany; Donets River, Ukraine; Kaolin, China.

Identification: Kaolinite minerals are typically white, soft (if scratched) and powdery.

The mineral kaolinite gives its name to the group of minerals from which kaolin (see China clay, right) comes and includes nacrite, dickite and halloysite as well as kaolinite – all with identical chemistry but differing crystal form. Most clay contains at least some kaolinite, and some clay beds are entirely kaolinite. Indeed, kaolinite is found almost everywhere – in soils, in stream beds, in rocks and many other places. Some kaolinites form as aluminium silicates in rocks and soils that are weathered and broken down. Others are formed in rocks such as granites and pegmatites when feldspars are altered by hot solutions. Kaolin formed from feldspar is often extracted from pegmatites, while in the china clay quarries of Cornwall, England, it comes from kaolinite formed from potassium feldspar in granite.

Kaolinized granite

Talc

Mg₃Si₄O₁₀(OH)₂ Magnesium silicate hydroxide

Talc is the softest of all minerals, rating 1 on the Mohs scale. The softness, whiteness and ability of pure talc to hold fragrances has long made it popular when ground into powder for cosmetics and babies. Talc is also used as a filler in paints, in rubber and in plastic. It typically occurs in magnesium-rich rocks affected by low-grade metamorphism – especially rocks such as peridotites and gabbros. It forms when magnesium silicate minerals such as olivine and pyroxene are altered by hydrothermal fluids forced through cracks in the rocks, often along fault lines. Talc also occurs in schists made from metamorphosed magnesian limestone. It typically occurs in association with serpentine, and may be compacted into a mass called soapstone, used since ancient times for carving and, because it becomes hard when heated, for heat-resistant utensils and insulators. Steatite is a particularly dense soapstone made from almost pure, white talc.

Talc

Magnesite

Identification: Talc is best identified by its extreme softness – you can scratch it with a fingernail – its soapy feel and its white or light grey or almond colour. It is often mixed with green chlorite.

Crystal system: Monoclinic.
Crystal habit: Crystals rare; usually forms grains of flaky masses.
Colour: White, greenish, grey.
Lustre: Dull, pearly, greasy.
Streak: White.
Hardness: 1.
Cleavage: Perfect in one direction, basal.
Fracture: Uneven to lameller (sheetlike).
Specific gravity: 2.7–2.8.
Other characteristics: Soapy texture to surface.
Notable locations: Shetland, Scotland; Florence, Italy; Tyrol, Austria; Transvaal, South Africa; Appalachian Mountains, Vermont; Connecticut; New York; Virginia; California; Texas.

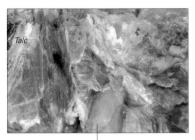

China clay
Although it is often taken for granted, kaolin, or china clay, is one of the most beautiful and useful of mineral products. It gets its name from Kao-ling hill in China, where it has been quarried for 1,400 years to make porcelain. It is a soft, white, very fine-grained clay made mainly from the mineral kaolinite. It forms from feldspars when soft soda feldspar-rich granites are broken down, and also when gneisses and porphyritic rocks crumble. In its natural state it is often stained yellow by iron hydroxides, and contaminated by micas and quartz. These contaminants can be washed out, and the kaolin bleached pure, shining white. It is this whiteness that makes it the perfect base material for paper, tied together with cellulose fibres from wood. When it is mixed with a quarter to a third water, it becomes 'plastic' – that is, mouldable – and once moulded and fired it retains its shape and white colour. This is what makes it so good for porcelainware. Besides China, the best known sources are Cornwall in England (pictured above; mining began here in the 18th century); Saxony, Germany; Sévres, France; and Georgia, USA.

Serpentine

Mg₃Si₂O₅(OH)₄ Magnesium iron silicate hydroxide (Lizardite)

Named after the green flecks said to make it look like snakeskin, serpentine forms when magnesium silicates in rocks like peridotite and dolomites are altered by hot fluids. The same process makes talc, veins of which are often found in serpentine. A rock entirely made from serpentine is called serpentinite. Serpentine also forms spidery veins in other minerals when they are altered by moisture, a process called serpentinization. Serpentine is not a single mineral but a group of minerals including chrysotile, antigorite and lizardite. Chrysotile is a fibrous, asbestos mineral, like crocidolite and anthophyllite. Antigorite is a corrugated kind of serpentine. Lizardite, named after the Lizard peninsula in Cornwall, England, is a fine-grained, platy variety.

Lizardite
Crystal system: Trigonal and hexagonal; (Antigorite, Clinocrysotile monoclinic; Ortho- and Parachrysotile orthorhombic).
Crystal habit: Usually forms fine-grained platy masses.
Colour: Green, blue, yellow, white.
Lustre: Waxy or greasy.
Streak: White.
Hardness: 2.5.
Cleavage: Perfect, basal.
Fracture: Conchoidal (Lizardite, Antigorite); Splintery (Crysotiles).
Specific gravity: 2.5–2.6.
Notable locations: Lizard (Cornwall), England; Salzburg, Austria; Liguria, Italy; Auvergne, France; Urals, Russia; Asbestos, Quebec; Hoboken, New Jersey; Eden Mills, Vermont.

Identification: Lizardite can be identified by its yellow-green colour, softness (as hard as fingernail), platy texture and association with talc in rocks like peridotite and dolomite.

MICAS

Micas are among the most instantly recognizable of minerals, with their flaky, almost transparent layers. These aluminium silicates are also among the most common of all the rock-forming minerals and major ingredients in all three rock types. There are 30 or so different kinds of mica altogether, but the most important ones are biotite, muscovite, phlogopite and lepidolite. Other important micas are glauconite and paragonite.

Biotite

K(Fe,Mg)₃AlSi₃O₁₀(OH,F)₂ Potassium iron magnesium aluminium silicate hydroxide fluoride

Biotite is a black, iron-rich mica. It is named after Jean Baptiste Biot, the French physicist who first described the optical effects of micas. It occurs in varying percentages in almost every kind of igneous and metamorphic rock, but is especially prominent in granites, diorites and andesites and schists, gneisses and hornfels. It adds glitter to schists, the black 'pepper' grains in granite and darkness to sandstones. Its iron content makes it the darkest of the micas. It is also one of the softest – scratchable with a fingernail – which is why good crystals are rarely found. All the same, single large plates or 'books' of biotite can grow to quite a size, especially in granite pegmatites. Biotite is part of a series with phlogopite, with denser, black biotite at the iron-rich end, and less dense, brown phlogopite at the iron-poor end. Weathered crystals of biotite can turn golden yellow and sparkle, fooling people into thinking they are gold. Biotite is easily altered; in the presence of seawater, for instance, it quickly turns into the mica mineral glauconite.

Identification: Biotite is usually easy to recognize by its dark colour and its soft flakes. It can look like phlogopite but is darker and denser because of its iron content.

 Crystal system: Monoclinic.
Crystal habit: Hexagonal tablet-shaped crystals; more often flakes, often in 'books', or grains in rock.
Colour: Black to brown and yellow with weathering.
Lustre: Vitreous to pearly.
Streak: White.
Hardness: 2.5.
Cleavage: Perfect in one direction, giving thin sheets or flakes.
Fracture: Not often visible but uneven when seen.
Specific gravity: 2.9–3.4.
Other characteristics: Flakes bend and spring back flat.
Notable locations: Norway; Monte Somma (Mount Vesuvius), Italy; Sicily; Russia; Bancroft, Ontario, and many other locations.

Phlogopite

KMg₃AlSi₃O₁₀(OH)₂ Potassium magnesium aluminium silicate hydroxide

Phlogopite is very similar to biotite, but contains little if any iron and so is lighter in colour, and less dense. It can sometimes have the red-brown hue that earned its name from the Greek *phlogopos*, meaning 'fire-like'. It often occurs with biotite, but whereas biotite is common in granite and other acidic igneous rocks, phlogopite is common only in ultramafic rocks such as pyroxenites and peridotites, and in metamorphosed limestones, notably magnesium-rich marbles. Its low iron content means phlogopite is a good electrical insulator, and is so valued by the electrical industry that there have been a number of attempts to synthesize. Synthetic phlogopite has so far proved too costly, however. When a thin flake of phlogopite is held up to the light, it will sometimes show asterism – a six-rayed star – due to small inclusions in the crystal.

Identification: Phlogopite is usually easy to recognize by its reddish brown colour and its soft flakes. It can look like biotite but is lighter in colour and less dense.

 Crystal system: Monoclinic.
Crystal habit: Hexagonal tablet-shaped crystals; more often flakes, often in 'books', or grains in rock.
Colour: Brown, reddish brown, maybe coppery.
Lustre: Vitreous to pearly.
Streak: White.
Hardness: 2.5–3.
Cleavage: Perfect in one direction, giving thin sheets or flakes.
Fracture: Uneven.
Specific gravity: approximately 2.9.
Other characteristics: Flakes bend and spring back flat.
Notable locations: Norway; Kovdor (Kola Peninsula), Russia; Madagascar; Bancroft, Ontario.

Muscovite

KAl₂(Si₃Al)O₁₀(OH,F)₂ Potassium aluminium silicate hydroxide fluoride

Muscovite is the most common mica, and is found nearly everywhere that intrusive igneous and metamorphic rocks occur. It is particularly abundant in granites and pegmatites, gneisses, schists and phyllites. It also forms when feldspars are altered. Microscopic-grained muscovite or sericite is what gives phyllite its silky sheen. In pegmatites, muscovite often occurs in giant sheets which are commercially valuable, and also in cavities, where the best crystals occur. Most muscovite in igneous rocks forms late as the magma solidifies. It is resistant to weathering, so is abundant in soils over muscovite-rich rocks, and in sands formed from them. Like all micas, muscovite breaks into flakes and can be almost transparent. In old Russia, sheets of muscovite were used for windows, being known as Muscovy glass, after the old name for Russia. It is very heat-resistant and it is still used for the windows in stoves. It is also a superb electrical insulator, and is used as artificial snow for Christmas trees.

Identification: Muscovite's thin flakes identify it clearly as a mica. It is much lighter in colour than either biotite or phlogopite mica.

Intergrowth crystals of muscovite

Crystal system: Monoclinic.
Crystal habit: Hexagonal tablet-shaped crystals; more often flakes, often in 'books', or grains in rock.
Colour: White, silver, yellow, green and brown.
Lustre: Vitreous to pearly.
Streak: White.
Hardness: 2–2.5.
Cleavage: Perfect in one direction, giving thin sheets or flakes.
Specific gravity: 2.8.
Other characteristics: Flakes bend and spring back flat.
Notable locations: Mursinka (Urals), Russia; Inikurti (Nellore), India; Brazil; Amelia County, Virginia, and many other locations.

Marvellous mica
Micas occur in two forms: flake mica, which occurs naturally in small flakes, and sheet mica which occurs in large sheets that can be cut into particular shapes. Sheet mica is much rarer, but large sheets of muscovite can occur in pegmatites, and these were used for the Muscovy glass once used in Russian windows. Because it is quite rare, sheet mica is often synthesized from flake or scrap mica. The high electrical and heat resistance of sheet mica means it is perfect for use as electrical insulators, for heat insulation, for the windows in stoves, and much more besides. Phlogopite is used as the insulation between the copper and steel in electric motors because it wears at exactly the same rate as the copper. Ground-up mica is also useful, and is added to plasterboard as a filler that prevents cracking. It is also used in paint to make it dry smoothly, and in oil well drills as a lubricant.

Lepidolite

K(Li,Al)₃(Si,Al)₄O₁₀(F,OH)₂ Potassium lithium aluminium silicate hydroxide fluoride

Pinky purple in colour, with a vitreous to pearly lustre, lepidolite is perhaps the most attractive of all the micas. However, it is relatively rare and forms only in granite pegmatites. Indeed, it pretty much only forms in complex pegmatites where there is plenty of lithium present, and where minerals have been replaced with other minerals repeatedly through time. It sometimes forms when lithium minerals alter muscovite, creating beautiful lace-like rims. It was once used as an ore of lithium, but the lithium content varies so much that most lithium is now taken from alkali lake brines. Some specimens are triboluminescent – that is, they will flash colours when pressed.

Crystal system: Monoclinic.
Crystal habit: Hexagonal tablet-shaped crystals; more often flakes, often in 'books', or grains in rock.
Colour: Violet to pale pink or white and rarely grey or yellow.
Lustre: Vitreous to pearly.
Streak: White.
Hardness: 2.5.
Cleavage: Perfect in one direction, giving thin sheets or flakes.
Specific gravity: approximately 2.8+ (average).
Other characteristics: Flakes bend and spring back flat. May show triboluminescence.
Notable locations: Varuträsk, Sweden; Penig (Saxony), Germany; Urals, Russia; Alto Ligonha (Zambesi), Mozambique; Madagascar; Pakistan; Londonderry, Western Australia; Minas Gerais, Brazil; Maine; Portland, Connecticut; San Diego County, California.

Identification: Flakes clearly identify lepidolite as a mica, but it is only likely to be lepidolite if found in complex dike rock along with lithium minerals.

MINERALOIDS

A few solid substances that occur naturally in the Earth do not quite conform with the basic properties of minerals. They fit into none of the chemical families, rarely form crystals and are typically organic in origin, evolving from fossilized or compacted living matter. Such substances are called mineraloids. Amber is fossilized pine tree resin. Jet, like coal, forms from the remains of trees. Pearl is formed by certain shellfish as a coating around irritant debris inside their shells. Whewellite forms from organic acids.

Amber

Approximately $C_{10}H_{16}O$ Succinic acid

Although it is often regarded as a gem, amber is not a mineral at all but a solid organic material. It is a form of tree resin that was exuded to protect the tree against disease and insect infestation. This then hardened and was preserved for millions of years. Amber slowly oxidizes and degrades when exposed to oxygen, so it survives only under special conditions. It is almost always found in dense, wet sediments, such as clay and sand formed in ancient lagoon or river delta beds. It is typically found embedded in shale or washed up on beaches. Most amber deposits contain only fragments of amber, but a few contain enough to make it worth mining, such as those found along the shores of Baltic seas where amber formed in sands 40–60 million years ago, and those from the Dominican Republic. Amber can form nodules, rods and droplets in various shades of orange and brown. Milky-white varieties are called bone amber. Modern analysis techniques are beginning to identify the detailed composition of ambers and link them to modern resin-making trees. Mexican amber, for instance, is linked to the *Hymenea* tree.

Crystal system: Usually amorphous.
Crystal habit: Nodules, rods, droplets.
Colour: Amber, brown, yellow.
Lustre: Resinous.
Streak: White.
Hardness: 2+.
Cleavage: None.
Specific gravity: 1.1 (will float in salty water).
Other characteristics: Can be burned, fluorescent.
Notable locations: Kaliningrad, Russia; Lithuania; Latvia; Estonia; Poland; Romania; Germany; Lebanon; Sicily; Dominican Republic; Mexico; Canada.

Identification: Amber is easy to identify from its amber colour, resinous lustre and smooth, round shape.

Jet

C Carbon

Like amber, jet comes from ancient trees and forests that grew hundreds of millions of years ago. It is sometimes thought of as a type of shiny black lignite coal, because it forms in a similar way, from compacted remains of trees. (Lignite is a less compressed brown coal which retains the structure of the original wood.) However, while coal formed in fresh water from swampy tropical forests that grew in the Carboniferous Age (354–290 million years ago), hard jet probably formed in many periods from logs that floated down river and out to sea to sink on to the sea bed. Despite its black colour, jet is not as rich in carbon as either anthracite or bituminous coal – as its brown streak reveals. The famous Whitby jet from the beaches of Whitby, in Yorkshire, England, has been used in jewellery since prehistoric times. It is found as lenses embedded in hard shales known as jet-rock.

Identification: Jet is easily identified by its jet-black colour. It can look like vulcanite, obsidian and black onyx, but has a brown streak.

Crystal system: Does not form crystals.
Crystal habit: Lumps with microscopic wood texture.
Colour: Black.
Lustre: Glassy.
Streak: Brown.
Hardness: 2–2.5.
Specific gravity: 1.1–1.3.
Other characteristics: Breaks with conchoidal fracture.
Notable locations: Whitby (North Yorkshire), England; Württemberg, Germany; Asturias, Spain; Aude, France; Pictou, Nova Scotia; Wayne Co, Garfield Co, Utah; San Juan Co, Guadalupe Co, Cibola Co, New Mexico; Custer Co, Colorado.

Pearl

Over 6,000 years ago, people around the Persian Gulf were buried with a pierced pearl in their right hand, and pearls featured in all the major cultures of the ancient world. Despite their beauty, pearls are not gemstones: they are organic materials formed by bivalve molluscs such as oysters and mussels. All these shellfish build up their shells with a hard, brown, hornlike protein called conchiolin, then line it with a smooth, shiny mother-of-pearl. Mother-of-pearl, a substance secreted from the outer membrane of the mollusc, is made from a material called nacre, built up from thin alternating layers of colourless conchiolin and the mineral aragonite. Pearls form when an irritant gets in the shell, such as a food particle. To stop it irritating, the mollusc gradually coats it in layers of mother-of-pearl. The longer this process goes on, the bigger the pearl will be. The biggest pearls come from marine oysters, but freshwater molluscs also make pearls. When pearls are cultured on oyster farms, an irritant is deliberately introduced into the oyster to stimulate the growth of a pearl.

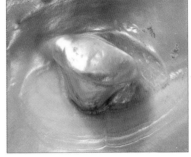

Crystal system: Does not form crystals.
Crystal habit: Forms microscopic alternating layers of aragonite and conchiolin.
Colour: Iridescent pearly white.
Lustre: Pearly.
Streak: White.
Hardness: 3.5–4.
Cleavage: None.
Specific gravity: 2.9–3.
Notable locations: Freshwater pearls: rivers across northern Europe, Asia and North America (notably Muscatine, Iowa, on the Mississippi River) Marine pearls: Venezuela, Persian Gulf.

Identification: The white aragonite in mother-of-pearl is visible through the translucent conchiolin, which gives the lining of the shell, and the resulting pearl, its famous lustre.

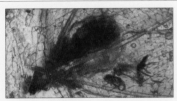

Amber time capsules
Perhaps the most amazing thing about amber is the way it perfectly preserves organisms, such as insects, trapped in it when it was liquid resin. Compounds called terpenes link up as the resin hardens, dehydrating the organisms and killing off bacteria that would make them decay. Amazingly, dehydration does not make the organism shrink, so its cell structure is preserved intact. Amber from the Cretaceous period (65–140 million years ago) offers a unique glimpse of insect life at a time when the dinosaurs flourished, not preserved in any other way. The oldest known bee, *Trigona prisca*, for instance, was found in 65–80 million-year-old amber from New Jersey. The oldest known mushroom, 90–94 million-year-old *Archaeo-marasmius*, was also found here in amber. Amber preserves things so perfectly that scientists still hold out the possibility suggested in the film *Jurassic Park* of recreating dinosaurs from DNA in blood sucked from them by mosquitoes preserved in amber. The best Cretaceous amber comes from northern Russia, but the oldest is from the Middle East. Early forms of animal life are also trapped in copal, a less mature resin often mistaken for amber.

Whewellite

CaC$_2$O$_4$.H$_2$O Hydrated calcium oxalate

Whewellite is named after the famous Victorian geologist William Whewell (1794–1866). Whewellite forms stones in human kidneys and mineralogists argue over whether it is a mineral or not. It is calcium oxalate, an organic chemical, and it forms from oxalic acid derived from organic sources such as coal and rotting vegetation. In the Arizona desert, whewellite forms on dead agave plants, which has lead some mineralogists to say that it is not a mineral. However, it forms crystals, concretions and crusts, and occurs in septarian nodules. Moreover, oxalic acid forms whewellite by reacting with calcium hydroxide, which occurs in ordinary groundwater and hydrothermal fluids. Whewellite typically forms where these fluids meet carbon-rich rocks, such as graphitic schists and anthracite coal.

Crystal system: Monoclinic.
Crystal habit: Tiny needles (slender prisms) and crusts.
Colour: Colourless, white, yellow, brown.
Lustre: Vitreous to pearly.
Streak: White.
Hardness: 2.5–3.
Cleavage: Good in three directions.
Specific gravity: 2.2.
Other characteristics: Very brittle, breaking in conchoidal fragments.
Notable locations: Freital, Burgk (Saxony), Gera (Thuringia), Germany; Alsace, Rhône-Alps, France; Kladno, Czech Republic; Yakutia, Russia; San Juan Co, Arizona; Havre (Hill Co), Montana.

Identification: Whewellite is best identified by the environment in which it forms, its tiny needle-like crystals and its very low density.

PRECIOUS METALS

Although traces of valuable metals may be found in a number of minerals, only those containing a relatively substantial amount – ore minerals – are of interest to prospectors. Silver, scandium and members of the platinum group of metals such as platinum and osmium are rare and precious metals, found in only a small number of places around the world. Where they are discovered, it is usually very expensive to extract them; however, their value and usefulness make the effort and cost worthwhile.

Silver ore: Proustite

Ag_3AsS_3 Silver arsenic sulphide

Silver is a rare and precious metal that sometimes occurs in pure form as a native element. It also occurs in no less than 248 other mineral species, such as argentite, eskimoite and uchucchacuaite, but only a few are common. Most silver tends to be recovered as a by-product from galena (lead ore), or tetrahedrite and chalcopyrite (copper ores). Silver compounds typically form near the Earth's surface when complex compounds lose their sulphur. Deeper down, silver occurs where hydrothermal veins create sulphides, antimonides and arsenides. Proustite is one of these silver sulphides. Named after the French chemist J L Proust (1755–1826), it is one of the few sulphides that is neither metallic nor opaque, and has beautiful red crystals that are highly sought after by collectors. However, the crystals quickly darken when exposed to light. Massive proustite is a commercial silver ore called ruby silver, but it occurs on only a small scale. Proustite is closely related to pyrargyrite, which substitutes antimony in place of proustite's arsenic.

Dark red proustite

Crystal system: Trigonal.
Crystal habit: Prism-shaped with rhombohedral or scalenohedral ends, resembling calcite dog-tooth spar. Also forms masses.
Colour: Scarlet to vermillion.
Lustre: Adamantine.
Streak: Scarlet.
Hardness: 2–2.5.
Cleavage: Distinct forming rhombohedrons.
Fracture: Conchoidal.
Specific gravity: 5.6.
Notable locations: Markirch (Alsace), France; Freiburg, Marienberg (Saxony), Germany; Joachimstal (Bohemia), Czech Republic; Chanarcillo, Chile; Chihuahua, Mexico; Lorrain, Ontario; Poorman Mine, Idaho.

Identification: Proustite's red colour and red streak help to identify it. It can be mistaken for its close relative pyrargyrite, but is generally lighter in colour.

Scandium ore: Thortveitite

$(Sc,Y)_2Si_2O_7$ Scandium yttrium silicate

Crystal system: Monoclinic.
Crystal habit: Typically prism-shaped crystals but also occurs in massive and granular forms.
Colour: Brown, greyish black, greyish green.
Lustre: Vitreous – adamantine.
Streak: Grey.
Hardness: 6.5.
Cleavage: Perfect.
Fracture: Conchoidal.
Specific gravity: 3.5.
Notable locations: Setesdalen (Aust-Agder), Norway; Ytterby, Sweden; Saxony, Germany; Ankazone, Madagascar; Kinki, Japan; Sterling, New Jersey.

Scandium is a rare metal first identified in the minerals euxenite and gadolinite in 1876, but not properly prepared until 1937. It is a soft, light, blue-white metal. Scandium oxide is used in high-intensity light, because it has a high melting point, and scandium iodide is added to the mercury vapour lights used for night-time filming. In the Sun, it is the twenty-third most common element, but on Earth, it ranks fiftieth. Traces of scandium can be found in about 800 minerals, but only wolframite, wiikite, bazzite and thortveitite are useful sources. Thortveitite is a rare silicate mineral found mainly in Scandinavia and Madagascar, typically with uranium ores in granite pegmatites. Scandium is often a by-product of uranium ore processing.

Identification: Thortveitite is best identified by its brown colour and its association with rare earth minerals in granite pegmatites. A thortveitite crystal in feldspar is shown above.

Osmium ore: Iridosmine

Os,Ir Osmium, Iridium

Discovered by English chemist Smithson Tennant in 1804 after dissolving platinum in acid, osmium is the densest naturally occurring element. It is the hardest of all the metals, and stands pressure better than diamond. It also has the highest melting point of all the platinum metals. Pure osmium never occurs naturally. Instead, it is found in natural alloys with iridium in siserskite (about four-fifths osmium), iridosmine (about a third osmium) and aurosmiridium (about a quarter osmium). Iridosmine is the most common of these, but is as rare as gold. It occurs in gravels and sands left by the weathering of platinum-rich ultrabasic rocks, such as in the Urals or in gold-bearing conglomerates, as in Witwatersrand, South Africa. This tough, non-corrosive alloy is used as it is, or synthesized to make pen nibs, surgical needles and spark points for cars. However, a significant amount of osmium is derived from other ores, such as the nickel ores of Sudbury, Ontario.

Identification: Iridosmine is best identified by its steel-grey colour. Its association with other platinum group metals or with gold is also a key indicator. Shown here are small grains.

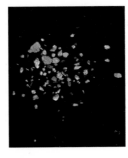

Crystal system: Hexagonal.
Crystal habit: Typically tiny crystals in matrix, but forms tablet-shaped hexagonal crystals.
Colour: Steel grey.
Lustre: Metallic.
Streak: Grey.
Hardness: 6–7.
Cleavage: Perfect.
Fracture: Uneven.
Specific gravity: 19–21.
Notable locations: Bulgaria; Sverdlovskaya (Urals), Kamchatka, Russia; Hunan, China; Witwatersrand, South Africa; Goodnews Bay, Alaska.
Caution: Handle iridosmine with care and keep in an airtight container. Osmium powder and vapour are very poisonous.

The platinum group
The platinum group of elements represents some of the rarest elements in the Earth's crust – ruthenium, rhodium, palladium, osmium and iridium. Together they are known as the platinum group metals, or PGM. Only two of them, platinum and palladium, occur naturally in pure form, as fine grains or flakes scattered throughout dark silicate rocks rich in iron and magnesium. The remainder occur as natural alloys with gold and platinum. Even platinum usually occurs mixed in with iridium. Most PGMs come from the Urals in Russia and the Bushveld complex in South Africa. All the platinum group metals are very dense with high melting points, which makes them extremely useful despite their rarity. Most of the commercially mined platinum group metals are used in catalytic car exhaust converters to cut down on the emission of smog-creating gases. They work by lowering the energy needed to convert these gases to harmless nitrogen and oxygen. Platinum group metals also have a number of medical uses. For example, they are used in dentistry, in chemotherapy to fight leukaemia, and in heart replacement valves.

Platinum ore: Sperrylite

PtAs₂ Platinum arsenide

Platinum is usually found as a native element, alloyed to similar metals – iridium, osmium, palladium, rhodium and ruthenium. Sperrylite (platinum arsenide), is the only known naturally occurring compound of platinum and the only ore. It was identified in 1889 and named after the American chemist F L Sperry, who found samples of it near Sudbury in Ontario, Canada. Sudbury remains the only major source of sperrylite, although good specimens have been found in places such as the Tarnak deposit near Yakut and in Kamchatka, Russia. It has also been found in the Bushveld complex in South Africa. It is a pyrite group mineral of basic pegmatites, often found concentrated in placer deposits.

Identification: Sperrylite has a similar crystal shape to pyrite, but is tin white to pyrite's gold. Sperrylite can also be identified by its association with chalcopyrite, pyrrhotite and pentlandite.

Crystal system: Isometric.
Crystal habit: Typically tiny crystals in matrix. Crystals rounded with mix of cubic, octahedral and other shapes.
Colour: Tin white.
Lustre: Metallic.
Streak: Black.
Hardness: 6–7.
Cleavage: Indistinct.
Fracture: Conchoidal.
Specific gravity: 10.6.
Other characteristics: Brittle.
Notable locations: Finnmark, Norway; Sakha, Kamchatka, Russia; Bushveld complex, South Africa; Sichuan, China; Sudbury, Ontario; Goodnews Bay, Alaska.

IRON ORES

Iron is the most common element in the Earth, making up a third of its mass. Most of this iron is buried deep in the Earth's core, but it is still abundant in rocks near the surface – here it is found mixed with other minerals in the form of ores – hematite, magnetite, siderite, goethite and limonite. It is only rarely found near the surface in pure form, such as in the basaltic rocks of Greenland.

Hematite

Fe_2O_3 Iron oxide

Massive hematite

Hematite is the most important ore of iron, containing 70 per cent of the metal. It gets its name from the Ancient Greek word for 'blood', because it gives rocks a reddish tinge. If soils or sedimentary rocks have a reddish colour, it is usually due to hematite, and this is what makes the planet Mars red. It occurs in other colours such as grey, brown and orange, but red is typical. Hematite often occurs in an earthy form called red ochre, but it can also occur as steel-grey crystals, in kidney-shaped lumps called kidney ore, and in pockets called iron roses. Most industrial sources of hematite come from massive layers in sedimentary rocks, such as those near Lake Superior in North America, or as beds of 'Clinton-type' ooliths laid down on the bed of shallow seas long ago. Sometimes it occurs in metamorphosed sediments such as those at Minas Gerais in Brazil.

Identification: The best clue to hematite's identity is its distinctive red streak.

Quartz
Hematite crystals

Massive hematite

Crystal system: Trigonal.
Crystal habit: Typically earth or sheetlike masses. Also reniform (kidney shapes), or tablet-shaped crystals.
Colour: Steel grey to black as crystals and red to brown in earthy and massive forms.
Lustre: Metallic, or dull in earthy forms.
Streak: Red.
Hardness: 5–6.
Cleavage: None.
Fracture: Uneven.
Specific gravity: 5.3.
Notable locations: Cumbria, England; Gotthard Massif, Switzerland; Elba, Italy; Australia; Minas Gerais, Brazil; Lake Superior, Canada.

Magnetite

$FeFe_2O_4$ Iron oxide

Crystal system: Isometric.
Crystal habit: Usually massive or grains. Crystals typically octahedral and dodecahedral.
Colour: Black.
Lustre: Metallic to dull.
Streak: Black.
Hardness: 5.5–6.5.
Cleavage: None.
Fracture: Conchoidal.
Specific gravity: 5.1.
Other characteristics: Magnetism is strong in massive specimens.
Notable locations: Binnatal, Zermatt, Switzerland; Zillertal, Austria; Traversella, Italy; Nordmark, Sweden; Russia; South Africa; Magnet Cove, Arkansas; Brewster, New York.

Magnetite is one of the few naturally magnetic minerals. Its name is thought to come from the region of Magnesia in Ancient Macedonia, where, according to legend, the shepherd Magnes discovered magnetism when the iron nails in his shoes stuck to the rock. Some specimens, called lodestones, are magnetic enough to pick up iron. Ships carried lodestones to magnetize their compass needles until the 18th century. Nowadays, geologists are able to trace the past movement of continents using magnetite crystals frozen in rock in alignment with the North Pole as they formed, a technique called paleomagnetism. Magnetite has over 72 per cent iron content, and is the most important iron ore after hematite. It occurs as grains in a huge variety of rocks, but especially igneous rocks and the sands that form from them as they are weathered. These are the main ores. But magnetite may also form crystals in hydrothermal veins and alpine fissures. In sands, magnetite is often associated with gold, but also forms black beaches of sand in its own right.

Identification: Magnetite's dark colour, black streak and magnetism will distinguish it clearly from most other iron-bearing minerals.

Quartz
Massive magnetite

Siderite

FeCO₃ Iron carbonate

Curved blades of siderite

Siderite gets its name from *sideros*, the Ancient Greek for 'iron'. It is similar to calcite, but with iron replacing calcium, often when limestone is altered by iron-bearing solutions. Siderite is widespread in sedimentary rocks, and forms iron-rich clay beds known as clay ironstone. Clay ironstone can form nodules with a nucleus of hematite, and a surface altered to limonite or goethite. Fossils of plants, millipedes or clams are sometimes found in the nodules. Siderite also forms in cool hydrothermal veins. The great iron deposits that gave birth to the European and North American iron industry were the 'Minette' ores, made of siderite ooliths laid down on seabeds hundreds of millions of years ago. Crystals of siderite are attractive but rare.

Identification: Siderite's softness (the ease with which it can be scratched) indicates that it is a carbonate. Its dark brown surface colour and white streak show it to be iron carbonate.

Crystal system: Trigonal.
Crystal habit: Crystals curved blades. More often earthy masses or nodules.
Colour: Dark brown, grey; surface may be iridescent goethite.
Lustre: Vitreous.
Streak: White.
Hardness: 3.5–4.5.
Cleavage: Perfect in three directions forming rhombs.
Fracture: Conchoidal, uneven.
Specific gravity: 3.9+.
Other characteristics: Magnetic when heated.
Notable locations: Cornwall, England; Allevard, France; Erzberg, Austria; Panasquiera, Portugal; Minas Gerais, Brazil.

Banded iron formations
Some of the world's largest iron ore deposits are remarkable formations of sedimentary rock called banded iron formations (BIFs). BIFs are rock beds some 50–600m/164–1,970ft thick consisting of minutely alternating layers of iron minerals, chert and jasper, each no more than 1cm/½in or so thick. Nearly all of these formations are at least 1.7 billion years old, and some are much older. It is thought that early in Earth's history, huge amounts of iron were dissolved in the oceans – so conditions must have been very different, since iron will not dissolve easily in water now. The remarkable theory is that as primitive plantlike bacteria in the oceans released oxygen, the oxygen combined with the iron and sank to the sea bed as insoluble iron oxides. The banding is thought to mark seasonal peaks in oxygen production. The most famous BIFs are near Lake Superior, USA and in Western Australia's Hammersley Trough.

Limonite and goethite

FeO(OH).nH₂O (Limonite) HFeO(OH) (Goethite) Hydrated iron oxide

Kidney-form goethite

Limonite and goethite are basically the same mineral, hydrated iron oxide, but goethite forms crystals, often silky and fibrous in appearance, while limonite is basically amorphous, forming earthy masses and nodules. Named after the German poet Goethe, goethite forms, along with hematite, fluorite and barite, when iron minerals are altered in hydrothermal veins, or as 'bog iron ore' made in lakes and marshes by iron-depositing bacteria. It contains 63 per cent iron and is the second most important iron ore after hematite. Limonite is essentially natural rust, and gives many soils and rocks their yellow-brown colour. It also stains many agates and jaspers. Limonite forms when the surface of iron minerals is altered by weathering and itself alters to hematite as it dries out. It is one of the oldest known pigments, called yellow ochre, and was used in many prehistoric cave paintings.

Crystal system: None for limonite; goethite is orthorhombic.
Crystal habit: Limonite forms earthy masses or kidney-shaped crusts; goethite forms prismatic and platy crystals.
Colour: Yellow, brown.
Lustre: Earthy to dull.
Streak: Brown to yellow.
Hardness: 4–5.5.
Cleavage: None.
Fracture: Crumbly (Limonite); splintery (Goethite).
Specific gravity: 2.9–4.3.
Notable locations: Limonite is found widely; goethite vein specimens common in England, France, Germany; pegmatite goethite crystals found in Florissant, Colorado; goethite fibres are found in the iron mines of Michigan and Minnesota.

Identification: Limonite can be identified by a yellow stain on a darker brown mass, and its yellowish streak. Fibrous-looking masses are probably goethite.

ALUMINIUM AND MANGANESE

Aluminium is the most abundant metal in the Earth's crust, and makes up 8 per cent of the crust by weight. Its compounds are spread throughout almost every rock, plant and animal on the planet. It is primarily extracted from bauxite. Although not quite so widespread as aluminium, manganese is also common and is recovered not just from native manganese, but from ores such as manganite.

Aluminium ore: Bauxite

Al(OH)₃ Aluminium hydroxide (Gibbsite)

Bauxite (below)

Named after the French town of Les Baux, where it was first discovered in 1821, bauxite is found on just about every continent in the world. It is typically buff or white, but can be red, yellow, pink or brown, or a mix of all of them. It can be as hard as rock or soft as clay, and can occur as compacted earth, small balls called pisolites, or hollow, twig-like tubules. Bauxite ore is bauxite with high enough levels of alumina (aluminium oxide) and low enough levels of iron oxide and silica to be worth extracting.

Gibbsite (below)

Bauxite is a mixture of three aluminium hydroxide minerals – gibbsite, diaspore and boehmite. Diaspore and boehmite have the same composition, but diaspore is denser and harder. Boehmite is abundant in European bauxites, but rare in America. Diaspore is common in American bauxites. Gibbsite is the dominant mineral in tropical bauxites.

Brown limonite coating

Gibbsite

Remarkably, despite its abundance, aluminium was not discovered until 1808. This is because it is so reactive that it almost never occurs as a pure native element. The only exceptions are microscopic inclusions and as nodules in volcanic mud. It can make gems and corundum, and related rubies and sapphires are all aluminium oxides. Topaz, garnet and chrysoberyl also all contain aluminium. Typically, aluminium occurs in igneous rocks in the form of aluminosilicate minerals such as feldspars, feldspathoids and micas. It also occurs in clay soils that are made from the breakdown of these igneous rocks. However, the most significant occurrence of aluminium is in bauxite. Bauxite is not a mineral, nor even a solid rock like some ores, but a kind of laterite. Laterite is a loose weathered material that forms deep layers in the tropics and subtropics. As rocks rich in aluminosilicates are weathered in the warm, wet climates typical of tropical rainforests, only iron and aluminium oxides and hydroxides are left behind, and everything else is washed away. Bauxite is the residue.

Bauxite is the source of 99 per cent of the world's aluminium, and is excavated in gigantic quarries. Over 40 per cent comes from the huge Australian reserves dug out in places such as Weipa (Queensland), Gove (Northern Territory) and the Darling Range (Western Australia). But Guinea, Brazil, Jamaica and India also have substantial reserves. Aluminium could be obtained from aluminous shales and slates, aluminium phosphate rocks and high alumina clays, but there is so much easily extractable bauxite, it makes no economic sense yet to use these sources.

Native Aluminium
Crystal system: Isometric.
Crystal habit: Microscopic inclusions and as nodules in volcanic muds.
Colour: Silvery white.
Lustre: Metallic.
Streak: White.
Hardness: 1.5.
Cleavage: Absent.
Fracture: Jagged.
Specific gravity: 2.72.
Notable locations: Russia; DR Congo; Baku, Azerbaijan.

Diaspore
Diaspore gets its name from the Greek for 'to scatter' because it splits violently when heated.

Gibbsite
Crystal system: Monoclinic.
Crystal habit: Typically plates or sheets, or grains. Maybe tablet-shaped crystals.
Colour: White, greenish grey, greyish brown, colourless.
Lustre: Vitreous to pearly.
Streak: White.
Hardness: 6.5–7.
Cleavage: Perfect in one direction, forming plates.
Fracture: Brittle, conchoidal.
Specific gravity: 3.4.
Other characteristics: When heated it splits violently into white pearly scales.
Notable locations: Les Baux, France; Larvik, Norway; Tyrol, Austria; Northern Cape, South Africa; Arizona; Nevada.

Manganese ore: Manganite, Psilomelane, Pyrolusite

MnO(OH) Manganese oxide hydroxide (Manganite)

A hard, brittle, grey-white metal, manganese gets its name from the same root as magnetite, because when alloyed with other metals such as copper, antimony and aluminium it can be magnetic. It was discovered in 1774 by the Swedish scientist Johan Gahn while heating the mineral pyrolusite in a charcoal fire. In fact, pyrolusite is manganese oxide, and it simply lost its oxygen to the charcoal to leave the metal. Tiny traces of manganese are vital for health, helping the body absorb Vitamin B1 and promoting the action of enzymes. But too much manganese can be toxic. Nowadays, manganese is used for making dry-cell batteries, and for adding to steel to make it harder. Manganese has never been found in pure form, but it combines readily with other elements and is a common but minor ingredient of many rocks.

Manganite (below)
Manganite is the mineral richest in manganese, and was once a major manganese ore. It is now too rare to be a significant source, but is often sought after for its distinctive crystals, which resemble only a few other metallic minerals such as enargite. It forms bundles of prismatic crystals or fibrous masses deposited from circulating waters and in low-temperature hydrothermal veins.

Manganite
Crystal system: Monoclinic.
Crystal habit: Prisms in bundles, or fibrous masses.
Colour: Black to steel grey.
Lustre: Submetallic to dull.
Streak: Reddish to black.
Hardness: 4.
Cleavage: Perfect lengthwise.
Fracture: Uneven.
Specific gravity: 4.2–4.4.
Other characteristics: Alters to give dull coat of pyrolusite fibres.
Notable locations: Cornwall, England; Ilfeld (Harz Mountains), Germany; Ukraine; Negaunee, Michigan.

Deep sea manganese
On the surface of the ocean bed, nodules of manganese and other metals are scattered across the ooze like countless marbles. They form when the hot waters from black smokers (undersea volcanic hot springs) meet the cold, deep ocean water. Because the waters are rich in manganese, manganese is deposited in nodules around the smokers, often accumulating on the surface of a grain such as a tiny piece of bone. Once started, each nodule grows slowly, by 1cm/½in in diameter in about 1 million years. As the ocean floor spreads wider, so these nodules are scattered over a vast area. Though rich in manganese, these nodules are too deep in the ocean and too costly to gather at the moment. But some fields of nodules, particularly in the Eastern Pacific, may one day be 'harvested' (mined) commercially as land sources diminish and cost-effective deep-sea mining methods are developed. The photo above was taken at a depth of 5,350m/17,550ft in the Atlantic Oceans.

It is the manganese oxides pyrolusite and psilomelane and the hydroxide manganite that provide most of the ores for manganese. Manganese silicates such as rhodonite and braunite are less important. Psilomelane gets its name from the Greek *psilos* for 'bald' and *melas* for 'black' because of the smooth, black, botryoidal masses it can form. It is now a general name for all the hard, dark, barium-rich manganese oxides, such as hollandite and romanechite. Like pyrolusite, it typically forms when rocks are weathered, and can often form large masses when other weathered materials are washed away. Wad, or 'bog manganese', is a soft, earthy mixture of manganese ores such as pyrolusite and psilomelane with water. It is often found around marshes and springs.

Psilomelane group
Crystal system: Monoclinic.
Crystal habit: Massive botryoidal, stalactites.
Colour: Black to dark grey.
Lustre: Submetallic.
Streak: Brown to black.
Hardness: 5–7.
Cleavage: None.
Fracture: Conchoidal to uneven.
Specific gravity: 3.3–4.7.
Notable locations: Cornwall, England; Schneeburg, Germany; Ouro Prêto (Minas Gerais), Brazil; Tucson, Arizona; Austinville, Virginia; Michigan.

Wad (right)
Wad is a shapeless, dark earthy mass of wet manganese ores that often forms in marshes, which is why it is also called bog manganese.

MOLYBDENUM AND TUNGSTEN

Molybdenum and tungsten melt only at very high temperatures. Molybdenum melts at 2,610°C/4,730°F, 1,000 degrees above most steels. Tungsten's melting point is even higher at 3,410°C/6,170°F. Both metals are used where heat-resistance is essential. Molybdenite is mainly extracted from molybdenite ores, but also from wulfenite and powellite; tungsten comes from the wolframite minerals, and from scheelite.

Molybdenum ore: Molybdenite

MoS₂ Molybdenum sulphide

Crystal system: Hexagonal.
Crystal habit: Rare crystals are thin hexagonal tablets. Typically forms soft sheetlike masses and crusts.
Colour: Bluish lead grey.
Lustre: Metallic.
Streak: Greenish grey.
Hardness: 1–1.5.
Cleavage: Perfect, basal, forming thin sheets.
Fracture: Flaky.
Specific gravity: 4.6–4.8.
Other characteristics: Leaves marks on fingers.
Notable locations: Cornwall, England; Raade, Norway; Hirase (Honshu), Japan; Kingsgate and Deepwater, New South Wales; Pontiac County; Quebec; Wilberforce, Ontario; Climax, Colorado; Lake Chelan, Washington.

Molybdenite gets its name from the ancient Greek word for 'lead', because its softness and dark grey colour made people wrongly think it contained lead. Molybdenite is widely distributed but rarely occurs in large quantities. There are an estimated 12 million tonnes of molybdenum in molybdenite around the world. Its molybdenum content means it forms mostly in high-temperature environments, typically in granite pegmatites and quartz veins. It also occurs in contact metamorphic deposits and skarns along with scheelite, pyrite and wolframite. It is one of the softest of all minerals, and is rather like graphite. As with graphite, it is soft because it has a sheetlike structure, with the molybdenum and sulphur in alternating layers that slide easily over one another. It also has a greasy feel and leaves dark marks on your fingers in the same way that graphite does. In fact, molybdenite and graphite are often hard to tell apart.

Molybdenite

Feldspar

Hexagonal molybdenite crystal in quartz

Identification: Soft, flexible flakes of molybdenum can only really be confused with graphite. Graphite leaves darker marks.

Molybdenum ore: Powellite

CaMoO₄ Calcium molybdate

Powellite is named after the American geologist John Wesley Powell (1834–1902), who led the first expedition down the Colorado River through the Grand Canyon. It is one of the few fairly common molybdenum minerals. Most powellite probably forms when hot hydrothermal solutions interact with molybdenite, so it occurs in many of the same places. In fact, powellite often mimics the shape of molybdenite, forming pseudomorphs as the atoms of molybdenite are replaced one by one. Powellite also forms directly in quartz veins. Good crystals are rare, and most come from the Deccan basalts of India, particularly from Nasik in Maharashtra. Powellite forms part of a series with scheelite, in which molybdenum replaces scheelite's tungsten. Both minerals are fluorescent, but scheelite glows bluish white while powellite glows golden yellow.

Identification: Powellite is identified by its colour, pyramidal cleavage, association with molybdenite and yellow fluorescent glow under UV.

Crystal system: Tetragonal.
Crystal habit: Rare, small, four-sided pyramids, or crusts on molybdenite.
Colour: White, yellowish brown, blue.
Lustre: Adamantine to greasy.
Streak: White.
Hardness: 3.5–4.
Cleavage: Pyramidal.
Fracture: Uneven.
Specific gravity: 4.2.
Other characteristics: Brittle, fluoresces golden yellow.
Notable locations: Cumbria, England; Telemark, Norway; Black Forest, Germany; Altai Mountains, Russia; Maharashtra, India; Sonora, Mexico; Arizona; Nevada; Seven Devils, Idaho; Keewenaw Peninsula, Michigan.

Tungsten ore: Scheelite

CaWO₄ Calcium tungstate

Crystal system:
Tetragonal.
Crystal habit:
Double pyramids
that look octahedral.
Masses, grains.
Colour: White, clove brown,
greenish grey (cuproscheelite).
Lustre: Adamantine.
Streak: White.
Hardness: 4.5–5.
Cleavage: Pyramidal.
Fracture: Conchoidal.
Specific gravity: 5.9–6.1.
Other characteristics:
Fluoresces blue-white.
Notable locations: Cumbria,
England; Saxony, Germany;
Slavkov, Czech Republic; Tong
Wha, South Korea; Mill City,
Nevada; Atolia, California;
Cochise County, Arizona.

Named after Karl Scheele, the 18th-century Swedish scientist who discovered tungsten, scheelite is an important ore of tungsten. Although most tungsten comes from wolframite found in Russia and China, scheelite provides the United States with most of its home-produced tungsten. It is found most often in skarns called tactites, where granitic magma intrudes into limestone. Scheelite forms here in association with garnet, epidote, vesuvianite and wolframite. Scheelite can also form in high-temperature, quartz-rich hydrothermal veins, along with cassiterite, topaz, fluorite, apatite and wolframite. Collectors treasure scheelite crystals for their distinctive double-pyramid crystals, with side facets that make them look octahedral. They are also cherished for their bright blue-white fluorescent glow, which can help miners discover the crystals in the dark with ultraviolet lamps. Scheelite is occasionally used as a gem.

Identification: Scheelite can be recognized by its distinctive eight-sided double-pyramid crystals and blue-white fluorescent glow in shortwave UV light.

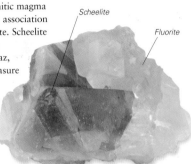

Scheelite

Fluorite

Tungsten ore: Wolframite

(Fe,Mn)WO₄ Iron manganese tungstate

Crystal system:
Monoclinic.
Crystal habit: Flat
tablets forming
bladed, sheetlike groups. Also
masses, grains.
Colour: Grey-black to
brownish black.
Lustre: Submetallic.
Streak: Brownish black.
Hardness: 5–5.5.
Cleavage: Perfect in
one direction.
Fracture: Uneven.
Specific gravity: 7–7.5.
Notable locations: Cornwall,
England; Erzebirge, Harz Mts,
Germany; Panasquiera,
Portugal; Caucasus Mts,
Lake Baikal, Russia; Tong
Wha, South Korea;
Nanling Mts, China;
Llallagua, Bolivia;
Northwest
Territories;
Boulder Co,
Colorado.

Legend has it that miners in medieval Saxony gave this mineral the name of 'the wolf' because it interfered with tin smelting. Now it is recognized as the major ore of tungsten. Wolframite is not a mineral in its own right but a mix of varying proportions of iron and manganese tungstate minerals. The iron-rich variety is called ferberite; the manganese-rich variety is hübnerite. Wolframite occurs typically in high-temperature quartz veins, pegmatites in granitelike rocks and skarns, often associated with cassiterite, galena and scheelite. It is also found concentrated in alluvial placer deposits. The mineral ore is extracted from opencast mines in Australia and Northwest Territories, Canada, but the largest deposits are in the Nanling Mountains, China, which need to be deep mined. Tungsten is used as a filament in lightbulbs and as a carbide in drilling equipment.

Identification: The wolframite minerals can best be identified by their black-brown colour and their distinctive sheetlike groups of flat crystals.

WSkarns
When a limestone or marble is invaded by hot magma, it can create remarkable deposits called skarns, which are sources of some of the world's most valuable ores. Hot, slightly acid solutions circulate through the rocks and react with the carbonates in the limestone to create a host of new minerals. The limestone supplies calcium, magnesium and carbon dioxide; the magma supplies silicon, aluminium, iron, sodium, potassium and other elements. The magma's silicon and iron, for instance, combine with the limestone's calcium and magnesium to form silicate minerals such as diopside, tremolite, wollastonite (as in the skarn shown above) and andradite. The hydrothermal solutions may also deposit ore minerals of iron, copper, zinc, tungsten or molybdenum. Tungsten skarns supply much of the world's tungsten from deposits such as those at Sangdong in Korea, King Island, off Tasmania in Australia, and Pine Creek in California, USA.

NICKEL

Nickel is a shiny, silvery white metal that is slightly magnetic. It is alloyed to iron in both the Earth's core and in many meteorites and was clearly an important element in the early solar system. It is rarely found in pure form in the Earth's crust, and industry extracts it either from the igneous sulphide ores pentlandite, nickeline and chloanthite or from ores in laterites such as garnierite.

Nickel sulphide ore: Nickeline (Niccolite)

NiAs Nickel arsenide

Crystal system: Hexagonal.
Crystal habit: Usually massive, though also forms in columnlike groups.
Colour: Copper red.
Lustre: Metallic.
Streak: Pale brownish black.
Hardness: 5–5.5.
Cleavage: Poor.
Fracture: Uneven.
Specific gravity: 7.8.
Notable locations: Black Forest, Harz Mountains, Germany; Jachymov, Czech Republic; Anarak, Iran; Natsume, Japan; Great Slave Lake, Northwest Territories; Sudbury, Cobalt, Ontario; Franklin, New Jersey.

The mineral nickeline, or niccolite, was known long before anyone identified the element nickel. Copper miners during the Middle Ages in Germany found copper-coloured niccolite deposits when they were looking for copper, but on processing it found only what looked like a white slag. Thinking it bewitched, they named the mineral *Kupfernickel* (copper nickel) after the Old Nick's (the Devil's) impish 'nickels' (little helpers) underground. It was in nickeline that Swedish chemist Axel Cronstedt finally identified nickel in 1751. Nickeline is not common, but it is mined together with other more important nickel and cobalt sulphide ores. It typically occurs in igneous rocks such as norite and gabbro, in association with pyrrhotite and chalcopyrite as well as pentlandite and nickel skutterudite. It also forms in hydrothermal veins along with silver, arsenic and cobalt minerals. On exposure to air, it alters to pale green annabergite, or 'nickel bloom'.

Identification: Nickeline's coppery colour is very distinctive and only really looks like breithauptite (nickel antimony).

Nickeline

Nickel sulphide ore: Pentlandite

(Fe,Ni)₉S₈ Iron nickel sulphide

The main nickel ore, pentlandite, is usually found intertwined with pyrrhotite (iron sulphide). The two can rarely be physically separated, though pyrrhotite is magnetic while pentlandite is not. They tend to form together when sulphides separate out of molten ultramafic magmas, notably norite and rare komatites. As the magma cools, metal sulphides crystallize and fall to the base of the magma to collect in a rich mass. But the famous nickel sulphide deposits at Sudbury, Ontario, Canada, were probably concentrated by the impact of a giant meteorite. Pentlandite – named after J B Pentland, who discovered it – is here embedded along with sperrylite and chalcopyrite in a huge mass of pyrrhotite in a trough of norite and gabbro rock 60km/ 40 miles long. Pentlandite has also been found in nickel-iron meteorites. Wherever it forms, pentlandite rarely makes good crystals, and is usually found only in the massive form.

Crystal system: Isometric.
Crystal habit: Mostly minute grains and masses.
Colour: Brassy yellow.
Lustre: Metallic.
Streak: Tan.
Hardness: 3.5–4.
Cleavage: None.
Fracture: Conchoidal.
Specific gravity: 4.6–5.
Notable locations: Iveland, Norway; Styria, Austria; Noril'sk-Talnakh, Russia; Bushveld complex, S Africa; Kambalda, Western Australia; Lynn Lake, Manitoba; Malartic, Quebec; Sudbury, Ontario; Ducktown, Tennessee; San Diego Co, California.

Identification: Pentlandite is best identified by its brassy colour and its association with pyrrhotite and chalcopyrite.

Nickel sulphide ore: Chloanthite (Nickel skutterudite)

NiAs$_{2-3}$ Nickel arsenide

Skutterudite, chloanthite, safflorite and rammelsbergite are rare nickel and cobalt arsenides that merge so closely they are hard to tell apart, and mineralogists argue over which is a proper mineral species. All are tin white, and have a bright metallic lustre and black streak. All were once nickel and cobalt ores, but the good veins are now all but used up. They are basically a pair of solid solution series: one running from skutterudite to chloanthite, the other from safflorite to rammelsbergite. Skutterudite and safflorite are at the cobalt-rich end and chloanthite and rammelsbergite at the nickel-rich end. Sometimes exposure to air helps distinguish them, as the cobalt-rich minerals get coated in red erythrite and the nickel-rich minerals get coated in green annabergite. In fact, *chloanthes* is Greek for 'green flowering'. Many mineralogists no longer recognize chloanthite as a separate species. They now prefer to call the nickel-rich end of the skutterudite-chloanthite series 'nickel skutterudite' and describe chloanthite simply as an arsenic-deficient version of this.

Native arsenic

Crystal system: Isometric.
Crystal habit: Massive, grains.
Colour: Tin white.
Lustre: Metallic.
Streak: Yellowish white.
Hardness: 5.5–6.
Cleavage: None, crumbly.
Fracture: Very brittle, conchoidal.
Specific gravity: 6.5.
Notable locations: Annaberg, Schneeberg (Saxony), Andreasberg (Harz Mountains), Germany; Cobalt, Ontario.

Identification: Chloanthite and other cobalt and nickel arsenides all look tin white and are hard to tell apart, except when they are coated green with annabergite.

Nickel laterite ore: Garnierite, Népouite

(Ni,Mg)$_3$Si$_2$O$_5$.(OH)$_4$ Hydrated nickel magnesium silicate

Laterites: tropical metal banks

Rocks weather quickly in warm, wet tropical climates to create incredibly deep soil-like layers called laterites. There are deep laterites beneath the Amazon basin, for instance. There are also ancient laterites in North America, and Europe, formed when they were in the tropics long ago. Yet although laterites are deep, they are not good soils, for heavy tropical rains ensure that just about all the useful minerals are quickly washed out. There isn't even much organic matter, since it decays rapidly in the tropical heat. Rainforest trees grow by continually recycling nutrients; when forests are cleared for farming, the soils lose all fertility in just a few years. But though laterites make poor soils, they do hold on to those iron, aluminium and nickel minerals that won't dissolve, and these often become concentrated by percolating water into valuable mineral deposits. Iron-rich laterites such as limonite are not exploited much, because banded iron formations provide a better source of iron. But aluminium-rich laterites are bauxites, the main sources of aluminium, and nickel-rich laterites are garnierites, which can yield nickel.

Garnierite is not a mineral species, but a blanket term for a range of different earthy nickel minerals, including népouite, pimelite and willemsite. Garnierite forms as gabbro, and peridotite rock rich in nickel is deeply weathered in warm climates to form a laterite. Weathering removes much of the original rock, but dissolved nickel may percolate down and collect in sufficient concentrations in the weathered material to create an ore. Garnierites are the 'nickeliferous' laterites richest in nickel, but limonites can contain a fair amount of nickel, too, and are far more widespread. Some garnierite laterites can contain almost 10 per cent nickel, such as in the népouite first excavated at Népoui in New Caledonia; other garnierite laterites have much less.

Crystal habit: Orthorhombic.
Crystal habit: Massive, earthy.
Colour: Pale green.
Lustre: Vitreous, pearly.
Streak: Greenish white.
Hardness: 2–2.5.
Cleavage: None.
Fracture: Uneven.
Specific gravity: 4.6.
Notable locations: Limpopo, Mpumalanga, South Africa; Népoui, New Caledonia; Australia; Cuba; Washington; Nevada.

Identification: Garnierite can be identified by its green colour and its earthy appearance in lateritic material.

CHROMIUM, COBALT, VANADIUM AND TITANIUM

Chromium and titanium are both tough, light metals that have found an increasing range of uses in the modern world, especially when alloyed with metals such as steel. Cobalt and vanadium, though soft as pure metals, also really toughen up steel and other metals when alloyed with them.

Chromium ore: Chromite

$FeCr_2O_4$ Iron chromium oxide

Crystal habit: Isometric.
Crystal habit: Crystals rare, octahedral. Mostly grainy masses.
Colour: Black, brownish black.
Lustre: Metallic.
Streak: Brown.
Hardness: 5.5.
Cleavage: None.
Fracture: Conchoidal.
Specific gravity: 4.1–5.1.
Notable locations: Outokumpu, Finland; Kempirsay Massif, Kazakhstan; Sarany (Urals), Russia; Guleman, Turkey; Luzon, Philippines; Andhra Pradesh, India; Camagüey, Cuba; Maryland.

Chromium derives its name from the Greek *chroma* for 'colour', because many minerals get their special colour from its presence, such as green emeralds. Yet although there are small amounts of chromium in many minerals – notably crocoite, in which it was discovered in 1797 – there is only one real ore: chromite. Chromite crystals are sometimes found in veins and scattered through serpentine. But most ore chromite forms as lens-shaped masses in ultramafic magmas, such as peridotites. Chromite has a high melting point and is one of the few ores to form directly from liquid magma. It crystallizes early in the cooling magma, then sinks to create these lenses. It may then stay unaltered when the host rocks are metamorphosed to serpentinites. Most commercial chromite comes from South Africa, Russia, Albania, the Philippines, Zimbabwe, Turkey, Brazil, Cuba, India and Finland.

Identification: Chromite is best identified by its association with serpentines. It looks like magnetite but is not magnetic. In the specimen below, all the black is chromite.

Cobalt ore: Cobaltite

CoAsS Cobalt arsenic sulphide

Skutterudite, a minor cobalt ore

The Ancient civilizations of Egypt and Mesopotamia used cobalt minerals to make deep blue glass, but it was not until 1735 that it was identified as an element. Now it is known as a widespread natural element found in small quantities in many different minerals. Most commercially mined cobalt is taken from ores of copper, nickel or copper-nickel, where cobalt occurs in sulphide minerals such as carrollite, linnaeite and slegenite, oxides such as heterogenite and absolite and the carbonate sphaerocobaltite. Much of the world's cobalt comes from copper ores in the Congo and Russia, nickel ores from New Caledonia, Cuba and Celebes, and copper-nickel sulphide ores in Canada, Australia and Russia. Very few ores are mined purely for their cobalt, except notably in Morocco. These include skutterudite, smaltite and cobaltite, noted for its remarkably pyrite-like crystals.

Identification: Cobaltite is usually grainy masses but can form either cubes, like galena, or pyrite-shape crystals, but cobaltite's silver white colour usually makes its identity clear.

Crystal habit: Isometric.
Crystal habit: Crystals are typically cubes and pyritohedrons (pyrite shapes), but typically grainy masses.
Colour: Silver-white with red tinge.
Lustre: Metallic.
Streak: Grey-black.
Hardness: 5.5.
Cleavage: Perfect like cubes.
Cleavage: Uneven to subconchoidal.
Specific gravity: 6–6.3.
Other characteristics: On cube faces, there are often striations.
Notable locations: Cumbria, England; Tunaberg, Sweden; Skutterud, Norway; Siegerland, Germany; DR Congo; Sonora, Mexico; Cobalt, Ontario; Boulder, Colorado.

Vanadium ore; Vanadinite

Pb₅(VO₄)₃Cl Lead chlorovanadate

Vanadium was named by 19th-century Swedish chemist Nils Seftsrom after Vanadis, the Norse goddess of beauty, because of the colours it forms in solutions, and vanadinite also often lives up to the name. Even in crust form, as seen here, it is an attractive deep red, but when it forms crystals it can be striking, although they tend to fade with time. Vanadinite crystals can even be multicoloured, like tourmaline. Vanadinite is almost always found where lead sulphide ores such as galena are weathered in arid climates, along with minerals such as pyromorphite, mimetite, wulfenite, cerussite and descloizite. Although vanadinite is the mineral with the highest vanadium content, it is too rare to be the only ore. Roscoelite, patronite and carnotite all provide ores, but most vanadium is recovered as a by-product of processing other ores, notably titanium-rich magnetite. It is also found in the ash from ships' smokestacks.

Identification: With its sharp, hexagonal prism crystals, vanadinite looks a little like pyromorphite and mimetite, but its blood-orange colour marks it out clearly.

Crystal system: Hexagonal.
Crystal habit: Small prisms.
Colour: Red, brown, yellow or orange.
Lustre: Resinous, adamantine.
Streak: Brownish yellow.
Hardness: 3–4.
Cleavage: None.
Fracture: Conchoidal.
Specific gravity: 6.7–7.1.
Notable locations: Dumfries and Galloway, Scotland; Carpathia, Austria; Urals, Russia; Mibladen, Morocco; Otavi, Namibia; Marico, South Africa; Chihuahua, Mexico; Tucson, Arizona; Sierra County, New Mexico.

Minerals in iron and steel

Extracting iron out of ore means smelting it in a blast furnace – heating it until the molten metal runs out, leaving the other minerals behind. Yet even then the iron is not pure. Pig iron from the blast furnace is only 93 per cent iron, with 4 per cent carbon, and traces of numerous other substances. Further refining turns it into cast iron, with 2–3 per cent carbon and 1–3 per cent silicon. All the carbon and silicon make iron too brittle to be beaten into shape, though it can be cast (poured into sand moulds). Wrought iron is almost pure iron with most of the carbon taken out so it can be bent and shaped into things like railings. Interestingly, though, pure iron can be toughened by adding the right 'impurities', including carbon, to make steel, a material first discovered in India 2,000 years ago. The most widely used kind of steel is carbon steel, which is 1 per cent carbon. Mild steel for car bodies may be 0.25 per cent carbon. High carbon steel for tools is 1.2 per cent carbon. Other steel alloys may be made by adding traces of other metals to give particular qualities. Chromium makes stainless steel (which clads the famous Gateway Arch of St Louis in the USA, above). Manganese, cobalt and vanadium give strength; molybdenum heat resistance and nickel corrosion resistance.

Titanium ore: Ilmenite, rutile

FeTiO₃ Iron titanium trioxide (Ilmenite)

Titanium was discovered first in black magnetic sands, then in rutile by German chemist M H Klaproth who named it after the Titans, the giants of Greek mythology, because of its strength. Titanium metal is used as a very strong lightweight material, but titanium dioxides such as rutile are used as the basic white pigment in paints, replacing toxic lead. Titanium is sometimes thought of as an exotic metal, but it is the fourth most abundant after aluminium, iron and magnesium, and ilmenite and rutile ores are abundant. Grains of ilmenite occur in rocks such as gabbro and diorite, and as the rocks ares weathered these grains can become concentrated in mineable sands, or as waterworn pebbles. Occasionally, good crystals form in pegmatites.

Crystal system: Trigonal.
Crystal habit: Thick tablet-shaped crystals, but usually grains and masses.
Colour: Black.
Lustre: Submetallic.
Streak: Dark brownish red.
Hardness: 5–6.
Cleavage: None.
Fracture: Conchoidal or uneven.
Specific gravity: 4.5–5.
Notable locations: Kragerø, Norway; Sweden; Lake Ilmen (Urals), Russia; Gilgit, Pakistan; Sri Lanka; Australia; Brazil; Lake Allard, Quebec; Bancroft, Ontario; Orange County, New York.

Ilmenite

Identification: Ilmenite looks like many black sulphosalts, but is harder. It can be easily distinguished from hematite by its streak and from magnetite by its lack of magnetism.

COPPER

As well as being one of the most distinctive metals, with its red-gold colour, copper is also one of the most widely available, and has long played a part in human culture. It is found worldwide as both a pure native element in basaltic lavas and also in ores such as chalcocite, chalcopyrite, bornite, cuprite, enargite and covellite. It can even be found in seaweed ash, sea corals, molluscs and the human liver.

Chalcopyrite

$CuFeS_2$ Copper iron sulphide

Crystal system: Tetragonal.
Crystal habit: Crystals appear tetrahedral; usually massive.
Colour: Brassy yellow, tarnishes to iridescent blues, greens and purples.
Lustre: Metallic.
Streak: Dark green.
Hardness: 3.5–4.
Cleavage: Poor.
Fracture: Conchoidal, brittle.
Specific gravity: 4.2.
Notable locations: Cornwall, England; Russia; Kinshasa, DR Congo; Zambia; Ugo, Japan; Olympic Dam, South Australia; Chihuahua, Naica, Mexico; El Teniente, Chile; Bingham, Utah; Ely, Nevada; Ajo, Arizona; Joplin, Missouri.

Most of the world's big copper deposits are 'porphyry' coppers, in which the copper minerals are scattered in veinlets evenly throughout the rock, which is typically porphyritic diorite, or schist. The copper ores of the gigantic Chilean deposits are porphyry coppers, as are those of the American south-west. Massive and vein deposits, while more concentrated, are far less extensive. Typically, the upper deposits in porphyries are copper oxides such as cuprite; the deeper deposits are sulphides such as chalcopyrite. Containing less than a third copper, chalcopyrite is the main source of copper, because it is found in such large quantities. It gets its name because it is the same gold colour as pyrite, although slightly yellower. Like pyrite, it has been mistaken for gold. Its surface is often coated with a dark greeny, purplish iridescent tarnish, earning it the name 'peacock copper'. As well as porphyries, it forms in hydrothermal veins and pegmatites, often with pyrite, sphalerite and galena.

Identification: Chalcopyrite looks like pyrite, but is more yellow. Like pyrite it also looks like gold, but is brittle and harder.

Chalcopyrite

Calcite

Cuprite

Cu_2O Copper oxide

Identification: A deep red mass with a coat altered to bright green malachite is likely to be cuprite. It is harder than cinnabar.

Cuprite is over two-thirds copper and is a widespread ore that forms by the alteration of other copper minerals. It tends to occur near the surface wherever there are copper sulphide deposits lower down, and forms as they are oxidized. It is typically found mixed in with iron oxides in earthy, porous masses. Some giant cuprite-dominated masses weighing several tonnes have been found. But cuprite can also form wonderful crystals in association with pure native copper and an array of other secondary copper minerals such as malachite, brochantite and azurite. Cuprite crystals range in colour from cochineal red through ruby red to purplish and even black. Fine hairlike masses of cuprite are called chalcotrichite. Tile ore is an earthy mix of the three minerals cuprite, limonite and hematite.

Cuprite

Crystal system: Isometric.
Crystal habit: Crystals usually octahedral, but can be other shapes; also acicular, massive, granular.
Colour: Red to very dark red.
Lustre: Adamantine to dull.
Streak: Brick red.
Hardness: 3.5–4.
Cleavage: Fair in four directions forming octahedrons.
Fracture: Conchoidal.
Specific gravity: Approximately 6.
Notable locations: Wheal Gorland (Cornwall), England; Chessy, France; Bogoslovsk (Urals), Russia; Ongonja, Namibia; Broken Hill (NSW), Wallaroo (SA), Australia; Bisbee, Arizona.

Chalcocite

Cu₂S Copper sulphide

Covellite is a copper sulphide with a deep indigo iridescence.

Chalcocite is the most copper-rich of all the copper ores, with a copper content of up to 80 per cent by weight. It is only less important as an ore than chalcopyrite because it is scarce, and most good deposits have been mined out. It can form as a primary mineral direct from magma as 'digenite'. But chalcocite more often forms by a process called secondary, or 'supergene' enrichment. In this, copper sulphide minerals that have been altered by oxidation in surface layers are washed downwards. As coppery solutions percolate down to the lower primary copper ores, they dissolve iron away. Iron-rich chalcopyrite ores are altered to copper-rich chalcocite along the levels of water tables in layers called chalcocite blankets. Typically, as iron is lost, the chalcopyrite alters first to bornite, then to covellite and finally to chalcocite.

Identification: Chalcocite is almost always associated with other copper sulphide minerals such as covellite and chalcopyrite. It can be distinguished from them by its dark grey colour.

Chalcocite coating

Crystal system: Orthorhombic below 105°C/ 221°F and hexagonal above 105°C.
Crystal habit: Prismatic or tabular crystals, rare; usually massive, or as powdery coat.
Colour: Dark grey to black.
Lustre: Metallic.
Streak: Shiny black to grey.
Hardness: 2.5–3.
Cleavage: Prismatic, indistinct.
Fracture: Conchoidal.
Specific gravity: 5.5–5.8.
Notable locations: Cornwall, England; Isère, Belfort, Alsace, France; Rio Tinto, Spain; Sardinia; Messina, South Africa; Butte, Montana; Bristol, Connecticut; Bisbee (Cochise Co), Arizona.

Arizona Copper

Arizona is sometimes known as 'The Copper State' because of the huge deposits of copper found there in beds of porphyry copper ore. The copper ores here originated 200–100 million years ago when intrusions of porphyry copper ore pushed up into the country rock, and were then concentrated by supergene enrichment (see Chalcocite above). What makes the deposits especially valuable is the presence of many other ores, such as uranium and vanadium. Two-thirds of all America's copper is hacked out of the ground in Arizona – a landscape shaped by strip (or surface) mining is shown above. If Arizona were a country, it would be the world's second largest producer of copper, after Chile. Towns such as Bisbee, Jerome, Globe and Clifton live in the shadow of vast plateaux of tailings piled up here since mining began in earnest in the late 19th century. The industry is now in decline, with copper reserves partly exhausted, and prices around the world falling. But the mines produced an astonishing amount in their heyday. Bisbee alone produced over 4 million tonnes of copper ores between 1877, when copper was first found here, and 1975, when the Phelps Dodge company ceased mining.

Bornite

Cu₂FeS₄ Copper iron sulphide

Enargite: Copper arsenic sulphide copper ore

Like chalcopyrite, bornite is often known as 'peacock ore' or 'peacock copper'. This is because its surface is often coated with a dark, greeny, purplish, iridescent tarnish made up of the oxides and hydroxides of copper. Peacock ore sold to collectors as bornite is often chalcopyrite with its iridescence enhanced by treatment with acid. To distinguish it from peacock chalcopyrite, you will need to scratch away the tarnish to reveal bornite's true pink copper colour. This pink gives bornite another, slightly less appealing, name: 'horseflesh ore'. Bornite occurs widely as masses with chalcopyrite in copper porphyry deposits, but it can also form fine-grained crystals within pegmatites and mineral veins.

Bornite

Crystal system: Isometric.
Crystal habit: Crystals rare but rough, pseudo-cubic and rhomb-dodecahedral; usually massive.
Colour: Reddish bronze with iridescent peacock tarnish.
Lustre: Metallic.
Streak: Grey-black.
Hardness: 3.
Cleavage: Very poor, octahedral.
Fracture: Conchoidal.
Specific gravity: 4.9–5.3.
Other characteristics: Tarnishes in hours.
Notable locations: Cornwall, England; Dzhezkazgan, Kazakhstan; Texada Island, British Columbia; Butte, Montana; Arizona.

Identification: Bornite tarnishes in hours to peacock ore. To tell it from peacock chalcopyrite, you have to scratch the surface to reveal the true bruised pink colour of bornite beneath.

LEAD

Easily shaped and resistant to corrosion, lead has been used since antiquity for items such as pipes and roofing, although, in recent years, the discovery of its toxicity has reduced its use. It is rarely found as a native element, but occurs in compound form in more than 60 minerals, including its main ores galena, cerussite and anglesite, as well as a number of minor ores such as mimetite, pyromorphite and minium.

Galena

PbS Lead sulphide

Made of 86.6 per cent lead, galena, or lead glance, has been the main lead ore since ancient times, described by Aristotle as the 'Itmid' stone. Its cube-shaped crystals and high density make it among the most distinctive of all minerals. It typically forms in hydrothermal veins, along with sphalerite, pyrite and chalcopyrite, as well as unwanted gangue minerals such as quartz, calcite and barite. Sometimes the hot fluids that deposit galena are trapped in branching cracks beneath the surface; in limestones, these fluids may ooze into cavities to create rich but patchy replacement deposits. Fluids may also well up in undersea volcanic activity to create 'volcanogenic' galena. Ores are often found when partially exposed at the surface by erosion, as in the famous Broken Hill and Mount Isa deposits of Australia. As these surface deposits are exhausted, mining companies are having to probe deeper. Galena usually contains only traces of silver, but so much galena is mined that it is the world's main ore of silver as well as lead.

Cubic crystals of galena

Identification: Galena crystals are easy to identify by their cubic shape, dark grey colour and high density. Its metallic lustre is usually hidden by a dull film, formed on contact with air.

Galena

Massive quartz

> **Crystal system**: Isometric.
> **Crystal habit**: Crystals typically cubic. Also forms masses and grains.
> **Colour**: Dark grey sometimes with a bluish tinge.
> **Lustre**: Metallic to dull.
> **Streak**: Lead grey.
> **Hardness**: 2.5+.
> **Cleavage**: Perfect in four directions forming cubes.
> **Fracture**: Uneven.
> **Specific gravity**: 7.5–7.6.
> **Notable locations**: Weardale, England; Black Forest, Harz Mountains, Germany; Sardinia; Trepca, Kosovo; Broken Hill (NSW), Mt Isa (QLD), Australia; Naica, Mexico; Tristate, Kansas-Missouri-Oklahoma.

Mimetite

Pb₅(AsO₄)₃Cl Lead chloroarsenate

Small wulfenite crystal

Mimetite

Identification: Slender yellow mimetite crystals and cauliflower crusts are usually easy to identify by their associations with lead minerals.

Mimetite's name comes from the Greek for 'imitator', as it resembles pyromorphite. Like pyromorphite, it occurs in association with galena, anglesite and hemimorphite, and is one of various lead ores that develop where galena is altered by exposure to air. Mimetite forms where arsenic is present too. It is quite rare and striking crystals like those from Tsumeb, Namibia, are rarer still. It is typically found as botryoidal or cauliflower-like crusts. It can also form barrel-shaped crystals called 'campylite'. Mimetite forms a series with pyromorphite, in which phosphorus replaces mimetite's arsenic, and another with vanadinite, in which arsenic is replaced by vanadium. Vandinite and mimetite look very different, but green mimetite crystals look like pyromorphite.

> **Crystal system**: Hexagonal.
> **Crystal habit**: Slender hexagonal prism-shaped crystals or barrel-shapes (campylite); usually botryoidal or 'cauliflower' crusts.
> **Colour**: Yellow, brown, green.
> **Lustre**: Resinous.
> **Streak**: Off-white.
> **Hardness**: 3.5–4.
> **Cleavage**: Rarely noticed.
> **Fracture**: Subconchoidal.
> **Specific gravity**: 7.1.
> **Other characteristics**: Garlic smell when heated.
> **Notable locations**: Johanngeorgenstadt (Saxony), Germany; Pribam, Czech Republic; Tsumeb, Namibia; Mount Bonnie (Northern Territory), Australia; Durango, Sonora, Chihuahua, Mexico; Gila County, Maricopa County, Arizona.

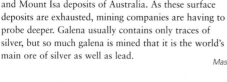

Pyromorphite

Pb₅(PO₄)₃Cl Lead chlorophosphate

Pyromorphite gets its name from the Greek for 'fire form' because of the way its crystals reform when heated and left to cool. It is a secondary mineral that forms where lead ores are oxidized by exposure to air, typically where there is phosphorus available in the apatite of local rocks. It is much less abundant than galena, cerussite or anglesite, but because it typically occurs in the surface layers, it was often used as an ore of lead in the past, notably at Broken Hill in Australia, Caldbeck Fells (Cumbria) in England and Leadville in Colorado, USA. Although it typically forms crusts and masses, pyromorphite is best known for its striking slender, hollow green crystals. It is often indistinguishable from mimetite, and like mimetite can sometimes form barrel-shaped crystals called campylite. Pyromorphite is much more common than mimetite, and good crystals are frequently found, such as in Spain, Idaho, France and now China.

Identification: Pyromorphite is best identified by its yellow-green colour, associations with other lead minerals and slender, hollow-ended crystals.

Crystal system: Hexagonal.
Crystal habit: Slender prism-shaped, often hollow-ended, crystals or barrel-shapes (campylite); also crusts and masses.
Colour: Green, yellow, brown.
Lustre: Resinous to greasy.
Streak: Off-white.
Hardness: 3.5–4.
Cleavage: Imperfect.
Fracture: Uneven.
Specific gravity: 7+.
Notable locations: Cumbria, England; Auvergne, France; El Nassau, Germany; El Horcajo, Spain; Guilin, China; Broken Hill (NSW), Australia; Durango, Mexico; Chester Co, Pennsylvania.

Bindheimite

Pb₂Sb₂O₆(O,OH) Lead antimony oxide hydroxide

Bindheimite is a minor ore of lead and antimony. It is an earthy yellow mineral that forms as lead and antimony minerals such as jamesonite and boulangerite are altered. As it forms, it often adopts the form of the original mineral and so becomes its pseudomorph. Bindheimite was named after the German mineralogist J J Bindheim (1750–1825) and was officially recognized as a mineral in 1868. However, it was known long before that. In fact, the Ancient Egyptians used it to give coloured glass a brilliant yellow hue. They mixed bindheimite with melted silica and soda ash to create yellow lead pyroantimonite which spread throughout the glass. This has since been used widely as a pigment, sometimes known as 'Naples yellow', in either synthetic or natural form.

Crystal system: Isometric.
Crystal habit: Typically cryptocrystalline masses or crusts but also pseudomorphs of lead antimony sulphides such as jamesonite.
Colour: Yellow to red-brown or greenish to white.
Lustre: Light greenish-yellow to brown.
Hardness: 4–4.5.
Cleavage: None.
Fracture: Earthy, conchoidal.
Specific gravity: 7.3–7.5.
Notable locations: Cardigan, Wales; Cornwall, England; Saxony, Black Forest, Germany; Carinthia, Austria; Auvergne, France; Nerchinsk, Siberia; Australia; Cochise and Pima Counties, Arizona; Black Hills, South Dakota; San Bernardino County, California.

Identification: Bindheimite is best identified by its yellow colour, its earthy habit and its association with stibiconite, jamesonite, cerussite, tetrahedrite, lewisite and partzite.

Bindheimite

Quartz

Poisoned by lead
Lead has been used for at least 7,000 years. Water pipes in ancient Rome, some of which still carry water today, were made of lead. The English word 'plumbing' is from the Latin word for 'lead', *plumbum*. Because it is so widely available and so resistant to corrosion, people went on using lead for water pipes until quite recently, and until the 1970s most homes were painted with lead-based paint. Motor fuel was often rich in lead because it made fuel burn in a more controlled way, protecting engines. But lead has proved to be a quiet killer, inflicting its damage for thousands of years unnoticed, poisoning the digestive system and causing brain damage in children. Some people attribute the downfall of the Roman empire to slow poisoning by lead. England's 16th-century Queen Elizabeth I and many of her ladies had their faces ruined by painting them with 'white lead' in the fashion of the time. Lead pipes are being replaced, paint no longer contains lead and cars are now required to run on unleaded fuel (above), but there is still plenty of lead around.

ZINC

Used since ancient times with copper in the alloy brass, zinc is a blue-grey, quite brittle metal, and is commonly used to coat iron and steel to stop them rusting in a process called galvanizing. Zinc is rarely found in pure native form, but it is widespread as a compound in rocks, and is extracted from a small number of ores. The chief ore is sphalerite, but smithsonite, zincite and hemimorphite are also important.

Smithsonite

$ZnCO_3$ Zinc carbonate

Named after James Smithson, the English benefactor who founded the Smithsonian Institution, smithsonite is still sometimes known, incorrectly, as calamine (see hemimorphite). It is a secondary mineral that forms in surface layers when zinc minerals like sphalerite are altered by weathering, particularly in limestones, which provide the carbonate. Until the 1880s, when it was replaced by sphalerite, it was the main source of zinc. Smithsonite is a surprisingly attractive mineral. Although it doesn't often form sparkling crystals, it instead forms grapelike or teardrop-shape masses with an extraordinary pearly lustre in a range of colours. Traces of copper give green to blue colours; cobalt, pink to purple; cadmium, yellow; iron, brown to reddish brown. The best known colour is apple-green, but the most sought after are the rarer lavenders. The best specimens come from Tsumeb, Namibia, from Broken Hill in Zambia, and the Kelly Mine in Magdalena, New Mexico.

Yellow smithsonite is sometimes called turkey fat ore

Identification: Smithsonite's pearly, waxlike lustre is very distinctive and only really looks like hemimorphite – but unlike hemimorphite, it looks like plastic at the broken edges.

Smithsonite

Crystal system: Trigonal.
Crystal habit: Crystals rare; usually botryoidal or globular crusts; also massive.
Colour: Green, purple, yellow, white, brown, blue, orange, peach, colourless, grey, pink or red.
Lustre: Pearly to vitreous.
Streak: White.
Hardness: 4–4.5.
Cleavage: Perfect, forming rhombohedrons.
Fracture: Uneven.
Specific gravity: 4.4.
Notable locations: Cumbria, England; Moresnet, Belgium; Bytom, Poland; Sardinia; Santander, Spain; Lavrion, Greece; Tsumeb, Namibia; Broken Hill, Zambia; Kelly Mine (Magdalena District, Socorro County), New Mexico; Leadville, Colorado; Idaho; Arizona.

Sphalerite

(Zn,Fe)S Zinc iron sulphide

Until the 19th century, sphalerite could not be smelted, so smithsonite was the main zinc ore. Now sphalerite is the most important zinc ore by far. It gets its name from the Greek for 'treacherous' because it is so easily mistaken for other minerals, especially galena and siderite. One of the problems with identification is that sphalerite tends to occur in exactly the same hydrothermal veins as galena, and the two are often intergrown. The German miners of old called sphalerite *blende*, which means 'blind', because it looked like galena but did not yield lead. Iron-rich varieties of sphalerite are known as marmatite, reddish ones as ruby-blende or ruby jack, and massive banded ones as schalenblende, in which the sphalerite is often intergrown with galena. Almost uniquely, sphalerite will cleave in six directions, although a single crystal rarely does.

Identification: In banded schalenblende, sphalerite is often intergrown with wurtzite and the two are always hard to tell apart. The best clue is in sphalerite's buff-coloured streak.

Crystal system: Isometric.
Crystal habit: Include tetrahedrons or dodecahedrons with cube faces. Often twinned. Also granular, fibrous, botryoidal.
Colour: Black, brown, yellow, reddish, green or white.
Lustre: Adamantine, resinous.
Streak: Buff.
Hardness: 3.5–4.
Cleavage: Very good parallel to faces of rhomb dodecahedrons.
Fracture: Conchoidal rarely seen.
Specific gravity: 4.0.
Other characteristics: Triboluminescent.
Notable locations: Cumbria, England; Trepca, Kosovo; Binnetal, Switzerland; Santander, Spain; Broken Hill (NSW), Australia; St Lawrence County, New York; Tristate, Kansas-Missouri-Oklahoma.

Zincite

ZnO Zinc oxide

Zincite is a rare secondary mineral that develops in surface layers of rock when zinc minerals such as sphalerite are weathered and oxidized. It is usually found in platy or granular masses, typically in association with smithsonite, franklinite, gahnite, willemite and calcite. There is only one place in the world where it is found in any quantity – the famous Franklin and Sterling Hill mines in New Jersey. It is found here as red grains and masses in a white, highly fluorescent calcite, typically associated with black franklinite and green willemite. There is so much zincite at Franklin and Sterling Hill that it is a major zinc ore. Good crystals, though, are exceptionally rare. When they are found, typically lying sideways in a calcite vein, they are hemimorphic – that is, they are different shapes at each end with a flat base and a hexagonal pyramid top.

Identification: Zincite is usually identified as in New Jersey, as a red mineral in association with white calcite, black franklinite and green willemite.

Franklinite (black)

Zincite (red)

Crystal system: Hexagonal.
Crystal habit: Usually massive, foliated, granular.
Colour: Orange-yellow to deep red or brown.
Lustre: Adamantine.
Streak: Orange-yellow.
Hardness: 4.
Cleavage: Perfect, basal.
Fracture: Conchoidal.
Specific gravity: 5.4–5.7.
Notable locations: Tuscany, Italy; Sardinia; Siegerland, Germany; Tasmania; Franklin and Sterling Hill, New Jersey.

Old Brass

Long before zinc was identified as an element, it was used with copper to make brass. At least 3,000 years ago, people in the Middle East discovered how to make brass by heating native copper and calamine (smithsonite) with glowing charcoal in a clay crucible. This was the method used to make brass right up until the 19th century, and 'calamine brass' is still often considered superior to other kinds. The great thing about brass was that it not only looked like gold, but was tougher, lighter, and cheaper – and corroded much less than any pure metal known except gold. It could also be beaten into shape as well as moulded (unlike bronze). The higher the zinc content, though, the less malleable brass is. Brass has a long history of continuous use in India. In Europe, it was used widely by the Romans for lamps, bowls, plates and other household items – a practice which continued for many centuries. In the Middle Ages, monumental brass plates were used to commemorate the dead. This important alloy has also been used throughout history on a vast number of items connected with time, navigation and observation, such as clocks, compasses, sundials and nautical instruments such as sextants. The latter is a small telescope mounted on an arc, which can help to determine latitude and longitude in relation to the position of celestial bodies.

Hemimorphite

Zn₄Si₂O₇(OH)₂.H₂O Hydrated zinc silicate hydroxide

Like smithsonite, hemimorphite was once called calamine and used as a powder in a lotion to soothe skin irritations. In America, calamine was hemimorphite; in Europe it was smithsonite. To avoid confusion, mineralogists no longer use the term calamine. The name hemimorphite means 'half form' – an illusion to its crystals, which means they are not symmetrical but blunt at one end and pointed like a pyramid at the other. Sometimes, though, in clusters, this hemimorphism is not apparent, because the crystals are attached at the base, hiding the blunt end. Occasionally, hemimorphite is found as clear, bladed crystals arranged in fan shapes; sometimes, it appears as a botryoidal crust similar to smithsonite but duller in lustre and a more acid green.

Crystal system: Orthorhombic.
Crystal habit: Hemimorphic flat based pyramids or arranged in fans; or botryoidal crusts.
Colour: Blue-green, green, white, clear, brown and yellow.
Lustre: Vitreous crystals, dull crusts.
Streak: White.
Hardness: 4.5–5.
Cleavage: Perfect in one direction.
Fracture: Conchoidal to subconchoidal.
Specific gravity: 3.4.
Other characteristics: Strongly pyroelectric and piezoelectric.
Notable locations: Black Forest, Germany; Sardinia; Broken Hill, Zambia; Durango, Chihuahua, Mexico; Leadville, Colorado; Franklin, New Jersey.

Identification: Hemimorphite forms knobbly crusts like smithsonite but these are a bluer blue-green and duller. Crystals are flat at one end and pointed at the other.

TIN, BISMUTH AND MERCURY

Tin, bismuth and mercury are all useful metals known since ancient times. Tin comes from the ores cassiterite and stannite, found mainly in river gravels and veins around granite intrusions. Bismuth comes mainly from bismuthinite, as well as bismite and bismutite, which are found in many of the same places as tin ores. Mercury comes almost entirely from the striking red mineral cinnabar.

Tin ore: Cassiterite

SnO_2 Tin oxide

Cassiterite gets its name from the Greek word for 'tin', which in turn came from the Sanskrit word *kastira*. Although a small amount of tin comes from sulphide minerals such as stannite, cassiterite is the only significant tin ore. It forms deposits mostly in granites and pegmatites, and in veins and skarns around granite intrusions where tin-bearing fluids from the granite magma have penetrated. Because it is very tough, it survives weathering to collect in river gravels, as gold does, in rounded grains (pebble tin). It was in these placer deposits that people first found tin thousands of years ago, and they still provide 80 per cent of all tin ore. Over half the world's tin now comes from placer deposits formed in river beds that have been submerged under the sea off the coasts of Malaysia, Indonesia and Thailand and are mined by dredging the seabed. Cassiterite has also long been mined from veins in Saxony and Bohemia, and, most famously, Cornwall, where tin mines date back to the Bronze Age. Cornish tin is now uneconomic to mine and most mined ore comes from Tasmania and Bolivia. Wood tin is cassiterite that looks like wood and is formed in high-temperature veins.

Identification: Cassiterite is best identified by its black colour, its hardness and its prism or pyramid-shaped crystals. Its high refractive index also gives crystals a tremendous adamantine lustre.

Crystal system: Tetragonal.
Crystal habit: Crystals pyramids or short prisms; also massive, granular.
Colour: Black or reddish brown or yellow.
Lustre: Adamantine or greasy.
Streak: White, or brownish.
Hardness: 6–7.
Cleavage: Good in two directions; indistinct in another.
Fracture: Irregular.
Specific gravity: 6.6–7.0.
Other characteristics: High refractive index – approximately 2.0.
Notable locations: Cornwall, England; Erzgebirge (Saxony), Germany; Zinnwald (Bohemia), Czech Republic; Spain; Fundào, Portugal; Tasmania, Greenbushes (WA), Australia; China; Malaysia; Indonesia; Thailand; Lallagua (Potosi), Araca, Oruro, Bolivia; Durango, Mexico (wood tin).

Tin ore: Stannite

Cu_2FeSnS_4 Copper iron tin sulphide

Identification: Stannite is best identified by its iron-black colour, with tinges of bronze yellow, and its association with sulphides.

Stannite is a comparatively rare tin sulphide mineral that also contains copper and iron, and was one of the ores once mined in Cornwall and Saxony. It forms grains in granite, or crystals and masses in hydrothermal veins near granite in association with other sulphide minerals. Its pure colour is iron black, but it is often turned bronze yellow by tarnishing or the presence of chalcopyrite. This is why miners sometimes call it bell-metal ore, or tin pyrites. In Bolivia, stannite occurs with silver ores; in Bohemia, it occurs with the minerals galena and sphalerite. The stannite group is also the name of a group of copper sulphides which includes luzonite, kësterite and velikite as well as stannite itself.

Stannite

Wolfram

Crystal habit: Tetragonal.
Crystal habit: Typically occurs as grains or masses in granite, but can form minute pyramid-shaped crystals.
Colour: Grey, black.
Lustre: Metallic.
Streak: Black.
Hardness: 4.
Cleavage: Imperfect.
Fracture: Uneven.
Specific gravity: 4.3–4.5.
Notable locations: Cornwall, England; Auvergne, Brittany, France; Erzgebirge (Saxony), Germany; Carinthia, Austria; Zinnwald (Bohemia), Czech Republic; Kamchatcka, Russia; Zeehan, Tasmania; Oruro, Llallagua (Potosi), Bolivia; Nova Scotia.

Bismuth ore: Bismuthinite

Bi₂S₃ Bismuth sulphide

Bismuth does occur naturally in native form but is only about as common as native silver. As a result, bismuth is often extracted from bismuthinite, and occasionally the oxide bismite and carbonate bismutite, which often forms pseudomorphs of bismuthinite when it is exposed to the air. Bismuthinite, or bismuth glance, typically forms when native bismuth is altered by hot fluids in hydrothermal veins and tourmaline-rich pegmatites. Minute, ribbonlike crystals have been found around fumaroles in the Lipari Islands, Italy. When it forms larger crystals in veins, they are often radiating sprays like those of stibnite (and its pseudomorph stibiconite), and the two can be hard to tell apart. Bismuthinite is heavier than stibnite, and its crystals have straighter, flatter sides. Bismuthinite is also commonly associated with tin ores, while stibnite is usually found with antimony and arsenic minerals.

Identification: Bismuthinite can usually be identified by its yellowish, slightly iridescent tarnish. Long crystals can be identified by bending them slightly.

Bismuthinite

Quartz

Crystal system: Orthorhombic.
Crystal habit: Usually massive, fibrous; rarely as sprays of long thin crystals.
Colour: Steel grey to white.
Lustre: Metallic.
Streak: Grey.
Hardness: 2.
Cleavage: Perfect in one direction.
Fracture: Uneven.
Specific gravity: 6.8–7.2.
Other characteristics: Thin crystals bend a little. Crystals may have a slight yellow or iridescent tarnish.
Notable locations: Cornwall, England; Siegerland, Vogtland, Germany; Lipari Island, Italy; Lllallagua (Potosi), Huanani (Oruro), Bolivia; Kingsgate (New South Wales), Australia; Guanajuato, Mexico; Timiskaming, Ontario; Haddam, Connecticut; Beaver County, Utah.

Bronze Age
Bronze is an alloy, usually of copper and tin. The discovery of this alloy in the Middle East about 5,300 years ago was one of the great milestones in the human story. People had been fashioning native metals such as copper and gold into ornaments and even tools and weapons since about 8000BC. But copper is too soft to keep its shape and stay sharp when made into blades. Yet by adding just a little tin (later found to be ideally 10 per cent), copper could be made into bronze. Bronze was not only more malleable than copper, it was much tougher. The first swords were bronze. So were the first suits of armour, the first metal ploughs and the first metal cooking pots. The illustration of bronze age implements above indicates how the alloy might be adapted to meet differing requirements, both in the prehistoric home and on the battlefield. It is hard to overestimate the impact of bronze on ancient technology. When the Ancient Greeks attacked Troy, they probably did so because it was a focus of the bronze trade. Initially, bronze was brought to Europe by traders from Mycenae. But to supply the smithies that sprang up, ore-hunters ranged across Europe. From 2300BC, tin mining and bronzemaking spread west from Unetice in Hungary and Mittelburg in Germany to Almaden in Spain and Cornwall in England.

Mercury ore: Cinnabar

HgS Mercury sulphide

Getting its name from the ancient Persian for 'dragon's blood', cinnabar is one of the most striking of all minerals. It is a brilliant scarlet colour and was once powdered to make the coveted paint pigment vermillion. Because mercury is liquid at room temperature, it is rarely found in nature except in tiny blobs in crevices, and cinnabar is the main source of mercury, as it has been since Roman times – the cinnabar mines at Almaden in Spain date back 2,500 years. It crystallizes in hydrothermal veins in sedimentary rocks only once the temperature has dropped quite low, so it is typically found fairly near the surface, or even around hot springs.

Crystal system: Trigonal.
Crystal habit: Crystals rhombohedral or thick tabular, sometimes short prisms or needles. Also grains and masses.
Colour: Scarlet to brick red. Darkens on exposure to light.
Lustre: Adamantine to submetallic.
Streak: Red.
Hardness: 2–2.5.
Cleavage: Perfect in three directions, forming prisms.
Fracture: Uneven to splintery.
Specific gravity: 8.1.
Notable locations: Almaden, Spain; Mount Avala (Idria), Serbia; Slovenia; Hunan, China; San Luis Potosi, Mexico; Cahill Mine (Humboldt Co), Nevada; Sonoma Co, San Benito Co, Santa Clara Co, California; Red Devil Mine (Sleetmute), Alaska.
Caution: Wash your hands carefully after handling cinnabar; mercury is poisonous.

Identification: Cinnabar is identifiable by its bright red colour.

ARSENIC AND ANTIMONY

Arsenic and antimony minerals are often found together along with silver. Although both are poisonous, arsenic is still used in making some kinds of bronze, in alloys designed to resist high temperatures and pyrotechnics, while antimony is used as a flame retardant in everything from plastics to textiles, added to tin to make pewter, and as a yellow pigment. The main arsenic ores are orpiment and realgar; ores for antimony include stibnite and bournonite.

Arsenic ore: Realgar

AsS Arsenic sulphide

Realgar got its name in the days of alchemy from the Arabic *rahj al ghar*, which means 'powder of the mine'. The name is apt, because it is liable to crumble as soon as it is exposed to light, slowly but surely.

Löllingite: A minor ore of arsenic, iron arsenide

Attracted by its startling orange-red colour, the Ancient Chinese used it to make carvings – and these carvings are now badly deteriorated. As it crumbles, realgar changes to yellow orange pararealgar. It was once used as a red paint pigment, and in many old paintings the original red has decayed to yellow or orange. Collectors are advised to keep realgar specimens in complete darkness. Realgar usually forms late in the cooler parts of hydrothermal veins and around hot springs and fumaroles, nearly always in association with orpiment, and calcite, and also with the mercury ore cinnabar and the antimony ore stibnite. Good crystals can be found in druses and cavities in China and Nevada.

Identification: Realgar is easily identified by its orange-red colour, softness and association with orpiment, calcite and stibnite.

Caution: Wash your hands carefully after handling realgar; arsenic is poisonous.

Crystal system: Monoclinic.
Crystal habit: Short, striated prisms, terminated by a wedge-like dome. Also found as grains, crusts and earthy masses.
Colour: Orange to red.
Lustre: Resinous, adamantine to submetallic.
Streak: Orange to orange-yellow.
Hardness: 1.5–2.
Cleavage: Good in one direction.
Fracture: Subconchoidal.
Specific gravity: 3.5–3.6.
Other characteristics: Unstable in light.
Notable locations: Carrara, Italy; Binnental, Switzerland; Tajowa, Hungary; Kresevo, Bosnia; Alsar, Macedonia; Transylvania, Romania; Turkey; Hunan, China; Mercur, Utah; Getchell, Nevada; King County, Washington.

Orpiment

As₂S₃ Arsenic sulphide

Orpiment got its name from a corruption of the Latin *auripigmentum*, 'golden coloured', and was long ago used as a pigment called 'king's yellow' and as a cosmetic by women unaware of its dangers. In Ancient China, it was used for gilding silk. Like realgar, orpiment usually forms late in the cooler parts of hydrothermal veins and around hot springs and fumaroles, nearly always in association with realgar and calcite, and also with cinnabar and stibnite. It is often found in mica-like foliated masses, but many fine crystals have been found in Hunan in China, Quiruvilca in Peru and, more recently, at the Twin Peaks gold mine in Humboldt County, Nevada. Like realgar, only less so, orpiment deteriorates on exposure to light, developing a white surface film.

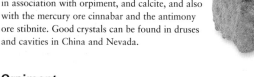

Caution: Wash your hands carefully after handling orpiment; arsenic is poisonous.

Identification: Orpiment is easily recognized by its golden-yellow colour, softness, garlic smell and often its mica-like flakes.

Crystal system: Monoclinic.
Crystal habit: Usually foliated or columnlike masses and crusts, also tiny tablet-shaped 'orthorhombic' crystals.
Colour: Orange-yellow to yellow.
Lustre: Resinous to pearly.
Streak: Yellow.
Hardness: 1.5–2.
Cleavage: Perfect in one direction giving bendable flakes.
Fracture: Flaky.
Specific gravity: 3.5.
Other characteristics: Unstable in light. Arsenic gives it a garlic smell.
Notable locations: Tuscany, Campi Fliegri, Italy; Binnental, Switzerland; Maramures County, Romania; Alsar, Macedonia; Hunan, China; Quiruvilca, Peru; Bolivia; Getchell Mine, Twin Peaks Mine, Nevada; Mercur, Utah; Green River Gorge, Washington.

Bournonite

CuPbSbS₃ Copper lead antimony sulphide

Named after the French mineralogist Count J L de Bournon (1751–1825), who made the first complete chemical analysis of it in 1804, bournonite is one of the most common sulphosalts. It is an ore of antimony that forms in hydrothermal veins, and also as a secondary mineral associated with copper ores. It is about 42 per cent lead, 24 per cent antimony, 21 per cent sulphur and 13 per cent copper. There are often vugs (open cavities) in the veins where good crystals of bournonite often form. Bournonite is famous for its twinned crystals, especially the groups of repeated twins that look rather like a chunky cogwheel. Many of these 'cog-wheel ores', or *Radelerz*, have been found in mines at Neudorf and Andreasberg in Germany and Cavnic and Baia Mare in Romania. The most spectacular specimens have come from the Herodsfoot mine near Liskeard in Cornwall, England, which was dug for its silver-bearing lead ores.

Identification: Bournonite is very easy to identify when the crystals are twinned, especially in the cogwheel pattern; otherwise it can easily be mistaken for many other dark, metallic minerals.

Caution: Wash hands after handling bournonite; antimony is poisonous.

Bournonite

Pyrite

Crystal system: Orthorhombic.
Crystal habit: Tablet- to prism-shaped crystals. Twins common, often repeated in shapes like cogwheels. Also grains and masses.
Colour: Silver-grey or black.
Lustre: Metallic.
Streak: Black.
Hardness: 2.5–3.
Cleavage: Poor.
Fracture: Subconchoidal.
Specific gravity: 5.7–5.9.
Other characteristics: Often gets a dull tarnish.
Notable locations: Endellion, Herodsfoot (Cornwall), England; Neudorf, Andreasburg, Germany; Baia Mare, Cavnic, Romania; Pribram, Czech Republic; Chichibu (Saitama), Japan; Quiruvilca, Peru; Park City, Utah.

Arsenic poisoning
Arsenic was used with copper to make the first bronze, but its toxicity as a poison was learned quite early on and it has been a popular poison for murderers for thousands of years. In low doses, though, it was used to treat diseases such as syphilis. In recent years, however, scientists have begun to appreciate that long-term exposure to even low doses may be dangerous – and arsenic is so common in the environment that many people are unwittingly ingesting arsenic in drinking water. Arsenic poisoning may manifest on the hands and feet as keratosis: hard, cornlike spots and callouses (above). Severe lethargy is another common symptom of drinking contaminated water. Some scientists are talking of potential catastrophes in certain parts of the world. Some of the danger is from the leaking of arsenic from mining waste into drinking water, as in parts of Australia and the UK. But the worst danger may be where wells have been sunk into groundwater contaminated by naturally occurring arsenic minerals, especially in Bangladesh, West Bengal and the Ganges plain. Some experts now fear that 330 million people in India and 150 million people in Bangladesh may be at serious risk.

Stibnite

Sb₂S₃ Antimony sulphide

Stibnite gets its name from the Latin for the metal antimony for which it is the main ore. Legend has it that antimony got its name when a monk called Valentinus put stibnite in his fellow monks' food to fatten them up – but, of course, it killed them, so was called *anti-monachium*, 'against monks'. Stibnite typically occurs along with quartz in low-temperature hydrothermal veins, or as replacements in limestone, and around hot springs. It is famous for its radiating sprays of long, thin, bladed or needlelike crystals. The planes within the crystal structure slide over each other, allowing the crystals to bend and droop, without fracturing.

Crystal system: Orthorhombic.
Crystal habit: Forms sprays of bladed or needle-like crystals often bent or curved. Also grains or masses.
Colour: Steel grey to silver.
Lustre: Metallic.
Streak: Dark grey.
Hardness: 2.
Cleavage: Perfect lengthways.
Fracture: Irregular.
Specific gravity: 4.5–4.6.
Other characteristics: Striated lengthwise; crystals slightly flexible.
Notable locations: Tuscany, Italy; Zajaca, Serbia; Alcar, Macedonia; Baie Mare, Cavnic, Romania; Xikuangshan (Hunan), China; Ichinokawa (Shikoku), Japan; Huaras, Peru; Nye County, Nevada.
Caution: Wash your hands carefully after handling stibnite; antimony is poisonous.

Identification: Stibnite is best identified by its very distinctive spray of long, thin crystals – similar only to bismuthinite and stibiconite.

PART TWO
FOSSILS

Fossils are the preserved remains of ancient life forms, from the simplest fungi and microbes to the first reptiles, plants and mammals, including the wonderful world of long-extinct species, such as the fascinating dinosaurs. The study of fossils can help us unravel the secrets of evolution itself, making it an endlessly captivating subject. This section explains what fossils are, where to find them and the best way to hunt for them. It also features an extensive directory containing over 375 key fossil discoveries.

Fossilized crinoids (sea lilies) found in Ontario, Canada. These marine animals were much more abundant and varied in the distant past than they are today.

INTRODUCING FOSSILS

Fossils have been treasured since ancient times. Long ago, people gathered these strangely shaped lumps of stone for many purposes – as objects of worship, signs of their worldly wealth and power, evidence of their god's handiwork, or simply as beautiful and elegant items to display and admire.

Fossil collectors of the ancient world could not have understood how their curiosities were formed. They could not know the enormous contribution of fossils to the development of modern scientific knowledge, about the Earth and the living things that have populated it through time. Had they understood the significance of fossils, they might well have been even more amazed at what they had collected.

More than 2,000 years ago in Ancient Greece, natural philosophers such as Aristotle mused on fossils and their origins. Some explanations were inorganic – fossils were nature's sculptures, fashioned by wind, water, sun, ice and other non-living forces. Some explanations were supernatural – fossils were the work of mythical beings and gods, placed on Earth as examples of their omnipotent powers. Aristotle himself described how fossils developed or grew naturally within the rocks, from some form of seed which he called 'organic essence'.

Below: Reconstructions from fossils of dinosaurs such as the famous Tyrannosaurus rex *continue to fascinate.*

In the 1500s, firmer ideas took root. In 1517 Italian physician-scientist Girolamo Frascatoro was one of the first on record to suggest that fossils were the actual remains of plants, creatures and other organisms or living things. In 1546 German geologist Georgius Agricola coined the word 'fossil', but not as we now understand it. His 'fossils' were almost anything dug from the ground, including coal, ores for metals and minerals, and what he believed to be rocks that just happened to be shaped like bones, teeth, shells and skulls. In 1565 Swiss naturalist Konrad von Gesner's works contained some of the first studied drawings of fossils. However, like Agricola, von Gesner believed that they were stones which, by chance, resembled parts of living things.

A fashion for fossils began in the 1700s in Europe. Wealthy folk established home museums where they displayed fossils, stuffed birds and mammals, pinned-out insects, pressed flowers and so on. Around 1800, scientists began to ponder more deeply on the true origins of fossils. It was suggested that they may indeed be the remains of once-living things. But as all life had been made by God, and

Below: Palaeontologists from the Natural History Museum, London, excavate 380 million-year-old tetrapod fossils in Latvia.

*Above: The rotting carcass of a spotted hyaena (*Crocuta crocuta*). Fossils begin like this – as the parts of living things that endure and are preserved in sediments.*

Below: This fossilized lower jaw of an extinct hyaena species is recognizably dog-like in character.

God would not allow any of his creations to perish, fossils could not actually be from extinct organisms. They must be from the types of living things which are still alive today.

Famed French biologist Georges Cuvier broke ranks. He studied the huge, well-preserved skull of a 'sea lizard' that looked like no living reptile, which he named *Mosasaurus*.

Above: Museum collections and exhibitions, especially well-designed displays like this one at the National Museum of Wales in Cardiff, are an excellent place to learn more about fossils.

Cuvier also examined fossil mammoth bones and saw their differences from those of living elephants. His new explanation was that these fossils were the remains of creatures that had indeed become extinct – in the Great Flood of Noah's time, as described in the Bible. As more and more varied fossils were dug, from deeper rock layers, the explanation grew into the story of Seven Floods. Each deluge saw a great extinction of living things. Then God re-populated the Earth with a new, improved set of organisms.

In 1859 English naturalist Charles Darwin's epic work *On The Origin Of Species* brought the idea of evolution to the fore. Fossils fitted perfectly into this scientific framework. Indeed they were used as major evidence to support it. In the struggle for survival, species less able to adapt to the current conditions died out, while better-adapted types took over. However the process never reaches steady state or an end point, since conditions or environments change through time – forcing the organisms to evolve with them. This explains not only the presence of fossils in the rocks, but also why these fossils exhibit changes over time.

Today the origins of fossils are well documented. And fossils are central to the story of Life on Earth – from the earliest blobs of jelly in primeval seas, to shelled creatures that swarmed in the oceans, enormous prehistoric sharks, the first tentative steps on land, the time of the giant dinosaurs, and the rigours of the ice ages with their woolly mammoths and sabre-toothed cats. A recent spate of discoveries now allows us to trace the origins of our own kind, back to ape-like creatures in Africa several million years ago. Yet, in addition to this scientific basis, we can still regard fossils as ancient people did – with wonder and appreciation of their natural beauty, while holding a part of Earth's history in our hands.

Left: The idea that fossils were the remains of long-extinct, fantastical creatures – such as this giant sea serpent in Charles Gould's Mythical Monsters, *published in 1886 – sparked some enduring myths.*

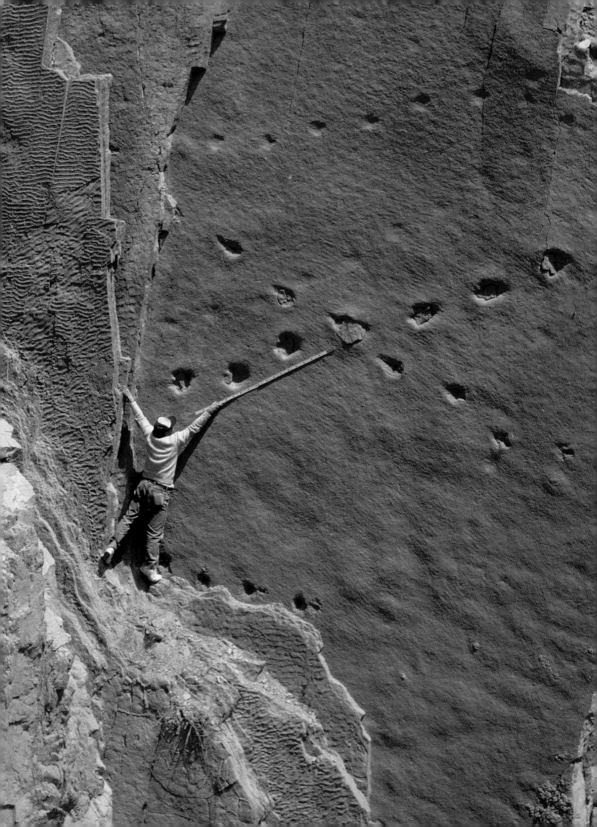

FOSSILS AND FOSSIL-HUNTING

Fossils are usually described as parts of living things which died long ago, and were preserved in the rocks, turned to stone. The processes of fossilization can happen in different ways, as described over the following pages. But they all share three main features.

First, the original objects preserved are from all kinds of living organisms – not only animals, but plants, fungi and even microbes. Second, the process takes a very long time – at least many thousands of years. Indeed, the majority of fossils are millions of years old. Third, most fossils are not the actual substances and materials from those once-living organisms. Fossilization gradually replaces living or organic matter with inorganic minerals and crystals. So, in general, a fossil is a lump of stone or rock. However its shape is taken from that of the original organism. This is summarized in the phrase 'bone to stone'.

Fossils themselves are fascinating, but many people gain even greater enjoyment by searching for them. It involves the thrill of the chase, and anticipation of an exciting new discovery. Fossil-hunting is an active outdoor pursuit, in the fresh air amid often beautiful scenery. It allows us to expand our basic understanding of geology and biology, from the way landscapes form and erode, to the shapes of snail shells and sharks' teeth.

Fossil-hunting always produces results. For some, the greatest satisfaction lies in creating a well-thought-out display. For others, it's about making a contribution to local or even national museums and to existing paleontological knowledge. For the few who come across a spectacular find, the attraction may even lie in fame and fortune.

Left: Rocks within the Earth's crust are subject to dramatic movements over a very long period of time, crumbling and slipping into the oceans or being raised into towering mountain ranges. These footprints were originally made by a carnivorous dinosaur upon an ancient Cretaceous shoreline, but have since been lifted, over millions of years, to form part of a near-vertical fault in the Andes. The palaeontologist measuring the stride of the creature has had to overcome some difficulty in access to record this information.

FOSSILS THROUGH TIME

Fossils trace the evolution of life from the tiniest, simplest scraps of living matter, to huge and complex creatures like dinosaurs, whales and mammoths. It is a very long story, spanning billions of years, with many surprising events along the way.

The oldest known fossils are microscopic shapes in rocks dating back more than 3,500 million (3.5 billion) years. That's 3,500,000,000 years – a very long time. In fact it's more than two-thirds of the time that our planet has been in existence.

Like the other planets of the Solar System, the Earth probably formed from a cloud of dust and rocks whirling in space around the embryo Sun, some 4,600 million (4.6 billion) years ago. About 1,000 million years later, signs of life were appearing as microscopic organisms in the 'primeval soup' of the early oceans. However for more than 2,000 million years after these beginnings, life stayed very small and very simple.

Changing Earth

The earliest living things probably resembled the organisms called bacteria and cyanobacteria (blue-green algae) today. They are the simplest forms of life and make up a kingdom

Below: As the present day continents began to take form during the Jurassic Period, reptiles dominated all habitats, with dinosaurs ruling the land and crocodiles presiding over freshwater lakes and rivers.

Above: A well-preserved trilobite fossil (Calymene blumenbachii) from Silurian times.

of organisms known as the Prokaryota. It's believed that their activities helped to change the nature of early Earth. At first the atmosphere was rich in poisonous gases such as methane. However the biochemical activities of the first life-forms pumped increasing amounts of oxygen into the air. This was a source of ozone which began to shield the surface from the Sun's harmful ultra-violet rays (as it still does today). These changing conditions allowed the next steps in evolution – see Burgess Shales.

Jelly to shells

From 2,000 million years ago, microorganisms grew in layered mounds of minerals in shallow water. These became well preserved as the objects known as stromatolites (shown in later pages of this book). Some stromatolites have been through the process of silicification, to form the kind of rock known as stromatolitic chert. Within these can be found amazingly preserved remains of the microbes. From about 1,000 to 800 million years ago there are increasing and tantalizing signs of multicellular organisms, perhaps resembling simple jellyfish: these might be called the first true 'animals'.

Fossils start to appear in appreciable numbers in rocks formed after about 540 million years ago. This was the

time that shelled creatures evolved in the seas. Shells, being hard and resistant to decay, fossilize fairly well. Living things before this time, from about 1,000 to 540 million years ago, had evolved greater size and complexity. But they were mostly jelly-like. Their soft bodies fragmented and rotted before preservation could occur. Only a few rare and precious fossils give glimpses into this misty phase of life's history.

From water to land

At first, life flourished only in the seas. Fossils form well in water, as organisms die and sink, and are

Origins of names

The names of the geological time spans called periods were mostly established in the 1800s. Some originate from the area where rocks of a particular age were first surveyed and studied. The Cambrian Period (540–500 million years ago) takes its name from Cambria, an old Roman-Latin term for Wales. Geologist Adam Sedgwick introduced the name in 1835 after describing rocks from North Wales. Other period names have their origins in the main types of rocks formed at the time. For example, the Cretaceous Period (135–65 million years ago) is so called from the huge thicknesses of chalk laid down as it progressed – *creta* ('kreta') is an old Latin term for 'chalk'.

Below: The ammonite Mantelliceras *is preserved in chalk typical of the Cretaceous Period.*

covered by silt, mud, sand and other sediments on the bottom. From the sea, plants moved into freshwater, and animals followed them. Their fossils show an increasing variety of forms. By 400 million years ago another major habitat was being conquered – the land. Until that time the ground had been barren, with no covering of soil as we recognize it. Pioneering plants and animals spread over the mud, sand and rocks, leaving their fossils as they colonized new areas. Their remains show that plants needed stiff stems to hold themselves up, since they were no longer buoyed by water. Animals could wriggle or squirm. But the more complex types changed their fins for limbs, and took their early steps over the ground.

Reptile domination

By 350 million years ago a new group of creatures had appeared, with scaly skin and waterproof-shelled eggs. These were the reptiles, and their fossils trace their climb to dominance.

Above: An early Palaeozoic plant, Moresnetia zalesskyii.

Below: Mesozoic brittlestars, Sinosura sp., from the late Jurassic Period.

By 200 million years ago they had given rise to the major large animals in all three major habitats – on land, as the dinosaurs; in the air, as the pterosaurs; and in the sea, as plesiosaurs and ichthyosaurs. Startling fossils of these great beasts, including their eggs and babies, and the animals and plants that shared their time, show how they lived and died.

Above: Early Cenozoic (Palaeocene) seedling Jaffrea speirsii, a relative of the living katsura tree in the magnolia group.

Below: Cenozoic mammoth hair dating from the Quaternary Period.

Above: The remains of a woolly mammoth and straight-tusked elephant being excavated from a clay pit in Sussex, England, in 1964. These large herbivores existed within the last 1.75 million years.

Towards the end of the 'Age of Reptiles', plants underwent a revolution. Conifers, ginkgos and similar trees had dominated for much of the era. But by 100 million years ago a new group, the angiosperms, were becoming established. These include the dominant plants we see around us today – flowers, herbs and broad-leaved bushes and trees.

Mass deaths

About 65 million years ago there was a great change in the types and varieties of living things. Dinosaurs, pterosaurs, plesiosaurs and many other groups of animals, and many plants too, all abruptly disappear. This is known as a mass extinction. It marked the end of the great time span known as the Mesozoic Era, 'Middle Life'. This era had begun with an even greater loss of life 250 million years ago, which had in turn marked the end of the previous era, the Palaeozoic or 'Ancient Life'. Following the Mesozoic came our own era, the Cenozoic or Cainozoic, 'Recent Life'. Each of these huge phases of life is documented by the types of fossils in the rocks, and the make-up of the rocks themselves.

THE GEOLOGICAL TIME SCALE

Fossils are records of life in the past. But when, exactly – how many thousands or millions of years ago? Palaeontologists and other scientists use a dating system known as geological time to put the various time spans of the past into perspective.

Rocks and the fossils they contain are dated in various ways, to find out how old they are in millions of years. These dates are fitted into a framework known as the geological time system. Its divisions are based on significant changes in the rocks and the types of fossils within them, as various kinds of life came and went throughout evolutionary history.

Eras and periods

As described on the previous page, the major divisions of geological time are termed eras. First is the Precambrian, when living things were mostly specks, although larger and more complex creatures resembling jellyfish and worms appeared towards the end. The Precambrian was followed by the Palaeozoic, Mesozoic and Cenozoic Eras – Ancient, Middle and Recent Life respectively. In turn, each of these great eras is divided into spans called periods. Most of these periods lasted a few tens of millions of years. Again, the transition from one to the next is marked by important changes in fossils, as extinctions of previously dominant organisms occur and new types take over.

Epochs

As we progress from the past towards the present, the amounts of fossils, and their details of preservation, generally increase – because earth movements, erosion and other forces have had less time to destroy them. Also, the technologies used for finding the age of a rock or fossil are more accurate with younger specimens. This allows increasing accuracy of dating in the Cenozoic Era. So the next-to-last and current periods, the Tertiary and Quaternary, are subdivided into shorter time spans known as epochs. Further subdivisions do exist, but are often used to date rocks on a more local scale.

Useful shorthand
The names of the eras, periods and epochs act as 'shorthand labels' to enable palaeontologists to define and communicate a fossil's place in the past.

Precambrian Era
4,600–540 million years ago

Palaeozoic Era
540–250 million years ago
• Cambrian Period
540–500 million years ago
• Ordovician Period
500–435 million years ago
• Silurian Period
435–410 million years ago
• Devonian Period
410–355 million years ago
• Carboniferous Period
355–295 million years ago
• Permian Period
295–250 million years ago

Mesozoic Era
250–65 million years ago

• Triassic Period
250–203 million years ago
• Jurassic Period
203–135 million years ago
• Cretaceous Period
135–65 million years ago

Cenozoic Era
• Tertiary Period
65–1.75 million years ago
  • Palaeocene Epoch
  65–53 million years ago
  • Eocene Epoch
  53–33 million years ago
  • Oligocene Epoch
  33–23 million years ago
  • Miocene Epoch
  23–5.3 million years ago
  • Pliocene Epoch
  5.3–1.75 million years ago
• Quaternary Period
1.75 million years ago – present
  • Pleistocene Epoch
  1.75–0.01 million years ago
  • Holocene Epoch
  0.01 million years ago (10,000 years ago) – present

Variations in dating
There are several variants of the geological time scale, used in different fields such as basic palaeontology, mainstream geology, geological surveying, and prospecting for fossil fuels. For example, in some systems the Cambrian Period starts 570, 560 or 543 million years ago, and the Jurassic Period begins 144 million years ago.

Also, in the US the Carboniferous Period is sometimes known by two other names, the Mississippian Period or System followed by the Pennsylvanian, with the split at about 320 million years ago. However the sequences of the names for the eras, periods and epochs remain the same in all of these time scales.

Right: Exposed rock in the Ouachita Mountains, Missouri. The mountains formed in what US geologists term the Pennsylvanian.

Palaeogene and Neogene

In recent years, some amendments have been suggested to the traditional dating system outlined above. These affect the Tertiary Period (65–1.75 million years ago) and its various epochs, which together account for almost all of the Cenozoic Era. Traditionally, the Tertiary was followed by the Quaternary Period – the tail-end of the Cenozoic – which ran to the present day. In the revised system, the Cenozoic Era is divided into two more equal time spans, the

Palaeogene and Neogene Periods. The former takes in the Palaeocene to Oligocene Epochs, which remain unchanged. The Neogene includes the Miocene and later epochs to the present, which again remain unaltered. So in the scheme depicted below, the Palaeogene Period would run from 65 to 23 million years ago, and the Neogene from 23 million years ago to the present. Thus the familiar labels 'Tertiary' and 'Quaternary' are superseded. However, many areas of palaeontology move rather slowly, and

numerous exhibitions, museums and collections still exhibit Tertiary specimens. For this reason, we have added 'Tertiary' and 'Quaternary' to the chart below, as they remain a point of reference for fossil enthusiasts.

Further amendments also affect the boundaries between recent time spans, such as shifting the end of the Eocene Epoch and start of the Oligocene from 33 to 34 million years ago, and likewise the end of the Pliocene Epoch and start of the Pleistocene from 1.75 to 1.8 million years ago.

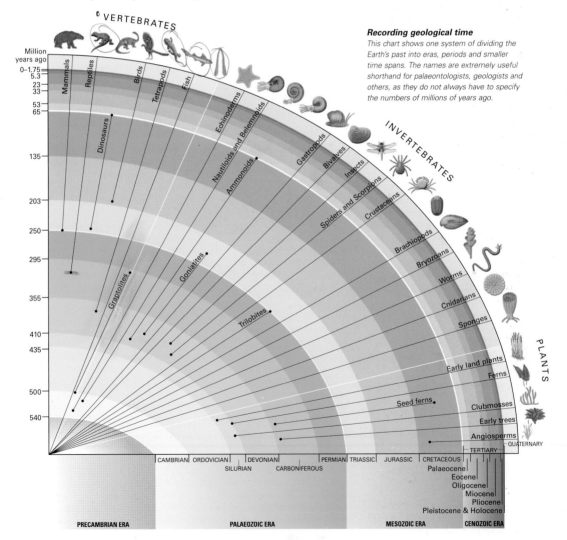

Recording geological time
This chart shows one system of dividing the Earth's past into eras, periods and smaller time spans. The names are extremely useful shorthand for palaeontologists, geologists and others, as they do not always have to specify the numbers of millions of years ago.

MAJOR FOSSIL SITES – THE AMERICAS

Some of the greatest fossil discoveries have been in the 'Badlands' of the North American Midwest – especially Alberta (Canada), and Montana, Wyoming, Colorado and Utah (USA). This is typical fossil-hunting country with bare rocks swept of soil by winds and rain, so that remains are easily spotted.

The landscape of the Badlands is subject to hot sun in summer, and freezing rain in winter, meaning plenty of erosion in the rocks as they weather, crack, peel and split, and fragments tumble down steep slopes or are washed away by flash floods. This ongoing geological activity continually exposes new layers with fresh fossils.

An embarrassing error

Perhaps the biggest 'fossil gold-rush' occurred in the Badlands region towards the end of the 19th century. It involved dinosaurs, of course, as well as many other reptiles, and plenty of other creatures, plus plants too. It was partly a result of bitter rivalry between two eminent palaeontologists, Edward Drinker Cope and Othniel Charles Marsh, and their respective teams of prospectors, workmen, restorers and benefactors. Their conflict began in about 1870, as an early wave of 'dinosaur-mania' was sweeping North America. Cope invited Marsh to view the remains of a long-necked, tubby-bodied, four-flippered, sea-going reptile, a plesiosaur called *Elasmosaurus*. Marsh noticed that Cope's reconstruction had the skull on the tail rather than the neck! From this time the two men were locked in competition to find the best fossils.

Below: Excavating dinosaur fossils on the Patagonian plains.

The first fossil gold rush

The intense antagonism of Cope and Marsh continued until their deaths in the late 1890s. Between them they named about 130 new kinds of dinosaurs and well over 1,000 other animals, from sharks to mammoths. Some of these finds have not stood the test of time, due to the race, rush and confusion of the so-called 'Bone Wars'. However the energy of the two men enthused many fellow fossil-hunters, and today North America vies with Asia (mainly China) to have the most fossil plants and animals described and catalogued.

The Burgess Shales

In 1909 eminent American palaeontologist Charles Walcott explored a quarry high in the Rocky Mountains near the town of Field, British Columbia. He came upon an astonishing treasure trove of fossils now know by the general name of 'the Burgess Shales'. The remains dated back to the Cambrian Period, some 530 million years ago. Walcott was astounded at the detailed state of preservation, which had occurred so fast that it entombed not only hard-shelled animals but also soft-bodied creatures like worms before they had time to rot away.

The Cambrian explosion

However, Walcott was less able to accept the incredible diversity of living things at such an early time – worms, sponges, jellyfish, arthropods related to crabs, and many more. According to beliefs of the time, creatures from that long ago should not be so diverse or varied. Life should have had started small and simple, and gradually worked up to the complexity and variety we see today. This idea is embodied in the concept of the 'evolutionary tree of life'. The tree

Above: A fossilized bee, more than 35 million years old, found in Colorado, USA.

began, long ago, with a trunk and a few branches, then expanded and diversified to the myriad twig ends of the present time. But finds such as the Burgess Shales show that this was not the case. Rather than a tree, an unkempt hedge might be a more apt comparison, with many branches appearing here and there throughout evolutionary time, some dying out rapidly but others enduring for much longer. The Burgess Shales represent one of these early bursts of evolution, which included the first shelled creatures. The time is often called the 'Cambrian explosion'.

South America

The number of significant fossil finds in South America has increased considerably since the 1960s. In particular, this continent lays claim to possibly the earliest, one of the biggest meat-eating, and the biggest of all dinosaurs – *Eoraptor*, *Giganotosaurus* and *Argentinosaurus*. Many other fascinating finds have come to light, often as a result of modern mining operations – in particular in Bolivia. Specimens include insects, frogs, small mammals and other creatures from the Tertiary Period, encased in amber.

Right: Map of the Americas, showing some of the major fossil discovery sites.

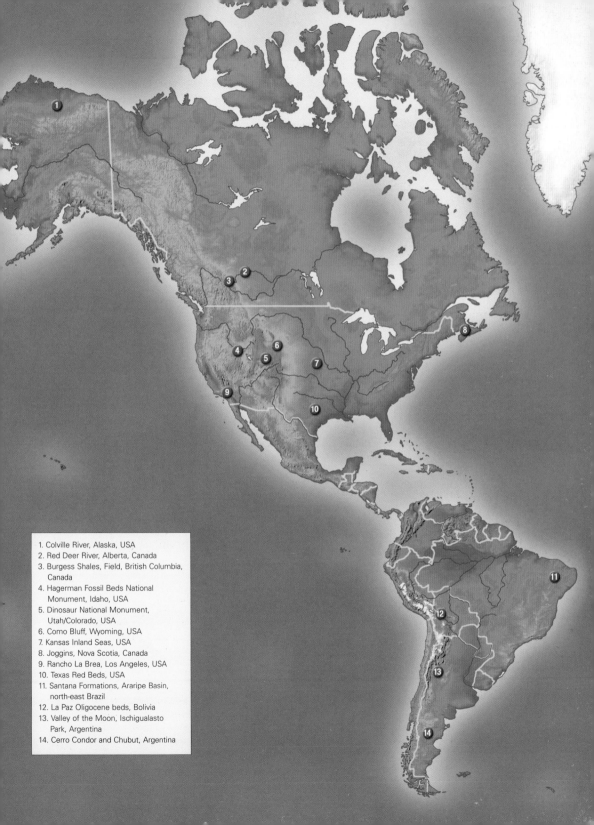

1. Colville River, Alaska, USA
2. Red Deer River, Alberta, Canada
3. Burgess Shales, Field, British Columbia, Canada
4. Hagerman Fossil Beds National Monument, Idaho, USA
5. Dinosaur National Monument, Utah/Colorado, USA
6. Como Bluff, Wyoming, USA
7. Kansas Inland Seas, USA
8. Joggins, Nova Scotia, Canada
9. Rancho La Brea, Los Angeles, USA
10. Texas Red Beds, USA
11. Santana Formations, Araripe Basin, north-east Brazil
12. La Paz Oligocene beds, Bolivia
13. Valley of the Moon, Ischigualasto Park, Argentina
14. Cerro Condor and Chubut, Argentina

MAJOR FOSSIL SITES – EUROPE, AFRICA

Collecting and displaying fossils as objects of beauty and intrigue – 'nature's sculptures' – gained popularity in Europe in the 1700s. So fossil-rich sites became well known even before people appreciated that these items were the remains of once-living things.

One of the first people known to make a living from fossil-hunting was a young woman, Mary Anning (1799–1847). Mary lived and worked in Lyme Regis, Dorset, South West England. Here, the coast is continually being eroded as waves eat into the cliffs. The region is sometimes referred to as the Jurassic Coast; many of its rocks were formed on the sea bed during this period, as countless marine creatures died and sank and were subsequently buried. The curly patterns of preserved ammonite shells are especially plentiful. Mary was one of those people who easily developed a 'good eye' for the best fossils, spotting tell-tale signs that enabled her to quickly distinguish fossilized creatures from ordinary rocks.

Fossil fame

Mary Anning spent her days on the Lyme Regis foreshore with her geological hammer and basket, gathering impressive specimens which were sold to private collectors and museums. She has a remarkable record of finds, including one of the first ichthyosaurs (dolphin-shaped sea reptiles) to be unearthed, in 1814; a virtually complete plesiosaur

Below: The 'Jurassic coast' of Dorset, UK.

(tubby-bodied, long-necked sea reptile) in 1824; and probably the first discovery in Britain of a flying 'pterodactyl' or pterosaur, in 1828. The shorelines near Lyme Regis still throng with fossil-hunters today.

Exquisite quality

Some of the world's most well known and valuable fossils come from the Solnhofen quarries of Bavaria, Germany. The limestone here is very fine-grained, known as 'lithographic' limestone since it was formerly quarried for use in printing. The seven or so preserved specimens of the first known bird, *Archaeopteryx*, were discovered in the area. So were the fossils of one of the smallest dinosaurs, *Compsognathus*. The amazing detail and quality of preservation allows study of their minutest features, including the feather structures of *Archaeopteryx*. These suggest that this bird, rather than being a clumsy glider, could truly fly – although probably not as aerobatically as most birds today.

The Karoo

Great Karoo Basin, centred around Victoria West in south-central South Africa, houses one of the largest and most important sets of fossil deposits in the world. Most of the rocks date

Above: After World War I, a team of scientists from the British Museum travelled to Tendaguru, Tanzania, to study and remove dinosaur fossils, including Brachiosaurus. The main excavations had been done by German palaeontologists prior to this.

from 270 to 230 million years ago – Permian to Early Triassic Periods. The rocks are divided into layers or series, which are (from oldest to youngest) the Dwyka, Ecca, Beaufort and Stormberg series. Long ago the climate was much damper than today. Ancient reptiles, in particular, roamed the luxuriant shrublands and swamps, and included mammal-like reptiles, pareiasaurs, titanosuchids and the smaller gorgonopsians.

Cradle of humankind

The vast continent of Africa is still relatively unexplored by fossil experts, but many palaeontologists accept that humans evolved here. Fossil remains of several kinds of prehistoric humans have been recorded. They range from smallish ape-like types who walked almost upright some five million years ago, to the first of our own kind, *Homo sapiens sapiens* or anatomically modern people, beginning around 150,000 years ago. Some experts recognize five or six main kinds of ancient human in Africa, while others contend there were twice as many.

Right: Map of Europe and Africa, showing some of the major fossil discovery sites.

1. Elgin, Scotland
2. Isle of Wight, England
3. Bernissart, Belgium
4. Neander Valley, Germany
5. Messel Quarries, Germany
6. Solnhofen, Bavaria, Germany
7. Beziers, France
8. Teruel, Spain
9. Tenere, Niger
10. Hadar, Ethiopia
11. Lake Turkana (Rudolf), Kenya
12. Olduvai Gorge, Tanzania
13. Tendaguru, Tanzania
14. Mafetang, Lesotho
15. Karoo Basin, South Africa

MAJOR FOSSIL SITES – ASIA, AUSTRALASIA

The tropical zones of the world, especially southern and South-east Asia, tend to be covered with lush vegetation and so have yielded few fossils. However, arid regions such as the Gobi and Australian deserts, with their exposed, eroded landscapes, are much more productive.

The vast arid stretches of Mongolia's Gobi Desert were first explored for fossils in the 1920s. Expeditions organized by the American Museum of Natural History visited the area, in the quest for the 'missing link' between apes and humans. One of the group's leaders was Roy Chapman Andrews, a larger-than-life taxidermist-turned-adventurer who always wore a ranger hat and carried a gun. The expedition sent back regular reports of exciting, perilous encounters with bandits and sandstorms – rather different from the usual staid world of palaeontology. Andrews is often identified, having been transplanted from palaeontology to archaeology, as the inspiration for the fictional movie hero Indiana Jones.

Remarkable finds
The Gobi rocks explored by the American expedition were far too old for human remains. Some dated to the Oligocene Epoch, 30–25 million years ago. Others were even more ancient, having formed in the late Cretaceous Period, some 70 million years ago, towards the end of the Age of Dinosaurs. However the expedition did

Above: Cambrian sponge fossil found in Archaeocyathid limestone, South Australia.

score some tremendous finds. One was the first known nest of dinosaur eggs, associated with the pig-sized, plant-eating horned dinosaur *Protoceratops*. One of the richest fossil sites in the Gobi became known as the Flaming Cliffs and continues to reveal major finds. Another exceptional find at Hsanda Gol was the remains of what is still the largest known mammal to have dwelt on land, *Paraceratherium* (formerly called *Indricotherium* and *Baluchitherium*). This vast Oligocene beast stood almost five metres tall at the shoulder and weighed 20 tonnes.

Scales and feathers
Some of the most astonishing fossil finds in recent years are from China's north-eastern province of Liaoning. Many are from the Early to Middle Cretaceous Period, 130 to 90 million years ago. They include some of the strangest dinosaurs ever found, and an incredible diversity of birds. Many of these creatures were covered in feathers, including some of the dinosaurs. The small predatory dinosaur *Microraptor gui* had feathers on all four limbs – the only known animal to exhibit such a feature. Exceptional preservation conditions in fine-grained rocks have helped to reveal the secrets of how and why feathers evolved.

Below: Dinosaur egg fossil discovered in Mongolia – the first such egg to be found.

More surprises
In addition, Liaoning has yielded fossils of mammals larger than any others dating from the Cretaceous Period. It was long believed that no mammal from the Mesozoic Era, the Age of Dinosaurs, was much larger than a rat. *Repenomamus gigantus* was dog-sized, resembled a small bear, and weighed in at about 15 kilograms, thereby overturning the accepted idea. Fossils of its cousin, the possum-sized *Repenomamus robustus*, reveal the remains of a small young dinosaur called a psittacosaur inside its body region – presumably its last supper.

Ediacara
In 1946–47 Australian geologist Reg Sprigg was surveying the Flinders Range of mountains in South Australia, seeking new deposits of the valuable metal ore, uranium. In the Ediacara Hills, Sprigg began to dig up strange fossils, the like of which had never been seen before. They dated back to Precambrian times, some 575–560 million years ago. They were soft-bodied life-forms, so unlike anything else in the fossil record, or alive today, that it's still not clear if some are plants or animals. Only a chance event meant they were buried rapidly and preserved swiftly, and then survived more than half a billion years to the present. These organisms, from a time before limbs or shells or fins, are now known by the general name of Vendian life. In contrast, the Riversleigh Quarries of Queensland have only been studied since the 1980s. They contain one of the world's densest concentrations of fossils, from the Late Tertiary Period, including kangaroos and marsupial predators resembling big cats.

Right: Map of Asia and Australasia, showing some of the major fossil discovery sites.

1. Hsanda Gol (Shand Gol),
   Mongolia
2. Nemegt Basin, Mongolia
3. Beipiao, Liaoning, China
4. Chaoyang, Liaoning, China
5. Umrer, India
6. Lufeng, China
7. Guangxi, China
8. Riversleigh Quarries, QLD,
   Australia
9. Muttaburra, QLD, Australia
10. Roma, QLD, Australia
11. Ediacara Hills, SA, Australia
12. Lightning Ridge, NSW, Australia
13. Dinosaur Cove, VIC, Australia
14. Mangahouanga, New Zealand

WHAT BECOMES A FOSSIL?

In theory, almost anything that was once alive could become a fossil – even a scrap of the jelly-like tissue from the soft body of a sea creature. In practice, most fossils are formed from hard parts of animals and plants, such as teeth, bones and shells, or tree bark and cones.

There are a number of popular misconceptions about what makes a fossil. Some people automatically think of dinosaurs, and ignore everything else. At the other end of the spectrum, the term 'fossil' once referred to any natural mineral object like a stone, rock or pebble. Today the accepted definition of a fossil involves an organic origin – an object that was once living. This includes all organisms or life-forms and parts thereof, from microbial organisms, tiny plants and

Above: The formidable killing claw of Dromaeosaurus albertensis was carried on the second toe of each foot.

Above: A Tyrannosaurus tooth dating from the Cretaceous Period, recovered near the Red Deer River, Alberta, Canada.

insects such as ants and gnats, to the greatest trees, sharks, dinosaurs, whales and mammoths.

Disappeared without trace

When most living things perish, they begin to rot and decay. Their remains may be decomposed by fungi and bacteria, and torn up, burrowed into or crunched to pieces by all manner of scavengers, from maggots to hyaenas. What is left weathers into fragments by the action of sun, rain, wind, ice and other elements of weather. In the water similar decay occurs, aided by waves and water currents, and the decomposing actions of fungi, worms, bacteria and others. So the vast majority of living things die and disappear without trace. They are recycled by natural processes back into the ground, lake or river bottom, or sea bed. They leave no fossils.

Above: In modern terminology, all 'fossils' have originated from once living matter.

Below: The scales, fins and tail of this fish, Notaeus laticaudatus, have been particularly well preserved in marl.

The harder, the better

Fossilization usually takes a long, long time, and is a chance-ridden process. As a result, usually only the harder parts of living things are preserved as fossils. These are the parts that resist decay, rot and scavenging, and which persist long enough for preservation to begin. Again, popular myth says that only bones form fossils, and perhaps teeth too. But there is a long list of other parts which are prime candidates for fossilization, as follows.

Plant parts

• The ribs or veins of plant leaves. (These are sometimes seen in winter, when the fleshy areas of the leaf have disintegrated to leave a 'skeleton' of harder, woodier veins.)
• The bark and wood of trees and other tall plants with trunk-like main stems.
• Roots, which are often among the hardest parts of a plant, and are usually underground, away from rotters and scavengers.
• Plant seeds, many of which are designed to resist harsh conditions such as winter, and then spring to life as they germinate when their surroundings improve.
• The woody cones of trees such as firs, pines and spruces.
• The thin, stiff, needle-like leaves of these various conifer trees.
• Pollen grains, which are tiny but tough, and resistant to most physical damage. Many are microscopic but

Above: The fossilized shells of bivalves.

they are produced in countless quantities, which improves the chances that some will be preserved.

Animal parts

• Shells or hard body cases, especially of molluscs such as snails and in the past, ammonoids and nautiloids. This includes the outer body casings or exoskeletons of insects, although these are usually only thick and resistant enough in certain examples, such as the hardened wing-cases, known as elytra, of beetles.

Spines

This very variable category of parts that might fossilize includes any long, narrow, sharp body projections. Spines occur on creatures as distantly related as sea urchins, spiny sharks (ancient fish), hedgehogs and porcupines. Often they fall from the main body after death and are preserved singly or in jumbled heaps.

Below: Whereas the spines of smaller creatures such as sea urchins may be relatively small, others – such as the ancient spiny shark Ctenacanthus – boasted pretty sizeable protrusions. The specimen below is a preserved spine (or ray) from one of Ctenacanthus's enormous fins, and it measures some 19cm (7.6in) across.

• Bones. Only the vertebrate or backboned animals possess true bones, which make up a skeleton. But then, not all vertebrates have bones – a shark's skeleton is made of cartilage or gristle, which is softer, bendier and more likely than bone to disintegrate.
• Scales, especially of fish and certain types of reptiles. In particular the large, hardened, bony scales, known as scutes, which covered certain dinosaurs, crocodiles, and fish such as sturgeons, are usually good candidates for fossilization.
• Teeth. These are possessed by most vertebrates, and certain other creatures like sea urchins also have tooth-like structures. For instance, in some snail-like molluscs the tongue is known as the radula and has hundreds of micro-teeth for scraping and rasping food. At the other end of the size range, the tusks of mammoths and other elephants were giant overgrown teeth.
• Horns. Usually these have a bony inner core covered with the hard, tough but light substance which itself is called horn (or keratin – see below). Various prehistoric mammals such as antelopes, cattle and rhinos had horns. There were also horned dinosaurs, or ceratopsians, such as *Triceratops*.
• Antlers, which are possessed chiefly by male deer and which are different from horns. They are formed of skin-covered bone and are shed each year to regrow the next, unlike horns, which are generally permanent and grow through life.

Above: Unearthing the remains of a 40 million-year-old giant whale, now in the desert at the Wadi Hitan 'Whale Valley' reserve, 150km/ nearly 100 miles south-west of Cairo, Egypt.

• Claws, nails and hooves, which are all formed from the same horny material as horns, known as keratin. Cats and birds of prey have sharp claws, the hoofed mammals or ungulates possess hoof-tipped toes, and most monkeys and apes have flattened nails similar to our own.

Above: The veins of a leaf Delesserites from the Palaeocene Epoch.

Below: Fossilized animal bones in a cave.

HOW FOSSILS FORM

Fossilization is a very variable process. It depends not only on the physical nature and chemical composition of the parts being preserved, but also on the nature of the sediments which bury them, water content, temperature, pressure, mineral availability and many other factors.

Fossils form in several ways. The 'normal' method is described here, and variations are included over the following pages, according to different habitats and conditions. The standard process occurs when a whole organism – or more often, parts of it – are buried. This usually happens in the sea, when a dead animal sinks to the bottom. It is covered by mud, silt or other small bits known as sedimentary particles. These are washed along by water currents or float down from above as the fine 'rain' of tiny fragments that settles continuously on the ocean floor. In general, the faster this initial burial, the better the state of preservation will be, since the animal or plant has less time subjected to decay, scavenging and disintegration.

Below: How a fossil forms, stage by stage.

Sediments and minerals

More sediments build up on top of the buried parts. This increases the thickness of the layers and the depth to which the parts are covered, and so increases the pressure on them. Gradually the pressure and squeezing begin to force together the particles, compressing the sediments harder and harder. Gradually, too, water seeps or percolates through these sediments, carrying with it natural minerals. These minerals slowly replace the organic substances or materials in the buried parts. The ongoing process of mineral replacement is known as permineralization.

Turned to stone

The original parts may be squashed, distorted and shattered during this phase, in which case the fossils are destroyed. In other cases the parts

Above: Soft-bodied, brittle-shelled creatures such as trilobites are often found fossilized in a squashed shape. These, of the genus Angelina, were recovered from the Late Cambrian Tremadoc Beds of Merioneths, Wales.

retain their original size and shape, and even their detailed inner organization, as their organic molecules and microstructures are replaced, one by one, by the inorganic minerals of rock. If the process continues to completion, then the original parts become completely

Stage 1: A hard part of a once-living thing, here symbolized by a single animal bone, resists weathering, scavenging and decay.

Stage 2: The part is washed into a river and sedimentary particles such as sand, mud or silt slowly settle and cover it.

Stage 3: Over time, more sedimentary layers collect on top. The part and its layer are squashed by the increasing weight.

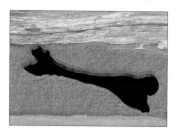

Stage 4: Minerals in the original part are dissolved and washed away, leaving an empty space or hollow chamber – a mould.

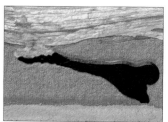

Stage 5: The space is filled by further minerals which trickle or percolate in from the surrounding rocks, forming a cast.

Stage 6: Earth movements, erosion and other processes gradually remove the upper layers, so that the fossil is eventually exposed.

Above: Rock strata are beautiful exposed in the iconic, partially eroded sedimentary landscape of western North America's canyons. The geology of these canyons is rich in the fossilized remnants of plant life and vertebrate animals.

inorganic or rocky. This is called petrification and is summarized in the handy but limited phrase 'bone to stone'. Meanwhile the rock around has been buried and compressed, and its grains and particles are cemented together by natural minerals too. As a result, the original parts of the plant or animal have become rock encased in rock.

Rocks that contain fossils

Because of the way fossils form, they are found only in certain types of rocks, known as the sedimentary group because they began as layers of sedimentary particles. Sedimentary rocks include sandstones, limestones, siltstones, mudstones and shales. In addition to sedimentary rocks, there are two other major groups of rocks: igneous and metamorphic. Igneous rocks form when original rocky material melts under great pressure and temperature and then cools again. Molten lava that flows from volcanoes hardens into igneous rock. The original rocky material may be sedimentary rock containing fossils, but melting in the igneous phase destroys them. Likewise, metamorphic rocks form when original rocky material is transformed by tremendous pressure, heat and chemical changes, but without actually melting. Again, if the original material had fossils, these are usually destroyed during the metamorphosis. The practical result of all this means that it's no use looking for fossils in igneous rocks, like basalts, granites or rhyolites, or in metamorphic rocks, like marbles, schists or gneisses. Usually, only sedimentary rocks will do, as shown on geological maps.

Moulds and casts

On occasion, as sedimentary rock forms around a piece of an animal or plant, the percolating mineral-laden water dissolves the part and washes it away. The result is a hole, space or chamber in the rock, of the same size and shape as the original part. This is a 'negative imprint' or impression, which paleontologists often refer to as a mould fossil. At a later time, this space may itself be filled by more minerals seeping in from the water and rock around. Now the result is a cast fossil, again of the same size and shape as the original organism part, but composed of a different selection of minerals and crystals.

Tiny part of a tiny part

It should be emphasized that at any time, in any of these processes, pressure and temperature and general earth movements can warp, twist, fragment and destroy fossils. Only a tiny proportion of plants and animals begin the preservation process. Only a tiny proportion of these end up as still-recognizable fossils. And only a tiny proportion of these fossils actually see the light of day, when we find them – the rest lie deep and undisturbed.

Trace fossils

Fossils which are not preserved actual parts of an animal or plant, but objects or signs made by them, or left by them, are known as trace fossils. Examples are footprints, burrows, nests, feeding marks and tail-drags. Perhaps some of the most unlikely are coprolites – fossilized droppings, lumps of dung and similar animal excreta. In their original state, coprolites would be squishy and smelly. But the preservation process turns them into hard objects made of rocky minerals.

Below: Fossilized fish coprolites originating from the Triassic Period.

FOSSILS IN THE SEA

Seas have covered most of the Earth's surface for most of its history. Sediments may be washed from land into the sea, where they settle on the bottom. The result is that the majority of fossils are from sea creatures, plus a few marine plants too.

It may seem strange that fossils found on high ground, such as the Burgess Shales in the Rocky Mountains, or the Vendian creatures of Australia's Ediacara Hills, came originally from life at the bottom of the sea. The answer lies in 'dynamic Earth' and its geology. The hard outer shell or crust of our planet is never still. Some of its activities are sudden and awesome, like a volcanic eruption, or an earthquake that sets off landslides, or an underwater earthquake that triggers tsumanis (giant fast-moving waves). But many other activities are too slow for us to see. Huge jagged sections or plates of the Earth's crust slide and drift, push into each other, or grind

under or over their neighbours. Where plates collide, the rocks fold and buckle to build mountains. The forces of erosion such as sun, wind, rain and ice counteract by wearing away rocks. These changes may occur at the rate of just a few millimetres each year. But they have continued over the vast span of geological time.

Ancient seas on dry land

As a result rocks are continually on the move, forming and drifting, distorting and twisting, and being worn away or destroyed by the forces of erosion. This is how sediments that were once on the sea bed may be uplifted into mountains, and then eroded to expose their fossils. The layers or strata of sedimentary rocks always form horizontally, with level surfaces as particles settle under the force of gravity. But later earth movements can ripple and fold these rock layers like

*Above: These 'chain coral' of the Silurian period (*Halysites* genus) are shaped like the links in a bicycle chain.*

paper, and even twist and turn them upside down. Usually the youngest or most recently formed sedimentary rocks are nearer the surface, so that as we excavate down to deeper strata, we 'dig back in time'. Inverted layers can confusingly reverse this sequence.

The most fossils

As previously described, conditions for fossilization are most likely in the seas, because of the way sedimentary rocks form. Also, compared to land, seas and oceans have always covered much more of our planet. (Today, the proportion is about 71% for seas and oceans.) Both these factors mean that the vast majority of fossils are from sea creatures – especially hard-shelled ones. Indeed, some sedimentary rocks are almost entirely composed of fossils. They began as piles of dead shelly sea organisms. The fossil remains may be recognizable items or fragments of the organisms, such as the shells of limpets and sea-snails. These rocks are sometimes known as fossiliferous limestones. There are three main types:
• Biohermal limestones – formed by living things growing in situ, such as the tiny stony cup-like skeletons of corals, or accumulations of creatures anchored to the sea bed, such as crinoids (sea-lilies) or mussels.

Left: Parkinsonia *ammonite recovered from Dorset, UK.*

Below: A stunning example of fossiliferous limestone rich in the remains of ammonites.

Above: Shells are usually smashed by wave action, but in sheltered bays they are often more likely to be buried intact.

Right: Very muddy seashores may discourage scavengers, which also boosts the chances of marine life being preserved.

• Biostromal limestones – mainly shells and other fragments of dead animals washed about and generally piled up, such as brachiopods, trilobites, oysters and other bivalve molluscs, and sea-snails like whelks.

Ediacara mudslide

The amazing fossils of Ediacara, South Australia, are from soft-bodied organisms, which usually rot away far too quickly for preservation. Scientists have concluded that this Precambrian community must have been surprised by a sudden mudslide, perhaps triggered by an earth tremor. Suffocating mud covered the area so fast, and there was no escape for its inhabitants. The blanket of mud kept away scavengers, and the low-oxygen conditions within it prevented the action of bacteria and other microbes which normally cause soft-tissue decomposition.

Below: The puzzling fossil Mawsonites *was one of those recovered from the Ediacara Hills, northern Flinders range.*

• Pelagic limestones – mainly tiny floating life-forms whose individual shells are often too small to make out. These die in their billions and sink to the sea bed, forming the greater proportion of ocean-floor ooze, which gradually hardens into sedimentary rock. As a result, pelagic fossiliferous limestones usually have a smoother, fine-grained appearance compared to the lumpy texture of the other types.

Bigger sea creatures

Most fossils are bits and pieces of organisms. Rarely a plant or animal is preserved whole, with its parts still together, in the positions they were in life. For a vertebrate animal like a fish, this is known as an articulated specimen. Stunning articulated fossils of fish show us their exact size and shape when alive. Sometimes even the positions of their fins are visible. Even the remains of what they last ate may be preserved in the stomach and gut region. In general, the bigger the organism, the more likely it is to come apart during preservation. So complete articulated skeletons of sea reptiles like ichthyosaurs and plesiosaurs, perhaps metres in length, are truly exceptional finds. It's likely that the original animals died in calm waters and their carcasses settled quietly on the sea bed, to be covered quickly but gently by sediments.

Below: Whole specimens such as this spiny-finned fish Diplomystus dentatus, *recovered from Wyoming, USA, and dating to either the Palaeocene or Eocene Epochs, can tell us much about the habits of the creature. This lake species was a relative of the herrings, and its upturned mouth suggests that it fed on insects at the water's surface.*

FOSSILS IN SWAMPS

Swamps, marshes and bogs are good preservers. Their still waters are usually low in oxygen, which inhibits microscopic decay by organisms such as bacteria. Also, mud and quicksand are excellent traps for unwary animals, which are 'swallowed' whole by the mire, ripe for preservation.

At times in the Earth's history, more than half the world's land area was freshwater wetlands. Mosaics of swamps, marshes and bogs were interspersed with pools, creeks and islands of drier land, constantly changing their pattern and shape as water levels fluctuated with the seasons and passing centuries. Perhaps the best known time for such conditions was the Carboniferous Period, 355 to 295 million years ago. The climate was also much warmer then. Tropical floodplains were home to vast forests of lush vegetation, where huge insects dwelt, and giant amphibians lurked in the steamy dark pools.

Dominant plants

There were no broad-leaved trees in Carboniferous times, as there are in tropical jungles today. The dominant Carboniferous plants were huge clubmosses and horsetails. Both of these groups still survive, but now they are much smaller in size and numbers. The enormous Carboniferous clubmoss *Lepidodendron* grew to 40m/130ft – as

Above: Coal being mined in China. This rapidly industrializing nation depends on fossilized carboniferous plants for much of its energy.

tall as the typical modern rainforest tree. Fossils of *Lepidodendron* are common worldwide and often show the characteristic 'scaly' bark with its kite- or diamond-shaped old leaf attachments. One of the largest horsetails was *Calamites*, at 15 and sometimes 20m/50–65ft tall, also with a woody strengthened stem. Its fossils are found mainly in Europe and North America.

Above: One of numerous fern impressions recovered from the Late Carboniferous Coal Measures, Illinois, USA.

Below: The 'coal swamps' formed during the Carboniferous supported giant insects and early tetrapods (four-legged vertebrates).

Above: Swamps remain productive habitats today. Since Mesozoic times, alligators have been among their most dominant carnivores – as here, in the Cypress National Reserve, USA.

The coal forests

These vast green swamps of clubmosses, horsetails, ferns, seed-ferns, and the newer groups of coniferous trees and cycads, produced vast amounts of vegetative growth during the Carboniferous. As the plants lived and died, their remains toppled and sank into the shallow bogs and marshes. Decomposition was slow, since warm water is low in oxygen, and the organisms of decay soon used up what was left of the oxygen for their life processes. So the slow-rotting plant matter piled up thickly, and was compressed under the weight of more layers above. Today's results of these 'Carboniferous coal forests' are various forms of coal and similar fossils fuels. They took tens of millions of years to form. We mine them with eye-blinking speed. When we burn these fuels, we are actually releasing the Sun's light energy, which those plants trapped more than 300 million years ago.

Giants of the swamps

Lumps of coal sometimes split open to reveal patterns of leaves such as ferns and clubmosses, and occasionally, animal remains. These were from creatures trapped in the semi-rotting vegetation and turned to fossils. The Carboniferous swamps were home to the biggest flying insect ever, the dragonfly *Meganeura*, with a tip-to-tip wingspan of 70cm/28in. It was probably snapped up occasionally by a new and enlarging group of four-legged, land-dwelling 'amphibians' such as temnospondyls and anthracosaurs. The fossils of these are also numerous in some regions. A well-known temnospondyl was *Eryops* of North America. At 2m/7ft long, it was one of the early large predators, resembling a huge and fierce salamander. Also in the Carboniferous, the first reptiles appeared as small, slim creatures outwardly resembling modern lizards.

Very special fossils

In rare cases, the fleshy body parts or soft tissues of a living thing are preserved as carbon-film fossils. The organic molecules of the ex-living tissue decay very slowly in an unusual way, into a thin dark film of carbon-rich substances. The resulting fossils themselves may look like a smear of oil or a scrape of soot in the rock – and that is approximately what they are. Carbon-film preservation can show the outlines of whole soft-bodied creatures, and the shapes of tentacles, flaps and fins around harder fossilized parts of animals such as fish, amphibians and reptiles.

Below: A carbon film impression of a Redwood tree leaf (Sequoia), from the natural history archives of Humboldt State University, California, USA.

Slow decay

The same processes that occurred most abundantly during the Carboniferous Period have happened in other times and places, and continue today. The Jurassic Period, in the middle of the Mesozoic Era, also saw a warm, damp climate and luxuriant plant growth, which fed the enormous dinosaurs of the time. Their legacy is some of the largest single fossils ever produced – huge limb bones and vertebrae (backbones) bigger than a person.

After the mass extinction at the end of the Cretaceous Period, 65 million years ago, the first phase of the Tertiary Period – the Palaeocene Epoch – was another span of moist and temperate conditions. This may have been one factor in the burst of evolution which saw new kinds of mammals and tall, flightless, predatory birds rise to prominence on land, in place of the once-ruling dinosaur-like reptiles.

Left: A Lyrocephaliscus euri frog skull recovered from Spitsbergen, in the Arctic Circle – a land once part of the tropics and rich in swamp life and lush vegetation.

FOSSILS IN FRESHWATER

Rivers and lakes support a variety of freshwater and semiaquatic life, ranging from fish to snails, worms, amphibians and reptiles. Yet they are also magnets for thirsty land animals. Heavy rain and floods wash remains from the land into these bodies of water, making preservation more likely.

It's a quiet day in the prehistoric world. Creatures have come down the steep riverbank to drink, bathe and cool down in the placid waters. Suddenly the river surges – a flash flood from torrential rains upstream crashes around the bend and sweeps all before it. The fitter, more agile animals can leap up the steep bank to safety. The young, old, sick and infirm have no chance, and the rushing torrent soon carries away their struggling bodies. Hours later, when the main flood has eased, the carcasses lap onto a sandbank downstream. Scavengers gather and the rot sets. A few days later, another flash flood covers the partly bared bones with sandy sediments. The process of preservation begins – and we dig up the resulting fossils millions of years later.

The science of taphonomy

It's exciting to speculate how ancient animals and plants met their ends, especially when large groups of remains are all preserved together. This applies through all eras and periods of geological time. Land animals need to drink, and so pools and rivers have always acted as magnets for wildlife. How did a herd of prehistoric mammoths or a flock of ancient birds all end up in the same place, entombed together? The science of taphonomy delves into the gap between death and rediscovery. It examines the processes by which dead organisms become fossilized, and how the circumstances, surroundings and other chance events all contribute to the groups or assemblages of fossils which we eventually uncover.

Above: Stegoceras *dinosaur vertebra of the Pale Beds, Alberta, Canada.*

Below: Freshwater habitats have supported animal life for millions of years. In some freshwater habitats, the remains of creatures were preserved in the base rock beneath the substratum, or lake bed.

Above: An amalgamation of pond snail fossils (Paludina) preserved in rock.

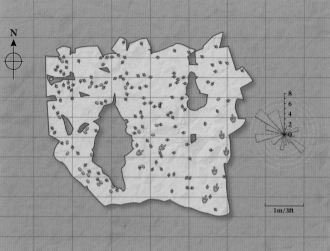

Prints on the bank

Footprints or paw tracks have created some of the most valuable trace fossils. They often occurred after animals had walked along riverbanks or lakesides, or across floodplains. The creatures made impressions with their feet, paws, hooves, claws, nails, tails and other parts, in the damp, soft clay, silt or mud. The indents were then dried and baked hard in the hot Sun. As the rains returned, another flood brought further sediments which filled the indentations, and so the fossilization process could begin. Many trackways of larger animals, such as dinosaurs, mammal-like reptiles and mammals, have been uncovered. In an area near Winton in Queensland, Australia, some 130 dinosaurs have left more than 3,200 prints in one rocky slab as they moved swiftly across the territory.

• The spacing of the prints (stride length) can suggest walking and running speeds.

• Print depth related to the firmness of the underlying sand or mud, and speed of movement, give valuable clues to body weight.

• Prints of many animals of the same kind and size, jumbled but all facing the same way, indicate a loose herd or family group on the move.

• Prints of the same kind but different sizes usually show a mixed-age group, with youngsters and adults.

The matrix

Close examination of fossils and the rocky material in which they are encased, called the matrix, reveals many clues. If the surrounding particles of sediments in the matrix are all aligned in a similar direction, and the fossils are a dense amalgamation of remains from just one kind of animal or plant, this may be due to flash flooding and powerful currents which swept away and drowned a herd or group. More jumbled and randomly oriented particles, with a mixed assemblage of fossils from various animals, could indicate a slower, more gradual process of accumulation.

Freshwater life

In lakes and rivers, fossils usually formed as described previously, when remains were covered by sediments. There are fossils of freshwater fish, amphibians such as frogs and salamanders, and aquatic insects like diving beetles and dragonfly nymphs. As usual, hard-shelled creatures are among the most numerous fossils, such as pond snails and freshwater mussels. Apart from the details of the animal remains themselves, plant fossils preserved with them can help to distinguish between the freshwater of inland sites and the saltwater habitat of seas and oceans. Preserved stems of

Above: Plotting footprints onto a map offers a clearer perspective of the movements of the animal, and points towards behaviour – perhaps that it is seeking food or trying to escape from a predator.

reeds and rushes, leaf 'skeletons', and hard seeds and cones all indicate freshwater origin. Flat, lobed blades or laminae, and grasping finger-like holdfasts, are distinguishing features of seaweeds.

Below: Fossilized footprints often tell us whether animals travelled in groups, such as this herd of sauropod dinosaurs.

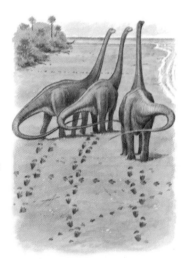

FOSSILS ON LAND

Most terrestrial (land-living) animals and plants which have left fossils were actually preserved in water – usually along the banks of rivers and lakes, or on seashores. Windblown sand or similar particles can also act as sediment to bury organisms in arid habitats.

Examples abound of creatures such as dinosaurs or prehistoric mammals being caught in a flash flood, drowned and buried under sediments on the river bed or lake bottom. Fossilization on dry land, without the role of water in any form, is rare by comparison. However, preservation may take place if the conditions are right.

Mummification

In very dry conditions, an animal carcass or dead plant may lose all its natural moisture and body fluids quite quickly, before decay can progress far. The micro-organisms and chemical processes that cause decomposition rely on the presence of moisture. So, once dried out, the remains may not disintegrate any further, or at least, do so very slowly. Plant scientists and florists use this technique for pressing flowers or making dried versions. This severe drying out of organisms is known as mummification.

Mummification can happen in deserts and drought-ridden conditions, and also in very dry caves. It's usually necessary for the remains to have some kind of physical protection against

Above: The dry, windswept, slowly-eroded Gobi Desert is a rich source of fossils dating especially from the Late Cretaceous Period, such as these dinosaur eggs.

wind, large swings in temperature, scavengers, moist air carrying microbes, and other agents which would cause them to degrade. This may happen when a dead animal is covered by sand in a desert storm, or is lost in a deep cave or tunnel.

A prelude to fossilization

Mummified remains of animals which became extinct relatively recently, like moas, mammoths, giant sloths and cave bears, are well known. However,

Above: The part-mummified flipper of a crabeater seal, preserved in the Dry Valleys of the Transantarctic Mountains, Antartica – a continent which is largely an 'ice desert'!

these specimens would not survive the rigours of geological time without further preservation. Mummification is usually seen as a helpful prelude to proper fossilization, with its entombment by sediments and mineral replacement. In such cases the initial drying out can help to preserve soft tissues such as skin and other features that would normally be lost. The giant tyrannosaur known as 'Sue' was partly mummified before true fossilization took place. So were some of the

Left: The remnants of plants, trees and even animal carcasses may be subject to limited decomposition, or 'mummification', in very dry or barren conditions.

Below: Petrified wood from a coniferous forest of Arizona, USA, from the Triassic Period.

Above: Insect droppings preserved in Dominican amber, from the Late Miocene.

Trapped in tar

Here and there, natural deposits of tar or asphalt ooze to the surface and form pools. If these are covered by rain water, they may look just like waterholes, inviting animals to come and drink. But as soon as the thirsty creatures wade in, they become stuck and start to sink, gripped by the thick, gooey tar. Their struggles may attract predators, who try to approach – and they too are trapped. The animals sink and are preserved by the chemicals in the tar, often in great detail. One of the best-known sites for tar pit preservation is La Brea, in the suburbs of Los Angeles, USA. More than 600 species were preserved from 40,000 to 10,000 years ago, in a cluster of about 100 tar pits. They range from mammoths, mastodons, dire wolves, sabre-toothed cats, bison, giant sloths and short-faced bears to rabbits, rats, birds, lizards, insects and various plants.

Below: The Page Museum at La Brea holds an annual excavation of specimens, where visitors can observe professionals, and volunteers like the two people shown, recovering the bones of animals that drowned in the ancient tar pits here.

remains of dinosaurs such as *Protoceratops*, and their nests and eggs, which were preserved in the dry scrub of what is now the Gobi Desert, some 70 million years ago.

In the freezer

Another temporary mode of preservation is deep-freezing. Some of the best-known examples are mammoths, woolly rhinoceroses, giant deer and other large mammals trapped in the snow and ice of the far north – the tundra and permafrost across Northern Europe and Asia. Specimens are regularly exposed from the ice, especially by mining and exploration for gas, oil and mineral ore deposits across the vast wastes of Siberia. But as soon as the thaw sets in, so does decomposition. It's necessary to melt the remains very slowly and treat with chemical preservatives at once, in an ongoing process, to have any hope of stabilizing them over the long term.

Amber

This substance is the hardened, fossilized resin, gum or sap of ancient plants, especially conifer trees such as pines, firs and spruces. The trees make this resin partly as a form of sealant to cover any cracks, breaks, gashes or 'wounds' in their bark – just as our blood clots to seal a cut in the skin. Sometimes smaller creatures passing by, such as flies and other insects,

Above: 'Dima' the baby mammoth was found in permafrost on the banks of the Kolyma River, Siberia, in 1977. She was believed to be between 6–8 months old at the time of her death, some 40,000 years ago.

spiders, centipedes, even small frogs and lizards and mammals, become stuck in the adhesive, glutinous secretions. So do plant items like windblown pollen, seeds, petals and fragments of leaves or twigs. If the resin or sap continues to ooze fairly rapidly, it may cover and encase the specimen, preserving every tiny detail. Amber contains some of the most beautiful and spectacular of all fossils, giving valuable insight into the evolution of small and delicate creatures that are hardly ever preserved in the usual way. A few specimens of amber date as far back as the Early Cretaceous, but most are from the Tertiary Period. Famous deposits occur in the Dominican Republic, Caribbean and the Baltic region of Europe.

Below: Cold and frozen conditions drastically slow decay and allow time for preservation.

GREAT FOSSIL-HUNTERS

In 1770 Dutch chalk miners unearthed what looked like the huge skull, jaws and teeth of a weird and wonderful creature. It was nicknamed the 'Meuse monster', after the region in which it was discovered. The great find prompted a heated debate about the creature's origins, and captured the public's imagination.

Theories about the Meuse monster's origins prompted more and more speculation. Were there more 'Meuse monsters' still alive in the depths of the sea? Was the beast a victim of a biblical catastrophe? Was the find a fake, put there by pranksters – or even by the Almighty, in order to test people's religious faith?

Monsters and myths

The 'Meuse monster' fossil made headline news around Europe. All of a sudden, ordinary folk began to imagine an exciting long-gone world populated by huge beasts. The specimen was examined by Georges Cuvier (1769–1832), who acknowledged that its kind were probably long extinct, and who named it *Mosasaurus*. Cuvier was the most famous and respected biologist of the time. He began the work of studying, naming and classifying fossils according to the same principles used for living things – a novel and slightly heretical idea at the time. Cuvier was based at the Paris Museum of Natural History, which was then the biggest scientific organization of its kind in the world. His specimens included many fossil reptiles and mammals from the Tertiary rocks of the Paris Basin.

The country doctor

In the 1830s tales began to surface about more great beasts from long ago, which had left fossil remains such as teeth, jaws and bones. Gideon

Above: Dr Gideon Algernon Mantell was a country doctor as well as a palaeontologist. However, his greatest 'discovery', the teeth of the dinosaur Iguanodon *(see below), was probably made by his wife, Mary Ann.*

Mantell (1790–1852) was a doctor from Sussex, England. He consulted Cuvier about some teeth he had found, about 1820, which resembled the teeth of a living iguana lizard but which were much larger. Mantell was a keen collector of fossils, and travelled widely to examine and obtain specimens. He also had a home museum where he proudly displayed his best finds. In 1825 Mantell decided to call his new find *Iguanodon*, meaning 'iguana tooth'. He imagined it as a massive, sprawling, plant-eating lizard.

Above: Sir Richard Owen was responsible for the naming of the fossil group Dinosauria. He is pictured here holding the enormous leg bones of a moa, a flightless New Zealand bird that was ultimately hunted to extinction.

Fossils become fashion

In 1842, Richard Owen (1804–92) produced a report on fossils of great reptiles. Owen was an expert anatomist and would become Superintendent of the British Museum of Natural History – now the Natural History Museum, London. His report suggested that certain reptiles known from fossils should be included in a new group, the Dinosauria, 'terrible lizards'. Its members included Mantell's plant-eating *Iguanodon*, and also *Megalosaurus*. This was named in 1824 from a fossil jawbone with large sharp teeth, by William Buckland (1784–1856), Professor of Geology at Oxford University.

The general public became more interested in dinosaurs as giant fearsome beasts from long, long ago. Their fascination increased sharply when Owen joined forces with sculptor Waterhouse Hawkins to produce life-sized models of the great creatures.

Above: British doctor Gideon Algernon Mantell named one of the first dinosaurs, Iguanodon. *Mantell's fossils consisted only of teeth, and it was more than 50 years before complete skeletons were found in Belgium. The creature is now believed to have been a four-footed herbivore.*

Above: The French palaeontologist and anatomist Georges Cuvier did much to propagate the principles of functional comparative anatomy, and courted controversy by classifying fossils by criteria usually reserved for living things.

These were displayed in the gardens of the Crystal Palace exhibition centre in Sydenham, south-east London from about 1854. People travelled from all around to gather open-mouthed around the reconstructions. Dinosaurs hit the news and fossil-hunting became the latest fashionable pastime among gentlefolk of the period.

Below: The decision to mount life-sized dinosaur models at Crystal Palace, south-east London, was due in part to a surge of public interest in prehistoric creatures.

Fossils go west

During the mid-1800s 'fossil fever' spread to North America. In 1868, life-sized models of dinosaurs such as *Hadrosaurus* were displayed in New York's Central Park. By the 1870s Cope and Marsh (see Major Fossil Sites – The Americas) were racing to see who could recover the most fossils. In the wake of the hunt for dinosaur remains, many other kinds of fossils were also found, studied and named. Some of the greatest contributors to the breadth of information that resulted are as follows.

• **Richard Lydekker** (1849–1915) published a vast 10-volume catalogue of fossils at the British Museum of Natural History (now The Natural History Museum), in 1891. His book *A Manual of Palaeontology* (1889) was for many years the 'bible' of fossil collectors and restorers.

• **Barnum Brown** (1879–1968) was perhaps the greatest collector of fossils in the early 20th century. Working for the American Museum of Natural History, he led explorations to the Red Deer River region of Alberta, Canada from about 1910 to 1915, and further excavations at Howe Ranch, Wyoming in the 1930s. He is perhaps best remembered for finding the first *Tyrannosaurus rex* specimen in 1902.

• **Eugene Dubois** (1858–1940), a Dutch anatomist, shook the world when he

Above: Richard Lydekker's cataloguing of specimens for A Manual of Palaeontology, published in the late 19th century, created an inspirational resource for fossil-hunters seeking a means of identifying and categorizing their finds.

announced his discovery of 'missing link' fossils in Trinil, Java, in 1891. He named the remains as 'upright ape man' *Pithecanthropus erectus*. Today these fossils are included in the profile of the species *Homo erectus*.

The Sternbergs

As families of fossil enthusiasts go, there are few to match the Sternbergs of North America. They were father Charles H (1850–1943) and sons George, Charles M and Levi Sternberg. As well as actively collecting specimens, they assisted both the science and hobby by devising many on-site fossil-hunting techniques, such as encasing fragile specimens in protective jackets or casts, as well as ways of copying fossils using latex rubber.

Below: Charles Sternberg Senior reflects on some of his collected fossils. He was one of the USA's most pioneering palaeontologists and fossil-hunters.

GREAT FOSSIL-HUNTERS (CONTINUED)

If the previous centuries saw the rise of palaeontology as a charismatic science, modern times have certainly kept the pace. During the 20th century, some staggering new finds reshaped many assumptions about the prehistoric world, and also marked a greater involvement for the amateur fossil sleuth.

During the 20th century, fossil-hunting and collecting spread around the world as a popular pastime for many and a profitable profession for a few. Some dipped into it as a hobby, while others retained a lifelong interest. And some known for contributions in other fields have also held a fascination for fossils.

Little-known fossil sleuth

Marie Stopes (1880–1958) is best known as a pioneer and writer on sex education and women's health issues. From 1921 she established Britain's first family planning or 'birth control' clinics. But in her early career Marie studied plants and fossils. In 1905 she became the youngest DSc (Doctor of Science) in Britain, and first female science lecturer at Manchester University. Marie studied cycads and other ancient plants and showed how their remains could decompose and fossilize to form coal.

Lucky find

Rarely, amateur fossil-hunters strike it lucky. Most of these instances involve dinosaurs. In 1983 William Walker was searching a clay pit, as he often

Below: Dr Marie Stopes, palaeontologist and later – more famously – social reformer.

did. A plumber and part-time quarry worker, at a pit in Surrey he noticed a claw-like fossil – a gigantic one that measured more than 30cm/12in. Local experts informed London's Natural History Museum, and staff soon determined that plenty more remains were at the site. The result was *Baryonyx walkeri*, 'Walker's heavy claw', a meat-eating dinosaur some 10m/33ft long. *Baryonyx* was a different kind of carnivore from others known at the time, and so of tremendous scientific value.

Time and experience

Ten years after Walker's discovery, another part-time palaeontologist entered the spotlight. Car mechanic Ruben Carolini was a fossil enthusiast who spent much of his spare time out in the field. While visiting the 'Valley of the Dinosaurs' in Neuquen province, Argentina, Carolini spotted some exciting-looking bones and called in the experts, led by Rodolfo Coria from Carmen Funes Museum. The result – another great meat-eating dinosaur – was named *Giganotosaurus carolinii*, or 'Carolini's giant southern reptile'. It quickly became regarded as one of the biggest terrestrial carnivores – larger even than *Tyrannosaurus rex*. However, a 2006 review of already-known *Spinosaurus* fossils from Africa deduced that this spinosaurid was the world's all-time biggest land predator.

Right: William Walker's story is cheering to all amateur fossil collectors. His famous Baryonyx claw, discovered in 1983, provided important information about the spinosaurids, which were huge predatory dinosaurs. Walker unearthed his 'lucky find' in a clay pit when he was not actively looking for fossils. A reconstruction of Baryonyx appears below.

The top full-timers

At the other end of the scale to these chance finds by amateurs, are renowned professionals who regularly turn up breathtaking discoveries. Most work for museums, universities or similar institutions. They have teams of surveyors, diggers and other staff, and resources to plan, equip and finance excavations in great detail. Among the leading professionals of the past few decades are Paul Sereno (1957–) and Dong Zhiming (1937–). Sereno is attached to the University of Chicago and has made discoveries of immense importance in North America, South America and Africa. They included record-breaking dinosaurs such as *Eoraptor* and *Herrerasaurus*, which were among the earliest of the dinosaur group.

China's 'Mr Fossil-finder'

Chinese palaeontologist Dong Zhiming has led expeditions to many sites, including Yunnan Province and the Gobi Desert. He and his teams have unearthed, studied and named hundreds of new species, not only dinosaurs but mammals and other creatures. One of his prize finds was a quarry near Dashanpu, in Sichuan

Unsung heroes

Fossils of dinosaurs and other great beasts usually grab the limelight. But all kinds of fossils are important, and many palaeontologists have contributed enormously to our knowledge of the prehistoric world while remaining relatively unknown. Dorothy Hill (1907–97) was based in Queensland, Australia. She first studied Mesozoic rocks, but then began to analyze fossil corals from the Palaeozoic Era. Corals and their reefs are hugely important in dating remains, as index or indicator fossils (see Dating Fossils, later in this section). Her three-volume *International Treatise on Invertebrates* is still admired by palaeontologists worldwide. Dorothy Hill also became the first woman to be made a professor at an Australian university.

Below: Professor Dorothy Hill is credited with making scientific studies more accessible to women in Australia.

Above: Chicago-based palaeontologist Paul Sereno, who identified a new species of dinosaur from bones collected in India, displays a model of the assembled skull of the 67-million-year-old carnivorous predator at an exhibition in Bombay. Working with Indian students, Sereno pieced together bones collected over a 20-year period from the western and central regions of the country.

Below: Giganotosaurus carolinii has been regarded as the largest known carnivore to roam the earth – a plaudit since taken by Spinosaurus. Amateur collector Ruben Carolini discovered the fossilized bones of this mighty meat-eater in Patagonia, Argentina in 1993, and a subsequent excavation unearthed a skull that was larger than any existing Tyrannosaurus. The reconstructed skeleton of this theropod is about 70% complete. Parts of a larger specimen have since been found, and indicate it was 8–10% greater in size.

(Szechwan) Province. This 'dinosaur graveyard' and has yielded thousands of fossils. At the museum on the site visitors can see large areas of fossils still partly embedded in the rocks.

Apart from dinosaurs, finds of early human fossils, and their relatives and ancestors called hominids, often grab the public's imagination (see also Famous Fossils, later in this section). Louis Leakey (1903–1972) and wife Mary Leakey (1913–1996) did much to establish East Africa as a centre for hominid evolution. In 1964 Louis and associates proposed the name *Homo habilis* ('handy person') for remains found associated with simple stone tools at Olduvai Gorge, Tanzania, from 1960. Mary and her team came upon the famous Laetoli hominid footprints in Tanzania, in 1978. The prints are dated to 3.7 million years ago and show a hominid with a true bipedal gait.

WHO OWNS FOSSILS?

Rules and regulations covering fossil-hunting are often quite vague – and, in any case, vary greatly around the world. This is rarely of consequence if people hunt fossils for pleasure, for their own personal collections and displays, cause minimal damage on site, and make no monetary gain.

Occasionally, there are problems when an unexpected find turns out to be very valuable or scientifically important. For example, assume a marvellous fossil is discovered, which is potentially worth a small fortune. Who gets the recognition and/or the money? In most cases, the actual finder, whether professional or amateur fossil sleuth, receives the recognition. Photographs of the specimen in situ with the person who spotted it are invaluable evidence in this respect, to ensure that credit is given where due. But the actual ownership of the fossil, and any profit arising from it, could be claimed by several parties, singly or in combination. They include:

• The landowner. This may be a private individual, a national or state organization like a heritage trust, or a business or commercial outfit such as a mining company.

• The benefactor or financier who has put up the funds to support the dig and pay for equipment, expenses and other costs.

Below: Fossil experts who lead walks and digs can advise on safety and legality, as can the local museum or palaeontological society.

• The leader, organizer or site manager of the dig, who takes ultimate responsibility for decisions about which fossils are worth excavating.

• The organization or institution which has set up and probably funded the excavation, such as a university, museum or palaeontological society.

• A commercial company which buys and sells fossils, and which may not be involved in a direct or practical way, but which has 'bought the rights' to any or all specimens yielded by the field trip.

Famous fossil battle

For day-to-day finds, such a list may seem fanciful. But there were months of discussions over who owned, and so had the right sell at auction, the enormous *Tyrannosaurus* 'Sue', worth US $8 million. This incredible specimen, the largest and most complete *T. rex* fossil ever discovered, was found in 1990 by dedicated fossil hunter Sue Hendrickson. A volunteer with the Black Hills Institute of Geological Research, South Dakota, Hendrickson was one of a party digging for fossils on land owned by

Above: Although educational trips to fossil sites are generally encouraged, schools should be careful to seek permission and ensure pupils are properly supervised.

Sioux leader Maurice Williams. Peter Larson, president of the Black Hills Institute, bought the rights to excavate the tyrannosaur for a comparatively small sum, but the validity of this sale was later contested by the US Federal Government. While her ownership was under question, 'Sue' spent time in transit under conditions that many palaeontologists argued would have a detrimental effect upon her bones. She was eventually sold in New York by Sotheby's, for the famous seven-figure sum, in October 1997, seven years after her discovery.

Public access

Some fossil sites have open access and can be enjoyed by virtually anyone at any time. The main examples are coasts and shorelines. Specimens can usually be picked up and removed without problems. On areas of 'common land' it may be possible to pick up a loose specimen lying at the surface. But any use of hammers or other tools, to chip and release an embedded fossil, is not advised. It could be viewed as illegal damage to the site, and legal claims of trespass and criminal damage could be brought.

For almost anywhere else, it is wise to obtain written permission from the landowner. This may vary from a

quarrying company or private estate owner to a highways authority or district council. The local palaeontology or geology society can help with this, or it may require a visit to the local council or Land Registry office. It is one of many reasons why novice fossil-collectors are urged to join a local club or society.

Many owners of fossil-rich sites have standard agreements ready to sign, date and exchange. Part of the agreement is to absolve the landowner from any risk, and make the fossil-hunter solely responsible for accidents and injuries. Fossil sites may well be steep and rocky, with loose stones – and claims for injury compensation are an all-too-familiar part of modern life, so do observe the general Countryside Code at all times when out and about.

Fossil-rustling

Taking a fossil from a site or collection without proper permission is, in effect, stealing. The recent explosion in prices paid for the best fossil specimens has meant that cases of 'fossil-rustling' have tripled in the past 20 years. One of the most infamous is the so-called Maxberg torso specimen of *Archaeopteryx*, one of only seven or so fossils of this earliest known bird. The only specimen to be held as part of a private collection (in Pappenheim, Germany), the Maxberg mysteriously disappeared following the death of its owner Eduard Opitsch in 1991. Despite police investigations it has not been found and is now believed to have been privately sold.

Below: Other Archaeopteryx *fossils form part of collections in Germany, and at The Natural History Museum, London (shown).*

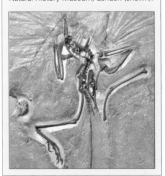

Special cases

Many valuable fossil sites, and other areas of natural beauty and special interest, have been made into national parks, wildlife refuges or heritage reserves. It is usually against the regulations to take any fossils or other items from these places. Some ill-informed amateurs may believe that a small fossil slipped into the pocket could not be traced. But an expert would soon analyse the specimen, its mineral make-up and matrix, and so pinpoint its place of origin.

Above: The gigantic Tyrannosaurus rex *fossil known as 'Sue' is now housed at the Chicago Field Museum, USA. The Museum famously won the bid for the bones, paying US $8.4m – the largest sum ever exchanged for a fossil, making 'Sue' a record breaker in every sense. 'Sue' is not a Jurassic specimen as many assume: rather, she is believed to have died at the end of the Cretaceous Period, possibly the victim of another tyrannosaur aggressor. Her discovery has greatly influenced the way in which palaeontologists perceive the movement and behaviour of these beasts. The fossils of 'Sue' were the subject of a protracted legal ownership battle between the landowner, the expedition sponsors, and the leading fossil-diggers.*

Countryside Code

In addition to various fossil-collecting codes, as detailed on the following pages, everyone is encouraged to observe the general rules of the Countryside Code when accessing rural land, whatever the reasons for doing so. In 2004, following new legislation concerning countryside access and the public's 'right to roam', the code for England and Wales was revised and updated. The rules focus on personal safety and the protection of the land.
• Be safe - plan ahead and follow any signs.
• Leave gates and property as you find them.
• Protect plants and animals, and take all litter away with you.
• Keep dogs under close control.
• Consider other people.
In addition:
• Beware of fire risks.
• Avoid damaging crops or disturbing animals.
• Respect the work of the countryside.

WHERE TO LOOK FOR FOSSILS

It sounds a little obvious, but the best place to look when starting to collect fossils is indeed a map. More precisely, several maps, especially geological maps showing rock types and ages, as well as guidebooks, and perhaps a visit to the local museum, fossil shop or antiquities centre.

The usual countryside map is a helpful starter, showing roads, towns, railways, parks, simple topography and similar features. This is useful for planning the trip and accessing the site. A geological map is also needed. This is available for purchase from specialist map suppliers, or perhaps for loan or hire from a local club or society. Geological maps show the general types and ages of rocks present or outcropping at the surface, as areas of lines and colours. Experts also employ satellite images and remote sensing to locate and survey new sites.

The right rocks

As described earlier, fossils are nearly always found in sedimentary rocks. So regions of igneous or metamorphic rocks are of little interest to fossil hunters. The age of sedimentary rocks is also important. If you have become

Below: The first geological map of Great Britain, drawn by William Smith (1769–1839). Many of his topographical studies were centred around the spa town of Bath in England, where he lived for much of his life.

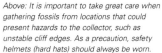

Above: It is important to take great care when gathering fossils from locations that could present hazards to the collector, such as unstable cliff edges. As a precaution, safety helmets (hard hats) should always be worn.

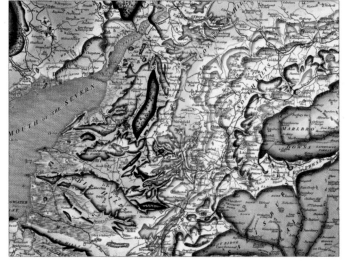

interested in, say, fossil mammals, then it's little use targeting rocks from the Palaeozoic Era. Likewise if you are looking to build a collection of trilobites or ammonites, then Tertiary rocks are too recent and will not contain them.

Also, the names of rock and mineral types can be bewildering to anyone unfamiliar with them. So it's advisable to read the maps in conjunction with a good rock-and-mineral guide book or identifier. In addition, a general guide book for the locality should describe areas of public access, private land,

Fossil-collector's Quarry Code

• One individual or party leader should obtain prior permission to visit a quarry or mine.
• Visitors should be familiar with the current state of the quarry, and consult the manager for areas which have access or hazards.
• Arrival and departure must be reported to the quarry office on every visit.
• Safety hats are usually compulsory; stout boots are also essential.
• Visitors should keep away from vehicles and machinery.
• All blast signals and warning procedures must be understood.
• Quarry faces and rockpiles are highly dangerous and liable to collapse without warning – stay away.
• Beware of wet sand and lagoons of spoil or sludge.
(See also the general Collector's Code in this section.)

Right: Forbes Quarry was the site of Gibraltar's first Neanderthal human find, in 1848. Five sites for Neanderthal finds have now been identified on 'The Rock'.

nature reserves and national parks and similar. This helps to decide when permission may be needed.

Types of sites

In many places, surface rocks are covered up: by soil, trees, crops, roads, buildings, golf courses and many other features. Fossils are most easily identified where rocks are bare and preferably in a continual process of being freshly exposed. This can happen by natural processes such as weathering and erosion, or by a range of human activities.

Coasts and shores

Waves, winds, tides, undercut cliffs, rolling boulders and rockfalls mean that seashores and coastlines are some of the best of all fossil sites. A visit after a storm may reveal marvellous new exposures. However the very agents that erode the rocks can present great danger too. So extra care is needed, especially with an eye on the tides to ensure that you are not trapped or cut off.

Below: The cave of Jebel Qafzeh, Israel, has yielded the remains of several Homo sapiens *up to 100,000 years old.*

Inland cliffs and bluffs

Outcrops of bare rock occur in many drier regions of scrub and desert. Hot days and cold nights alternate to expand and cool the rock, so that it cracks and flakes. Winds may pick up sand or dust and blast the exposed surfaces. Also the winds and occasional flash floods wash away particles to prevent soil formation that would cover the bare ground. So fresh fossils appear regularly at the surface.

River valleys

In some regions, rivers gradually cut their way downwards, exposing layers of rock as they go. Sometimes the layers or strata are clear enough to be 'read like a book', yielding different kinds of fossils from their various ages of formation. However eroded, rocky riverbanks can be steep and slippery, with difficult access.

Quarries and mines

Some of the best fossil finds have come from mining and quarrying sites. Vast amounts of ores, minerals and rocks are worked, exposed and removed. However the site owners may have strict rules about fossil collecting, so that it does not interfere with the operation of what is an extremely expensive business (see panel).

Cuttings

Artificial 'valleys' cut for roads, railways, canals and even pipelines can present many fossil-hunting opportunities. However the land is owned by someone, and permission should be sought. The noise and activity of a busy road or railway line nearby not only presents hazards such as fast-moving vehicles, but can also be very tiring and distracting.

Caves

Fossil-hunting underground, in caves and tunnels, is a specialist activity that needs expert help. There are many dangers such as lack of light for working, seeping gases, and possible cave-ins or floods. Some professional excavations of caves have unearthed important ancient human remains.

PLANNING THE TRIP

Nearly every region has a geology or palaeontology society, natural history club, local museum and history group, rock-and-mineral association, or similar organization. Both careful consideration and expert advice will help to ensure that the trip does not become a wasted opportunity.

Regional and local groups undoubtedly offer enormous benefits to the keen but inexperienced fossil hunter. They can advise on planning and equipment, likely finds, and perhaps loan items of equipment such as tools and containers. There are also the benefits of shared experience and social interaction. A visit to the local museums also helps, to see the types of fossils found in the locality.

Another way of gaining experience is as a volunteer on an organized dig by a museum, university or similar institution. This can vary from a day or two at a local site, to a longer period in a more exotic location. Some holiday companies specialize in these types of breaks and vacations. It can be an invaluable experience to mix with the experts, and learn from their knowledge with 'on-the-job training'.

Equipment

Proper preparation and equipment make fossil-hunting infinitely more pleasurable and rewarding than when without these items. A collection of tools, bags and other kit can be built up over time. It need not be expensive

Below: The fossil collector's kit should contain items both for personal safety as well as for exposing specimens on site.

Above: With field gear safely tucked away in portable rucksacks, you are free to make and record observations about what you see.

– ordinary household items or do-it-yourself tools will often do, until the proper versions are affordable.

Clothing and safety

Warm, comfortable clothes, that fit loosely and do not chafe, are a must. People today are much less accustomed to spending all day outdoors, with the sun – or rain – beating down. Several layers of shirts and jumpers, plus a hooded waterproof jacket or coat, are ideal. So are stout shoes or boots.

Rocks are usually hard and sharp, so gloves are very important. A safety helmet or 'hard hat' is obligatory in some situations, and always helpful near tall rock faces. Sunglasses or a peaked cap help to shield the eyes from glare, especially from light-coloured shiny rocks. Goggles or a visor protect the eyes from flying shards of rock while hammering. And a mobile phone is an excellent aid in case of accident or injury.

Location and recording

Maps, local guide books, and fossil or mineral field identifiers are invaluable, usually in compact pocket form. A magnetic compass helps to locate the site in detail and orientate the finds, for example, the direction of fossilized footprints. A field notebook with pencil (and sharpener) or waterproof-ink pen is necessary for notes and sketched records. Modern lightweight cameras are a boon for 'point-and-shoot' photographs of fossils.

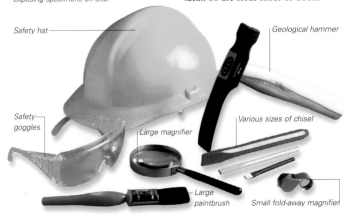

Safety hat

Geological hammer

Safety goggles

Large magnifier

Various sizes of chisel

Large paintbrush

Small fold-away magnifier

Felt marker pen

Clear plastic bags

Tweezers

Notebook for recording finds

Camera

Map

Kitchen paper

Selection of writing tools

Compass

Above: The above items are essential for recording the details of finds both in situ and partly removed, and should be present on site.

Field tools

The main hand tools for fieldwork are:
• Geological hammer, with one blunt face and one pick-end.

Construction sites

As builders and construction companies dig footings and foundations, they may expose fossil-containing rocks. However, they may not always welcome fossil hunters. Having extra people on site can cause problems with safety, insurance and the use of machinery. Also a large, urgent construction project is unlikely to pause while a few fossils are recovered. However, it's always worth asking …

Below: An informal chat with construction staff on site may quickly reveal whether or not the type of items being uncovered are likely to be of interest.

• Guarded chisels, usually 'cold chisels' with hardened blades and plastic hand grips which protect against hammer blows to the hand. (Wood chisels are not suitable.)
• A mallet for use with the chisels, although the geological hammer may suffice.
• Trowels, large and small, for lifting and picking at loose items.
• Brushes, generally a small and large paintbrush, for sweeping away fine loose debris. Shaving brushes are excellent.
• Hand lens or magnifier, to study small details.
• Sieves, to separate actual specimens from loose sediments such as fine gravel, sand or silt. These can be lightweight plastic, of rigid or collapsible foldaway design.

Containers

A selection of plastic or canvas bags is ideal for carrying fossils. Newspaper, tissues or kitchen paper can be used to protect them in transit. Smaller specimens are safest packed into lightweight, rigid, plastic lidded containers such as sandwich boxes. See also equipment for plastering or jacketing fossils.

All of this, and more, is best carried in a sturdy backpack or knapsack. This leaves both hands free, is comfortable to wear rather than hold, especially when supporting packages of heavy fossils, and is easier to balance when walking in rough country with uneven terrain underfoot.

Photography and sketching

Some fossil-hunters take a few quick snaps of their finds in situ, but are later disappointed at the lack of clarity in the images. The undulating, random surfaces of rocks, and their often varied colours, mean that it can be difficult to make out shapes and objects. It is advised to take some photographs from directly above, and others from much lower viewpoints and from several angles, so that features can be identified by comparing these images. Shadows can be troublesome if they are from harsh flash or a low sun, but equally, lack of shadows when the sun is overhead also results in lack of definition.

Below: Many local museums display fossils excavated from the surrounding region. Staff are often willing to advise on the origins of particular specimens, which can help in the planning of a field trip to the local area.

DIGGING UP FOSSILS

Before setting out on a day's fossil-hunting, remember to tell someone reliable about where you are going and when you expect to return, supplying the relevant map references where possible. Also check the weather forecast, and revise the plan if it looks bad – digging really does require dry weather.

Think about personal necessities and sustenance before the trip – take food as necessary, and plenty of fluids. Being out in the sun and wind all day is thirsty work. On arrival at the site, carry out a quick survey and decide which areas look best for excavation, by a few short 'test digs' or blows with the hammer. Remains may already be 'eroding out' in some places. Look for shapes and textures that suggest fossils, like smooth curves of bone ends or regular lines and striations on shells, which distinguish them from the surrounding rock. See also the Collector's Code, later in this section.

Above: A pair of protective gloves, that still allow sufficient grip and manoeuvrability, is important when working in the field.

Above: This conifer wood bored by Teredo bivalve (shipworm) is an Eocene specimen from the Isle of Sheppey. It is a good example of a specimen likely to be tough to extract from rock without losing much of the detail.

Big to small

When digging for fossils, there are several basic guidelines.

• Go easy and take care. A too-hard hammer blow could shatter a prize specimen. If time is short or the weather turns foul, you may simply have to leave the fossil where it is, partly excavated. It will still be there tomorrow or the day after.

• Work from big to little, tough to delicate. Use the larger tools to remove unwanted rock, then work more sensitively and in greater detail as the specimen is exposed.

• Work with the rock rather than against it, using its natural splitting or bedding planes to your advantage.

• Remove a block containing the fossil, rather than the specimen itself. Leave detailed exposure and cleaning until later, back at the workroom, where it is more comfortable and proper equipment is at hand.

• Don't forget to wear safety equipment such as goggles and gloves.

• Record each stage of the excavation with a simple map, a few sketches or (if you have brought a camera) photographs, and plenty of notes. The latter should include map reference and the surrounding rock types.

Initial stages

Large quantities of material covering a fossil, known as the overburden, may be loose gravel, sediments or rockfall. It can be removed with a spade or perhaps a pick-axe. (Professionals sometimes use road-drills, jack-hammers, construction diggers or even explosives, but these are beyond most part-timers.) Then hammer and chisels can be employed. Keep an eye on the rock fragments as they are removed. If the fossil is well embedded in matrix, try to assess its overall size and position, and work around this, leaving plenty of room for error. Try a trial few blows on a rock of the same type which does not contain a fossil, to see how the rock reacts and splits or shatters. Do not chip or scratch the fossil itself, which could spoil its appearance and considerably reduce its scientific value.

Later stages

As the fossil is uncovered, switch to gentler methods. A small trowel or dental pick can lever out detached

Left: At a communal site, be careful not to stray onto a neighbouring patch, or damage delicate fossils with boots or piles of debris.

Shall I, shan't I ...?

Faced by a jumble of rocks and fossils at a new site, one of the most vexing questions for the novice fossil hunter is: Which specimen shall I choose? In other words, which ones will be worth the time and effort to excavate? Or even: Is it a fossil or just a weirdly shaped rock? For beginners, the answer is usually quite simple: Choose almost any one, since it will be valuable practice that brings precious hands-on experience.

Also consider: Would I wish to spend time cleaning up this specimen for my collection and perhaps display? It's wise to spend some time giving the whole site a brief survey, rather than stopping at the first attractive fossil. There might be a better-preserved, more complete version just a few metres away.

Below: Use of a geological hammer and chisel will preserve the fossilized specimen while separating it from the rock (and it would be safer wearing gloves).

Above: When uncovering larger fossils, an excavator carefully works around the specimen to remove 'overburden' and detach it from the surrounding matrix.

Left: Both a trowel and metal tray are useful items when sifting through sediment.

fragments. For very fragile fossils, a paintbrush, bradawl or dental-type pick may be suitable. It's best to chisel out a larger lump of rock around the fossil and lift out the whole item. Very fragile or crumbly specimens can be strengthened or stabilized as described later in the section.

Notes

Your notebook should contain plenty of jottings that record details such as locality, date, type of rock, position of fossil specimen, its size, notes on initial identification and other relevant information. All these details help to confirm formal identification later. A sketch can help to pinpoint an unusual feature, such as a spine or claw, that may not be easily visible on a photograph of the complete specimen.

Below: When labelling new specimens, for example by using corrective fluid and a ballpoint pen, be sure to mark the least important area, so that the fossil itself (like this small graptolite, to the right of the label) is not obscured.

COLLECTOR'S CODE AND FOSSILS

Each fossil presents a unique challenge for those tasked with removing it from the site and transporting it back to the workroom. Not least, some fossils are extremely heavy! As when digging up specimens, moving fossils could potentially disturb the local environment and should be conducted with great care.

Palaeontology and geology associations offer codes of behaviour when fossil-hunting. These are based on respect for the countryside and its inhabitants, as well as the need for caution and safety.

Field Collector's Code

1 Obey the Country Code and observe local byelaws. Remember to shut gates and leave no litter.
2 Always seek permission before entering onto private land.
3 Don't interfere with machinery.
4 Don't litter fields or roads with rock fragments that could injure animals or be a hazard to people or vehicles.
5 Avoid disturbing wildlife.
6 On coastal sections, check tides or local hazards such as unstable cliffs, if necessary with the Coastguard.
7 In mountains or remote areas, follow advice given to mountaineers and trekkers. In particular, inform someone of where you are going and when you are due back. Do not go alone.
8 Do not venture underground except with expert help.
9 Never take risks on cliffs or rock faces. Take care not to dislodge rock: others may be below.
10 Be considerate and do not leave hazards or dangers for those who come after you.

Below: Palaeontologists in Nigeria strengthen the weaker areas of a sauropod dinosaur bone before applying a hessian bandage.

Great weight

Most fossils are solid rock. So they are heavy. A hand-sized specimen is already a considerable weight to carry, especially when trekking on foot. A dinosaur leg bone may weigh several tonnes. So it's vital to be realistic about what can be transported back from a collecting trip. Much depends on whether vehicles can get near to the dig site. If not, it is best to budget for two or three trips on foot, from the site back to the car or truck, carrying sets of specimens each time. Although

Above: A palaeontologist examining a six million year old fossil in Abu Dhabi. The hot and arid conditions, and the sheer size of the fossil, make this a very challenging excavation.

you may be tempted to try and save time, if you try to carry lots of heavy specimens in one journey, you risk breaking them – or yourself.

Fragile fossils

Sometimes fossils start to crack or crumble as they are being excavated. The first response is: Is it worth continuing? Is the specimen valuable

Below: Further plaster is then applied to the bandaged specimen to complete the 'jacket' before the fossil is moved.

Below: Once the plaster has hardened the bone must be turned over so that the under surface can be fully encased within the 'jacket'.

Maps and grids

On larger sites, or when many fossils are preserved together, it's important to know their positions in relation to each other. This can help to reconstruct how the original organisms died and were preserved. Sometimes the site is covered by a grid of strings or wires, set at regular intervals such as one metre. Then the position and orientation of each fossil can be measured and plotted. This process continues with the excavation, so that the depth of specimens is also known in order to complete a three-dimensional location.

Below: This excavation of dinosaur bones in Nigeria illustrates how palaeontologists may be compelled to work within a tight, clearly-defined space, even when tackling the largest of creatures. The grid markers help the excavators to record the position of each part of the fossil.

Above: Excavation of a large dinosaur skeleton at Dinosaur Monument, Utah, Colorado, USA. As different sections of the specimen are uncovered, it is vital that the findings of each part are carefully recorded, to reconstruct accurately the creature's dimensions.

enough to warrant extra time and care? Perhaps there's another one nearby which is tougher and easier to remove. If not, then a weak or fragile specimen can be stabilized and protected in various ways, for example, by spray-on or brush-on adhesive or stabilizing compound, DIY spray-glue or purpose-made chemicals used by masonry restorers. This must be allowed to dry or set, so schedule time into the day to carry out other tasks during this period.

Jackets

Larger specimens that might crack or shatter can be protected and strengthened by jacketing. There are several common methods, usually involving plaster bandaging (plaster-of-Paris) or some type of glass-fibre resin or compound. One of the simplest ways is to use sheets of material or first aid bandages and ordinary plaster-

of Paris. An alternative is ready-made plaster-impregnated bandages as used to encase broken limbs, which simply need moistening. Wrap and smooth the jacket over the fossil and beyond, and allow plenty of time for it to set, before chipping the whole section free of the surroundings rock.

Wraps and containers

Each fossil should be put into a separate plastic bag – small transparent freezer-bags used for food are ideal. Tear off a piece of paper, write on the fossil's initial identification, as coded into your main notebook, and include this in the bag too. Smaller fossils can

be put into rigid plastic boxes, but pad them to avoid rubbing and chipping. If there are no padding materials such as newspaper available, some dry vegetation like old leaves or grass stems will suffice.

Below: Uneven or largely-absent roads, and a shortage of road signs to indicate the direction of travel, are just two of the challenges that might confront a team of palaeontologists transporting fossils from remote locations.

CLEANING FOSSILS

Rarely, some fossils fall out of the rocks in almost perfect condition. All they need for display is buffing with a soft cloth, and a discreet label. But in most cases, specimens benefit from some judicious preparation and sympathetic cleaning.

Before rushing to clean and prepare a specimen, there are some questions to consider.

• Is it really worth a place in the displayed collection? What seemed like a fine example when you were removing it, out in the field some time ago, may not now seem quite so marvellous. It may have been eclipsed by specimens you found subsequently. On closer examination, it could have small flaws or chips. It might be very similar to a specimen you have already prepared or obtained. In these cases it might be worth saving for your general 'background' or reference collection, but without too much time spent on preparation since it will probably not go on display.

• How much should it be cleaned? Leaving some surrounding rock or matrix in place may add artistic merit and 'character' to a fossil. Take some time to look at real specimens in

Above: Some kind of magnification tool may be required in order to work out where the fossil ends and the matrix begins.

museums and exhibitions, and of photographs of specimens in books and on websites. See how sometimes only a portion has been exposed and cleaned up. The result can be more dramatic, as the fossil 'appears' mysteriously from its background, while at the same time providing enough scientific information for firm identification.

The work space

For cleaning and preparation, you need a clear working space with a hard non-slippery surface, such as wood, which does not matter if it chips or dents. Beware of fragments of rock flying away from the specimen as you work. Some preparators arrange series of long, low, upright panels around the work surface, like fencing, to stop this from happening.

Below: Items such as safety goggles and a face mask are a useful safeguard against dust thrown up when cleaning specimens. Before beginning to clean, wash your work space with a clean cloth or sponge, and keep a pair of protective gloves on hand, particularly if handling chemicals such as dilute acid.

Always do a 'test run'

Whichever technique you use to clean a fossil, try it first on an unwanted fragment of rock, or on one of the less important specimens, or on a less important surface of the main specimen which will not be visible on display. If things go wrong, you have not lost your most prized example. Or at least, the problem will be 'round the back' when the item is displayed.

Below: Dilute acid is carefully applied to a disposable area to see if the specimen contains certain reactive minerals, some of which may help to dissolve the matrix.

Above: For specimens without much adherent matrix, simple household items such as cotton buds (swabs) or small brushes can be useful when cleaning specimens. These can also be used to remove small patches of sediment once more rigorous methods have been tried.

Above: Similarly, a soft paintbrush can push away friable matrix in small amounts if used delicately – caution when applying strokes to the surface of the specimen is the main thing to remember. The brush can be washed and dried regularly to prevent clogging.

Above: A miniature sandblaster, powered by compressed air, helps to remove loose sediment without damaging the specimen.

Above: A vibrating pen, or 'vibrotool', is powered in a similar fashion, and can help to define the shape of a specimen.

Plenty of light from a couple of desk lamps, one on either side to minimize shadows, is important. Try not to place the light sources on the opposite side of the specimen from yourself, otherwise the areas you are working on will be shaded. Goggles for eye protection against flying shards are also strongly advised.

Experts usually clean smaller fossils while looking through a magnifier lens on a stand, or through a binocular (two-eye) microscope. This helps to protect your eyes and prevent eye strain as well as giving a closer view.

A fishing-tackle box or toolkit case, with small drawers and compartments, makes a good container for tools and equipment. Similar many-compartment cases intended for items such as screws and bolts, available from do-it-yourself

and hardware outlets, are very handy to store your specimens while they are being prepared.

It helps if you have a room or corner to leave this equipment laid out in place, rather than having to set it all out and then pack it away again after each session. If tools and specimens can be left out, covering them with a cloth or sheet will prevent dust gathering, which tends to dull the colour of fossils and obscure detail.

Cleaning with chemicals

Often a fossil is different in mineral composition from the matrix around it. If the fossil is much harder, it may be possible to brush away soft matrix, perhaps after softening or soaking it with water. You could test-soak unwanted pieces in water or perhaps

a dilute solution of vinegar, which may soften or dissolve calcium-rich matrix while leaving the fossil untouched. Fossils from the seashore may be soaked for a few hours in a watery, weak solution of bleach. This helps to remove the salt. Again, try these techniques on an unwanted piece first.

Another chemical method is acid etching. Kits of acid-etch equipment and solutions can be purchased from specialist fossil suppliers. This is a fairly complex, time-consuming technique and involves handling potentially harmful chemicals. Good ventilation is essential, and other safety precautions are also necessary. Sometimes local palaeontology club members band together to research, purchase and set up such equipment, so that it can be used during carefully monitored communal sessions.

Physical methods

When removing rock and matrix by physical means, it is advised to study the hidden shape and contours of the fossil, and where the matrix probably ends and the specimen begins. This can be done by consulting guide books and fossil identifiers. When chipping carefully with a lightweight hammer and narrow chisel, making sure the specimen is well supported so it cannot fly away or slide off the worktop. More delicate tools include awls, bradawls, dental picks and mounted needles for removing smaller and smaller particles. Even general household items such as cotton buds (swabs) and paintbrushes can be used.

Power tools

Powered engravers, drills or sanders can save much time. Dental drills and modelling power tools often have tiny circular-saw blades and fine file-like abraders for gradually and carefully removing adherent matrix. With all these methods, pause now and again to clear the dust. Take a wider view, from several angles, to check progress of the work. Power tools throw up a lot of dust and debris, so wear a face mask to filter the air, as well as the usual protective goggles.

IDENTIFYING AND STORING FOSSILS

After spending hours out in the field collecting fossils, and in the workroom cleaning them – what next?
Most collectors like to make a display to show off their prize specimens. But there is some forethought
and preparation to do first, such as setting the fossils in context.

Fossils can be admired for their simple shapes, colours and beauty. But this is like admiring a painting without the insight and knowledge of the painter, which period it dates from, the circumstances of the subject or scene, and how the work fits into the history of art. Likewise, identifying fossils and finding out which organisms left them – when and how and where – adds vastly to the interest and pleasure of owning and displaying them.

Identification

There are many ways to pin down the identity of a fossil. For some people a simple label such as 'ammonite' may suffice. But there are thousands of kinds of ammonites known from fossils, so it's worth delving more

Above: Record as much information as possible about a specimen, though do keep to a consistent and acceptable labelling system for ease of reference.

Above: In addition to actually labelling the specimens, it is important to record corresponding information about your collection in a log book.

Below: Identification guides, augmented by further information gleaned from the internet or museum displays, can help to pinpoint the origins of a fossil.

deeply into the science of taxonomy and the system of classifying living things. The traditional system starts with large groups, phyla, which are subdivided in turn into classes, orders, families, genera and species, as will be explained later. A name at the hierarchy level of family or genus is

usually sufficient for most amateur collectors, depending on the specimen's state of preservation and rarity. Identification is carried out by noting specific features, characteristics or traits in the specimen, as listed in classification keys, as well as by overall visual appearance, coupled with the age of the rocks and their general nature, such as sandstone or limestone, which indicates the original habitat of the organism.

Helpful aids to identification
• Pictures and notes in the later pages in this book.
• Similar pictures and lists of key features, in fossil guides and identification books.

Below: Specimens become dusty and details obscured if kept in open displays, so choose a suitable container for them.

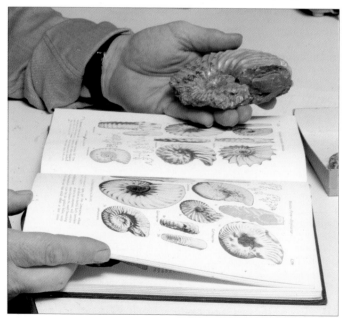

• The same material from internet sites, which are a fast-growing resource for palaeontologists.
• Expert help and advice from knowledgeable people.
• Similar specimens in other collections, such as in museums or palaeontology societies.

Storage

Fossils that are not for display are best stored in rigid containers such as sets of drawers or compartments, which are dry and dark. Moisture and bright light can degrade several kinds of minerals. Purpose-made cupboards or chests of drawers, often with the classic shallow trays as used in museums, can be picked up at auctions, sales or house clearances. Or sets of wooden or plastic storage cases can be pressed into use. Open shelves are less useful, since dust accumulates and gradually dulls and obscures a fossil's detailed surface features. Remember that fossils are stone, and so heavy. Any storage system should be able to cope with the weight. The drawers, compartments or boxes should be clearly labelled with their contents, and arranged in some sort of logical order, for example, by time period or major groups of living things.

Below: Soft padding protects the fossils from scratches or cracks if the container is jolted. Do, however, avoid cotton wool or similar materials where fibres are likely to dislodge and cling to the specimen.

Labels, notes and indexing

Each fossil in a collection should be given a brief identification name and a unique code or log number. This keys into an index for the collection, which can be kept as cards in a box, in a loose-leaf folder or on computer disc. Information from the field notebook is transcribed onto a list for each specimen, along the lines of:
• Unique code, index or log number.
• Name, for example, to family, genus or even species level.
• Approximate date of preservation, from a period or era of geological time such as Carboniferous or Eocene, to much narrower or shorter units such

Left: Parts or fragments of one fossil item may be assembled like a jigsaw puzzle into the more complete version, simply by trial-and-error fitting.

as ages or zones of time. This is linked to the type of rock in which the fossil was preserved, known as the lithological unit.
• Location and date of find, with map reference if possible.
• Nature of the matrix and features of surrounding rock.
• Special notes about the site, methods used for extraction or cleaning, glues and preservatives, and similar data.
• General information such as notes on palaeo-ecology (see next page), behaviour or lifestyle: 'active carnivore in shallow Silurian seas' or similar.
• To tie the fossil to its information, include a short label in its storage compartment. Or write the index code in indelible ink on an unimportant part of the fossil. Take fossils from their storage places one at a time, to avoid mixing and misplacing.

Below: Accessible museum displays may provide clues as to the identify of your fossils. Staff may allow you to handle them (always check), or the displays may be interactive, with information given at the touch of a button.

DISPLAYING FOSSILS

One of the greatest satisfactions of fossil-hunting is to build up a well-presented and informative display of the best specimens, so that other people can appreciate and understand these fascinating objects. This can be done by following a few very simple, practical tips.

Fossil displays can be a permanent installation, or 'collapsible' and brought out for special occasions. It's well worth putting some thought into what you would like to convey in showing a range of specimens together – the most interesting displays usually have some sort of linking theme or topic, for example:

• Varied fossils from the same locality at different times, illustrating the range of life which has populated the area through the ages.

• Fossils of the same geological age and site, revealing the variety of plants and creatures that lived together in the same place. This can allow reconstruction of a whole habitat and how the animals and plants lived together and interacted, an area of science known as palaeo-ecology.

• A group of organisms or similar fossils through time, such as sea-snails, petrified bark or fish teeth. These can be compared and contrasted to see how they evolved.

Below: Include notes on possible origins wherever you can. For example, it might be conjectured that the death of an entire shoal of fish (the species Gosiutichthys parvus, shown below) was caused by poisonous gas in the water.

A practical display

Almost any size of display can be made attractive and informative, from a shoebox to a huge wall cabinet or stand-alone glass case. Ensure the case or container is presentable, perhaps with a coat of paint or paper lining that contrasts with the samples. Choose only the best specimens available – do not 'dilute' them with poorly preserved or broken examples. As your collection builds, you can transplant new and better finds to replace older, less impressive ones. Position and support each specimen at its best angle to show off the most informative side or face.

Above: It makes sense to display fossils together which have similar origins. Notable differences between the members of a group may indicate stages in growth or evolution. Clockwise from top-left: seeds of the Carboniferous plant Trigonocarpus; four echinoid, or sea urchin, specimens; the teeth of the huge shark Carcharodon megalodon.

You may need to trim off unwanted edges to make a pleasing shape and set the fossil to its best advantage. Include neatly printed labels or cards, both for individual specimens and groups of them. Or label a diagram or photograph of the collection as a key.

Selection

If you have a container or case for a collection, try various options of fossils before deciding on the final selection. Move them about and group them in different ways, to show similarities or contrasts. It's not unlike trying furniture in different places in a room, before deciding on the best layout.

Left: The cut of the matrix around this attractive Late Cretaceous fossil (Ctenothrissa radians) is perfectly suited to an eyecatching stand-alone display.

Sections and infills

Sometimes the broken surface of a fossil can be used to advantage. It may be sawn or cut off flat and then sanded and polished to reveal the inner structure of the specimen. This may be very beautiful, with colourful swirls or patterns of mineral crystals adding to the effect. Another option is to 'fill in' the missing part, to make the specimen whole again. This is usually done with some type of plaster or resin, but in a colour that contrasts with the actual fossil. The size, shape and contours of the infill can be gleaned from pictures of whole specimens. Attempts to blend the infill portion so that it looks like the rest of the fossil need great skill, and usually end up detracting from the overall effect. A faintly tinted clear plastic resin is another option for infills. It is usually better, more honest, and also scientifically acceptable, to make the infill very obvious.

Above: Casts made of glass reinforced plastic are laid out prior to the mounting of the dinosaur Baryonyx *on exhibition panelling.*

Varnish or not?

Some collectors like to varnish their specimens. This may help to seal and stabilize the minerals. It may also deepen hues and bring out textures and colours. However over the long term, certain varnishes can degrade the minerals of the fossils. Also a shiny varnish means light glints and reflects off the surface, which can be irritating and obscure detail. However, some fossils are made by a process called pyritization. The original material of the organism is replaced by mineral iron pyrites. In moist air this is unstable and gradually disintegrates. Such fossils are best dried thoroughly and varnished to seal them from moisture in the air.

Below: Following expert guidance, museum staff may decide to add certain finishing touches to enhance or protect the surface of the reconstructed skeleton.

Preserving

Sometimes a broken fossil can be mended with clear glue. As already advised, try the glue on an unwanted piece or surface first, to check it adheres and does not discolour. Fossils that are weak or crumbly can be stabilized with chemicals called consolidants, hardeners or protective resins, specially made for the purpose. Again, try them on unimportant areas first, in case they cause discoloration or a chemical reaction which may degrade the minerals.

Below: The tail of this Diplodocus *replica at the Natural History Museum, London, was temporarily replaced with a cardboard one while the original was being reconstructed.*

RECONSTRUCTING FROM FOSSILS

For certain people, finely preserved and well-displayed fossils, with a wealth of information about their origins, is not quite the end of the process. Some collectors like to take the next step, of trying to reconstruct or 'rebuild' an extinct plant or creature as it would appear in life.

Lifelike reconstructions range from sketches on paper, to illustrations 'borrowed' from books or websites for display, to simple scale models or even complex and detailed full-sized renderings. Most of the fossils in the later pages of this book have illustrations showing the appearance of the living plant or animal, as it looked millions of years ago. How do we know it is accurate?

Composites

As we have seen, many fossils are just pieces or fragments of whole living things. Not only this – the parts can be squashed and twisted out of shape. One way around the problem is to assemble a composite version, using pieces from another specimen identified as the same species. There are many limitations on this method of 'borrowing' from some fossils to fill in missing sections of others. But the composite can help to give an general view of shape, size and features.

Above: This atmospheric diorama combines life-sized model dinosaurs with: their habitat; the fossil skeletons used as the basis of the reconstructions; and an informative text panel with illustrations.

Below: Artists' reconstructions of prehistoric creatures, such as Pteranodon shown here, may show them feeding, fighting, breeding or in the company of similar species. Concurrent flora and fauna should also be depicted.

Above: Examining the surface of fossilized body parts, such as limbs, for scars and indentations made by muscle and other body tissues helps with reconstruction. Both of the above specimens originate from dinosaurs of the Cretaceous. Top is a toe bone of the plant-eating Maiasaura. Below it is a hoof bone of the big-horned Styracosaurus albertensis.

What colour were they in real life?

This question often plagues those making reconstructions of fossilized animals and plants. The short answer is – no one knows. A glance through a selection of fossil books will probably show the same famous creature, such as the earliest known bird *Archaeopteryx*, in a startling variety of plumages. Fossils are not original copies of the once-living matter. They are mineral replacement versions made of rock or stone. So they are the colour and pattern of their constituent minerals. Was a long-extinct fish or shellfish dull green or brown for camouflage, or brightly striped with pink and yellow to advertize its venomous sting or bite? Take your pick. When scientists, artists and modelmakers reconstruct extinct organisms, they have to make intelligent guesses as to the patterns and hues visible on the outside. Their informed guesswork is often based on the features exhibited by similar organisms, or living relatives.

Above: A safe option is to reconstruct an image of Archaeopteryx using little colouring, or even in black and white.

Above: A more adventurous rendering paints Archaeopteryx in slate-blue shades. But even this might be a conservative treatment.

Comparisons

There are also comparisons with living organisms. The science of comparative anatomy has a long and honourable tradition in palaeontology. For example, molluscs such as ammonoids no longer exist. But their modern cousins do, including the cephalopod mollusc group such as octopus, squid and especially the deep-sea, curly-shelled nautilus. Comparisons of ammonoid fossils with the equivalent parts of these living relatives helps to reconstruct the ammonoid's soft parts. Similar principles are used for all manner of ancient life, from simple plants to sharks, reptiles and whales.

A note of caution involves the process called convergent evolution. Organisms of different kinds come to resemble each other, generally or in specific features, since they are adapted to the same mode of life. A simple example involves the body shapes of certain fish like sharks, the extinct swimming reptiles known as ichthyosaurs, and modern dolphins. They have all evolved a similar streamlined body shape, with fins and tail, as a result of adaptation to speedy swimming, rather than because of any shared ancestry.

Flesh on bones

Some animals and plants are defined in size, shape and external appearance by their strong outer casing, such as a tree's bark or a trilobite's shell. This helps greatly to recreate their appearance in life. However most vertebrate animals – those with backbones – have an inner skeleton which may fossilize, covered with muscles and other soft tissues which hardly ever do. How are these fibres reconstructed by scientists?

Soft tissues can be built up stage by stage. Many fossil bones show roughened areas on their surfaces called muscle scars. These are patches where muscles were attached to the bones, and they may have flanges, ridges or crests on the bone to secure their fixings. Often they correspond to the muscle anchorage sites in modern relatives of the creature. The size and extent of a muscle scar indicates the size and power of the muscle which anchored there. So a fossil skeleton can be partly clothed in muscles. In turn, the musculature often defines the shape and contours of the whole animal, since muscles are generally clothed only by skin.

Right: A museum reconstruction of a chalk sea floor of the Cretaceous Period.

Below: A fossil of a sea floor at Kings Canyon, Northern Territory, Australia. The ripples in the rock were created by water movement.

DATING FOSSILS

How old is a fossil? There are two main methods of dating or aging either the rocks themselves or their preserved remains, as used by geologists and palaeontologists. These are comparative or relative dating, and absolute or chronometric dating.

Palaeontologists often use comparative or relative dating, which deduces the age of a fossil by comparing it to the rock around it, especially the other fossils contained within this rock, and to the rocks and fossils in the layers above and below it. Because of the way sedimentary rocks form, the younger layers or strata usually overlie the older ones. So deeper fossil are older. Also the same sequence or series of sedimentary layers often crops up at many sites, with the same sequence of fossils within it. These sequences have been correlated and matched or combined to give an overall progression of rock types and fossils through the ages all around the world. This is known as general stratigraphic history. If a fossil comes from an isolated layer of rock, the features of this layer can be used to place it into the overall progression or stratigraphic history, and so the age is revealed.

Index fossils
Some types of organisms are very widespread as fossils, and they have also changed in a characteristic and

recognizable way through time. These evolutionary sequences are very useful for dating. The well-known fossils in them are known as index, marker or dating fossils. They help to pinpoint the time origin of more obscure fossils found with them. Common examples of index fossils include shelled marine animals, which formed plentiful remains in marine sediments and are often worldwide in distribution. (All are included on later pages in this book.) For instance:

• Trilobites are useful for dating Cambrian rocks.
• Graptolites are common and widespread index fossils for the Ordovician and Silurian Periods.
• Belemnites and ammonites show characteristic evolutionary sequences through the Jurassic and Cretaceous Periods.
• Foraminiferans, which are tiny and simple shelled organisms of the plankton, are used through the Tertiary Period.

Left: Successively aging rock strata are clearly visible in the steep walls of a canyon.

Above: A member of an excavation team from The Natural History Museum, London, uses a device known as a compass-clinometer. This measures the incline, or 'dip', of the bedding plane – the boundary separating one layer of rock from the layer above or below – at various magnetic compass orientations. Assessing the nature, formation and subsequent movement of each rock stratum will help the team to assess, from the position of the now-exposed fossil, its approximate age.

Absolute dating
Also known as chronometric dating, this method is based principally on the physical process called radioactive decay. Some minerals of the Earth's crust contain pure chemical elements that are radioactive (usually in a very minor, harmless way). Through time after their formation, these elements give off radiation energy (radioactivity), at a regular and constant rate. As they do so, they change from one form or isotope of a chemical element into another, or even from one element into another. Scientists can measure the relative proportions of different isotopes or elements in certain rocks, to deduce when the mineral formed.

Each isotope or element series has its own fixed rate of decay. Carbon is relatively fast, and is used in 'radio-carbon dating' for specimens up to about 70,000 years ago. Uranium decays very slowly and can date the oldest rocks, from billions of years ago. Potassium-argon decay is also widespread and commonly used for times spans greater than 100,000 years ago, mainly where there has been local volcanic activity and the rocks are rich in potassium. However, measuring radioactive decay involves complex and expensive scientific equipment such as mass spectrometers, which are beyond the finances of most amateurs!

In practice, absolute dating pinpoints the actual age in numbers of years, for a limited number of suitable rocks. There is usually a margin of error given. For example, rocks from the Cretaceous-Tertiary (Palaeogene) transition may be dated to 65.3 million years ago ± (+/- or plus-or-minus) 0.4 million years. The results from

Above: Samples containing fossil fungal spores being collected by scientists at Butterloch Canyon in Italy. The Permian-Triassic boundary between these rocks is visible as the dark sediment line running around the cliff area from upper left and to the right.

absolute dating are then fed across to comparative dating. Once certain fossils and their rocks are aged in this way, the dating can be used for similar specimens in other locations and for smaller timescales on a day-to-day basis.

Above: Trilobite fossils, such as this specimen found in Montana, USA, are useful for dating the Cambrian rocks prominent in some of the geology of this state.

Palaeomagnetism

As some types of rocks formed, tiny iron-rich particles in their minerals were affected by the Earth's natural magnetic field. The field made the particles line up, like miniature compass needles. It is known that the planet's magnetic field has varied through time, both in strength and orientation, as the North and South Poles 'wandered' (and still do today). Sometimes the field even flipped or reversed as North became South and vice versa. All these changes are preserved in certain rocks, usually of volcanic origin. The sequences of changes are known, and fossils found in sedimentary rocks above or below the volcanic ones can be matched into the sequence. This gives another form of dating, known as palaeomagnetism.

Below: Iron ore displaying black areas of massive magnetite, and silvery quartz.

Below: The limestone and shale cliffs at Gros Morne National Park, Newfoundland, Canada, define the boundary between the Cambrian and Ordovician Periods.

FOSSILS AND EVOLUTION

The general notion of evolution has been around for more than two thousand years. Fossils are one of the key areas of evidence supporting the theory of evolution, which is a cornerstone of the modern life sciences.

'The crust of the Earth with its imbedded remains must not be looked at as a well-filled museum, but as a poor collection made at hazard and at rare intervals.' Charles Darwin, Conclusion to *On The Origin of Species by Means of Natural Selection* (1859).

Aristotle (384–322 BC) of Ancient Greece vaguely mentioned what he called a hierarchy or ladder of life, with Man (humans) at the top. In these writings he perhaps foresaw some kind of evolutionary process. Many other natural philosophers mused on the idea of a time before humans, when different kinds of animals and plants lived. Some even suggested that kinds or species or living things were not fixed and unchanging – that is, they evolved. The notion gained ground in the early 1800s when experts such as Georges Cuvier began to accept that there had been previous extinctions. However at this time species were still

Below: Darwin's revolutionary theories about evolution were subject to both outrage and ridicule in certain quarters upon going into print. This cartoon shows a gorilla protesting of Darwin (shown right, holding his book On The Origin of Species*): "That man wants to claim my pedigree. He says he is one of my descendants." Henry Bergh, co-founder of the Society for the Prevention of Cruelty to Animals (shown middle), replies: "Now, Mr Darwin, how could you insult him so?"*

Above: German biologist Ernst Haeckel (1834–1919). Haeckel was greatly influenced by Darwin's work on evolution, although he remained suspicious about natural selection as a means of determining survival.

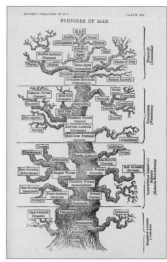

Above: An extract from Haeckel's work The Evolution of Man: *a popular exposition of the principal points of human ontogeny and phylogeny, published in 1879. Haeckel believed in a link between phylogeny – the concurrent changes demonstrated by a group of organisms – and biological development. However, his geneaological trees, such as the one shown above, often represented older ideas about the 'scale of being'. The above illustration shows the evolution of the human species, the 'pedigree of man', with the trunk representing a complete and linear line of progress from higher to lower forms.*

seen as never-changing between creation and extinction, according to teachings derived from the Bible.

The 'great book'

The idea of evolution smashed onto the scene in 1859 when English naturalist Charles Darwin published *On The Origin of Species by Means of Natural Selection.* Darwin argued that most living things produce too many offspring for all to survive the struggle for existence. Organisms struggle against the weather and physical conditions, against enemies and predators and similar problems, and against the limits of vital needs such as food, shelter or living space. The offspring which survive are those best suited or adapted to conditions of the time. This happens because not all offspring are identical. They naturally vary slightly. The variation usually

comes from genes, which are units of inheritance, like 'instructions' for how a living thing develops and functions. In sexual reproduction, genes from the parents are shuffled into different combinations for different offspring. Some of the offspring possess certain features or characters which other offspring lack.

Sexual reproduction and genetic variation provide the 'raw material' on which evolution acts. If a certain organism's features are an aid to survival, and are also inherited by its offspring, then the offspring are better adapted to the prevailing conditions. They have a better chance of survival. And so on, generation after generation.

Change with time

Darwin proposed that over very long time periods, as conditions and environments altered, certain kinds or species of living things gradually changed or evolved into different ones. Two chapters of *The Origin* dealt with fossils. Darwin realized that fossils represented excellent evidence for his ideas. They were remains of living things which had lost the struggle for survival. They died out as conditions changed, to be replaced by newer, better adapted forms. And this is a key point. Sooner or later, environmental conditions, such as the climate, begin to change. So living things are always playing 'catch-up' by evolving to fit the new conditions, even as that environment continues to alter.

Above: Scientists at Beijing University attempt to extract fragments of DNA from a large fossilized dinosaur egg.

A long long time

In his great book, Darwin was cautious about time spans. Many people of his period believed the literal truth of the Bible, which led to the calculation that Earth was created in the year 4004 BC. Darwin saw that the formation of fossils according to the theory of evolution would need much longer. He estimated Earth's age at up to 400 million years. He was also extremely aware of the chance-ridden nature of fossilization, and the random nature and huge gaps in the fossil record. So he warned that fossils were unlikely to show complete and gradual evolutionary sequences – there would be many gaps and jumps.

Below: Charles Darwin (1809–1882).

Left: Organisms trapped in amber, like this spider (left), are helping palaeontologists in their endeavour to extend the DNA database from living to extinct organisms. The idea is that this will help them to clarify evolutionary relationships between genera and species.

In the early 1970s, Stephen Jay Gould (1941–2002) and Niles Eldridge (1943–) proposed a modified version of Darwin's original idea. Based on the fossil record, they suggested that many organisms go through long periods of stability or stasis, when conditions remain relatively constant. Then there is a brief period of change, when new species rapidly appear, before stability again. This twist on Darwinism is known as punctuated equilibrium, or 'evolution by jerks', in comparison to the traditional 'evolution by creeps'.

Acceptance

Many ordinary people were outraged by Darwin's ideas, which seemed to fly in the face of religious ethic. But most scientists quickly saw how his proposals fitted with many strands of evidence. These included the similarities between various groups of living things, leading to how they are grouped or classified, and the evidence of changing fossils through time. Since Darwin's day many more strands of evidence have supported the idea of evolution. They include the basic nature of genes, and the structure of the chemical DNA that forms them, which is shared by nearly all living things. Some genes that control basic biochemical pathways, such as breaking apart energy-rich food substances to obtain energy for life processes, are found in identical form throughout vast numbers of hugely varied organisms. This is powerful evidence for the evolutionary process and the relatedness of living things.

CLASSIFICATION OF FOSSILS

For a long time, naming and classifying living things, and their fossils, was based on a system devised in the mid-18th century. However recent years have seen the rise of a newer system, which its supporters contend is based more on logic and less on judgement, known as cladistical analysis.

Most fossil-hunters name their finds according to a system devised in 1758 by Swedish-born plant biologist Carolus Linnaeus (Carl von Linné, 1707–1778). In his vast work *Systema Naturae* Linnaeus proposed that each kind or species of living thing should have two names. The first was its genus name – a genus being a group of similar or closely related species. The second name was unique to the species. From about 1780 the Linnaean system was gradually applied to fossils as well as living or extant species. So one of the genus names for certain kinds of shellfish called brachiopods (lampshells) which are common as fossils is *Lingula*. Individual species include *Lingula gredneri*, *L. anatina*, *L. cornea*, and dozens of others. Genus and species names are usually printed in *italics*, while names of other groups, like classes or orders, are not.

The origins of names

Usually the scientific names for living things and fossils are derived from Latin or another ancient language. They are international and recognized

Below: Carolus Linnaeus (1707–1778), from a portrait completed after the death of the 'Founder of Taxonomy'.

Above: For many years experts have debated whether modern-day birds evolved from dinosaurs. In recent years the evidence has supported the answer 'Yes', and has also shown that feathers are not the unique possession of birds, as was once believed. Fossils of tyrannosaurs recently discovered in China show that these great predators had a partial covering of fine, downy feathers. A much smaller feathered dinosaur, Microraptor gui, is shown right.

by scientists all around the world, no matter what their own spoken or written languages. The meaning of a name may be based on a particular feature of the genus or species concerned, or perhaps where it is found, or who first discovered it.

Taxonomy

Linnaeus' binomial nomenclature (giving two names as a unique identification) has been extended and refined into a hierarchy of classification for all living things. The science of classifying organisms, living or extinct, is known as taxonomy, and its basic unit or 'building block' is the

species. The species included in the same genus are selected by studying similarities and differences. The same is done with genera when they are grouped into a family, and so on, into larger groups. The biggest or topmost groups are kingdoms. Working from the top down:

- Kingdom
- Phylum (animals) / Division (plants)
- Class
- Order
- Family
- Genus
- Species

Levels of identification

The level of identification for a fossil, within the hierarchy of groups, varies hugely – according to practical considerations, amount of expertise, resources available, and scientific needs. Some amateur palaeontologists consider the label 'trilobite' is enough for a particular specimen. This is the everyday version of the scientific name for the class Trilobita. However other enthusiasts with more time and resources may wish to attempt a more accurate identification, perhaps even down to species level, such as:

Kingdom Animalia
Phylum Arthropoda
Class Trilobita
Order Phacopida
Family Calymenidae
Genus *Calymene*
Species *Calymene blumenbachii*

Below: The name of the trilobite Calymene blumenbachii may be rather a mouthful, so in England it has the nickname of the 'Dudley bug' after the Midlands region where this fossil species is especially common.

Traditional classification

Birds are a separate and higher rank (a Class) than dinosaurs (an Order).

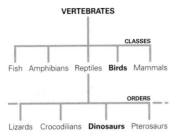

VERTEBRATES

CLASSES

Fish Amphibians Reptiles **Birds** Mammals

ORDERS

Lizards Crocodilians **Dinosaurs** Pterosaurs

Above: Taking the classification of dinosaurs and birds as an example, we can see how the organization of animal groups differs radically in the traditional and cladistic methods. In the former, Birds are given the rank of class, equivalent to that of Reptiles. Dinosaurs are a subgroup of Reptiles, generally with the rank of order (Dinosauria). In cladistical analysis (shown right), Birds have a much lower rank than Dinosaurs, and are in fact part of the dinosaur group or clade.

Cladistics

From the 1960s, a different method of grouping living and extinct organisms has gradually been taking over from the traditional Linnaean version, especially among professional biologists, palaeontologists and life scientists. It is known as the cladistic method of phylogenetic systematics. It was devised by German biologist and expert on flies, Willi Hennig (1913–1976).

In the Linnaean system, the features or traits used to group organisms are often a matter of discussion or opinion. In cladistics, these features are specially identified as synapomorphies or 'unique derived characters'. Each unique derived character of feature arose only once. Descendants of the ancestor in which the character arose all share it. They, and only they, form a group known as a clade. This is an evolution-based group – all members of a clade are descended from the same common ancestor. As evolution continues, different derived characteristics evolve, so clades split

Left: Computers are massively powerful aids for comparing the features and relatedness of both living organisms and fossils, especially using the cladistic approach.

Cladistic view

Birds appear as a sub-group within the dinosaurs clade.

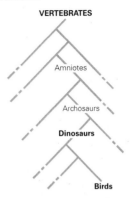

VERTEBRATES

Amniotes

Archosaurs

Dinosaurs

Birds

into subclades, and so on, and on. This creates the family tree type format shown above.

Cladograms

For example, assume that birds evolved from small, meat-eating dinosaurs. Then birds should be included within the dinosaur clade, as part of it. In the Linnaean system, birds are a separate and in fact larger group than dinosaurs. Birds are the class Aves, while dinosaurs are the order Dinosauria. Cladistics uses branching diagrams known as cladograms to show the evolutionary relationships between groups of organisms. The diagrams show branching points where new derived characteristics, and therefore new groupings, appear. This testable system is less prone to opinions and errors than the Linnaean version, and its administrative logic can be applied to other areas of life besides science. However for most practical and everyday purposes, the overall groupings of the Linnaean system are still in use.

Right: Limpets are included within the gastropod group, and given the genus name Patella, due to their similarity in shape to the human kneecap bone of this name.

FAMOUS FOSSILS

Fossils become famous for various reasons, such as scientific significance, scarcity or size. On the negative side, some specimens are more infamous or notorious, with links to tales of skulduggery, fossil-rustling and similar underhand tactics.

What makes a fossil famous? Usually, a mixture of reasons. The specimen might open up a whole new area of science, for example, by establishing a major and previously unsuspected group of extinct organisms. Its discovery could be a tale of romance and adventure, with fossil-hunters braving danger and taking risks to bring home their prize. The find might set a new record as the biggest, smallest or oldest of its kind. Or the fossil could turn up in a very unexpected place, chanced upon by a keen amateur who is suddenly catapulted into the media spotlight.

Conodonts

Tiny fossils commonly known as Conodonts or 'cone-teeth' resemble various tooth-like structures – long single fangs; many teeth in a line, like a crocodile's jaw; rows of spikes like teeth on a comb, either curved or straight; and other variations. Many are the size of sand grains and their wonderful shapes and arrangements can only be appreciated through a powerful hand lens or binocular microscope. These items are a common find in marine sediments worldwide during most of the Palaeozoic Era. They remained a mystery until the 1980s, when whole animals were

Above: Tooth-like conodont fossils from the Silurian Period, discovered in Russia. The real nature of these tiny yet intricate objects remained a mystery for centuries.

found which probably possessed them. A conodont creature resembled a long, slim fish or eel, roughly finger-sized or smaller, with two large eyes and blocks of muscles along the body. The conodont 'teeth' were arranged in the throat or neck region, but not in the jaws. Conodont creatures seem unrelated to any other groups of marine animals, living or extinct.

Hallucigenia

One of the best-known yet most puzzling creatures from the fossils in the 530 million-year-old Burgess Shales is *Hallucigenia*. It was described in 1977 as a surreal creature from a dreamworld – a long and vaguely worm-like body, blob-like head and narrow curved tail (or the reverse), seven pairs of stiff spines below for walking, and a row of about seven wavy tentacles above, each possibly tipped with a pincer-like 'mouth' for

feeding. However further finds shed more light on *Hallucigenia*. In 1991 a revised reconstruction showed the original version had been upside down. The twin rows of spines pointed upwards for protection. There were twin rows of tentacles too, not one. These pointed downwards and the 'mouths' were in fact feet used for walking. Hallucigenia may have been a very early type of onychophoran or velvet-worm, a group which includes *Peripatus* and still survives today, mainly in tropical rainforests.

Dimetrodon

Names are important for fossil-hunters. *Dimetrodon* was a three-metre-long predator from the Early Permian Period, some 280 million years ago. It had long fang-like teeth and a 'sail' of skin rising from its back, held up by long rods of bone. Its remains are plentiful in the fossil-rich 'Red Beds' of Texas, and have also been found in Oklahoma and in Europe. This powerful and impressive predator is often called a 'dinosaur'.

Below: The fossil Hallucigenia sparsa, *now held to have been a type of worm, found in The Burgess Shales, British Columbia.*

But it was not. It lived 50 million years before the first known dinosaurs. It belonged to a reptile group known as the pelycosaurs, which were the dominant large land animals of the time. The pelycosaurs were in turn part of the group known as synapsids or 'mammal-like reptiles'.

Mosasaurus

This fossil is mentioned several times in this book. It comprises parts of the skull, jaws and teeth of a large marine reptile. The remains were found in a chalk mine near Maastricht, Netherlands in 1770. In 1808 they were called *Mosasaurus* by Georges Cuvier, after the local region of the River Meuse. They led the renowned and influential Cuvier to a new proposal for the time, that many creatures had become extinct in previous catastrophes. Following many more similar finds, mosasaurs are now regarded as a group of large predatory sea reptiles from the Cretaceous Period. Some reached 15m/49ft in length. Their closest living relatives

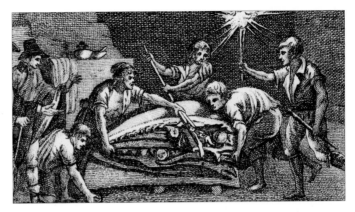

may be the big lizards known as monitors, such as the Nile monitor and Komodo dragon.

Forgeries

In 1912 remains of a human-like skull and ape-like jawbone were found at Piltdown, East Sussex, England. Experts hailed a marvellous new species of early human, *Eoanthropus dawsoni*. In 1953 careful scientific

Above: An engraving depicting the excavation of Mosasaurus *near Maastricht, published in a natural history study of the area in 1799.*

analysis revealed 'Piltdown Man' as a hoax. The 'fossils' were part of a amalgamation of a mediaeval human skull and an old orang-utan jawbone and chimpanzee tooth stained to match in colour. Piltdown still serves as a stark warning to over-zealous fossil sleuths everywhere.

Evidence of parental care

The parrot-beaked dinosaur *Psittacosaurus* from the Early Cretaceous Period is well known from many fossil finds in Asia. In 2004 a new discovery from Liaoning, China showed an adult *Psittacosaurus* surrounded by up to 34 part-grown youngsters. The amazing conclusion is that young from several parents formed a 'creche' that was looked after by one adult, perhaps while the others were feeding. These types of exciting discoveries allow us to reconstruct the behaviour of extinct animals with greater detail, showing that some dinosaurs were far from slow, stupid, simple beasts.

Below: Psittacosaurus had a hooked beak for snipping off vegetation and rows of crushing cheek teeth.

*Above and below: Two modern-day reptiles, the Nile monitor (*Varanus niloticus, *below left) and Komodo dragon (*Varanus komodoensis,*

below right), may be closely related to prehistoric reptilian mosasaurs such as the species Platycarpus ictericus *(above).*

LIVING FOSSILS

Fossils cannot come to life, of course. But the close similarity of extant organisms to extinct ones has led to the concept of the 'living fossil'. This applies to plants and animals, but caution is needed when drawing conclusions from what may be misleading superficial comparisons.

The term 'living fossil' is something of a misnomer. Its usual implication is that exactly the same kind of plant or animal which was alive millions of years ago is still alive today. This is only partly true. Usually it is members of the same group which survive, such as a family or genus, rather than the actual species. However the survivors are largely unchanged from their cousins of millions of years ago.

Horsetails

The simple, flowerless plants called horsetails include the genus *Equisetum*, with about 20 species around the world. They are common 'weeds' of disturbed ground, each with a tall stem bearing whorls of

Left and below: Pictures from fossil horsetail specimens (left) display remarkable similarity with living species (below), even to the number of leaf-ribs in each whorl.

slim leaves like the ribs on an upside-down umbrella. Horsetails belong to the plant group called sphenopsids. They were far larger and more common in ancient times, with types such as *Equisetites*, *Asterophyllites* and *Calamites*. Their growths formed much of the vast Carboniferous forests and became preserved in coal.

Maidenhairs (ginkgos)

The maidenhair tree *Ginkgo biloba* is the only living species of a large group that flourished from the Early Permian through to the Tertiary Periods. Ginkgoes were especially common during the Jurassic, when giant dinosaurs walked among them and browsed their leaves. Today's maidenhair tree grows to 30m/130ft in height and has the characteristic fan-shaped leaves of its group. It was cultivated in the gardens of palaces and temples of Ancient China, its geographical home. In the early 18th century it was 'discovered' by European plant collectors and is now an ornamental park and street tree in many parts of the world.

Magnolias

The group known as angiosperms or flowering plants, which today includes familiar flowers, herbs, blossoms and broadleaved or deciduous trees, arose in the Cretaceous Period. Among the early types were magnolias, dating back more than 110 million years. Today's main ornamental types, with their large white and perhaps pink-tinged early blooms, are derived from Ancient Chinese species. Other flowers and trees which have changed little since Cretaceous times include waterlilies, planes and laurels.

Above: A reconstruction of a prehistoric brachiopod (of the Terebratula genus). Modern day relatives of these mollusc-like creatures look very similar to fossils (see opposite).

Lampshells (brachiopods)

From the outside, a lampshell or brachiopod ('arm-foot') looks similar to a bivalve mollusc such as a mussel, clam or oyster. However the two are very different. Lampshells have been

Above and below: Reconstructions of Cretaceous magnolias (below) show the same flower arrangement as the modern day Magnolia species, such as this Magnolia reliata (above).

Above: This Jurassic brachiopod fossil (Lingula beani) exhibits a shiny lustre created by its mineral content (calcium phosphate). It has a characteristic lampshell shape.

around since the Cambrian Period, more than 500 million years ago. Their two shells are usually very similar but one is larger than the other, and each shell is symmetrical from side to side (unlike most bivalves). More than 30,000 kinds of fossil lampshells are known. There are only 350 living species and most, such as *Lingula*, are filter-feeders in the deep sea.

Nautiluses

The few living species of pearly nautilus (including *Nautilus pompilius*) are the only remnants of a vast group of molluscs that were

very common predators in Palaeozoic seas. Today's living nautiluses are large-eyed, curly-shelled, many-tentacled hunters of fish and other prey. They dwell in the deep sea. But there were more than 2,500 prehistoric types including straight-shelled nautiluses more than five metres long. Nautiloids are sometimes confused with ammonoids and belemnoids, which were also cephalopod molluscs and also extinct. Their living cousins are octopuses, squid and cuttlefish.

Tuataras

The tuatara resembles an iguana lizard. But it belongs to a very different group of reptiles called rhynchocephalians, which thrived at the time of the early dinosaurs, some 250–200 million years ago in the Triassic Period. Rhynchocephalians like *Scaphonyx* of the Middle Triassic were bulky plant-eaters almost two metres long, weighing nearly 100kg/220lb. Only two tuatara species now survive, on rocky islands off the coast of New Zealand.

They grow to about 60cm/24in long and come out of their burrows at night to feed on worms, insects and spiders.

Above and below: The 'relict reptile', the tuatara (Sphenodon, above), gives a rare and valuable insight into the evolution of its group, the rhynchocephalians (below) that were common during the Triassic Period.

Left: Millions of fossil nautiloid specimens (like the example left) confirm the close relationship with the living representatives of the deep sea, such as Nautilus pompilius (below).

An extinct species returns to life

The coelacanths are bony fish which have 'lobe-fins' (Osteichthyes) – a fleshy muscular base or stalk to each fin. They belong to the group Sarcopterygii, in contrast to the Actinopterygii or 'ray fins', the group that contains the vast majority of bony fish. Coelacanths first appeared way back in the Devonian Period. They reached 3m/10.5ft in length during the Early Cretaceous Period but faded away. Thought to be extinct from about 70 million years ago, a living coelacanth was caught off the coast of East Africa in

1938. Several specimens have been obtained since, and a second species, the Indonesian coelacanth, was discovered in the late 1990s. These fish grow to about 1.8m/6.5ft long, live in deep water, eat crabs and similar prey, and have a curious three-lobed tail.

Below: A reconstruction of an ancient (left) and modern day (right) coelacanth. These fish have many intriguing features, including a small tassel-like addition to the two main lobes of the caudal (tail) fin.

FOSSILS AND US

Since ancient times, people have speculated that we, as humans, evolved from some type of ape-like ancestors. The search for fossils of our ancestors and extinct cousins is one of the most exciting and controversial areas of palaeontology.

More than 2,000 years ago Aristotle of Ancient Greece wrote that 'some animals share the properties of Man and the quadrupeds, as the ape, the monkey and the baboon'. When Charles Darwin proposed the theory of evolution by natural selection in 1859, he was vilified for daring to suggest that our ancestors were some kinds of monkeys or apes. In fact, in *On The Origin of Species*, his only comment about human evolution was that 'Much light will be thrown on the origin of man and his history'.

The quest for 'missing links'

In 1856 some human-like bones were discovered by workmen mining limestone in a cave in the Neander Valley near Dusseldorf, Germany. The skull has a low brow, very thick bones and heavy jaw. Some experts dismissed it as 'an individual affected with idiocy and rickets'. But others began to think seriously that remains of ancient humans, including our ancestors, were preserved in rocks. The search for 'missing link' fossils had begun.

Below: Donald Johanson, discoverer of 'Lucy', with a plaster cast of her skull.

Above: Cast of an original skull of the 'Taung child', Australopithecus africanus, discovered by Raymond Dart in Taung, Cape Province, South Africa.

Gathering evidence

One of the first major discoveries was made by Dutch scientist Eugene Dubois near Trinil, Java in 1890–92. It consisted of part of an upper skull, a piece of jawbone with tooth, and a thigh bone (femur), and soon became known as 'Java Man'. Dubois named it *Pithecanthropus erectus*, and stated: 'I consider it a link connecting apes and men.' In 1927 a team led by Canadian anatomist Davidson Black excavated 'Dragon Bone Hill' at Zhoukoudian (Chou'kou'tien) near Beijing, China. They found many bones of another missing link which they named *Sinanthropus pekinensis*, 'Peking Man'. With continual revision of fossils and their names, both of these discoveries are now included within the species *Homo erectus*. Meanwhile in 1925 Australian anatomist Raymond Dart named a much more ape-like skull from Southern Africa, informally known as the 'Taung child', as *Australopithecus africanus*.

No longer missing

Gradually more fossils came to light of prehistoric humans. One of the most celebrated finds was 'Lucy' – the two-fifths complete skeleton of a species known as *Australopithecus afarensis*, found in Hadar, Ethiopia. It was

discovered in 1974 by Donald Johanson and represented a small 1m/3.5ft-tall ape-like creature from more than three million years ago, which could walk upright almost like us. Many experts now believe that 'Lucy' was in fact a male and have renamed him 'Lucifer'. Further finds in the 1990s and 2000s include even more ape-like creatures from even longer ago. They include *Ardipthecus ramidus* from Aramis, Ethiopia, whose fossils are almost four and a half million years old, and *Orrorin tugenensis*, a more controversial discovery from Kenya, which is some six million years old. Ethiopia was again the site of a remarkable discovery in September 2006: the skeleton of a 3 year-old child, more or less complete, was uncovered by scientists based at the Max Planck Institute for Evolutionary Anthropology, Leipzig, Germany. Although older than 'Lucy', the child was also a member of *A. afarensis*, and her lower limbs again suggested a

Below: Dr G.H.R. von Koenigswald examining the upper jaw of Pithecanthropus robustus, once known as Java Man. This species was later assigned to the genus Homo as an example of Homo erectus.

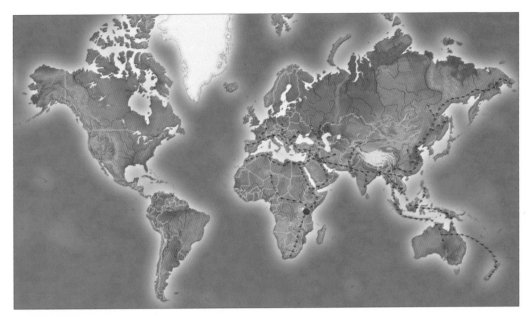

Above: Our own species, Homo sapiens, *is thought to have appeared in East Africa less than 200,000 years ago (as shown by nucleus on map), and then spread mainly to the north and east (indicated by direction of arrows).*

The 'hobbit humans'

In 2004 an astonishing find on the island of Flores, Indonesia, overturned many ideas about recent human evolution. It was believed that since the extinction of the last Neanderthal and *Homo erectus* people by 25,000 years ago, our own species *Homo sapiens* was the only hominid on the planet. However it seems that as recently as 13,000 years ago, diminutive humans lived on Flores. Only one metre tall, they are known as *Homo floresiensis*. The skulls of these mini-people are hardly the size of grapefruits.

Yet relatively sophisticated stone tools and the remains of charred animal bones have also been found at the site. This raises huge questions about brain size related to the intelligence needed for making tools and using fire. The size reduction of species when confined to an island habitat – a phenomenon known as 'island dwarfism' – appears many times in the fossil record, and can be seen in mammoths and elephants. However the conclusion that the Flores remains represent an entirely separate species of human is disputed until, as palaeontologists tend to say in furtherance of their profession, 'further evidence is forthcoming'.

Below: Debate surrounds the Flores hominids. Was the main species, reconstructed below, small but otherwise normal, or suffering from some kind of growth malformation?

capacity to walk upright. Her ape-like arms and shoulders, however, indicated that she may also have spent much of her life sheltering in trees.

'Found links'

Apart from fossils, the evidence of genes, DNA and other molecules within the bodies of ourselves and chimpanzees show that they are our closest living relatives. It is estimated that humans and chimps split from a common ancestor some seven or eight million years ago. Our 'branch' of the evolutionary tree makes up the family Hominidae. Some palaeontologists contend that it includes 20 or more species, many overlapping in time and place, especially in Africa. Others propose as few as eight species. It seems that the 'cradle of human evolution' was indeed Africa, that upright walking began several million years ago, that tool use dates back more than two million years ago, and that brain size has tended to increase through all this time.

DIRECTORY OF FOSSILS

No single bookbound collection of fossil images could ever incorporate every group and individual kind of organism that has left its record in the rocks. However, one can strive to create a soundly representative compilation of the common and recognizable forms – importantly, from different times and places around the world – plus a few glimpses of the most rare and precious of all fossils.

The following pages include many groups of organisms which are familiar to hordes of amateur fossil enthusiasts, such as ferns, corals, marine molluscs and echinoids (sea urchins), but which, nevertheless, retain their own beauty, fascination, and importance, especially in the context of a local collection. There are no apologies for including a few tantalizing examples of momentous finds that would make any palaeontologist gasp with awe and glow with a sense of achievement.

The directory is organized into three major traditional groupings: plants, invertebrates (mostly non-chordate) animals, and vertebrate (most chordate) animals. Each grouping is generally subdivided into the conventional taxonomic assemblages, such as, taking members of the phylum Mollusca as an example, the bivalves, gastropods, ammonites and others. However, as noted earlier in this book, there are disagreements among experts about the extent or separation of certain groups. And the ways in which these groups are ranked are being continually altered and refined with the aid of scientific scrutiny and the logical framework provided by the system of cladistical taxonomy (see Classification of Fossils in the previous chapter).

Left: The Dinosaur Provincial Park in Alberta, Canada, a designated World Heritage Site, demonstrates the 'pulling power' of important fossils.

HOW TO USE THE DIRECTORY OF FOSSILS

The information in the directory is easy to locate, thanks to its logical and consistent framework. The featured specimens range from fossils of simple organisms to the more complex and evolutionarily advanced, and are grouped as plants, invertebrate animals and vertebrate animals.

The following pages provide easily discerned and digested information on many hundreds of fossil specimens: through the use of photographs of the fossils themselves, often annotated to assist identification; through drawings, and of course through the arrangement of the accompanying text. The name of each specimen is usually given at the level of genus, or species 'group' – that is, above the rank of species and below the rank of family. Conventionally, the names of genera are italicized in the running text, but not always when standing alone, such as in headings.

Occasionally a full species is identified, which then has two Latin names, such as *Homo neanderthalensis*. This is the standard binomial method of classifying and naming species. The first name is the genus; this always begins with an upper-case letter. The second name is for the species itself, and all letters here are lower-case.

Descriptions and terminology
The text attempts to describe each specimen in a straightforward way. However some technical terms are included as important 'shorthand',

especially for anatomical features used in identification. For a group of related organisms, these terms are usually explained in at least one of the specimen descriptions, but not all, to avoid unnecessary repetition. If you cannot immediately find the meaning of a term, it is worth checking adjacent specimen entries for explanations. There is an extensive glossary of terms near the end of the book. All of the individual specimen names, in their common and scientific forms, and key palaeontological terms, are included in an extensive index.

Main headings
These identify the major group of fossil organisms covered in that section (such as Molluscs), and, where applicable, subgroups (such as Ammonites). The major groups are often (but not always) equivalent to the classification levels known as divisions in plants and phyla in animals.

Introductory text
This introduces the group, with its main characteristics such as anatomical features and geological time span. Where a large group or even subgroup (such as Ammonites, shown here) extends over a number of pages, the continuation is indicated in the main heading.

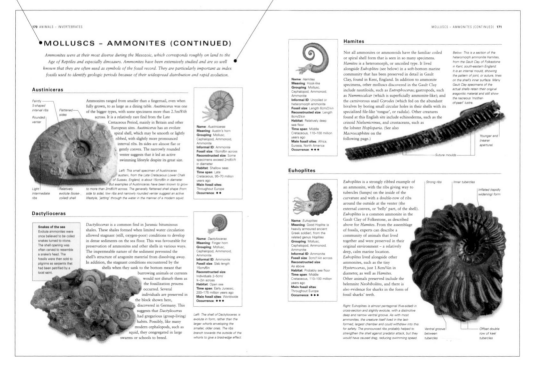

Scientific name
The internationally accepted scientific name for the organism is usually based on ancient Latin or Greek, or commemorates its finder or discovery site. Names are for genera, or species.

General description
The main text supplies useful information about the organism, such as its growth habits and ecology, its grouping and relationships, and methods of preservation where significant.

Factfile data
This panel summarizes the main categories of data for the specimen, as applicable and as available. It is shown below left in expanded form. (A? indicates uncertain data.)

Reconstruction
Illustrations recreate organisms as they may have appeared in life. However some fossils provide limited insight, so it should be borne in mind that reconstructions are conjectural.

• Dactylioceras

Snakes of the sea
Evolute ammonites were once believed to be coiled snakes turned to stone. The shell opening was often carved to resemble a snake's head. The fossils were then sold to pilgrims as serpents that had been petrified by a local saint.

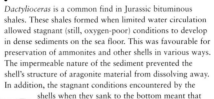

Dactylioceras is a common find in Jurassic bituminous shales. These shales formed when limited water circulation allowed stagnant (still, oxygen-poor) conditions to develop in dense sediments on the sea floor. This was favourable for preservation of ammonites and other shells in various ways. The impermeable nature of the sediment prevented the shell's structure of aragonite material from dissolving away. In addition, the stagnant conditions encountered by the shells when they sank to the bottom meant that burrowing animals or currents would not disturb them as the fossilization process occurred. Several individuals are preserved in the block shown here, discovered in Germany. This suggests that *Dactylioceras* had gregarious (group-living) habits. Possibly, like many modern cephalopods, such as squid, they congregated in large swarms or schools to breed.

Name: *Dactylioceras*
Meaning: Finger horn
Grouping: Mollusc, Cephalopod, Ammonoid, Ammonite
Informal ID: Ammonite
Fossil size: Slab length 15cm/6in
Reconstructed size: Individuals 2–5cm/ ¾–2in across
Habitat: Open sea
Time span: Early Jurassic, 200–175 million years ago
Main fossil sites: Worldwide
Occurrence: ◆ ◆ ◆

Left: The shell of Dactylioceras is evolute in form, rather than the larger whorls enveloping the smaller, older ones. The ribs branch towards the outside of the whorls to give a braid-edge effect.

Tinted panel
Occasional panels provide background or allied information, for example, about living relatives or about the period in which the specimen existed.

Specimen image
The photograph has been taken from a view that provides maximum visual detail. Many specimens include labels drawing out important features.

Caption
This text points out diagnostic or important features in the specimen image, and adds details that may not be visible in the view.

Name
The specimen's scientific name, usually given at the genus level.

Grouping
The list of names descends through the taxa, that is, from larger to smaller groups.

Fossil size
Approximate dimensions of the specimen itself or the slab in which it is embedded.

Habitat
The broad habitat or environment of the organism when alive.

Main fossil sites
Broad geographical range of the sites of modern discoveries.

• **Name**: *Dactylioceras*
Meaning: Finger horn •
• **Grouping**: Mollusc, Cephalopod, Ammonoid, Ammonite
Informal ID: Ammonite •
• **Fossil size**: Slab length 15cm/6in
Reconstructed size: •
Individuals 2–5cm/ ¾–2in across
• **Habitat**: Open sea
Time span: Early Jurassic, 200–175 million years ago •
• **Main fossil sites**: Worldwide
Occurrence: ◆ ◆ ◆ •

Meaning
Most scientific names allude to features, people, places or events.

Informal ID
The casual or everyday name by which the specimen is known.

Reconstructed size
Dimensions of organism, within a range or up to a likely maximum.

Time period
Usually expressed in geological periods, with 'millions of years ago' to localize within that period.

Occurrence
A sliding scale, from ◆ for rare to ◆ ◆ ◆ ◆ for very common.

Geological time systems
There are several systems for organizing the Earth's history into time spans, though most have the same eras, periods and epochs or stages/ages.
• The terms Early, Middle (or Mid) and Late differ between schemes, with some periods lacking a Middle.
• Nomenclature may differ between regions. The Early Triassic period may consist of the Griesbachian, Dienerian, Smithian and Spathian, or (covering the same times) the Induan and Olenekian. Some regions term the latter two of these the Feixianguanian and Yongningzhenian, respectively.
• The Cenozoic Era may consist of the Palaeogene (65–23mya) and Neogene (23–1.8mya) Periods, and in some schemes the Quaternary Period (1.8mya to today).

MICROFOSSILS

Many types of life have left microfossils – remains that are too small to see or identify individually with the naked eye. These include single-celled life-forms, such as bacteria, cyanobacteria or 'blue-green algae', protists or protoctists (protozoa and protophyta), and also micro-sized products of larger organisms, such as the spores of fungi, mosses and ferns, and the pollen grains of conifers and angiosperms.

Bacteria

Below: This computer-coloured SEM (scanning electron micrograph) shows fossilized bacteria as red sausage- or rod-like blobs. They are calcified on the sculptured shell surface of a single-celled life form called a foraminiferan. 'Forams' have beautifully geometric chambered shells, or tests, sometimes resembling large seashells, and are themselves preserved in vast numbers. Foram shells form an important component of chalk and many deep-sea muds and oozes.

Bacteria are generally regarded as being some of the earliest forms of life to have taken hold on the planet. They appeared in the 'primeval soup' of organic compounds in Earth's ancient seas more than 3,000 million years ago, and since then have been evolving, diversifying and dying out – just like bigger plants and animals, but in their trillions and in all habitats. Bacterial microfossils may be ball-shaped (cocci), rod-like (bacilli), corkscrew-shaped (spirilli) or take on many other forms. These early bacteria began to break apart the chemicals and minerals in the water in order to gain energy for life, growth and reproduction, and are known as chemosynthetic organisms. Some bacteria thrive in much the same way today in the mineral-rich, superheated water of deep-sea hot springs known as hydrothermal vents.

Name: Bacteria (various kinds).
Meaning: Little sticks.
Grouping: Monerans (single-celled life forms having no proper nucleus or 'control centre' in the cell).
Informal ID: Bugs, bacteria, microbes, germs (if harmful).
Fossil size: Most types 0.2–20µm (⅕,₀₀₀–¹⁄₅₀mm/ ¹⁄₁₂₅,₀₀₀–¹⁄₁,₂₅₀in).
Reconstructed size: As above.
Habitat: Warm, shallow seas at first, gradually spreading to every habitat on Earth.
Time span: Early Precambrian (Archean), more than 3,000 million years ago, to today.
Main fossil sites: Worldwide.
Occurrence: ◆ ◆

Cyanobacteria

Below: This light micrograph shows fossilized Chlorellopsis, a freshwater cyanobacterium in Eocene rocks (55–33 million years ago) from Wyoming, USA. Each individual globular cell is about 100µm (0.1mm/¹⁄₂₅₀in) across. Cyanobacteria may be solitary or colonial, spherical, cylindrical or polygonal, and form clumped masses or long, hair-like filaments.

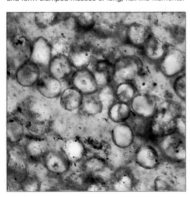

Cyanobacteria, like bacteria, are monerans – single cells lacking the membrane-contained nucleus, or 'control centre', found in cells of more complex organisms. Cyanobacteria are sometimes called blue-green algae. They are not true algae, but they are greenish, blue-green or red-brown and live by trapping light energy using photosynthesis. Cyanobacteria live in all aquatic habitats and are familiar as the 'scum' on ponds and also as 'toxic tides' when they bloom and release harmful byproducts. As they evolved and became more numerous, from about 3,000 million years ago, their photosynthetic activities produced oxygen, which accumulated in the water and then the atmosphere of the early Earth. The oxygen changed the previously poisonous air by, for example, screening out more of the sun's harmful ultraviolet rays, and such atmospheric changes allowed more complex oxygen-using life-forms to evolve.

Name: Cyanobacteria (various kinds).
Meaning: Blue little sticks (see Bacteria).
Grouping: Monerans (single-celled life forms having no proper nucleus or 'control' centre' in the cell).
Informal ID: Blue-green algae, microbes.
Fossil size: Most individuals 0.01–1mm/¹⁄₂,₅₀₀–¹⁄₂₅in.
Reconstructed size: As above.
Habitat: Warm, shallow seas at first, gradually spreading to most aquatic habitats and damp land habitats.
Time span: Early Precambrian (Archean), more than 3,000 million years ago, to today.
Main fossil sites: Worldwide.
Occurrence: ◆ ◆

Radiolarians

Name: Radiolarians.
Meaning: Little radii (circular, wheel- or spoke-like patterns).
Grouping: Protists, Sarcodinans, Actinopods.
Informal ID: Bugs, microbes, microplankton, zooplankton.
Fossil size: Average 0.1–0.5mm/$\frac{1}{250}$–$\frac{1}{50}$in diameter, some species more than 1mm/$\frac{1}{25}$in.
Reconstructed size: As above.
Habitat: Warm shallow seas at first, gradually spreading to most marine habitats.
Time span: Very Late Precambrian, about 550 million years ago, to today.
Main fossil sites: Worldwide.
Occurrence: ◆ ◆

Radiolarians are protists or protoctists – single-celled, animal-like life-forms, often called protozoa. Each individual 'body' is made of soft jelly similar to that of the well-known pond-mud protist amoeba. It is two-part, with the cell's main control centre, or nucleus, and inner jelly-like endoplasm within a porous container or shelly capsule, and more jelly-like ectoplasm, plus perhaps spines and long, thin filaments, around this. Characteristic of radiolarians after death are the beautiful and intricate 'skeletons', or shells, which are usually spherical, conical or cylindrical with many faces and pores (gaps or holes). They have been likened to chandeliers, and since in most radiolarians they are made from silicate (siliceous) minerals, which is the basic material of glass, these tiny organisms are said to live in 'glass houses'.

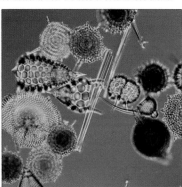

Below: Several different forms of fossilized radiolarians are shown in this light microscope photograph of material collected at Mount Hillaby, Barbados. Living radiolarians occur in marine plankton in countless trillions and are important constituents near the base of oceanic food chains.

Seeing microfossils

Palaeontologists will often use the following when studying microscopic detail in fossils:
• A hand lens – single-lens magnifier or 'magnifying glass' – is always a useful tool for studying small details of bigger fossils as well as for examining large microfossils. The lens power is usually on the order of 5 to 20 times magnification.
• A stereo microscope on a stand, with plenty of bright light falling on the specimen, is useful for smaller items. Most stereo microscopes, which have two complete sets of lenses, work in a similar way to binoculars and have magnifying powers of up to 50. For the enthusiastic fossil-lover this is a worthwhile investment since it has many other 'close-up' uses.
• A standard light microscope (LM, transmission light microscope) can enlarge a few hundred times, with a practical limit in most models of 2,000. The specimen may be lit from below so this microscope is usually suited to very thin, ground-down sections of fossils, through which light can pass. There are monocular (one eyepiece lens) versions and, to avoid squinting, binocular, or 'two-eyed', versions. Unlike a stereo microscope, they do not give a three-dimensional image with 'depth'.
• When higher magnifications are called for, the equipment becomes far more specialized, and expensive. The main viewing tools are the scanning electron microscope (SEM) and scanning tunnelling electron microscope (STEM/STM), which give realistic three-dimensional images that are magnified hundreds to thousands of times.

Coccoliths

Coccoliths are microscopic scale-like plates of calcium-rich or calcite minerals that were once the outer 'skeletons' (shells or tests) of single-celled, plant-like algal organisms known as coccolithophores. Each of these organisms secreted a protective chamber around itself, the coccosphere, composed of perhaps 30 or more coccoliths. At death, the coccolith plates of the casing separate, sink to the sea bed mud and, being formed of calcium carbonate, may become fossilized in chalky or limestone deposits. Some coccolithophores continually shed plates when alive, as they grow and multiply. The plates occur in a bewildering diversity of shapes and sizes, from crystal-like rods and blocks to rings and 'hubcaps'.

Name: Coccoliths.
Meaning: Berry stones.
Grouping: Protoctists (Protists), Haptophytes.
Informal ID: Plant microplankton, phytoplankton, calcareous nannoplankton.
Fossil size: Individual plates vary but average 1–10μm (up to 0.01mm/$\frac{1}{2,500}$in), many at 2–4μm ($\frac{1}{7,500}$in).
Reconstructed size: Individual whole organism 10–50μm ($\frac{1}{2,500}$–$\frac{1}{500}$in).
Habitat: Marine.
Time span: Triassic, about 225 million years ago, to today.
Main fossil sites: Worldwide.
Occurrence: ◆ ◆

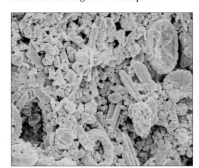

Left: This SEM (scanning electron micrograph) reveals a diversity of coccoliths from the famous Cretaceous chalk 'White Cliffs of Dover' near Folkestone, south-east England. Names such as 'calcareous nannoplankton' and 'calcareous microplankton' are used for coccoliths and similar fossils.

PLANTS

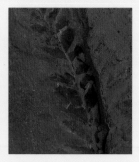

Most biologists offer the defining feature of the kingdom Plantae as photoautotrophism, or making one's own food with light – absorbing minerals and other raw materials from the surroundings, and using captured sunlight energy to convert these to sugars and new tissues. Fossils cannot show any of this directly. But the anatomy of fossil plants is often immediately comparable to their living cousins. This section aims to trace the appearance of the main plant groups through time, charting increasingly complex structures such as roots, stems thickened with woody fibres, and fronds or leaves. Much of the classification is based on modes of reproduction, from spores in the earlier groups, such as mosses and ferns, to the cones of gymnosperms. The angiosperms, broadly meaning 'enclosed seeds' or 'seeds in receptacles', were the last major plant group to appear, in the Cretaceous. They now dominate as flowers, herbs and broad-leaved, deciduous bushes and trees.

Above, from left: Fern imprint, Archaeopteris (the world's first tree), Sassafras.

Right: The Triassic 'petrified forests' of Arizona, USA, are literally trees turned to stone. Woody tissues reinforced by lignin are more resistant to decay, and specimens of the extinct conifer Araucarioxylon arizonicum were buried by stream sediments or volcanic ash before they could rot away, then mineralized as quartz.

EARLY PLANTS

The very first life on Earth was mainly microscopic. The first multicelled plants appeared in the sea, perhaps around one billion years ago. They are usually known as algae – the simplest category of plants, lacking true roots and leaves and reproducing by simple spores instead of flowers and seed-containing fruits. Well-known algae today include sea-lettuces, wracks, kelps, oarweeds and dulses.

Mixed algal–Chaetetes deposit

Below: This specimen, which incorporates algae and Chaetetes depressus, *is from varied Late Carboniferous limestones that were found on inland cliff deposits exposed at the River Avon Gorge in Clifton, Bristol, south-west England.*

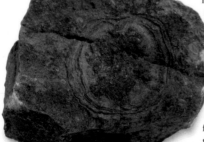

Chaetetes has long been something of a mystery. It has been interpreted as a coral of the tabulate (flat-topped or table-like) group, or an encrusting sponge (poriferan) of the coralline demosponge type. Specimens are often combined with encrusting algae rich in the mineral calcium carbonate, making varied forms of limestone. Similar fossil structures have been called stromatoporoids, or 'layered pores', and these, in turn, show similarities to a relatively newly identified group of living sponges – the encrusting sclerosponges. So *Chaetetes* may well have been a sponge rather than a coral. However, in fossils it is often difficult to differentiate between the algal, poriferan and coral elements. Similar algal–*Chaetetes* carbonate accumulations from the Ely Basin, dating to the Mid to Late Carboniferous, are found along the western edge of North America. 'Chaetetes Bands' are also common in northern England and several European regions, including Germany and Poland.

Name: Algal–Chaetetes limestone.
Meaning: —
Grouping: Encrusting alga (plants) and *Chaetetes* (probably poriferan or sponge).
Informal ID: Algal-sponge-coral accumulation, 'sea floor stone'.
Fossil size: Specimen 6cm/2¼in across.
Reconstructed size: Some formations cover many square kilometres.
Habitat: Warm, shallow seas.
Time span: Mainly Mesozoic, 250–65 million years ago.
Main fossil sites: Europe, North America.
Occurrence: ◆ ◆

Carboniferous algal deposit

Below: This Carboniferous structure may be a cyst – a tough-walled container enclosing the thallus, or main body, of the alga, possibly a frond. What looks like the 'leaf' of a seaweed-type alga is known as the frond, the 'stem' is termed the stipe, and some types have 'root'-like structures, known as holdfasts, to anchor them to rocks.

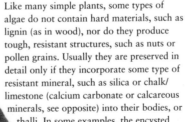

Like many simple plants, some types of algae do not contain hard materials, such as lignin (as in wood), nor do they produce tough, resistant structures, such as nuts or pollen grains. Usually they are preserved in detail only if they incorporate some type of resistant mineral, such as silica or chalk/limestone (calcium carbonate or calcareous minerals, see opposite) into their bodies, or thalli. In some examples, the encysted form shows a tough wall developed to resist drying or similar adverse conditions (see left). However, algal fossils have an immense time span extending back to the Precambrian Period, more than 540 million years ago, and have been used as index or indicator fossils to date marine deposits, especially in the search for petroleum oil.

Name: Carboniferous algal deposit.
Meaning: —
Grouping: Alga, Cyanophyte.
Informal ID: Seaweed.
Fossil size: 5cm/2in across.
Reconstructed size: Whole plant up to 1m/3¼ft.
Habitat: Warm seashores and shallow seas.
Time span: Precambrian, before 540 million years ago, to today.
Main fossil sites: Worldwide.
Occurrence: ◆ ◆

Coelosphaeridium

Name: Coelosphaeridium.
Meaning: Little hollow spheres.
Grouping: Alga, green alga.
Informal ID: Ball-shaped stony seaweed.
Fossil size: Individual specimen 7cm/2¾in across.
Reconstructed size: As above, forming large beds of many square metres.
Habitat: Warm seashores and shallow seas.
Time period: Mainly Palaeozoic, 540–250 million years ago.
Main fossil sites: Northern Europe.
Occurrence: ◆ ◆

Calcareous alga such as *Coelosphaeridium* lay down deposits of chalky or limestone minerals (calcium carbonate) and jelly-like substances within and sometimes around their tissues. This gives the plant a stiff, stony feel and aids preservation. *Coelosphaeridium* is common in certain parts of northern Europe, especially Scandinavia. It forms mixed beds along with other algae, such as *Mastopora* and *Cyclocrinus* (named from its original identification as a sea-lily or crinoid, a member of the echinoderm group, but now regarded by some authorities as a siphonean alga), and with various encrusting animal invertebrates, such as sponges and corals.

Below: These fossilized remains of the algal 'cells' of Coelosphaeridium are from Ringsaker, Norway, where they occur in large accumulations. They date back to the Ordovician Period, about 450 million years ago. The outer wall resembles bark with a pattern of radiating spoke-like elements enclosing a central chamber, where the radial structure is less defined.

Outer wall

Radiating spokes

Central chamber

Algal stromatolitic limestone (mixed composition)

Name: Algal stromatolitic limestone.
Meaning: Stromatolite = 'layer stone'.
Grouping: Algae.
Informal ID: As above, 'seaweed rock'.
Fossil size: Specimen 19cm/7½in across.
Reconstructed size: Large deposits can extend for kilometres.
Habitat: Warm seashores and shallow seas.
Time period: Precambrian, before 540 million years ago, to today.
Main fossil sites: Worldwide.
Occurrence: ◆ ◆

Fossilized stromatolites, or 'layered stones', were once thought to be produced by tiny animals known as protozoans, or by inorganic (non-living) processes of mineral deposition. However, comparison with rocky deposits found along many warm seashores today, famously Shark Bay in Western Australia, reveal that these humped or mound-like structures, generally varying in size from that of a tennis ball up to that of a family car, usually have an organic origin. Tiny threads of green algae and blue-green 'algae' (see Cyanobacteria, discussed earlier) thrive in tangled, low mats, covering themselves with slime and jelly for protection. These growths trap sand, silt and other sediments, as well as fragments of shell and other organic matter. New algal mats grow on top as the ones below harden, and slowly the mound builds up as a multi-layered stromatolite. In calm waters, the stromatolite shape is often rounded, resembling a burger bun, while strong currents cause them to elongate, like French baguettes (sticks), lying parallel to the current's direction.

Below: This specimen dates from the Late Carboniferous Period and is from Blaenavon, in Monmouth, Wales. The laminated, or layered, structure is typical of stromatolitic formations.

EARLY LAND PLANTS

Before the evolution of primitive land plants, some 440 million years ago, the land was rocky and barren except for mats of algae and mosses along the edge of the water. By about 400 million years ago vascular plants, such as the Rhyniales and Zosterophyllales, began to spread. These early plants reproduced by unleashing clouds of spores, in the manner of simpler plants, such as mosses and ferns, today.

Cooksonia

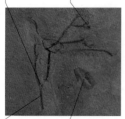

Y-shaped branching point | Terminal spore capsule

Stem | Eurypterid

Above: The presence of a eurypterid (sea-scorpion) in this sample, from the Upper Olney Limestone of Onondaga County, New York, USA, suggests that the rock was formed on the sea bed. So the plant had been transported and washed out to sea, perhaps by heavy rains.

One of the earliest and best-known vascular land plants appeared late in the Silurian Period. *Cooksonia* was a small and simple plant, yet it achieved worldwide distribution – although it is dominantly known from Eurasia and, especially, Great Britain. Its success may be largely due to its vascular system (see *Rhynia*, overleaf), which allowed water, minerals and sugars to be distributed throughout its entire body. *Cooksonia* had no leaves like modern plants, but it was probably green and photosynthesized (captured the sun's light energy) over its whole stem surface. It is recognizable by its characteristic smooth, Y-branching stems that end in spore capsules shaped like kidney beans.

Right: Features of Cooksonia *include the Y-shaped branching stem and the spore capsules (sporangia) borne singly at the end of each stem. Some specimens have just one branch forming a Y; others have five or more levels. The stems are smooth, rarely carrying surface features.*

Name: *Cooksonia*.
Meaning: Named for Australian palaeobotanist Isabel Cookson.
Grouping: Rhyniale, Rhyniacean.
Informal ID: Simple vascular plant, early land plant.
Fossil size: Smaller slab 4cm/1½in across.
Reconstructed size: Entire height usually less than 10cm/4in.
Habitat: Shores near rivers, lakes.
Time period: Late Silurian to Devonian, 410–380 million years ago.
Main fossil sites: Worldwide, especially Eurasia and Britain.
Occurrence: ◆ ◆

Zosterophyllum

Branching point | Laterally clustered sporangia

A late Silurian vascular plant, like *Cooksonia* (above), *Zosterophyllum* is distinguished by having Y-like and also H-shaped branches, where each branch is divided into two equal stems. Each stem was smooth, and the spore capsules were clustered in the manner of a 'flower spike' towards the end of the stem but on the sides, rarely the tip. The capsules split along the sides, unfurling like a fern to release their spores. Nutrients were shared between stems by a network of underground root-like rhizomes. Plants such as *Zosterophyllum* were probably ancestral to the giant lycopods, or clubmosses, that dominated the coal swamps of the Carboniferous Period.

Left: Zosterophyllum *had its oval- or kidney-shaped sporangia (spore capsules) on short stalks, clustered in long arrays along the sides of the terminal stem, rather than at its end (as in Cooksonia).*

Name: *Zosterophyllum*.
Meaning: Girded leaf.
Grouping: Zosterophyllale, Zosterophyllacean.
Informal ID: Simple vascular plant, early land plant.
Fossil size: Specimen length (height) 4.5cm/1¾in.
Reconstructed size: Entire height 25–30cm/10–12in.
Habitat: Edges of lakes, slow rivers.
Time period: Late Silurian to Middle Devonian, 410–370 million years ago.
Main fossil sites: Worldwide.
Occurrence: ◆ ◆

Gosslingia

Name:
Gosslingia.
Meaning: For
Gosling.
Grouping:
Zosterophyllale,
Gosslingian.
Informal ID:
Simple vascular
plant, early land plant.
Fossil size: Slab 4cm/
1½in across.
Reconstructed size: Height
up to 50cm/20in.
Habitat: Edges of lakes.
Time period: Early Devonian,
400 million years ago.
Main fossil sites:
Throughout Europe.
Occurrence: ◆ ◆

This fossilized remains of this early land plant show it to be smooth-stemmed with a distinguishing Y-shaped branching pattern. The sporangia (spore capsules) are scattered along the length of the stem on small side branches, rather than clustered at or near the end. In addition, some stem tips are coiled, much like those of a modern fern, in a form known as circinate. Many of these plants were washed downstream by floods and deposited in oxygen-deficient mud. These anaerobic conditions prevented decay and allowed the plants to fossilize as pyrite, or 'fool's gold' (known as pyritization), preserving in microscopic detail the cellular structure of the original plant as it was in life.

Below: Gosslingia shows the typical Y-shaped branching pattern of many early plants. However, in this specimen, small lateral or side branches are also evident, some bearing sporangia.

Sporangia (spore capsules)

Lateral branch

Main stem

Compsocradus

This genus of plant forms part of a group known as the iridopteridaleans, from the Mid to Late Devonian Period in Venezuela and the Carboniferous of China. The branches are whorled (grouped like the ribs of an umbrella) and vascularized for photosynthesis. Some of the uppermost branches divide up to six times before ending in curled tips, while others have paired spore capsules at their tips.

Right: Laterally compressed (squashed flat from side to side), this fossilized specimen of Sawdonia shows its spiny nature. In life, the stems may have bunched together, possibly to form a prickly thicket that gave mutual support and protection and allowed the plants to reach greater heights.

Sawdonia

Sawdonia is distinctly different from other early land plants in being covered with saw-toothed, spiny or scale-like flaps of tissue along its stems. However, these flaps of tissue were not vascularized – in other words, did not have fluid-transporting pipe-like vessels – and therefore cannot be considered as true leaves. Their function is not clear. They may have been spiky defences against early insect-like land animals. The presence on the flaps of stomata – tiny holes allowing the exchange of gases and water vapour between the inside of the plant and the surrounding air – suggests that they increased the photosynthetic surface area of the plant for greater capture of light energy. The sporangia of *Sawdonia* are found laterally (along the sides) and the branch tips are circinate, uncoiling like the head of a fern.

Name:
Sawdonia.
Meaning:
Saw-tooth.
Grouping:
Zosterophyllale,
Sawdonian.
Informal ID:
Simple vascular
plant, early land plant.
Fossil size: 23cm/9in.
Reconstructed size: Height
up to 30cm/12in.
Habitat: Shores.
Time period: Early Devonian,
400 million years ago.
Main fossil sites: Worldwide.
Occurrence: ◆ ◆

Stem branching

Saw-toothed tissue

RHYNIOPHYTES, HORSETAILS

A great advance in plants was vascularization, which consists of networks of tiny vessels – pipes or tubes
– within the plant. There are two main types: xylem vessels transport water and minerals; phloem carry
sugar-rich sap around the plant as its 'energy food'. Xylem vessels, in particular, became stronger and
stiffer in early plants. This gave rigidity to the stem, enabling these plants to grow taller and stronger.

Rhynia

One of the most famous early plants,
Rhynia is named after the village of Rhynie
in Aberdeenshire, Scotland. This area was
once marshy or boggy, probably with hot
springs that gushed water rich in silica
minerals. These conditions have preserved
various organisms, including *Rhynia* and
the insect-like creature *Rhyniella*, in
amazing detail in rocks known as Rhynie
cherts. *Rhynia* had horizontal growths,
or rhizomal axes, bearing hair-like fringes,
which probably had stabilizing and water-
absorbing functions, but were not true
roots. The horizontal growths bent to form
upright, gradually tapering leafless stems, or
aerial axes. These branched successively in a
V-like pattern, each with a spore capsule at
its tip. In the vascular system, each 'micro-
pipe' is made from a series of individual
tube-shaped cells attached end to end, with
their adjoining walls degenerated to leave a
gap – like a jointed series of short lengths of
hosepipe. In *Rhynia* the xylem vessels were
clustered in the centre with phloem vessels
around, forming a vascular bundle. Similar
arrangements are seen in vascular bundles in
many other plants, including flowers and
trees living today. The vascular bundles
branch out of the stem and into the leaves,
forming the thickened leaf 'veins'.

Below: This much enlarged view (25-30 times life
size) shows complete soft-tissue preservation, down
to the level of individual cells. The view is a transverse
section through the stem of a plant – much as a tree
trunk could be sliced across in order to reveal its
growth rings (see panel below). The bundles of tiny
tubes or vessels, phloem and xylem, are the key
feature of vascular plants. Xylem vessels conduct
water and minerals, which are usually absorbed from
the ground, to the leaves of the plant to sustain their
growth, while phloem vessels transport energy-rich
sap, the product of photosynthesis, away from the
leaves and around the rest of the plant.

Name: *Rhynia*.
Meaning: After
the location,
Rhynie (see
text).
Grouping:
Vascular plant,
Rhyniophyte.
Informal ID: Early land plant.
Fossil size: Stem actual size
2–3mm/¹⁄₁₆–¹⁄₈in across.
Reconstructed size: Whole
plant height up to 20cm/8in,
rarely 40cm/16in.
Habitat: Various, from sandy
to marshy.
Time period: Early Devonian,
400 million years ago.
Main fossil sites: Europe
(Scotland).
Occurrence: ◆

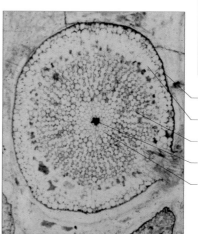

— Cuticle (outer 'skin')

— Outer cortex

— Inner cortex

— Xylem vessels

— Phloem vessels

Stronger stems
As plants became taller, in the evolutionary race to
outshade their competitors and receive their full
quota of sunlight, their stems not only needed
greater width but also increased stiffness or
rigidity. This is achieved by laying down woody
fibres, usually of the substance lignin, as seen in
today's bushes, shrubs and trees. In a cut tree
trunk or large branch, the annual growth rings
show where each season's set of new vessels
develops. As more rings are added on the outside,
the inner ones become more fibrous and stiff, as
sapwood. The oldest rings towards the centre of
the trunk, forming the heartwood, play almost no
role in transporting water, minerals or sap.

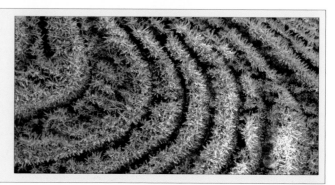

Calamites (Annularia, Asterophyllites)

Name: *Calamites*.
Meaning: Reed.
Grouping:
Sphenopsid,
Equisetale.
Informal ID: Giant
horsetail.
Fossil size: Stem
length 30cm/12in,
Annularia slab
15cm/6in across.
Reconstructed size: Height
up to 30m/100ft.
Habitat: Shifting sands by
rivers and lakes.
Time period: Carboniferous,
300 million years ago.
Main fossil sites: Worldwide.
Occurrence: ◆ ◆ ◆

During the Carboniferous Period, some 300 million years ago, vast forests of giant tree-like plants thrived in the warm, damp swamps and marshes that covered much of the land at this time. One of these early giants was *Calamites*, which could grow to reach heights of 30m/100ft. It was a member of the group called horsetails, scouring rushes, sphenopsids or sphenophytes. These are still seen today, although they are much smaller in stature – such as *Equisetum*, the common horsetail, which springs up from freshly disturbed soil, such as a dug garden. The general form of horsetails is a stiff central stem or trunk with whorls of upswept leaves (like inverted umbrellas) at regular intervals. *Calamites* was successful worldwide, particularly along the banks of rivers, where it was often buried by the shifting sands even as it continued to grow. This adaptation to its environment actually helped to support the plant, preventing it from falling over during times of local flooding. *Calamites* formed spore cones and, much like the earliest plants, released spores from these as its method of reproduction. In addition, it also had an underground system of runner-like rhizomes that could produce new individuals asexually (as in ferns, such as bracken, today), to colonize new environments rapidly. The names *Annularia* and *Asterophyllites* are both applied to the leaves of *Calamites* and similar sphenopsids.

Below: Asterophyllites *leaves were rigid and sword-like, arching away from their node (the point where they grew from the stem) to touch the whorl growing above them. In some types of plant, the stems branched off the main trunk in pairs and the reproductive spore cones grew at their tips.*

| Stem

Above: Annularia *is the name that is given to certain types of leaves from* Calamites-*type sphenopsids or, in some cases, to the whole plant itself. The leaves were softish and lance- or blade-like in shape, and they were arranged in whorls, or rings (annuli), looking like the spokes of an umbrella.*

Whorl of
leaves

Main stem

Transverse bands
showing old
whorl leaf
segments

| Branch

Fine vertical
ribbing

Left: The term Calamites *was originally applied only to the stem of this tree-sized horsetail plant, recognizable by its distinctive vertical ribbed pattern. After burial, probably by the banks of a river or the shores of a lake, the plant's thick trunk – up to 30cm/12in in diameter – rotted away. The resulting cavity was quickly filled by sediment creating a cast of the inner pith cavity, as in this specimen, which was flattened during preservation.*

Living representatives
The main extant horsetail genus *Equisetum* is represented by about three dozen species and hybrids, found in most corners of the world. The water or swamp horsetail *E. fluviatile* (here shown with a few leaves of marsh yellowcress, *Rorippa palustris*) grows rapidly in shallow water and can be a troublesome aquatic weed. The hollow stem is rich in silica minerals, which give a gritty or abrasive texture – hence the old-time uses of these plants as abraders and polishers, with the common name of scouring rushes.

HORSETAILS (CONTINUED), LIVERWORTS AND FERNS

Liverworts are non-vascular plants usually with a low, lobe-like body, or thallus. They grow mainly in damp places such as near streams. Ferns, or monilophytes (pteridophytes), are common today and were especially so throughout prehistory, some growing as large as trees and known appropriately as tree-ferns.

Sphenophyllum

Below: The whorled leaves of many specimens are larger at the bottom than farther up. This suggests some sort of light-partitioning system to ensure maximum productivity, allowing lower whorls to receive light energy that bypassed the upper ones.

A creeping or climbing plant with an exceptionally long time range, *Sphenophyllum* appeared in the Devonian and survived into the Triassic. Like many of its modern horsetail relatives, it favoured damp conditions, growing on floodplains or along the banks of rivers and lakes. It may have formed a tangled understorey in lycopod (clubmoss) forests, draping itself over the large trunks in the manner of vines and creepers in rainforests today. The stems were long and thin, and branches rare. The slightly fan-like leaves were distinctive, forming whorls in threes or multiples of three. Another distinctive feature of the plant was that its vascular system comprised three bundles of vessels arranged in a triangular shape. Fertile (reproductive) branches can be recognized by the spore cones they carry.

Name: *Sphenophyllum*.
Meaning: Wedge leaf.
Grouping: Sphenopsid.
Informal ID: Climbing horsetail.
Fossil size: Slab size 12cm/4¾in across.
Reconstructed size: Trailing stems for many metres, height depending on available support.
Habitat: Damp places, floodplains, around the edges of streams and lakes.
Time span: Carboniferous to Early Triassic, 350–240 million years ago.
Main fossil sites: Worldwide.
Occurrence: ◆ ◆

Hepaticites

Below: Hepaticites (including some species formerly classified as Pallavicinites) was one of the first liverworts, from Late Devonian times, and survived through to the Carboniferous Period, and beyond in some regions. Its fossils are known from most continents; this specimen is from Jurassic rocks in Yorkshire, northern England. Thallose liverworts have lobed or ribbon-like branching bodies with ribs that resemble the veins of true leaves, but which lack conducting vessels inside.

A member of the liverwort group, *Hepaticites* was low growing, fleshy and non-vascular – it lacked the network of water- and sap-carrying tubes found in most plants. The liverworts, or hepatophytes (Hepaticae or Hepatopsida, with more than 8,000 living species), along with the mosses (Musci), make up the major plant group known as the Bryophyta. They have no proper roots, stems, leaves or flowers. The main part of the plant is known as the thallus, and while there may be root-like rhizoids, these are only for anchorage. The thallus absorbs moisture directly from its surroundings, which is why many bryophytes were and are confined to damp, shady places. Bryophytes appeared early in the fossil record, but their soft tissues, without woody parts, mean they were rarely preserved. The common name 'liverwort' refers to the often fleshy shape, which is reminiscent of an animal's liver.

Name: *Hepaticites*.
Meaning: Of the liver, liver-like.
Grouping: Bryophyte, Hepaticean, Metzgerialean.
Informal ID: Liverwort.
Fossil size: 4.1cm/1¾in.
Reconstructed size: As above.
Habitat: Damp places, floodplains, around the edges of streams and lakes.
Time span: Late Devonian to Cretaceous, 360–100 million years ago.
Main fossil sites: Northern Hemisphere.
Occurrence: ◆

Phlebopteris

Name: *Phlebopteris*.
Meaning: Fleshy wing.
Grouping: Monilophyte, Matoniale.
Informal ID: Fern.
Fossil size: Slab 10cm/ 4in across.
Reconstructed size: Height 1–2m/3¼–6½ft.
Habitat: Marsh edges, levees and flood plains.
Time span: Triassic to Early Cretaceous, 230–120 million years ago.
Main fossil sites: North America, Europe.
Occurrence: ◆ ◆ ◆

This genus of fern first appeared during the Early Triassic Period and persisted at least into the early part of the Cretaceous – an immense time span of more than 100 million years. The leaves, or fronds, followed one of the typical fern patterns of paired, branching leaflets. However, the leaves appear to have formed a radial pattern, something like the spokes of a wheel, around the terminal part of the stalk or stem – the remainder of which was bare. Like most later ferns (and many ferns living today) *Phlebopteris* grew in drier areas, such as those around the edges of marshes and swamps, and on top of river levees (banks). This specimen is of the species *Phlebopteris smithii* from the Triassic deposits known as the Chinle Formation, in New Mexico, USA. They are also found in the nearby famous Petrified Forest National Park of Arizona.

Above: The main frond had up to 14 pinnae, or leaflets, and these divide into oblong, narrow pinnules or subleaflets with rounded ends (for a definition of this terminology see the following page). Each pinnule bears a strong central ridge, the main vein or midrib, which extends along most of its length and then divides into several smaller veins that run to the edge.

Cladophlebis

Name: *Cladophlebis*.
Meaning: Branching flesh.
Grouping: Monilophyte, Filicale.
Informal ID: Fern.
Fossil size: Slab 10cm/ 4in long.
Reconstructed size: Whole leaf frond 60cm/24in long.
Habitat: Marsh edges, levees and floodplains.
Time span: Triassic to Cretaceous, 230–120 million years ago.
Main fossil sites: North America.
Occurrence: ◆ ◆ ◆

This Mesozoic fern, like *Phlebopteris* above, also survived from the Triassic Period through to the Cretaceous Period. It is one of the most common fern-like leaves found in the famed location of the Petrified Forest National Park, Arizona, USA. The pinnules, or subleaflets, were small – just a couple of millimetres wide and twice that in length. Each pinnule bears a midrib (the central main vein), from which thinner side, or lateral, veins branch in an alternating pattern. The pinnae, or leaflets, branch off in pairs from the main stem. Some species of *Cladophlebis* have small, saw-like serrations on the margins of their leaflets. This is not, however, a distinctive character as other *Cladophlebis* species lack them.

Ferns and coal

Fern fossils and impressions are common in lumps of coal formed during the Carboniferous (Mississippian-Pennsylvanian) Period, when great 'coal measure' steaming tropical swamps covered much of the Earth's land. Sometimes, a piece of coal splits, or cleaves, easily along its natural bedding plane to reveal beautiful fern shapes. These may be indistinct in colour due to the coal's blackness. But in soft coals they can, with care, be turned into dark, rubbed images on white or pale paper, in the way that brass rubbings are created.

Terminal edge

Central stem or rachis

Broad, angular leaflets

Above: The pairs of broad, angular leaflets are preserved in this fossil along with the depression of the central stem. The frond was probably longer than that shown here, and it is the missing part that would have attached to the main stem.

FERNS (CONTINUED), SEED FERNS

There are more than 11,000 species of fern today, making them the largest main group of plants after the flowering plants, or angiosperms. Like many other types of simpler plants, they reproduce by spores. Ferns first appeared in the Devonian, thrived through the Carboniferous, became less common during the Late Permian and the Middle Cretaceous, but underwent a resurgence during the Tertiary.

Psaronius, Pecopteris

Fern sporangia

A fern's spore containers, or capsules, known as sporangia, sometimes look like rows of 'buttons' or 'kidneys' along the underside of the fern frond. These are the sporangia of the extant male fern *Dryopteris filix-mas*, a common woodland species. The protective scale which forms the outside of each sporangium is known as the indusium. This becomes grey as the spores ripen ready for release into the wind.

Main leaflet (pinna) composed of paired subleaflets (pinnules)

Fragmented leaflets

Central stem or rachis

The identification and grouping of fern fossils is a very tricky subject, as exemplified by the preserved remains of fronds called *Pecopteris*. It was eventually discovered that these grew on tree ferns that had already received their own names, such as *Psaronius*, from the fossils of their strong, thick trunks. In addition, other parts of these plants, such as the rhizomes ('roots'), have also received yet more names. (The same problems have also arisen with seed ferns, see overleaf.) *Psaronius* was one of the most common tree ferns of Late Carboniferous and Early Permian times. It would have superficially resembled a modern palm tree, growing with a tall, straight, unbranched trunk and an umbrella-like crown of fronds. Small roots branching from the lower main stem, called adventitious roots, caused thickening of the lower trunk, which was known as the root mantle. This arrangement helped to stabilize the whole tree fern. *Psaronius* is thought to have grown on elevated and drier areas around the swamplands. Its dominance towards the end of the Carboniferous Period is associated with climate warming and the disappearance of flooded wetlands. It faded later in the Permian, when the climate became drier and hotter. To add to the confusion, some specimens of *Pecopteris* have been reassigned to the genus *Lobatopteris*, which is not a true fern but a member of a different plant group – the seed ferns (see opposite and overleaf).

Left: The genus Pecopteris *is plentiful and widespread in the fossil record. Some specimens are in excellent condition, but others are only fragments, making identification difficult. This has caused considerable confusion with several hundred species assigned to the genus in some listings. In most specimens, the leaflets had two or three subdivisions (see opposite) and were shaped like rectangles with rounded ends. The central vein, or midrib, gave off smaller side veins at right angles.*

Name: *Psaronius*.
Meaning: Of grey, in grey.
Grouping: Monilophyte, Marattiale.
Informal ID: Tree fern.
Fossil size: —
Reconstructed size: Height up to 10m/33ft.
Habitat: Elevated, drier areas of swampland.
Time span: Late Carboniferous to Mid Permian, 310–270 million years ago.
Main fossil sites: Worldwide.
Occurrence: ◆ ◆ ◆

Name: *Pecopteris*.
Meaning: Comb wing.
Grouping: Monilophyte, Marattiale.
Informal ID: Frond of tree fern such as *Psaronius*.
Fossil size: Slab 10cm/ 4in long.
Reconstructed size: Frond length to 1.5m/5ft.
Habitat: See *Psaronius*.
Time span: See *Psaronius*.
Main fossil sites: Worldwide.
Occurrence: ◆ ◆ ◆

Lobatopteris

Name: *Lobatopteris*.
Meaning: Lobed wing.
Grouping: Pteridosperm.
Informal ID: Tree-like
seed fern.
Fossil size: Slab 10cm/
4in long.
Reconstructed size: Height
5m/16½ft.
Habitat: Drier parts of
swampland.
Time span: Carboniferous to
Permian, 320–270 million
years ago.
Main fossil sites:
Northern Hemisphere.
Occurrence: ◆ ◆

Part of the famous Mazon Creek fossils
of central North America, *Lobatopteris*
is an example of a common problem that
occurs in palaeontology. Often, a genus
name has to be changed at some later date,
because new discoveries about its true
nature and classification come to light
as further finds are made. As a result,
some species of *Pecopteris*, from the
true ferns (see opposite), have now been
renamed in the genus *Lobatopteris*, and
reclassified as seed ferns, or
pteridosperms (see overleaf). The whole
plant would have resembled a sizable tree,
growing several metres in height, and living
in the drier, probably more elevated
areas around the swamplands.
Pteridosperms mostly had fern-like
foliage, but they produce real seeds, as
opposed to spores as in true ferns.

*Right: This fossilized imprint shows a typical
fern-like leaf, or frond, preserving the individual
leaflets in detail. The central stem, or rachis, while
not visible, is represented by a depression running
along the middle.*

Subleaflets

Leaflet

Central stem
depression

Ferns, fronds, leaves and leaflets

The parts of a typical fern commonly called leaves are also known
as fronds. These may be smooth-edged and undivided, as in today's
hart's-tongue fern. Alternatively, they may be deeply divided, or
'dissected', into many smaller sections, which are themselves
subdivided, and so on, with a repeating pattern. The terminology
associated with all these parts is complex, but in general:
• The frond is comparable to the whole leaf with its stalk, which
grows up from the root area.
• The stalk of the frond is called the stipe.
• The rachis is the main stalk, or stem, that runs along the middle
of the frond, from which the other parts branch to each side, often
in simultaneous pairs or alternately.

Below: The unfurling frond or 'fiddlehead' of a male fern.

• The pinna, or leaflet, is one of these branches from the main stalk,
or rachis, of the frond.
• Pinnules, or subleaflets, may branch from the stalk of the pinna.
This is called a twice-divided, or twice-cut, frond.
• A pinnule may, in turn, have a lobed structure, giving a thrice-
divided, or thrice-cut, frond.
• When the fronds are young, they are curled up in a spiral form
and as they unfurl they are known as the fiddle-head.
    With their changing shapes and detailed structures, ferns are often
important as index, or marker fossils, helping to date the rock layers
in which they are found.

*Below: Oak ferns jostle for space with a familiar angiosperm,
the violet (the genus* Viola).

SEED FERNS (CONTINUED)

The seed ferns or pteridosperms were once grouped with the true ferns. However, discoveries during the last century showed that these plants reproduced by seeds, which form when female and male structures come together at fertilization, as in the conifers and flowering plants on later pages. This was an advance on more primitive spore-bearing and other methods used by horsetails, ferns and other simpler plants.

Neuropteris, Alethopteris, Medullosa

The confusion that has arisen over the years from giving different names to the different parts of the same plant, such as the tree ferns on the previous pages, is all too familiar when we come to the seed ferns. Examples of this include the leafy foliage known as *Alethopteris*, *Neuropteris* and *Macroneuropteris* (see opposite), which are regarded by many experts as being borne on smaller branch-like leaf stalks, or petioles, known as *Myeloxodon*, which in turn grow from a large, thick, trunk-like main stem known as *Medullosa*. The whole formed a tree-like plant that grew up to about 5m/17ft tall, which is sometimes also known by the name *Medullosa*, and generally classified as a seed fern. To add to the confusion, its seeds are given yet another name, *Trigonocarpus* (see overleaf). However, this is a generalization, and not all species of the above-mentioned genera may have been part of the same plant in this way. In addition, some authorities continue to use names such as *Neuropteris* or *Trigonocarpus* when referring to the whole plant. Researchers continue to discover new associations as they investigate fossil specimens and work out the relationships between them.

Above: The leaves of the genus Alethopteris were highly variable in shape. Some specimens show that mature leaves, or fronds, sprout subdivisions, pinnae (leaflets) and pinnules (subleaflets, see previous page), only after reaching a certain size. In some cases, division continues to produce ever-increasing multi-leaflet fronds. In other forms, the pinnae were smooth-edged. Fossil slab 21cm/8¼in across.

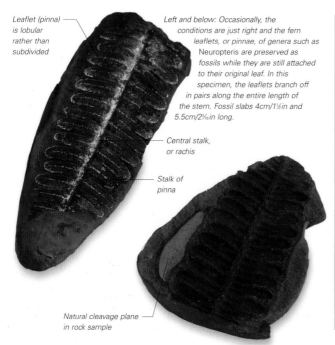

Leaflet (pinna) is lobular rather than subdivided

Left and below: Occasionally, the conditions are just right and the fern leaflets, or pinnae, of genera such as Neuropteris are preserved as fossils while they are still attached to their original leaf. In this specimen, the leaflets branch off in pairs along the entire length of the stem. Fossil slabs 4cm/1½in and 5.5cm/2³⁄₁₆in long.

Central stalk, or rachis

Stalk of pinna

Natural cleavage plane in rock sample

Medullosa growth

Several other tree-like seed ferns, such as *Sutcliffia*, had a similar overall shape to *Medullosa*. Distinguishing them by using features such as their leaf bases is difficult. Each of these grew wrapped like a sheath around the main central stem (trunk) and then angled away to bear the leaf-like frond. As the plant enlarged, new leaf bases grew above and within the previous ones. The older fronds fell away as the fresh ones unfurled towards the stem's upper end, causing the plant to gain height. In this way, the leaf bases built up like scales along the trunk. A similar growth pattern occurs today in the sago palm or cycad.

Living cycads show the general shape and growth habit of the extinct seed-ferns.

Name: *Medullosa* (used for main trunk and also whole plant); *Alethopteris* (foliage), *Neuropteris* (foliage).
Meaning: Marrow, inner layer (*Medullosa*); lacy wing (*Neuropteris*).
Grouping: Pteridosperm, Medullosale.
Informal ID: Tree-like seed fern.
Fossil size: Slab length 10cm/4in.
Reconstructed size: Height of whole plant up to 5m/16½ft.
Habitat: Swamps, damp regions.
Time span: Late Carboniferous to Early Permian, 310–270 million years ago.
Main fossil sites: Worldwide.
Occurrence: ◆ ◆

The detailed study of leaves such as *Neuropteris* and *Alethopteris* has shown that adaptations have occurred that are normally associated with low light levels, especially in the former. This is evidence to suggest that the Carboniferous forest canopy was layered, and that these tree-like seed ferns may have been found in the understorey where light would have been restricted. In addition to the foliage shown here (right), other fern-type leaves that are often assigned to *Medullosa* or related tree ferns include *Odontopteris*, *Callipteridium* and *Mixoneura*.

Right: This is one of the detached leaves of a seed fern that has been assigned to the genus Neuropteris. *The specimen of the rock type, called siderite concretion, is from Braidwood, Mazon Creek, Illinois, USA. Fossil slab size is 10cm/4in long.*

Lacy pattern
In the *Neuropteris* and *Macroneuropteris* series of seed fern foliage, the thickened vein-like areas branch in a curved pattern from the central stem to produce a lacy, fan-like, filamentous effect, reminiscent of the branching nerve system within the human body. A similar effect is seen in the wings of the predatory insect called the lacewing, in the group Neuroptera.

Terminal point —

Stalk attachment —

— Fine, lacy surface pattern

Macroneuropteris

Name: *Macroneuropteris*.
Meaning: Big lacy wing.
Grouping: Pteridosperm, Medullosale.
Informal ID: Leaf of tree fern eaten by insect.
Fossil size: Slab width 4cm/1½in.
Reconstructed size: See *Medullosa*, opposite.
Habitat: Floodplain deposits.
Time span: Late Carboniferous, 300 million years ago.
Main fossil sites: Worldwide.
Occurrence: ◆

Macroneuropteris follows the typical fern growth pattern of having individual leaflets (or pinnae) branching alternately from either side of a single stem. The leaflets in this fern are characteristically long and fat with rounded tips. This type of leaf was born on the tree-like seed fern plant that is sometimes referred to as *Medullosa*, as described opposite. Arthropod predation of early plants – such as spore-eating as well as the piercing and sucking of fluid from the stems and leaves of plants – probably originated as far back as the Devonian Period, some 400–360 million years ago, when land animals made their first appearance. Direct evidence for leaf grazing, as in the example shown here (right), becomes common during the Carboniferous, due to a number of factors, including the increasing volume of plant growth and the rapidly diversifying insect groups. It is the generally good state of fossil preservation from the Carboniferous period that allows us to study the evidence.

— Leaflet

— Central stalk left uneaten

— Grazing trace

— Grazed area

— Grazed area

Above: The distinctive semicircular grazing patterns that are evident on certain fossil leaves give a direct insight into the feeding behaviour of herbivorous arthropods that lived millions of years ago. The consumer of this specimen has nibbled away successive arcs of leaf flesh, in a way similar to today's caterpillars.

SEED FERNS (CONTINUED)

The leaves, or fronds, of seed ferns are commonly found in 'Coal Age' Carboniferous deposits, mainly across the Northern Hemisphere. As explained on previous pages, the different parts of what is now known to be a single plant have often received different names. Seed ferns combined features of true ferns with characteristics of the more advanced trees known as cycads.

Trigonocarpus

The seeds of tree ferns are known as *Trigonocarpus* or, sometimes, *Pachytesta*, depending on the way they have been preserved. Such seeds may represent one of the first steps in animal–plant relationships – animal dispersal of seeds. The seeds themselves are large, up to 6cm/2⅜in long, avocado-like and with a fleshy layer beneath a tough outer coat. This size alone would have made these wingless seeds difficult to transport far from the parent plant unless they were first eaten by animals and then transported to new growing sites.

Below: Isolated seeds such as these Trigonocarpus, *each embedded in a rocky nodule, come from seed ferns. Although commonly found, they are very rarely preserved in situ, attached to the parent plant. Each seed is strongly ribbed and divided into three equal sections, with several longitudinal fibrous bands acting to toughen the seed wall.*

Name: *Trigonocarpus*.
Meaning: Three ribbed cup.
Grouping: Pteridosperm, Medullosale.
Informal ID: Ancient seed of seed fern.
Fossil size: Seed length 1–1.5cm/⅜–½in.
Reconstructed size: Seed length varies from 5mm–6cm/³⁄₁₆–2⅜in.
Habitat: Floodplains, swamps, marshy ground.
Time span: Mostly Late Carboniferous, 300 million years ago.
Main fossil sites: Worldwide.
Occurrence: ◆ ◆

— *Fleshy layer*
— *Tough outer coat (test)*
— *Stone-like seed*

Fossil counterpart of specimen on left

Mariopteris

Smaller, younger leaflets

Central stem

The presence of 'climbing hooks' suggests that this seed fern may have been a climbing, trailing or scrambling plant, similar to a modern vine, in swampy or waterlogged habitats. The plant itself may have grown up to heights of around 5m/16½ft. However, it is the leaflets that best identify this genus. Unlike many other seed ferns, the leaflets of this plant all have the same shape, with stems originating alternately from a central branch, and decreasing in size with growth. The distinctive four-pronged shape of the leaflet is most obvious towards the base of the stem. It becomes sharper and more shovel-like towards the terminal edge of the branch.

Older, larger leaflets at base

Lateral stem

Left: A late Cretaceous swamp plant, the seed fern Mariopteris had a distinctive leaf, or frond, shape, where each leaflet was part of a four-pronged frond. These leaflets had a central vein or midrib from which secondary veins sprouted. The specimen shows part of one frond, from the Late Carboniferous layers of Caerphilly, Wales.

Name: *Mariopteris*.
Meaning: Sea wing.
Grouping: Pteridosperm, Medullosale.
Informal ID: Trailing seed fern.
Fossil size: Slab 10cm/ 4in across.
Reconstructed size: Up to 5m/16½ft.
Habitat: Swamps.
Time span: Late Carboniferous to Early Permian, 310–270 million years ago.
Main fossil sites: Worldwide.
Occurrence: ◆ ◆ ◆

Glossopteris

Name: *Glossopteris*.
Meaning: Tongue wing.
Grouping: Pteridosperm, Glossopteridale.
Informal ID: Gondwana tree, tree-like seed fern.
Fossil size: Leaf length could exceed 10cm/4in.
Reconstructed size: Whole plant height 6m/19½ft, rarely 8m/26¼ft.
Habitat: Warm, damp lowlands.
Time span: Permian to Triassic, 270–210 million years ago.
Main fossil sites: Southern Hemisphere.
Occurrence: ◆ ◆ ◆

This is one of the largest and best known of the seed ferns, and it has given its name to its own subgroup – the Glossopteridales. Although a large plant, managing to reach heights of up to 8m/26ft, precisely what it looked like in life is still much discussed. For many years it was reconstructed as a bush or shrub, but it is now thought to have been more tree-like in appearance, with one main stem or trunk. *Glossopteris* originated in the great southern supercontinent of Gondwana during the Permian Period and rapidly spread across the rest of the Southern Hemisphere. The plant was unusual in many respects. Its leaves, or fronds, were narrow in shape and elongated, in some cases lanceolate – shaped like the head of a lance or spear. The roots of the plant had a characteristic pattern of regularly spaced partitions, giving it the appearance of a vertebrate backbone (like our own spinal column) and earning isolated specimens of it a separate name, *Vertebraria*. Lobed wood is also present at the centre of these roots, with internal spaces between the lobes. The seeds appear to have developed on the undersides of the leaves, while the pollen was produced in anther structures found elsewhere on the plant. The rise and spread of *Glossopteris*, as shown by its fossil distribution, have both helped to map the drifting continents and the extent of Gondwana. At the end of the Permian this seed fern's domination, combined with the characteristic vegetation of the region, is known as the '*Glossopteris* fauna' and the plant itself has the informal name of the 'Gondwana tree'.

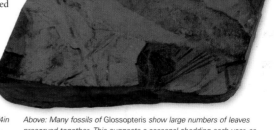

Below: Some mature leaves of Glossopteris grew in excess of 10cm/4in in length, as shown by this Triassic specimen from the Upper Karroo region of South Africa. The well-developed central vein was actually configured from many small veins. Minor veins branched from this and spread to form a characteristic network enclosing small polygonal flatter regions, looking like tiny crazy-paving.

Above: Many fossils of Glossopteris show large numbers of leaves preserved together. This suggests a seasonal shedding each year, as with most modern deciduous trees. This specimen is Permian and from Mount Sirius in the Central Transantarctic Mountains of Antarctica. It bears different-shaped leaves from two species, Glossopteris indica and G. browniana.

Central vein

Numerous lateral veins

Polygonal 'crazy paving' network effect

The Gondwana Tree

During the Permian Period the world's land masses were linked into the greatest supercontinent, Pangaea. The masses were subgrouped into northern and southern clusters. From the Triassic and through the Jurassic, these would pull apart as northern Laurasia and southern Gondwana (Gondwanaland). Fossils of *Glossopteris* have been found in swathes of sites that cut across South America, Africa, Madagascar, India, Antarctica and Australia. This provides major evidence for the arrangement of these land masses at the time. Since Permian times India has drifted far northwards and collided with southern Asia, while the other continents have moved apart but mainly stayed in the Southern Hemisphere. In late 1912 Antarctic explorers found *Glossopteris* among the specimens collected on Captain Robert Scott's ill-fated trip to the South Pole earlier the same year. In 1915 Alfred Wegener proposed early ideas about continental drift, using *Glossopteris* to suggest the existence of Pangaea.

Laurasia

Gondwana

CLUBMOSSES (LYCOPODS)

The swamps of the Carboniferous represented perhaps the greatest accumulation of biomass (living matter)
ever. We exploit these fossilized remains when burning coal and natural gas. Much of this wetland
vegetation consisted of plants called lycopods (lycopsids), or clubmosses – some as tall as 50m/165ft.
Clubmosses were not closely related to true mosses (Bryophytes, Musci) or to early ferns and seed ferns.

Baragwanathia

Name: *Baragwanathia*.
Meaning: After Australian geologist William Baragwanath.
Grouping: Lycopod, Baragwanathiale.
Informal ID: Clubmoss.
Fossil size: Slab height 10cm/4in.
Reconstructed size: Plant height up to 25cm/10in.
Habitat: Damp lowland areas, floodplains.
Time span: Late Silurian to Early Devonian, 420–400 million years ago.
Main fossil sites: Australia.
Occurrence: ◆

This curious and controversial plant is named in honour of William Baragwanath, who was the Director of the Geological Survey in the southern state of Victoria, Australia, from 1922 to 1943. William Baragwanath was born in the gold-mining town of Ballarat in 1878, the son of Cornish immigrants from the west of Britain lured to Australia in the hope of striking it rich in the booming gold fields of the time. The plant that bears his name is regarded by many experts as being a clubmoss, but from a very early time, more than 400 million years ago. This was when very few other land plants are known, and those that were present – such as *Cooksonia* (detailed earlier) – were much more primitive. In this context, 'primitive' means that it had fewer specialized features. *Baragwanathia* was first identified as imprints in rocks from the well-known Yea site in Victoria, Australia, and has since been discovered in other Australian localities. Dating methods put the remains at the Ludlovian Age of the Late Silurian Period, some 420 million years ago. The dating uses the fossils of graptolites, which are commonly found in sediments deposited in deep-water settings, and has been challenged by some authorities. However, it is possible that those plants of the great southern supercontinent of Gondwana were more advanced than their northern contemporaries.

Living clubmosses

There are more than 1,000 species of extant (still living) clubmosses on the planet today. However, these are mostly smallish, low-growing, creeping plants, generally simple in form, and are most often to be found growing in the undergrowth, and they are merely a shadowy relict of their former glory in the Carboniferous Period when they could attain heights of up to 50m/165ft. Today's clubmosses resemble tough, tall mosses, but they possess the distinguishing feature of having a simple yet definite vascular system of vessels (tubes) to distribute water and nutrients around the plant. The spores are shed from spore-bearing leaves known as sporophylls. The species shown below is *Lycopodium annotinum*, commonly known as the 'stiff clubmoss'. It has narrow shoots comprised of tightly overlapping pointed leaves. When mature, it may grow to 30cm/1ft in height. Its favoured habitat is moist forest and thickets, and this species is particularly common in the foothills of Alberta, Canada.

Central stem

Clothing strip-like leaves give 'brush' effect

Leaves were borne at upright angle

Left: The stems of Baragwanathia *were clothed in what are regarded as true leaves – which resembled small strips, narrow ribbons or spines – each up to 1cm/⅜in in length. The growth habit included branching horizontal stems, which also branched to give upright stems, growing to 10–25cm/4–10in in height. There were spore capsules in the axils – where the leaves joined the stem.*

Drepanophycus

Name: *Drepanophycus*.
Meaning: Sickle plant.
Grouping: Lycopod, Baragwanathiale.
Informal ID: Clubmoss.
Fossil size: Fossil slab 14cm/5½in long.
Reconstructed size: Height 50cm/20in, rarely 1–2m/3¼–6½ft.
Habitat: Moist settings, river and lake banks.
Time span: Early Devonian to Carboniferous, 390–300 million years ago.
Main fossil sites: Eurasia.
Occurrence: ◆ ◆

The clubmoss *Drepanophycus* had a lowish 'creeping' form, evolved sometime in the Early Devonian Period and persisted through most of the Carboniferous, although as a genus it appears to have been restricted to the Northern Hemisphere. Some of these plants could attain several metres in height, although 50cm/20in was more usual. Each upright stem forked into two at an acute angle, perhaps more than once. The stem was covered with spiny, thorny or scale-like leaves, probably as a means of defence against insect predation. The leaves arose from the stem in a spiral pattern, although this was variable and at times they could appear whorled (emerging like the spokes of a wheel).

Right: The fossil shows the impression of Drepanophycus *along with its spiny leaves. A characteristic feature of the genus is its thickened stem, which can reach several centimetres in diameter. This specimen is from the Devonian Old Red Sandstone of Rhineland, Germany.*

Spiny leaves —

Side fork —

Thickened stem

Sigillaria

Name: *Sigillaria*.
Meaning: Seal-like.
Grouping: Lycopod, Lepidodendrale.
Informal ID: Tree-like clubmoss.
Fossil size: Slab size could not be confirmed.
Reconstructed size: Height 20m/66ft, rarely 30m/100ft plus.
Habitat: Moist settings, marshy forests.
Time span: Early Carboniferous to Permian, 340–260 million years ago.
Main fossil sites: North America, Europe.
Occurrence: ◆ ◆

Sigillaria is a large, well-known clubmoss, known mainly from the Late Carboniferous Period. The genus varies in form, but most types were fairly tall, standing around 20m/66ft in height, yet also very sturdy – as much as 2m/6½ft across at the base. The main trunk tapered slowly with height and did not divide profusely, with many specimens sending out only a few arm-like side branches. The upper regions of the plant were covered with long, slim leaves, resembling blades of freshly growing grass, that grew directly from the main stem and fell away to leave behind characteristic leaf scars between the strengthening vertical ribs. The leaves were long and narrow – up to 1m/3¼ft long and only about 1cm/⅛in across. Some of the fossils known as *Stigmaria* may be preserved roots of *Sigillaria*.

Right: This section of Sigillaria *trunk bears the distinctive 'seal'- or 'crab'-like leaf scars, from where the leaves grew straight out. Older leaves fell from the main trunk as the plant gained height, so most foliage was found in the upper region. Some types of* Sigillaria *are thought to have reached heights of 35m/115ft. This specimen comes from Carboniferous rocks near Barnsley, England.*

Strengthening vertical ribs —

Leaf scars —

CLUBMOSSES (CONTINUED)

Clubmosses are one of the earliest major plant groups. Their history stretches back to the Late Silurian, more than 400 million years ago, and there are about 700 species still living today – including Lycopodium, Phylloglossum and Selaginella. Most are smallish plants creeping on the ground or up larger plants, such as trees. They are small shadows of their Carboniferous cousins, such as Lepidodendron.

Lepidodendron

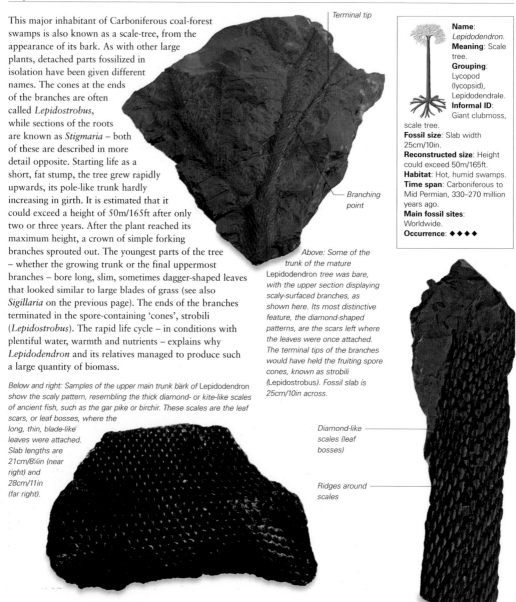

This major inhabitant of Carboniferous coal-forest swamps is also known as a scale-tree, from the appearance of its bark. As with other large plants, detached parts fossilized in isolation have been given different names. The cones at the ends of the branches are often called *Lepidostrobus*, while sections of the roots are known as *Stigmaria* – both of these are described in more detail opposite. Starting life as a short, fat stump, the tree grew rapidly upwards, its pole-like trunk hardly increasing in girth. It is estimated that it could exceed a height of 50m/165ft after only two or three years. After the plant reached its maximum height, a crown of simple forking branches sprouted out. The youngest parts of the tree – whether the growing trunk or the final uppermost branches – bore long, slim, sometimes dagger-shaped leaves that looked similar to large blades of grass (see also *Sigillaria* on the previous page). The ends of the branches terminated in the spore-containing 'cones', strobili (*Lepidostrobus*). The rapid life cycle – in conditions with plentiful water, warmth and nutrients – explains why *Lepidodendron* and its relatives managed to produce such a large quantity of biomass.

Below and right: Samples of the upper main trunk bark of Lepidodendron show the scaly pattern, resembling the thick diamond- or kite-like scales of ancient fish, such as the gar pike or birchir. These scales are the leaf scars, or leaf bosses, where the long, thin, blade-like leaves were attached. Slab lengths are 21cm/8¼in (near right) and 28cm/11in (far right).

Terminal tip

Branching point

Above: Some of the trunk of the mature Lepidodendron tree was bare, with the upper section displaying scaly-surfaced branches, as shown here. Its most distinctive feature, the diamond-shaped patterns, are the scars left where the leaves were once attached. The terminal tips of the branches would have held the fruiting spore cones, known as strobili (Lepidostrobus). Fossil slab is 25cm/10in across.

Diamond-like scales (leaf bosses)

Ridges around scales

Name: *Lepidodendron.*
Meaning: Scale tree.
Grouping: Lycopod (lycopsid), Lepidodendrale.
Informal ID: Giant clubmoss, scale tree.
Fossil size: Slab width 25cm/10in.
Reconstructed size: Height could exceed 50m/165ft.
Habitat: Hot, humid swamps.
Time span: Carboniferous to Mid Permian, 330–270 million years ago.
Main fossil sites: Worldwide.
Occurrence: ◆ ◆ ◆ ◆

Lepidostrobus

The genus name *Lepidostrobus* is assigned to the reproductive parts – variously called fruiting bodies, cones, spore capsules, sporangia or strobili – of the huge, tree-like clubmoss *Lepidodendron* (opposite). Some specimens of *Lepidostrobus* are smallish, only about finger-size, while others are considerably larger and can exceed 30cm/12in in length. Most of them were elongated in shape, resembling cigars, cylinders, eggs or pears in outline. These spore-bearing cones were made up of sporophylls with sporangia on the upper surface of the base. The cones are always found intact as fossils, proving that they were integrated structures rather than aggregate clumps of sporangia. On the living tree they would be sited at the ends of the branches at the top of the plant.

Left: The fruiting body, or spore-bearing cone, of the Lepidodendron *tree is known by the name* Lepidostrobus. *These structures housed the spores, which were tiny (microsporous, male), large (megasporous, female) or both, depending on the species, and there are an estimated 100-plus species in the genus. This is* Lepidostrobus variabilis *from Carboniferous rocks in Staffordshire, England.*

Name: *Lepidostrobus*.
Meaning: Twisted scale.
Grouping: Lycopsid, Lepidodendrale.
Informal ID: Spore cone or fruiting body of *Lepidodendron* and similar giant clubmosses.
Fossil size: Length 8cm/3¼in.
Reconstructed size: As above (can be up to 30cm/12in).
Habitat: As *Lepidodendron*, see opposite.
Time span: As *Lepidodendron*, see opposite.
Main fossil sites: As *Lepidodendron*, see opposite.
Occurrence: ◆ ◆

Stigmaria

Name: *Stigmaria*.
Meaning: Holed, pricked, marked.
Grouping: Lycopsid, Lepidodendrale.
Informal ID: Root system, or rhizome, of *Lepidodendron* and similar giant clubmosses.
Fossil size: See captions for slab widths of both specimens.
Reconstructed size: Length up to 5m/16½ft.
Habitat: As *Lepidodendron*, see opposite.
Time span: As *Lepidodendron*, see opposite.
Main fossil sites: As *Lepidodendron*, see opposite.
Occurrence: ◆ ◆

Right: The straight depressions running away from the main root represent the areas where the branching rootlets once existed. These often resemble black ribbons in the grey shale rocks. Fossil slab 36cm/14in across.

Stigmaria is the genus name for the detached rhizomes, or root structures, of large tree-like lycopods, such as *Lepidodendron* (opposite) and *Sigillaria* (previous pages). These strong, woody, branching rootstock systems had a number of functions, such as stabilizing the tall, slender tree-like structure above, in the manner of prop roots, and taking up nutrients from the watery sediment. A typical *Stigmaria* branched into two stems close to the base of the tree, and these stems tended to run almost horizontally within the sediment or soil for as far as several metres. Along this length, branches, variously called rootlets or 'giant root hairs', emerged at right angles, helping with anchorage and absorbing water and nutrients.

Below: In this impression of a Stigmaria *root, the texturing of circular scars preserved along the length of the fossil represents the scars where fine rootlets once attached to the main root during life. Fossil slab 25cm/10in across.*

Circular scars

Rootlets ('giant root hairs')

Main root

Rootlets branch at right angles

EARLY TREES – CYCADS

Plants can be divided into two groups. Sporophytes reproduce by spores (simple cells with no food store or protective coat). In spermatophytes, male and female cells come together and the resulting fertilized cells develop into seeds. Spermatophyes are divided into gymnosperms ('naked seeds') – including cycads, cycadeoids or bennettitales, cordaitales, ginkgos and conifers – and the flowering plants.

Williamsonia

Below: The fronds of Williamsonia are characteristically slender. Like all cycadeoids (the group Bennettitales), they have many small leaflets, or pinnae, alternating on either side of the main stem, or midrib. This pattern is known as the Ptilophyllum type, and differs from cycadales in which the pinnae usually occur in opposite pairs.

Cycads were among the dominant large plants found in the Mesozoic Era. The cycadales include both fossil and living forms, with about a hundred extant species in Mexico, the West Indies, Australia and South Africa, while the cycadeoids (bennettitales) are all extinct. *Williamsonia* is the best-known cycadeoid. The trunk was long and slender (although it could be short and bulbous in other cycadeoids), and the leaves narrow and frond-like. Fossil cycadeoids and cycadales can be distinguished if the leaf cuticle is exceptionally well preserved and cell walls can be seen under the microscope. They are quite different in their reproductive structures. Cycadales (including all living species) produce conifer-like cones, while cycadeoids had flower-like structures that suggest they could have played a part in the origin of true flowering plants, the angiosperms.

Name: *Williamsonia*.
Meaning: Honouring W C Williamson (1816–95), surgeon and naturalist.
Grouping: Gymnosperm, Cycadophyte, Cycadeoid.
Informal ID: Gymnosperm, cycad, cycadeoid.
Fossil size: Slab 29cm/11½in across.
Reconstructed size: Frond width 6–7cm/2½–2¾in; height of whole plant 2m/6½ft.
Habitat: Dry uplands.
Time span: Early Triassic to Cretaceous, 245–120 million years ago.
Main fossil sites: Western Europe, Asia (India).
Occurrence: ◆ ◆

Cycadeoidea

Below: This section of fossilized Cycadeoidea trunk distinctly shows the reproductive structures (or 'buds') embedded just under the surface. These were originally thought to have been precursors to true flower structures, but they are now known to have remained closed in life and so were probably self-pollinated.

Cycadeoidea was one of the most common North American cycadeoids. Its fossils are usually of the short, almost spherical or barrel-shaped trunk that was topped by a crown of fronds. When these trunks are well preserved, several 'buds' can be seen embedded just below the surface. These buds bear many small ovules (containing egg cells) in a central structure, surrounded by filaments bearing pollen sacs with the male cells. It was thought that these buds would later emerge as flower-like structures, but they are now known to have remained closed, and were probably self-pollinated. *Cycadeoidea* was a very common plant in the Cretaceous Period of North America, but it is rarely found elsewhere. The leaf bases are usually preserved covering the trunk, making many of the fossils resemble petrified pineapples.

Name: *Cycadeoidea*.
Meaning: Cycad-like.
Grouping: Gymnosperm, Cycad, Cycadeoid.
Informal ID: Cycadeoid, bennettitale.
Fossil size: Slab 15cm/6in high.
Reconstructed size: Whole plant height up to 1m/3¼ft.
Habitat: Dry uplands.
Time span: Jurassic to Cretaceous, 170–110 million years ago.
Main fossil sites: North America, Asia (India).
Occurrence: ◆ ◆ ◆ (North America only).

Cycadites

Name: *Cycadites*.
Meaning: Related to cycads.
Grouping: Gymnosperm, Cycad, possibly Cycadale.
Informal ID: Cycad.
Fossil size: Slab 8cm/3¼in wide.
Reconstructed size: Whole fronds may exceed 20cm/8in in length.
Habitat: Moist to dry forests.
Time span: Jurassic to Cretaceous, 180–100 million years ago.
Main fossil sites: Worldwide.
Occurrence: ◆ ◆ ◆

The fossil leaf *Cycadites* is found in many parts of the world, from the Indian subcontinent through to Scandinavia. Despite its abundance, however, comparatively little is known of *Cycadites* in life. This is because well-preserved fossils of the whole plant are exceptionally rare. The plant seems to be more closely related to the cycadales than to the cycadeoids (bennettitales), as evidenced by the leaflets (pinnae) emerging in opposing pairs from the frond's main stem, or midrib. *Cycadites* is an example of a form-genus – a scientific name that is assigned to a certain part of a plant. As with examples such as *Psaronius* and *Medullosa*, which have been detailed on previous pages, this has resulted in numerous difficulties for researchers and scientists in identifying and naming specimens. Other cycad fragments similar to *Cycadites* include *Pagiophyllum* and *Otozamites* (*Otopteris*).

Right: This fragment of Cycadites *frond has been preserved in red (oxidized) mudstone. The fragment is approximately 12cm/4¾in long. Note the leaflets (pinnae) in opposite pairs. Some species of* Cycadites *are known from Permian times, but most thrived during the Jurassic and Cretaceous Periods.*

Midrib | Pinnae

Zamites

Name: *Zamites*.
Meaning: From *Zamia*.
Grouping: Gymnosperm, Cycad, Cycadale.
Informal ID: Cycad.
Fossil size: Fronds up to 20cm/8in across.
Reconstructed size: Height up to 3m/10ft.
Habitat: Wide range of terrestrial habitats.
Time span: Tertiary, less than 65 million years ago.
Main fossil sites: Worldwide.
Occurrence: ◆ ◆ ◆

Right: Here a fragment of Zamites *leaf, or frond, has been preserved in red (oxidized) mudstone. The species is* Zamites buchianus *from the Fairlight Clays of Sussex, England. The living* Zamia *has a short, wide, pithy stem topped by leathery fronds.*

Zamites is, like *Cycadites* (above), a form-genus – and its fossil leaf is extremely similar in form to the living cycad genus *Zamia*. Today, *Zamia* is found exclusively in the Americas, its habitat ranging from Georgia, USA, south to Bolivia. *Zamites* and similar forms have been found in places as far apart as France, Alaska and Australia, a pattern which suggests that *Zamia*, or its ancestors, had a much wider geographic distribution in the past. This makes *Zamites* very interesting from the point of view of biogeography, which is the study of the distribution of animals and plants across the planet. The oldest cycad fossils are from the Early Permian of China, some 280 million years ago, and show that cycads have remained virtually unchanged through their long history. In general, a typical cycad looks outwardly like a palm tree or tree fern, with a trunk-like stem that can be either short and bulbous (resembling a pine cone or pineapple) or tall and columnar, crowned by an 'umbrella' of arching evergreen fronds. Cycads have ranged widely in size: some ground-hugging at just 10cm/4in tall; others more tree-like at almost 20m/66ft. Their average height, however, is about 2m/6½ft. Some types are informally called sago palms, although they are not true palms (which are flowering plants, Angiosperma).

Strap-like leathery fronds

Central frond stalk (midrib)

EARLY TREES – PROGYMNOSPERMS, GINKGOS

The first woody trees appeared around Middle Devonian times, about 375 million years ago. These trees were progymnosperms such as Archaeopteris. *They shared many features with gymnosperms, such as leaves, trunks and cones, but they shed their spores into the air, like ferns. In contrast, gymnosperms such as the ginkgos retain their large female spores on the plant and these become seeds when fertilized.*

Archaeopteris (includes Callixylon)

Archaeopteris, from the Late Devonian Period, has been called the 'world's first tree' and formed the Earth's early forests, becoming the dominant land plant worldwide. *Archaeopteris* was originally named from its leaf impressions. In the 1960s, palaeontologist Charles Beck showed that *Callixylon*, a fossil known only as mineralized tree trunks, and *Archaeopteris* were actually different parts of the same plant. *Archaeopteris* is an important fossil in that it combines spore-releasing reproductive organs similar to ferns with an anatomy, particularly in its wood structure, that resembles the conifers. Some experts believe that *Archaeopteris* is an evolutionary link between the two groups.

Left: This frond, or leaf, of Archaeopteris hibernica is from the Old Red Sandstone (approximately 375 million years old) Devonian rocks of Ireland. The leaflets, or pinnules, overlap one another and within the genus the leaflet shape ranges from almost circular to nearly triangular, rhomboid or wedge-shaped. Fertile branches had spore capsules rather than leaves.

Name: *Archaeopteris.*
Meaning: Ancient leaf.
Grouping: Progymnosperm, Archaeopteridale.
Informal ID: *Archaeopteris*, 'world's first tree'.
Fossil size: Slab 20cm/8in long.
Reconstructed size: Whole plant height 10m/33ft.
Habitat: Seasonal floodplains.
Time span: Late Devonian to Early Carboniferous, 370–350 million years ago.
Main fossil sites: Worldwide.
Occurrence: ◆ ◆

Cordaites

Below: This leaf fragment from a Cordaites tree shows the distinctive parallel vein pattern known as linear venation. The great size of what would be a whole leaf is evident when compared with the Archaeopteris fronds (see above) preserved alongside it.

With fossil trunks measuring up to 1m/3¼ft in diameter and 30m/100ft in height, the long-extinct cordaitales were some of the tallest trees growing in the Carboniferous Period. Their large strap-shaped leaves were similar in shape to the living kauri pines, *Agathis*, of the Chilean pine, or monkey-puzzle, group, and were arranged in a spiralling fashion along slender, long branches. The veining pattern is distinctive, with long parallel veins but no midrib in most species. The root system was extensive, often consisting of lateral (side) roots forming arch-like clusters on only one side of the main root. In living trees, this pattern is typical of species with 'stilt roots' inhabiting mangrove swamps, which might provide a clue to the environment in which *Cordaites* grew. From the structure of the loosely clumped cones and other reproductive parts, the cordaitales are considered to be ancestors of the true conifers, Coniferales.

Part of Cordaites leaf | Parallel veins

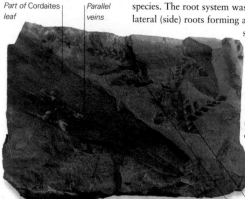

Archaeopteris *fronds*

Name: *Cordaites.*
Meaning: Heart-like.
Grouping: Gymnosperm, Cordaitale.
Informal ID: *Cordaites*, Conifer ancestor.
Fossil size: Leaf length up to 1m/3¼ft; width up to 15cm/6in.
Reconstructed size: Whole tree height 10–15m/33–50ft; some forms up to 30m/100ft.
Habitat: Swamps.
Time span: Early Carboniferous to Permian, 340–260 million years ago.
Main fossil sites: Worldwide.
Occurrence: ◆ ◆

Ginkgo

Name: *Ginkgo*.
Meaning: Traditional name for the tree (see panel below).
Grouping: Gymnosperm, Ginkgophyte.
Informal ID: Ginkgo, maidenhair tree.
Fossil size: Leaf widths 5cm/2in, 8cm/3⅛in.
Reconstructed size: Living tree height up to 40m/130ft.
Habitat: Wide-ranging.
Time span: Early Jurassic, 180 million years ago to today.
Main fossil sites: Worldwide.
Occurrence: ◆ ◆

The living maidenhair, or ginkgo tree, *Ginkgo biloba*, is a popular ornamental plant grown today, and is often seen in city parks and avenues in temperate regions. The fact that *Ginkgo* is a common sight belies its unique status as one of the plant world's genuine 'living fossils'. Fossilized preserved remains virtually identical to the living species are known from the Early Jurassic Period, about 180 million years ago. As a group, the ginkgophytes had their heyday between the Triassic and Cretaceous Periods, a span of about 165 million years – the 'Age of Dinosaurs' – when they extended their range throughout Laurasia (the Mesozoic northern supercontinent that included Europe, North America and most of Asia) and diversified into at least 16 genera. The end of the Cretaceous Period saw all but one of these genera, *Ginkgo* itself, become extinct. Although *Ginkgo* managed to retain a wide geographical range during most of the Cenozoic Era, reaching as far north as Scotland, by about 2 million years ago, it had disappeared from the fossil record virtually everywhere in the world apart from a small area of China. It was from this relict population (a species surviving as a remnant) that Chinese people first took seeds for cultivation in the eleventh century. *Ginkgo* became a traditional feature of temple gardens, and spread along with Buddhism to Japan and Korea. From there it was taken around the world by European traders. It is somewhat ironic that this ancient Mesozoic survivor is one of today's most urban-tolerant trees, growing in conditions that are too harsh for many other species to survive.

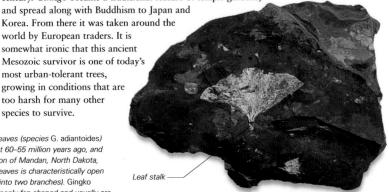

Right and below: These two Ginkgo *leaves (species* G. adiantoides*) are from the Palaeocene Epoch, about 60–55 million years ago, and were found in the Fort Union Formation of Mandan, North Dakota, USA. The pattern of the veins in the leaves is characteristically open and dichotomous (divides repeatedly into two branches). Gingko leaves, symbols of long life, are commonly fan-shaped and usually are partially split into two lobes. However, there is considerable variation in the degree of lobing and splitting displayed, even among the leaves from a single tree.*

Leaf stalk

Fan-like margin | Branching veins

Lone survivor
The extant ginkgo tree grows in a relatively upright habit when young, but branches out in the upper regions with age. Some specimens reach 30m/100ft in height and 30m/100ft across the broadest part of the crown. The leaves are bright green but not always obviously bilobed, since the central division halfway along the 'fan' may be almost non-existent, or it may be accompanied by other subdivisions in each of the halves. Fossil specimens show similar variation. The name 'ginkgo' is apparently derived from the Chinese term *yin kuo*, which refers to the silvery fruits. The hard fruit kernels can be roasted as a hangover cure.

CONIFERS (GYMNOSPERMS)

The conifers, Coniferales (Coniferophytes), are gymnosperms ('naked-seed' plants) that include forest trees like pines, firs, cedars and redwoods. They also include yews and monkey-puzzle trees (araucarians). Most bear woody cones in which the seeds ripen. Although a few conifers are medium-sized shrubs, the majority are tall with straight trunks and represent the highest and most massive trees on Earth.

Araucaria

Araucaria cones

Araucarias are dioecious – that is, they produce male (pollen) cones and female (seed) cones on separate trees. When the smaller male cones are mature, the woody scales open to allow the tiny pollen to blow away and reach the female cones. Later, the female cones open to let their ripe seeds disperse in the wind. These fossil cones of the species *Pararaucaria patagonia* are from the Cerro Cuadrado region of Argentina.

Mature (open) male cone

Immature (closed) male cone

Right: *These fossil twigs of Araucaria sternbergii from the Eocene rocks of Germany (55 million years ago) are very similar to the living forms, and show the characteristic covering of small, sharp leaves.*

A 'living fossil', remains of *Araucaria* date back to the Triassic, about 230 million years ago. Today it survives in the wild in isolated areas of the Southern Hemisphere, although it formerly had a worldwide distribution. However, it is widely planted as the ornamental species *Araucaria araucana*, or the monkey-puzzle tree. Tall and elegant, it bears arc-like horizontal branches covered in small, tough, spiny, scale-like leaves.

Needle-like leaves on twig

Separated leaves

Name: *Araucaria*.
Meaning: After the Arauco Indians, Chile.
Grouping: Gymnosperm, Coniferale.
Informal ID: Monkey-puzzle tree, Chile pine.
Fossil size: Twig length 5cm/2in.
Reconstructed size: Tree height typically 30m/100ft plus.
Habitat: Subtropical forests.
Time span: Triassic, 230 million years ago, to today.
Main fossil sites: Worldwide.
Occurrence: ◆ ◆ ◆

Taxodites

This 20-million-year-old fossil is of a twig bearing very small lanceolate (spear-point shaped) leaves, similar to leaves seen today in living cypresses, such as the swamp cypresses (*Taxodium*) and the Chinese swamp cypress (*Glyptostrobus*). However, because there is insufficient detail that can be seen – the specimen lacks clearly visible reproductive features – it cannot confidently be assigned to either group. The main difference between the groups is the shape of the seeds – winged in *Glyptostrobus* and not in *Taxodium*. Because, as their common names imply, both groups favour a swampy habitat, *Taxodites* is a good indicator fossil for its environmental conditions.

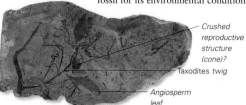

Crushed reproductive structure (cone)?

Taxodites twig

Angiosperm leaf

Left: *In this Miocene specimen from Germany, the smaller twigs are Taxodites while the larger leaves belong to an angiosperm (flowering plant). The rounded shape in the centre may be the reproductive structure, or cone, of Taxodites, but it is too damaged for identification.*

Name: *Taxodites*.
Meaning: Related to *Taxus* (yew).
Grouping: Gymnosperm, Coniferale.
Informal ID: Cypress-type twig.
Fossil size: Twig diameter less than 1cm/½in.
Reconstructed size: Whole tree height up to 30m/100ft.
Habitat: Swampy forests.
Time span: Miocene, 20–10 million years ago.
Main fossil sites: Worldwide.
Occurrence: ◆ ◆ ◆

Sequoia, Metasequoia

Name: *Sequoia, Metasequoia.*
Meaning: After Sequoyah, a native American (Cherokee) from the early 19th century.
Grouping: Gymnosperm, Coniferale.
Informal ID: Sequoia, redwood.
Fossil size: See text.
Reconstructed size: Whole tree height up to 70m/230ft.
Habitat: Damp mountain slopes.
Time span: *Sequoia* – Jurassic, 200 million years ago, to today; *Metasequoia* – Cretaceous, 110 million years ago, to today.
Main fossil sites: *Sequoia*, Worldwide; *Metasequoia*, Northern Hemisphere.
Occurrence: ◆ ◆

The sequoias, or redwoods, are conifers that include the largest organisms ever to have lived on the planet. Members of the cypress family Taxodiaceae (see opposite), they were an important part of the extensive Mesozoic forests that were roamed by the dinosaurs, although today their distribution in the wild is much restricted. In fact, *Metasequoia*, the dawn redwood or 'water fir', was known as Mesozoic fossils from China, until a small stand of living trees was discovered in an isolated valley in Szechwan province, China, in 1941. For fossil plant experts at the time, it was like discovering a living dinosaur! The cones are small and ball-like. During the past two centuries, American sequoias have been planted in parks and gardens around the world.

Rght: The three or so living species of sequoias are the coastal redwood (S. sempervirens, see below), the dawn redwood (Metasequoia glyptostroboides) and the giant or Sierra redwood, or 'big tree' (Sequoiadendron giganteum), shown here. It can grow to almost 100m/330ft in height and can live for several thousand years.

Conifer wood

Conifer wood lacks hard fibres and is composed almost exclusively (about 90 per cent) of types of tube-like cells called tracheids. These are elongated woody cells that transport water (containing dissolved minerals) around the tree. As a result, this made the wood permeable, allowing a fossilization process by which waterlogged wood was replaced with minerals dissolved in the water, called permineralization. This occurs when water-borne minerals deposit around the structure of the wood itself, and this happened more frequently in areas that were prone to seasonal flooding. The famous 'petrified forest' of Arizona, USA is made up of permineralized conifers.

Left: This Sequoia affinis *twig is from Miocene rocks associated with the Florissant Beds of Colorado, USA. Today, the genus* Sequoia *in North America is restricted to* Sequoia sempervirens, *the coast redwood, which grows wild only in damp mountain forests along the western coastline. Slab 7cm/2¾in across.*

Collections of small, flattened, scale-like needles (leaves)

Main twig

Right: This open Metasequoia *cone, typically globular, is from the famous Hell Creek Formation of Late Cretaceous rocks in South Dakota, USA. Today, the genus* Metasequoia *grows wild only in isolated valleys of China. Cone about 2cm/¾in across.*

Seed recess

Scales of cone

FLOWERING PLANTS (ANGIOSPERMS)

The angiosperms include most flowers, herbs, shrubs, vines, grasses and blossom-producing broadleaved trees. The presence of flowers, and seeds enclosed in fruits (angiosperm means 'seed in receptacle'), differentiate them from gymnosperms, such as conifers. There are two great groups of angiosperms: monocotyledons and dicotyledons. 'Monocot' seeds have one cotyledon, or seed-leaf, in the seed.

Phragmites (reed)

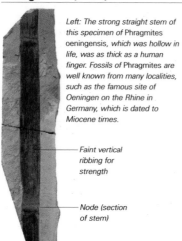

Left: The strong straight stem of this specimen of Phragmites oeningensis, which was hollow in life, was as thick as a human finger. Fossils of Phragmites are well known from many localities, such as the famous site of Oeningen on the Rhine in Germany, which is dated to Miocene times.

— Faint vertical ribbing for strength

— Node (section of stem)

The genus *Phragmites* is well known today around the world as the common reed or marsh grass *P. australis*. Tall and tough, in some regions it grows higher than 5m/16½ft and its dense, tangled rootstock spreads rapidly. *Phragmites* fringes not only lakes and slow rivers but also brackish water, and is regarded in many areas as an invasive pest. It has stout unbranched stems and long, slim leaves with bases that sheath the stem as it grows. The soft, feathery flower-heads are a shiny purple-brown colour. The dried stems are cut for many purposes, including roof thatching. *Phragmites* fossils are among the most common plant remains in some localities, especially from the Miocene Epoch to recent times. Reeds, grasses and rushes are placed in the large monocotyledon family Gramineae.

Name: *Phragmites*.
Meaning: Fence-like growth.
Grouping: Angiosperm, Monocotyledon, Graminean.
Informal ID: Reed.
Fossil size: Specimen height 18cm/7in.
Reconstructed size: Plant height 4m/13ft or more.
Habitat: Fringing bodies of fresh and brackish water.
Time span: Mostly Tertiary, 60 million years ago, to today.
Main fossil sites: Worldwide.
Occurrence: ◆ ◆

Sabal (palm)

Below: This specimen of a fan-like leaf pattern from Aix-en-Provence, France, dates to the Eocene Epoch. Palm leaves are generally tough and durable (as known from their uses today, such as thatching, wrapping and cooking). They have thickened veins that run side by side, known as parallel venation, which is characteristic of all monocotyledons.

Palm trees (family Palmae or Arecaceae) are familiar in many warmer regions, and are grown for their wood, frond-like leaves and nut-like or juicy seeds that produce oil, starch and other useful materials, as well as providing edible dates and coconuts. A typical palm has an unbranched, almost straight trunk covered with ring-, arc- or scale-like scars where leaf bases were once attached. The trunk cannot grow thicker like most other trees but remains almost the same diameter to the crown of frond-like leaves, which may be fan-like, feathery or fern-like. Palm fossils of genera such as *Sabal* and *Palmoxylon* are known from the Late Cretaceous Period, 80-plus million years ago. By the Early Tertiary Period, 60 million years ago, they were evolving fast, and living genera such as *Phoenix* (date palms) and *Nypa* (mangrove palms) had appeared. There are hundreds of extinct species and more than 2,700 living ones.

Name: *Sabal*
Meaning: Food
Grouping: Angiosperm, Monocotyledon, Palmaean
Informal ID: Palm tree
Fossil size: Specimen length 32cm/12½in
Reconstructed size: Tree height 20–30m/66–100ft
Habitat: Tropics
Time span: Eocene, 50 million years ago
Main fossil sites: Worldwide
Occurrence: ◆ ◆

Bevhalstia

Name: *Bevhalstia*.
Meaning: For palaeontologist L Beverly Halstead.
Grouping: Angiosperm.
Informal ID: 'Wealden weed'.
Fossil size: Stems and shoots up to 15cm/6in; flower-like structure 5mm/³⁄₁₆in across.
Reconstructed size: Whole plant height up to 25cm/10in.
Habitat: Freshwater swamps.
Time span: Early Cretaceous, 130–125 million years ago.
Main fossil sites: Europe (Southern England).
Occurrence: ◆ ◆

Bevhalstia was possibly one of the first flowering plants to have appeared on the Earth and it dates back to the Early Cretaceous Period, some 130–125 million years ago. The plant had an enigmatic combination of features, with leaves that had a vascular system (tube-like network) similar to those of ferns and mosses, together with angiosperm-like buds and flower-like features. *Bevhalstia* was likely to have been a delicate herbaceous (non-woody) plant. Despite its fragile nature, however, it has been found in abundance in the fossil record, suggesting that it grew very near to its burial sites, which were possibly quiet lakes or swamp bottoms. If this is accurate, then *Bevhalstia* probably had an aquatic way of life, perhaps similar to today's *Cabomba* or fanwort pond and aquarium plants. Some of the first specimens of *Bevhalstia* were collected in the 1990s as part of the studies of the fish-eating dinosaur *Baryonyx* being carried out at London's Natural History Museum.

Above: In this fragment of Bevhalstia the delicate nature of the stem is a clue to the plant's probable way of life, part-submerged in the manner of modern pondweeds.

Right: The flower-like structure of Bevhalstia is one of the first to appear in the fossil record, in the Early Cretaceous, and it already shows what could be interpreted as 'petals'.

Pabiana

Name: *Pabiana*.
Meaning: After Pabian (see text).
Grouping: Angiosperm, Dicotyledon, Magnoliacean.
Informal ID: Magnolia leaf.
Fossil size: Leaf length 5–7cm/2–2¾in.
Reconstructed size: Tree height 10m/33ft plus.
Habitat: Warmer forests.
Time span: Middle Cretaceous, 100 million years ago.
Main fossil sites: North America.
Occurrence: ◆ ◆

Pabiana is usually regarded as a Cretaceous member of the magnolias, family Magnoliaceae. These were among the first flowering trees to evolve, in the Early to Middle Cretaceous Period. The group still has about 80–120 living species, mostly distributed in warmer parts of the Americas and Asia. Although many similar flowers are usually associated with pollination by bees, magnolias evolved their large blossoms well before the appearance of bees, and instead they are pollinated by beetles. Because of this, the bloom is quite tough and fossilizes fairly well – that is, compared with other flowers. A primitive aspect of magnolias is their lack of – or combination of, depending on the point of view – distinct petals or sepals. The name 'tepal' has been coined to describe these intermediate structures in magnolias, which look like the petals of other flowers. The original specimens of *Pabiana* were discovered at the Rose Creek Quarry near Fairbury, Nebraska, USA, in 1968 by a team including American palaeontologist Roger K Pabian, who was actually searching for fossil invertebrates. The genus was named after him in 1990.

Right: Leaves of Pabiana were lanceolate (lance-shaped) and were very similar to those of living magnolias. This specimen is from Middle Cretaceous rocks found in Nebraska, USA.

FLOWERING PLANTS (CONTINUED)

Most of the 250,000 living angiosperm, or flowering plant, species are 'dicots' – their seeds have two cotyledons, or seed-leaves. An enormous area of interest in both gymnosperms and angiosperms centres around the study of pollen. Pollen grains are tough, resistant, incredibly numerous and mostly microscopic. Their preserved remains give vast amounts of data on the origins and ranges of plant species.

Typha

Below: Bulrush leaves are long and strap- or blade-like, with virtually parallel sides and prominent, almost parallel veins, held nearly upright on the unbranched stem. In some species of Typha, the leaves are more than 50cm/20in in length. This species is Typha latissima from Oligocene rocks of the Isle of Wight, southern England.

The monocotyledonous genus *Typha* is known to some people as bulrushes and to others as reedmace. The confusion is usually traced back to a painting of the Bible's account of Moses in the Bulrushes, *Finding Moses*, in 1904 by Sir Lawrence Alma-Tadema. He substituted the more impressive plant with the then-common name of reedmace for the plant that was then known as bulrush, which has much more delicate, branching flower spikes. Other common names for *Typha* include false-bulrush, cattail and cat's-tail. The scientific name *Typha* cannot be confused, however. Its brown, sausage-shaped female flower head is closely packed with many tiny flowers. The fluffy, feathery male flower head is on the same stem above the female one. For almost 100 million years bulrushes (reedmace) have had thick underwater roots and been invasive, quickly choking waterways such as brooks and streams, and diminishing the open water of lakes and slow rivers.

Name: *Typha*.
Meaning: Bog, marsh.
Grouping: Angiosperm, Monocotyledon, Typhacean.
Informal ID: Reedmace, bulrush.
Fossil size: Slab 13cm/5in across.
Reconstructed size: Plant height 2m/6½ft or more.
Habitat: Margins of still or slow fresh water.
Time span: Middle Cretaceous, 100 million years ago, to today.
Main fossil sites: Northern Hemisphere.
Occurrence: ◆ ◆

Populus

The poplars, members of the willow group, are well known for their long-stalked 'trembling' leaves, large and colourful hanging catkins, and rapid growth, which makes them useful as seasonal windbreaks. The genus *Populus* includes aspens and cottonwoods and consists of more than 30 living species, mainly in the Northern Hemisphere. These are grouped as balsams, white poplars, aspens and black poplars. *Populus* is well known from several Eocene Epoch sites, such as the Florissant Fossil Beds National Monument in Colorado, USA, where they thrived in the dampness along stream banks and adjacent to lakes. Leaf shape generally varies from rounded to heart-like, with smooth or slightly toothed edges. Some species have conspicuous glands on the upper side of the leaf where it joins its stalk.

Left: Many species of Populus have the characteristically long leaf stalk, or petiole, which, combined with the alternate and well-spaced pattern of leaves along the twig, allows them to rustle in the breeze. Populus glandulifera may be the ancestor of today's black poplar group. This specimen is of Miocene origins from Oeningen, Baden, Germany.

Name: *Populus*.
Meaning: People, crowd.
Grouping: Angiosperm, Dicotyledon, Salicacean.
Informal ID: Poplar tree.
Fossil size: Leaf width 8cm/3in.
Reconstructed size: Tree height 30m/100ft.
Habitat: Temperate mixed woodland.
Time span: Eocene, 50 million years ago, to today.
Main fossil sites: Northern Hemisphere.
Occurrence: ◆

Fruits, nuts and seeds

The terminology of plant reproductive structures is famously complex. In general, a fruit is the ripe or mature ovary or seed-container of a plant, together with its contents. In fruits such as apples (known technically as pomes) the fruit is strictly only the apple core, while the fleshy part is derived from structures around the ovary – in this case the floral receptacle. Usually only hard fruits, such as nuts, form fossils. The kernels, or 'nuts', themselves are the seeds, while the surrounding container (pericarp, exo/endocarp and/or cupule) is known as a shell or husk, especially when it is dry and fibrous.

Walnuts are fruits known botanically as drupes, with an outer fleshy exocarp, a hard endocarp (shell) and the woody, wrinkled nuts within.

Carya

The genus *Carya* includes about 22 species of living trees, known variously as hickories and walnuts, that are mainly native to North America. They are usually placed in the walnut family, known as the juglands group. Hickory wood, being straight-grained and strong yet elastic, is often compared with ash, and it is used to make tool handles, oars, wheel spokes and similar items where shock absorbency is important. The tough-walled nuts fossilize well and can usually be distinguished by their similarity to a walnut with one end more pointed, generally flat and angular sides, and four lengthwise grooves or narrow, wing-like extensions.

Right: This preserved seed of Carya ventricosa from Hungary is dated within the Eocene Epoch, some 45 million years ago. Inside the tough husk the seeds would have been pale and edible.

— Nut

Name: *Carya*.
Meaning: From the Greek *karya*, walnut (formerly applied to many similar nuts).
Grouping: Angiosperm, Dicotyledon, Juglandacean.
Informal ID: Hickory tree.
Fossil size: Seed length 4cm/1½in.
Reconstructed size: Tree height 20–30m/65–100ft.
Habitat: Temperate mixed woodland.
Time span: Palaeocene, 60 million years ago, to today.
Main fossil sites: Chiefly Northern Hemisphere.
Occurrence: ◆ ◆

Fagus/Berryophyllum

Name: *Berryophyllum*.
Meaning: Fagus – beech; Berryophyllum – berry leaf, rounded leaf.
Grouping: Angiosperm, Dicotyledon, Fagacean.
Informal ID: Beech tree.
Fossil size: Leaf length 9cm/3½in.
Reconstructed size: Tree height 20–30m/65–100ft.
Habitat: Temperate woodland regions.
Time span: Eocene, 45 million years ago.
Main fossil sites: North America.
Occurrence: ◆

Beech trees, genus *Fagus*, are familiar plants growing in northern temperate countries today. One species is found in North America and one grows in Europe, with another 10 or so species native to East Asia (principally China). With colloquial identities such as 'Queen of the Forest' these stately, massive trees provide homes and food for myriad wildlife. In addition, they produce vivid autumn leaf colours, small nuts called beechmast in spiny or prickly husks (cupules) and valuable timber for a variety of uses. The fossil history of *Fagus* is complex and has been much discussed. The genus may date back to Palaeocene times and was far more widespread than it is today, growing in huge areas around almost all of the Northern Hemisphere. The preserved three-angled nuts and paired husks, either together or separated, have been given various names. So have the preserved fossils of the characteristic smooth bark and also isolated leaves, which are alternate on the stem, typically ovate to lanceolate in shape with a range of tooth- or saw-like serrated edges. Since the leaves, husks and nuts of living beech are variable, not only within one species but even on the same tree, the fossil record of the genus will take much unravelling.

Below: Mystery surrounds the leaves known as Berryophyllum. They may come from an early type of beech tree. The central vein, or midrib, is prominent, as are the parallel lateral or secondary veins. Similar leaves are known as Castenophyllum. This specimen comes from the Eocene of Tennessee, USA, and is associated with 'frilly' beech-like husks.

FLOWERING PLANTS (CONTINUED)

The earliest angiosperm fossils are generally Early Cretaceous, about 130 million years ago – so in terms of plant evolution they are relative latecomers. They could, however, have evolved much earlier in dry uplands where fossilization was unlikely. Or their evolution could have been triggered by interaction with animal groups such as browsing dinosaurs or possibly new types of pollinating insects.

Castanea

Below: Sweet chestnut leaves are simple (that is, they are not divided into lobes or leaflets), and are usually arranged alternately along the twig. Their shape is long or lanceolate and with serrations, crenations, or fine teeth, although some species are smoother-edged. This specimen from southern Germany, dated to the Eocene Epoch, shows the typical parallel lateral veins branching from the main vein, or midrib.

Midrib — Fine serrations

Sweet chestnut trees and chinkapins, with about eight living species in the genus *Castanea*, are members of the beech and oak family, Fagaceae. (They should not be confused with horse chestnuts or buckeyes, *Aesculus*, which form their own family, Hippocastanaceae.) Sweet chestnuts, including the Spanish and American types, are known for their highly edible fruits, which are sweet-tasting, smooth-surfaced nuts, usually two or three together in a fleshy, prickly or thorny husk. The Spanish chestnut bark is also characteristic when fossilized, being smooth in younger trees but developing a pattern of heavy, hard ridges arranged in a right-handed spiral pattern up the main trunk as the tree ages.

Name: *Castanea*.
Meaning: From the Greek *kastanon*, chestnut.
Grouping: Angiosperm, Dicotyledon, Fagacean.
Informal ID: Sweet chestnut tree.
Fossil size: Leaf width 3cm/1in.
Reconstructed size: Tree height 30m/100ft or more.
Habitat: Mixed warm-temperate woodland.
Time span: Late Cretaceous, 70 million years ago, to today.
Main fossil sites: Worldwide.
Occurrence: ◆ ◆

Myrica

Below: Most Myrica leaves are simple in outline with a tough, shiny surface in life, and possibly with finely toothed or crinkled edges. The leaves are arranged spirally along the stem and their shape is known as oblanceolate (tapered to a point at both ends but broader towards the tip). Lateral veins branch off the midrib in a 'herringbone' pattern.

Lateral veins | Midrib (main vein) | Leaf base

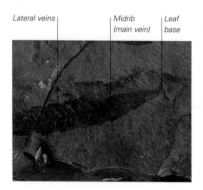

A genus of the beech and oak order, Fagales, *Myrica* (or myrtles) includes between 35 and 50 species of plants living today, varying from small-growing bushes or shrubs to tall trees reaching up to about 20m/66ft in height. They are mostly evergreen plants and are widely distributed throughout the African, Asian, North American, South American and European continents – the only exception being Australasia. Some species are well known in ornamental woods, parks and gardens, and are planted for their aromatic qualities. They include bayberry, sweet gale or bog myrtle, candleberry and wax-myrtle. The flowers are arranged as catkins, and the berry-type fruits (known as drupes) often have a thick, waxy coating. At one time, this wax was collected for making into candles, leading to such common names as candleberry, wax-myrtle and tallow-myrtle.

Name: *Myrica*.
Meaning: Myrtle.
Grouping: Angiosperm, Dicotyledon, Myricacean.
Informal ID: Myrtle.
Fossil size: Leaf length 7cm/2¾in (incomplete).
Reconstructed size: Shrub height 10m/33ft.
Habitat: Warm temperate woods, including peat bogs.
Time span: Oligocene, 30 million years ago, to today.
Main fossil sites: Northern Hemisphere.
Occurrence: ◆

Zelkova

Name: Zelkova.
Meaning: From the local Caucasian name.
Grouping: Angiosperm, Dicotyledon, Ulmacean.
Informal ID: Zelkova, Asian elm, Japanese elm and others (see text).
Fossil size: Leaf length 4cm/1½in.
Reconstructed size: Tree height up to 35m/115ft.
Habitat: Mixed temperate woodland.
Time span: Palaeocene, 55 million years ago, to today.
Main fossil sites: Europe, Asia.
Occurrence: ◆

There are about six species in the genus *Zelkova* (*Zelcova*) living today, varying from small shrubs to tall trees. In the wild, they are found in Southern Europe (on the Mediterranean islands of Crete and Sicily) and they are also native to Central to East Asia. The genus is classified within the elm family but it is separate from the true elms. This has led to confusion with common names such as Caucasian elm and Japanese elm being applied to the Caucasian zelkova *Z. caprinifolia* and the Japanese zelkova or keaki *Z. serrata*, which is commonly planted as a street tree in North America; it not only produces good shade and attractive autumn colour, but is also tolerant of wind, heat, drought and the pollution associated with urban conditions. Evidence suggests that the genus may have arisen in Palaeocene times, more than 50 million years ago, possibly in the lands around the North Pacific – there are fossils both from Asia and from North America to support this, as in the specimen here, which is from the Eocene Epoch in Utah, USA. Various forms of *Zelkova* continued to thrive throughout North America and Northern Europe, but these disappeared with the Pleistocene glaciations. Fossils from Europe, dated to the Miocene and Pliocene Epochs, resemble the modern *Z. carpinifolia* and *Z. serrata*, as mentioned above.

Above: Most Zelkova leaves have saw- or tooth-like edges, with the size of the teeth varying from fine to coarse, and their tips ranging from almost blunt to very finely sharp-pointed. The overall leaf shape of early Tertiary types is long and narrow, as here, with some modern species being more rounded or oval. This specimen is part of the same item as Cardiospermum (see following pages).

Sassafras

Name: Sassafras.
Meaning: From Native American 'green stick'.
Grouping: Angiosperm, Dicotyledon, Lauracean.
Informal ID: Sassafras tree.
Fossil size: Leaf width 8cm/3⅛in.
Reconstructed size: Tree height 20–30m/66–100ft.
Habitat: Temperate mixed woodland.
Time span: Middle Cretaceous, 100 million years ago, to today.
Main fossil sites: Scattered, mainly North America.
Occurrence: ◆

Sassafras – which is both its scientific and its common name – is familiar in some parts of the world as a tree planted for the aromatic qualities of its oil, bark and roots. In some areas it is also planted in order to help repel mosquitoes and other insects. Also called the root beer tree, ague tree and saloo, its root essences were used as a central flavouring in root beers and teas, and are also incorporated into cosmetics, ice-creams, salads, soups, jellies, perfumes and many other food and drink items. There are three living species of *Sassafras*, the chief one known in the wild from eastern North America. The genus is classified as part of the laurel family, and is closely related to the 'true' laurels. The whole tree is roughly cone-shaped and is admired for its blazing red and gold autumn leaf colours. The leaves, both fossilized and from living specimens, are famous for their heterophylly, or variation in shape – they range from roughly oval (unlobed), bilobate (mitten shaped) and trilobate (three pronged), even on the same branch of the same tree.

Below: The leaves of Sassafras are commonly three-lobed, or trilobate, with three equal lobes. However, they may also be trilobate with a larger central lobe, or two-lobed (bilobate) with equal or unequal lobes, or even elliptical with almost no lobing at all. Each lobe has a central midrib with branching lateral veins. This specimen is from the Dakota Group of Central Kansas, USA.

Regular three-lobed form

Tripartite vein branching

Leaf stalk (petiole)

FLOWERING PLANTS (CONTINUED)

Angiosperms, or flowering plants, range from delicate flowers and soft herbs to shrubs, bushes and trees.
Fossilized remains of tender, soon-shrivelled parts, such as flower petals, are rare. Most specimens are
of bark, woody stems, roots, hard seeds such as nuts, and leaves which have fallen into mud
or similar sediments with good preserving qualities.

Liquidambar

Below: Liquidambar leaves are typically palmate with three (rarely), five (more commonly) or seven (rarely) sharply pointed lobes. The leaves resemble those of Platanus (see below) in shape, and are borne alternately, spiralling along the stem. This specimen of Liquidambar europaeum is from the Miocene Epoch of Germany.

Trees of the genus *Liquidambar* are known commonly as sweetgums. The gum, sometimes called storax or styrex, is used in herbal medicine and the fruits are burr-like capsules ('gumballs'). Formerly, *Liquidambar* was included in the family Hamamelidaceae, the witch-hazels. Newer schemes place it in the family Altingiaceae, along with the altingia trees of South-east Asia. There are about six living species of *Liquidambar*, most in East Asia, with the American sweetgum in eastern North America down to northern Central America. The genus is a relict, being much more widespread in the Middle Tertiary Period. The numerous fossil species include *Liquidambar europaeum*, shown here. However, the European representatives disappeared during the recent ice ages, and climate changes caused it to fade also in western North America.

Name: *Liquidambar*.
Meaning: From the runny gold-coloured gum.
Grouping: Angiosperm, Dicotyledon, Altingiacean.
Informal ID: Sweetgum tree
Fossil size: Leaf width 9cm/3½in.
Reconstructed size: Tree height to 30m/100ft.
Habitat: Mixed woodland.
Time span: Eocene, 40 million years ago, to today.
Main fossil sites: Worldwide.
Occurrence: ◆

Platanus

Below: This Eocene specimen of the species Platanus wyomingensis is from Utah, USA. The fine, well-spaced points, or 'teeth', sometimes termed multi-serrate or multi-denticulate, are visible chiefly on the three central lobes.

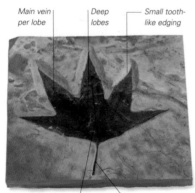

Main vein per lobe | Deep lobes | Small tooth-like edging

Palmate central area | Leaf stalk (petiole)

Considerable confusion surrounds the common names of species in the genus *Platanus*. In general, they are known as planes in areas such as Great Britain, but sycamores in North America. (The sycamore in Britain is a member of the genus *Acer*, otherwise called the sycamore maple.) The American sycamore is also called the buttonwood tree, and the London plane is hardy and resistant to air pollution. Plane leaves can be mistaken for those of the maples, with their palmate form (lobes coming from a central area, like the fingers of a palm), and often deep and sharp-pointed lobes, sometimes bearing toothed edges. However, *Platanus* leaves are arranged alternately along the twig, while *Acer* leaves grow in opposite pairs. The group has one of the longest fossil records of any well-known tree, and the six living species of *Platanus* are all found in the Northern Hemisphere.

Name: *Platanus*.
Meaning: Broad, flat.
Grouping: Angiosperm, Dicotyledon, Platanacean.
Informal ID: Plane or sycamore tree.
Fossil size: Leaf width 19cm/7½in.
Reconstructed size: Tree height up to 40m/130ft.
Habitat: Mixed woodland.
Time span: Early to Middle Cretaceous, 115 million years ago, to today.
Main fossil sites: Northern Hemisphere.
Occurrence: ◆ ◆

Acer

Name: *Acer.*
Meaning: Maple.
Grouping: Angiosperm, Dicotyledon, Aceracean.
Informal ID: Maple tree.
Fossil size: Leaf width 4cm/1½in.
Reconstructed size: Tree height 20–35m/66–115ft.
Habitat: Temperate mixed woodland.
Time span: Oligocene, 30 million years ago, to today.
Main fossil sites: Mainly Northern Hemisphere.
Occurrence: ◆ ◆

The maple family, Aceraceae, includes about 150 living species of maples and sycamores, and possibly as many fossil species. The species living today include both trees and shrubs and are renowned for their often spectacular autumn colour. One of their most characteristic features is the seeds, each with a single membraneous 'wing' that causes the seed to twirl around like a mini-helicopter as it is shed from the parent tree. The seeds are borne in pairs, with the angle between the pair varying from almost a straight line to a very sharp V, according to the species. These seeds are fairly hardy and make easily identifiable fossils. The leaves are usually long-stalked and distinctively three-lobed, with each lobe toothed to a varying degree, again depending on the species concerned. The sugar maple is the state tree of both New York and West Virginia, USA. Most maples, both living and fossilized, are from the Northern Hemisphere, with the majority in East Asia.

Above: The three-lobed maple leaf is an internationally recognized symbol of the North America region, especially of the latitudes around the border between Canada and the USA, and it is also the central motif of the Canadian flag.
This French specimen in loosely cemented limestone from Quaternary rocks has minimal toothing – that is, it has a rounded outline.

Cardiospermum

Name: *Cardiospermum.*
Meaning: Heart seed.
Grouping: Angiosperm, Dicotyledon, Sapindacean.
Informal ID: Balloon vine (also see text).
Fossil size: Leaf length 5cm/2in.
Reconstructed size: Height to 3–4m/10–13ft.
Habitat: Mixed temperate woodland.
Time span: Eocene, 40 million years ago.
Main fossil sites: North America.
Occurrence: ◆ ◆

The genus *Cardiospermum* includes living woody vines and shrubby plants with a trailing tendency, mostly from the tropical Americas. It is part of the Sapindaceae, or soapberry, family with more than 1,300 species worldwide, many of which are toxic. A well-known living species is *Cardiospermum halicacabum*, the balloon vine, which has been commonly planted and in some areas has escaped to become a problem weed. The flowers are small and white, but the plant is known for its fruits, which are pale, inflated-looking, membraneous 'balloons', each about 3cm/1¼in across, containing three black seeds. Each seed has a white heart-shaped scar – hence the genus name, which means 'heart seed'. Another common name for this species is love-in-a-puff, which belies its status as a noxious weed and plant pest in some regions. The species *Cardiospermum coloradensis* is well known from the Eocene Epoch rocks of the Green River Formation, Utah, USA, where its leaves form some of the most distinctive fossils. This extinct species appears to have been more of a shrub than a trailing vine.

Above: The leaves of the Colorado balloon vine, Cardiospermum coloradensis, are typically trilobate, although some specimens show four or five rounded lobes. The leaf stalk, or petiole, can be considerably longer than that seen on this specimen.

Below: The balloon vine, originally from Central and South America, is now a weed in North America, Australia, South Africa and other regions.

ANIMALS – INVERTEBRATES

Vertebrae are commonly known as backbones – we have them in our own 'backbone' or spinal column. This feature is often used to divide the kingdom Animalia into two great groups – those with vertebrae and those without. (The accurate biological situation is slightly more complicated, as explained on the opening pages of the next chapter, Animals – Vertebrates).

Invertebrate or 'spineless' animals range from the simplest sponges, the poriferans, which are devoid of nerves or muscles or brains, to complex and highly developed cephalopod molluscs such as the octopus and squid, with sophisticated and intelligent behaviour. The molluscs are one of the best-represented invertebrate groups in the fossil record, since their hard shells end up in the high-probability preservation conditions of the sea bed.

One of the largest invertebrate groups is the arthropods or 'joint-legs'. It takes in the extinct trilobites, myriad crustaceans such as barnacles and crabs, and the land-dwelling insects and arachnids.

Above from left: the coral Meandrina, *trilobite* Trinucleus *and mollusc* Dactylioceras.

Right: Ammonites are one of the most recognizable and evocative symbols of the prehistoric world. This huge group of molluscs came to prominence in the Devonian Period and faded during the Cretaceous, along with the dinosaurs. These specimens are Promicoceras planicosta *from the Jurassic Period.*

EARLY INVERTEBRATES

The most ancient animal fossils, such as those from Precambrian Ediacara in Australia or the Middle Cambrian Burgess Shale in North America, have received varying interpretations over the years. Some of those life-forms were so different from any others, living or extinct, that the term 'Problematica' is used when precise taxonomic position is uncertain.

Mawsonites

Hand-sized *Mawsonites* are one of the most difficult of the Ediacaran life-forms to interpret with any degree of certainty. With its multi-lobed, expanding radial symmetry – that is, a wheel- or petal-like structure – it has been called almost everything from a flower to an aberrant sea lily (crinoid). However, it comes from a time well before flowering plants had yet evolved. One of the more established theories regarding its origins is that it was some kind of scyphozoan – that is, a jellyfish from the cnidarian group. Its unique features, however, make it difficult to assign to any

of the known jellyfish groupings, either those living or extinct. In addition, its surface topography, which has some fairly sharp and well-defined ridges, is not reminiscent of a floppy, jelly-like organism. Another possibility regarding its origins is that the fossils are the remains of a radial burrow system made by some type of creature, perhaps a worm, tunnelling and looking for food in the sandy, silty sea bed of the seas and oceans of the time. This would mean it is a trace fossil – traces left by an organism, rather than preserved parts of the actual organism itself.

Name: *Mawsonites*.
Meaning: Mawson's animal (after Australian Antarctic explorer Sir Douglas Mawson).
Grouping: Animal, Cnidarian, Scyphozoan.
Informal ID: Ediacara 'jellyfish' or trace fossil.
Fossil size: Overall width 12–14cm/4¾–5½in.
Reconstructed size: Unknown.
Habitat: Sea bed.
Time span: Precambrian, about 570 million years ago.
Main fossil sites: Australia.
Occurrence: ◆

Central disc

Innermost set of concentric rings

Outermost lobes

Outer margin

Left: Resembling multiple, expanding whorls of flower petals, the reddish sandstone impression known as Mawsonites has an overall circular outline composed of curved lobes. The central disc is well defined, with concentric sets of roughly circular raised 'rings' that become more straight-sided towards the outer margin.

Naraoia

This thumb-size arthropod was at first believed to be some kind of branchiopod crustacean (a cousin of today's water-flea, *Daphnia*). The shiny, two-part covering 'shield', or carapace, bears a central groove from head to tail and a division from side to side into two valves, front and rear. Around the edge of the carapace is a fringe of limb and appendage endings. Dissection into the thickness of the fossil, down through the carapace, has revealed the limbs and appendages in more detail, showing how and where they join to the main body underneath. The results indicated, very unexpectedly, that *Naraoia* was some type of trilobite. However it is not tri-lobed, with left, middle and right sections to the carapace, but rather bi-lobed, with just left and right. Similarly, it has not three divisions from head to tail, but two.

Below: Preservation of Burgess Shale trilobites is exceptional, as some of these are entire animals rather than just shed body casings. Naraoia is an early and unusual trilobite, with front and rear sections to the carapace and one central furrow from head to tail. Comparison of specimens shows that Naraoia may have retained immature features into mature adulthood, a phenomenon known as neoteny.

Name: *Naraoia*.
Meaning: From the nearby locally named Narao lakes.
Grouping: Arthropod, Trilobite.
Informal ID: Early trilobite.
Fossil size: Carapace front–rear about 3cm/1⅛in.
Reconstructed size: Width 2.5cm/1in, including limbs.
Habitat: Sea bed.
Time span: Middle Cambrian, 530 million years ago.
Main fossil sites: North America.
Occurrence: ◆

Burgessia

Named after the shale rocks of its Burgess Pass discovery region in Canada, *Burgessia* was probably some type of bottom-dwelling arthropod. The creature is known from thousands of specimens discovered at the locality. *Burgessia* probably walked, burrowed or swam weakly across the sea floor, and fed mainly by filtering tiny edible particles of food from the general ooze on the ocean floor. Its protective, convex (domed) shield-like carapace was about the size of a fingernail and it covered the softer, vulnerable parts beneath, so that only the ends of the limbs, two antennae-like feelers, which were directed forwards, and the long tail spine were visible from above. In the carapace was a branching set of canals, or grooves, which may have been part of the digestive system. In some specimens the tail spine is twice as long as the body.

Hurdia

The Burgess fossils show the first large-scale evolution of a mineralized, or chitinous, exoskeleton – a hard outer casing, usually found over the top of the body and used for protection as well as muscle anchorage. *Hurdia* has been included in the group known as anomalocarids. Its mouthparts may have had an extra set of teeth within, forming a so-called 'pharyngeal mill' that would be able to grind up other harder-bodied victims – showing that predators were already adapting to well-protected prey.

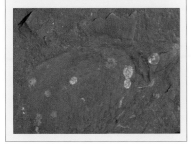

Name: *Burgessia*.
Meaning: Of Burgess.
Grouping: Arthropod, otherwise uncertain.
Informal ID: Burgess arthropod.
Fossil size: 1cm/⅜in across.
Reconstructed size: Head–tail length 2.5cm/1in.
Habitat: Sea bed.
Time span: Middle Cambrian, 530 million years ago.
Main fossil sites: North America.
Occurrence: ◆

Tail section | Main carapace

Tail spine | Head end

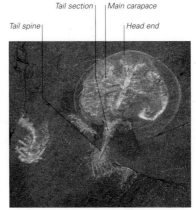

Left: Detailed study of Burgessia *reveals that the body has a cephalic (head) region, about nine main body sections or segments, a telson (tail section) and a long spiny tail. The two antennae were directed so that they pointed forwards. Piecing together the remains of many specimens shows that the creature's 9 or 10 pairs of legs were uniramous (unbranched) and probably bore gills for obtaining oxygen from the water.*

SPONGES

The sponges, phylum Porifera, are strange animals that live permanently rooted to the rocks or mud of the sea bed. They flush water through their porous bodies and filter out tiny food particles. Sponges have an extremely simple anatomy, lacking specialized organs, nerves and muscles and possessing only a few basic cell types. From Cambrian times, most sponges have left plentiful fossils of their mineralized bodies.

Doryderma

Name: *Doryderma*.
Meaning: Porous-skinned.
Grouping: Poriferan,
Demosponge, Lithistid,
Megamorinan, Dorydermatid.
Informal ID: *Doryderma*
sponge.
Fossil size: Specimen
7cm/2¾in high.
Reconstructed size: 50cm/
20in plus.
Habitat: Sea floor.
Time span: Carboniferous
to Late Cretaceous, 350–85
million years ago.
Main fossil sites:
Throughout Europe.
Occurrence: ◆ ◆

Doryderma is an example of the Demospongea class of sponges – a group that is distinguished by building their skeletons out of both silica spicules and/or a network of tough, cartilage-like fibres of the substance spongin, which is a nitrogenous, hornlike material. Where spicules are present in this class, they are typically both highly complex in shape and capable of meshing together to form a network of significant strength – the skeletons of living demosponges have been observed to hold together long after the death of the animal itself. *Doryderma* is a reasonably common fossil in European marine rocks. However, the atypical branching form that makes it so attractive also makes it prone to damage and fragmentation.

Exhalant openings
(oscula)

Branching
points

*Left: The
Cretaceous genus
Doryderma has an
unusual branching form.
The ends of the branches
bear multiple openings
that are exhalant: that is,
from which water would have
been expelled from them after
having been filtered for food
inside the main body cavity.*

Sponge shapes and anatomy

There are something in the region of 10,000 species of sponges living in the world today, and many thousands of them are also known from the fossil record. They have existed since Cambrian times, more than 500 million years ago, and the abundance of individual specimens as well as the number of different species provide useful marker, or index, fossils for dating rocks. Common sponge shapes include vases, tubes, mushrooms and funnels. Some types of sponge branch either regularly (as illustrated here) or follow a more random pattern, but they all share the same major feature – a cavity inside where the water is sieved and the food removed. In addition, a growing sponge may be constrained by its locality, such as a cleft in the sea bed rock, or be nibbled by predators, and both of these factors can have an effect on its overall form and size. The three main types of sponge are calcareous (calcisponges) with spicules (small, pointed structures) of hard, calcium-containing minerals; siliceous sponges, in which the spicules are based on the mineral silica and have a glass-like quality; and horny sponges, in which the skeleton is composed of protein-based or similar substances such as spongin.

Spongia

Name: *Spongia*.
Meaning: Sponge.
Classification: Poriferan, Demosponge, Keratosid, Spongiid.
Informal ID: Bath sponge.
Fossil size: 8cm/3in across.
Reconstructed size: Up to 20cm/8in across.
Habitat: Sea floor.
Time span: Carboniferous, 350 million years ago, to today.
Main fossil sites: Worldwide.
Occurrence: ◆ ◆

The Spongiidae family includes excellent examples of demosponges, which possess a skeleton composed entirely of branching and interwoven fibres made of spongin. This makes it highly flexible even when dry. The much harder, mineralized, spicule-type elements of the skeleton are almost completely absent. Modern forms of the genus *Spongia* are probably most familiar as the traditional 'bath sponges' of recent years – which are actually the dried corpses of this most 'spongy' of sponges, used for their mild exfoliating properties and ability to absorb water. The fossil record of the Spongiidae shows that these sponges have been a feature of the sea floor at varied depths since the beginnings of the Carboniferous Period.

Right: Spongia *is a fossil from the same group as the modern 'bath sponge'. It possessed a soft, highly flexible skeleton – a feature that often resulted in poor preservation as a fossil. The example shown here is from the Cretaceous Red Chalk of Hunstanton in Norfolk, England.*

Verruculina

Verruculina is a genus of demosponge that lived from the Mid Cretaceous to the Tertiary, chiefly in Europe. The genus possessed a network of spongin reinforced by small, simple spicules embedded in the elastic material (in a similar manner to fibreglass). This gave the whole sponge a skeleton midway in rigidity between those of *Doryderma* and *Spongia*. *Verruculina* did not have an almost fully enclosed central space in the manner of most sponges – with many small, inhalant pores for drawing in water, and one large exhalant pore (the osculum) for pushing the water out. Instead, it possessed a broad, squat body that unfolded at the top, resembling a sprouting leaf or a bracket fungus.

Below: Verruculina *was a distinctive demosponge and it resembles the bracket fungus, which grows on tree trunks. It was once common in the Cretaceous seas that covered Europe – this specimen is Late Cretaceous chalk – but it steadily declined during the Tertiary Period before its eventual extinction.*

Name: *Verruculina*.
Meaning: Wart-like.
Grouping: Poriferan, Demonsponge, Lithistid, Leiodorellid.
Informal ID: Bracket-fungus sponge.
Fossil size: 7cm/2¾in across.
Reconstructed size: Typically 10cm/4in diameter.
Habitat: Sea floor.
Time span: Middle Cretaceous to Tertiary, 110–2 million years ago.
Main fossil sites: Throughout Europe.
Occurrence: ◆ ◆ ◆

SPONGES (CONTINUED)

Most sponges, past and present, lived in the seas, and typically had a hollow, porous body attached to the sea bed at one end. Water is drawn through the body wall, where feeding cells lining the interior extract food particles. The internal skeleton can consist of horny material; hard, mineralized shards or spicules; or both horny and mineralized elements, embedded in the body wall.

Porosphaera

Porosphaera is an example of the calcarean sponges, or calcisponges. This is a group that builds skeletons made entirely of spicules made from calcite, with no spongy material or silica. Each of the hundreds of tiny spicules that form the skeleton often resembles a Y or tuning fork in shape, allowing them to mesh together and form a strong, rigid network that is often easily fossilized. *Porosphaera* possessed a roughly grape-size, spherical body, made rigid by interlocking spicules and also covered in tiny spines, presumably for defence against predators. These could range from fish to starfish and sea urchins. It is a relatively common fossil from the Cretaceous marine rocks of Europe, particularly chalk deposits.

Below: Tiny Porosphaera *was a Cretaceous calcisponge that was relatively common in European rocks, especially – as with this specimen – from the Upper Chalk beds of Sussex, England. This view shows the large exhalant opening, or osculum, from which water left the chamber within, and which probably faced directly upwards in life.*

— Globular body

— Exhalant opening (osculum)

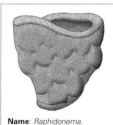

Name: *Porosphaera*.
Meaning: Porous sphere.
Grouping: Poriferan, Calcarean, Pharetronid, Porosphaeridan.
Informal ID: Spiny globe sponge.
Fossil size: Generally less than 1cm/⅖in but some specimens over 2cm/¾in.
Reconstructed size: As above.
Habitat: Sea floor.
Time span: Mainly Cretaceous, 145–65 million years ago.
Main fossil sites: Throughout Europe.
Occurrence: ◆ ◆ ◆

Raphidonema

The calcisponge *Raphidonema* was basically elongated and cup-shaped, but individuals grew in a great many variations on this simple theme. The thick walls tapered downward from a broad exhalant opening at the top, often giving a form similar to a vase or funnel. The walls themselves were well perforated with small canals, and the outer surface was frequently covered by knobbly or nodular protrusions characteristic of this genus, while the inner walls lining the interior chamber were much smoother. *Raphidonema* was widespread in the warm, shallow seas of the Cretaceous Period.

Left: Fossils of Raphidonema *are famously common in the Farringdon Sponge Gravels of Oxfordshire, England, where colonies of individuals formed enormous, coral-like structures known as sponge-beds.*

Above: Raphidonema, *a vase-shaped Cretaceous sponge, can be easily recognized by the lumpy, nodule-like ornamentation on the outside surface. In contrast, the interior of the sponge is smooth. The simple vase- or cup-like shape was often bent, curved, squashed and distorted during preservation.*

Name: *Raphidonema*.
Meaning: Seamed one.
Grouping: Poriferan, Calcarean, Pharentronid, Lelapiid.
Informal ID: English vase sponge or vase 'coral'.
Fossil size: 7cm/2¾in.
Reconstructed size: Variable, height typically up to 10cm/4in.
Habitat: Sea floor.
Time span: Triassic to Cretaceous, 250–65 million years ago.
Main fossil sites: Throughout Europe.
Occurrence: ◆ ◆ ◆

Ventriculites

Ventriculites belongs to the Hexactinellidae, which is a class of sponges that form skeletons of glassy mineralized silica spicules only, without networks of the more flexible spongin material. Hexactinellid spicules occur in both small and large forms, a novel feature that allows them to interlock more tightly and so form a strong skeleton.

This sponge was an important element of the Santonian Micraster Chalk Communities found on the coast of Kent, in the south of England. (*Micraster* was an echinoid, or sea urchin, and appears later in this chapter.) These sea-bed assemblages of urchins, starfish and other echinoderms, plus molluscan shellfish, probably relied on the large sponges for shelter, both from currents and predators – a role more usually held by corals. Remains of *Ventriculites* are often found in association with the fossils of huge, coiled-shelled ammonoids, up to 2m/6½ft across. It is possible that the sponges provided shelter and protection for the young of these predatory molluscs.

Left: The attractive Ventriculites *from the Cretaceous Period of Europe grew in a funnel shape, tapering from a very broad exhalant opening at the summit to a surprisingly narrow base, which was anchored to the shallow sea bed.*

Name: *Ventriculites*.
Meaning: Stomach stone.
Grouping: Poriferan, Hexactinellid, Lychniscosan, Ventriculitid.
Informal ID: Funnel sponge, funnel 'coral'.
Fossil size: Height 15cm/6in.
Reconstructed size: Height up to 30cm/12in.
Habitat: Sea floor.
Time span: Middle to Late Cretaceous, 110–65 million years ago.
Main fossil sites: Throughout Europe.
Occurrence: ◆ ◆

Archaeocyatha

Name: *Archaeocyatha*.
Meaning: Ancient cup
Grouping: Uncertain – classified on their own or within the phylum Porifera.
Informal ID: Cambrian cup 'sponge'.
Fossil size: Longest side 10cm/4in.
Reconstructed size: Height up to 25cm/10in.
Habitat: Sea floor.
Time span: Cambrian, 550–500 million years ago.
Main fossil sites: Worldwide.
Occurrence: ◆ ◆

The archaeocyathans (archaeocyathids) are a very ancient 'anomaly' from the Cambrian Period. They have many similarities to sponges and also to certain corals. Some experts consider them to be sponges, others as allied to corals, and still others as not closely related to sponges or corals at all. They are sometimes called the 'first reef-builders', but their actual importance in the earliest reefs is not certain. Archaeocyathans were solitary organisms, each inhabiting a double-walled, porous skeletal structure that resembled a tall cup – their name is derived from the old Greek for 'ancient cup'. The cup was attached to the sea floor by thick roots, but the form and lifestyle of the animal that lived within remains a mystery. The strange structure and lack of soft tissue anatomy has frustrated attempts to define their relationships to other groups of sponges and similar simple creatures.

Below: The mysterious archaeocyathans are sometimes found in great concentrations that resemble primitive reefs. This assemblage is from Cambrian rocks of the Flinders Ranges, Australia. In some Cambrian rocks found in Russia, the fossils are numerous enough to be used as indicator or marker fossils for dating layers.

CNIDARIANS – JELLYFISH, CORALS AND RELATIVES

Cnidarians are often regarded as the most basic forms of true animals. The soft, sac-like body has a single opening, which doubles as both mouth and anus. The life cycle of most cnidarians includes a free-swimming, jellyfish-like stage, or medusa, and an anemone-like stage, or polyp, attached to the sea bed.

Essexella

Jellyfish, known as scyphozoans, spend most of their lives as large, free swimming medusae. Their soft bodies have bequeathed them a generally poor fossil record, restricted to areas where an unlikely combination of circumstances have combined to preserve exceptionally detailed fossils. *Essexella*, from the Upper Carboniferous rocks of North America, is one of the better-known fossil jellyfish, due to mass discoveries of them in several quarries within the state of Illinois, USA. The fine

preservation shows an unusual anatomy with a large, mushroom-like body, the bell, trailing an apron- or sheet-like membrane rather than the usual dangling mass of tentacles. This strange feature has earned it, and similar forms, the nickname 'hooded jellyfish'.

Below: This specimen of Essexella had its umbrella-like body preserved in semi-3D, as part and counterpart in two halves of a split rock. The sheet-like membrane hangs below the bell, which may have pulsated gently as the jellyfish drifted with the currents.

— Bell

'Hood' (apron-like sheet)

Name: *Essexella*.
Meaning: From the estuarine Essex fauna of Mazon Creek.
Grouping: Cnidarian, Scyphozoan, Rhizostoman.
Informal ID: Hooded jellyfish.
Fossil size: Bell 5cm/2in across.
Reconstructed size: Bell width up to 10cm/4in.
Habitat: Open sea.
Time span: Late Carboniferous, 300 million years ago.
Main fossil sites: North America.
Occurrence: ◆ ◆

Ediacaran medusoid

Above and right: Ediacaran medusoids are a well-known form among the rare and beautiful fossils of that time. Despite their name, however, it is likely that they are unrelated to modern jellyfish, and were probably incapable of swimming. The fossilized ripple effect shows that the creature was preserved on the sea bed.

The incredible fossils of Ediacara offer a rare glimpse into the long-vanished world of the Precambrian Era, some 570 million years ago. They are the faint, compressed outlines of strange, soft-bodied animals that bathed in shallow seas about 50 million years before the dawn of the Cambrian 'explosion' of more modern life-forms. Some experts contend that they were a 'failed experiment' – long-vanished organisms that came and went, were fundamentally different to all modern creatures, and left no descendants. Some are termed vendobionts, or 'gas creatures', because of their inflated, sac-like bodies.

Name: Ediacaran medusoid.
Meaning: Jellyfish-like creature from Ediacara.
Grouping: Unknown.
Informal ID: Ediacaran 'jellyfish'.
Fossil size: Up to 10cm/4in across.
Reconstructed size: As above.
Habitat: Sea floor.
Time span: Late Precambrian, 570 million years ago.
Main fossil sites: Australia.
Occurrence: ◆

Zaphrentoides

Zaphrentoides was a solitary form of rugose coral from the Early or Lower Carboniferous Period (for general information on corals, see next page). It had a form typical of the solitary rugose corals – a curved, horn-shaped structure subdivided by many radiating internal ribs, known as septa. Rugose corals have long disappeared, seemingly wiped out in the truly catastrophic series of extinctions that marked the end of the Permian Period, 250 million years ago. Unlike their tabulate and scleractinian cousins, the solitary rugose corals did not solidly attach themselves to the seabed, and probably did not form a large part of Palaeozoic reef communities. *Zaphrentoides* is one of the less common forms. Known informally as a horn coral, *Zaphrentoides* would have lived half-buried in the soft mud. Being overturned was a common danger, as many fossil specimens are found on their sides or even upside down.

Below: The horn-like structure of Zaphrentoides is shown cut through to reveal the flanges, or ribs, arranged in a radiating pattern, like the spokes of a wheel. This specimen is from County Fermanagh, in Northern Ireland.

Septa

Horn-like body shape

Narrow end of horn

Name: *Zaphrentoides*.
Meaning: Tube connections.
Grouping: Cnidarian, Anthozoan, Rugosan, Hapsiphyllid.
Informal ID: Horn coral.
Fossil size: Width 2cm/⅜in.
Reconstructed size: Typical width 2cm/⅜in, height 10cm/4in.
Habitat: Sea floor.
Time span: Early Carboniferous, 350 million years ago.
Main fossil sites: Europe, North America.
Occurrence: ◆ ◆

Lithostrotion

Name: *Lithostrotion*.
Meaning: Star stone, tread rock.
Grouping: Cnidarian, Anthozoan, Rugosan, Lithostrotionid.
Informal ID: Spaghetti rock.
Fossil size: Individual 'threads' 1cm/⅜in thick.
Reconstructed size: Colonies commonly 10m/ 33ft across.
Habitat: Sea floor.
Time span: Early to Middle Carboniferous, 350–320 million years ago.
Main fossil sites: Worldwide, especially Central Europe and Asia.
Occurrence: ◆ ◆ ◆

A colonial (group-living) form of rugose coral, *Lithostrotion* remains are usually composed of thousands of individual polyp skeletons consolidated into a single, tight-knit structure. (For general information on corals, see next page.) It occurs in Early Carboniferous rocks on all continents, and especially in Middle Carboniferous limestone rocks in Central Europe and Asia. In some localities in Scotland, deposits are so common that blocks measuring metres across are nearly entirely composed of these fossils. *Lithostrotion* forms a major part of the Carboniferous 'coral-calcarenite' fossil assemblages, such as those commonly found in limestones of northern England dated to Asbian (Visean) times, 340 million years ago. These fossilized sea-bed communities of corals, brachiopods and brittlestars were an important haven for marine life in the Mid-Palaeozoic seas.

Right: Lithostrotion, a colonial rugose coral, can be recognized by its structure of thick, thread-like strands, which to some people resemble over-cooked pasta. This has earned it the nickname 'spaghetti rock'.

CORALS

Corals (zoantharia) are cnidarians that have discarded the free-swimming stage and live as attached polyps, either singly or in colonies. They are common fossils due to their mineralized, often elaborate skeletons, which generally resemble a cup surrounding the anemone-like creature. Coral reefs have occurred for much of the last 500 million years and are useful indicators or markers for dating rocks.

Favosites

Below: Favosites *is recognized by its characteristic hexagonal column structures, which have earned it the nickname 'honeycomb coral'.*

The tabulate coral *Favosites* was one of the longest-surviving members of this now-extinct group. Tabulates are named for the plate-like structures, tabulae, that divide them horizontally, so that each individual when cross-sectioned resembles a packet of sweets (candies) cut lengthways. *Favosites* colonies formed medium-size, dome-like structures that were an important element of the tabulate reef communities during the Silurian and the Devonian Periods, around 400 million years ago. The remains of *Favosites* are often found in association with fellow reef-builders *Heliolites* and *Halysites*. After their heyday, which was from the Ordovician to the Devonian Periods, they became less common, although in some areas they are known to have persisted until the Permian Period.

Name: *Favosites*.
Meaning: Favoured form (used in ancient Greece for a hexagon).
Grouping: Cnidarian, Anthozoan, Tabulatan, Favositinan.
Informal ID: Honeycomb coral.
Fossil size: 5cm/2in across.
Reconstructed size: Variable, up to 40cm/16in.
Habitat: Sea floor.
Time span: Ordovician to Permian, 500–250 million years ago, gradually becoming less common.
Main fossil sites: Worldwide.
Occurrence: ◆ ◆ ◆

Heliolites

Close examination of the remarkable *Heliolites*, or 'sun stone' fossil, shows that it appears to be covered with circles surrounded by radiating lines, bearing a striking resemblance to tiny, rayed suns. In this genus of extinct tabulate corals (see *Favosites* above), the tiny 'suns' are, in fact, formed by the skeletal ribs, or septa, in the chambers inhabited by the polyps that built the coral skeleton. *Heliolites* formed a major part of Mid-Palaeozoic tabulate reef communities, and it is often found in association with *Favosites* and *Halysites*.

Right: Heliolites *formed large, pillow-like structures that contributed greatly to the reefs that existed 400 million years ago. These would have provided shelter for a variety of different primitive fish, molluscs such as ammonoids, and other creatures, as well as food, as predators tried to nibble the soft-bodied coral polyps from their stony chambers.*

Tiny rayed 'suns' dot the surface

Name: *Heliolites*.
Meaning: Sun stonel.
Grouping: Cnidarian, Anthozoan, Tabulatan, Heliolitid.
Informal ID: Sun coral.
Fossil size: 7cm/2¾in across.
Reconstructed size: Colonies 1m/3¼ft or more across.
Habitat: Sea floor.
Time span: Late Silurian to Middle Devonian, 415–380 million years ago.
Main fossil sites: Worldwide.
Occurrence: ◆ ◆ ◆

Halysites

Halysites, a colonial tabulate coral (see *Favosites*), is relatively common in Silurian rocks. It can be recognized by the distinctive pattern of winding, beaded lines covering its surface, rather like a conglomeration of pearl necklaces. This feature has earned it the informal name of 'chain coral'. Like the other specimens on these two pages, *Halysites* was an important Mid-Palaeozoic reef builder. It formed large, upright colonies that resembled a curved or folded sheet. Its remains are often found in association with those of *Heliolites* and *Favosites*.

Name: *Halysites*.
Meaning: River stone.
Grouping: Cnidarian, Anthozoan, Zoantharian, Tabulatan, Halysitid.
Informal ID: Chain coral.
Fossil size: 5cm/2in.
Reconstructed size: Typically 10–20cm/4–8in across.
Habitat: Sea floor.
Time span: Ordovician to Silurian, 500–410 million years ago.
Main fossil sites: Worldwide.
Occurrence: ◆ ◆ ◆

Above: The skeletons of colonial Halysites *polyps linked together laterally (that is, side to side), as can be seen in this three-quarter view. In this way, their fossils formed sheet-like folds of mineralized material.*

Left: When viewed from above, Halysites *has a distinctive beaded, winding pattern, earning it the name 'chain coral'.*

Syringopora

Name: *Syringopora*.
Meaning: Porous tube.
Grouping: Cnidarian, Anthozoan, Zoantharian, Tabulatan, Auloporid.
Informal ID: Organ pipe coral.
Fossil size: 6cm/2½in across at top.
Reconstructed size: Average colony 5–10cm/2–4in; columns about 5mm/⅛in across.
Habitat: Sea floor.
Time span: Silurian to Late Carboniferous, 440–310 million years ago.
Main fossil sites: Most continents.
Occurrence: ◆ ◆

Syringopora is a relatively common Mid to Late Palaeozoic genus of colonial tabulate coral (see *Favosites*). It is often found in association with the rugose colonial coral *Lithostrotion* as part of the Carboniferous 'coral-calcarenite' fossil assemblages, where the smaller *Syringopora* probably grew in the shelter provided by the larger corals. *Syringopora* consists of a loose fabric of tubes, which can be distinguished from similar corals by the presence of cross-links joining together neighbouring columns. It displays a very similar shape to the unrelated modern genus *Tubipora*, suggesting that it occupied a similar ecological role in Palaeozoic seas. *Syringopora* is a relatively common fossil in marine rocks from the Silurian Period.

Right: Syringopora *colonies were bundles of loosely interconnected tubes, often called organ-pipe coral from their resemblance to the stack of pipes of a large concert organ.*

CORALS (CONTINUED)

The tabulate and rugose ('wrinkly') corals shown on the previous pages did not survive the major end-of-Permian mass extinction event. All post-Triassic corals, such as those shown on these two pages, and including living types, are known as scleractinians. They superficially resemble rugose corals, but they can be distinguished by their internal ribs, or septa, which are always arranged in multiples of six.

Montlivaltia

This solitary scleractinian coral, squat and cup-shaped, colonized 'hardground' areas of the sea floor during much of the Mesozoic Era. These were areas where the sediment had become saturated with minerals produced by the decomposition of shelly material, resulting in a solid, concrete-like seabed.

Montlivaltia probably established itself in such an environment by inhabiting holes eroded by current- or wave-borne pebbles or abandoned mollusc borings. It is a moderately common fossil of Triassic and Cretaceous rocks, and it is found on all continents. Like all coral polyps, it would extend its stinging tentacles from its surrounding skeleton.

Left: Montlivaltia was a small, solitary form, and one of the few corals capable of colonizing hardened areas of sea floor. This specimen is from the Lower Lias of Walterston, near Cardiff, Wales.

Name: *Montlivaltia*.
Meaning: From a French site.
Grouping: Cnidarian, Anthozoan, Zoantharian, Scleractinian, Montlivaltid.
Informal ID: Hardground cup coral.
Fossil size: Slab 6cm/2½in across.
Reconstructed size: Individuals 1–5cm/½–2in across.
Habitat: Sea floor.
Time span: Middle Triassic to Late Cretaceous, 240–70 million years ago.
Main fossil sites: Worldwide.
Occurrence: ◆ ◆

Thamnasteria

This Middle Mesozoic genus is an example of a colonial scleractinian coral – the group that forms the largest and most elaborate colonies known. *Thamnasteria* formed medium-size, domed colonies, and is a relatively common fossil in marine deposits. As a colonial coral, it is more colonial than most. In fact, *Thamnasteria* is the model genus for a type of structural organization among corals known as 'thamnasteroid'. In this growth pattern, the skeletons of individual polyps overlap and merge to form a cohesive whole. This further blurs the boundaries between individuals of the colony, making it hard to distinguish where one polyp ends and another begins.

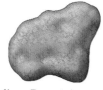

Individual polyp chambers

Adjoining wall

Radiating ribs, or septa

Right: Thamnasteria inhabited Mesozoic environments known as 'patch reefs'. These were discontinuous reefs formed in warm, shallow waters, where large corals provided habitats for a diverse range of molluscs, crustaceans and fish. This is a Jurassic specimen.

Name: *Thamnasteria*.
Meaning: Bushy star.
Grouping: Cnidarian, Anthozoan, Zoantharian, Scleractinian, Thamnasteriid.
Informal ID: Colonial scleractinian coral with polygonal cups.
Fossil size: 7cm/2¾in across.
Reconstructed size: Individuals 5–10mm/⅕–⅖in across.
Habitat: Sea floor.
Time span: Middle Triassic to Mid Cretaceous, 240–100 million years ago.
Main fossils sites: Most continents.
Occurrence: ◆ ◆ ◆

Meandrina

This genus takes its name from the meandering, wandering ridges and furrows that cover its surface. It is an excellent example of the so-called 'brain corals', an informal term that refers to a variety of colonial scleractinian corals that are not closely related. In the brain corals, the skeletons form bulbous domes covered in convoluted, lobe-like folds, the whole structure bearing an uncanny resemblance to the human brain. Modern *Meandrina* corals form large groups of colonies in the Atlantic coastal waters of southern North America, Brazil, the West Indies and Europe. Fossil forms can be found from about 55 million years ago, especially in Europe, the West Indies and in Brazil.

Name: *Meandrina*.
Meaning: Meandering coral.
Grouping: Cnidarian, Anthozoan, Zoantharian, Scleractinian, Meandrinid.
Informal ID: Maze coral, brain coral.
Fossil size: Length 8cm/3in.
Reconstructed size: Up to 30cm/12in.
Habitat: Sea floor.
Time span: Eocene, 55 million years ago, to today.
Main fossil sites: North and South America, Europe.
Occurrence: ◆ ◆ ◆

Left: Meandrina, *a brain or maze coral, gets its latter name from the maze-like surface network of grooves. Discoveries of fossil* Meandrina *show that this coral has been an important part of Atlantic sea-bed communities for nearly 55 million years. This specimen from Florida, USA, is about 5 million years old (Late Miocene to Early Pliocene).*

Flabellum

A genus of solitary scleractinian coral, *Flabellum* is today commonly found living attached to boulders and other hard substrates, in the seas around Hawaii and the Azores islands. The coral can be readily recognized by its fan-shaped body, which is typically a flattened circle in cross section and gives *Flabellum* a superficial resemblance to solitary forms of the extinct rugose corals. *Flabellum* fossils are found in deposits from the Eocene Epoch to the present day in rocks all around the world, although it is particularly abundant in Middle Tertiary rocks of the Apennine Mountains in Italy.

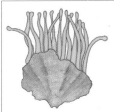

Name: *Flabellum*.
Meaning: Fan-shaped.
Grouping: Cnidarian, Anthozoan, Zoantharian, Scleractinian, Flabellid.
Informal ID: Fan coral.
Fossil size: Individual structures 1.2–1.5cm/ ½–⅝in wide.
Reconstructed size: Height up to 10cm/4in.
Habitat: Sea floor.
Time span: Eocene, 55 million years ago, to today.
Main fossil sites: Most continents.
Occurrence: ◆ ◆ ◆

Oppelismilia

The solitary scleractinian coral *Oppelismilia*, up to 5cm/2in across, is a relatively rare fossil from rocks of Upper Triassic to Jurassic times in the Americas and Eurasia. This small, conical coral grew in soft, muddy conditions that resulted in frequent uprooting. Thus, it was never able to establish firm anchorage and develop substantial reef communities.

Above: The fan coral Flabellum *is today known to survive at spectacular depths – the deepest recorded colonies are more than 3,000m/9,840ft below the surface. Fossil finds show that this coral has been present in sea-bed communities since the Early Eocene Epoch. The flange-like septa within the outer wall and the elliptical cross-sectional shape are clearly visible.*

WORMS

Worms are entirely soft-bodied animals, which means they are rarely found as fossils except where there is an exceptional level of preservation. Worms are found in the Burgess Shale Formations of North America preserved as carbon films, and in the Chengjiang Formations of China, where they have been mineralized with pyrites. Their trace fossils, such as furrows and burrows, are found worldwide.

Ottoia

A common marine worm from the Burgess Shale fossils of British Columbia, Canada, *Ottoia* is from the worm group known as priapulids or proboscis worms, which appeared in Cambrian times, some 530 million years ago, but which has only about 10 living species. *Ottoia* was about 8cm/3in long with a spiny, bulbous, moveable 'snout', or proboscis. The spines on the proboscis were probably used like stabbing teeth to capture prey. How *Ottoia* lived is uncertain, but it may have been an active tunneller living in a U-shaped burrow, in the manner of today's lugworm. At Burgess, its fossils have been found with soft tissue preserved showing both muscle and gut material. Analysis of the gut contents shows a diet of mainly molluscs – although there is also evidence of cannibalism.

Below: Ottoia is commonly found in the Cambrian Burgess Shale. Its fossils are most often preserved as a thin film on the fine-grained shale. This specimen has its gut preserved.

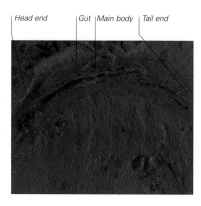

Head end | Gut | Main body | Tail end

Name: *Ottoia*.
Meaning: Of Otto.
Grouping: Priapulid.
Informal ID: Priapulid worm, proboscis worm.
Fossil size: Length 7cm/2¾in.
Reconstructed size: Length 8cm/3in.
Habitat: Shallow sea floor.
Time span: Cambrian, 530 million years ago.
Main fossil sites: North America.
Occurrence: ◆ ◆

Serpulid worm casts

Below: These serpulid worm casts are preserved in chalk from the Late Cretaceous rocks of Suffolk, England. They are not equivalent to the squiggly 'worm casts' familiar on lawns, which are the eaten and excreted soil of earthworms. They are calcareous tubes that the serpulid worms constructed and inhabited. These remains show the irregular spiralled shape typical of the casts.

Serpulid worms are members of the Annelida or segmented worms. It is the most familiar worm group known today and it includes both earthworms and seashore ragworms. Modern serpulids, also called tubeworms or plume worms, each live in a calcareous tube that they construct around themselves. This is attached to a hard surface, such as a seashore rock, mussel or crab shell. Serpulids have structures called radioles, made of rows of micro-hairs called cilia, which are used for respiration and feeding. Modern serpulid worms are very diverse, but their fossil record is limited. The main evidence for them includes trails, burrows and casts – their stony living tubes – but these cannot be linked with a specific species. The earliest serpulid worm tubes date from the Silurian Period, although serpulids become abundant only from the Jurassic onwards.

Name: Serpulid worm.
Meaning: Spiral.
Grouping: Annelid, Polychaete, Serpulid.
Informal ID: Tube worm or plume worm.
Fossil size: Each coiled tube 2cm/¾in across.
Reconstructed size: Length up to 10cm/4in.
Habitat: Wide range of marine environments.
Time span: Arose in the Silurian, 420 million years ago, but common from the Jurassic, to today.
Main fossil sites: Worldwide.
Occurrence: ◆ ◆

Nereites

Nereites is a fossilized trace or trail, probably left by a wandering annelid worm, consisting of a middle furrow with tightly spaced lobes on either side. *Nereites* facies (a group of rocks with this trace fossil as the common element) are used to represent deep ocean environments that are subjected to turbidity currents. This means that *Nereites* can be used as a palaeo-environmental and palaeodepth indicator – that is, it lived in the very deep sea and so, therefore, did other creatures whose fossils are found associated with it. Geologist Roderick Murchison named this particular trace fossil *Nereites cambrensis* in 1839 – the species name of *cambrensis* for its discovery site of Wales. It is the holotype – the specimen used for the original description of the species, and with which all subsequently discovered specimens are then compared.

Name: *Nereites*.
Meaning: From Nereus, the all-wise son of Gaia, a Roman sea god.
Grouping: Annelid, Pascichnian.
Informal ID: Worm feeding or grazing trail.
Fossil size: Slab width 5cm/2⅜in; length 6cm/2⅜in.
Reconstructed size: Worms reached tens of centimetres in length.
Habit: Deep sea bed.
Time span: Specimen is from the Silurian Period (about 430 million years ago).
Main fossil sites: Worldwide.
Occurrence: ◆ ◆

Left: The trace represents the feeding trail made by the worm as it meandered across the sea-bed sediment looking for food. The lobed depressions may have been left by the parapodia – oar-like flaps along the sides of the worm's body, which are used for 'walking' and swimming, as in the modern ragworm.

Spirorbis worm casts

Name: *Spirorbis*.
Meaning: Spiralled coil.
Grouping: Annelid, Polychaete, Sabellid.
Informal ID: Coiled tubeworm.
Fossil size: Individual coils less than 1cm/⅜in across.
Reconstructed size: Range from a few millimetres across the coil to 2cm/¾in.
Habitat: Varied marine environments.
Time span: Late Ordovician (450 million years ago), to today.
Main fossil sites: Worldwide.
Occurrence: ◆ ◆

Spirorbis worms are related to serpulid worms (see opposite) and are commonly known as coiled tubeworms. Like the serpulids, *Spirorbis* builds itself a white calcareous tube in which to live. It pokes a 'plume' of tiny greenish tentacles from the tube's open end in order to catch any floating particles of food or micro-organisms from the plankton. Today's *Spirorbis* tubes are a common sight on seaside rocks, large seaweeds, such as kelps, and also shellfish. Each coil is just a few millimetres in diameter and resembles a miniature white snail. Fossils show that *Spirorbis* lived like this in the past, too, attaching itself to hard surfaces such as the shells of other organisms, including ammonites and crabs.

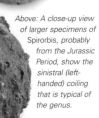

Above: A close-up view of larger specimens of Spirorbis, *probably from the Jurassic Period, show the sinistral (left-handed) coiling that is typical of the genus.*

Right: Spirorbis worm casts are the living tubes of tiny worms related to fanworms and other tubeworms. In this specimen, the tubes have a loosely spiralled form and are preserved in sandstone.

BRYOZOANS

Bryozoans are groups or colonies of zooids, which superficially resemble miniature sea anemones. They construct calcareous cups, tubes, bowls, boxes or other containers around themselves, known as zooecia. As they multiply, the colony grows in a certain pattern according to the genus, from an encrusting 'carpet' over the rocks, to upright delicate fan-, lace- or bush-like structures. Today there are about 4,300 species.

Fenestella

A marine bryozoan, *Fenestella* was particularly common during the Early Carboniferous Period (350–320 million years ago), but it became extinct at the end of the Permian. Sometimes called a 'lace-coral', it is not a colony of coral polyps, but tiny bryozoan zooids as explained above. Its network of interconnected branches often preserve in a pale colour compared with the substrate, making its fossils look like pieces of delicately woven lacework. *Fenestella* colonies fed by producing water currents that flowed through holes in the colony, where the feeding zooids filtered out suspended food particles.

Name: *Fenestella*.
Meaning: Small window.
Grouping: Bryozoan, Fenestellid, Fenestratan.
Informal ID: Lace 'coral'.
Fossil size: Slab width 5cm/2in.
Reconstructed size: Colonies range in size from less than 1cm/⅖in to 50cm/20in across.
Habitat: Hard substrate on the sea floor.
Time span: Silurian to Early Permian, 420–280 million years ago.
Main fossil sites: Worldwide.
Occurrence: ◆ (◆ ◆ in the Early Carboniferous)

Left: This example of Fenestella plebia from a road cutting in Oklahoma, USA clearly shows the lace-like branching structure typical of the genus. The net-like colony would have been erect or vertical in life. The crosswise connecting bars, called dissepiments, lacked the tiny zooid creatures, while the upright branches each had a double row of zooids.

Archimedes

Archimedes is an extinct bryozoan that had a central 'screw' structure to support the colony. Often the branches and the tubes secreted by the zooids are not preserved, since they are very delicate, so all that remains is the more robust central support. The individual zooids had a flat, table-like shape. *Archimedes* is most commonly found in fine-grained sediments laid down in calmer areas of the sea, such as sheltered bays. New colonies formed by branching off a parent colony before growing a new screw. In some instances, thousands of colonies can be traced back to a single parent branch.

Name: *Archimedes*.
Meaning: After Archimedes, inventor of the screw device for lifting water.
Grouping: Bryozoan, Fenestrid, Fenestellid.
Informal ID: Archimedes screw, screw bryozoan.
Fossil size: Specimens about 3cm/1in long.
Reconstructed size: Length up to 50cm/20in.
Habitat: Calm, sheltered marine environments.
Time span: Carboniferous to Permian, 350–270 million years ago.
Main fossil sites: Worldwide.
Occurrence: ◆ ◆

Below: Four specimens of Archimedes show the central 'screw'. The outer edges of each show traces of the branching structure of the bryozoan animals themselves, which attached to this central support structure.

Chasmatopora

One of the bryozoans that formed colonies in curtain-like curved sheets, *Chasmatopora* (sometimes known as *Subretepora*) formed its fronds by branching off the main stems from the narrow base or 'root' anchored to the substrate. Each of the stems or strips is about six zooecia wide – that is, formed from six rows of tiny zooid animals in their box-like containers. These sets of parallel rows of tiny containers, known as zooecia, branched and then joined to form an enlarging net riddled with a regular pattern of holes. More than a dozen species of *Chasmatopora* have been identified, most from Eastern Europe (Estonia, Russia) and China, with a few from North America.

Right: This Chasmatopora *specimen is dated to the Middle Ordovician Period, about 465 million years ago, and comes from Kuckersits, Estonia. The original wavy curtain-shaped colony probably broke into several pieces due to a storm that caused an underwater sediment slump.*

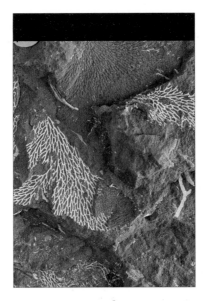

Name: *Chasmatopora.*
Meaning: Ravine or gorge holes.
Grouping: Bryozoan, Fenestrid, Phylloporinin.
Informal ID: Sea fan, lacy sea mat.
Fossil size: Colonies about 10cm/4in across.
Reconstructed size: As above.
Habitat: Sea floor.
Time span: Ordovician to Silurian, 480–420 million years ago.
Main fossil sites: Europe, Asia, North America.
Occurrence: ◆ ◆

Ptilodyctia

In this bryozoan, the colony grows in a slightly curved or an almost straight, single-branch structure. It consists of a middle wall with rows of box-shaped zooecia (containers) along either side, like tiny shoe boxes arranged in a row. In each zooecium lived a zooid, which typically had a pear- or stalk-like body and a crown of tentacles (the lophophore), for catching tiny particles of suspended food and also for obtaining

oxygen. However, in some bryozoan colonies different types of zooids specialized in feeding, respiration, protection, cleaning or reproduction – which was by asexual 'budding', where a new individual grew as a plant-like bud from an older one. Just above the example shown below is a fossil brachiopod, or lampshell (see next page). The brachiopods are considered to be the bryozoans' closest relatives.

Name: *Ptilodyctia.*
Meaning: Soft or downy finger.
Grouping: Bryozoan, Cryptostomatid, Ptilodyctyid.
Informal ID: Bryozoan, sea-comb.
Fossil size: Length 23cm/9in.
Reconstructed size: Length up to 30cm/12in.
Habitat: Sea floor.
Time span: Ordovician to Devonian, 450–370 million years ago.
Main fossil sites: Worldwide.
Occurrence: ◆

Left: From the Silurian Wenlock Limestones of Dudley in the English West Midlands, this specimen of Ptilodyctia *would have been attached to the sea bed at its narrow end. This had a rounded or cone-shaped end that fitted into a similar-shaped depression in the base, which was attached to the sea bed.*

BRACHIOPODS

Brachiopods, or lampshells, are soft-bodied marine creatures that live in a shell consisting of two parts called valves. Superficially they look like bivalve molluscs, such as oysters and mussels, but brachiopods form a separate main grouping, or phylum. One of the key differences between the shells of brachiopods and bivalves concerns the line of symmetry. Brachiopods often have different-sized valves and are symmetrical along the midline, whereas the valves of bivalves are the same size: that is, the symmetry runs between the two valves. Brachiopods originated in the Cambrian and became very common and useful as indicator fossils, but their diversity was much reduced by the end-of-Permian mass extinction. event.

Orthid brachiopod

Name: Orthid brachiopod.
Meaning: Straight arm-foot.
Grouping: Brachiopod, Orthid.
Informal ID: Brachiopod, lampshell.
Fossil size: Slab length 10cm/4in.
Reconstructed size: Individuals 2cm/¾in wide (grew up to 5cm/2in).
Habitat: Sea floor.
Time span: Cambrian to Permian, 500–250 million years ago.
Main fossil sites: Worldwide.
Occurrence: ◆ ◆ ◆

The orthids were an early and important group of articulate brachiopods (see later in this chapter). They first appeared in the Early Cambrian Period and were extremely diverse by the Ordovician Period, some 450 million years ago. The brachiopods in this group are sub-circular to elliptical in form, with generally biconvex (or outward-bulging) valves, although one valve is flatter and so bulges less than the other. The hinge line of this brachiopod is normally strophic, or straight, giving rise to the name 'orthid'. Other common features of this brachiopod group include ribs radiating outwards in the shape of a fan from the area of the hinge, and the presence of a concave sulcus (a depression or groove) on one valve and a corresponding fold in the opposite valve.

Plectothyris
Plectothyris is a brachiopod from the Jurassic Period. It is biconvex (both valves bulge outwards) with a non-strophic, or curved, hinge line. It has no sulcus, or fold. The line along which the two valves meet, called the commissure, is zigzagged, with the zigzag becoming more defined towards the midline. Each valve is ribbed towards the edge. Also growth lines are visible towards the centre of the valve. They formed as the valves became bigger, probably due to seasonal growth, similar to the growth rings of a tree trunk. Details such as these help to distinguish many hundreds of types of brachiopod. Their sea-bed habitat and their very hard and resistant shells mean that brachiopods were preserved in huge numbers and are very useful for dating rock layers.

Left: A mass accumulation is a large number of the same organisms found preserved in one place. Such accumulations usually form under extreme conditions – for example, a large influx of sediment causing the organisms to be rapidly buried, or a similar sudden change to a hostile environment. These particular orthids may have been buried by an underwater avalanche of mud or silt.

Pedicle area

Convex pedicle valve

Radiating ribs on shell

Brachial valve (non-pedicle 'shell')

Strophomenid brachiopod

Name: Strophomenid brachiopod.
Meaning: Twisted or malformed arm-foot.
Grouping: Brachiopod, Strophomenid.
Informal ID: Brachiopod, lampshell.
Fossil size: 4.5cm/1¾in wide along the hinge line.
Reconstructed size: As above.
Habitat: Sea floor.
Time span: Ordovician to Triassic, 450–210 million years ago.
Main fossil sites: Worldwide.
Occurrence: ◆ ◆

Strophomenid brachiopods are more diverse in form than any other brachiopod group. They are normally plano-convex, which means that one valve is convex (bulging out), while the other is flat; or concavo-convex, meaning that one valve bulges out, while the other curves inwards. In addition, strophomenids have no pedicle opening – the pedicle is a muscular structure, like a stalk or the 'foot' of a mollusc, that secures the brachiopod to a hard surface. So strophomenids probably remained unattached on the sea floor. The subgroup known as true strophomenids were wider than they were long, and had a very small body cavity between the valves. They were important in Early Palaeozoic times, but faded by the Jurassic Period. Other groups of strophomenids became abundant during the Late Palaeozoic.

Above right: This view shows the valve of a strophomenid brachiopod. The valves are very unornamented and there is a long strophic (straight) hinge line. This specimen has an impression of another brachiopod near its hinge line.

Valve margin

Hinge line

Second brachiopod impression

Pentamerus

Name: *Pentamerus*.
Meaning: Five parts or portions.
Grouping: Brachiopod, Pentamerid.
Informal ID: Brachiopod, lampshell.
Fossil size: 5cm/2in wide.
Reconstructed size: As above.
Habitat: Sea floor.
Time span: Late Cambrian to Devonian, 500–365 million years ago.
Main fossil sites: Worldwide.
Occurrence: ◆ ◆

Pentamerus is an articulate brachiopod, explained in more detail on the next page, but differs from other articulate brachiopods in that it has a spoon-shaped structure, the spondylium, on the rear portion of the pedicle valve, which is the valve bearing the muscular stalk-like pedicle that usually faced downwards. The hinge line is short and non-strophic (not straight). In addition, the fold and sulcus (depression) are opposite to those found on most brachiopods, with the fold on the pedicle valve. *Pentamerus* arose in Mid-Cambrian times and became very common in the Ordovician Period. During the Silurian Period, it lived in colonies with its pedicle buried in the sea floor. In some species within the genus, there was no pedicle and the colony was self-supporting. When displayed in museums the colonies are often exhibited upside down! The name comes from its five-fold symmetry, which, when the fossil is seen in section, can often resemble an arrow head.

Right: Both valves in this species of Pentamerus *are ribbed in shape. The downwards-facing pedicle valve is larger than the opposing valve, known as the brachial valve, which usually faced upwards. The shell looks roughly five-sided, or pentagonal, in outline – hence its name.*

Brachial valve

Pedicle foramen

Hinge line

Pedicle valve

Vaguely five-sided valve margin

Radiating ribs

BRACHIOPODS (CONTINUED)

Most brachiopods were connected to a hard surface by a pedicle – a muscular stalk-like structure. They
were filter-feeders, using a structure called a lophophore to extract food from the water. The two main
groups are the Articulata, which has a socket that hinges the two valves, and the hingeless Inarticulata.
More than 3,000 fossil genera are known, but only 350 species survive today – all in the sea.

Spirifer

Spiriferids made up a diverse group of brachiopods that had
an enormous range of shell shapes. *Spirifer* had a spiralled
brachidium – the support arrangement for the feeding
structure, or lophophore, within the shell. Some experts
regard this as a uniting feature of the spiriferids, although
others disagree. The valves are biconvex (both bulge
outwards), usually with a long, straight hinge line, and
typically with ribs. Some species are 'winged', but these
wings are often missing in the fossil record.

Below: External (top) and internal
(bottom) views of Spirifer. The
internal view shows the spiral
brachidium after which the
genus is named.

Name: *Spirifer.*
Meaning: Bearing or carrying
a spiral (from the spiralled
brachidium).
Grouping: Brachiopod,
Spiriferid.
Informal ID: Brachiopod,
lampshell.
Fossil size: Width 3cm/1¼in.
Reconstructed size: Less
than 10cm/4in.
Habitat: Sea floor.
Time span: Ordovician to
Permian, 480–250 million
years ago.
Main fossil sites: Worldwide.
Occurrence: ◆ ◆

Below: Internal view of one valve
of Spirifer, which is broad, ribbed
and has a strophic (straight)
hinge line.

Brachidial region

Torquirhynchia

Torquirhynchia is an exception to the rule that states
brachiopods are bilaterally symmetrical. Its shell is fairly
large and has the outline of a wide triangle. It is also
ornamented with thick ribs, and the two valves are slightly
out of line with each other. The hinge line is non-strophic
(not straight) and the commissure – the line along which
the two valves meet – is jagged or zigzagged.

Below: Two specimens of
Torquirhynchia, both showing
strong ribbing, the non-strophic
hinge line and the 'nut shape'
that is typical of its subgroup,
called the rhynchonellids. In
these examples, the two valves
are slightly opened.

Name: *Torquirhynchia.*
Meaning: Twisted beak,
twisted nose.
Grouping: Brachiopod,
Rhynchonellid.
Informal ID: Brachiopod,
lampshell.
Fossil size: Width 3cm/1¼in.
Reconstructed size: As above.
Habitat: Sea floor.
Time span: Late Jurassic,
150 million years ago.
Main fossil sites:
Throughout Europe.
Occurrence: ◆ ◆

Zigzagged commissure

Non-strophic hinge line

Strong ribbing

Lingulid brachiopod

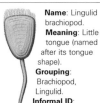

Name: Lingulid brachiopod.
Meaning: Little tongue (named after its tongue shape).
Grouping: Brachiopod, Lingulid.
Informal ID: Brachiopod, lampshell.
Fossil size: 1cm/⅖in.
Reconstructed size: Up to a few cm/about 1in.
Habitat: Inter-tidal zones.
Time span: Cambrian, more than 500 million years ago, to the present.
Main fossil sites: Worldwide.

Lingulid brachiopods have an extremely long fossil record, which stretches back into the Cambrian Period, more than 500 million ago, and belong to one of the oldest families of multi-cellular animals known to science. Since their origins, these creatures have remained fairly unchanged, and they still survive today in the deep sea, earning them the nickname 'living fossils'. Lingulids became fairly diverse during the Cambrian Period but went into decline during the following Ordovician Period and have never since been a varied group. *Lingula* itself is a relatively small inarticulate brachiopod (one that lacks a hinge to hold the two valves together). Its valves are thin, biconvex (both bulge outwards) and tongue-shaped in outline, lacking strong ornamentation. Today's *Lingula* lives buried in sediments along the shore's inter-tidal zone, and is eaten as a delicacy in some parts of South East Asia. The lingulid shell is made of protein, chitin and calcium phosphate.

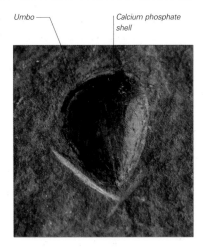

Umbo — Calcium phosphate shell

Above: This fossilized specimen of a lingulid brachiopod shows the typical tongue shape that gives the genus its name. The specimen also shows the characteristic lack of ornamentation, although some growth lines are visible.

Actinoconchus

Name: *Actinoconchus*.
Meaning: Rayed shell.
Grouping: Brachiopod, Spiriferid.
Informal ID: Brachiopod, lampshell.
Fossil size: Length 2.5cm/1in.
Reconstructed size: Length up to 4.5cm/1¾in.
Habitat: Sea floor.
Time span: Carboniferous, 300 million years ago.
Main fossil sites: Worldwide.
Occurrence: ◆ ◆

In this distinctively shaped brachiopod, both of the valves are very convex in outline and are almost the same overall size, thus creating a nearly circular outline. Thin but broad expansions of the shell can be seen along the ridged growth lines. The valves are finely ribbed, with the ribs diverging towards the front. Living during the Carboniferous Period, *Actinoconchus* attached itself by a short, stalk-like pedicle to hard surfaces. The shape of the brachial (upper) valve is similar to that of the hard surface to which the brachiopod was attached.

Below: Seen upside down here, this pedicle (lower) valve of Actinoconchus shows the expansions along the growth lines, but the fine ribbing is not clearly visible in this specimen. These layered expansions helped to prevent Actinoconchus sinking into the muddy sea floor it inhabited, and also acted as camouflage, protecting it from the attentions of predators. These creatures fed by filtering out tiny particles of food from the surrounding water.

———— Umbo

Expanded shell along growth lines

ARTHROPODS – TRILOBITES

Arthropods include millions of species that are alive today: insects, arachnids or spiders and scorpions, millipedes or diplopodans, centipedes or chilopodans, and crabs, shrimps and other crustaceans. These are covered in the following pages, starting with the long-extinct trilobites. The uniting feature of arthropods is their hard outer casing and usually numerous jointed limbs. The name means 'joint-foot'.

Anomalocaris

Most creatures of Cambrian times, as revealed by Burgess Shale fossils, were small – generally one to a few centimetres. The arthropods known as anomalocarids, however, commonly grew to 50cm/20in, and some reached almost 2m/6½ft. They were the largest animals of their time. *Anomalocaris* had a disc-like mouth that was originally interpreted as being a separate jellyfish-like animal, known as *Peytoia*. It also had large, strong, forward-facing appendages, which in some species bore sharp spines. These powerful spiked 'limbs', along with large eyes and strong swimming lobes, like paddles along the sides of the body, suggest that *Anomalocaris* was the top predator of its age. It probably fed on soft-bodied animals, such as worms, and perhaps even on trilobites. The relationship of anomalocarids to other arthropods is not yet fully understood (see also *Hurdia*).

Below: This lateral (side) view shows one of the two multi-segmented frontal appendages. Such specimens were originally interpreted as separate shrimp-like creatures, until experts eventually realized that they were detached parts of a larger animal. The small spines on each segment may have assisted in capturing their prey.

Name: *Anomalocaris*.
Meaning: Anomalous shrimp.
Grouping: Arthropod, Anomalocarid.
Informal ID: Anomalocarid, Burgess 'supershrimp'.
Fossil size: 7.5cm/3in.
Reconstructed size: Total length, including front appendages, 60cm/24in.
Habitat: Shallow seas.
Time span: Cambrian, 530 million years ago.
Main fossil sites: North America (Burgess), Asia (China).
Occurrence: ◆

Cruziana

The name *Cruziana* is used for a trackway, or 'footprints', made by an arthropod filter-feeding in the muddy sea bed. The chevron-like indentations were created as the animal used its limbs to plough into the ooze, stirring the sediment into suspension (floating in the water). While moving the sediment towards its rear end, the animal could filter out edible particles and move them forward towards its mouth. As a trace fossil, or ichnogenus, *Cruziana* refers to a particular type of behaviour and the evidence that results, but not the type of animal that created it. However, it is commonly assumed that some types of trilobites were responsible. In some examples, specimens of *Calymene* (see opposite) have been found at the end of the trackway. But any arthropod feeding or moving in this manner could conceivably be the trackmaker.

Name: *Cruziana*.
Meaning: Of Cruz (in honour of General Santa Cruz of Bolivia).
Grouping: Arthropod, trilobite.
Informal ID: Cruziana, trilobite trackway.
Fossil size: Track width 3cm/1in.
Reconstructed size: As above.
Habitat: Muddy sea floor.
Time span: Cambrian to Permian, 540–250 million years ago.
Main fossil sites: Europe, North America, South America, Australia.
Occurrence: ◆ ◆ ◆

Left: It is most likely that the trace fossils known as Cruziana were made by trilobite-like creatures. The trackmaker's direction of travel is towards the open end of the V-shapes, left by appendage movement.

Calymene

Name: *Calymene*.
Meaning: Stony crescent.
Grouping: Arthropod, Trilobite, Phacopid.
Informal ID: Trilobite.
Fossil size: 3.2cm/1¼in.
Reconstructed size: Unrolled head–tail length 4cm/1½in.
Habitat: Shallow seas.
Time span: Silurian to Devonian, 430–360 million years ago.
Main fossil sites: Europe, North America.
Occurrence: ◆ ◆ ◆ ◆

Calymene was a medium-size trilobite. It was probably a predator and was found in shallow Silurian seas, usually in lagoons or reefs. It had a semicircular-shaped cephalon (head shield) with three or four distinct 'lobes' that ran laterally along its central head section, known as the glabella. The glabella itself was bell-shaped, containing small eyes. There was a variable number of segments (or tergites) in the middle region, the thorax, depending on the species concerned – British specimens, for example, may have up to 19 segments, while those from North American species have only 13. Its tough exoskeleton, like that found on many post-Cambrian trilobites, was a good defence against predation and also increased its own chances of being preserved in the fossil record. This has helped to make *Calymene* one of the most commonly found trilobite genera from the Middle Ordovician Period.

Above: Viewed from above, Calymene shows its segmented main body structure and its almost non-existent pygidium, or tail.

Right: This is a three-quarter front view of a partially enrolled Calymene. The ability of many trilobites to roll up is thought to be a defence against predators attacking the unprotected underside.

Right: In front view, the cephalon of Calymene blumenbachii was more triangular, compared with the rounded cephalon of most other trilobite groups.

Diacalymene

Name: *Diacalymene*.
Meaning: Through Calymene.
Grouping: Arthropod, Trilobite, Phacopid.
Informal ID: Trilobite.
Fossil size: Head–tail length 5cm/2in.
Reconstructed size: As above.
Habitat: Shallow seas.
Time span: Silurian to Devonian, 430–360 million years ago.
Main fossil sites: Europe, North America.
Occurrence: ◆ ◆ ◆ ◆

Diacalymene is closely related to *Calymene*, to the point where some palaeontologists have suggested that they should be regarded as the same genus. However, there are some differences between the two. For a start, *Diacalymene* has a distinct ridge along the anterior of the cephalon, as well as a narrower glabella (central head region) than *Calymene*. In addition, *Diacalymene* also has a slightly more triangular-shaped cephalon (head shield), and it also tended to live in muddier sediments than did *Calymene*. Like *Calymene*, however, *Diacalymene* could roll up tightly for protection against threats, such as predators and storms.

Right: This large and complete fossilized specimen of Diacalymene was discovered in a trilobite-rich formation in an area known as the Laurence Uplift, Oklahoma, USA.

ARTHROPODS – TRILOBITES (CONTINUED)

Almost all trilobites had a similar body: a cephalon (head end), thorax (middle section) and pygidium (tail). However, there were exceptions (see Naraoia). Note that the name 'trilobite' does not come from this three-section head-to-tail structure. It is derived from the three lobes seen from side to side, being the left, central and right lobes, which were formed by two furrows or divisions running from head to tail.

Ogygopsis

Ogygopsis was a medium-size trilobite, and was more highly evolved than its early Cambrian ancestors. The body was fairly wide, and although the cephalon (head end) lacked the complexity or ornamentation of other forms, Ogygopsis did have long genal spines projecting from the 'cheeks' or edges of its cephalon. These genal spines reached to about the middle of the thorax. The tail, or pygidium, was larger than in most other Cambrian trilobites, consisting of multiple thoracic segments fused together to form a single large plate. When the pygidium is almost as large as the cephalon, such as in Ogygopsis, this is known as an isopygous type of trilobite. However, loss of segments to the tail meant that there were only eight segments in the thorax.

Left: The lack of side-cheeks on the cephalon (head region) of this specimen indicates that this is a moulted or cast-off body casing, which happened as the animal grew (as in crabs and insects today), rather than a whole dead animal.

Name: *Ogygopsis.*
Meaning: Eye of the ancient King Ogyges.
Grouping: Arthropod, Trilobite, Corynexochid.
Informal ID: Trilobite.
Fossil size: Head–tail length 4cm/1½in.
Reconstructed size: As above.
Habitat: Muddy sea floor.
Time span: Middle Cambrian to Early Ordovician, 520–470 million years ago.
Main fossil sites: Worldwide.
Occurrence: ◆ ◆ ◆

Phacops

Phacops was a medium-to-large trilobite that lived in shallow water. The creature had 11 segments in its thorax section, and the pygidium (tail unit) was small and semi-circular. Like other trilobites, Phacops was capable of rolling itself into a tight ball when under threat of predation. The distinctly round glabella ('forehead') expands forward and is covered in rounded lumps, known as tubercles. Another distinctive feature of Phacops was its eyes, which were schizochroal, or multi-faceted. This means that each eye had many discrete, individual lenses, each with its own covering, or cornea, and separated from neighbouring lens units by areas called sclerae. These large eyes could swivel, and would have allowed an almost all-round field of vision. This indicates that Phacops was most likely a predator, perhaps hunting in low light levels.

Below: A view from above shows the large eyes on either side of the head region, or cephalon.

Left: This specimen shows how tightly enrolled Phacops could become. All the antennae and appendages would have been protected within this tight exoskeletal (outer-cased) ball.

Name: *Phacops.*
Meaning: Shiny eye.
Grouping: Arthropod, Trilobite, Phacopid.
Informal ID: Trilobite.
Fossil size: Head–tail length 4cm/1½in.
Reconstructed size: As above.
Habitat: Muddy sea floor.
Time span: Silurian to Devonian, 430–360 million years ago.
Main fossil sites: North America, Europe.
Occurrence: ◆ ◆ ◆ ◆

Encrinurus

Name: *Encrinurus*.
Meaning: In hair.
Grouping: Arthropod,
Trilobite, Phacopid.
Informal ID: Strawberry-
headed trilobite, stalk-eyed
trilobite.
Fossil size: Head–tail length
2.5cm/1in.
Reconstructed size:
Head–tail length 4.5cm/1¾in.
Habitat: Shallow sea floor.
Time span: Middle
Ordovician to Silurian,
470–410 million years ago.
Main fossil sites: Worldwide.
Occurrence: ◆ ◆ ◆

Encrinurus was a large type of trilobite with a forward-facing, rounded glabella (or forehead), and pimple-like lumps, known as tubercles, all over its cephalon (head). These distinctive characterisitics have led to the creature's informal name of the 'strawberry-headed trilobite'. Leading away from the outer corners, or cheeks, of the cephalon were short genal spines. The thorax had 11 or 12 segments. *Encrinurus*'s pygiduim was fairly long and unusual in construction in that there were many more segments in the central, or axial, portion than on the two lateral portions, called pleural ribs. These pleural ribs curved towards the rear. However the most unusual feature was the stalked eyes prominently protruding from the side-cheeks, which accounts for its other informal name of the 'stalk-eyed trilobite'.

Above right: The ornamental 'pimpled' cephalon and the unique stalked eyes can be seen clearly in this specimen. The pygiduim, or tail section, can also be seen under the forehead-like glabella.

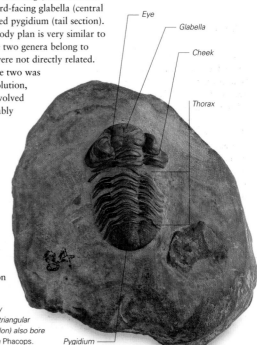

Longitudinal division or furrow

Thoracic carapace of central lobe

Thoracic carapace of left lobe

Cheek

Eye

Tubercles

Glabella

Pygidium

Acaste

Name: *Acaste*.
Meaning: After the mythical
Greek goddess.
Grouping: Arthropod,
Trilobite, Phacopid.
Informal ID: Trilobite.
Fossil size: Head–tail length
2cm/¾in.
Reconstructed size:
As above.
Habitat: Shallow sea bed.
Time span: Silurian to Early
Devonian, 430–390 million
years ago.
Main fossil sites: Worldwide.
Occurrence: ◆ ◆

Acaste is a medium-size trilobite with large, schizochroal (multi-lens) eyes, a round, forward-facing glabella (central head section), and a small, tapered pygidium (tail section). Although its overall shape and body plan is very similar to that of *Phacops* (opposite), these two genera belong to different trilobite families, and were not directly related. The physical likeness between the two was achieved through convergent evolution, which means that both groups evolved similar features separately, probably due to living in a similar environment and following a similar lifestyle. In both types of creature, natural selection favoured certain adaptations, such as schizochroal eyes to see well, and so the two types end up looking very similar. Indeed, it has been suggested that its visual acuity was more than the creature actually required. The two differ in some features, however, including ornamentation at the head end.

Right: Acaste had a more tapered body shape than Phacops, and also a more triangular pygidium (tail). Its head section (cephalon) also bore more complex ridges and furrows than Phacops.

Eye

Glabella

Cheek

Thorax

Pygidium

ARTHROPODS – TRILOBITES (CONTINUED)

Trilobites appeared in Cambrian times as some of the earliest of the complex, hard-cased creatures. They were adapted to a wide range of habitats, from shallow reefs to deep ocean floors. More than 5,000 genera and 15,000 species have been identified, with more discovered every year. Their abundance and distinctive features, which changed through time, make them useful indicator fossils for dating rock layers.

Ogygiocaris

Name: Ogygiocaris.
Meaning: Shrimp of the ancient King Ogyges.
Grouping: Arthropod, Trilobite, Asaphid.
Informal ID: Trilobite.
Fossil size: 20cm/8in.
Reconstructed size: Head–tail length 13cm/5in.
Habitat: Deep ocean floor.
Time span: Early to Middle Ordovician, 490–470 million years ago.
Main fossil sites: Worldwide.
Occurrence: ◆ ◆

Ogygiocaris was a large, wide trilobite with a flattened body margin (edge) that probably helped it stay on top of the muddy sea floor rather than sinking in. Its cephalon (head section) was round and had short genal spines leading away from the corners. The pygidium, or tail, was also round and was about the same size as the head. As the tail was so large, there are a reduced number of segments in the thorax – only eight. The multiple rear-facing lateral (side) segments of the tail, known as pleural spines, are also a distinct feature.

Right: This specimen of Ogygiocaris *became disarticulated (its main sections came apart) at some point prior to fossilization. However, the large cephalon (head shield) and pygidium (tail), with its curved pleural spines, can be clearly seen.*

Ogyginus
Similar and closely related to *Ogygiocaris*, *Ogyginus* had small genal spines on the corners of its cephalon. Like *Ogygiocaris*, it probably lived in muddy marine environments, feeding on small edible particles swept up from the substrate with its appendages. Note the smaller juvenile forms near the larger adult.

Agnostus

Agnostus was a blind, simply shaped trilobite reaching the size of a little fingernail. The cephalon, or head end, lacked the usual grooves or sutures that separate its central glabella from the side-cheeks. The pygidium, or tail, was rounded and smooth. In addition, *Agnostus* had only two segments in its thorax. This was the result of a pattern of thoracic segment reduction that began in the Early Cambrian and reached its peak at the end of that period. *Agnostus* probably lived on the deep ocean floor, using its short antennae to sweep up food particles from the sediment.

Left: The presence of different types of agnostid trilobites – seen here in a specimen from Sweden – is used to help demarcate rocks of specific dates within the Cambrian Period.

Above: In this close-up view of the specimen seen on the left, the isopygous pygidium (which means it was about the same size as the head, or cephalon) and the fact that the two are plain and very similar in outline, make it difficult to tell the head of Agnostus *from its tail.*

Name: Agnostus.
Meaning: Ignorant, not known.
Grouping: Arthropod, Trilobite, Agnostid.
Informal ID: Trilobite.
Fossil size: Slab length 12cm/4¾in.
Reconstructed size: Individual head–tail length 1cm/⅜in.
Habitat: Deep sea floor.
Time span: Late Cambrian, 510–500 million years ago.
Main fossil sites: Europe, North America, East Asia, Australia.
Occurrence: ◆ ◆ ◆

Ellipsocephalus

Name: *Ellipsocephalus*.
Meaning: Ellipse
(oval) head.
Grouping: Arthropod,
Trilobite, Ptychopariidid.
Informal ID: Blind trilobite.
Fossil size: Fossil slab
length 18cm/7in.
Reconstructed size:
Head–tail length 2cm/¾in.
Habitat: Muddy sea floor.
Time span: Middle
Cambrian, 520–510 million
years ago.
Main fossil sites: Worldwide.
Occurrence: ◆ ◆

This small, eyeless trilobite was micropygous: that is, it had a very small pygidium, or tail section. This allowed it to tuck not only its pygidium but also some thoracic segments under the front edge of its cephalon (head shield) when enrolled. *Ellipsocephalus* lacked any complex features on its cephalon and a suggested reason for this is that, because it could not see, it did not select a mate of its own kind based on visual features that distinguished it from other trilobite types. The lack of eyes is a secondary condition, since from their beginning most trilobites were sighted – in fact, they were the first creatures known to have large, complex eyes. The ancestors of *Ellipsocephalus* had them, but as part of its evolution when living in the dark depths, the eyes gradually became reduced as they were no longer necessary, and eventually sealed over.

Below: The fossilized remains of Ellipsocephalus *are often found crowded together, either as moulted, or cast-off, body casings, or as whole-animal death assemblages, as is the case in the specimen seen here. Some experts consider that this is evidence for some form of social behavior among trilobites. They may, for example, have come together in sizeable groups when migrating or mating.*

Trinucleus

Trinucleus was a small, blind trilobite with several unique and distinguishing features. The glabella, or forehead, and side-cheeks on the cephalon, or head shield, form raised elevations, giving *Trinucleus* its name, which means 'three centres'. The cephalon is also very large in comparison with the rest of the body, accounting for almost half the animal's total length. Projecting from either edge of the cephalon were genal spines that were longer than the body itself. In addition, along the anterior margin, or front edge, of the cephalon was a fringe that contained rows of small pits. This wide fringe, combined with the long genal spines, may have helped *Trinucleus* to stay on top of the muddy substrate it lived in without sinking down, working in a similar way as a snowshoe to spread the animal's weight. There was a reduced number of thoracic segments, and the pygidium, or tail section, was also very short and circular.

Right: The long genal spines sticking out from the head end, which would face the rear along the sides of the body, are missing from this specimen. However, the unique cephalic fringe is clearly visible.

Name: *Trinucleus*.
Meaning: Three nuclei,
three centres.
Grouping: Arthropod,
Trilobite.
Informal ID: Trilobite.
Fossil size: Head–tail length
2cm/¾in.
Reconstructed size:
As above.
Habitat: Muddy deep-
sea floor.
Time span: Middle to Late
Ordovician, 470–440 million
years ago.
Main fossil sites: Western
Europe, North Africa.
Occurrence: ◆

ARTHROPODS – CRUSTACEANS

Crustaceans are the largest group of sea-dwelling arthropods, occupying a vast variety of environments and with more than 40,000 living species, including some in freshwater and on land. They include crabs, lobsters, shrimps, prawns and the shrimp-like krill in the seas, as well as pillbugs and woodlice on land, water-fleas in ponds, and barnacles along seashores.

Beaconites

Regular infills (menisci)

Usually straight burrow

Branching points are less common

Beaconites is the name given to the trace fossils of burrow or tunnel networks that were first seen in the Cambrian Period more than 500 million years ago. These burrows tend to be small in scale, cylindrical and unbranched, and they may have been constructed either in a straight line or in a more curving, sinuous shape. The burrows often display regularly organized infill layers, called menisci, that were produced behind the animal as it moved forward. The burrows themselves typically have rounded ends. *Beaconites* is often found in sediment that has been heavily disrupted by various different feeding and burrowing organisms, known as bioturbation. As is the case with most ichnogenera, it is difficult to specify precisely which type of animal left these traces behind. A variety of arthropods, including crustaceans, may have been responsible for making them.

Name: *Beaconites*.
Meaning: From the Beacon Heights Mountain, Antarctica.
Grouping: Probably Arthropod, possibly Crustacean.
Informal ID: Fossilized dwelling/feeding tunnel system.
Fossil size: Slab length 50cm/20in.
Reconstructed size: Burrow width rarely up to 20cm/8in.
Habitat: Shallow seas and lakes.
Time span: Cambrian, 540 million years ago, to today.
Main fossil sites: Antarctica, Europe, North America.
Occurrence: ◆ ◆ ◆

Left: The direction of travel of the animal that made this burrow – most likely an arthropod of some description – was opposite to the concave (dished) infillings within the burrow itself. This specimen is from the Wealden Lower Cretaceous beds in Surrey, England.

Barnacles

Also known as cirripedes, barnacles are ancient creatures. They are found way back in the fossil record, some 540 million years ago, yet they still exist in huge numbers in the seas and oceans of today. Although they are sometimes confused with molluscs, barnacles are in fact a distinct group of crustaceans that have lost most of their crustacean features, as these are unnecessary in their sessile (meaning settled or stationary) lifestyle. Some types of barnacle live in the deep oceans, but many more thrive on the narrow intertidal band of rocky shores, where they have evolved special adaptations to surviving periods of low tide when the water retreats and they are exposed to the air. At this time, the biggest threat to the creatures' survival is water loss, so barnacles have special plates that seal them within their own calcareous 'shell' until the tide rises once more. There are two main groups of barnacle. Acorn barnacles cement their cone- or volcano-like shells directly to an object, while goose barnacles attach with a long stalk.

Six sections or plates form a cone-like mound

Opening sealed with plates while exposed to air

Hard surface

Balanus

Name: *Balanus*.
Meaning: Acorn.
Grouping: Arthropod, Crustacean, Cirriped.
Informal ID: Acorn barnacle.
Fossil size: Each individual 3.2cm/1¼in high.
Reconstructed size: Height 4cm/1½in, including feathery limbs.
Habitat: Intertidal rocks or shallow sea floor.
Time span: Middle Cretaceous, 125–100 million years ago, to today.
Main fossil sites: Worldwide.
Occurrence: ◆ ◆ ◆

Balanus is a common and widespread genus of medium-to-large barnacles, some growing to a sizeable 10cm/4in across. Today they are found living on most rocky shorelines and they are extremely abundant worldwide. *Balanus*, like most other crustaceans, begins life as a tiny planktonic larva (young form). However, during this juvenile period, it loses most of its typical crustacean features, such as a long body with numerous walking limbs. The larva spends a month or more floating through the ocean, moulting several times before settling onto a firm base. Here, it secretes six calcareous plates that eventually grow to form a cone-like shell. Four further plates seal the opening of the shell, except when the barnacle extends its feather-like limbs. These curl through the water, like a grasping hand, in order to catch any passing planktonic food.

Below: The thick, calcareous shell of Balanus *is robust, meaning that it is more likely to be preserved in the fossil record than many other thinner-shelled crustaceans. This view, which was taken from above, shows four individuals that have settled and grown together.*

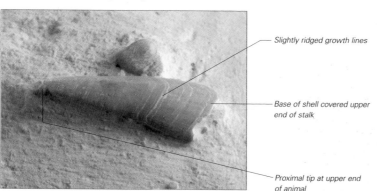

Cretiscalpellum

Cretiscalpellum is an extinct member of a group of barnacles known as goose or gooseneck barnacles, many of which still thrive today. The name derives from the long, flexible, extendable muscular stalk, like a goose's neck, that attaches to the substrate. Five bivalve-like calcareous plates protect the soft body. Goose barnacles live in groups attached to a variety of both floating and stationary objects, from pieces of driftwood and great whales to solid rocks, molluscs and other barnacles. As in acorn barnacles, the large, feathery limbs, or cirri, extend through the opened shell to filter small, floating edible particles from the water.

Below left: Cretiscalpellum *is the most abundant type of barnacle found in British Chalk beds. Pictured here are parts of the outer casing or shell, which, in life, resembled the covering of a mussel.* Cretiscalpellum *beds coincided with a period of worldwide chalk deposition that occurred during the Late Cretaceous Period, around 100 million years ago.*

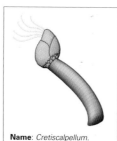

Name: *Cretiscalpellum*.
Meaning: Small sharp chalk.
Grouping: Arthropod, Crustacean, Cirriped.
Informal ID: Goose or gooseneck barnacle.
Fossil size: Plates 1.2cm/½in long.
Reconstructed size: Main body 4.5cm/1¾in long, plus length of stalk.
Habitat: Warm seas.
Time span: Middle Cretaceous, 125–100 million years ago, to today.
Main fossil sites: Worldwide.
Occurrence: ◆ ◆ ◆

Slightly ridged growth lines

Base of shell covered upper end of stalk

Proximal tip at upper end of animal

ARTHROPODS – CRUSTACEANS (CONTINUED)

Crustaceans are found in the fossil record from as early as the Cambrian Period. But apart from barnacles, with their strong calcareous plates, they are generally rare due to their lack of truly hard body parts. It was not until the appearance of decapods, or 'ten-limbs' – crabs, lobsters, crayfish, shrimps and prawns – with their thick exoskeletons, that crustaceans began to fossilize with some regularity.

Cyamocypris

Countless tiny crustacean 'shellfish' crop up in the fossil record – then, as now, vital links in the food chain for larger creatures. One group is the ostracods, also called mussel-shrimps or seed-shrimps. They look superficially like very small bivalve shellfish, but they are not molluscs or shrimps – they form their own major group, Ostracoda, within the Crustacea. Mostly smaller than a fingernail, their two-part body casings, or carapaces, are familiar components of mud and ooze in today's oceans, lakes and ponds, and also in preserved sediments from these habitats. Ostracods have thrived since Ordovician times, almost 500 million years ago, at first in the sea, and then from the Carboniferous Period in fresh waters, with genera such as *Darwinula*.

Below: The Jurassic freshwater Cyamocypris was approximately the size of a large grain of rice. As with many ostracods, the left and right parts of the valves of its carapace had a rounded outline, with a small pointed 'beak' on the underside. The left valve was larger than the right and overlapped it.

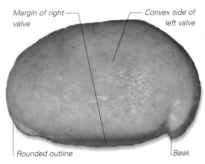

Margin of right valve — — Convex side of left valve

Rounded outline — Beak

Name: Cyamocypris.
Meaning: Bone shrimp.
Grouping: Arthropod, Crustacean, Ostracod, Podocopid.
Informal ID: Ostracod, 'tiny shellfish'.
Fossil size: Length 5mm/³⁄₁₆in.
Reconstructed size: As above.
Habitat: Fresh water.
Time span: Early Cretaceous, 100 million years ago.
Main fossil sites: Europe.
Occurrence: ◆ ◆

Palaeastacus

Below: Although one of the more common crustaceans from the Late Cretaceous Period, whole Palaeastacus specimens are rarely found intact. The enlarged first appendages with pincer ends, known as chelae, are the most common parts found fossilized. The exoskeleton (outer casing) of these appendages was thicker and stronger than in other areas of the body, and so was more likely to remain intact after death. Palaeastacus may have lurked in a crevice or burrow, emerging only to grab victims or scavenge.

Astacus is one of the main genera of today's crayfish. *Palaeastacus* is its fossil equivalent as a lobster, most commonly found in the Late Cretaceous Chalk rocks of England and Germany. Its features are very similar to those of modern lobsters and crayfish. The long, thin exoskeletal covering over most of the head and thorax is known as a carapace. There are also large, powerful chelae or pincers at the front and an elongated abdomen at the rear containing powerful swimming muscles. At the end of the abdomen was a flattened, fan-like tail, the telson, used to propel *Palaeastacus* rapidly backwards, away from advancing predators.

Name: Palaeastacus.
Meaning: Ancient lobster/crayfish.
Grouping: Arthropod, Crustacean, Decapod.
Informal ID: Lobster/crayfish.
Fossil size: Whole specimen length 19cm/7½in.
Reconstructed size: Pincer–tail length 35cm/14in.
Habitat: Shallow seas.
Time span: Early Jurassic to Late Cretaceous, 190–65 million years ago.
Main fossil sites: Europe, North America, Australia, Antarctica.
Occurrence: ◆ ◆

Main body of chela —

Joint with body —

— Tips of chela (claw or pincer)

— Hinge area of chela

Enoploclytia

Name: *Enoploclytia*.
Meaning: Loosely folded.
Grouping: Arthropod, Crustacean, Decapod.
Informal ID: Lobster.
Fossil size: Pincer-appendage length 12cm/4¾in.
Reconstructed size: Pincer–tail length 32cm/12½in.
Habitat: Shallow seas.
Time span: Late Cretaceous to Palaeocene, 95–55 million years ago.
Main fossil sites: Throughout Europe.
Occurrence: ◆

Enoploclytia was a type of lobster found in Upper Cretaceous Chalk beds, very much like *Palaeastacus* (opposite, below). It displayed the characteristic section of exoskeleton, known as the carapace, that covered both the creature's head and thorax, with an extended abdomen containing powerful swimming muscles. Species of this creature tend to have robust exoskeletons (outer casings) with coarse nodes or spines. *Enoploclytia* is generally rare in the fossil record, and is usually found only in Cretaceous beds. Some species, however, have been found in rocks dating to the Palaeocene. This means that it managed to survive the catastrophic end-of-Cretaceous, or K/T, extinction that occurred 65 million years ago. This wiped out a large proportion of the world's marine and terrestrial life and brought about the end of the dinosaurs.

Below: Like modern lobsters, Enoploclytia *had a pair of large pincers. These looked similar to a pair of needle-nose pliers, being thinner and more elongated than those of Palaeastacus. They were perhaps used for probing into crevices to search for food rather than for simple crushing.*

Joint with body

Hinged tip of chela

Main body of left chela (claw or pincer)

Fixed tip of chela

Dromiopsis

Name: *Dromiopsis*.
Meaning: Running face.
Grouping: Arthropod, Crustacean, Decapod.
Informal ID: Crab.
Fossil size: Carapace width 2cm/¾in.
Reconstructed size: Width across legs 4cm/1¼in.
Habitat: Shallow marine environments.
Time span: Late Cretaceous and Early Tertiary, 95–55 million years ago.
Main fossil sites: Throughout Europe.
Occurrence: ◆

This relatively recently extinct crab was very similar in appearance to modern crabs. It had a thick hexagonal (six-sided) exoskeletal shell, known as a carapace, covering the head and thorax. The abdomen was greatly reduced and tucked or folded under the carapace. Crabs are one of the younger crustacean groups in the fossil record, with less time for major change between the first representatives and today's living creatures. Unlike other decapods (meaning ten-limbs), whose tail extends behind the body, the tail of *Dromiopsis* and other crabs is usually only extended for the purposes of mating.

Below: Although only the exoskeleton, or carapace, of Dromiopsis *can be seen in this specimen, the distinct hexagonal shape and surface design make it clearly identifiable as a crab. This Early Tertiary specimen is from Denmark and has an attached spirorbid worm tube.*

Head end

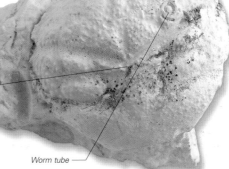

Rear of carapace is broken off

Worm tube

ARTHROPODS – CRUSTACEANS (CONTINUED)

The group of crustaceans we are most familiar with is the decapods. This term refers to the 'ten feet' or, rather, limbs of these types of crustacean, being the two front pincers and usually four pairs of walking legs. Decapods include more than 20,000 living species, which encompass shrimps and prawns – two common names that have no strict scientific basis – along with all types of crayfish, lobsters and crabs, as also shown on the previous pages.

Eryma

Appearing in the Jurassic Period, *Eryma* represents the oldest known 'true' lobster. *Eryma* is characterized by having typical decapod-like features, which include multiple pairs of appendages, such as antennae and mouthparts, as well as the ten main limbs. In addition, it had an elongated, segmented body, tough exoskeletal armour (carapace) that covers the thorax and head, and a fan-like tail, or telson. Like modern lobsters, *Eryma* most likely swam backwards, using its large tail muscles to flick its telson down and forwards and so propel itself quickly backwards, away from any predators. The pair of long pincers (chelae) could also be used for defence as well as for feeding. Compared with thicker-shelled crabs, however, *Eryma* and other lobsters and shrimp have a relatively thin carapace. This means that only rarely is the entire body preserved completely intact, usually in cases of Konservat-Lagerstatten – a German term for a site of 'exceptional preservation'.

Name: *Eryma*.
Meaning: Fence, barrier or guard.
Grouping: Arthropod, Crustacean, Decapod.
Informal ID: Lobster.
Fossil size: Large slab length 8cm/3in.
Reconstructed size: Pincer–tail length up to 9cm/3½in.
Habitat: Shallow seas.
Time span: Late Jurassic, 150–140 million years ago.
Main fossil sites: Throughout Europe.
Occurrence: ◆

Left: The Late Jurassic Solnhofen Limestones of Bavaria, southern Germany, where this specimen originates, constitute a site of exceptional preservation. The limestones were deposited in the calm marine environment between a reef and land. A wide variety of animals, including small dinosaurs and the earliest-known bird, Archaeopteryx, became preserved after their bodies were quickly buried in the hypersaline (extra-salty) bottom waters.

Right: More than 600 fossil species have been preserved in the Solnhofen Limestones in Germany. Among the crustaceans found here, decapods are the most abundant group. Soft-tissue preservation is so good in some specimens that even the eyes may remain intact. This individual is about 3.5cm/1½in long from its claw-tips to the end of its tail.

Below: Modern lobsters like Homarus have one larger, more robust claw or chela for crushing, and one narrower, slimmer claw for cutting and picking.

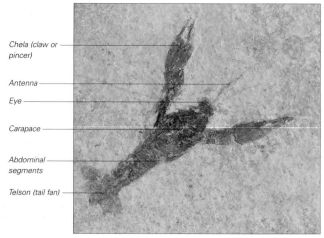

Chela (claw or pincer)

Antenna

Eye

Carapace

Abdominal segments

Telson (tail fan)

Thalassinoides

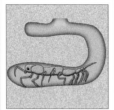

Name: *Thalassinoides*.
Meaning: Of the sea.
Grouping: Arthropod, Crustacean, probably Decapod.
Informal ID: Fossilized feeding tunnel system.
Fossil size: Entire specimen 25cm/10in across.
Reconstructed size: Burrow width average 4–5cm/1½–2in.
Habitat: Fairly calm deeper sea floor.
Time span: Jurassic, 190 million years ago.
Main fossil sites: Worldwide.
Occurrence: ◆ ◆ ◆

Thalassinoides is a trace fossil tunnel system most likely made by a decapod crustacean, such as a shrimp, while feeding. The tunnels range from between 2 and 6cm/¾ and 2⅓in in diameter and are usually found in rocks that were deposited in calm waters, below the turbulent waves and currents of the tidal zone. Because wave and current action were not dominant here (unlike the settings for *Ophiomorpha*, see below), animals tended to feed closer to the substrate. This often resulted in branching Y- or T-shaped tunnel networks, as with *Thalassinoides*. Again, unlike *Ophiomorpha*, the tunnels are usually smooth-sided. *Thalassinoides* burrows were often infilled by sediment after the inhabitant had vacated them. Following erosion, the network may have become separated from the original substrate, transported and then reburied in younger sediments. This makes it unusual, being one of the few trace fossils that may not have been made at its discovery site.

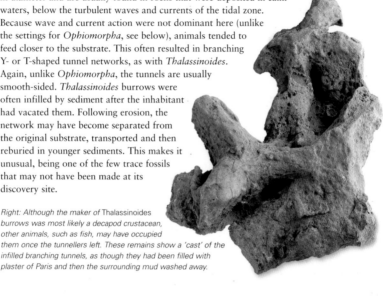

Right: Although the maker of Thalassinoides *burrows was most likely a decapod crustacean, other animals, such as fish, may have occupied them once the tunnellers left. These remains show a 'cast' of the infilled branching tunnels, as though they had been filled with plaster of Paris and then the surrounding mud washed away.*

Ophiomorpha

Ophiomorpha, like *Thalassinoides* above, is a trace fossil tunnel network most likely made by a decapod crustacean, such as a shrimp. It is usually found in rocks deposited under shallow, high-energy settings where wave action and tidal currents were present. In this type of environment, the majority of organisms live within the sediment (known as infaunal). *Ophiomorpha* tends to be more vertically orientated, as these tunnels were probably a dwelling, rather than just a feeding burrow. This can be deduced from the pellet-like ornamentations along the tunnel walls, which may have been faecal material (droppings) from the animal living there, or possibly lumps of sand grains. Often *Ophiomorpha* tunnels become infilled and eroded, as described above for *Thalassinoides*. This would result in an isolated 'cast' of the tunnel network that could be transported and redeposited somewhere else.

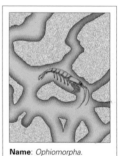

Name: *Ophiomorpha*.
Meaning: Serpent shape.
Grouping: Arthropod, Crustacean, probably Decapod.
Informal ID: Fossilized dwelling tunnel system.
Fossil size: Longest item 9cm/3½in.
Reconstructed size: Burrow width average 1.5–2cm/½–¾in.
Habitat: Shallow sea floor.
Time span: Permian to Late Tertiary, 270–3 million years ago.
Main fossil sites: Worldwide.
Occurrence: ◆ ◆ ◆ ◆

Left: These fossilized infilled specimens of Ophiomorpha *were discovered in Eocene rocks in Berkshire, southern England. They were eroded out of the sediment where they were originally dug in the Jurassic Period, most likely by a decapod crustacean, revealing the internal shape of the construction. The burrow system, which may have been used as a dwelling by the tunneller, was lined with nodules of sand.*

ARTHROPODS – EURYPTERIDS, XIPHOSURANS

The eurypterids were predatory arthropods that appeared in the Ordovician Period. Also called 'sea scorpions', some ventured into fresh water, and a few could survive on land for a time. They were chelicerates, the same group as spiders and scorpions, with large claw-like front limbs, or chelicerae.

Erieopterus

Eye — Carapace

Opisthosoma (abdomen) | Sixth prosomal appendage, or 'swimming paddle' | Telson (tail spine)

Erieopterus was one of the most widespread genera of eurypterids. Unlike some of its group – who could walk on land for short periods of time – it was fully aquatic. It possessed a pair of well-developed, paddle-like sixth appendages, behind feeding claws and four pairs of walking limbs. This characteristic of eurypterids gave them their name, from the Greek for 'broad wing'. These rearmost legs were used as oars for swimming. The relatively small eyes and leg spines of *Erieopterus* suggest that, unlike many others in its group, it was not a top predator of its shallow seas, rivers and estuaries. Instead, it was probably an unspecified feeder, taking primarily small invertebrates.

Above left: Erieopterus would have used its paddle-like sixth limbs to swim actively over the sand and mud of both salty and fresh water, looking for small invertebrate prey, such as worms and molluscs.

Name: *Erieopterus.*
Meaning: Erie wing (from the North American Great Lake).
Grouping: Arthropod, Chelicerate, Eurypterid.
Informal ID: Sea scorpion.
Fossil size: Head-tail length 13cm/5in.
Reconstructed size: Whole specimens reached 20cm/8in in length; fragmentary remains suggest much larger individuals.
Habitat: Inshore sea coasts, tidal estuaries.
Time span: Silurian to Devonian, 410–370 million years ago.
Main fossil sites: North America.
Occurrence: ◆ ◆

Eurypterus

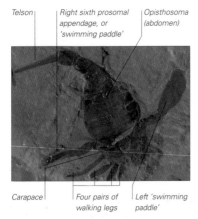

Telson | Right sixth prosomal appendage, or 'swimming paddle' | Opisthosoma (abdomen)

Carapace | Four pairs of walking legs | Left 'swimming paddle'

Above: This Silurian specimen from New York State shows the different specializations of Eurypterus's appendages. The relative positions of the left and right 'swimming paddles' highlight their wide range of movements.

The genus that gave the eurypterid group its name was relatively small and unspecialized. (Its close cousin, the Devonian *Pterygotus*, was one of the largest arthropods of all time at 2.5m/8¼ft long.) However, *Eurypterus* was relatively widespread and common, considering the eurypterids' overall rarity. Its fossils are usually found in fine-grained dolostone (waterlime) containing 'salt hopper' structures, which indicate hypersaline (extra-salty) conditions. However, the fossils seem to have been transported to the area before burial. Many specimens are incomplete and most are the remains not of the animals, but of moulted, or cast-off, body casings. Many individuals of *Eurypterus* seem to have congregated in the area to moult and mate – giving them safety in numbers and from the high salinity of the water, which would deter other creatures – before dispersing.

Name: *Eurypterus.*
Meaning: Broad wing.
Grouping: Arthropod, Chelicerate, Eurypterid.
Informal ID: Sea scorpion.
Fossil size: Head-tail length 8cm/3in.
Reconstructed size: Total length up to 40cm/16in.
Habitat: Shallow coasts, hypersaline lagoons.
Time span: Silurian, 430–410 million years ago.
Main Fossil Sites: North America.
Occurrence: ◆ ◆

Belinurus

Name: *Belinurus*.
Meaning: Dart or needle tail.
Grouping: Arthropod, Chelicerate, Xiphosuran, Limulid.
Informal ID: Extinct horseshoe crab.
Fossil size: Length, including tail, 5cm/2in.
Reconstructed size: As above.
Habitat: Freshwater and brackish (part-salty) swamps.
Time span: Devonian to Late Carboniferous, 360–300 million years ago.
Main fossil sites: Throughout Europe.
Occurrence: ◆

Belinurus was another aquatic chelicerate. It was not a eurypterid, but a member of the group called xiphosurans ('sword tails'), which includes the horseshoe or king crabs of today. (These are not true crabs from the crustacean group, but are more closely related to scorpions and spiders.) *Belinurus* is one of the first known xiphosurans and already possessed the large spine- or dagger-like telson, or tail. Other features that are characteristic of the group include the large, hoof-shaped prosoma, or head, bearing the mouth, eyes and legs, and a shorter abdomen. The abdomen was still segmented in *Belinurus*. Unlike its living marine relatives, *Belinurus* was a freshwater, swamp-dwelling animal. However, it probably fed in a similar fashion, by walking over soft mud or sand as it searched for worms and other small invertebrate creatures to eat.

Below: Belinurus remains are mostly found associated with plants of the Carboniferous 'Coal Forests', which grew mainly in warm, tropical freshwater to brackish swamps.

Mesolimulus

Mesolimulus was almost identical to today's horseshoe crab, *Limulus*, which is why the latter is often dubbed a 'living fossil'. *Mesolimulus* probably had a similar lifestyle, crawling on the soft bottom of shallow coastal marine waters, feeding on small invertebrates and perhaps scavenging dead fish. Its fossils are famously known from the German Solnhofen Limestones. Some specimens are found at the centre of spiraling 'death march' tracks, left as the dying animal was succumbing to the toxic waters at the bottom of the lagoon. Paradoxically, this suggests that *Mesolimulus* could tolerate a wide range of conditions – most other Solnhofen creatures were already dead by the time they settled on the bottom. Other preserved tracks show that *Mesolimulus* probably swam like the modern *Limulus* – upside down. The traces show evidence of the animal turning itself over after 'touchdown' from swimming, before walking off.

Name: *Mesolimulus*.
Meaning: Mesozoic (Middle Life) equivalent of *Limulus* (horseshoe crab).
Grouping: Arthropod, Chelicerate, Xiphosuran, Limulid.
Informal ID: Extinct horseshoe crab.
Fossil size: Specimen head–tail length 9cm/3½in.
Reconstructed size: Average head–tail length 12cm/4¾in (however, most specimens are interpreted as being juveniles).
Habitat: Warm, shallow coastal waters, brackish estuaries.
Time span: Jurassic to Cretaceous, 170–100 million years ago.
Main Fossil Sites: Europe, Middle East.
Occurrence: ◆ ◆

— Prosoma (head and thorax)

— Eyes

— Genal spines

— Fused abdomen

— Long, sword-like telson (tail spine)

Left: Many complete and exquisitely preserved specimens of Mesolimulus are known from the fine-grained Late Jurassic Solnhofen Limestones. This horseshoe crab may have used the curved front edge of its prosoma, or head, to plough into the soft sea bed, either to burrow and hide, or to disturb possible food.

ARTHROPODS – SPIDERS, SCORPIONS

The arthropods include several kinds of chelicerates – spiders, scorpions, xiphosurans (horseshoe crabs) and eurypterids (sea scorpions). These all possess chelicerae, powerful front appendages variously modified as pincers, claws or fangs. Scorpions and spiders are known as arachnids, characterized by four pairs of walking limbs. Scorpions were among the first wave of land animals more than 380 million years ago.

Leptotarbus

Head end | Mouthparts | Abdomen

Above: This Late Carboniferous specimen of Leptotarbus *lacks its walking legs, but illustrates the wide 'waist' and overall oval shape typical of a phalangiotarbid or living harvestman.*

Leptotarbus belongs to an obscure group of arachnids, the Phalangiotarbida. It resembled a spider, having four pairs of walking legs, multiple pairs of eyes (*Leptotarbus* had three, contrasting with the four pairs of most modern spiders), a combined head and middle body section, the cephalothorax or prosoma, and a large abdomen. Overall it resembles the living long-legged arachnids called harvestmen, such as the modern genus *Phalangio*, which are not classed as true spiders. The body is flattened and oval in shape, with a wide connection between prosoma and abdomen, instead of the narrow 'wasp waist' of true spiders. It is difficult to interpret their lifestyles, considering their extinction and the rarity and poor condition of their fossils. Their air-breathing organs suggest they were terrestrial, and their mouthparts were weak.

Name: *Leptotarbus*.
Meaning: Slender terror.
Grouping: Arthropod, Chelicerate, Arachnid, Phalangiotarbid.
Informal ID: Extinct spider-like arachnid or harvestman.
Fossil size: Head–body length 1.5cm/½in.
Reconstructed size: Total width across legs 4cm/1½in.
Habitat: Unknown, possibly dense undergrowth.
Time span: Carboniferous, 300 million years ago.
Main Fossil Sites: Throughout Europe.
Occurrence: ◆

Maiocercus

Maiocercus was a relatively large spider for its group, the Trigonotarbida, and lived in the humid, tropical forests that covered Europe in the Carboniferous Period. Although superficially similar to the 'true' spiders, or araneans, *Maiocercus* lacked the 'wasp waist' and perhaps also silk glands. It possessed a thick, plated, chitinous outer body casing, or exoskeleton. However, it was a close relative of araneans and probably had a similar lifestyle to today's wolf-spiders, hunting smaller invertebrates by ambushing them in thick forest undergrowth or by chasing or cornering them. This interpretation is strengthened by the discovery of certain specimens with four pairs of strong walking legs. Trigonotarbid spiders represent the oldest known land arachnids, appearing in the Late Silurian Period. They became most diverse in the Late Carboniferous, and had faded by the early Permian Period.

Below: This Maiocercus *abdomen was recovered in coal from the Rhondda region of South Wales. It represents the holotype of the genus – the one that is used for the original description, and against which all subsequent specimens are compared, to identify them as* Maiocercus.

Name: *Maiocercus*.
Meaning: Mother (original) cercus (tail-like appendage).
Grouping: Arthropod, Chelicerate, Arachnid, Trigonotarbid.
Informal ID: Spider.
Fossil size: 1–2cm/⅖–¾in.
Reconstructed size: Head–body length 2–3cm/¾–1in.
Habitat: Warm, dense, humid forest.
Time span: Carboniferous, 310 million years ago.
Main fossil sites: Throughout Europe.
Occurrence: ◆

Fossil scorpions

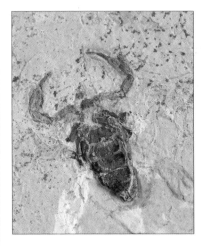

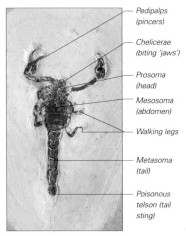

Pedipalps (pincers)

Chelicerae (biting 'jaws')

Prosoma (head)

Mesosoma (abdomen)

Walking legs

Metasoma (tail)

Poisonous telson (tail sting)

Name: Scorpions.
Meaning: From *skorpios* (Greek).
Grouping: Arthropod, Chelicerate, Arachnid, Scorpion.
Informal ID: Scorpion.
Fossil size: More complete specimen (far right), head–tail length 1.4cm/½in.
Reconstructed size: The group varies from head–tail length less than 1cm/⅜in to almost 1m/3¼ft – larger forms were aquatic.
Habitats: Variable, generally warm, including deserts, forests, grasslands, marine intertidal; many Palaeozoic types were aquatic.
Time span: Middle Silurian, 420 million years ago, to today.
Main fossil sites: Worldwide.
Occurrence: ◆

Above: Protoischnurus, from Brazil's Crato Formation, lacks its walking legs and tail. However, its remains show well-developed pincers.

Above: This scorpion was also found in the Crato Formation of north-east Brazil – a location known for the remarkable level of detail preserved in its fossils.

True scorpions appeared in the Middle Silurian Period, some 420 million years ago, and most early types were, in fact, aquatic creatures. Land-living (or terrestrial) forms of scorpions are known from the fossil record of the Early Devonian Period, while the last aquatic types survived into the Jurassic Period. They are thought to be cousins or descendants of eurypterids (sea scorpions), from which they inherited their overall body plan. This comprised a large mesosoma (or abdomen) of seven segments, and an elongated metasoma (tail section) of five slender segments, ending with a telson or tail spine. The telson of true scorpions is a poisonous sting.

Origins of spiders

The true spiders, Araneae, have an obscure ancestry. Specimens are known throughout the Late Palaeozoic Era, especially from the Carboniferous Period. A few have been found from the Mesozoic Era, and many more from the Tertiary, often preserved in amber. Many of the living types of spider can be traced back to the Mesozoic Era. For example, a tarantula relative is known from Late Triassic rocks (dating back 210 million years) in France, and relatives of web-spinning species are known from the Jurassic Period. In particular, a Lower Cretaceous specimen from Spain shows three claws on the end of each leg, which modern spiders use for walking on their webs. Defining characteristics of spiders include the appendages known as chelicerae modified into poisonous fangs, a thin 'waist', and a system of glands and appendages (spinnerets) that produces silk. The example here is one of the Saticidae (jumping spiders), entombed in Baltic amber dating back to the Late Eocene, some 40 million years ago. Typical victims of the sticky tree sap before it fossilized include midges, cockroaches, nymphs and ants, as well as jumping spiders.

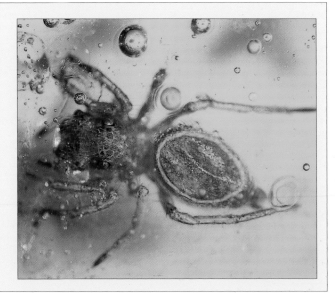

ARTHROPODS – INSECTS

In the fossil record, the vast arthropod or 'jointed-leg' group is dominated by aquatics, such as the still-thriving crustaceans and long-gone trilobites. In comparison, there are few remains of the main arthropod group we encounter nearly every day – insects. This is because most were (and are) small, relatively fragile and often consumed by predators. If not, they died in places such as moist forests, where decay was swift.

Proeuthemis

The very first insects were tiny and wingless, and were among the early land animals of the Devonian Period. By the next period, the Carboniferous, they had evolved into several groups and greatly increased in size – and some had developed wings to become the first creatures to take to the air. Indeed, the steamy swamps of the Late Carboniferous, 300 million years ago, saw the ancestors of dragonflies, known as protodonatans, which were the largest flying insects ever. The famed protodonatan 'dragonfly' *Meganeura* had wings spanning 70cm/28in. The true dragonflies and their smaller cousins, damselflies, in the group Odonata, follow their ancestors' lifestyle as fast, darting aerial predators. (See also *Libellula*, below.)

Main vein
Root of wing

Subsidiary vein
Wing tip

Above: This Proeuthemis wing shows the dragonfly pattern of veins. In life they were tubes taking blood to the wing membrane and also helping to stiffen the wing. Dragonflies and their ancestors are four-winged insects.

Name: *Proeuthemis*.
Meaning: Before *Euthemis* (a similar dragonfly genus).
Grouping: Arthropod, Insect, Odonatan.
Informal ID: Dragonfly.
Fossil size: Wing length 2.5cm/1in.
Reconstructed size: Wingspan 5cm/2in.
Habitat: Swamps, moist woodlands.
Time span: Early Cretaceous, 120 million years ago.
Main fossil sites: Throughout Europe.
Occurrence: ◆

Libellula

Below: Identified as Libellula ceres, *this Miocene specimen, possibly showing segments of the abdomen, is from Rott, near Siegburg, in the Rhine-Sieg region of Germany. Rott is known for its brown coal or lignite formations, which have also preserved bees, flies and mites.*

Dragonflies and damselflies (Odonata), like mayflies (Ephemeroptera), have the primitive winged-insect feature known as palaeopteran – they cannot fold their wings back to lie along the body. Instead they project sideways at rest. With about 5,000 living species, the dragonflies and damselflies are relatively well represented as fossils. They occur in rocks that were originally formed in swamp and marsh habitats – where dragonflies pursue small flying insects, such as midges, gnats and moths, to eat. Libellulids survive today as the dragonflies known as darters or skimmers, most of which have a broad body flattened from top to bottom. Their typical behaviour is to rest on a waterside perch, watch for victims with their domed eyes – their vision is among the sharpest in the insect world – and then dart out to catch the prey in a 'basket' formed by their six dangling legs.

Name: *Libellula*.
Meaning: Leaflet, booklet
Grouping: Arthropod, Insect, Odonatan.
Informal ID: Darter dragonfly.
Fossil size: Fossil slab 6cm/2⅜in across.
Reconstructed size: Wingspan up to 15cm/6in.
Habitat: Marshes, bogs.
Time span: Late Miocene, 8 million years ago.
Main fossil sites: Northern Hemisphere.
Occurrence: ◆

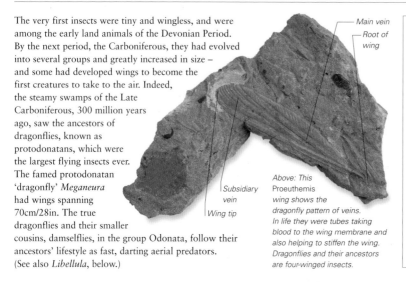

Palaeodictyopteran

Below: A typical palaeodictyopteran had not only two pairs of dragonfly-like wings, but also a smaller pair of 'winglets' in front of the foremost main wings. It also had a beak-like mouth for tearing and sucking. In some types, the bold-patterned markings on the wings are clearly visible in the fossil.

Name: Palaeodictyopteran.
Meaning: Ancient net wing.
Grouping: Arthropod, Insect, Palaeodictyopteran.
Informal ID:
Palaeodictyopteran, extinct dragonfly cousin.
Fossil size: Wingspan 7cm/2¾in.
Reconstructed size:
Wingspan of some types exceeded 50cm/20in.
Habitat: Forests.
Time span: Most types Late Carboniferous to Late Permian, 320–255 million years ago.
Main fossil sites:
Europe, Asia.
Occurrence: ◆ ◆

'Dictyopteran' was an older name for a main group, or order, of insects that has now, in most modern classifications, been split into two orders: the cockroaches, Blattodea, and the mantids (preying mantises), Mantodea. The palaeodictyopterans, all extinct since the Palaeozoic Era, were medium-to-large insects with an outward similarity to odonatans, such as dragonflies (opposite). They had a palaeopterous wing structure, which did not allow the wings to be folded back along the body, but only held out to the sides or above. The two pairs of wings were similar in size and shape. Palaeodictyopterans are not now regarded as being one evolutionary group with a single ancestor, but an assemblage of primitive insect groups. They lived from the Middle Carboniferous to the Late Permian, about 320 to 255 million years ago. In some evolutionary schemes, certain of the Palaeodictyoptera are regarded as ancestors of the Protodonata (see *Proeuthemis*, opposite).

Blattodean (cockroach)

Name: Blattodean.
Meaning: From Greek *blatta* for cockroach.
Grouping: Arthropod, Insect, Blattodean.
Informal ID: Cockroach.
Fossil size: Overall length 2.5cm/1in.
Reconstructed size:
As above.
Habitat: Damp places, leaf litter, forests.
Time span: Carboniferous, 340 million years ago, to today.
Main fossil sites: Worldwide.
Occurrence: ◆ ◆ ◆

Cockroaches are infamous as hardy survivors in almost every habitat, including urban settings where they infest houses, food stores and other buildings. However fewer than 25 of the 4,000 or so living species are serious pests. These flattened, tough-bodied, fast-scuttling lovers of dark, damp warmth are also well known as some of the earliest insects, dating from Carboniferous times. Their remains are common from the Coal Forest swamps, where they lived among damp vegetation and were readily buried by mud and preserved. Most had generalized mouthparts and were omnivores, eating a wide range of foods. Their body shape and design has changed little since, with a low profile that allows them to hide in cracks and crevices. Some types of cockroach have large wings extending over the abdomen, while others have much reduced or even absent wings.

Above: A cockroach has the usual three insect body parts of head, thorax and abdomen. However the head may be obscured by a shield-like pronotum, and the tough wings lie over the abdomen. These Late Carboniferous specimens are from the Coal Measure rocks of Avon, southwest England.

ARTHROPODS – INSECTS (CONTINUED)

A typical insect has three principal body parts. These are the head, usually bearing antennae or feelers, eyes and mouthparts; the thorax, carrying the wings and legs; and the abdomen, containing the insect's digestive, waste-removal and reproductive organs. Typically, an adult insect has six legs and four wings, although true flies (Diptera) have just one pair of wings.

Cratoelcana

Below: Amazing detail can be seen in this Mesozoic cricket, which is some 120 million years old – yet its features are very similar to living types. The rear leaping legs are almost fully extended, as they would be in life as the animal jumped. The wings are folded back in the resting position. They could also be extended as the animal leaped, to allow a fluttering flight, usually to escape from predators.

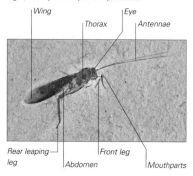

Wing | Eye | Thorax | Antennae | Rear leaping leg | Front leg | Abdomen | Mouthparts

The famed Early Cretaceous rocks of the Crato Formation, in the northeast region of Brazil, have yielded many amazingly detailed fossils, ranging from small scorpions through to huge pterosaurs, which were the first vertebrate animals to evolve true flight, and fish such as *Dastilbe*. Among the most represented insect groups found at the site are the orthopterans, which today include more than 20,000 living species of grasshoppers, crickets and katydids. *Cratoelcana* is a type of ensiferan – the subgroup including bush crickets, which are sometimes called long-horned grasshoppers, also field and cave crickets, and katydids. The very long 'horns' referred to are, in fact, antennae or feelers. The genus name refers to the Crato Formation and also to the primitive cricket family Elcanidae, which is now extinct.

Name: *Cratoelcana*.
Meaning: Elcanid of Crato.
Grouping: Insect, Orthopteran, Ensiferan.
Informal ID: Cricket.
Fossil size: Total length including antennae 5cm/2in.
Reconstructed size: As above.
Habitat: Woods, forests, shores.
Time period: Early Cretaceous, 120 million years ago.
Main fossil sites: South America.
Occurrence: ◆ ◆

Panorpidium

A member of the Ensifera, or long-horned grasshoppers and bush crickets (see *Cratoelcana*, above), *Panorpidium* dates from the Late Cretaceous Period, between 80 and 65 million years ago. Orthopterans as a group are far older, stretching back to the Triassic Period more than 220 million years ago. However, there could be no 'grasshoppers' at that time, since grasses would not evolve for almost another 200 million years. The early orthopterans presumably lived among the conifers, cycads and similar dominant plants of the time. The rear pair of orthopteran wings are enlarged for flight. The front pair, if present, are small and hardened flaps that help to protect the rear pair. The antennae vary from short to very long, depending on the species involved, and the hind legs are enlarged and especially modified for jumping.

Below: The rear wing of this Panorpidium *has an area of damage towards the lower tip that could have been an injury in life. The box-like network of main longitudinal veins and cross-connecting transverse veins typical of the orthopterans is clearly visible. The specimen is from the Late Cretaceous Wealden Formation of Surrey, southern England.*

Area of damage | Veined network | Wing root (joins to body)

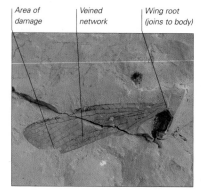

Name: *Panorpidium*.
Meaning: Not known.
Grouping: Insect, Orthopteran, Ensiferan.
Informal ID: Bush cricket.
Fossil size: Wing length 1.5cm/⅝in.
Reconstructed size: Total length (excluding antennae) 2cm/¾in.
Habitat: Undergrowth.
Time period: Late Cretaceous, 80–65 million years ago.
Main fossil sites: Europe.
Occurrence: ◆

Trichopterans (caddisflies)

Caddisflies are not true flies (which belong to the insect group Diptera), but form their own group, Trichoptera. This has more than 8,000 living species and is most closely related to butterflies and moths (Lepidoptera). Fossils from before the Mesozoic Era show that caddisflies have been following a similar way of life for more than 250 million years. The soft-bodied larvae dwell in fresh water and most build tube-like homes from sand, pebbles, leaf fragments or pieces of twig, depending on the species and material available. They emerge from the water and moult into the adults, which vaguely resemble a combination of moth and lacewing, with two pairs of long wings, long, slim legs and long antennae.

Right: These specimens are from the Yixian Formation near Beipiao, northern China. They are similar in many details to today's caddisflies. The Early Cretaceous Yixian rocks are perhaps more famous for amazingly preserved dinosaurs, many with feathery or filamentous coverings. Specimen total length 1.2cm/½in (right) and 2.5cm/1in (far right).

Above: This Tertiary larval case was fashioned from tiny shards of rock and shells, similar to the 'homes' built today by Mystacides, *the genus whose adults are known by freshwater anglers as the 'silverhorn' caddis. Each time the larva shed its soft skin, it would construct a new, larger case. The length of this example is 2cm/¾in.*

 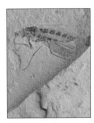

Name: Trichopteran.
Meaning: Hairy wing.
Grouping: Arthropod, Insect, Trichopteran.
Informal ID: Caddisfly.
Fossil sizes: See text.
Reconstructed size: Most types 1–5cm/ ½–2in long.
Habitat: All habitats in or close to fresh water.
Time period: From Late Permian, 250 million years ago, to today.
Main fossil sites: Worldwide.
Occurrence: ◆

Trapped in amber

Small insects and other arthropods were often trapped in the resin oozing from pines and other conifers, which hardened over millions of years into amber. Some amber is so transparent that every detail can be seen of the creature within. These specimens are two dipterans, or true flies, one resembling a gnat (below left), the other a midge (below right), along with a small moth or lepidopteran (bottom). They are all dated to the Late Eocene Epoch, about 35 million years ago.

Cupedid (reticulated) beetle

Beetles, Coleoptera, are the largest subgroup of insects, with more than 370,000 living species. One of their distinguishing features is that the front of the two pairs of wings has usually thickened and hardened into a pair of wing cases called elytra. In flight, the elytra pivot to the side and upwards, and the main flying wings unfold. The family Cupedidae is widely known from the Mesozoic Era, and in some regions forms almost a third of all beetle fossils during the Late Triassic Period. But their numbers fell away drastically by the Early Cretaceous. Today, they are regarded as a primitive and relict family (showing ancient features), being much less common than previously.

Name: Cupedid beetle.
Meaning: Cup, dome, turret.
Grouping: Arthropod, Insect, Coleopteran.
Informal ID: Beetle.
Fossil size: Total length 1.5cm/⅝in.
Reconstructed size: As above.
Habitat: Varied.
Time period: From Early Permian, 280 million years ago, to today.
Main fossil sites: Northern Hemisphere.
Occurrence: ◆ ◆

Left: Cupedid beetles are also called reticulated or net beetles due to the pattern on the elytra, which variously resembles a chessboard, net or rows of small depressions. This specimen is from China and is about 125 million years old.

ARTHROPODS – INSECTS (CONTINUED), MYRIAPODS

Myriapods, 'myraid legs', is a general name used for the arthropods most people know as centipedes and millipedes, with the scientific names Chilopoda and Diplopoda, respectively. Millipede-like creatures were among the earliest-known land animals in Silurian times.

Trematothorax

Wing | Thorax

Eyes | Leg

The Hymenoptera group of insects includes some 200,000 living species of bees, wasps, ants and sawflies. Like many major insect groups, it appeared in the Permian–Triassic Periods, some 250 million years ago, and then diversified rapidly with the spread of flowering plants (angiosperms) from the Early Cretaceous. Most types of bee and wasp have two pairs of fairly long, narrow wings. The front and rear wings on each side are joined by tiny hooks so that they effectively beat as one. Another feature of modern hymenopterans is the 'wasp waist', the narrow constriction between the thorax and abdomen. The egg-laying tube at the rear of the abdomen is modified into a sting.

Left: Trematothorax is a genus of the sepulcid group, which in turn is more closely related to the sawflies called stem sawflies than to typical wasps. Sawflies are more primitive, have wider 'waists' and do not sting. The eyes, thorax, legs and wings, with the characteristic pattern of veins, are clearly seen here.

Name: *Trematothorax*.
Meaning: Three-clubbed or three-rod chest.
Grouping: Arthropod, Insect, Hymenopteran, Sepulcid.
Informal ID: Wasp, sawfly.
Fossil size: Total length 1cm/⅖in.
Reconstructed size: As above.
Habitat: Woods, scrubland.
Time span: (*Hymenoptera* genus) Triassic, 220 million years ago, to today.
Main fossil sites: Worldwide.
Occurrence: ◆ ◆

Tiphid ant

Below: This specimen shows the characteristic ant features – long, powerful running legs, fairly large eyes and crooked, or 'elbowed', antennae (feelers). The fossil comes from the famous Crato Formation of Brazil (see Cratoelcana), and shows the amazing detail of preservation in these rocks, known as Lagerstatten. It is dated to the Early Cretaceous, about 125–120 million years ago.

Ants, which belong to the subgroup of Hymenoptera known as formicids, have about 9,000 species alive today. The other major subgroups are apids (including honey bees and bumblebees) and vespids (including hornets and wasps). Most ants are less than 1cm/⅖in long. The tiphid ants still thrive in warmer parts of the world, such as the genus *Thipia* in the Caribbean region. They are active creatures during the summer months and are predators of ground beetle larvae, or 'grubs'. Some female tiphids dig into the soil to look for the larvae. For many years ants were thought to have appeared in the Tertiary Period, dating from about 45 million years ago. However, more recent discoveries have pushed back their origins to Cretaceous times, some 125 million years ago.

Name: Tiphid ant.
Meaning: Swamp/marsh ant.
Grouping: Arthropod, Insect, Hymenopteran, Formicid.
Informal ID: Ant.
Fossil size: Total length 1cm/⅖in.
Reconstructed size: As above.
Habitat: Woods, forests.
Time span: From Early Cretaceous,125 million years ago, to today.
Main fossil sites: Worldwide.
Occurrence: ◆ ◆

Diplopodan (millipede)

Name: *Diplopodan*.
Meaning: Double footed.
Grouping: Arthropod,
Diplopodan.
Informal ID: Millipede
('thousand feet').
Fossil size: 7.3cm/3in.
Reconstructed size: This
specimen (unstraightened)
6.3cm/2½in; largest species
exceeded 2m/6ft in length.
Habitat: Moist forest floor.
Time span: Ordovician, over
435 million years ago, to today.
Main fossil sites: Worldwide.
Occurrence: ◆

Preserved remains of millipedes date back to the Silurian Period, and some of their tracks or traces may be even earlier (see panel below). This suggests that they originated in aquatic habitats but soon moved onto land. Millipedes are also known as diplopodans, reflecting the feature of two pairs of legs per body section or segment (centipedes have one pair). Most of the 10,000 living species chew vegetable matter, from juicy leaves to dead wood. The first millipedes probably led a similar lifestyle. They were among the first of all land creatures, perhaps capitalizing on the largely uncontested decaying remains of early terrestrial plants. *Arthropleura*, a millipede of the Late Carboniferous, possibly exceeded 2m/6ft in length, making it the biggest known land invertebrate of all time.

Above: An unidentified diplopodan preserved as part and counterpart in a Carboniferous siltstone nodule of Yorkshire, north-east England. Most millipedes have always lived in damp habitats, such as forest floors, and their exoskeletons are not mineralized, making their fossils scarce.

Chilopodan (centipede)

A centipede has one pair of legs per body section, or segment (millipedes have two pairs, see above), and usually more than 15 segments. Most of the 3,000 living species are predators of small worms, insects and similar creatures. What is effectively the first pair of walking legs has become a pair of curved, fang-like claws called forcipules. These seize prey and inject poison. Like millipedes, most centipedes frequent moist places, such as woods and leaf litter, where they would soon rot after death, and so their fossils are scarce. Centipedes also often consume their own shed body casings as they grow and moult and, presuming ancient

Name: Chilopodan.
Meaning: Poison feet.
Grouping: Arthropod,
Chilopodan.
Informal ID: Centipede.
Fossil size: Centipede length
14mm/½in.
Reconstructed size: Large
species up to 30cm/12in.
Habitat: Moist forests.
Time span: This specimen,
Late Eocene, 40–35 million
years ago.
Main fossil sites: Worldwide.
Occurrence: ◆

centipedes did the same, this would further reduce the material available for fossilization. The fossil record for centipedes may stretch back to the Devonian or even Late Silurian, with relatively plentiful numbers from the Carboniferous, but finds from the Mesozoic Era are rare.

Millipede trace fossils

These tracks (footprints) were probably made by a millipede-like creature scurrying across a silty underwater surface. As the water current pushed it one way, the animal placed its legs out to one side in order to stay balanced, to keep itself in touch with the substrate and to keep itself going in the desired direction. Its feet slid sideways in the soft silt, leaving slim furrows. The fossil is Devonian, from the Old Red Sandstone near Brecon, Wales.

Right: In this piece of Baltic amber, dating from the Late Eocene Epoch, about 40–35 million years ago, the trapped centipede is in the centre. Its long horn-like antennae protrude at the head (left) with the first few pairs of walking legs just visible as the body curves away, terminating in the tail (upper-right). Some small flies and other insects were also entrapped.

Head

Tail

MOLLUSCS

The phylum (major animal group) Mollusca is one of the largest in the animal kingdom, with more than 100,000 living species. They range from gastropods – slugs, sea slugs (nudibranchs), land snails and sea snails, such as whelks and limpets – and bivalves, which are shellfish with a two-part shell, such as oysters, clams and mussels, to the many-tentacled cephalopods – octopus, squid and cuttlefish. The ancient representatives of creatures such as these, as well as many thousands of their long-extinct molluscan cousins, such as the ammonites and belemnites, form a significant portion of the fossil record. The various subgroups of the Mollusca are covered on the pages that follow.

Dentalium

Scaphopods are distinctive molluscs with a long, slim, tapering, tubular shell that is open at both ends. Its resemblance to an elephant's tusk has earned the group the common name of 'tusk shells'. In the living animal, the fleshy, finger-like 'foot' and head extend from the wider, lower end for feeding. Water is drawn into the narrow upper end for obtaining oxygen. Unlike most molluscs, scaphopods lack gills and, instead, absorb oxygen over the folded fleshy surfaces within the shell. Scaphopods live part-buried in mud or similar sediments on the sea floor and use the foot to construct a feeding cavity. Most types consume microorganisms, such as foraminifera. One subgroup, the dentaliids, have rigid sets of teeth, which they use to grind small prey. In another subgroup, the gadilids, the teeth are more flexible and function to chop and crush. Scaphopods occur in the fossil record from the Ordovician Period, more than 450 million years ago. They are extremely common in certain places and their preserved shells range in size from larger than a forefinger to as small as a matchstick.

Below: Dentalium striatum is named after the striations, or stripes, along its length, which show patterns of growth rings (like a tree) as the shell lengthens with age. This specimen, from the Late Eocene Epoch, is from the Barton Beds of Hampshire, England.

Name: *Dentalium.*
Meaning: Small tooth.
Grouping: Mollusc, Scaphopod.
Informal ID: Tusk shell.
Fossil size: *D. striatum* shell length 4.5cm/1¾in.
Reconstructed size: Most specimens 1–10cm/¾–4in
Habitat: Sea floor.
Time span: Ordovician, over 450 million years ago, to today.
Main fossil sites: Worldwide.
Occurrence: ◆ ◆ ◆ ◆

Bands of striations

Faint longitudinal faces give hexagonal cross-section

Narrow upper (posterior) opening for respiratory water current

Wide lower (anterior) opening where main body protruded in life

Left: Scaphopods show little variety of form, most having the curved, tapering shape of an elephant's tusk. The shell surface ornamentation is either absent or comprised of raised ridges or ribs, as here in Dentalium neohexagonum, *which is from the Pleistocene Palos Verdes Sand of San Pedro, Los Angeles, California, USA.*

Helminthochiton

Name: *Helminthochiton*.
Meaning: Leaf shield.
Grouping: Mollusc,
Polyplacophoran.
Informal ID: Chiton,
coat-of-mail shell.
Fossil size: Total length
8cm/3⅛in.
Reconstructed size: Total
length 10cm/4in.
Habitat: Sea floor.
Time span: Ordovician to
Carboniferous, 500–300
million years ago.
Main fossil sites: North
America, Europe.
Occurrence: ◆

Chitons are known scientifically as polyplacophorans, and informally as coat-of-mail shells. A chiton has a slug-like body shielded beneath eight jointed plates or valves, and flanked by an oval spiny girdle. When threatened, the chiton can roll up for protection in the manner of a woodlouse. Unlike most molluscs, chitons possess multiple pairs of gills for obtaining oxygen from water. Some scientists interpret this as a primitive feature inherited from a segmented ancestor, but others view it as an evolutionary adaptation for better respiration. Chitons are common today in rocky marine environments where they cling limpet-like to hard surfaces. Most are fingernail-sized but the giant chiton is as large as a shoe. However, they are scarce in the fossil record, usually occurring as isolated valves from the Cambrian Period, more than 500 million years ago.

Below: A chiton's valves usually separate soon after death to form isolated fossils. This specimen has been reconstructed with all its main valves articulated (positioned as in life), although the flange-like girdle, which would have joined to their edge all around, like a 'skirt', is missing.

Second valve | *Head end*

Eighth valve

Proplina

Name: *Proplina*.
Meaning: Early pot.
Grouping: Mollusc,
Monoplacophoran.
Informal ID:
Monoplacophoran.
Fossil size: 3cm/1in.
Reconstructed size:
As above.
Habitat: Sea floor.
Time span: Late Cambrian to
Early Ordovician, 510–470
million years ago.
Main fossil sites: North
America, Northeast Asia.
Occurrence: ◆ ◆

Monoplacophoran molluscs are termed 'living fossils'. These primitive marine molluscs were presumed extinct from the Devonian Period, more than 300 million years ago, until live specimens were collected in 1952 from the Pacific deep-sea floor. They are considered similar in form to the ancestral or 'original' mollusc, comprising a simple conical shell and a fleshy foot for locomotion. They also possess multiple gills, like the chitons. Superficially they resemble limpets, but the latter belong in a different group, the gastropods or snails, due to the internal twisting of the body. Monoplacophorans appeared during the Cambrian Period and they were common in the shallow seas of the Early Palaeozoic Era.

Margin (lower edge) of shell | *Apex (upper tip) of shell pitched forward over anterior (head) end*

Right: The spoon-shaped monoplacophoran Proplina *is found in various rocks, such as cherts from eastern and southeastern North America. This specimen is from Missouri, USA. In well-preserved specimens, eight muscle 'scars' can be seen on the shell, arranged in a horseshoe pattern. These 'scars' are where the foot-retracting muscles were anchored in life.*

Spiral-shelled mollusc

Posterior end

MOLLUSCS – BIVALVES

In general the mollusc group have left plentiful fossils and they are especially common in rocks from the Cambrian Period onwards. The bivalves are perhaps the most familiar molluscs and include the clams, oysters, cockles, scallops and mussels. The shell has two parts or valves. The group seems to have appeared in the Cambrian Period and become well established by the Ordovician.

Inoceramus

The hard parts of a bivalve consist of a pair of calcareous valves. Broadly speaking, the two valves of a bivalve are mirror images of each other, which distinguishes them from the brachiopods. The valves close to protect the soft parts, with an articulation or hinge composed of teeth and sockets. Ligaments on one side of the hinge-line hold the valves open, while powerful muscles, called adductors, pull the valves tightly shut. Typically, each bivalve has a pair of adductor muscles that attach to the inside of the valves, leaving distinctive impressions called muscle scars. The exact structure of the hinge area and the arrangement of the muscle scars are key features when it comes to distinguishing the various bivalve groups.

Left: The two valves of this specimen of Inoceramus lamarckii, *a type of mussel (see* Sphenoceramus, *below), have been preserved together, still articulated. The decaying fleshy remains within the shell acted as a site for silica precipitation within the sediment, and the valves are now partially engulfed within a flint (silica) concretion.*

Name: *Inoceramus*.
Meaning: Strong pot.
Grouping: Mollusc, Bivalve.
Informal ID: Mussel.
Fossil size: 10cm/4in.
Reconstructed size: Most specimens up to 20cm/8in.
Habitat: Sea floor.
Time span: Late Cretaceous, 90 million years ago.
Main fossil sites: Worldwide.
Occurrence: ◆ ◆ ◆

Sphenoceramus

Right: This elaborate specimen of Sphenoceramus, *from the Late Cretaceous Chalks of Sussex, England, has sets of radiating ridges arranged in growth arcs, leading to its species name of* S. pinniformis *(meaning 'in the form of a leaf').*

Pseudomytiloides
Freak currents have swept these shells along the sea floor and banked them against a log of driftwood. During the fossilization process the shells have incorporated the iron mineral pyrite, or 'fool's gold', giving them an unusual and attractive golden coloration.

Sphenoceramus and *Inoceramus* (above) belong to a group known as the inoceramids. They were particularly successful during the Late Cretaceous, where many grew to a great size – individuals more than 1m/3¼ft in length were not uncommon. Most lived reclining on the sea floor and acted as oases of hard substrate for small communities of oysters, sponges, corals, barnacles, brachiopods, worms and bryozoans. Some shells contain the remains of schools of fish, which apparently lived within them and were trapped when the shell died and clamped shut.

Name: *Sphenoceramus*.
Meaning: Wedge pot.
Grouping: Mollusc, Bivalve.
Informal ID: Mussel.
Fossil size: 17cm/6¾in in length.
Reconstructed size: Some individuals more than 1m/3¼ft in length.
Habitat: Sea floor.
Time span: Late Cretaceous Period, 85 million years ago.
Main fossil sites: Worldwide.
Occurrence: ◆ ◆ ◆

Spondylus

Name: Spondylus.
Meaning: Vertebra
(backbone).
Grouping: Mollusc, Bivalve.
Informal ID: Thorny or
spiny oyster.
Fossil size: 4cm/1½in.
Reconstructed size: Up to
15cm/6in across, including
the spines.
Habitat: Sea floor.
Time span: Jurassic, 190
million years ago, to today.
Main fossil sites:
Worldwide.
Occurrence: ◆ ◆ ◆

*Right: The upper valve is largely
free of spines, while the lower
valve is densely armoured. Only
the spine bases have been
retained in this specimen.*

Some bivalves have shells ornamented with tall, sharp ridges, and there are even spines protruding from along the edge or margin of one or both valves. These ribs and spines can provide strength, protection and stability, as well as encouraging other organisms, such as algae (seaweeds), to grow on top to act as camouflage. Modern forms of *Spondylus* are called spiny or thorny oysters, although they are not actually true oysters. In many, the lower valve becomes encrusted to a hard substrate, such as a rock. The Cretaceous species *S. spinosus* was not encrusted but free-living. In life, it possessed a dense array of particularly elongated and slender spines projecting from around the valve edge, which possibly prevented the animal from sinking into soft sediment.

Growth rings | Scalloped valve margin | Curled, pointed 'beak'

Above: The ligaments that hold the creature's valves open are more resistant to decay than the adductor muscles that pull the valves shut, so the shell often gapes after death. These two specimens were probably washed together at the same stage of decay.

Radial ridges — Spine base

Neithea

Name: Neithea.
Meaning: Reliquary,
strongbox.
Grouping: Mollusc, Bivalve.
Informal ID: Clam, scallop.
Fossil size: 3cm/1in across.
Reconstructed size: Up to
10cm/4in across.
Habitat: Sea floor.
Time span: Cretaceous,
130–70 million years ago.
Main fossil sites:
Worldwide.
Occurrence: ◆ ◆ ◆

Neithea was a scallop-like bivalve in the group called pectens (see following page), with one flat valve and one convex (outward-bulging) valve. Presumably it lived reclined on the surface of the sea floor with the convex valve lowermost. Like all bivalves, it lacked a defined 'head'. Instead, the shell gaped slightly so that water could pass over the sheet-like gills hanging within the body cavity, moving as a current generated by the gill lining of microscopic, whip-like structures known as cilia. In addition to taking in oxygen, the gills act as sieves, straining the water for tiny food particles, typically microscopic organisms of the plankton. These are gathered in sticky mucus (slime) and passed to the mouth.

Right: Neithea had about five large radiating ridges or ribs interspersed with groups of four or so narrower, lower ribs. This design, like the wavy corrugations in metal sheet, provided strength and stiffness for minimum material usage and weight.

Convex valve | Overhanging 'beak' of hinge region

Main ribs | Secondary ribs | Flattened valve

MOLLUSCS – BIVALVES (CONTINUED)

During the Early Palaeozoic, bivalves were successful in aquatic environments where the water was shallow and salty – conditions shunned by their main rivals, the brachiopods (lampshells). Through the Devonian and Carboniferous, bivalves became more common in lagoons, estuaries and fresh water. Today, bivalves dominate many shallow marine habitats while brachiopods have become relatively scarce.

Aviculopecten

Below: Aviculopecten had the typical scallop, fan-like structure of ribs and furrows radiating out to the valve margins (edges). The 'ears' on either side of the hinge area are asymmetrical. The front, or anterior, ear has an in-curved notch (more pronounced in the right valve).

The typical behaviour of the scallops, or pecten bivalves, is to rest on the sea floor. Some types anchor themselves to a hard substrate, such as rock, by tough stringy structures called byssus threads (see opposite), in the manner of mussels. Others are able to 'swim', as described below, but take flight only when they are threatened by predators or when the environmental conditions become unfavourable for some reason. *Aviculopecten* was probably anchored by threads to a hard surface and so could not swim freely. Bivalves are often brightly coloured in life, but colour pigments quickly break down after death and are preserved in fossils only in exceptional circumstances. The specimen shown here has retained its striking pattern of zigzag banding, but the original colours have been lost.

Name: *Aviculopecten.*
Meaning: Bird/winged comb.
Grouping: Mollusc, Bivalve.
Informal ID: Scallop.
Fossil size: 3cm/1in.
Reconstructed size: Most specimens less than 10cm/4in.
Habitat: Sea floor.
Time span: Devonian to Permian, 360–260 million years ago.
Main fossil sites: Worldwide.
Occurrence: ◆ ◆ ◆

Streblopteria

Scallops include forms with the ability, unique among bivalves, to 'swim'. This is achieved by repeated powerful contractions of the adductor muscles, which cause the valves to 'clap'. Jets of water are expelled through openings either side of the hinge area and thus propel the animal across the sea bed and even up into mid water. In life, scallops possess a feature unique to bivalves – rows of tiny iridescent light sensors, or 'eyes', fringing the opening of the shell. These are relatively simple, but are able to sense light and movement and can be regrown if damaged or lost. The development of sight among scallops complements their swimming ability, and allows them to orientate themselves and swim purposefully in a particular direction.

Left: The preservation of fossil patternation is generally very rare. It is well known in bivalves from the Carboniferous limestones of Castleton in Derbyshire, England, however, where this specimen and Aviculopecten (above) were found. The direction of patternation follows, but does not coincide exactly with, the radiating ribs that strengthen the valves.

Name: *Streblopteria.*
Meaning: Twisted/rolled wing.
Grouping: Mollusc, Bivalve.
Informal ID: Scallop.
Fossil size: 3cm/1in.
Reconstructed size: As above.
Habitat: Sea floor.
Time span: Carboniferous to Permian, 350–260 million years ago.
Main fossil sites: Worldwide.
Occurrence: ◆ ◆ ◆

Plagiostoma

Name: *Plagiostoma*.
Meaning: Slanting mouth, sideways opening.
Grouping: Mollusc, Bivalve.
Informal ID: Clam.
Fossil size: 10cm/4in.
Reconstructed size: As above, reaching 20cm/8in.
Habitat: Sea floor.
Time span: Triassic to Cretaceous, 230–65 million years ago.
Main fossil sites: Worldwide.
Occurrence: ◆ ◆ ◆

Extinct bivalves such as *Plagiostoma* probably produced a unique substance known as byssus, as used by many types of living bivalve today. Byssus is a form of fine thread that the animal uses to attach and anchor itself to hard substrates, such as rocks. (The common mussel, for example, attaches itself to rocks and breakwaters using byssal threads.) Byssus is produced by the bivalve's foot as a fluid that solidifies once it is in contact with water. The foot draws the material out into threads that it then attaches to the substrate. It is not necessarily a permanent connection, and it is possible for the animal to break free if threatened or if it needs to relocate. (Fabric made from byssus was prized in ancient times and referred to as sea-silk. The fabled Golden Fleece of ancient myth was perhaps made of bivalve byssus.)

Right: Plagiostoma *is believed to have attached itself to rocks or large pieces of driftwood using byssal threads. This species,* P. giganteum, *is a common fossil from the Liassic rocks of the Jurassic Period. It grew to a large size and can be found preserved in the position it would have adopted in life, orientated vertically.*

Myophorella

Myophorella belongs to a group of bivalves known as trigoniids – types of clam characterized partly by hinge structure. The right valve part of the hinge has two broad, ribbed teeth forming a V shape. The teeth fit into corresponding indentations in the left valve. The thick shell of *Myophorella* is partly covered by curving lines of knobs or tubercles, which would have aided its passage through and purchase within the sediment when burrowing. The area lacking tubercles would have projected above the sediment, with the line where the ornamentation is lost marking the surface of the sediment. Trigoniids often occur in dense associations within Mesozoic marine limestones, but they are rare today.

Name: *Myophorella*.
Meaning: Little muscle bearer.
Grouping: Mollusc, Bivalve.
Informal ID: Clam, brooch clam.
Fossil size: 5cm/2in across.
Reconstructed size: As above.
Habitat: Sea floor.
Time span: Early Jurassic to Early Cretaceous, 190–130 million years ago.
Main fossil sites: Worldwide.
Occurrence: ◆ ◆ ◆

Laevitrigonia

If the shell of a fossil such as a bivalve decays, it leaves an internal mould shape in the rock, as shown here. This is referred to by the German word *steinkern*, meaning 'stone seed'. The steinkerns of *Laevitrigonia* are referred to in folklore as 'osses 'eds'. When viewed from a certain angle, their shape can be likened to the head of a horse. The shell itself, before it decayed to leave the mould, was quite similar in general form to that of *Myophorella* (see right).

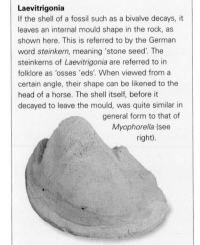

Below: Myophorella *has an almost half-circle shell shape partially ornamented by rows of low bumps or pimples termed tubercles. The rear, or posterior, part of each valve is plain and would have projected above the sediment in life. Fossils associated with this bivalve suggest that it lived in deep-sea sands, muds and similar types of sediment.*

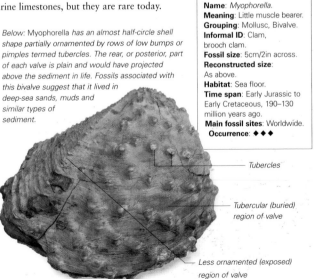

Tubercles

Tubercular (buried) region of valve

Less ornamented (exposed) region of valve

MOLLUSCS – BIVALVES (CONTINUED)

The bivalve molluscs known as oysters are characterized by their tendency to encrust rocks and other shells. Oysters are useful 'way-up' indicators. This is because they lie encrusted on one valve, so the other exposed, upward-facing valve acts as a hard surface for its own smaller encrustations such as barnacles. This indicates which way up the oyster – and often fossils associated with it – were in life.

Lopha

The lifestyle and habitat of ancient bivalves are usually derived from analogy to their living relations. The shell shape and form of oysters such as *Lopha*, *Ostrea* and *Rastellum* are often highly irregular and reflect the shape of the surface on which that particular individual grew. In many cases the external features of oyster shells are so variable that recognizing the various species is very difficult, especially among mixed assemblages. As a general rule, internal features, such as muscle scars, are more reliable for identification. Oysters possess only one adductor muscle, which produces large and distinct muscle scars where it attaches to the inside of the valves. A much smaller pair of muscles, the Quenstedt muscles, attach to the gills and leave small muscle scars.

Above: Jurassic Lopha marshi *(Ostrea marshii) grew in a variety of shapes, partly determined by the room available. The curled shape and prominent V-like ridges have led to the common name of coxcomb oyster.*

Left: Lopha marshi *often lived among hardground communities where ageing, lithified sediments were exposed on the sea floor, giving a weathered, cement-like substrate.*

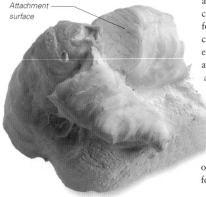

Name: *Lopha*.
Meaning: Crested, peaked.
Grouping: Mollusc, Bivalve.
Informal ID: Oyster, coxcomb (cock's comb) oyster.
Fossil size: 6–7cm/2½in.
Reconstructed size: Less than 10cm/4in.
Habitat: Shallow sea floor.
Time span: Late Jurassic to Late Cretaceous, 160–70 million years ago.
Main fossil sites: Europe, Asia.
Occurrence: ◆ ◆

Gryphaeostrea

Below: The flattened attachment surface present on one of these oysters suggests that they were attached to a large aragonite bivalve, which has not been preserved.

Attachment surface

Most mollusc shells are calcareous, which means that they are built from calcium carbonate minerals. There are two main forms of calcium carbonate: calcite and aragonite. Aragonite is less robust than calcite, and often dissolves away during fossilization. Oysters form their shells from calcite, but many of the groups that they encrust, such as cephalopods, gastropods and other bivalves, have shells formed of aragonite. The result is that fossil oysters are often found apparently unattached, because the shells they were encrusting have not been preserved. However, the encrusting surface of the oyster represents a very faithful mould showing the shape of the surface it was once attached to – often sufficiently detailed for the former host to be identified.

Name: *Gryphaeostrea*.
Meaning: Grabber oyster.
Grouping: Mollusc, Bivalve.
Informal ID: Oyster.
Fossil size: 4cm/1½in.
Reconstructed size: Up to 10cm/4in.
Habitat: Hard objects, such as shell or bone, found on the sea floor.
Time span: Cretaceous to Miocene, 115–10 million years ago.
Main fossil sites: Worldwide.
Occurrence: ◆ ◆ ◆

Gryphea

Name: Gryphea.
Meaning: Grabber.
Grouping: Mollusc, Bivalve.
Informal ID: Oyster, 'devil's toenails'.
Fossil size: Specimen lengths 7–10cm/2¾–4in.
Reconstructed size: As above.
Habitat: Sea floor.
Time span: Late Triassic to Late Jurassic, 220–140 million years ago.
Main fossil sites: Worldwide.
Occurrence: ◆ ◆ ◆ ◆

Various forms of *Gryphea* were among the first true oysters, appearing in the Triassic Period. The tiny free-moving larva, or 'spat', attached to a small particle (rather than a large rock) as its initial hard substrate. It soon outgrew this, however, and became essentially free-living, reclined on the sea floor. Like other bivalves, the valves of the oyster shell are on either side of the animal (even though one usually faces up and the other down) and can be referred to as the left and right valves. For true oysters which encrust, the left valve attaches to the hard substrate. In the free-living adult *Gryphea* the left valve is large and coiled while the right valve is more like a small cap. In life, the animal would have rested on the left valve with the right valve facing upwards. The thick plug of calcite around the umbo (the original, first-formed, beak-like part) of the left valve acted as a counterbalance to keep the opening between the valves raised above the sediment. Many types of *Gryphea* were very successful through the Mesozoic Era.

Pearls
Oysters are famed for containing pearls. These are generated as a defensive substance to 'wall off' particles that enter the shell and irritate the animal. This process can be simulated artificially by inserting shell fragments into farmed oysters. Pearls are sometimes found preserved in fossil oysters. However, the usual mineral replacement, which is part of fossilization, has taken place, and so the pearly lustrous appearance has long been lost.

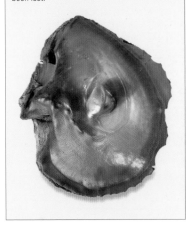

Thick inrolled umbo (first-formed part of valve), acting as a counterbalance

Cap-like right valve

Left: Gryphea *oysters usually show pronounced growth ridges, as the valves of the shell grew seasonally at their widening edges, or margins. Fossils of these early oysters are common enough finds to have entered into folklore – their talon-like shape has earned them the nickname of 'devil's toenails'.*

Growth ridges

Large coiled left valve

Umbo

Area of cap valve

Right: Gryphea *of the Early Jurassic display changes in time whereby the left, or coiled, valve becomes broader and more flattened, while the right, or cap, valve becomes larger and more concave. This gave the animals an overall bowl-like appearance, perhaps representing an adaptation to the sediment on which they were living.*

MOLLUSCS – BIVALVES (CONTINUED)

The morphology (shape and form) of a bivalve shell can say a lot about the owner's mode of life, its orientation within the sediment, depth of burial, the strength of the currents and the layout of the soft body parts within. Many burrowing bivalves have a well-developed, muscular foot. It can extend between the valves, expand at the tip to anchor itself, and then shorten to pull the valves towards it.

Tellina

Tellins are mostly smallish bivalves with a thin-walled, symmetrical, plain shell. They live along sandy and muddy coasts and estuaries today. Many tellins are brightly coloured, although this characteristic is seldom seen in life since they remain buried just under the surface. Like other burrowing bivalves, such as cockles and clams, the tellin extends two muscular tubes, called siphons, which project just above the surface. One takes in water, which is filtered for food and from which oxygen is absorbed. The water then exits through the other siphon. Tellins are sometimes found packed in dense groups with many hundreds in one square metre/yard.

Name: *Tellina*.
Meaning: Of the earth.
Grouping: Mollusc, Bivalve.
Informal ID: Freshwater tellin.
Fossil size: 4cm/1½in.
Reconstructed size:
As above.
Habitat: Rivers, lakes.
Time span: Mainly Early
Tertiary, 50 million years ago,
to today.
Main fossil sites: Worldwide.
Occurrence: ◆ ◆

Left: This specimen of freshwater tellin, Tellina aquitanica, is from Pliocene rocks of Asti, Italy. It shows the growth rings typical of many bivalves. These reflect different rates of shell enlargement through the seasons, with cold weather slowing growth. Tellin valves typically have smooth, 'clean' lines, with little ornamentation. However, their thin and relatively delicate shell structure means that they are among the less common bivalves found as complete fossils.

Modiolus

Life within sediments, such as sand, silt and mud, is known as infaunal. Members of the genus *Modiolus*, which still thrive along shorelines today, are known as horse mussels. When alive, they tend to lie part-buried in the sediment with their generally oblong, slightly lop-sided valves lying parallel to the sediment surface, and with their byssal threads (see *Plagiostoma*) attached to the relatively firm substrate. *Modiolus* tends to grow larger than common mussels (*Mytilus*) and has a thicker, tougher shell. This forms a hard surface for small encrusting organisms such as barnacles and tubeworms, which are themselves sometimes preserved on the mussel.

Name: *Modiolus*.
Meaning: Washer, collar.
Grouping: Mollusc, Bivalve.
Informal ID: Horse mussel.
Fossil size: 7cm/2¾in.
Reconstructed size: As
above, but some specimens
grow to 20cm/8in or more.
Habitat: Sea floor.
Time span: Devonian, 400
million years ago, to today.
Main fossil sites: Worldwide.
Occurrence: ◆ ◆ ◆

Growth
rings

Long hinge line

Tumid central region

Left: Modiolus, the horse mussel, has altered very little in the course of its evolutionary history. Its slipper-shaped valves show growth lines and are unornamented, except for very fine ribs in some specimens that spread out like the ribs of a fan. The overall shell shape is bulging or swollen, mainly in the central region, known technically as a tumid form.

Pholadomya

Name: Pholadomya.
Meaning: Piddock clam.
Grouping: Mollusca, Bivalve.
Informal ID: Gaper,
gaper clam.
Fossil size: 5cm/2in.
Reconstructed size: Up
to 10cm/4in.
Habitat: Sea floor.
Time span: Triassic,
220 million years
ago, to today.
Main fossil sites: Worldwide.
Occurrence: ◆ ◆ ◆

Pholadomya belonged to a group of bivalves called gapers, gaper clams or soft-shelled clams, which still thrive on the sea floor today. They burrow deep into the sediment in order to gain protection and to avoid unfavourable conditions, such as a withdrawing tide. Their shell does not fully close and so is of lesser protective importance. This allows for greatly enlarged siphons. The siphons are paired and part-fused tubes that extend to the surface of the sediment, draw in particles of food and water, and expel the water, after absorbing oxygen, along with any wastes. A large gap between the 'closed' valves shows this deep-burrowing mode of life, and can be seen in fossils. A broad but shallow furrow on the inner surface of the shell, known as the pallial sinus, is also a sign of the enlarged siphons, which, when fully extended, could be up to twice the length of the shell.

*Below: The posterior of
Pholadomya's shell is drawn
out and expanded so that it
can accommodate the greatly
enlarged siphons. In life, this
end would have been orientated
upwards, towards the surface
of the sediment, several
centimetres above.*

Umbo (original growing
point of shell)

Collar-like gap
for siphons

Growth lines

Posterior
(rear) of shell

Teredo

Name: Teredo.
Meaning: Grub.
Grouping: Mollusc, Bivalve.
Informal ID: Shipworm.
Fossil size: Wood specimen
length 12cm/4¾in.
Reconstructed size: Shells
about 1cm/⅖in across; whole
animal may be 50cm/20in
plus in length.
Habitat: Driftwood.
Time span: Eocene, 50
million years ago, to today.
Main fossil sites: Worldwide.
Occurrence: ◆ ◆

Many bivalves have developed the capacity to bore into hard substrates, such as wood or rock, where they are largely protected from predators as well as any hostile elements. The bivalves can employ both physical and chemical means to achieve this goal. The piddock, which is familiar from most rocky shores, rotates its abrasive shell to bore into stone. *Teredo*, erroneously known as the 'shipworm', is a bivalve mollusc that has greatly modified its body for a life within driftwood. Its trace fossils of burrows are familiar in preserved wood. The shell has been reduced to two small rasping plates that abrade the tunnel, and the worm-like body then fills the tube and lines it with calcareous cement. The entrance to the tube has two small trapdoors, termed pallets, that can close to seal off completely the outside world. The shipworm siphons water containing suspended nutrients into its burrow, and it is also able to digest the wood as it bores.

*Below: This piece of fossilized conifer driftwood has
been water-worn to expose the Teredo tubes within,
lined with stony calcium-rich minerals. This example
is from the Eocene deposits of the Isle of Sheppey,
in south-east England, where such specimens are
commonly found.*

Teredo *tunnels*

MOLLUSCS – GASTROPODS

Gastropods include very familiar creatures, such as garden snails and slugs, along with pond and water snails, a huge variety of marine shelled forms, such as seashore winkles, periwinkles, whelks, limpets and abalones, and the sea-slugs. The name 'gastropod' means 'stomach-foot', and comes from the early observation that these animals seem to slide along on their bellies. The group first appeared in the seas towards the end of the Cambrian Period, more than 500 million years ago. Since then, they have spread to freshwater and terrestrial habitats. Their shells are not especially tough but have left huge numbers of fossils, mainly during the Cenozoic Era (from 65 million years ago to the present). For details of the gastropod shell and body, and gastropod classifications, see the following pages.

Athleta

Typically, it is only the shell of a gastropod that is preserved and comes to us in the form of a fossil. The shell is constructed mainly from a calcareous mineral called aragonite, which is unstable over geological time, and is often either replaced by calcite or dissolved away, leaving a void in which a mould fossil may form. *Athleta* was similar in overall shape to the living oyster-drill or sting-winkle, *Ocenebra*. This creature preys on living bivalve molluscs, such as oysters and mussels, scraping out the soft flesh from within. *Athleta* may have had a broad foot, which it used to pull apart the two valves and so gain access to its food.

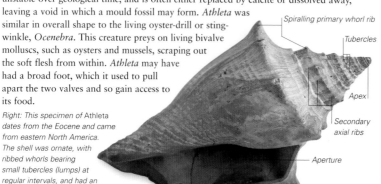

Spiralling primary whorl rib

Tubercles

Apex

Secondary axial ribs

Aperture

Outer lip

Right: This specimen of Athleta dates from the Eocene and came from eastern North America. The shell was ornate, with ribbed whorls bearing small tubercles (lumps) at regular intervals, and had an elongated aperture.

Name: *Athleta*.
Meaning: Contestant, athlete.
Grouping: Mollusc, Gastropod, Prosobranch, Neogastropod.
Informal ID: Sea snail.
Fossil size: Shell length 5cm/2in.
Reconstructed size: Shell length up to 12cm/4¾in.
Habitat: Marine.
Time span: Mainly Late Palaeocene and Early Eocene, 55–45 million years ago.
Main fossil sites: North America, Europe.
Occurrence: ◆ ◆

Potamaclis

Below: In this accumulation, the orientation of tiny Potamaclis specimens records the direction in which the water was flowing at the time of deposition – the tips of the spires point downstream.

Gastropods which have shells with tightly formed whorls that progress rapidly along the main lengthways axis as they grow are known by the common names of spireshells or turretshells. Many types live in estuarine mudflats today. Several species of *Potamaclis* inhabited freshwater environments with flowing water, such as rivers and streams. Their remains are particularly common in Oligocene rocks from the Isle of Wight and Hampshire, southern England. The most prevalent form is known as *Potamaclis turritissima*.

Name: *Potamaclis*.
Meaning: Closed/shut river.
Grouping: Mollusc, Gastropod, Prosobranch, Mesogastropod.
Informal ID: River spireshell, turretshell.
Fossil size: Slab width 16cm/6¼in.
Reconstructed size: Individual shell length up to 3cm/1in.
Habitat: Flowing fresh water.
Time span: Mainly Oligocene, 30–25 million years ago.
Main fossil sites: Europe.
Occurrence: ◆ ◆

Viviparus (including Paludina)

The common freshwater river snail *Viviparus* of today has changed little throughout its long fossil record. It has a thin-walled shell, which is quite brittle, and rounded, bulging whorls with shallow sutures giving a relatively smooth outline when viewed laterally. The shell surface is also smooth, with little in the way of spines or other ornamentation, although in life the young snail has small hair-like processes in bands around each whorl. The aperture shape is almost oval or perhaps slightly heart-like, and the aperture is at right angles to the main axis along which the shell lengthens. The operculum, or closure plate, for the aperture is horny and so rarely forms a fossil. These snails cannot tolerate much in the way of salty (saline) water and so their fossil presence indicates that a body of water was fresh rather than brackish or saline. Like many gastropods, *Viviparus* is mainly a herbivore, grazing on small plants. Most specimens previously identified as *Paludina* are usually now referred to as *Viviparus*: for example, *P. lenta* is synonymous with *V. lentus*, and *P. angulosa* with *V. angulosus*. (See also *Lymnaea* on the following pages, which is another common freshwater snail, usually called the pond snail.)

Apex — Suture — Main whorl

Above: Within the genus *Viviparus* are numerous species, each with a modification on the basic river snail shape. This specimen is the species *Viviparus lentus*, and it comes from Oligocene deposits on the Isle of Wight, southern England.

Right: This accumulation of Viviparus *may have been picked up by a flush of river water and deposited as the current slackened at the river estuary, where salty water would kill the snails.*

Name: *Viviparus*.
Meaning: Bearing live young.
Grouping: Mollusc,
Gastropod, Prosobranch,
Mesogastropod.
Informal ID: River snail, pond snail, live-bearing snail.
Fossil size: 2.5cm/1in.
Reconstructed size: Shell length usually up to 7cm/2¾in.
Habitat: Fresh water including ponds, lakes, slow rivers.
Time span: Jurassic, 200 million years ago, to today.
Main fossil sites: Worldwide.
Occurrence: ◆ ◆

Fossil operculum
Some gastropods in the prosobranch group have an operculum. This is a hard door-like flap or fingernail, which is used for defence after the animal has withdrawn its soft parts into the shell. In some kinds of gastropod, the operculum fits neatly into the aperture itself; in others, it is considerably smaller than the aperture and is drawn part way into the first whorl. When the gastropod emerges from its shell, the operculum usually lies out of the way on the upper surface of the rear foot. The operculum is horny or calcareous. Horny types soon decay leaving no trace, but the calcareous forms may be tougher than the shell and are often preserved as fossils, having characteristic shapes that allow identification.

Right: This turban or wavytop snail Lithopoma undosum *has withdrawn most of its body (partly visible upper right) into the shell. The door-like operculum is the central dark-edged surface with three pale curves; as the snail continues withdrawal, the operculum will fit snugly within the shell.*

MOLLUSCS – GASTROPODS (CONTINUED)

Gastropods are by far the most successful mollusc group, with perhaps 75,000 species alive today, and upwards of 15,000 types known from fossils. The body is soft, covered by a sheet- or cloak-like mantle, with a well-developed head bearing eyed tentacles and a mouth equipped with a file-like radula 'tongue' for rasping food. Characteristic of the group is a twisted body inside a spiral or helical shell.
The twisting, which happens early in development, is known as torsion. It means the mantle cavity and gills (for aquatic gastropods) are positioned above the head rather than behind, as in most other molluscs.

Neptunea

Below: The rounded whorls of Neptunea *are indented by deeper sutures to give a 'stepped spire' appearance. The final whorl where the animal lives, known as the body whorl, is typically convex, or dish-shaped. There are also visible transverse growth lines, usually caused by the seasonal speeding up and then slowing down of the rate at which material was added to the shell opening.*

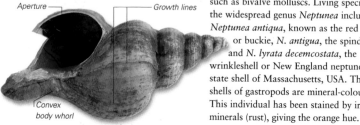

Aperture — — Growth lines

Convex
body whorl

The typical 'sea-shell' we pick up on the beach and hold to our ear, to listen to the sounds of the waves, is likely to be a whelk, or buccinid. This gastropod group includes mainly carnivorous sea snails that scrape up flesh, either scavenging on carcasses of dead crabs and fish or actively preying on victims such as bivalve molluscs. Living species of the widespread genus *Neptunea* include *Neptunea antiqua*, known as the red whelk or buckie, *N. antiqua*, the spindleshell, and *N. lyrata decemcostata*, the wrinkleshell or New England neptune – the state shell of Massachusetts, USA. The fossil shells of gastropods are mineral-coloured. This individual has been stained by iron minerals (rust), giving the orange hue.

Name: *Neptunea*.
Meaning: From the Greek god of the sea.
Grouping: Mollusc, Gastropod, Prosobranch, Neogastropod.
Informal ID: Whelk, sea snail.
Fossil size: Shell length 7cm/2¾in.
Reconstructed size: Shell length up to 10cm/4in.
Habitat: Marine.
Time span: Eocene, about 40 million years ago, to today.
Main fossil sites: Worldwide.
Occurrence: ◆ ◆

Vermetus

Name: *Vermetus*.
Meaning: Worm-like.
Grouping: Mollusc, Gastropod, Prosobranch, Mesogastropod.
Informal ID: Worm-shell, worm-snail.
Fossil size: Slab width 4cm/1½in.
Reconstructed size: Shell length up to 10cm/4in.
Habitat: Marine, mainly tropical and subtropical seas.
Time span: Early Tertiary, about 60 million years ago, to today.
Main fossil sites: Worldwide.
Occurrence: ◆ ◆

The vermetids were, and are, gastropods with a misleading appearance. They resemble the spiral tubes built on rocks and other hard surfaces by true worms (annelids), such as the spirorbid and serpulid worms, known as coiled tubeworms. In *Vermetus*, the whorls separate with growth, so that the inner surface of the later whorl does not contact the outer surface of the preceding one. In some species, the shell lies flat on the substrate; in others, it lengthens as it grows, producing a shape like an enlarging corkscrew. Most living vermetids occupy tropical waters, are cemented to a hard substrate, and secrete mucus that traps food items.

Below: This Vermetus *worm-shell specimen is from the Atherfield Clays at Atherfield Point, Isle of Wight, southern England. The typical vermetid shell gradually coils wider and straighter with age. Also its growth becomes more irregular, leading to many differently shaped individuals.*

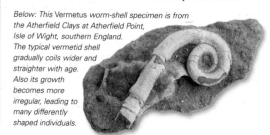

Parts of the gastropod shell
Typically, the gastropod shell is twisted, either in one plane, as a spiral (like a hosepipe coiled on the ground), or as a widening helix, as in the familiar snail. The turns of the shell are called whorls, and the junction lines between them are known as sutures.
The open end is the aperture, and the pointed end is the apex. In many types of gastropod, the whorls form a lengthening spire as they grow. The animal grows from the apex by adding more shell material to the aperture, which is formed by two lips – the outer or free lip and the inner lip based on the previous whorl.

Lymnaea (Limnaea)

Name: *Lymnaea*.
Meaning: Of the marsh.
Grouping: Mollusc,
Gastropod, Pulmonate.
Informal ID: Pond snail, great
pond snail.
Fossil size: Specimen shell
lengths 2–3cm/¾–1in.
Reconstructed size: Shell
length in some species up
to 6cm/2½in.
Habitat: Fresh water, usually
still or slow flowing.
Time span: Late Palaeocene,
55 million years ago,
to today.
Main fossil sites: Mainly
Northern Hemisphere.
Occurrence: ◆ ◆

The great pond snail *Lymnaea stagnalis* is a well-known inhabitant of still fresh waters and grows as big as a thumb. Its thin-walled shell is longer and more spire-like than that of *Viviparus*, which is another freshwater gastropod with the common name of pond or river snail (see previous pages). In particular, the apex of *Lymnaea* is relatively pointed or 'sharp'. This snail is a pulmonate gastropod, having adapted to life on land with a lung-like organ instead of gills, but then returning to the water. Its aperture is flared, especially in older specimens, which often exhibit growth lines. The last, or body, whorl of *Lymnaea* tends to look out of proportion and larger than it should, relative to the size increase of the other whorls.

Below: The genus Lymnaea *has a long fossil record, stretching back to the Palaeocene Epoch, and has been very widespread across northern continents. In these Oligocene specimens the apex characteristically narrows to a sharp point.*

Gastropod classification

For many years, gastropods have been classified according to the following categories:
Prosobranchia, with three main orders:
• Archaeogastropoda: these are sea-living forms such as limpets, abalones and ormers. The twisting is not clear in the adult form, which has a cone-like shape.
• Mesogastropoda: these mostly marine animals include towershells, worm-shells and periwinkles.
• Neogastropoda: this order includes predatory forms, such as whelks and coneshells, as well as tritons. Most fossils fall into the prosobranch group.
Opisthobranchia, the sea-slugs or nudibranchs, which are usually shell-less, soft-bodied and brightly coloured.
Pulmonata, familiar snails and slugs of the garden, which are mostly terrestrial with the gills modified as air-breathing lungs. Some pulmonates have gone back to fresh water and the lung has been further modified for breathing under the surface of the water.
Recent studies have shown that the above groupings do not reflect the evolution and true relationships of the gastropods. A newer cladistic system is being developed based around groups including the Orthogastropoda and Eogastropoda. Most modern systems of biological classification are now based on cladistic analysis. This system (as explained earlier in the book) uses evolutionary relationships and the possession of what are known as derived features, inherited from a common ancestor and which no other organisms possess, to form groups into units known as clades.

Planorbis

Several genera of freshwater snails are known as ram's-horns or ramshorns, from the flattened spiral shell shape that resembles the curled horns of a male sheep when viewed from the side. These genera include *Planorbis*, *Planorbarius* (*Planorbarius corneus* is known as the great ram's-horn), *Gyraulus* and *Anisus*. Today, these snails live mainly in shallow, weedy waters. Like *Lymnaea* (above), the ram's-horns are air-breathing pulmonates. At the surface, they take a bubble of air into the body cavity, which lasts a considerable time when submerged. Some species have sharp ridges or keels around the widest part of the whorl and transverse growth lines are often prominent.

Name: *Planorbis*.
Meaning: Flat spiral or circle.
Grouping: Mollusc,
Gastropod, Pulmonate.
Informal ID: Ram's-horn snail.
Fossil size: Shell diameter
2–4cm/¾–1½in.
Reconstructed size: Shell
diameter in some species up
to 8cm/3⅛in.
Habitat: Fresh water, usually
still or slow flowing.
Time span: Late Palaeocene,
55 million years ago, to today.
Main fossil sites: Worldwide.
Occurrence: ◆ ◆

Right: This species is identified as Planorbis euomphalus, *from Oligocene deposits on the Isle of Wight, southern England. It has a relatively square whorl cross-section showing little curvature on each side, with transverse growth ridges and grooves.*

Mineralized banding

Mould surface

Body whorl

MOLLUSCS – GONIATITES, CERATITES

The Ammonoidea were a huge group of molluscs, mostly with coiled shells, that lived in the seas and oceans. They started to establish themselves in the Early to Mid Devonian Period, from about 400–390 million years ago. Different subgroups came and went, some becoming incredibly numerous and widespread, until the end of the Cretaceous Period, some 330 million years later. Ammonoid subgroups included goniatites and ceratites shown here, and the ammonites. Together with belemnoids and nautiloids, as well as squid, octopuses and cuttlefish, the ammonoids are included in the major cephalopod, or 'head-foot', group of molluscs (see following pages).

Goniatites

The goniatite subgroup of ammonoids takes its name from the genus *Goniatites*, meaning 'angle stones'. This refers to the angled zig-zags in the sutures – the lines showing clearly at intervals on the outside of the shell. Sutures mark the sites where the coiled part of the shell is joined internally to a series of thin, dividing cross-wall partitions, known as septa. These septa separate the inside of the shell into a succession of compartments, which become larger as the animal grows. The creature itself lived in the last, latest compartment. The septa were rarely simple, flat cross-walls. They often had complex wave-like angles and undulations, like large pieces of paper pushed into the shell's interior, where they folded and crumpled as they became wedged in. The pattern of septal folding can be discerned in fossils broken open and its edging shows on the outside as the suture. This pattern is the same for all the septa within an individual shell and for all members of a species, but it differs between species and genera. So the patterns of both sutures and septa can be used to distinguish the thousands of ammonoid species.

The goniatite group, which were among the first ammonoids to appear, were present in the seas for 170 million years. However, they suffered a major decrease in numbers at the end of the Permian Period, 250 million years ago, with just a few types surviving through to the following Triassic Period. The specimen shown here, like many in the genus *Goniatites*, has a relatively wide shell when looked at 'end-on' – that is, directly into the opening (at right angles to the view in this photograph). This wide, fat shell suggests that the animal was a poor swimmer. Faster, more manoeuvrable ammonoids had a narrow shell when viewed end-on. *Goniatites* probably lived in groups or swarms on midwater reefs.

Name: *Goniatites*.
Meaning: Angle stone.
Grouping: Mollusc, Cephalopod, Ammonoid, Goniatite.
Informal ID: Goniatite.
Fossil size: 4.5cm/1¾in across.
Reconstructed size: Some species up to 15cm/6in across.
Habitat: Seas, probably reefs.
Time span: Early Carboniferous, 355–325 million years ago.
Main fossil sites: Worldwide, mainly Northern Hemisphere.
Occurrence: ◆ ◆ ◆

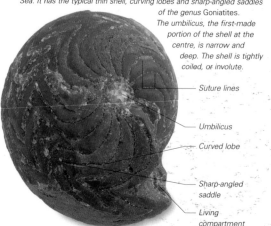

Below: This specimen of Goniatites crenistria *is from the Carboniferous Poyll Vaaish Limestones on the shores of the Isle of Man in the Irish Sea. It has the typical thin shell, curving lobes and sharp-angled saddles of the genus* Goniatites. *The umbilicus, the first-made portion of the shell at the centre, is narrow and deep. The shell is tightly coiled, or involute.*

— Suture lines

— Umbilicus

— Curved lobe

— Sharp-angled saddle

— Living compartment

The importance of sutures

Ammonoids are immensely important as marker or index fossils for dating Late Palaeozoic and Mesozoic rocks and characterizing various kinds of marine communities. Much of their identification relies on the patterns of sutures – the wavy or angled lines on the shell. For how these relate to the inner dividing walls, or septa, see the main text for *Goniatites*.
• The parts of the suture curving or angled towards the open, head end are known as lobes.
• Those directed towards the shell's diminishing centre or umbilicus are called saddles.
• The contours of the suture reflect the folding, waving and crumpling of the edges of the septum, the dividing partition within the shell.
• Goniatites, in general, were early ammonoids with relatively direct sutures, which consisted of curves and sharp angles, but neither lobes nor saddles were subdivided.
• Ceratites were slightly later and had subdivided lobes, but undivided saddles.
• Ammonites were the third major ammonoid group to appear. Their sutures are overall more complex, with both lobes and saddles subdivided in an enormous variety of patterns.

Cheiloceras

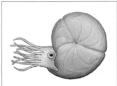

Name: Cheiloceras.
Meaning: Lip horn.
Grouping: Mollusc,
Cephalopod, Ammonoid,
Goniatite.
Informal ID: Goniatite.
Fossil size: 1.5cm/½in across.
Reconstructed size: As
above, some specimens
3–4cm/1–1½in across.
Habitat: Seas.
Time span: Late Devonian,
375–360 million years ago.
Main fossil sites: Worldwide.
Occurrence: ◆ ◆ ◆

*Below: These specimens of
Cheiloceras were preserved in iron
pyrite. They are small – each would fit
on a thumbnail – and show the widely
spaced sutures characteristic of
several species in the genus. They are
from Famennian Age deposits near
Gladbach, along the River Mulde in the
Sachsen region of eastern Germany.*

This well-known, small, geometrically shaped,
involute (tightly coiled), distinctive goniatite – one of the
earlier types in the group – has been found at several
locations, including Europe (Iberian Peninsula, France
and Germany), North Africa (Morocco), East Asia
(China) and Australia (Western Australia, New South
Wales). It is a useful marker or index fossil and lends its
name to the Cheiloceras beds or deposits typical of the
time span known as the Famennian Age of the Late
Devonian Period, which is dated to approximately
375–360 million years ago. In some localities it has
been suggested that *Cheiloceras* is also found in the
Late Frasian Age, the time span before the
Famennian. Cross-matching *Cheiloceras*
and other fossils
found with it
in these layers
has allowed
the detailed
correlation
of rocks
from around
the world.

Ceratites

The ceratites, the second main group of
ammonoids to evolve, appeared during the
Mississippian – the first phase of the
Carboniferous Period – about 350 million years
ago. They became much more numerous and
diverse during the Early Triassic, after the
goniatites had faded, but then went extinct
themselves by the end of that period.

*Hollandites was a Mid Triassic ceratite. This
specimen hails from the McLearn Toad
Formation along the Tetsa
River of northern British
Columbia, Canada.
Note the subdivided
lobes – the curved
parts of the sutures
facing towards the
main opening.
Specimen measures
3cm/1in across.*

*Right: Identified as Gastrioceras listeri, this fossil is
from the Carboniferous 'Coal Measures' of Yorkshire,
north-east England. The genus Gastrioceras was
common in the shallow, warm seas around the
tropical freshwater swamps and marshes of the
time. The umbilicus, or central, first-grown part of
the shell is wide and fairly deep. The venter –
meaning the ventral surface, which is exposed in the
final whorl – is wide and relatively flattened.*

Gastrioceras

This Carboniferous goniatite shows the
beginnings of shell ornamentation that
would become more complex among the
ceratites and reach elaborate forms in
the ammonites of the Mesozoic Era. The
main ornamentation of *Gastrioceras* consists
of lumps, or tubercles, on the side flanks,
towards the dorsal surface – the inward- or
centre-facing surface of an outer whorl (turn
of the shell), where it is fused to the ventral,
or outward-facing, surface of the whorl
within it. In some specimens, the tubercles
narrowed to form ribs
that passed around
the curve of the
shell to the
tubercles
situated
on the
opposite
side.

Name: Gastrioceras.
Meaning: Stomach horn.
Grouping: Mollusc,
Cephalopod, Ammonoid,
Goniatite.
Informal ID: Goniatite.
Fossil size: 3cm/1in across.
Reconstructed size: Up to
6cm/2½in across.
Habitat: Seas.
Time span: Carboniferous,
340–295 million years ago.
Main fossil sites: Worldwide,
mainly Northern Hemisphere.
Occurrence: ◆ ◆ ◆

Venter

Umbilicus

Tubercles

MOLLUSCS – AMMONITES

The ammonites' name comes from the shell's resemblance to a coiled ram's horn (the ram being the symbol of the Egyptian god Ammon). Like the extinct belemnites and the living nautiloids, octopuses, squid and cuttlefish, they belonged to the mollusc group known as the cephalopods. This name comes from the Greek 'kephale' (head) and 'podia' (foot), referring to the tentacles they have in the head region.

'Ammonite jaws'

Above: These 'ammonite jaws', or aptychi, date from the Late Jurassic Period. An aptychus is only rarely found alongside the ammonite of which it was part, and to which it can be attributed, so most specimens are usually given their own scientific names. This is Laevaptychus – *the 'smooth aptychus'.*

Palaeontologists do not agree on the true function of these common wing-shaped fossils, also known as aptychi (singular aptychus). They are usually interpreted as being part of the 'jaw'-like mouthparts of ammonites, other ammonoids and nautiloids.

Aptychi are abundant and often well preserved as fossils. But they are usually found separated from the main shells of their owners. For this reason, only rarely can palaeontologists match the aptychus with the ammonite species to which it belonged. Other theories regarding their identity have suggested that they were the preserved shells of creatures in their own right, perhaps the valves of bivalve molluscs, such as clams and oysters, or the operculum, or 'door', that closed the shell opening in many other kinds of molluscs, such as gastropods.

Name: *Laevaptychus.*
Meaning: Smooth fold.
Grouping: Mollusc, probably Cephalopod, Ammonoid, Ammonite.
Informal ID: Ammonite jaws.
Fossil size: Each item 3cm/1in across.
Reconstructed size: Unknown.
Habitat: Sea.
Time span: Mainly Mesozoic, 250–65 million years ago.
Main fossil sites: Worldwide.
Occurrence: ◆ ◆ ◆ ◆

Phylloceras

Phylloceras is one of the earlier ammonites and is moderately commonly found in Northern Hemisphere deposits dating from the Early Jurassic Period, some 185–180 million years ago. *Phylloceras* is thought to be ancestral to the psiloceratid ammonites typified by *Psiloceras* (see opposite), and thus to the later very successful and diverse Jurassic ammonites. It is an involute form – which means that the spiral is tightly coiled and wraps over itself – and of small-to-medium size. Because of its compressed, smooth profile and rounded venter (the external convex, or 'belly' part, of the shell), it is thought *Phylloceras* swam about with the aid of jet propulsion, in a manner perhaps similar to the living *Nautilus*.

Left: Phylloceras *gets its name 'leaf-horn' from the extremely complex and frilly looking lobed, leaf-like or spatulate sutures, or joint lines. This mould fossil formed when the inside of the shell was filled by sediments, and clearly shows the internal suture pattern.*

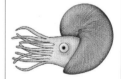

Name: *Phylloceras.*
Meaning: Leaf horn.
Grouping: Mollusc, Cephalopod, Ammonoid, Ammonite.
Informal ID: Ammonite.
Fossil size: 10cm/4in across.
Reconstructed size: As above.
Habitat: Shallow seas.
Time span: Late Triassic to Early Cretaceous, 210–120 million years ago.
Main fossil sites: Northern Hemisphere.
Occurrence: ◆ ◆ ◆

Psiloceras

Name: *Psiloceras*.
Meaning: Smooth horn.
Grouping: Mollusc,
Cephalopod, Ammonoid,
Ammonite.
Informal ID: Ammonite.
Fossil size: Slab 35cm/
13¾in long.
Reconstructed size:
Individuals up to 7cm/
2¾in across.
Habitat: Shallow seas.
Time span: Early Jurassic,
200 million years ago.
Main fossil sites: Worldwide.
Occurrence: ◆ ◆ ◆

As one of the first ammonites to appear in the shallow seas of the Jurassic Period, some 200 million years ago, *Psiloceras* was probably ancestral to most of the later Jurassic ammonites. *Psiloceras* is the first widely distributed ammonite and provides a valuable insight into the extent of those shallow sea areas during Early Jurassic times. The beautiful specimens shown here have been fossilized in the form of ammolite, which is an iridescent, opal-like gemstone. Certain mollusc shells in life are made of a type of calcium carbonate called aragonite. In most cases, this material is replaced by other minerals as part of the fossilization process. In this example, however, impermeable clay sediments must have covered the shells before the replacement process could begin, thus allowing the original aragonite to be preserved as ammolite.

Below: The shale that flattened these small ammonites also allowed for their beautiful preservation. Over time, the surrounding sediments impregnated the nacreous (iridescent 'mother-of-pearl') shells with trace elements, such as iron and magnesium, and this accounts for the bright red and green colours.

Stephanoceras

Name: *Stephanoceras*.
Meaning: Crown horn.
Grouping: Mollusc,
Cephalopod, Ammonoid,
Ammonite.
Informal ID: Ammonite.
Fossil size: 6cm/2⅓in across.
Reconstructed size:
As above.
Habitat: Shallow seas.
Time span: Middle Jurassic,
160 million years ago.
Main fossil sites: Worldwide.
Occurrence: ◆ ◆ ◆

Stephanoceras was an ammonite of the shallow seas of the Mid Jurassic Period, some 160 million years ago. It had a disc-shaped, regularly spiralled shell, with a strongly ribbed pattern. Deep ribs such as those shown by many ammonites may have contributed to the strength of the shell, thus providing them with protection against predators such as fish, marine reptiles and their larger cephalopod cousins. There are many species in the genus *Stephanoceras*, and it is also widely distributed, with fossils having been found around the world. Specimens have been discovered from Europe and North Africa to the Andes, and sometimes in great abundance.

Right: The shell of Stephanoceras has a rounded profile about the venter (the external convex, or 'belly' part, of the shell), rather than being drawn into a sharp keel, which suggests that it probably had a floating habit in life. The prominent ribs would have created drag, however, preventing it from being a fast swimmer. This specimen is from Sherbourne in Dorset, southern England.

Prominent ribs

Venter has rounded profile

Umbilicus (first-formed central part of shell)

Younger, wider whorls of shell overlap older, smaller ones

Aperture

MOLLUSCS – AMMONITES (CONTINUED)

Cephalopod molluscs, as well as being ecologically important today, also have a fossil record going back to the beginning of complex animal life during the Cambrian. Ammonites dominated most marine environments for a period of 300 million years, from the Late Silurian to the end of the Cretaceous about 65 million years ago, when they died out with many other groups in the mass extinction of the time.

Caloceras

Above: Caloceras was an iridescent ammonite that grew in a regular, evolute spiral (its whorls are exposed and not enveloped by later ones). It has prominent ribs and is a well-known zone fossil indicating the Early or Lower Jurassic Period.

Beautiful iridescent ammonite specimens, such as the one pictured here, are much sought after by fossil collectors. Unfortunately, this demand has had a negative effect, since unscrupulous individuals have been lured by the high prices that good specimens can command. The dark shale surrounding *Caloceras* accounts for its bright colours, as described for *Psiloceras* (see previous page). The shale covered the shell in an impermeable layer before the original aragonite (a carbonate mineral) was replaced by minerals in the water, which then eventually impregnated the shell with iron, magnesium and other trace elements present in the sediment. *Caloceras* is important as a zone fossil for the Hettangian Age of the Early Jurassic Period, dating back some 200 to 196 million years ago.

Name: *Caloceras*.
Meaning: Beautiful horn.
Grouping: Mollusc, Cephalopod, Ammonoid, Ammonite.
Informal ID: Ammonite.
Fossil size: 10cm/4in across.
Reconstructed size: As above.
Habitat: Shallow coastal areas.
Time span: Early Jurassic, 200–196 million years ago.
Main fossil sites: Throughout Europe.
Occurrence: ◆ ◆ ◆

Lytoceras

Lytoceras has an evolute growth pattern, meaning that all of its whorls are exposed and not enveloped or grown over by later, younger ones. The thin, often widely separated ribs are characteristic of its group, the Lytoceratida. By observing the flow of water around models of ammonite shells in flow tanks, researchers have shown that fine ribs such as these reduce drag – the resistance of water to objects moving through it. The ribs, therefore, make the animal's swimming motion more efficient and ease the effort of moving a bulky shell through the water.

Minimal contact between successive whorls

Left: In water tank tests, the low, thin ribs ornamenting the shell of Lytoceras have been shown to reduce drag and so increase swimming efficiency. The evolute growth pattern leaves almost all of the earlier, older whorls exposed. The name 'unbound horn' refers to specimens where the coils are hardly linked or joined to each other.

Reduced ribbing

Name: *Lytoceras*.
Meaning: Loose or unbound horn.
Grouping: Mollusc, Cephalopod, Ammonoid, Ammonite.
Informal ID: Ammonite.
Fossil size: 10cm/4in across.
Reconstructed size: As above.
Habitat: Deep water.
Time span: Early Jurassic to Late Cretaceous, 200–70 million years ago.
Main fossil sites: Worldwide.
Occurrence: ◆ ◆ ◆

Liparoceras

Name: *Liparoceras*.
Meaning: Smooth or oily horn.
Grouping: Mollusc, Cephalopod, Ammonoid, Ammonite.
Informal ID: Ammonite.
Fossil size: 10cm/ 4in across.
Reconstructed size: As above.
Habitat: Shallow seas.
Time span: Early Jurassic, 185 million years ago.
Main fossil sites: Europe.
Occurrence: ◆ ◆ ◆

The Liparoceratidae group is one of the best-studied examples of lineages, or lines of descent, among the ammonite fossils. This is because the members are very abundant and completely span the Pliensbachian Age of the Early Jurassic Period, usually dated from 192 or 190 to 183 million years ago. During this time, the typical shell shape changes from an evolute spiral (where whorls are loosely coiled and do not cover each other) to an involute spiral (where later whorls envelop, or grow over, the earlier ones). *Liparoceras* is also interesting in studies concerning evidence for ammonite sexual dimorphism. This is the difference in size and shape between males and females of the same species. In some ammonites, the female form is larger than the male, and it or its shell is termed the macroconch. Its fossils can be as much as a hundred times more abundant than the male form, or microconch, with its smaller shell.

Above: This specimen of Liparoceras *is* L. contractum, *from the coast of Charmouth, Dorset, in southern England. It has the involute shell pattern. The smaller male form, or microconch, of* Liparoceras *is sometimes called by a different scientific name, Aegoceras.*

Amaltheus

Name: *Amaltheus*.
Meaning: From Amalthea, ancient Greek symbol for 'horn of plenty'.
Grouping: Mollusc, Cephalopod, Ammonoid, Ammonite.
Informal ID: Ammonite.
Fossil size: 6cm/2½in across.
Reconstructed size: As above.
Habitat: Hardground sea bottoms.
Time span: Early Jurassic, 185 million years ago.
Main fossil sites: Throughout Europe.
Occurrence: ◆ ◆ ◆

Amaltheus is one of the few ammonites from the fossil record in which colour has occasionally been preserved, in the form of longitudinal brown stripes on a white background. The thin disc shape and narrow, sharp keel (the outer edge of the shell spiral) suggest that it was a fast swimmer. *Amaltheus* lived in 'hardground' marine environments – cemented (hardened) sea beds recognized by the presence of borings and incrustations. Cemented beds or bottoms came about when the settling or deposition of sediment occurred very slowly. Dissolving of the molluscan aragonite shells on the sea floor was followed by calcium carbonate coming out of solution into the sediment. This event acted as a cement to bind the general sea floor material into a type of concrete.

Right: Amaltheids are an extremely polymorphic group, meaning that the different species vary greatly in shape. This is Amaltheus subnodosus, *from Early Jurassic rocks near Goppingen in West Germany. It has almost straight ribs that branch near the outer part of the whorl to give a comb or rope-like effect to the keel.*

Sharp comb-effect keel   Ribs   Involute form

Coloured stripes

MOLLUSCS – AMMONITES (CONTINUED)

Ammonites were at their most diverse during the Mesozoic, which corresponds roughly on land to the Age of Reptiles and especially dinosaurs. Ammonites have been extensively studied and are so well known that they are often used as symbols of the fossil record. They are particularly important as index fossils used to identify geologic periods because of their widespread distribution and rapid evolution.

Austiniceras

Faintly S-shaped interval ribs

Flattened sides

Rounded venter

Ammonites ranged from smaller than a fingernail, even when fully grown, to as large as a dining table. *Austiniceras* was one of the bigger types, with some specimens more than 2.5m/8¼ft across. It is a relatively rare find from the Late Cretaceous Period, mainly in Britain and other European sites. *Austiniceras* has an evolute spiral shell, which may be smooth or lightly ribbed, with slightly more pronounced interval ribs. Its sides are almost flat or gently convex. The narrowly rounded venter suggests that it led an active swimming lifestyle despite its great size.

Light intermediate ribs

Relatively evolute (loose-coiled) shell

Left: This small specimen of Austiniceras austeni, from the Late Cretaceous Lower Chalk of Sussex, England, is about 15cm/6in in diameter. But examples of Austiniceras have been known to grow to more than 2m/6½ft across. The generally flattened shell shape (from side to side), low ribs and narrowly rounded venter suggest an active lifestyle, 'jetting' through the water in the manner of a modern squid.

Name: *Austiniceras*.
Meaning: Austin's horn.
Grouping: Mollusc, Cephalopod, Ammonoid, Ammonite.
Informal ID: Ammonite.
Fossil size: 15cm/6in across.
Reconstructed size: Some specimens exceed 2m/6½ft in diameter.
Habitat: Shallow seas.
Time span: Late Cretaceous, 95–70 million years ago.
Main fossil sites: Throughout Europe.
Occurrence: ◆ ◆

Dactylioceras

Snakes of the sea
Evolute ammonites were once believed to be coiled snakes turned to stone. The shell opening was often carved to resemble a snake's head. The fossils were then sold to pilgrims as serpents that had been petrified by a local saint.

Dactylioceras is a common find in Jurassic bituminous shales. These shales formed when limited water circulation allowed stagnant (still, oxygen-poor) conditions to develop in dense sediments on the sea floor. This was favourable for preservation of ammonites and other shells in various ways. The impermeable nature of the sediment prevented the shell's structure of aragonite material from dissolving away. In addition, the stagnant conditions encountered by the shells when they sank to the bottom meant that burrowing animals or currents would not disturb them as the fossilization process occurred. Several individuals are preserved in the block shown here, discovered in Germany. This suggests that *Dactylioceras* had gregarious (group-living) habits. Possibly, like many modern cephalopods, such as squid, they congregated in large swarms or schools to breed.

Name: *Dactylioceras*.
Meaning: Finger horn.
Grouping: Mollusc, Cephalopod, Ammonoid, Ammonite.
Informal ID: Ammonite.
Fossil size: Slab length 15cm/6in.
Reconstructed size: Individuals 2–5cm/ ¾–2in across.
Habitat: Open sea.
Time span: Early Jurassic, 200–175 million years ago.
Main fossil sites: Worldwide.
Occurrence: ◆ ◆ ◆

Left: The shell of Dactylioceras is evolute in form, rather than the larger whorls enveloping the smaller, older ones. The ribs branch towards the outside of the whorls to give a braid-edge effect.

Hamites

Name: Hamites.
Meaning: Hook-like.
Grouping: Mollusc, Cephalopod, Ammonoid, Ammonite.
Informal ID: Uncoiled or heteromorph ammonite.
Fossil size: Length 6cm/2⅓in.
Reconstructed size: Length 8cm/3⅛in.
Habitat: Relatively deep sea floor.
Time span: Middle Cretaceous, 110–100 million years ago.
Main fossil sites: Africa, Eurasia, North America.
Occurrence: ◆ ◆ ◆

Not all ammonites or ammonoids have the familiar coiled or spiral shell form that is seen in so many specimens. *Hamites* is a heteromorph, or uncoiled type. It lived alongside *Euhoplites* (see below) in a soft-bottom marine community that has been preserved in detail in Gault Clay, found in Kent, England. In addition to ammonite specimens, other molluscs discovered in the Gault Clay include nautiloids, such as *Eutrephoceras*; gastropods, such as *Nummocalcar* (which is superficially ammonite-like); and the carnivorous snail *Gyrodes* (which fed on the abundant bivalves by boring small circular holes in their shells with its specialized file-like 'tongue', or radula). Other creatures found at this English site include echinoderms, such as the crinoid *Nielsenicrinus*, and crustaceans, such as the lobster *Hoploparia*. (See also *Macroscaphites* on the following page.)

Below: This is a section of the heteromorph ammonite Hamites, from the Gault Clay of Folkestone in Kent, south-eastern England. It is an internal mould, showing the pattern of joint, or suture, lines on the shell's inner surface. Many Gault Clay specimens of the actual shells retain their original aragonitic material and still show the nacreous 'mother-of-pearl' lustre.

Younger end (nearer aperture)

Suture moulds

Euhoplites

Name: Euhoplites.
Meaning: Good Hoplite (a heavily armoured ancient Greek soldier), from the related genus *Hoplites*.
Grouping: Mollusc, Cephalopod, Ammonoid, Ammonite.
Informal ID: Ammonite.
Fossil size: 3cm/1¼in across.
Reconstructed size: As above.
Habitat: Probably sea floor.
Time span: Middle Cretaceous, 110–100 million years ago.
Main fossil sites: Throughout Europe.
Occurrence: ◆ ◆ ◆

Euhoplites is a strongly ribbed example of an ammonite, with the ribs giving way to tubercles (lumps) on the inside of the curvature and with a double-row of ribs around the outside at the venter (the external convex, or 'belly' part, of the shell). *Euhoplites* is a common ammonite in the Gault Clay of Folkestone, as described above for *Hamites*. From the assemblage of fossils, experts can describe a community of animals that lived together and were preserved in their original environment – a relatively deep, calm marine location. *Euhoplites* lived alongside other ammonites, such as the tiny *Hysteroceras*, just 1.8cm/¾in in diameter, as well as *Hamites*. Other animals preserved include the belemnite *Neohibolites*, and there is also evidence for sharks in the form of fossil sharks' teeth.

Right: Euhoplites is almost pentagonal (five-sided) in cross-section and slightly evolute, with a distinctive deep and narrow ventral groove. As with most ammonites, the creature itself lived in the last-formed, largest chamber and could withdraw into this for safety. The pronounced ribs probably helped to strengthen the shell against predator attack, but they would have caused drag, reducing swimming speed.

Strong ribs

Inner tubercles

Inflated (rapidly widening) form

Ventral groove between tubercles

Offset double row of keel tubercles

MOLLUSCS – AMMONITES (CONTINUED)

Ammonites are usually reconstructed with large eyes and long tentacles similar to a squid. Some are assumed to have been speedy, predatory swimmers. The thousands of types are distinguished by features such as the coiling of the shell – from evolute to involute – as well as the shell's width in its coils and umbilicus (the central first-formed region), zigzag lines, surface ribs and other ornamentations.

Hyphoplites

Umbilicus

Fine inner ribs

Broad outer ribs

Hyphoplites is a beautiful, elaborately sculptured ammonite from the Cenomanian Age of the Mid to Late Cretaceous Period, around 95 million years ago. Sculpturing like this probably contributed to strengthening of the shell. However, these deep ribs and tubercles (bumps), along with the flattened venter ('belly' region of the shell) suggest this ammonite was not an active swimmer. Instead, it probably led a more sluggish lifestyle, on or near the sandy sea bottom, where its remains were preserved. *Hyphoplites* is an important component of the sand communities of its time, along with echinoderms, bryozoans, other ammonites and the strange coiled oyster *Exogyra*.

Left: Hyphoplites *has a mix of pronounced, wide ribs with large and small tubercles (bumps) around the outer part of the whorl, and fine oblique (diagonal) ribs along the inner region of the whorl. Its umbilicus (central, oldest part of the shell) is wide and shallow.*

Name: *Hyphoplites*.
Meaning: Under Hoplite (a heavily armoured ancient Greek soldier), from the related genus *Hoplites*.
Grouping: Mollusc, Cephalopod, Ammonoid, Ammonite.
Informal ID: Ammonite.
Fossil size: 5cm/2in across.
Reconstructed size: Up to 10cm/4in across.
Habitat: Sandy sea floor.
Time span: Middle to Late Cretaceous, 95 million years ago.
Main fossil sites: Throughout Europe.
Occurrence: ◆ ◆

Macroscaphites

Macroscaphites was an unusual-looking heteromorph (not completely spiral-shaped) ammonite from the Early Cretaceous Period, mainly in Europe. Heteromorph ammonites (see also *Hamites* on the previous page) have long puzzled palaeontologists. Early researchers thought that the strange shapes were the result of 'evolutionary decadence', when features become more exaggerated and outlandish, usually just prior to extinction – which, for the ammonites, would have been at the end of the Cretaceous Period. However, it is now known that such ammonites had a long history that began at least in the Early Triassic and lasted until the very end of the Cretaceous. They coexisted with normally coiled ammonites and were even more numerous than coiled types in some habitats.

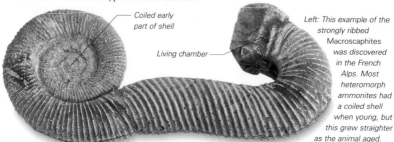

Coiled early part of shell

Living chamber

Left: This example of the strongly ribbed Macroscaphites *was discovered in the French Alps. Most heteromorph ammonites had a coiled shell when young, but this grew straighter as the animal aged.*

Name: *Macroscaphites*.
Meaning: Large hoe or shovel.
Grouping: Mollusc, Cephalopod, Ammonoid, Ammonite.
Informal ID: Uncoiled ammonite.
Fossil size: Length 10cm/4in.
Reconstructed size: Length as above.
Habitat: Calm seas.
Time span: Early Cretaceous, 135–95 million years ago.
Main fossil sites: Europe, North America.
Occurrence: ◆ ◆ ◆

Ludwigia

Name: *Ludwigia*.
Meaning: After Ludwig.
Grouping: Mollusc,
Cephalopod, Ammonoid,
Ammonite.
Informal ID: Ammonite.
Fossil size: 11cm/4⅓in across.
Reconstructed size:
As above.
Habitat: Shallow seas.
Time span: Middle Jurassic,
175–165 million years ago.
Main fossil sites: Worldwide.
Occurrence: ◆ ◆ ◆

Ludwigia was a stout-whorled, involute ammonite with a broad, strongly keeled venter ('belly' region of the shell). The genus *Ludwigia* is important in stratigraphy – the study of the deposition and age of sedimentary rock strata (layers or beds). Ammonites of this genus are found abundantly in the Murchison beds of the Swiss Jura Mountains – the location after which the Jurassic Period was originally named. Finding the same species in other locations helps to date and correlate the rock layers to the Aalenian Age of the Jurassic, about 175 million years ago.

Left: This specimen of Ludwigia murchisonae *is from Dorset, southern England. The ammonite displays low, S-shaped ribs and the keeled venter.* Ludwigia *is named in honour of Christian Gottlieb Ludwig (1709–1773), German professor of natural history and medicine.*

— Umbilicus

— Evolute form (younger, larger whorls overlap older, smaller ones)

— Keeled venter

— S-shaped ribs

Predator and prey

Active and fast-swimming ammonites such as *Macrocephalites* and *Cardioceras* (9cm/3½in across, in the specimen shown here) usually had a moderately compressed form – in other words, the shell was flattened from side to side. They also had a strongly carinated venter, which means that the 'belly' part of the shell was drawn into a ridged or sharpened keel. The ridges in the shell may have helped reduce drag in the water, making them more efficient swimmers. Fossils associated with *Cardioceras* show other, larger marine reptiles such as ichthyosaurs, plesiosaurs and marine crocodiles. Bite marks are sometimes found on preserved ammonite shells. The size, spacing and pattern of the indentations can sometimes be attributed to a specific predator.

Macrocephalites

Macrocephalites had quite a wide and bulky shell covered in thin ribs, which branched into two (bifurcated) around each side. As with other fine-ribbed ammonites, this ornamentation would reduce drag and help the animal move its cumbersome shell through the water. *Macrocephalites* has been found in fossil assemblages all over the world, including the Ellsworth Mountains in Antarctica. It is an important index fossil for the Callovian Age of the Middle Jurassic Period, especially in Northern Europe from Britain to southwest Germany. The Callovian Age is dated at about 160 million years ago.

Name: *Macrocephalites*.
Meaning: Like a large head.
Grouping: Mollusc,
Cephalopod, Ammonoid,
Ammonite.
Informal ID: Ammonite.
Fossil size: 6cm/2⅓in across.
Reconstructed size: Some
specimens up to 2m/6½ft in
diameter.
Habitat: Shallow seas.
Time span: Middle Jurassic,
165–160 million years ago.
Main fossil sites: Worldwide.
Occurrence: ◆ ◆ ◆

Right: This specimen of Macrocephalites, *from Porta Westfalica, near Hanover, Germany, could easily fit into the palm of the hand. The genus could, however, reach gigantic proportions – in fact, almost 2m/6½ft in diameter. The fine ribbing pattern probably aided in swimming. The shell is extremely involute, with each coil overlapping and obscuring the previous, older one.*

— Rib bifurcations

— Extreme involute form (outer whorls obscure preceding ones)

MOLLUSCS – NAUTILOIDS

The great cephalopod ('head-foot') group of molluscs encompasses ammonoids, actinoceratoids, bactritoids, belemnoids, the living squid and octopus and cuttlefish – and nautiloids, both living and extinct. The surviving genus Nautilus *includes the Indo-Pacific pearly nautilus,* Nautilus pompilius, *plus four or five other extant species in two genera,* Nautilus *and* Allonautilus. *They are the only living cephalopods with a true external shell, and sole remnants of a once enormous mollusc group. Knowledge of their soft-tissue anatomy and mode of life allows detailed reconstructions of fossil nautiloids and their marine communities.*

Orthoceras

Below: The shell chambers of Orthoceras are known as concavo-convex, having a concave, or in-curved, surface facing to the rear (older end of the shell) and a convex, or out-bulging, surface directed to the front (the living chamber end). The structure is often likened to a row of contact lenses. This species, identified as Orthoceras canaliculatum, *is from the Silurian Wenlock Beds of a quarry near Cardiff, Wales.*

This well-known nautiloid is extremely abundant in certain European localities, especially from the Middle to Late Ordovician Period. The first-formed portion of the shell was rounded. The shell then grew straight but gradually increased in diameter to form a gently widening cone shape. In nautiloids, belemnoids and similar cephalopods the chambered portion of the shell is termed the phragmocone, as opposed to the large living compartment where the main body of the creature is located and which ends at the aperture. The siphuncle (inner tube passing through the chambers) was centrally placed. In life, *Orthoceras* probably swam with its shell horizontal, parallel to the sea bed, as it chased small prey or scavenged across the ocean floor.

Name: *Orthoceras*.
Meaning: Straight horn.
Grouping: Mollusc, Cephalopod, Nautiloid.
Informal ID: Nautiloid.
Fossil size: Length 12cm/4¾in.
Reconstructed size: Length up to 20cm/8in.
Habitat: Seas.
Time span: Ordovician–Silurian, 460–430 million years ago.
Main fossil sites: Throughout Europe.
Occurrence: ◆ ◆ ◆

Eutrephoceras

Remains of this nautiloid come mainly from Late Cretaceous deposits found in the Midwest of the USA. At that time, the region was covered by a shallow sea, the Western Interior Seaway, which harboured not only nautiloids, but also a vast array of other marine life, including the ferocious sea-going reptiles known as mosasaurs (ancient relatives of today's monitor lizards). Many nautiloid and ammonoid shells bear grooves or cracks that correspond well with mosasaur tooth shape and spacing. The shell of *Eutrephoceras* was involute – that is, tightly coiled – and rapidly enlarging in cross-sectional area. The central, or first-formed, part of the shell known as the umbilicus, in the middle of the spiral, is likewise narrow.

Name: *Eutrephoceras*.
Meaning: Well fed horn.
Grouping: Mollusc, Cephalopod, Nautiloid.
Informal ID: Nautiloid.
Fossil size: Shell 7cm/2¾in across.
Reconstructed size: Diameter up to 12cm/4¾in.
Habitat: Shallow seas.
Time span: Late Cretaceous, 75–65 million years ago.
Main fossil sites: North America.
Occurrence: ◆ ◆

Umbilicus hidden within matrix

Living chamber

Venter (rounded ventral surface)

Left: Identified as Eutrephoceras dekayi, *this nautiloid from the Fox Hills Formation of South Dakota, USA, is dated to the Maastrichtian Age, about 70–65 million years ago – which was the last time span of the Cretaceous Period. It has a wide shell with a rounded shape, like a partial doughnut. The smaller scattered shells resembling mussels or clams are not molluscs but brachiopods.*

Cenoceras

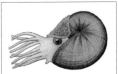

Name: Cenoceras.
Meaning: Recent horn.
Grouping: Mollusc,
Cephalopod, Nautiloid.
Informal ID: Nautiloid.
Fossil size: 5cm/2in across.
Reconstructed size: Up to
15cm/6in across.
Habitat: Seas.
Time span: From Late
Triassic to Middle Jurassic,
220–160 million years ago.
Main fossil sites: Worldwide.
Occurrence: ◆ ◆ ◆

Different species of *Cenoceras* show varying
contours on the shell, from simple, gently curving
sutures (as in this example), to narrow ribs following
the curling spiral of the shell. Characteristic of the
genus are an involute, or tightly coiled, shell form
and a very narrow umbilicus (the central, earliest-
grown region of the shell). Inside the shell, the
siphuncle (which is not visible in this
photograph) is not central within the dividing
walls, or septa, rather it is offset towards the
dorsal side, which is the inner portion of the
whorl. When the animal swam, the shell would
have been in much the position shown here, with
the eyes and tentacles at the shell opening facing
to the right. The inner surface of each whorl is the
upper or dorsal side of the animal, and the outer
surface the ventral side or belly.

Narrow umbilicus | Dorsal surface of opening

Sharp-rimmed sutures | Ventral surface of opening

Right: This species Cenoceras inornatum *dates from the Bajocian
Age of the Mid Jurassic Period, about 170 million years ago. It was
recovered from loose material in a road cutting in Dorset, southern
England. The sutures show a sharp-margined, slightly overlapping
pattern similar to that of fish scales.*

Septa, sutures, siphon and siphuncle
A nautiloid lives in the largest, last compartment
of its many-chambered shell. The chambers are
formed by dividing cross-walls, septa, usually
shaped like shallow bowls. Where their edges
join the main shell they form lines, sutures,
visible on the outside. Nautiloid sutures are
usually plain and ring-like, in contrast to the
elaborate sutures of ammonoids.

The nautiloid's rear body tapers into a long
fleshy stalk, the siphon. This passes though an
opening in the most recent septa, and then
though similar openings in each of the septa in
turn. Where the siphon passes through, it is
surrounded by a collar-like structure, the neck.
Successive necks are linked by tube-shaped
connecting rings, forming a long passageway
with the siphon inside. The series of necks and
rings is called the siphuncle. In many species it
was thickened with aragonite and preserved well.
Some nautiloids are identified by the shape and
arrangement of necks and rings in the siphuncle.

*Right: This specimen from north Wales shows the
inner shell region of Lituites, with the shell coiled in
a loose spiral – so loose, that successive whorls
were not in contact on their dorsal-ventral surfaces.
As the animal aged, the shell would straighten up as
it lengthened to give a walking stick or shepherd's
crook effect. The umbilicus at the centre, and the
close-set ridged sutures – indicating the concavo-
convex septa, or dividing walls, within the shell
(see above) – are clearly visible. The large, straight
living chamber is mostly missing.*

Lituites

An early and aberrant nautiloid – that is,
a departure from the standard type – was
Lituites from the Ordovician and Silurian
Periods. In its younger stages it grew as a
loosely coiled, or evolute, spiral. However,
with increasing age from juvenile to adult,
the shell straightened and grew as a slowly
enlarging straight or slightly curved tube,
which could achieve lengths of 30cm/12in
or more. The result looked similar to the
'walking stick' ammonite *Baculites*, which
could reach even greater lengths of up to
1m/3¼ft. (See also *Cyrtoceras*, page 431.)

Name: Lituites.
Meaning: Like a trumpet,
curved staff, crook.
Grouping: Mollusc,
Cephalopod, Nautiloid.
Informal ID: Nautiloid.
Fossil size: Slab 9cm/
3½in across.
Reconstructed size: Total
shell length 30cm/12in plus.
Habitat: Seas.
Time span: Ordovician to
Silurian, 460–420 million
years ago.
Main fossil sites: Asia
(China), northern Europe.
Occurrence: ◆ ◆

Broken-away area of shell

MOLLUSCS – NAUTILOIDS (CONTINUED), BELEMNOIDS

*Nautiloids are among the most primitive cephalopods, and one of the earliest groups to appear, with
Plectronoceras dating to the Late Cambrian Period, some 500 million years ago. Nautiloids dominated
the Palaeozoic seas, with many hundreds of well-described fossil genera. Their shell shapes vary from
straight and almost tube-like to conical, hooked, or spiralled as in the living species. Rare glimpses of
soft-tissue preservation suggest that some early nautiloids had fewer arms (tentacles), perhaps only ten or
so, compared with the living nautilus, which may possess up to 90. (Note that the two specimens shown
opposite are not nautiloids but belemnoids – see following pages.)*

Nautilus

Numerous species of the living genus *Nautilus* are known
from fossils dating back tens of millions of years. The gently
curved shell sutures are characteristic. The living nautilus
has a zebra-like pattern of brownish stripes on a pale
background and, in rare cases, marks on fossils suggest
extinct species could have been likewise patterned.
The living nautilus makes a new dividing wall, or septum,
about every two weeks as its aragonite-reinforced shell
gradually extends and widens at the open end, or
aperture. To move, the nautilus sucks in water and
squirts this out through a muscular funnel as a fast-
moving jet. (Squid have a similar system, but this
involves movements of the fleshy 'cloak', or mantle,
around the main body.) The fleshy siphon running through
the siphuncle within the shell can alter the amount of gases
inside the shell chambers and so control the nautilus's
buoyancy, allowing it to hang in mid-water, rise or sink.
The chambered part of the shell, or phragmocone, is gas-
filled and light, while the living compartment containing
the animal is relatively heavy. This is why the nautilus
floats with its body lowermost. The genus *Nautilus* is
known from Oligocene times (about 30 million years ago),
but its nautiloid subgroup, Nautilida, first appeared in the
Devonian Period some 400 million years ago.

*Above: This Cretaceous Nautilus
has a deep, narrow umbilicus (the
central, first-grown region of the
shell) and, as with other
members of its genus, is involute,
or tightly coiled, with younger,
later whorls partly obscuring
older, smaller ones. Specimen
3cm/1in in diameter.*

Umbilicus

Name: *Nautilus*.
Meaning: Sailor.
Grouping: Mollusc,
Cephalopod, Nautiloid.
Informal ID: Nautilus.
Fossil size: See text.
Reconstructed size: Some
species exceeded 50cm/20in.
Habitat: Seas.
Time span: Genus *Nautilus*,
Oligocene, 30 million years
ago, to today.
Main fossil sites: Worldwide.
Occurrence: ◆ ◆ ◆

*Left: In this specimen of
the species Nautilus
bellerophon, the hole
known as the siphuncle
(see previous page) is
visible in the middle of
the last dividing wall, or
septum. The main living
chamber has been lost.
The specimen is Cretaceous
and from Faxe, Denmark.
Specimen 2cm/¾in
in diameter.*

Sutures

Septum

Siphuncle

Nautiloid lifestyle

The living nautilus, a voracious,
nocturnal predator, suggests the
lifestyle of fossil species. By day it
rests on or near the sea bottom, at
depths down to 800m/2,600ft, perhaps
clinging to a rock with some of its
arms. A horny protective 'hood' is
drawn over most of its head, eyes and
tentacles. At night, the hood tilts back
and the animal becomes active, rising to around 200m/650ft deep
to search by smell, feel and perhaps sight. The nautilus darts
suddenly to surprise and grab victims with its tentacles, which are
ridged, but not hooked or suckered as in squid or octopus. The fish,
crab or other meal is ripped apart by the mouth, which is equipped
with a parrot-like beak at the centre of the tentacles.

Nautiloid 'hooks'

Cyrtoceras (*Meloceras*) was a finger-size, hook-like nautiloid from Silurian–Carboniferous times. It demonstrates the way that nautiloid evolution explored many different patterns of shell growth and shape (see also *Lituites*, shown previously, which changed from a youthful, loosely coiled spiral to become a straight or slightly curved adult). This specimen is from Silurian deposits in the Bohemian 'Cephalopod Quarry' region of Lochkow, Czech Republic. Fossil length 8cm/3¼in.

Gas-filled compartments

Oldest region of shell floated uppermost

Aperture

Living compartment

Hibolithes

One of the most classically shaped 'belemnite bullets', this fossil of the belemnoid *Hibolithes* is of the body part called the guard. (For general information on belemnoids, see next page.) The guard was the usually pointed structure within the rear end of the squid-like body. Since belemnoids, like squid, presumably often jetted backwards, the guard provided strong internal support when the animal bumped 'tail-first' into an obstacle while shooting through the water. The long, slender shape of the whole guard and its attached internal shell, or phragmocone, suggest that *Hibolithes* was a slim, streamlined belemnoid that could swim at speed. It has given its name to a family of belemnoids, the Hibolithidae.

Right: This specimen is the rearmost end part of the guard from the species Hibolithes jaculoides. *It dates from the Hauterivian–Barremian Ages of the Early Cretaceous, 135–125 million years ago. These fossils occur in abundance in the Speeton Clay of North Yorkshire, England. More complete specimens show that the whole guard was long and slender and would extend to the right in the view shown here.*

Name: Hibolithes.
Meaning: Stick stone.
Grouping: Mollusc, Cephalopod, Belemnoid.
Informal ID: Belemnoid.
Fossil size: Length 6cm/2⅜in.
Reconstructed size: Whole animal exceeded 30cm/12in.
Habitat: Seas.
Time span: Middle to Late Jurassic and Cretaceous, 160–70 million years ago.
Main fossil sites: Northern Hemisphere.
Occurrence: ◆

Smooth surface

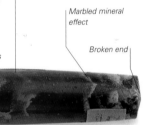

Marbled mineral effect

Broken end

Pachyteuthis

Name: Pachyteuthis.
Meaning: Thick squid.
Grouping: Mollusc, Cephalopod, Belemnoid.
Informal ID: Belemnoid.
Fossil size: Length 5cm/2in.
Reconstructed size: Whole animal up to 50cm/20in.
Habitat: Seas, usually deeper water.
Time span: Middle to Late Jurassic, 170–145 million years ago.
Main fossil sites: Worldwide.
Occurrence: ◆ ◆

Many belemnoids were finger-, hand- or foot-size but *Pachyteuthis*, at 50cm/20in or perhaps more, was almost arm-length. The proportions of the rear part of the guard show that the restored creature was a stoutly built heavyweight. This is reflected in its name, which means 'thick squid' – 'teuthis' is a common element of squid names, such as *Architeuthis*, the living giant squid. Fossils found with its remains indicate that it hunted or scavenged in relatively deep waters, probably using size and strength to overwhelm its prey, rather than surprise and sudden bursts of speed. *Pachyteuthis* belonged to the Cylindroteuthididae family of belemnoids. Specimens originally named as *Boreioteuthis* are generally reassigned to *Pachyteuthis*. For general information on belemnoids and *Cylindroteuthis*, see next page.

Below: This specimen of Pachyteuthis densus *is from the Sundance Formation of Late Jurassic rocks in Wyoming, USA. The whole guard would extend to the left, forwards, to surround the phragmocone – the belemnoid's chambered internal shell. *Pachyteuthis *is well known from various fossil sites, especially in the USA (the Dakotas, Wyoming, Montana) and in Europe (in particular, Germany). It was prominent during the late Mid Jurassic, from about 160–155 million years ago. Its remains are often mixed with those of the sea-going, dolphin-shaped reptiles known as ichthyosaurs, which may have dived deep to prey on these and other belemnoids.*

Relatively blunt posterior end

Broken surface

MOLLUSCS – BELEMNOIDS (CONTINUED)

The belemnoids – a name derived from the Greek for a thrown dart or javelin – were members of a vast group of cephalopod molluscs that began their rise in the Carboniferous, and swarmed in their millions in Jurassic and Cretaceous seas, but declined in the Late Cretaceous. Their kind did not survive the end-of-Cretaceous mass extinction of 65 million years ago, although a few squid-like types are claimed for the Eocene Epoch, 50 million years ago. Belemnoids belonged to the group of cephalopods known as coleoids. Also in this group are extinct and living teuthoids or squid, octopods or octopuses, and sepioids or cuttlefish. Coleoids have two gills, whereas other cephalopods, such as nautiloids and ammonoids, had four. (Two belemnoids are shown on the previous page.)

Belemnitella

The distinctive guard of this belemnoid – one of the last of the group, from the end of the Cretaceous Period – is nearly tube-shaped for much of its length. The rear end curves around as it narrows and the rearmost tip has a nipple-like structure, the mucron. A network of small, shallow grooves can be seen on the surface of many specimens. These may have been occupied by blood vessels and/or nerves in the living animal. Various species of *Belemnitella* are well known from locations across the Northern Hemisphere, especially eastern North America and Europe, where massive accumulations occur in some rocks. One member of the genus, *B. americana*, is the state fossil of Delaware, USA.

Below: This specimen of the species Belemnitella mucronata *is from Late Cretaceous deposits in Holland. The 'nipple', or mucron, is at the rear tip. The alveolus is the 'socket' that housed the rear part of the animal's chambered internal shell, the phragmocone. The broken surface indicates where the guard would have continued forward, becoming much thinner and more fragile as it surrounded the phragmocone.*

Name: *Belemnitella*.
Meaning: Small dart.
Grouping: Mollusc, Cephalopod, Belemnoid.
Informal ID: Belemnoid.
Fossil size: Length 8cm/3⅛in.
Reconstructed size: Whole animal up to 45cm/18in.
Habitat: Shallow seas.
Time span: Late Cretaceous, 80–65 million years ago.
Main fossil sites: Northern Hemisphere.
Occurrence: ◆ ◆ ◆ ◆

Main body of guard

Break line at alveolus

Mucron (tip)

Surface network of fine grooves

Curved taper

Belemnoid guards and phragmocones

The most common fossils of belemnoids are their guards (rostra). These are variously shaped like darts, bullets, cigars, swords and daggers, and are known in folklore as 'devil's fingers' or 'thunderbolts', from the belief that they were cast down by the gods during thunderstorms. The guard was in the belemnoid's rear body, with its tapering end or apex pointing backwards. At the front end in some belemnoids was a deep socket, the alveolus, into which fitted the phragmocone, an internal shell with a chambered structure, not unlike that of the nautiloid shell, but generally conical or funnel-shaped. The animal altered the balance of fluids and gases in the phragmocone to adjust its buoyancy. In other belemnoids the phragmocone simply butted on to the front end of the guard. Phragmocones were fragile, and therefore easily disrupted during preservation. In addition, many guards broke at the place where they became very thin as they arched around the narrow end of the phragmocone. So, in general, only the rear, solid part of the guard is found preserved as a fossil.

Above: Passaloteuthis, Early Jurassic, Dorset, southern England. Fossil length 10cm/4in.

Below: Hastites, Late Jurassic, Austria. Fossil length 9cm/3½in.

Above: Oxyteuthis, Early Cretaceous, Hanover, Germany. Fossil length 8cm/3⅛in.

Cylindroteuthis

Name: Cylindroteuthis.
Meaning: Cylinder squid.
Grouping: Mollusc, Cephalopod, Belemnoid.
Informal ID: Belemnoid.
Fossil size: Length 10cm/4in.
Reconstructed size: Whole animal up to 80cm/32in.
Habitat: Seas, usually deeper shelving areas.
Time span: Jurassic, 165–145 million years ago.
Main fossil sites: North America, Europe.
Occurrence: ◆ ◆ ◆

The main guard of *Cylindroteuthis* is long and pointed, like a poker, and is a very characteristic fossil. The rear end, or apex, tapers to a sharp tip. The front end expands at the alveolus, or socket, which can occupy up to one-quarter of the guard's total length. This housed the phragmocone – the animal's internal chambered shell – and the guard became very thin as it covered the rear portion of this structure. The phragmocone was broad and conical, expanding towards its front end, which was roughly in the centre of the animal's body. The guard length can reach 25cm/10in in this genus. Working on the assumption that the complete guard can occupy one-quarter to one-third of the animal, this gives a suggested head–body length of 70–80cm/ 28–32in, making *Cylindroteuthis* one of the larger belemnoids.

Below: This Cylindroteuthis specimen from near Chippenham, Wiltshire, west England, is dated to the Callovian Stage of the Late Middle Jurassic Period, about 155 million years ago. It was found in the Oxford Clays, which have provided many thousands of similar specimens.

Fractured remains of chambered phragmocone

Alveolus of guard

Posterior sharp tip of guard

Cylindrical main body of guard

Actinocamax

Life and death of belemnoids

Rare soft-tissue fossils, especially from Solnhofen, Germany, show that belemnoids were very similar to living squid. They had ten tentacles armed with small, strong hooks, a beak-like mouth at the centre of the tentacles, two large eyes, a torpedo-shaped body covered by a fleshy mantle tapering to a point at the 'tail' end, and two fleshy side fins near the rear. They swam by jet propulsion, taking water into the mantle and squirting it out through a funnel-like siphon. Some soft-tissue remains even show an ink sac, allowing the belemnoid to discharge a water-clouding pigment when in trouble. Huge accumulations of belemnoid guards occur in some localities. These 'belemnite graveyards' or 'battlefields' may be the result of a sudden mudslide burying a swarm of them, or semelparity – mass death following breeding – as in some types of squid today.

The guard of this belemnoid has a distinctively long groove in the ventral (lower) side. The guard also expands towards the rear end before curving in, a shape known as lanceolate. The apex, or rear, forms a nipple-like tip, the mucron. If a belemnoid guard is broken or sliced open, its inner structure is seen as concentric onion-type layers, light and dark, made of calcite crystals radiating outwards like spokes. The overall pattern is similar to the growth rings of a tree trunk. Assuming these guard layers were true growth rings, estimates show that a typical belemnoid lived for about four years. The solid, heavy, mineralized guard probably acted as a counterweight at the rear end of the animal, to the head and tentacles at the front. This allowed stable swimming as the chambered, partly gas-filled phragmocone in the middle was adjusted for buoyancy.

Name: Actinocamax.
Meaning: Not known, possibly radiating or spoke-like.
Grouping: Mollusc, Cephalopod, Belemnoid.
Informal ID: Belemnoid.
Fossil size: 9cm/3½in.
Reconstructed size: Whole animal 40cm/16in.
Habitat: Shallow seas.
Time span: Late Cretaceous, 90–70 million years ago.
Main fossil sites: Northern Hemisphere.
Occurrence: ◆ ◆

Below: This view of the underside of an Actinocamax guard shows the long ventral groove. The specimen comes from Late Cretaceous chalk deposits along the coat of Sussex, southern England. The Plenus Marls of the 'White Cliffs' along Kent and Sussex coasts are named from the many specimens of the species shown here, Actinocamax plenus.

ECHINODERMS – CRINOIDS (SEA LILIES, FEATHER STARS)

The crinoids are members of the phylum Echinodermata – creatures that have been common in the oceans since the Cambrian (500 million years ago). All echinoderms possess wheel-like radial symmetry usually based on the number five and a 'skeleton' of countless hard, small plates, termed ossicles.

Encrinus

Crinoids appeared in the oceans in the Early Ordovician Period, 480 million years ago, and thrived throughout the Palaeozoic Era. Their plant-like body plan has led to them being given the common name of sea lily, and consists of a crown of radiating arms mounted on an elongated stalk. In order to feed, the crown of the animal is raised above the surface on the stalk, and the arms of the sea lily open out to gather any food particles floating past. These are then passed along the arms towards the mouth by finger-like fleshy extensions of the water-filled hydrostatic system, known as tube feet. Crinoids still thrive in the oceans today, though mainly in the deep sea, and there are more than 600 living species known to science.

Below: Crinoids, such as Encrinus, have a tendency to close the arms of the crown after death, due to muscle contraction and decay of the water-pressurized hydrostatic system. This makes them resemble a flower with its petals folded up – as though just emerging from a bud. Crinoids are often preserved in this state.

Name: *Encrinus*.
Meaning: Lily-like.
Grouping: Echinodermata, Crinoidea.
Informal ID: Sea lily.
Fossil size: 13cm/5in.
Reconstructed size: Up to 30cm/12in.
Habitat: Sea floor.
Time span: Middle Triassic , 230 million years ago.
Main fossil sites: Throughout Europe.
Occurrence: ◆ ◆ ◆

Marsupites

Below: Complete cups of Marsupites are relatively common crinoid fossils, perhaps because they were almost ready-buried in the sediment when the animals were alive. The arms, however, are never retained intact.

Many crinoids looked like flowers on stalks. *Marsupites*, however, lacked a stalk and so is known as a stemless crinoid. This creature apparently lived with its bulbous, pouch-like cup, or calyx, planted on the sea floor and its elongate arms extended into the water above. The mouth faced upwards and was located at the centre of a sheath, or tegmen, stretched over the top of the cup. *Marsupites* flourished briefly in the Late Cretaceous and the distinctive polygonal (many-sided) plates of its cup act as a worldwide marker fossil for rocks of that age. The closely related *Uintacrinus* apparently lived as large colonies on the sea floor. Rock layers bearing hundreds of associated individuals occur in the Cretaceous Chalks of Kansas, USA.

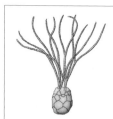

Name: *Marsupites*.
Meaning: Pouch-like.
Grouping: Echinodermata, Crinoidea.
Informal ID: Sea lily.
Fossil size: 2cm/¾in.
Reconstructed size: Calyx 3–5cm/1–2in; arms up to 1m/3¼ft.
Habitat: Sea floor.
Time span: Late Cretaceous, 85 million years ago.
Main fossil sites: Worldwide.
Occurrence: ◆ ◆ ◆ ◆

Eucalyptocrinites

The arms of the crinoid's crown attach to a plated cup-like main body, or calyx, which contains the bodily organs. These organs comprise mainly a looped stomach and intestine. The stem extends from the base of the cup and bears regular branching structures, termed cirri, which give the crinoid anchorage. A nervous system runs along the arms and stem and is controlled by a central 'brain' within the cup. Many modern crinoids live on the deep-sea floor, but most groups in the fossil record preferred shallow water. Well-preserved specimens of *Eucalyptocrinites* from the Waldron Shale rocks of Indiana, USA, show other creatures living on their stems, including brachiopods (lampshells), tabulate corals, tube or keel worms, such as *Spirorbis*, and bryozoans (sea-mats).

Below: The columnals of the crinoid stem, here of Eucalyptocrinites, are the skeletal plates or ossicles covering the stem or stalk. They are common fossils and come in all shapes and sizes. Many are flattened discs, some are five-pointed stars and others are elongate rods.

Name: *Eucalyptocrinus*.
Meaning: Covered cup lily.
Grouping: Echinodermata, Crinoidea.
Informal ID: Sea lily.
Fossil size: Ossicles each 2cm/¾in wide.
Reconstructed size: 40cm/16in across.
Habitat: Sea floor.
Time span: Middle Silurian to Middle Devonian, 420–390 million years ago.
Main fossil sites: Europe, North America, Siberia, Australia.
Occurrence: ◆ ◆ ◆

Saccocoma

Stalked crinoids were hugely successful in the Palaeozoic Era and their remains form thick limestone layers from the shallow sea beds of that time. During the Mesozoic Era stalked crinoids declined and stemless forms, known as feather stars or comatulids, began to dominate instead. These still survive and are free roaming, able to walk along the sea floor, climb corals and swim in the water. Although they have lost the stem, comatulids have retained a ring of hair-like cirri (tassel-like filaments) that enable them to anchor themselves once they have found a suitable location. *Saccocoma* has flared, paddle-like ossicles (skeletal plates) at the base of its arms and was evidently a swimmer. Specimens are commonly found in the famous Jurassic Lithographic Limestones of Solnhofen, Germany, which were laid down in a shallow lagoon. Evidently, individuals were periodically swept into the lagoon where they succumbed to the inhospitable, extremely salty waters.

Name: *Saccocoma*.
Meaning: Hairy pouch.
Grouping: Echinodermata, Crinoidea.
Informal ID: Feather star.
Fossil size: 3cm/1in.
Reconstructed size: Calyx less than 1cm/⅜in; arms less than 5cm/2in long.
Habitat: Various marine habitats.
Time span: Jurassic to Cretaceous, 160–70 million years ago.
Main fossil sites: Europe, North Africa.
Occurrence: ◆ ◆

Bourgueticrinus
These worm-like crinoids anchored themselves by a branching root network, known as a radix, at the base of their stem. The fossilized calyx and columnals are commonly found intact, though articulated material, such as arms, are rarely retained. Other crinoids have structures similar to the root bulbs of plants, coiled whips and grappling hooks.

Left: The skeletons of feather stars such as Saccocoma were not robust and so quickly disarticulated (fell apart or were pulled to pieces) after death. Whole specimens of these creatures are found only in unique circumstances – where the sea floor was free of disruptive scavengers, for example.

ECHINODERMS – ASTEROIDS (STARFISH), OPHIUROIDS (BRITTLESTARS)

Starfish and brittlestars appear as fossils in the Ordovician Period, 450 million years ago, and are presumably related. They are usually placed in a single group of echinoderms termed the Asterozoans, as they both share the familiar five-armed body plan, but the two groups are different structurally.

Crateraster

Below: The skeleton of a goniasterid starfish such as Crateraster *comprised a mosaic of small ossicles on the upper side and undersurface, and an outer edging or marginal frame of larger, blocky ossicles, which extend along the arms.*

Small inner ossicles

Marginal frame

Arm

Although asterozoans have been common since the Palaeozoic Era, they are very rarely found whole as fossils. The flexible skeleton is a jigsaw of countless small, often brick-like calcareous plates, or ossicles, that are weakly bound together in life, and so quickly fall apart after death. The arms of living individuals are easily detached, a feature that serves as a defence mechanism, and can be subsequently regenerated. *Crateraster* was a typical goniasterid starfish whose isolated ossicles are commonly found in Cretaceous chalky limestones. Articulated remains are very scarce, although bedding planes littered with complete individuals occur in the Austin Chalk of Texas, USA. Specimens are often found as pellets of regurgitated ossicles, spat out by a predator that was unable to digest them. Such predators included larger starfish.

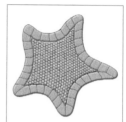

Name: *Crateraster.*
Meaning: Pitted star.
Grouping: Echinoderm, Asteroid.
Informal ID: Starfish.
Fossil size: 4cm/1½in.
Reconstructed size: 3–10cm/1–4in across.
Habitat: Sea floor.
Time span: Late Cretaceous, 90–70 million years ago.
Main fossil sites: Europe, North America.
Occurrence: ◆ ◆ ◆ ◆

Metopaster

The underside of a starfish has a central mouth and five radiating rows of suckers mounted on its fleshy tube-feet. These move the creature along the sea floor and also enable it to grip its prey. Starfish feed mainly on molluscs, including larger bivalves, such as mussels and clams, by pulling apart the valves (two halves of the shell) and then turning their stomach inside out in order to digest the soft parts. Many starfish differ from the familiar five-pointed-star body plan. For example, the modern crown-of-thorns starfish comprises a large, spiny disc with up to 20 arms, while cushion-stars are simple, rounded domes. *Metopaster* had highly reduced arms and complete fossilized specimens are biscuit-(cookie-)like, with a circular profile.

Right: The simple and robust marginal frame of Metopaster *was often relatively well preserved. However, the smaller ossicles (skeletal plates) of the upper and undersurfaces are typically lost.*

Name: *Metopaster.*
Meaning: Forehead star.
Grouping: Echinoderm, Asteroid.
Informal ID: Starfish.
Fossil size: 5cm/2in.
Reconstructed size: 3–10cm/1–4in across.
Habitat: Sea floor.
Time span: Late Cretaceous to Eocene, 90–35 million years ago.
Main fossil sites: Europe, North America.
Occurrence: ◆ ◆ ◆ ◆

Urasterella

Name: *Urasterella*.
Meaning: Tailed star.
Grouping: Echinoderm, Asteroid.
Informal ID: Starfish.
Fossil size: 10cm/4in.
Reconstructed size: Up to 15cm/6in across.
Habitat: Sea floor.
Time span: Ordovician to Permian, 450–250 million years ago.
Main fossil sites: Canada, Europe, Russia.
Occurrence: ◆ ◆

Right: The basic body plan of the asterozoans, with a central disc and radial arms, has remained unchanged since the Early Palaeozoic Era. This allows even ancient forms, such as Urasterella, to be easily recognized, and so they are rarely confused with other echinoderms or other creatures.

The seven modern orders (main groups) of starfish all arose in the Early Mesozoic Era and are viewed as distinct from those of the preceding Palaeozoic. The Devonian (400-million-year-old) Hunsrück Slate of Germany provides a rare window into the world of those Palaeozoic asterozoans. Scarce but spectacular echinoderm fossils preserve the finest details of the anatomy, including replacement of the soft, fleshy body parts by iron-sulphide minerals. This mode of preservation allows the delicate specimens to be analyzed by X-ray photography while they are still encased in their rock matrix.

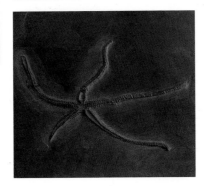

Sinosura
In exceptional circumstances, many asterozoans may be preserved together intact within a single bedding plane of rock. Such 'starfish beds', as they are known, usually result from the rapid burial of live individuals on the sea floor by an influx of gravity-driven sediment – in other words, an underwater landslide or submarine avalanche. The smooth arms of *Sinosura* distinguish them from specimens of *Ophiopetra*, which have distinctly spiny arms.

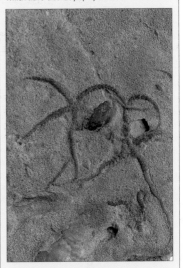

Palaeocoma

Name: *Palaeocoma*.
Meaning: Ancient hair.
Grouping: Echinoderm, Ophiuroid.
Informal ID: Brittlestar.
Fossil size: 7cm/3in.
Reconstructed size: Less than 10cm/4in.
Habitat: Sea floor.
Time span: Early to Middle Jurassic, 180–160 million years ago.
Main fossil sites: Throughout Europe.
Occurrence: ◆ ◆

The arms of brittlestars, or ophiuroids, are built around a core of ossicles, or skeletal plates, shaped like vertebrae (backbones). This allows the arms to move freely in almost any direction. Controlled bending or flexure of the arms enables the brittlestar to move rapidly across the sea floor, and also allows it to climb structures such as branching corals. Most brittlestars today, as in ancient times, are found in the deep sea. They exist either on or within the sea-floor sediments like sand or mud, and feed on any particles of nutrients that collect there. Their informal name is well deserved, since the numerous arms are brittle and are sometimes broken off by predators.

Right: The species Palaeocoma egertoni, *from the Early Jurassic beds of Eype Cliff, Dorset. The fossils here were captured in a freak sudden burial, whereby the sea floor was rapidly smothered by a great slump of sediment, burying the inhabitants alive.*

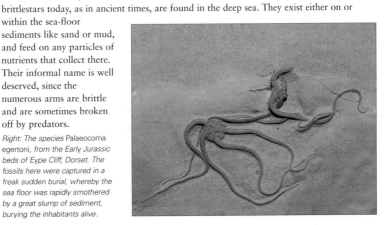

ECHINODERMS – ECHINOIDS (SEA URCHINS)

Sea urchins first appeared in the Late Ordovician, but only became successful after the end-of-Permian mass-extinction event, which eliminated many groups of plants and animals and 'remoulded' life on the sea floor. The echinoids thrived in this new world, and diversified into a variety of forms and lifestyles. They form robust fossils commonly found in shallow marine rocks of the Mesozoic and Cenozoic Eras.

Phalacrocidaris

Echinoids, or sea urchins, possess a globular or spherical shell, called the test, which is constructed from a series of interlocking plates made of calcite minerals. The outer surface of the body is also armoured, protected with a great number of spines, and in life it has flexible tube feet protruding through tiny holes in the test. 'Regular' echinoids, such as *Phalacrocidaris*, possess long spines ornamented with thorns, which they use like stilts to move along the sea floor. These spines also deter predators and often contain toxins. The area of attachment of each spine is protected by smaller spines. Most echinoids lose their spines after death as the tissues that attach them to the body quickly decay. However, rare specimens of *Phalacrocidaris* with its spines still attached are known from Cretaceous chalks around Europe. Even rarer specimens retain the five-part mouth 'jaws', which are called the Aristotle's lantern, in honour of the Greek naturalist and philosopher who first described it. The Aristotle's lantern is a highly complex beak-like structure, which extends through the mouth on the underside of the animal and can be used to scrape food off rocks or directly from the sea floor.

Name: *Phalacrocidaris*.
Meaning: Bald tiara.
Grouping: Echinodermata, Echinoidea.
Informal ID: Sea urchin.
Fossil sizes: 5 and 6cm/ 2 and 2¼in.
Reconstructed size: Body with spines up to 14cm/ 5½in across.
Habitat: Sea floor.
Time span: Late Cretaceous, 90 million years ago, to today.
Main fossil sites: Europe, Australia.
Occurrence: ◆ ◆

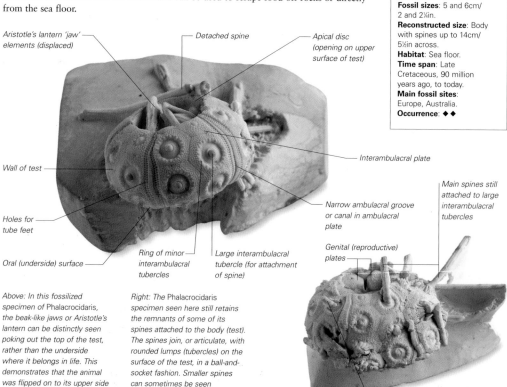

Aristotle's lantern 'jaw' elements (displaced)

Detached spine

Apical disc (opening on upper surface of test)

Wall of test

Holes for tube feet

Oral (underside) surface

Ring of minor interambulacral tubercles

Large interambulacral tubercle (for attachment of spine)

Interambulacral plate

Narrow ambulacral groove or canal in ambulacral plate

Main spines still attached to large interambulacral tubercles

Genital (reproductive) plates

Broken-off spine

Above: In this fossilized specimen of Phalacrocidaris, the beak-like jaws or Aristotle's lantern can be distinctly seen poking out the top of the test, rather than the underside where it belongs in life. This demonstrates that the animal was flipped on to its upper side before it became fossilized.

Right: The Phalacrocidaris specimen seen here still retains the remnants of some of its spines attached to the body (test). The spines join, or articulate, with rounded lumps (tubercles) on the surface of the test, in a ball-and-socket fashion. Smaller spines can sometimes be seen protecting these joints.

Hirudocidaris

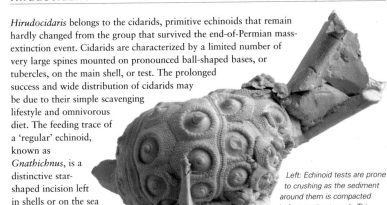

Hirudocidaris belongs to the cidarids, primitive echinoids that remain hardly changed from the group that survived the end-of-Permian mass-extinction event. Cidarids are characterized by a limited number of very large spines mounted on pronounced ball-shaped bases, or tubercles, on the main shell, or test. The prolonged success and wide distribution of cidarids may be due to their simple scavenging lifestyle and omnivorous diet. The feeding trace of a 'regular' echinoid, known as *Gnathichnus*, is a distinctive star-shaped incision left in shells or on the sea floor, produced by the scraping action of its five-part jaws – the Aristotle's lantern.

Left: Echinoid tests are prone to crushing as the sediment around them is compacted before turning to rock. This specimen of Hirudocidaris most likely fell on to its side and became squashed before it was fossilized.

Name: *Hirudocidaris*.
Meaning: Leech tiara.
Grouping: Echinodermata, Echinoidea.
Informal ID: Sea urchin.
Fossil size: 11cm/4⅓in.
Reconstructed size: Body with spines up to 16cm/6½in across.
Habitat: Sea floor.
Time span: Late Cretaceous, 100–70 million years ago.
Main fossil sites: Throughout Europe.
Occurrence: ◆ ◆ ◆

Paracidaris

Each echinoid spine, here of *Paracidaris*, is formed from a crystal of calcite and is highly resistant to weathering processes after death. For this reason, the spines are often much more common than the echinoid tests themselves. They can be an important component of sedimentary rocks.

Rows of small tubercles (pimples)

Socket at base

Tylocidaris

Echinoid spines come in all shapes and sizes, and they are often the most characteristic feature of the creature's skeleton. This is useful for fossils, since most spines are washed away from their owners before being preserved and have to be identified in isolation. *Tylocidaris* possessed a relatively small number of large and particularly distinctive club-shaped spines. The end of each spine furthest from the body is swollen, and ridges of low 'thorns' run along its length. Isolated *Tylocidaris* spines are common fossils in British Chalk, and with great luck complete individuals, with their spines still attached, can occasionally be found.

Name: *Tylocidaris*
Meaning: Knobbly tiara
Grouping: Echinodermata, Echinoidea
Informal ID: Sea urchin
Fossil size: 5cm/2in
Reconstructed size: Body with spines up to 11cm/4½in across
Habitat: Sea floor
Time span: Middle Cretaceous, 100 million years ago
Main fossil sites: Europe, North America
Occurrence: ◆ ◆ ◆

Left: This Tylocidaris specimen has been prepared using an air abrasive – a tool that 'hoses' abrasive powder over the fossil to remove the rock and expose the finest details. This preparation method works particularly well with chalk echinoids.

ECHINODERMS – ECHINOIDS (SEA URCHINS) (CONTINUED)

At the beginning of the Mesozoic Era, all echinoids had a 'regular', spherical body plan and radially arranged spines. This enabled them to roam the sea floor in search of food. However, in the Jurassic Period, many developed oblong, 'irregular' bodies with reduced spines, to help them burrow for nutrients in the sediment.

Diademopsis

Regular echinoids are classified in part by the detailed structure of their 'jaws', or mouthparts, known as the Aristotle's lantern. Four groups are recognized: cidaroid, aulodont (which includes *Diademopsis*, shown here), stirodont and camarodont. However the majority of irregular echinoids have evolved to lose their lantern. So another important feature used in echinoid classification is the structure of the ambulacra. These are five paired strip-like columns of porous plates, sometimes equally spaced, that run the vertical length of the echinoid test (body shell), in the manner of the segments of an orange. The tube feet are aligned in rows along the ambulacra, and connect to the internal water-pressurized, or hydrostatic, system through the holes, or pores, in the test. The tube feet of regular echinoids are often armed with suckers, which allow the animal to grasp food or cling to rocks.

Below: The test of Diademopsis *has broad and relatively subtle ambulacral columns. Decay has caused all of the spines to fall off, and water currents have separated them from the test.*

Name: *Diademopsis*.
Meaning: Crown appearance.
Grouping: Echinodermata, Echinoidea.
Informal ID: Sea urchin.
Fossil size: 2.5cm/1in.
Reconstructed size: Body with spines up to 12cm/5in.
Habitat: Sea floor.
Time span: Triassic to Jurassic, 230–180 million years ago.
Main fossil sites: Europe, North Africa, South America.
Occurrence: ◆ ◆ ◆

Phymosoma

Below: Phymosoma *had a flattened shape, somewhat like a doughnut, bearing rows of tubercules (small knobs) that curve around from bottom to top. In preserved remains, the spines sometimes associated with the test are smooth and tube-like. This specimen is seen from the underside, so the large central opening (the peristome) is for the mouth and lantern.*

The ball-shaped shell, or test, of an echinoid has two major openings: the peristome, for the mouth, on the underside and the periproct, for the anus, on the upper side. In 'regular' echinoids, the periproct is encircled by the apical system – a ring of plates associated with reproduction and regulating water intake. In 'irregular' echinoids, the anus is separate from the apical system. Both of these features help to group fossil urchins. The main internal organs inside the test are a large, coiled stomach and intestinal system, and the valves and tubes associated with the hydrostatic system. Each valve links through the ambulacral pore, or small hole, to a long, finger-like tube foot outside the test.

Name: *Phymosoma*.
Meaning: Swollen body.
Grouping: Echinodermata, Echinoidea.
Informal ID: Sea urchin.
Fossil size: 4cm/1½in.
Reconstructed size: Body with spines up to 10cm/4in across.
Habitat: Sea floor.
Time span: Middle Jurassic to Palaeocene, 165–60 million years ago.
Main fossil sites: Worldwide.
Occurrence: ◆ ◆ ◆

Conulus

The 'irregular' echinoids or sea urchins, such as *Conulus*, are characterized by spines that have been reduced into a dense coat of small, flexible 'hairs'. In some types of sea urchin this coat almost resembles fur and serves a number of specialist functions. Some of these spines maintain a burrow around the animal, while some push the animal through the sediment or beat to generate water currents within the burrow. Yet others secrete slimy sticky mucus to catch unwanted sediment. *Conulus* is a very common echinoid from the Late Cretaceous Period, but, like all irregulars, the hair-like spines are almost never found attached to the test.

Offaster

Echinoids, such as *Offaster*, often make good zone fossils for dating rocks. As they modified their tests to suit new life strategies and conditions, so they left behind an evolutionary lineage in the rock record, with each form representing a unique interval in time.

Name: *Conulus.*
Meaning: Conica.l
Grouping: Echinodermata, Echinoidea.
Informal ID: Sea potato.
Fossil size: 3–5cm/ 1–2in across.
Reconstructed size: As above.
Habitat: Sea floor.
Time span: Cretaceous, 135–65 million years ago.
Main fossil sites: Worldwide.
Occurrence: ◆ ◆ ◆

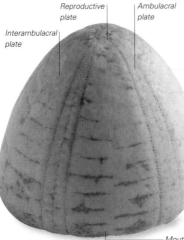

Interambulacral plate

Reproductive plate

Ambulacral plate

Mouth and anus on underside

Left: The hair-like, flexible spines of irregular echinoids are mounted on numerous microscopic tubercles (socket-like joints). Although densely covering the test, the spines are barely visible to the naked eye – but they do produce a rough feel when touched. The pairs of vertical bands, or ambulacra, are particularly prominent on this specimen. Each band represents the location of a row of tube feet in life. The sea urchin has five such ambulacra.

Micraster

Name: *Micraster.*
Meaning: Small star.
Grouping: Echinodermata, Echinoidea.
Informal ID: Sea potato, heart urchin.
Fossil size: 5cm/2in.
Reconstructed size: 3–9cm/ 1–3½in across.
Habitat: Sea floor.
Time span: Cretaceous to Palaeocene, 90–60 million years ago.
Main fossil sites: Europe, North Africa, Asia.
Occurrence: ◆ ◆ ◆

The heart urchins are almost unrecognizable from their regular sea urchin ancestors. Their bodies are highly modified for a life within the sediment. In addition to an upper and lower surface, they have a definite front and back end, rather than a simple, circular, same-all-around body plan like regular echinoids. The mouth has been drawn to the front of the body and modified into a scoop, while the anus has moved from the top of the animal to the rear. The tube feet have certain specialized functions, such as maintaining the passage of oxygenated water through the burrow and selecting food particles.

Below: The five 'petal' depressions on top of this echinoid are modified ambulacra. In life, they carried specialized tube feet. The animal's front is to the left, at the groove.

OTHER ECHINODERMS

After their appearance in the Palaeozoic Era, echinoderms quickly diversified. They remained highly varied throughout the Palaeozoic, and so abundant at times that their remains now form entire layers of rock. Many were stalked and crinoid-like, while the edrioasteroids were small domes that encrusted hard surfaces. Most Palaeozoic groups died out before the Mesozoic and so lack any modern relatives.

Cothurnocystis

Below: This mould fossil of Cothurnocystis is from the Late Ordovician Period. The side view shows the typical boot-like body shape of this genus, with its three-pronged cup or theca (body), tilted to the right. The stem (tail) points to the left and down and has wider stacked laminae (layers) at its upper end, where it bends towards the theca.

Echinoderms make up the invertebrate group that is most closely related to the vertebrates, or animals with backbones, and both are classified together within a larger group, the deuterostomes. It has been suggested that the curious echinoderms known as carpoids are, in fact, the ancestors of all backboned creatures, from the most primitive fish to birds and mammals. Carpoids appeared in the Cambrian Period, about 520 million years ago, and died out in the Devonian, some 370 million years ago. Their tadpole-shaped body consists of a plated body housing, or theca, with a tail-like stem and feeler-like appendages. Some experts suggest that the flattened theca could contain slits that might work as gills, both for respiration and for filtering small food particles, while the tail-like stem could have contained the rod-like notocord, which was the evolutionary forerunner of the vertebral column or backbone.

Name: *Cothurnocystis*.
Meaning: Boot pouch.
Grouping; Echinoderm, Carpoid.
Informal ID: Carpoid.
Fossil size: Slab 7cm/ 2¾in across.
Reconstructed size: Theca (body) 4cm/1½in across.
Habitat: Sea floor.
Time span: Ordovician Period, 450 million years ago.
Main fossil sites: Most continents worldwide.
Occurrence: ◆

Pseudocrinus

The echinoderms called cystoids are found worldwide in rocks dating from the Ordovician Period, 490 million years ago, to Permian times, about 290 million years ago. Although they are superficially similar to the carpoids (above), cystoids share anatomical details with the blastoids, such as *Pentremites* opposite. A characteristic feature of many cystoids is the possession of polygonal grille-like structures, termed pore-rhombs, variably arranged on the theca (main body housing), which is formed of skeletal plates. These connect with the inside of the theca and probably allow the passage of fluids and respiration. Pore-rhombs are located across the junction between adjoining plates of the theca, and may be hidden beneath a thin covering. Cystoids can be both bulbous and radially symmetrical, or more flattened and bilaterally symmetrical.

Left: The plated theca and stout stem of a typical cystoid were commonly preserved nearly intact. However, the delicate arms, used in life for gathering food, are usually absent.

Name: *Pseudocrinus*.
Meaning: False lily.
Grouping: Echinoderm, Cystoid.
Informal ID: Cystoid.
Fossil size: 4cm/1½in.
Reconstructed size: With complete stem and arms, about 10cm/4in.
Habitat: Sea floor.
Time span: Silurian, 400 million years ago.
Main fossil sites: Throughout Europe.
Occurrence: ◆ ◆ ◆

Pentremites

Name: Pentremites.
Meaning: Five petals or flaps.
Grouping: Echinoderm, Blastoid.
Informal ID: Blastoid, fossil sea bud or rose bud.
Fossil size: Height 4cm/1½in.
Reconstructed size: Height, including stalk and arms, about 30cm/12in.
Habitat: Sea floor.
Time span: Early Carboniferous, 330 million years ago.
Main fossil sites: Worldwide.
Occurrence: ◆ ◆

The blastoids are yet another group of stalked echinoderms from the Palaeozoic Era. Their robust body containers, or thecas, form resilient and distinctive fossils, variously referred to as 'sea buds', 'rose buds' or 'hickory nuts'.

Rows of pores radiate from the summit of the theca, and open into a feature within the theca that is unique among the blastoids, termed the hydrospire. This is a delicate, folded cavity that probably functioned in both respiration and reproduction. Water entering the hydrospire via the pores could exit the theca through holes, termed spiracles, at the top. Blastoids occurred from the Middle Ordovician to Late Permian times. The genus *Pentremites* is especially common in the Carboniferous (330-million-year-old) deposits of North America.

Right: Like most preserved blastoid specimens, this example of Pentremites, from Illinois, USA, has lost its fragile arms, known as brachioles. These projected from the ambulacra (the five rows of small holes that look like starfish arms). The brachioles in life would have gathered small items of food. A thin stem attached the base of the theca (main body casing) to seabed rock.

Achistrum

Holothuroideans, or sea cucumbers, are related to echinoids (sea urchins), but they are virtually unrecognizable as echinoderms. The body takes the form of a leathery tube, with the skeleton of mineral plates, or ossicles, reduced to microscopic ossicles embedded in the skin. These form distinctive microfossils, shaped like wagon-wheels, anchors and hooks, and are found in rocks from the Silurian Period onwards. Holothurians are still common today, with more than 1,000 species both in shallow tropical waters and the deep sea. Some creep along the sea floor, in a manner akin to a slug, while others are free-swimming or drift with the plankton. The mouth is surrounded by tube feet modified as branching tentacles for gathering food. Rows of regular tube feet armed with suckers run the length of the body and are used for locomotion and grip.

Name: Achistrum.
Meaning: Ancient swelling.
Grouping: Echinoderm, Holothuroidean (Holothurian).
Informal ID: Sea cucumber.
Fossil size: 7cm/3in.
Reconstructed size: Length up to 10cm/4in.
Habitat: Sea floor.
Time span: Carboniferous to Cretaceous, 300–100 million years ago.
Main fossil sites: Europe, Egypt, North America.
Occurrence: ◆

Tentacles at head end

Left: This fossil of Achistrum is typical of Mason Creek, Illinois, USA, which is one of the few localities in the world where whole holothuroidean remains can be found. Swift burial and the rapid formation of ironstone nodules within the sediment led to the preservation of the sea cucumber's soft parts.

GRAPTOLITES

The hemichordates are relatively simple animals that share features with the vertebrates, such as gill slits and a dorsal nerve cord. Modern members of the Hemichordata are the acorn worms and the tiny coral-like pterobranchs. A similar skeletal structure is found in the extinct graptolites. These graptolites were highly abundant globally in rocks from the Early to Mid Palaeozoic.

Desmograptus

Name: *Desmograptus*.
Meaning: Joined/linked marks in stone.
Grouping: Hemichordata, Graptolithina.
Informal ID: Graptolite.
Fossil size: 12cm/4¾in.
Reconstructed size: Up to 20cm/8in.
Habitat: Sea floor, open ocean.
Time span: Ordovician to Silurian, 500–410 million years ago.
Main fossil sites: Worldwide.
Occurrence: ◆ ◆

The first graptolites were bushy colonies that lived permanently attached to the sea floor. However, by the end of the Cambrian Period (some 500 million years ago) many had adopted a free-floating existence among the plankton. This lifestyle apparently favoured a simpler colony structure, and the general evolutionary trend was a reduction towards a stick-like form. Bushy, or dendroid, graptolites continued to exist, however, and in the end they outlived their stick-like, or graptoloid, relatives. The protein-rich graptolite 'skeleton' is most often preserved as a flattened carbon film or 'smear'. But rare three-dimensional specimens, which have been acid-etched from limestone, reveal the fine-scale structure of the colony. Evidently, the bushy dendroid graptolites had a conical form in life, as opposed to the flattened net- or leaf-like fans that they formed as fossils.

Right: Fossil graptolites such as Desmograptus resemble hieroglyphic markings in the rocks, hence the Greek origin of their name – graptos lithos literally means 'writing [marks] in stone'. Desmograptus can resemble the corals known as sea fans.

Network of branching stipes

Didymograptus

Below: The classic tuning-fork shape of Didymograptus is formed by two stipes, or branches, with inward-facing thecae (cup-like containers). It is unclear how the planktonic colony would have been orientated in life.

The colonial skeleton of a graptolite, referred to as the rhabdosome, is formed of branches (stipes) lined with inclined tubular cups called thecae. When flattened during preservation, these take on the appearance of a saw-blade. By analogy to their modern living relatives, the pterobranchs, each theca housed a minute graptolite animal called a zooid, which was perhaps similar to a tiny sea anemone. Each zooid was linked to the rest of the colony via a connective stalk. Zooids gathered food and nutrients from the water with a crown-like feeding structure of tiny tentacles, known as the lophophore. It is possible that planktonic graptolites were able actively to move up and down in the water as part of their feeding strategy.

Name: *Didymograptus*.
Meaning: Paired marks in stone.
Grouping: Hemichordata, Graptolithina.
Informal ID: Graptolite.
Fossil size: 1.5cm/½in.
Reconstructed size: Up to 15cm/6in.
Habitat: Open ocean.
Time span: Early to Mid Ordovician, 500–450 million years ago.
Main fossil sites: Worldwide.
Occurrence: ◆ ◆ ◆

Monograptus clingani

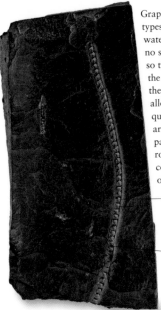

Graptolite fossils are most commonly found in the types of sediments that had been deposited in deep-water settings, such as dark shales and slates. Little or no sea-floor life was present in these environments and so the delicate graptolite skeletons that sank down from the surface waters above and settled could accumulate there undisturbed. The planktonic lifestyle of graptolites allowed them to be ocean-going and each new form quickly spread right around the globe. Their fossils are an invaluable resource for geologists and palaeontologists, since Early and Middle Palaeozoic rocks from around the world can be dated and correlated by the graptolites they contain as marker or index fossils.

Individual cup-like thecae (containers) along stipe

Stipe (branch) is straight and undivided

Left: Monograptids were among the most simply shaped graptoloids, with one stipe that varied in form, from almost straight to coiled like a spiral. They were the only graptolite group to survive into the Devonian Period, when they finally died out after 400 million years ago.

Name: *Monograptus clingani*.
Meaning: Clingan's single mark in stone.
Grouping: Hemichordata, Graptolithina.
Informal ID: Graptolite.
Fossil size: 7cm/2¾in.
Reconstructed size: Up to 15cm/6in.
Habitat: Open ocean.
Time span: Silurian to Devonian, 430–390 million years ago.
Main fossil sites: Worldwide.
Occurrence: ◆ ◆ ◆

Monograptus spirilis

Name: *Monograptus spirilis*.
Meaning: Spiral single mark in stone.
Grouping: Hemichordata, Graptolithina.
Informal ID: Graptolite.
Fossil size: 2cm/¾in.
Reconstructed size: Up to 10cm/4in.
Habitat: Open ocean.
Time span: Silurian, 435–410 million years ago.
Main fossil sites: Worldwide.
Occurrence: ◆ ◆ ◆ ◆

Planktonic graptolites may have used various mechanisms to remain afloat in the water. Some forms possessed a rod-like structure, referred to as the nema, which appears to have attached the colony to some kind of float. Early dendroid graptolites probably attached to naturally buoyant objects, such as floating seaweeds, but later forms may have generated their own float. Rare specimens show a number of graptolites united around a central disc, which could have been this buoyancy aid. Such associations are termed synrhabdosomes. Alternatively, colonies may have accumulated lightweight fats or gases within their own bodies to keep themselves buoyant.

Above: The spiral form of Monograptus would have been a coiled spire, which may have caused it to twist around, slowing its descent through the water.

Orthograptus

The action of occasional currents on the deep-sea floor often swept the remains of graptolites together into dense associations, as here with *Orthograptus*. The alignment of their stick-like skeletons records the direction of those ancient currents and helps with oceanographic studies.

ANIMALS – VERTEBRATES

Vertebrate creatures have a vertebral column or backbone, separating 'inverts' and 'verts'. However, a more scientific definition relies on the presence of a notochord – a stiffened, rod-like structure along the back, envisaged as an evolutionary forerunner of the vertebral column. The phylum (major animal group) Chordata includes vertebrates plus some vertebrateless but notochord-possessing creatures known as urochordates, hemichordates and cephalochordates (see previous chapter). Traditionally vertebrates have comprised five major classes: fish, amphibians, reptiles, birds and mammals. DNA and cladistic analysis have fragmented these into numerous subgroups. For example, fossil creatures once known as 'the first amphibians' are now more commonly called early tetrapods, 'four-legged' or 'four-footed'.

Above from left: Head of Hoplopteryx, detail from Hyracotherium tooth, the fossil of an ancient bat, Palaeochiropteryx, showing one of the wings folded over the backbone.

Right: Fossilized dinosaur tracks in a creek bed below the Black Mesa mountain, near Kenton, Oklahoma, USA. The mountain was named for the black lava rock which formed at the top, and is now part of a nature preserve.

FISH – SHARKS

The sharks were among the very earliest groups of fish, and their sleek, streamlined body design has changed little throughout their long history. Sharks and their close cousins the rays are together known as elasmobranchs and, like other fish, have an internal skeleton. This skeleton is unusual, however, as it is made of the tough, gristly substance known as cartilage, rather than bone, giving them and chimaeras, or ratfish, the name of cartilaginous fish (Chondrichthyes). Cartilage degrades more rapidly than bone after death, so most of our knowledge about prehistoric sharks comes from their well-preserved, abundant teeth and fin spines. These date back to the Early Silurian Period, more than 420 million years ago.

Hybodus

Name: *Hybodus*.
Meaning: Healthy, strong tooth.
Grouping: Fish, Chondrichthyean, Elasmobranch, Selachian, Hybodontiform.
Informal ID: Shark.
Fossil size: Length 16cm/6¼in (partly complete).
Reconstructed size: Head to tail length up to 2.5m/8¼ft.
Habitat: Oceans.
Time span: Late Permian to Cretaceous, 260–70 million years ago.
Main fossil sites: Worldwide.
Occurrence: ◆ ◆

Below: This dorsal fin spine is probably from Hybodus. The portion that attached to the body is on the right. It is likely that the spine was partly embedded in the flesh of the fin for about half of its length, with the pointed end projecting freely as a deterrent.

Hybodus is one of the best-known representatives of a group of smallish, highly successful sharks that ranged from the Late Permian and Early Triassic Periods, some 250 million years ago, to the great end-of-Cretaceous mass extinction 65 million years ago. *Hybodus* grew to more than 2m/6½ft in length and, compared with many of today's modern sharks, it had a bluntish snout. Apart from this, however, it was similar in shape to its living cousins and belonged to the modern shark group, Selachii. It may have had a varied diet, suggested by two types of teeth. The sharp teeth at the front of the mouth were for gripping and slicing slippery prey, such as fish and squid, while the larger, molar-type teeth at the rear of the jaws were perhaps designed for crushing hard-shelled food, such as shellfish. Attached to the front of each of its two dorsal (back) fins was a spine that may well have served as a defence to deter predators that were larger than itself. The spines are the body parts most commonly found as fossils.

Shark jaws

Like the rest of a shark's skeleton, the animal's cartilaginous jaws were rarely preserved. Decomposition of the springy cartilage occurred readily following death, but rare examples of jaw preservation, such as the example below, show teeth in their position in life. Teeth grew continuously behind those in use, and gradually moved to the edge of the jaw. As the teeth at the front were broken off or wore away, the younger teeth moved forward to replace them.

Below: Several rows of teeth formed an effective way of holding and slicing soft, slippery prey, such as fish and squid. Most shark teeth are either triangular in general shape, with a straight base fixed to the jaw, or have a tall, dagger-like central point growing from a low base. The tooth itself is slim, like a blade, and often has tiny saw-like serrations along its exposed cutting edges.

Oldest row of teeth in use | Length around curve of jaw 10cm/4in

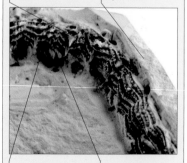

Central point of tooth | Base of tooth

Ptychodus

Name: *Ptychodus*.
Meaning: Folded tooth.
Grouping: Fish,
Chondrichthyean,
Elasmobranch, Selachian,
Hybodontiform.
Informal ID: Shark.
Fossil size: Single tooth
6cm/2⅜in across; slab
20cm/8in across.
Reconstructed size:
Head–tail length up to 3m/10ft.
Habitat: Shallow seas.
Time span: Cretaceous,
120–70 million years ago.
Main fossil sites: Worldwide,
especially North America.
Occurrence: ◆ ◆ ◆

Not all sharks had razor-sharp teeth like blades or daggers. *Ptychodus*, for example, possessed large, ribbed, flattened crushing teeth. Hundreds of these teeth were arranged in parallel interlocking rows to form a plate-like grinding surface in the mouth. This form of dentition suggests that such sharks fed chiefly on hard-shelled invertebrates, such as crustaceans and ammonites. Associated fossils show that *Ptychodus* probably lived in shallow marine conditions, and while it attained a worldwide distribution, its remains are particularly abundant in the states of Texas and Kansas, USA.

Below right: Identified as the species Ptychodus polygyrus, *this tooth close-up shows the slightly convex (domed) main biting surface as a row of ridges surrounded by lower lumps, or tubercles. The whole tooth is roughly rectangular in shape.*

Shark vertebrae

Vertebrae are sometimes called 'backbones', but in sharks they are cartilage like the rest of the skeleton. However, some shark vertebrae may have deposits of hardened, mineralized material similar to true bone, formed by a process known as ossification. Shark vertebrae can usually be recognized by the circular centrum, or main 'body', and the ringed appearance – almost like an animal version of a sectioned tree trunk. This example is 11cm/4⅜in in diameter.

Left: Following the animal's death, the tooth plate broke up, creating a jumble of separated, or disarticulated, elements. The different sizes of teeth came from different parts of the grinding tooth plate within the mouth. These teeth have been identified as the species Ptychodus mammillaris *from the Late Cretaceous Period.*

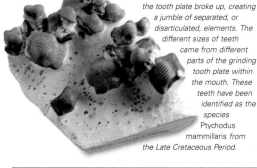

Margin of crown

Convex biting surface

Coarse ridges

Tubercles (pimple-like lumps)

Fish coprolites

Coprolites, or fossilized droppings (faeces), are important for determining the diet of extinct animals. The remains of creatures eaten by sharks, such as shell fragments, bones and scales, can be preserved within the coprolite, showing exactly what these prehistoric creatures were eating. Shark coprolites often take on the spiral or helical pattern of a part of the shark's digestive tract (intestine) called the spiral valve, making them look like pine cones or squat corkscrews.

Below: Shark coprolite, or enterospirae, Late Cretaceous, length 6cm/2½in.

Left: Fish coprolite, Triassic, length 2.5cm/1in.

FISH – LOBE-FINNED BONY FISH

Most fish species today, and those found in the fossil record, have skeletons made of bone, not cartilage, and belong to the fish group Osteichthyes. Devonian times saw a major group of these fish, known as lobe-fins, or sarcopterygians. Lobe-fins include lungfish and coelacanths, and also long-extinct fish closely related to the ancestors of tetrapods – amphibians and other four-legged vertebrates.

Osteolepis

The tetrapods – backboned animals with four limbs, including amphibians, reptiles, birds and mammals – almost certainly evolved from some form of Devonian lobe-finned fish or sarcopterygian. This would have possessed the same pattern of bones in its fin bases as found in limbs of terrestrial vertebrates. *Osteolepis* was a relative of such a fish. This is not only because of the structural similarity between its fin bones and a tetrapod's limb bones, but also because of other features found in the first amphibian-type tetrapods. These include an internal link between the nose and throat, termed the choanae, which most fish lack, but all tetrapods have. *Osteolepis* lived in shallow bodies of fresh water, feeding on any prey it could find, and was able to breathe air as well as obtaining oxygen through its gills.

Name: *Osteolepis*.
Meaning: Bony scaled.
Grouping: Fish, Osteichthyan, Crossopterygian, Rhipidistian, Osteolepiform, Osteolepid.
Informal ID: Lobe-finned Devonian fish.
Fossil size: Head–tail length 18cm/7in.
Reconstructed size: Head–tail length 20cm/8in.
Habitat: Shallow lakes and rivers.
Time span: Middle to Late Devonian, 380–360 million years ago.
Main fossil sites: Antarctica, Europe, Asia.
Occurrence: ◆ ◆

Rounded snout	Eye orbit (socket)	Fleshy-based pectoral fin	Caudal fin (tail)

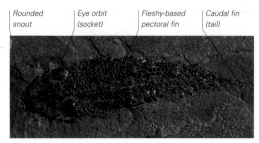

Left: Osteolepis was a slim-bodied, multi-finned rhipidistian, or fan-sail, and belonged to the subgroup of lobe-fins called crossopterygians, or tassel-fins, which also included coelacanths. It had large, shiny, squarish-shaped scales of the type known as cosmine, which had a surface layer of spongy bone. Cosmine also covered the bones of the head.

Coccoderma

Within the sarcopterygian (lobe-fin) group, coelacanths belong to the crossopterygians, or tassel fins. Coelacanths first appeared in the Middle Devonian Period and were believed to be extinct for some 70 million years until they were rediscovered in 1938. This specimen of the fossil coelacanth *Coccoderma* came from the famous Solnhofen limestone rocks of Bavaria, southern Germany, which have yielded many spectacular fossils, including the earliest known bird, *Archaeopteryx*, and the very small dinosaur *Compsognathus*.

Name: *Coccoderma*.
Meaning: Berry skin.
Grouping: Fish, Osteichthyan, Sarcopterygian, Crossopterygian, Coelacanth (Actinistian).
Informal ID: Coelacanth.
Fossil size: Head–tail length 32cm/12½in.
Reconstructed size: As above.
Habitat: Shallow marine lagoons.
Time span: Late Jurassic, 150 million years ago.
Main fossil sites: Throughout Europe.
Occurrence: ◆

Left: Coccoderma is smaller than, but otherwise very similar to, the living coelacanth Latimeria. It had a deep body and fleshy bases to its main fins. Visible as a diagnostic feature of coelacanths is the third lobe of the caudal fin, or tail – the smaller rearmost 'tassel', projecting backwards from between the tail's main upper and lower lobes. (This specimen was prepared using a technique called acid transfer and mounted on a polygonal glass-fibre block.)

Dipterus

Name: *Dipterus*.
Meaning: Two wings.
Grouping: Fish,
Osteichthyan, Sarcopterygian,
Dipneustian.
Informal ID: Lungfish.
Fossil size: 20cm/8in.
Reconstructed size:
Head–tail length 30cm/12in.
Habitat: Seas.
Time span: Devonian,
360 million years ago.
Main fossil sites: Europe,
North America.
Occurrence: ◆ ◆

*Right: This well-preserved fossil
of the species* Dipterus
valenciennesi, *from Scottish Old
Red Sandstone, shows clearly the
asymmetrical tail fin. Also visible
are the part-rounded hexagonal
scales, known as cosmoid scales,
characteristic of extinct lungfish
and the coelacanths.*

Lungfish, or dipneustians, first appeared in the Devonian Period, some 400 million years ago. *Dipterus* is one of the oldest and most primitive lungfish genera. It originated some 370 million years ago, obtained a worldwide distribution in the fossil record, and although it went extinct, it may have been an ancient ancestor or cousin of the living African lungfish, *Protopterus*. Like all but the very earliest lungfish, *Dipterus* possessed specialized teeth, which were fan-shaped and ridged, to grind down hard prey, such as molluscan and crustacean shellfish. Lungfish were among the first vertebrate animals to develop air-breathing organs, lungs, by modifying a type of swim bladder to allow the absorption of oxygen.

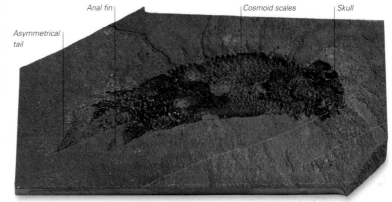

Anal fin | Cosmoid scales | Skull

Asymmetrical tail

Ceratodus

Name: *Ceratodus*.
Meaning: Horn tooth.
Grouping: Fish,
Osteichthyan, Sarcopterygian,
Dipneustian.
Informal ID: Lungfish.
Fossil size: Width 7cm/2¾in.
Reconstructed size:
Head–tail length 50cm/20in.
Habitat: Shallow seas.
Time span: Early Triassic to
Early Tertiary, 240–55 million
years ago.
Main fossil sites: Worldwide,
especially Europe and Africa.
Occurrence: ◆ ◆ ◆

The living Australian lungfish, *Neoceratodus*, is regarded as the most primitive existing member of the group – that is, it is least changed from the original lungfish body plan. It reaches 1.8m/6ft in length and weighs more than 40kg/88lb. *Ceratodus* is one of its ancient cousins, but it attained only about one-third the size. Like other lungfish, it could breathe air and had large, plate-like teeth for grinding shelled invertebrates. Early evidence of *Ceratodus* comes from the Triassic Period, some 240 million years ago. The genus went extinct in the Early Tertiary Period after a very long-lived existence of more than 180 million years.

Below: The genus Ceratodus *was named after its distinctive horn-like teeth, which were grouped to form two upper and two lower plates within the animal's mouth. This view shows a single tooth, which in life would have been growing in a row or stack with several others.*

Wave-like outer margin

Lateral margin

Convex shape to body of tooth

Base of tooth

FISH – PRIMITIVE AND RAY-FINNED BONY FISH

*Most fish today belong to the osteichthyan (bony fish) subgroup – ray-fins, or actinopterygians. In these,
the spine-like fin rays, which hold each fin open like the ribs of a fan, emerge directly from the body,
rather than from a fleshy or lobe-like base. However, before the ray-fins developed, there were earlier
groups, including the jawless fish, or agnathans, and the long-extinct placoderms, or 'flat-plated skins'.*

Cephalaspis

*Below: This Cephalaspis head shield shows the
animal's shovel-shaped head used for lying on and
ploughing into bottom sediments. The eyes are set
close together and face upwards. The pineal
opening, sometimes called a 'third eye', would have
been able to detect changes
of light levels but probably
not form clear images.*

The very first fish had no true fins and no
jaws either, and are known as agnathans.
Their history stretches back more than 500
million years into the Cambrian. Today
fewer than 100 species survive, as eel-like,
sucker-mouthed lampreys and hagfish.
Cephalaspis was an armoured agnathan
with a prominent flattened head shield,
suggesting it was a bottom-dweller. Its lack
of biting jaws meant a diet of soft-bodied
animals small enough to be swallowed
whole. *Cephalaspis* also lacked mobile true
fins and, instead, possessed fleshy 'flaps' for
clumsy swimming. Its skull shows slightly
depressed areas – sensory plates or fields.
 These suggest nerve endings in the cheek
and forehead that could detect water
currents, ripples from the movement of prey
or perhaps tiny electrical signals from the
active muscles of prey, as in sharks today.

Name: *Cephalaspis*.
Meaning: Head shield.
Grouping: Fish, Agnathan,
Osteostracan.
Informal ID: Armoured
jawless fish.
Fossil size: Head shield
width 12cm/4¾in.
Reconstructed size:
Head–tail length up to
30cm/20in.
Habitat: Rivers, estuaries,
perhaps also shallow
sea coasts.
Time span: Early Devonian,
400 million years ago.
Main fossil sites:
Throughout Europe.
Occurrence: ◆ ◆

Pterichthyodes

*Below: The body shield of Pterichthyodes dominates
the fossil, although the smaller head is also clearly
visible and would have formed an effective defence
against larger predators. The unusual wing-shaped,
arm-like appendages would provide additional
movement by pushing through the sediment.
This specimen comes from the famed Devonian
Period Old Red Sandstone, near
Caithness, in Scotland.*

— *Small head
shield*

— *Large body shield*

— *Wing-like appendage*

Pterichthyodes was one of a number of
heavily armoured fish called antiarchs,
which were a subgroup of the early fish
group called placoderms. Its entire head and
body were encased by large armoured plates
as a formidable defence against predation.
This fish was benthic – a bottom-dweller –
as shown by the eyes, which are located on
top of the head rather than to the sides.
The mouth acted as a shovel to churn up the
sediment in search of small prey. Placoderms
evolved rapidly in the Devonian to become
the dominant vertebrate predators. Some
were giants, with the Late Devonian
Dunkleosteus growing to more than
6m/20ft long – the largest predator of
its time. However, they had died out
by the Early Carboniferous Period.

Name: *Pterichthyodes*.
Meaning: Like a winged fish.
Grouping: Fish, Placoderm,
Antiarch.
Informal ID: Armoured or
tank fish.
Fossil size: Length 8cm/3¼in.
Reconstructed size: Head to
tail length up to 20cm/8in.
Habitat: Bottom of shallow
lakes.
Time span: Middle Devonian,
380 million years ago.
Main fossil sites:
Throughout Europe.
Occurrence: ◆ ◆ ◆

Palaeoniscus

Name: *Palaeoniscus*.
Meaning: Ancient small fossil.
Grouping: Fish,
Actinopterygian,
Palaeonisciform, Palaeoniscid.
Informal ID: Predatory
freshwater fish.
Fossil size: Length 23cm/
9¾in.
Reconstructed size: Head to
tail length 20–30cm/8–12in.
Habitat: Fresh water.
Time span: Middle Devonian
to Late Triassic, 370–210
million years ago (but most
abundant in Carboniferous.
Main fossil sites: Europe,
Greenland, North America.
Occurrence: ◆ ◆

The strong, fusiform (spindle-shaped) build of *Palaeoniscus* indicates that it was a powerful swimmer rather than a bottom-dweller. Its large, forward-placed eyes also suggest that vision was its most important sense. It had proportionately very large jaws, lined with rows of short, sharp teeth that continuously replaced themselves (as in sharks). *Palaeoniscus* is an early or basal actinopterygian (ray-fin). Its deeply forked, strongly heterocercal tail (the two lobes unequal in size) is entirely fleshy to the tip of its upper lobe – while, in more derived ray-fins, only the bony rod-like rays support the tail, which is often homocercal (with equal lobes). Additionally, *Palaeoniscus* regulated its buoyancy using a pair of air sacs connected to the throat, instead of the swim bladder as in most later ray-finned fish, which develops as an outgrowth of the digestive tract.

Dorsal (back) fin | Caudal fin

Body | Tail base

Above: Although anatomical details are obscured on this Palaeoniscus specimen, the animal has obviously been rapidly buried, preserving all features, down to each scale, in their precise life arrangement.

Lepidotes

Name: *Lepidotes*.
Meaning: Covered with
scales, scaly.
Grouping: Fish,
Actinopterygian,
Semionotiform, Semionotid.
Informal ID: Ray-finned
cosmopolitan fish.
Fossil size: Scale slab
25cm/10in long.
Reconstructed size:
Head–tail length up to
2m/6½ft.
Habitat: Lakes, shallow
coastal waters.
Time span: Middle Triassic to
Cretaceous, 230–100 million
years ago.
Main fossil sites: Worldwide.
Occurrence: ◆ ◆ ◆

An extremely successful Triassic-to-Cretaceous genus of ray-finned fish, the *Lepidotes* group was related to the modern gars. Around a hundred species have been listed from Mesozoic sediments laid down around the world, although some badly preserved specimens have probably been mistakenly identified as entirely new species. Different forms of *Lepidotes* have been found in a wide variety of environments, including rivers, lakes and shallow coastal sea areas. The various species range in length from 30cm/12in up to 2m/6½ft or more. The jaws had developed a new mechanism for shaping themselves into a tube that could suck in prey from a distance, and its flatenned or peg-like teeth were capable of crushing molluscan prey. Some of the recognizably thick scales of *Lepidotes* have been found in the fossilized chest region of the large dinosaur *Baryonyx*; the fish-eater perhaps scooped them up with its large claws, like a bear.

Below: Lepidotes scales were about thumbnail-sized, rhomboid or diamond-shaped and arranged in rows. Each scale was composed of layers of bone, dentine and enamel (as in our own teeth). The enamel often gives its fossils the shiny, jet-black appearance as seen here. These scales were probably from the side, or flank, of the body.

Rhomboid (diamond or kite) shape | Dorsal scales

Preserved black enamel

Ventral (belly) scales

FISH – TELEOSTS

Most of today's fish are in the ray-fin (actinopterygian) group called teleosts, or 'complete bone', where few, if any, parts of the skeleton are made of cartilage (gristle). The teleosts include the vast majority of living fish, with more than 20,000 species. They arose in the Triassic Period and underwent rapid evolution in the Middle Cretaceous, when many of the modern families became established.

Leptolepis

Above: This specimen, from the famed Late Jurassic limestone of Solnhofen, Germany, shows the protractible mouth, which had shorter jaws than earlier fish, but could gape more widely, forming a tube-like structure to suck in small food. The vertebrae of the backbone, as well as the deeply forked homocercal (equally lobed) tail, are visible.

Leptolepis is usually classed as an early teleost. In particular, this herring-like swimmer possessed a fully ossified (bony) vertebral column, or backbone. In addition, its scales lacked a heavy enamel coating and were much thinner and lighter than in more primitive fish, as well as being more flexible and more cycloidal, or rounded, in shape. *Leptolepis* remains are most abundant in Jurassic rocks, where entire fossilized schools of hundreds of individuals can be found on single slabs. Other, more fragmentary remains indicate that the genus persisted throughout almost the whole Mesozoic Era. These relatively small, warm-sea fish fed on the small planktonic organisms.

Name: *Leptolepis*.
Meaning: Thin scale.
Grouping: Fish, Actinopterygian, Teleost, Leptolepiform, Leptolepid.
Informal ID: Extinct 'herring'.
Fossil size: Head–tail length 8cm/3¼in.
Reconstructed size: Head–tail length up to 25cm/10in.
Habitat: Warm coastal seas.
Time span: Middle Triassic to Late Cretaceous, 230–80 million years ago.
Main fossil sites: Europe, Africa, North America, Australia.
Occurrence: ◆ ◆ ◆

Eurypholis

Below: This fossilized specimen of Eurypholis illustrates the species' large head, the narrow, streamlined body and the long, fine, ragged-looking teeth. Several of the fins are also visible in this specimen, including the front paired fins (or pectorals), the rear paired fins (or pelvics), and the unpaired back (or dorsal) fin.

Eurypholis – also called the 'viper fish' – was a Cretaceous euteleost with no living close relatives, whose fossils come chiefly from the region of Hadjoula, Lebanon. Its common name reflects its long, thin, 'ragged' teeth, reminiscent of the living sand tiger shark. The scientific name refers to the row of three large, bony scales superimposed behind its skull on its back. The dentition, wide eyes and very large mouth give *Eurypholis* a fearsome appearance and indicate its predatory lifestyle. *Eurypholis* fossils have been recovered with other fish in their stomach region – signs of their last meal. These fish lived in the warm, shallow seas of the Mesozoic Era. Individuals are found alone in 'mass mortality layers' known from Lebanon, indicating that they were solitary rather than shoaling fish.

Name: *Eurypholis*.
Meaning: Broad scale.
Grouping: Fish, Actinopterygian, Euteleost, Aulopiform, Eurypholid.
Informal ID: Extinct viper fish.
Fossil size: Head–tail length 12cm/4¾in.
Reconstructed size: Head–tail length up to 20cm/8in.
Habitat: Shallow coastal seas, lagoons.
Time span: Middle to Late Cretaceous, 100–70 million years ago.
Main fossil sites: Europe, Middle East, North Africa (especially Lebanon).
Occurrence: ◆

'Snaggle-tooth' dentition

Dorsal fin

Vertebral column (backbone)

Hoplopteryx

Name: *Hoplopteryx*.
Meaning: Armoured wing.
Grouping: Fish,
Actinopterygian, Teleost,
Beryciform, Trachichthyid.
Informal ID: Sawbelly,
slimehead.
Fossil size: Each specimen
about 13cm/5in long.
Reconstructed size:
Head–tail length 30cm/12in.
Habitat: Warm, shallow
coastal seas.
Time span: Late Cretaceous,
80–70 million years ago.
Main fossil sites:
Northern Hemisphere.
Occurrence: ◆ ◆

A relative of modern perch-like fish,
Hoplopteryx is from the Late Cretaceous
Period and it showed many anatomical
characteristics that are seen today in the
living ray-finned fish group Beryciformes,
which includes roughies, squirrelfish,
pinecone fish and the unpleasantly named
slimehead fish. The protractible mouth of
Hoplopteryx, which was lined with
minuscule teeth, would suck in a small
volume of water when it was speedily
opened, thereby capturing prey from a
distance. The upturned mouth is typical
of surface feeders, making it a probable
predator of small creatures in shallow
coastal waters. It had large eyes, small
pectoral fins, forward-placed pelvic fins,
a homocercal tail and a narrow body.

Beryx

Beryx superbus, like *Hoplopteryx* (see left), was
a Late Cretaceous member of the Beryciformes
group. Its genus is still represented by two
species found almost worldwide today – the
common alfonsino (*B. decadactylus*) and
the splendid alfonsino (*B. splendens*). *Beryx*
are relatively large fish, up to 70cm/27½in in
length, and feed on small fish, cephalopods
and crustaceans on the coastal sea floor.

*Below: This specimen shows the skull and thoracic
(chest) region and the vertebral column (backbone)
with the vertebral spines projecting above it.*

Gill cover region
(operculum)

*Right: This spectacular
specimen is beautifully
preserved in three dimensions
within the chalk deposits typical
of Late Cretaceous marine rocks.*

Mouth

Eye orbit (socket)

Belonostomus

Name: *Belonostomus*.
Meaning: Arrow mouth.
Grouping: Fish,
Actinopterygian, Teleost,
Aspidorhynciform,
Aspidorhynchid.
Informal ID: Needlefish.
Fossil size: Jaw 16.5cm/6½in.
Reconstructed size: Head–tail
length up to 40cm/16in.
Habitat: Warm coastal seas,
possible also brackish and
fresh water.
Time span: Late Jurassic to
Late Cretaceous, 150–65
million years ago.
Main fossil sites: Europe,
Middle East, South and
North America.
Occurrence: ◆ ◆

Superficially similar to the living needlefish (*Belone belone*),
Belonostomus was a long, slender, ray-finned fish of the
latter half of the Mesozoic Era. It had a long, pointed
rostrum (or snout) packed with closely spaced teeth.
Its dorsal and anal fins were positioned far back along the
body, and it possessed recognizably deep, elongated flank
scales. Most fossils of this fish have been found in rocks
that were deposited in shallow coastal marine areas.
However, some specimens from North America are known
from what were freshwater habitats, and although most
remains are Mesozoic, one jaw fragment is known from
the Early Tertiary Period.

*Below: This is a fossil of the long,
pointed, upper jaw (or rostrum)
of Belonostomus, showing some
of the teeth sockets. This fish
was probably a speedy predator
of small surface fish, such as
herring (clupeomorphs). Fins set
near the rear of the body signify a
fast-accelerating fish.*

Anterior (front) end
of jaw bone

Teeth

Tooth
sockets

FISH – TELEOSTS (CONTINUED)

Among the modern ray-finned bony fish, teleosts, the largest subgroup, include the perciforms, or perch-like fish, such as Mioplosus. Together with their close relatives, teleosts make up the Acanthopterygii, which includes more than half of all kinds of living fish. Their characteristic features include stiff bony spines in or near the front dorsal fin, giving the whole group the common name of spiny-rayed fish.

Mioplosus

Below: The fine-grained calcite limestones in which fossils such as this specimen have been found suggest that Mioplosus frequented areas similar to those of perches today – mainly upper and middle lake zones. This specimen is from Wyoming, USA, its main area of occurrence.

The spiny-fin, or perciform, *Mioplosus* has been found, although relatively rarely, in the exceptionally well-preserved Green River Formation rocks of Wyoming, USA. Its two dorsal (back) fins, and the comparable size of the second (rearmost) of these to the anal fin, which is directly beneath, are identifying features. The body is quite variable in proportions, but its length and strong build give it an overall perch-like appearance. Specimens of *Mioplosus* often seem to have died while trying to swallow, and choking on, fish up to half their own size. This suggests *Mioplosus* was a voracious predator, while the isolated nature of the finds indicate that it was solitary rather than shoaling.

Second dorsal fin

First dorsal fin

Mouth line

Anal fin

Name: *Mioplosus*.
Meaning: Small plosus.
Grouping: Fish, Actinopterygian, Teleost, Acanthopterygian, Perciform.
Informal ID: Extinct perch.
Fossil size: Head–tail length 28cm/11in.
Reconstructed size: Head–tail length up to 40cm/16in, and rarely up to 50cm/20in.
Habitat: Warm temperate to subtropical freshwater lakes.
Time span: Eocene, 50 million years ago.
Main fossil sites: North America (especially Wyoming, USA).
Occurrence: ◆

Diplomystus

This member of the herring family, Clupeidae, is recognized by its anal fin extending to its tail, and its deep body. *Diplomystus* also possessed two rows of scutes (bony plates) behind its head, dorsally (on its back) and ventrally (on the belly), extending to its paired fins. It lived from the Cretaceous Period through to the Early Tertiary Period, therefore surviving the mass extinction of 65 million years ago. Its fossils have been found worldwide, in both fresh water and marine habitats. It is especially abundant in the Green River Formation rocks of Wyoming, USA (see *Mioplosus*, above), where it grew to lengths in excess of 60cm/24in. *Diplomystus* fed on surface-water fish, as indicated by specimens fossilized with their prey still in their mouths or guts.

Name: *Diplomystus*.
Meaning: Double recess, twice hidden.
Grouping: Fish, Actinopterygian, Teleost, Clupeiform, Clupeid.
Informal ID: Extinct herring.
Fossil size: Head–tail length 43cm/17in.
Reconstructed size: Head–tail length up to 65cm/26in.
Habitat: Varied, seas and lakes.
Time span: Mid to Late Cretaceous to Early Tertiary, 80–50 million years ago.
Main fossil sites: Europe, Middle East, Africa, North and South America.
Occurrence: ◆ ◆ ◆

Left: Diplomystus had a drastically upturned mouth, a characteristic typical of surface-feeding fish. The extremely long anal fin on the underside of the rear body extends right to the deeply forked, homocercal (equal-lobed) tail fin.

Knightia

Name: Knightia.
Meaning: After Knight
(Wilbur Clinton Knight,
Wyoming's first state
geologist).
Grouping: Fish,
Actinopterygian, Teleost,
Clupeiform, Clupeid.
Informal ID: Extinct herring.
Fossil size: Head–tail length
9cm/3½in.
Reconstructed size:
Head–tail average length
10cm/4in, up to 25cm/10in.
Habitat: Warm temperate to
subtropical fresh water.
Time span: Mainly Eocene,
50 million years ago.
Main fossil sites: North and
South America (especially
Wyoming, USA).
Occurrence: ◆ ◆ ◆

Knightia has the distinction of being the vertebrate fossil most often found completely articulated – that is, with its structural parts still attached or aligned as in life. A herring and relative of *Diplomystus* (see opposite), it can be distinguished from the latter by its smaller size, shorter anal fin and slender body. *Knightia* is frequently found in mass mortality layers, in which thousands of individuals have died virtually simultaneously. This clearly suggests shoaling or schooling, but often the cause of death remains unclear – perhaps sudden temperature changes, or water stagnation with falling oxygen content, or rising levels of toxins due to algal blooms. Its small size, and common presence in the jaws or guts of larger fish, suggest that *Knightia* was near the beginning of the food chains and probably fed on plankton. Its phenomenal abundance in the Green River Formation rocks (see opposite) has led to it being appointed as Wyoming's state fossil.

Right: Soft-tissue preservation and perfect articulation of Knightia *fossils, as seen here, is sometimes observed in thousands of specimens packed densely together, with up to several hundred fish per square metre.*

Deeply forked homocercal tail · Orbit · Mouth

Vertebral column

Short anal fin

Pelvic fin

Pectoral fin

Dastilbe

Name: Dastilbe.
Meaning: Not known.
Grouping: Fish,
Actinopterygian, Teleost,
Ostariophysian,
Gonorynchiform.
Informal ID: Extinct milkfish.
Fossil size: Head–tail length
10.5cm/4in.
Reconstructed size:
Head–tail length of largest
individuals in excess of
20cm/8in, average 10cm/4in.
Habitat: Fresh to salty water,
including inland salt lagoons.
Time span: Early Cretaceous,
120 million years ago.
Main fossil sites: South
America (Brazil), Africa
(Equatorial Guinea).
Occurrence: ◆ ◆ ◆

Dastilbe was a relation of today's milkfish, a member of the major teleost group called the Ostariophysi, which includes more than 6,000 living species, such as catfish and carp. Their fossil history shows such features as modified hearing parts, termed Weberian ossicles, which link the swim bladder to the inner ear. This means that the swim bladder acts as a resonating chamber to improve underwater hearing. *Dastilbe* did not spread particularly far or last very long, but wherever it was present it was abundant, especially in the shallow tropical lakes of what is now Brazil. Fossil specimens are found with prey still in their mouths, indicating that *Dastilbe* was a predator – and when fully grown it was a cannibal, since some were eating young of their own kind.

Below: Fossilized Dastilbe *remains have been found in rocks from Brazil's Cretaceous Crato Formation, which were deposited as the beds of brackish lagoons. Most specimens were small juveniles, such as the example shown here, which could suggest that the adults were migratory and perhaps even marine. The young died when the water's salt level rose and became too high, and the fish were not sufficiently large to leave the protective waters of their nursery lagoon.*

Deeply forked homocercal tail

TETRAPODS – AMPHIBIANS

Tetrapods were and are vertebrates with four limbs – which today includes amphibians, reptiles, birds and mammals (although some, such as snakes, have lost limbs during their more recent evolution). The first air-breathing tetrapods probably evolved from fleshy-finned fish resembling Eusthenopteron *and* Panderichthys *into creatures such as* Ventastega *and* Acanthostega, *some 380–360 million years ago.*

Branchiosaurus

Below: The wedge-shaped skull, slim limbs (one front limb is missing due to slab fracture) and long tail are clearly visible in this example of Branchiosaurus from Rotleigend in northern Germany – an area famous for its Permian rocks, fossils and mineral reserves, such as natural gas. Total specimen length 9cm/3½in.

Tail — Rear limb

Front limb — Skull

Not to be confused with the massive dinosaur *Brachiosaurus*, which was some 26m/85ft long, *Branchiosaurus* belonged to the group of tetrapods known as temnospondyls. These appeared during the Early to Mid Carboniferous Period and included newt-like aquatic forms as well as large, powerful, land-going predators, such as *Eryops* (see overleaf) and *Mastodonsaurus*, which had scaly or bony plates on the skin and bore a vague outward resemblance to crocodiles. *Branchiosaurus* was a smallish, fully aquatic form, about 30cm/12in in length, that in some ways resembled the modern salamander-like axolotl. It had four weak limbs for walking in water, a long finned tail for swimming, and feathery external gills. Its family date from the Late Carboniferous Period through to the Permian. Usually only the young or larval forms of amphibians, known commonly as tadpoles, have external gills. ('Amphibian' is a commonly used though imprecise term usually applied to tetrapods that breathe using gills when young, and by gills and/or lungs when adult.) Keeping the larval gills as the rest of the body became sexually mature is an example of a well-known phenomenon in animal development termed paedomorphosis – retention of juvenile features into adult life – something that it shares with the Mexican axolotl. *Branchiosaurus'* way of life as a sharp-toothed aquatic predator was similar to that of *Micromelerpeton*, opposite.

Name: *Branchiosaurus*.
Meaning: Gill lizard.
Grouping: Vertebrate, Tetrapod, Temnospondyl, Dissorophoidean, Branchiosaurid.
Informal ID: Amphibian, prehistoric newt or salamander or axolotl.
Fossil size: See text.
Reconstructed size: Total length up to 30cm/12in.
Habitat: Tropical fresh water, swamps.
Time span: Permian, 290–260 million years ago.
Main fossil sites: Europe.
Occurrence: ◆ ◆

Below: This specimen of Branchiosaurus, also from Germany, measures 8cm/3⅛in in length. The front limbs are unclear. The gills of Branchiosaurus would have been very soft and fragile in life, and preserved only rarely. They would be positioned just behind the rear lateral angle, or backward-facing 'cheek', of the skull.

The Mexican axolotl
The rare Mexican axolotl salamander is said 'never to grow up'. This description comes about because it retains its feathery external gills, which most modern amphibians have only when they are larvae (immature young), throughout its adult life.

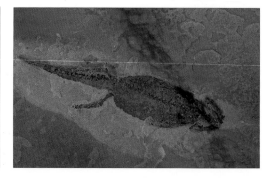

Micromelerpeton

Name: *Micromelerpeton*.
Meaning: Tiny *Melerpeton* (another proposed tetrapod genus).
Grouping: Vertebrate, Tetrapod, Temnospondyl, Dissorophoidean.
Informal ID: Amphibian, prehistoric newt or salamander or axolotl.
Fossil size: Total length 18cm/7in.
Reconstructed size: Total length up to 25cm/10in.
Habitat: Tropical fresh water, swamps.
Time span: Permian, 290–260 million years ago.
Main fossil sites: Europe.

In many ways similar to *Branchiosaurus* (see opposite), *Micromelerpeton* was a temnospondyl amphibian from the Permian Period. Its fossils are best known from Germany, such as in the Saar region, and France. Like *Branchiosaurus*, and modern salamanders and newts, it was an aquatic predator of small creatures, such as young fish, insect larvae and worms. It had sharp teeth not only in the jaws, but also on the palate (mouth roof). It, too, showed the phenomenon of paedomorphosis, retaining its larval feature of feathery external gills as the rest of its body metamorphosed into the sexually mature adult. At one time its remains were thought to be the larva of an unknown land amphibian. Some features of its shoulder girdle and other skeletal parts suggest that its ancestors may have been land-dwellers, after which *Micromelerpeton* returned to the water.

Below: This specimen of Micromelerpeton *distinctly shows the small, sharp teeth along the animal's jaw bone used for grabbing and impaling slippery prey, such as fish and worms. The rear limbs were slim and weak. They were not strong enough for walking on land, but they were suitable for manoeuvring among waterweeds or walking along the bed of a stream or lake. This species,* Micromelerpeton credneri, *is named in honour of German palaeontologist Hermann Credner (1841–1913).*

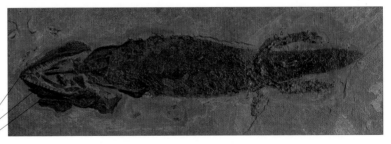

Snout
Teeth
Jaw bone

Diplocaulus

Name: *Diplocaulus*.
Meaning: Two-stemmed (double-stalked).
Grouping: Tetrapod, Lepospondyl, Nectridean.
Informal ID: Newt-like amphibian.
Fossil size: Width of skull 40cm/16in.
Reconstructed size: Head–tail length 1m/3¼ft.
Habitat: Streams and lakes.
Time span: Early Permian, 290–270 million years ago.
Main fossil sites: North America, also North Africa.
Occurrence: ◆ ◆

One of the most distinctive of all animal fossils is the skull of the 'giant Permian newt' *Diplocaulus*. The skull is immediately recognized from its boomerang-like shape, with two long, backswept, pointed 'horns', on either side, smallish eye sockets (orbits) set relatively close together near the front, and two much smaller nostrils just on top of the blunt snout. Associated backbones (vertebrae) and limb bones show that *Diplocaulus* was probably newt-like in shape and about 1m/3¼ft in length. The function of *Diplocaulus*'s extraordinary head shape is much debated. One suggestion is that it was a defensive adaptation – a predator would need a very large mouth gape to engulf the head. Or the skull may have been hydrodynamic, working like the wing of an aircraft, but in water, to provide a lifting force as *Diplocaulus* swam with its relatively weak limbs.

Right: This specimen of Diplocaulus *is from Baylor County, Texas, USA – the region of the famed Texan 'Red Beds', which have provided marvellous fossils of amphibians, reptiles and many other life-forms from Permian times. It shows the distinctive skull, most of the vertebral column and part of the shoulder girdle for the front left limb.*

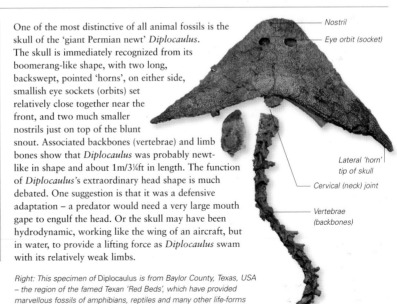

Nostril
Eye orbit (socket)
Lateral 'horn' tip of skull
Cervical (neck) joint
Vertebrae (backbones)

TETRAPODS – AMPHIBIANS (CONTINUED)

In the past, some types of amphibian were perfectly at home on dry land and, presumably, needed water only to lay their jelly-covered eggs in when spawning. Living amphibians are known in modern classification schemes as lissamphibians. The three main groups are frogs and toads, the tailless anurans; salamanders and newts, the tailed urodeles; and the legless, worm-like caecilians or apodans.

Eryops

A temnospondyl amphibian, like *Branchiosaurus*, *Eryops* was a strong, stoutly built and powerful predator. Its sturdy limbs probably allowed it to move reasonably well on dry land, and the tall spinal processes along the vertebrae of its backbone suggest effective back muscles that swished its body and tail from side to side when walking and swimming. Its way of life, however, was probably semi-aquatic. Both its eyes and nostrils were on top of the head, allowing it to lurk almost hidden in water yet still see above and breathe air. Or it may have skulked along the banks of rivers and lakes, on the lookout for likely prey. The temnospondyl group had died out by the Late Jurassic or perhaps the Early Cretaceous Period, more than 100 million years ago.

Name: *Eryops*.
Meaning: Long eye.
Grouping: Vertebrate, Tetrapod, Temnospondyl.
Informal ID: Giant prehistoric amphibian.
Fossil size: Total length 2m/6½ft.
Reconstructed size: As above.
Habitat: Rivers, lakes and swamps with dry ground.
Time span: Late Carboniferous to Permian, 300–280 million years ago.
Main fossil sites: North America.
Occurrence: ◆ ◆

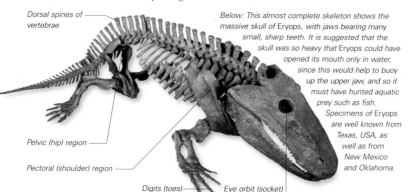

Dorsal spines of vertebrae

Below: This almost complete skeleton shows the massive skull of Eryops, with jaws bearing many small, sharp teeth. It is suggested that the skull was so heavy that Eryops could have opened its mouth only in water, since this would help to buoy up the upper jaw, and so it must have hunted aquatic prey such as fish. Specimens of Eryops are well known from Texas, USA, as well as from New Mexico and Oklahoma.

Pelvic (hip) region

Pectoral (shoulder) region

Digits (toes)

Eye orbit (socket)

Eye socket (orbit) | Skull | Ribs | Rear limb | Long tail
Front limb

This unidentified Miocene salamander shows the bony processes in the tail that held up a ridge of skin, for propulsion in the water.

Salamanders

The urodeles are commonly known as newts (mostly aquatic) and salamanders (mainly terrestrial). However, usage of these common names varies around the world, leading to potential confusion, and encompasses salamanders with names such as congo-eels, olms, axolotls and mudpuppies. They are mostly speedy swimmers and voracious hunters of varied small prey, from worms, grubs and insect larvae to fish and the tadpoles (larvae) of fellow amphibians. Although many invertebrates have the ability to regrow body structures, urodeles are the only examples of vertebrates that can completely regenerate limbs as adults. The fossil history of the salamander can be traced back to forms such as *Karaurus* from the Late Jurassic Period in Kazakhstan, Central Asia. Their design has changed little in almost 150 million years, with a long body and tail, and legs that project almost sideways. This specimen is dated to the Middle Miocene Epoch, about 10 million years ago.

Rana (R. pueyoi)

Name: *Rana pueyoi*.
Meaning: Frog of Pueyo.
Grouping: Vertebrate, Tetrapod, Lissamphibian, Anuran, Ranid.
Informal ID: Frog.
Fossil size: Head–body length 10cm/4in.
Reconstructed size: As above.
Habitat: Ponds, marshes and streams.
Time span: From Early Oligocene Epoch, 30 million years ago, to today.
Main fossil sites: Worldwide.
Occurrence: ◆ ◆

The huge genus *Rana* (typical or 'green' frogs) contains numerous species of living frogs, including the familiar common frog of Europe, *R. temporaria*, and the American bullfrog, *R. catesbeiana*. The fossil history of frogs and toads, the Anura, stretches back 200 million years, to the Early Jurassic Period. About 40 million years before this, the very frog-like *Triadobatrachus* is known from Early Triassic rocks of Madagascar. 'Modern' frogs of the genus *Rana*, with their relatively large, wide-mouthed, pointed skull, shortened, compact body, small shock-absorbing front legs for landing and large rear limbs and feet adapted for leaping, have remained essentially unchanged through most of the Tertiary.

Right: This specimen of Rana pueyoi *is from Late Miocene rocks (8 to 6 million years old) near Teruel in the mountainous, east-central region of Aragon, Spain – a locality also noted for its Early Cretaceous dinosaur fossils. It is very similar to living frogs of the same genus, with large eyes and an extensive eardrum behind each of these.*

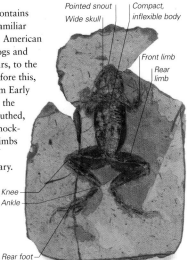

Pointed snout | Compact, inflexible body
Wide skull
Front limb
Rear limb
Knee
Ankle
Rear foot

Nanopus

Name: *Nanopus*.
Meaning: Tiny foot.
Grouping: Vertebrate, Tetrapod, possibly Temnospondyl.
Informal ID: Amphibian footprints.
Fossil size: Print size 1–2cm/½–¾in.
Reconstructed size: Animal length estimates 20–30cm/8–12in.
Habitat: Fresh water, marshes, rivers, lakes, ponds.
Time span: Late Carboniferous to Early Permian, 310–290 million years ago.
Main fossil sites: North America, Europe.
Occurrence: ◆ ◆

Nanopus is an ichnogenus – evidence for a creature based only on the traces it left, such as droppings, burrows or, in this example, footprints. The shape of the prints differ between the manus (the 'hand' of the front limb) and the pes (the 'foot' of the rear limb). Particularly well-known prints ascribed to the ichnogenera *Matthewichnus* and *Nanopus*, along with footprints and trackways made by many other vertebrates and invertebrates, are known from sites such as the Union Chapel mine of Walker County, Alabama, USA. These fossils are dated to the Westphalian (Moscovian) part of the Late Carboniferous Period, just more than 300 million years ago. Studies of these and other prints, including their spacing and progress pattern, the way the toes splay, and associated tail drags, give clues to the walking and swimming abilities of tetrapods such as *Nanopus*, *Limnopus* and *Matthewichnus* from Carboniferous times.

Above: The size of Nanopus and similar prints is generally on the order of 1–2cm/½–¾in. Some were left in the damp sand or mud where the currents did not disturb them. The water receded and the sediments were baked hard by the sun before fossilization. Expert analysis suggests several species of the genus Nanopus, including N. caudatus, N. obtusus, N. quadratus, N. quadrifidus and N. reidiae.

Left: These prints were made by the animal walking from right to left in soft sand. The placement can suggest if the animal was walking slowly or hurrying, as here. Blank areas between prints may suggest that the animal was partly floating – perhaps kicking its feet to move forward.

Heel area
Toe impression
Two prints close together

REPTILES

One of the most significant of evolutionary events was the amniotic egg – an egg sealed in an amniotic sac, usually enclosed in a tough outer shell. It first appeared with the tetrapods we call reptiles more than 300 million years ago. Its strong, encased structure allowed the reptiles to free themselves from the water, which the amphibians – dominant land vertebrates at the time – needed to lay their jelly-coated spawn.

Hylonomus

Often referred to as the 'first reptile', *Hylonomus* lived about 310 million years ago in the Late Carboniferous Period. Its remains come from the famed site of Joggins in Nova Scotia, Canada. *Hylonomus* outwardly resembled a lizard, but the lizard group of reptiles would not appear for many millions of years after this time. There has been much discussion about the circumstances of preservation of *Hylonomus*. Huge tree-like *Sigillaria* clubmosses lived and died in the area, snapping off their trunks low down as they fell, leaving stumps behind. Floods brought mud that built the ground up to the same level of the snapped-off *Sigillaria* stumps. These, in time, rotted from within to form hollows. Insects, millipedes and other small creatures fell into these 'pitfall traps'. Small predators, such as *Hylonomus*, followed them in to feed, but became trapped themselves and were buried in later sediments.

Below: The preservation detail of Hylonomus fossils is remarkable, with almost every body part present, even skin scales. The jaws shown here bear many conical teeth suited to snapping up small prey such as insects, grubs and worms.

Name: *Hylonomus*.
Meaning: Forest mouse.
Grouping: Reptile, Captorhinid.
Informal ID: First reptile, 'lizard'.
Fossil size: Jaw length 2cm/¾in.
Reconstructed size: Total length 20cm/8in.
Habitat: Swampy forests.
Time span: Late Carboniferous, 310 million years ago.
Main fossil sites: North America.
Occurrence: ◆

Elginia

Below: Elginia had an elaborate head shape with knobs, spikes and two long horns formed by outgrowths of the skull bones. This skull's larger side opening is the eye socket, the smaller one near the front is the nostril, and the central one on the forehead is the pineal, or 'third eye', which was possibly used to help regulate temperature.

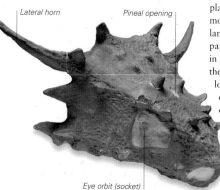

| Lateral horn | Pineal opening |

Eye orbit (socket)

The captorhinids were the earliest main reptile group to appear, led by *Hylonomus* (above). During the Middle Permian one of the subgroups, the pareiasaurs, began to spread and evolve into larger, more powerful forms. They were strongly built plant eaters and some reached the size of a modern buffalo, being among the largest land animals of the time. *Elginia* was a pareiasaur, but a dwarf form about 1.5m/5ft in length – one of the smallest and last of the group. It is named after its discovery location of Elgin, Scotland. Like many other pareiasaurs, it had bony growths over its head. Their function is not clear, but they may have been for defence against predators of the time, or for show and display to partners of their own kind when breeding. The pareiasaurs became all but extinct at the end of the Permian Period.

Name: *Elginia*.
Meaning: Of Elgin.
Grouping: Reptile, Captorhinid, Pareiasaur.
Informal ID: Pareiasaur, prehistoric reptile.
Fossil size: Skull length 25cm/10in.
Reconstructed size: Head–tail length 1.5m/5ft.
Habitat: Mixed scrub and plant growth.
Time span: Late Permian, 270–250 million years ago.
Main fossil sites: Europe.
Occurrence: ◆

Stereosternum

Name:
Stereosternum.
Meaning: Two-sided breastbone.
Grouping: Reptile, Captorhinid, Mesosaur.
Informal ID: Mesosaur
Fossil size: Slab 25cm/10in across.
Reconstructed size: Total length up to 30cm/12in.
Habitat: Shallow seas.
Time span: Early Permian, 285 million years ago.
Main fossil sites: South America, South Africa.
Occurrence: ◆

Soon after reptiles had developed their protective, tough-shelled amniotic egg and scaly skin to reduce water loss when on dry land, some types went back to the water. The first mesosaurs date from the Early Permian Period, approximately 290 million years ago, but the group did not last until the end of the period. Their fossils are known only from the Southern Hemisphere. Superficially resembling crocodiles, they are in a very different group of reptiles, usually linked to the 'primitive' captorhinids (see opposite). *Stereosternum* was a typical member of the mesosaur group and was adapted for an aquatic lifestyle. It had a long, slim neck, body and tail for sinuous swimming, and short limbs (rear ones larger than forelimbs), probably with webbed feet for paddling and steering.

Below: This beautifully preserved specimen from Brazil clearly shows the detailed skeleton of Stereosternum, *including the girdle bones, shoulder and hip, for both sets of the animal's limbs. Mesosaurs had long, thin, delicate teeth, which may have worked like a comb or sieve, to filter small creatures, such as shrimps or similar crustaceans that were plentiful at the time, from the ocean water.*

Hyperodapedon

Name: *Hyperodapedon*.
Meaning: Ground-hugging, short-legged reptile.
Grouping: Reptile, Archosauromorph, Rhynchosaur.
Informal ID: Rhynchosaur.
Fossil size: Skull length 22cm/8¾in.
Reconstructed size: Total length 1.4–1.8m/4½–6ft.
Habitat: Dry scrub with seed ferns.
Time span: Late Permian, 260–250 million years ago.
Main fossil sites: Europe, Asia.
Occurrence: ◆

The rhynchosaurs, or 'beaked lizards (reptiles)', were a spectacularly successful, although short-lived group from the Triassic Period. They were mostly stout animals with a barrel-shaped body with strong limbs, a hook-ended upper jaw and blunt teeth for crushing plant food. Their fossils are found so commonly in some localities, especially the southern continents, that they have been described as dotting the landscape and munching on plants like 'Triassic sheep'. Like most other rhynchosaurs, *Hyperodapedon* had several rows of teeth set in each side of the upper jaw, with a furrow or groove between them. The singe row of teeth on each side of the lower jaw fitted into this furrow to give the animal a powerful crush–shear action when chewing.

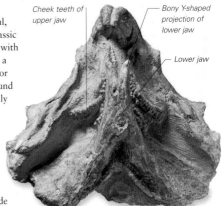

Cheek teeth of upper jaw

Bony Y-shaped projection of lower jaw

Lower jaw

Above: This view from below shows the tooth rows of Hyperodapedon in the upper jaw, and the much narrower lower jaw with its Y-shaped, or forked, tip.

Eye orbit (socket)

Snout

Down-curved 'beak' (rostrum) of upper jaw

Left: The distinctive shape made by the downward-curving tip of the upper jaw of rhynchosaurs such as Hyperodapedon may have been a food-gathering device. The creature might have used it to hook plant material towards its batteries of crushing teeth. The dished area to the upper middle-left is the eye socket, while the dished area to the upper right is a skull 'window', or fenestra. This may have allowed the jaw muscles to pass through or bulge easily as they contracted while chewing.

REPTILES – CHELONIANS, PHYTOSAURS

One of the most distinctive, long-lived reptiles groups is the chelonians – turtles, tortoises and terrapins. They appeared during the Triassic and have changed little in basic body form in more than 200 million years. The phytosaurs were also reptiles and from the same time. However, they are linked to a different group, the archosaurs ('ruling reptiles'), which included true crocodiles, dinosaurs and pterosaurs.

Emys

The family Emydidae of 'typical' freshwater and semi-aquatic turtles includes more than 90 living species, as well as many more extinct ones, and it makes up not only the largest, but also the most diverse of the turtle families, being found in both North and South America, Europe, Northern Africa and Asia. Although emydid species live mainly in fresh water, some species are also found in brackish waters or are terrestrial in habitat. The genus *Emys* is represented by well-known living species, including the European pond turtle (*Emys orbicularis*), the water terrapin (*E. elegans*) and the western pond turtle (*E. marmorata*). Early members of the *Emys* group are known in the fossil record from Nebraska, in the central USA, dating back to the Barstovian (Langhian) Age of the Middle Miocene Epoch – approximately 15–13 million years ago. However, other fossils, which have been found both in North America and Asia, and the geographic spread of these remains, suggest an origin for the group in the Late Eocene. For further details of turtle anatomy see below.

Name: *Emys*.
Meaning: Turtle.
Grouping: Reptile, Chelonian, Emydid.
Informal ID: Turtle.
Fossil size: Fragments 2–3cm/¾–1in across.
Reconstructed size: Total length 50cm/20in.
Habitat: Fresh water.
Time span: Late Eocene, 35 million years ago, to today.
Main fossil sites: Worldwide.
Occurrence: ◆ ◆

Broken edge of plate

Junction line between plates

Left: These fragments of shell are probably from a representative of the turtle genus Emys. *The dividing lines between the plates, where they are joined firmly together, form a characteristic 'paving' effect. Sometimes the vertebrae (backbones) and ribs are also preserved, firmly fixed to the inner side of the upper shell, or carapace.*

The turtle shell

Chelonian anatomy has changed little since the group appeared relatively suddenly in the Late Triassic Period. The skull of the animal is box-like, with large eye sockets (orbits) but no openings or 'windows' behind them – a distinctive skull structure known as anapsid. The jaws lack teeth and form a horny sharp-edged beak for chopping food. The shell consists of a dome-like carapace above and a flatter, shelf-like plastron on the underside. The vertebrae, ribs and limb girdle bones are all fused to the inside of the carapace. The shell, in fact, has two layers, each made up of flat sections fused together. The inner layer has bony plates and the outer one has horny scutes that form the visible observed pattern of the shell. The carapace pictured here, from the Purbeck Beds of Dorset, southern England, is dated to the Late Jurassic Period or Early Cretaceous Period, about 140–130 million years ago.

Vertebral column (backbone)

Main area of carapace

Lateral struts

Lateral plates form edging 'skirt'

Phytosaurus

Name: *Phytosaurus*.
Meaning: Plant lizard/reptile.
Grouping: Reptile, Archosaur, Crocodylotarsian, Phytosaur.
Informal ID: Phytosaur, crocodile-like extinct reptile.
Fossil size: Tooth length 4.5cm/1¾in.
Reconstructed size: Total length 5m/16½ft.
Habitat: Fresh water, rivers and lakes and swamps.
Time span: Late Triassic, 220–210 million years ago.
Main fossil sites: North America.
Occurrence: ◆ ◆

The phytosaurs (parasuchians) were one of several groups of crocodile-like reptiles whose fossil record dates them from the Triassic Period. They were among the main semi-aquatic predators found in the Northern Hemisphere at that time, while their similar cousins, the rauisuchians, prowled more on dry land, and a third related group, the aetosaurs, became plant-eaters. All three of these groups have been placed within the crocodylotarsians, or 'crocodile ankles'. This is because fossils of their limbs show that their feet could be turned so that the toes faced more to the side, rather than being restricted to just facing forwards. *Phytosaurus*, which gave the phytosaur group its name, was misleadingly named 'plant lizard (reptile)', since its sharp teeth clearly show that it was a carnivore. The name came from the parts that were first studied as fossils. These were cast fossils from the blunt-tipped empty tooth sockets that acted as moulds, and these were mistaken at the time for the crushing teeth typical of a herbivore. (See also *Rutiodon*, below.)

Below: The conical tooth of Phytosaurus *is recurved: that is, bent backwards towards the animal's throat, to prevent slippery prey such as fish from wriggling free. This specimen showing wear marks from struggles with victims is from Texas, USA.*

Base of tooth

Posterior (rear) surface

Wear marks

Anterior (front) surface

Rutiodon

A member of the phytosaur group (see *Phytosaurus*, above), *Rutiodon* – which may include specimens known as *Leptosuchus* – is known from many fossil sites across North America and Europe. Like most phytosaurs it was armoured over its body with bony plates known as scutes, similar to those of a crocodile, which form characteristic fossils even when they fall away from the decayed skin. Its snout was very long, low and narrow, similar to the modern-day crocodilian known as the gharial from the Indian region. This design may be adapted to swishing sideways through the water with minimal resistance when snapping at prey. The skulls of phytosaurs such as *Rutiodon* and *Phytosaurus* can be distinguished from true crocodiles especially by the presence of nostrils high up on the forehead, just in front of and between the eyes, rather than on the top of the tip of the snout as in crocodilians.

Name: *Rutiodon*.
Meaning: Wrinkled tooth.
Grouping: Reptile, Archosaur, Crocodylotarsian, Phytosaur, Angistorhininan.
Informal ID: Phytosaur, crocodile-like extinct reptile.
Fossil size: Tooth length 5cm/2in.
Reconstructed size: Total length 3m/10ft.
Habitat: Fresh water.
Time span: Late Triassic, 220–215 million years ago.
Main fossil sites: Europe, North America.
Occurrence: ◆ ◆

Below: The gharial is one of the largest living crocodilians, growing to lengths in excess of 6m/20ft. With weak legs and webbed feet, it is also one of the most aquatic, both catching and eating prey – chiefly fish – in the water.

Below: This specimen of a Rutiodon tooth from Arizona, USA, shows the characteristically prominent vertical grooves, or striations – the genus name means 'wrinkled tooth'. The root is flat and 'cut-off'. Like crocodilians, phytosaurs lost their teeth regularly and, as this happened, new ones grew from the jaw, so the mouth was usually filled with different-sized teeth of various ages.

Wrinkles or grooves

Base of tooth

Convex front surface

REPTILES – CROCODILIANS

Crocodiles, alligators, caimans and gharials are semi-aquatic predators around the warmer parts of the world. Yet the 23 or so living species of crocodilians are a small remnant of a once large and diverse group dating back more than 220 million years. They may have started as lizard-like terrestrial animals. Crocodilians are included with dinosaurs and pterosaurs as archosaurs, or 'ruling reptiles'.

Goniopholis

Supratemporal fenestra
(skull opening)

The typical body design of a crocodilian such as the Late Jurassic–Early Cretaceous *Goniopholis* is for ambush predation. The skull was adapted for lurking in water. The eyes and nostrils are on the upper surface, so that the animal can see, breathe and smell while floating almost submerged. (Crocodilian nostrils are at the tip of the snout, differing from the position in phytosaurs already covered.) The snout is long and low, the body thick-set and powerful and the legs usually project sideways for a sinuous, slithering gait. The tail is narrow and tall and provides the main thrust for fast swimming, when the limbs are usually held against the body; the webbed feet may be used as paddles for slower movement and manoeuvring in water. Crocodilians have protective bony slabs called scutes that grow embedded in the skin, especially along the upper surface and sides of the animal (see below), and provide plentiful fossils.

Right: Goniopholis *had a single nostril opening, an undivided naris, at the upper tip of the snout. The bone surface is ornamented with a waffle-like texture. The eye sockets, or orbits, are smaller than the gaps above them in the skull, known as the supratemporal fenestrae.*

Nostril

Name: *Goniopholis*.
Meaning: Angled scutes.
Grouping: Reptile, Archosaur, Crocodilian.
Informal ID: Crocodile.
Fossil size: Skull length 70cm/27½in.
Reconstructed size: Head–tail length 3–4m/ 10–13ft.
Habitat: Rivers, lakes and swamps.
Time span: Late Jurassic to Early Cretaceous, 150–130 million years ago.
Main fossil sites: Europe, North America.
Occurrence: ◆ ◆

Diplocynodon

Below: This is a preserved piece of the body armour of Diplocynodon, *from the bony plates called scutes (see* Goniopholis, *above). Each type of crocodilian has a characteristic pattern of scute sizes, shapes and surface textures – the one here being squarish and dimpled.*

Crocodiles can be distinguished from alligators principally by the structure of their teeth. In alligators, for example, the fourth tooth in the lower jaw fits into a corresponding socket, or pit, in the upper jaw, and so cannot be seen when the mouth is closed. In crocodiles, however, this tooth slides into a notch on the side of the jaw, and so the tooth is readily visible when the animal's mouth is closed. This key identifying feature is useful when unknown fossil material of the jaws and teeth are available, in which case the relevant part of the upper jaw can distinguish between them. In addition, the snout shape of an alligator such as *Diplocynodon* is generally broader, more foreshortened and more rounded at the tip than that of a crocodile. *Diplocynodon* remains are found especially from swampy and marshy habitats in Europe about 30–20 million years ago, during the Oligocene Epoch, although remains of *Diplocynodon* from the Bridger beds of Wyoming, USA, have also been discovered.

Name: *Diplocynodon*.
Meaning: Double dog tooth.
Grouping: Reptile, Archosaur, Crocodilian.
Informal ID: Alligator.
Fossil size: 4cm/1½in across.
Reconstructed size: Head–tail length 3m/10ft.
Habitat: Lakes, rivers and swamps.
Time span: Eocene to Pliocene, 45–5 million years ago.
Main fossil sites: North America, Europe.
Occurrence: ◆ ◆

Metriorhynchus

Name: *Metriorhynchus*.
Meaning: Moderate snout.
Grouping: Reptile, Archosaur, Crocodilian.
Informal ID: Marine crocodile.
Fossil size: 22cm/8⅔in.
Reconstructed size:
Head–tail length 3m/10ft.
Habitat: Shallow seas.
Time span: Middle Jurassic to Early Cretaceous, 160–120 million years ago.
Main fossil sites: South America, Europe
Occurrence: ◆ ◆

This crocodile was highly adapted for life at sea, and may have come ashore only to lay eggs, in much the same way that marine turtles do today. Its limbs were paddle-shaped with extensive toe webbing. The tail end of the backbone (vertebral column) angled downwards and supported a large fish-like fin that provided sideways, sweeping propulsion through the water. The snout was very slim, long and pointed, in the style of the living crocodilian known as the gharial (gavial). This species is mainly a fish-eater, swiping its snout sideways to catch slithering prey with its many sharp teeth; *Metriorhynchus* probably did the same. Although shorter than modern crocodiles, *Metriorhynchus* would have been a formidable hunter, with a powerful, streamlined body ideally suited for swimming down a variety of prey. An adaptation to its fast-swimming way of life meant that it lost most of its protective armour, leaving it vulnerable to attack from larger and more powerful reptiles.

Below: The vertebrae of Metriorhynchus, *seen here from above, had characteristic joints that distinguished modern crocodiles from more primitive ones. The large lateral processes projecting to either side would have anchored powerful back muscles along the sides of the body and tail, that arched them left and right when swimming. Specimen length 22cm/8⅔in.*

Right: The teeth of Metriorhynchus *were slimmer and more pointed than the sturdier, conical teeth of land-dwelling crocodilians. Tooth length 3cm/1in.*

Conical tip

Convex anterior (front) surface

Recurved posterior (rear) surface

Tooth base

Dorsal spines or processes (projections)

Transverse, or lateral, processes (projections)

Zygapophyseal prongs

Intervertebral joint

Centrum (body) of vertebra

Body of rib

Distorted fossil femur

These femurs (thigh bones) of *Metriorhynchus* contrast a healthy one (on the left) with a diseased, arthritic one (on the right). In a land crocodilian, where the walking weight is borne solely by the limbs, this disfiguring condition may have been a much more considerable handicap compared with the same condition in a swimming *Metriorhynchus*. The rear limbs of *Metriorhynchus* were larger than the front pair. Femur length 28cm/11in.

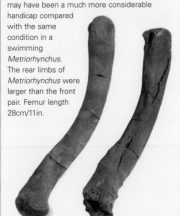

Above: The ribs of Metriorhynchus *were slim and lightweight. Despite this marine crocodile's highly modified adaptations, it remained an air-breather and so the ribcage and the lungs within would function in the normal reptilian way. Rib length 18cm/7in.*

Joint with thoracic vertebra (backbone in chest)

REPTILES – NOTHOSAURS, LIZARDS AND SNAKES

Several groups of reptiles took to the seas during the early Mesozoic, including the semi-aquatic placodonts, nothosaurs such as Lariosaurus, *their later cousins the plesiosaurs and also the ichthyosaurs. Meanwhile, on land the lizards had appeared, but snakes did not evolve until the end of the Mesozoic.*

Lariosaurus

Above: This fine juvenile specimen of Lariosaurus *from Como, Italy is dated to the Mid Triassic, about 225 million years ago. Its fossils are also known from Spain and other European locations. The large spread fingers may have been encased in tough skin and connective tissue in life, to form flipper-like structures. The skull is wedge-shaped and the neck is flexible but relatively short.*

The nothosaurs are named after their best-known member, the 3m/10ft *Nothosaurus* of Europe and West Asia. *Lariosaurus* was much smaller, just 60cm/24in long. Some dwarf types of nothosaurs were just 20cm/8in long. Nothosaurs had partly webbed feet. Larger nothosaurs probably lived in a similar way to seals today, swimming fast after prey such as fish, which they seized in their long, pointed, fang-like teeth. Then they would haul out to rest on land. *Lariosaurus* was less aquatic than other types. It may have foraged among rockpools or shallow seaweed beds, slithering and paddling after small victims such as young fish and shrimps.

Name: *Lariosaurus*.
Meaning: Lario (Lake Como) lizard or reptile.
Grouping: Reptile, Nothosaur, Nothosaurid.
Informal ID: Nothosaur.
Fossil size: Slab width 25cm/10in.
Reconstructed size: Head–tail length up to 60cm/24in.
Habitat: Coastal waters.
Time span: Middle Triassic, 225 million years ago.
Main fossil sites: Europe.
Occurrence: ◆ ◆

Kuehneosaurus

Vertebrae (3)

A type of 'flying lizard' or 'gliding lizard', *Kuehneosaurus* fossils come mainly from England. The ribs were long and hollow and directed sideways, being longest in the middle of each side. This produced a wing-like framework that in life was probably covered by tough skin, to give a gliding surface. The modern flying lizard of Southeast Asia, *Draco volans*, probably leads a similar lifestyle. It uses its gliding ability to swoop from the upper part of one tree to the lower part of a nearby one, in order to find fresh feeding areas and, especially, to escape from danger. However the ribs of the modern *Draco* are jointed with the vertebral column so the wings can be folded against the sides of the body. The 'wingspan' of *Kuehneosaurus* is estimated at about 30cm/12in, being half its total length.

Left: This specimen of Kuehneosaurus *shows mainly the long, slim, hollow, light ribs, plus a femur, or thigh bone. This genus is regarded as a primitive member of the group, not advanced enough to belong to the 'true' lizards.*

Femur

Ribs

Name: *Kuehneosaurus*.
Meaning: Kühne's (Kuehne's) reptile.
Grouping: Reptile, Squamate, Kuehneosaurid.
Informal ID: Flying lizard.
Fossil size: Slab 15cm/6in across.
Reconstructed size: Head–tail length 60cm/24in.
Habitat: Woodland, scrub.
Time span: Late Triassic, 210 million years ago.
Main fossil sites: Europe.
Occurrence: ◆

Ardeosaurus

Name: *Ardeosaurus*.
Meaning: Water lizard/reptile.
Grouping: Reptile,
Squamate, Lacertilid,
Ardeosaur.
Informal ID: Gecko.
Fossil size: Slab 9cm/
3½in across.
Reconstructed size: Total
length 20cm/8in.
Habitat: Woods, forests.
Time span: Late Jurassic,
150–140 million years ago.
Main fossil sites:
Throughout Europe.
Occurrence: ◆ ◆

The ardeosaurs are better known as geckos – lizards named from their sharp, bark-like calls. Today, the gecko family is huge, with approaching 700, mainly tropical, species around the world. The geckos were also one of the first modern lizard groups to appear, in the Late Jurassic Period, along with others, such as skinks, iguanas and monitor lizards, or varanids. However, lizard-like reptiles had been around since the Early Triassic Period and even the Late Permian, more than 250 million years ago. *Ardeosaurus* probably lived largely nocturnally, very much like its present-day equivalents. The large orbits (eye sockets) in the skull show that it had big eyes, probably to enable it to see well at night as it snapped up insects, spiders, grubs and similar small prey.

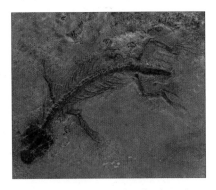

Above: This type specimen cast from Bavaria, southern Germany, is dated to the Late Jurassic Period, some 150 million years ago. It has been preserved in the very fine-grained 'lithographic' limestone that also trapped the early bird Archaeopteryx *and the tiny dinosaur* Compsognathus. *Most of the tail of this individual is missing.*

Columber

The snake group, Serpentes, is placed with lizards in the major reptile group Squamata. The fossil history of snakes is far shorter than that for lizards, stretching back to the Late Cretaceous Period about 80 million years ago. The group origins are obscure: two suggestions include evolution from some type of burrowing lizard, hence the loss of limbs; or evolution from a form of aquatic lizard. The snake skull is usually low and squat, with a flattened forehead and loose jaw articulation (joint). This allows a very wide gape and 'dislocation' of the jaws to permit swallowing large prey whole, since the teeth are slim and fang-like and cannot bite off lumps or chew.

Name: *Columber kargi*
Meaning: Kargi's whip-snake
Grouping: Reptile,
Squamate, Serpent
Informal ID: Whip-snake
Fossil size: Slab width
30cm/12in
Reconstructed size: Total
length 80–90cm/32–36in
Habitat: Forests
Time span: (This specimen)
Late Miocene, 8 million years
ago. Genus continues today.
Main fossil sites: Europe
Occurrence: ◆

Right: This specimen, Columber kargi, *is in Late Miocene rocks from Oenigen, Switzerland. The snake skeleton has typically just a skull and a very elongated spinal column of many vertebrae (backbones), sometimes more than 400, mostly with arched ribs.*

Ribs

Skull

Cervical (neck) vertebrae

Caudal (tail) vertebrae

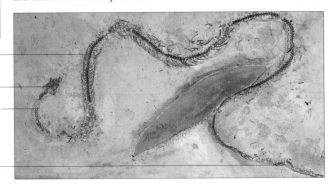

MARINE REPTILES – ICHTHYOSAURS

The ichthyosaurs, or 'fish-reptiles', were the reptiles most highly adapted to aquatic life. They appeared abruptly in the Early Triassic Period and by the Late Triassic to Late Jurassic they were among the dominant large predators in many seas. However, they declined during the Cretaceous and had all but disappeared by the end-of-Cretaceous (K-T) mass extinction that marked the end of the Mesozoic Era.

Ichthyosaurus

This group of marine reptiles is named after the medium-sized *Ichthyosaurus*, which grew up to 3m/10ft in length. Many species have been assigned to this genus, sometimes on the flimsiest of fossil evidence. The genus is said to have persisted in a variety of different forms from the Early Jurassic Period to the Early Cretaceous Period, a time span covering more than 80 million years. A 'typical' *Ichthyosaurus* was shaped like a fish or dolphin – making it an example of 'convergent evolution'. This is when different organisms come to resemble each other due to a similar habitat and/or lifestyle. The jaws of *Ichthyosaurus* were very long, slim and pointed – similar to a dolphin's beak, only more exaggerated. These predators were armed with many small, sharp teeth for impaling slippery prey such as fish and squid. The eyes of many types were large, indicating that they hunted at night or deep in the gloomy water where light penetration was poor. All four limbs were modified as paddles, with the front pair usually larger than the rear ones. These were used mainly for steering, manoeuvring and braking. A dolphin-like fin on the back prevented the animal leaning or rolling to the side. The caudal vertebrae (tail backbones) angled down at the end to support a large upright fin, similar to the tail of a fish. Swishing this from side to side provided the thrust for speedy swimming.

Name: Ichthyosaurus.
Meaning: Fish reptile (fish lizard).
Grouping: Reptile, Ichthyosaur.
Informal ID: Ichthyosaur.
Fossil size: Slab length 1m/3¼ft.
Reconstructed size: Total length up to 3m/10ft.
Habitat: Seas.
Time span: Chiefly Jurassic, 200–150 million years ago.
Main fossil sites: Worldwide.
Occurrence: ◆ ◆ ◆

Below: This Early Jurassic ichthyosaur is Stenopterygius, and it is from the Early Jurassic Posidonia Shale of Holzmaden, Germany. Some of the ring-like vertebrae have been displaced during preservation, as has the end of the rear (pelvic) flipper. Long, thin ribs have also been scattered across the main body region. The front flipper shows numerous digit bones set closely together, which in life would have been bound with muscle and connective tissue.

Left: Shonisaurus from the Late Triassic was one of the largest ichthyosaurs, reaching lengths of 15m/50ft – as big as some great whales of today.

| Long, slim jaws | Small sharp | Large orbit | Shoulder girdle | | Hip girdle | Displaced | Rear flipper |
| Pointed snout | teeth | (eye socket) | Front flipper | Ribs | | vertebrae | Angle in tail |

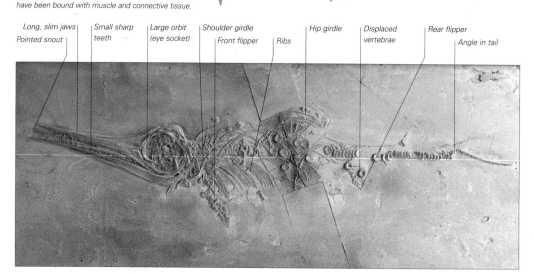

Ophthalmosaurus

One of the larger ichthyosaurs, *Ophthalmosaurus* was more than 3m/10ft long. (Some ichthyosaurs reached huge sizes, in excess of 15m/50ft in length.) The genus was named after its remarkable distinguishing feature – its huge eyes. Each eyeball was more than 10cm/4in across and was housed in a large orbit whose deepest recess almost touched that of the orbit in the other side of the animal's skull. Around the orbit was a ring arrangement of rectangular to wedge-shaped bony plates, known as the sclerotic ring (see also right). All ichthyosaurs had this structure, which probably protected the soft eyeball tissues from excessive water pressure at depth. However, the ring was very large and pronounced in *Ophthalmosaurus*.

Below: Ichthyosaur vertebrae (backbones) have a characteristically disc-like construction, with pronounced dished (concave) surfaces. The two larger fossil specimens shown here have a slight constriction around the side, producing an hour-glass shape, and are from Peterborough, England, and they are probably from Ophthalmosaurus itself. The smaller one for comparison is from an unidentified species of ichthyosaur.

Name: *Ophthalmosaurus*.
Meaning: Eye reptile (eye lizard).
Grouping: Reptile, Ichthyosaur, Ichthyosaurid.
Informal ID: Ichthyosaur.
Fossil size: Eye ring 16cm/6¼in across, largest vertebra 8cm/3¼in across.
Reconstructed size: Length 3.5m/11½ft.
Habitat: Seas.
Time span: Middle to Late Jurassic, 160–140 million years ago.
Main fossil sites: Chiefly North and South America, Europe.
Occurrence: ◆ ◆

Ichthyosaur details and lifestyle

Ichthyosaurs were air-breathers like other reptiles. Their nostrils were placed just in front of their eyes, and they would hold their breath while diving to locate and chase prey. Many details are known about their soft body tissues due to the remarkable preservation of specimens at sites such as Holzmaden, Germany. Slow decay has produced carbon film outlines, which delineate the shapes of the flippers and the outline of the dorsal (back) and caudal (tail) fins, all as they were in life. Preserved skin shows that it was smooth, not scaly. In addition there is evidence for live birth rather than egg-laying, with many fossils of young developing within their mother's body. There are even remarkable specimens of babies being born tail-first (as in modern whales and dolphins).

Above: Furrow-surfaced coprolites – fossilized droppings – of ichthyosaurs contain remains of fish scales and bones, as well as the hard parts such as the 'beaks' of molluscs such as squid.

Below: This ichthyosaur skull clearly shows the protective sclerotic ring around the eye (see above and right) and the very long and sharp-toothed jaws resembling needle pliers. Specimen length 50cm/20in.

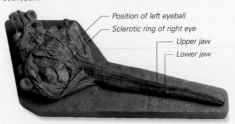

— Position of left eyeball
— Sclerotic ring of right eye
— Upper jaw
— Lower jaw

Below: The sclerotic ring of Ophthalmosaurus probably formed an anti-pressure guard around the exposed part of the eyeball. Such huge eyes were probably an adaptation for hunting prey either at night or at depth.

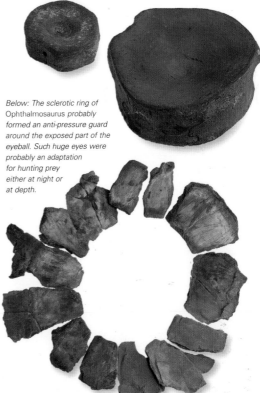

MARINE REPTILES – PLESIOSAURS AND THEIR KIN

Plesiosaurs were the greatest ocean-going reptiles of the Jurassic and Cretaceous seas at the time when dinosaurs ruled the land. There were two groups. The plesiosaurs were long-necked with smallish heads and hunted smaller prey than the pliosaurs. The pliosaurs had much larger heads and mouths armed with sharp teeth. All died out with the dinosaurs, some 65 million years ago.

Pliosaur

A typical plesiosaur or pliosaur had a tubby body, four paddle-like limbs, and a shortish, tapering tail. An Early Jurassic pliosaur was *Macroplata*, and a Later Jurassic one *Peloneustes*, both 3–4m/10–13ft in overall length. During the following Cretaceous Period, some pliosaurs reached massive proportions. One was *Liopleurodon*, whose size estimates range up to 15–20m/50–65ft plus from nose to tail, and its weight a staggering 50 or even 100 tonnes. If so, this would place it as the largest predator ever to live on Earth – larger even than today's record-holder, the sperm whale. Pliosaurs seized substantial prey, such as big ammonites and fish, as well as other marine reptiles, too, with their rows of sharp, back-curved fangs.

Above: A vertebra (backbone) from a large pliosaur, dated to the Middle Jurassic, showing the rounded cotton-reel-like shape that is common in many Mesozoic marine reptiles.

Name: Pliosaur (genus unknown).
Meaning: Further lizard.
Grouping: Reptilia, Plesiosaur, Pliosaur.
Informal ID: Pliosaur, short-necked plesiosaur.
Fossil size: Thigh bone 50cm/20in; tooth 10cm/4in.
Reconstructed size: Nose–tail length 5–10m/16½–33ft.
Habitat: Seas and oceans.
Time span: Much of Mesozoic, 250–65 million years ago.
Main fossil sites: Worldwide.
Occurrence: ◆ ◆ ◆

Plesiosaurs

These marine dinosaurs did not reach the massive size of pliosaurs. However, they did develop amazingly long necks. *Elasmosaurus* from eastern Asia and North America reached 14m/46ft in length, and more than half of this was its neck. This is a vertebra from the plesiosaur (long-necked type) *Plesiosaurus*, which grew to about 2.2m/7¼ft in total length. This is a similar size to many modern dolphins.

Above: Pliosaur teeth were huge and saw-edged, showing that this reptile was a savage predator. Wear facets indicate where the teeth situated on either side would have rubbed in life. This specimen may be from Liopleurodon.

Below: One of several possible paddle limb bones of a pliosaur from the Middle Jurassic Period. It may be the femur (equivalent to the thigh bone) at the base of the left rear paddle.

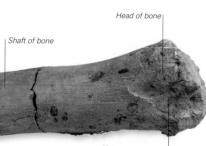

Head of bone

Shaft of bone

Articular (joint) surface

Coniasaurus

Coniasaurs were probably aquatic reptiles similar in overall appearance to a small mosasaur or lizard. Their relationships with the mosasaurs and aigialosaurs are not clear, but from various skull and skeletal features they are thought to be members of the larger squamate group, which included lizards and snakes as well as mosasaurs. Coniasaurs were probably predators on small fish, young ammonites and similar animals. Their remains were once known only from England, but from the 1980s they have been discovered in North America, in the Eagle Ford Group of Texas (Travis, Bell and McClennan counties in the Balcones Fault zone and farther north in Dallas County).

Name: *Coniasaurus*.
Meaning: Cretaceous reptile.
Grouping: Reptile,
Squamate, Coniasaur.
Informal ID: Coniasaur.
Fossil size: 10cm/4in.
Reconstructed size:
Nose–tail length 50cm/20in
when adult.
Habitat: Seas.
Time span: Late
Cretaceous, 85–65 million
years ago.
Main fossil sites: Europe,
North America.
Occurrence: ◆

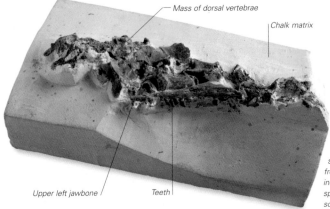

Mass of dorsal vertebrae

Chalk matrix

Upper left jawbone

Teeth

Left: The upper left maxilla (jawbone) of a juvenile of the best-studied (although still poorly known) species, Coniasaurus crassidens. The animal's upper front teeth are pointed, but those situated behind are increasingly rounded and blunt or bulbous. This specimen is from the Cretaceous chalk beds of Sussex, southern England.

Mosasaurus

Mosasaurs were large, four-flippered, slim, long-tailed, predatory marine lizards, named after the Meuse region of the Netherlands where the first scientifically described fossil specimen was dug from a chalk mine in the 1770s. They were close cousins of the varanids, or monitor lizards, of today, which include the Nile monitor and huge Komodo dragon. Mosasaurs, along with pliosaurs, dominated the Late Cretaceous seas, and although the two look similar, they were from different reptile groups.

Below: This is the lower jaw bone, with teeth in place, from the species Mosasaurus gracilis, meaning 'slender mosasaur'. Its teeth are well spaced, like a crocodile's today, but they are much sharper. The V-shaped patterns of mosasaur tooth marks appear on many prey creatures of the time, including the shells of ammonites.

Name: *Mosasaurus*.
Meaning: Meuse
lizard.
Grouping: Reptile,
Squamate, Lacertid, Varanid.
Informal ID: Mosasaur.
Fossil size: Lower jaw
38cm/15in long.
Reconstructed size:
Nose–tail length up to
10m/33ft.
Habitat: Seas.
Time span: Mid to Late
Cretaceous, 100–65 million
years ago.
Main fossil sites: Europe,
North America.
Occurrence: ◆ ◆

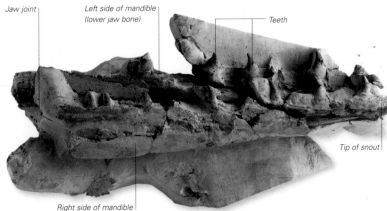

Jaw joint

Left side of mandible
(lower jaw bone)

Teeth

Tip of snout

Right side of mandible

DINOSAURS AND THEIR KIN

Dinosaurs were members of the reptile group called archosaurs and dominated the land from the Mid to Late Triassic Period, around 225 million years ago, to the sudden end-of-Cretaceous mass-extinction event of 65 million years ago. One of their key features was legs held directly below the body, as in birds and mammals, rather than angled out to the side, like other reptiles such as lizards and crocodiles.

Chirotherium

Chirotherium looked similar to a dinosaur, and probably belonged to the same major group of reptiles as the dinosaurs, called the archosaurs, although it lived just before the dinosaur group appeared. It had a large mouth with sharp teeth for catching varied prey, including smaller reptiles. *Chirotherium* usually walked on all fours, although its rear legs were larger and stronger than the front ones, with its tail held out behind, rather than dragging. Its legs were straight, in the upright posture of dinosaurs. Fossil bones of *Chirotherium* have not been found. Its restoration comes from the remains of *Ticinosuchus*, which had similar feet.

Right: The print is a mould fossil, where sediment filled in the impression and then hardened, and so is shown here effectively upside down. It is a rear right foot with four smaller claw-tipped toes. What looks like the big toe is the smallest one set at an angle.

Claw of outer digit
(smallest toe)

Claws of four
larger digits

Name: *Chirotherium.*
Meaning: Hand beast (footprint is similar to the impression of a human hand).
Grouping: Reptile, Archosaur, possibly Pseudosuchian.
Informal ID: Chirotherium, 'pre-dinosaur'.
Fossil size: Print about 15cm/6in across.
Reconstructed size: Nose to tail length 2–2.5m/6½–8¼ft.
Habitat: Dry scrub.
Time span: Triassic, 250–203 million years ago.
Main fossil sites: Throughout Europe.
Occurrence: ◆

Grallator (dinosaur footprint)

Below: This 10cm-/4in-long specimen is dated to 207–206 million years ago. The relative length of the toes suggests a Coelophysis-type carnivore, which would have had a smallish head, flexible neck, slim body, long rear legs and a very long tail.

Long central toe

Claw impression

Grallator is an ichnogenus – in other words, it is named from trace fossils rather than from the remains of the creature itself. The footprints known as *Grallator* were made by an unknown dinosaur. However, the size and the shape of the prints allow scientists to make certain assumptions, and indicate that the creature was probably an early meat-eating or theropod-type of dinosaur. It would have measured around 3m/10ft in length and weighed in at some 35–40kg/ 77–88lb, similar to the well-known *Coelophysis*, which lived about 20 million years earlier. The prints of several individual animals often occur in lines or trackways, showing that this was probably a group-dwelling or herding dinosaur.

Name: *Grallator* (print fossil).
Meaning: Stilt walker.
Grouping: Reptile, Dinosaur, Theropod.
Informal ID: Grallator.
Fossil size: Prints up to 17cm/6¾in long.
Reconstructed size: Nose–tail length 3m/10ft.
Habitat: Varied land habitats.
Time span: Late Triassic and Early Jurassic, around 200 million years ago.
Main fossil sites: North America, Europe.
Occurrence: ◆ ◆ ◆

Gastroliths

Gastroliths, or stomach stones, are from the digestive system of larger plant-eating dinosaurs, especially members of the sauropod group, such as *Diplodocus*, *Brachiosaurus* and *Apatosaurus* ('Brontosaurus'). Such dinosaurs had a very small head and weak, rake-like teeth that could only pull in vegetation, which was swallowed without chewing. Stones were swallowed, too. In the stomach (gizzard), powerful muscular writhing actions worked like a grinding mill to make the stones crush and pulp the food for better digestion. In the process, the stones became rounded and shiny. In some fairly complete dinosaur skeletons, a pile of gastroliths is located in the position where the stomach would have been in life. Some crocodilians also swallow stones, to alter their buoyancy.

Below: This gastrolith is from Early Cretaceous rock called Paluxy Sandstone, from Carter County, Oklahoma, USA. It was associated with dinosaur fossils and shows the typical polished surface with worn, rounded corners.

Name: Gastroliths (trace fossils).
Meaning: Stomach stones.
Grouping: Usually associated with large plant-eating sauropod dinosaurs.
Informal ID: Stomach stones.
Fossil size: From pea-sized to as large as soccer balls.
Reconstructed size: —
Habitat: Dinosaur gizzard/stomach.
Time span: Mainly Jurassic and Cretaceous, 200–65 million years ago.
Main fossil sites: Worldwide.
Occurrence: ◆ ◆ ◆

Diplodocus

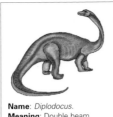

Name: *Diplodocus*.
Meaning: Double beam.
Grouping: Reptile, Dinosaur, Sauropod, Diplodocid.
Informal ID: Diplodocus.
Fossil size: Skull length 60cm/24in.
Reconstructed size: Nose–tail length 27m/88ft.
Habitat: Scattered woodlands, shrubs.
Time span: Late Jurassic, 150–140 million years ago.
Main fossil sites: North America.
Occurrence: ◆ ◆

The sauropod dinosaurs were mostly huge, with a small head, long neck, bulky body, four column-like legs and a long, whippy tail. *Diplodocus* was one of the slimmer types, weighing 'only' 15–20 tonnes. But it is also one of the longest dinosaurs for which relatively complete fossils are known. It lived in North America during the heyday of the sauropods, the Late Jurassic Period, around the same time as the mega-predator *Allosaurus*. Finds of numerous individuals preserved together suggest that sauropods lived and travelled in herds of young and adults. *Diplodocus* is named 'double beam' for the two ski-like chevrons on the undersides of the caudal vertebrae – the bones in the middle section of its tail.

Right: The skull of Diplodocus *is low and rather horse-like. Slender, peg- or pencil-like teeth formed a fringe at the front of the mouth, but there are no rear chewing teeth. The front teeth probably worked like a garden rake to strip leaves from branches.*

Nostrils on upper forehead

Eye socket

Jaw joint

Peg-like teeth fringe front of jaws

PREDATORY DINOSAURS

The dinosaurs shown here were all members of the theropod group, which was almost exclusively carnivorous, or meat-eating. Tyrannosaurus *had long enjoyed fame as the largest predator to walk the Earth. Recently, however,* Carcharodontosaurus *and the even bigger* Giganotosaurus *had taken this record. The dromaeosaurs were much smaller but their fossils suggest that they hunted in packs.*

Tyrannosaurus

Name: *Tyrannosaurus*.
Meaning: Tyrant lizard.
Grouping: Reptile, Dinosaur, Theropod, Tetanuran.
Informal ID: T-rex.
Fossil size: Phalanx 8cm/3¼in; tooth tip 5cm/2in; centrum 6cm/2⅜in.
Reconstructed size: Nose–tail length 12m/40ft.
Habitat: Varied land habitats.
Time span: Late Cretaceous, 75–65 million years ago.
Main fossil sites: North America.
Occurrence: ◆ ◆

The symbol of a mighty hunting dinosaur, *Tyrannosaurus* weighed 6 tonnes or more, yet could probably run faster than any human today, as suggested by the spacing of its preserved footprints. It was one of the very last dinosaurs, surviving to the great end-of-Cretaceous extinction 65 million years ago. Bite marks of its teeth have been found on the plant-eating dinosaur *Triceratops*. Its large, strong teeth and thick, powerful neck suggest that it tackled living prey, biting with incredible power and pulling or 'sawing' its head from side to side to slice out chunks of flesh.

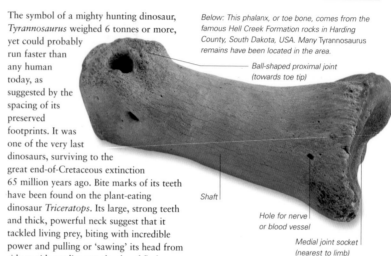

Below: This phalanx, or toe bone, comes from the famous Hell Creek Formation rocks in Harding County, South Dakota, USA. Many Tyrannosaurus remains have been located in the area.

Ball-shaped proximal joint (towards toe tip)

Shaft

Hole for nerve or blood vessel

Medial joint socket (nearest to limb)

Left: The centrum is the central part of a vertebra, or backbone. In many dinosaurs the end caudal, or tail, vertebra lacked the usual muscle-anchoring flanges of the other vertebrae within the main body, and were reduced to simpler, almost cylindrical shapes. This specimen is probably from a juvenile's tail.

Joint surface with neighbouring vertebra

Big teeth
The mouth of *Tyrannosaurus* probably contained anywhere between 50 and 60 teeth at any one time. The longest projected some 15cm/6in above the jaw, with the same length, or more, beneath, firmly anchored within the jaw bone itself. As the animal's old teeth wore down or were broken, new ones grew to replace them, giving the mouth something of a gappy appearance.

Broken lower tooth merged into root

Carcharodontosaurus

Even larger than *Tyrannosaurus* (opposite), this species is named *Carcharodontosaurus saharicus*, since its original fossils, found in 1927, come from the Sahara Desert. In 1996, larger and more complete remains were discovered in the Tegana Formation region of K'Sar-es-Souk Province, south of Taouz, Morocco. These specimens reveal a very large predator indeed, but from a slightly different meat-eater group, being a closer cousin of the North American, Late Jurassic carnivore *Allosaurus*.

Below: The tooth of Carcharodontosaurus is curved backwards (recurved), pointing to the rear of the mouth, to prevent struggling prey from slipping out. As with the Tyrannosaurus tooth, this is only the end of the visible portion, probably a shed (discarded) specimen.

Name: *Carcharodontosaurus*.
Meaning: Shark-tooth lizard.
Grouping: Reptile, Dinosaur, Theropod, Allosaur.
Informal ID: Carcharodontosaurus.
Fossil size: Tooth tip 6cm/2¼in.
Reconstructed size: Nose–tail length 14m/46ft.
Habitat: Floodplains, marshes, swamps.
Time span: Middle Cretaceous, 110–100 million years ago.
Main fossil sites: North Africa.
Occurrence: ◆

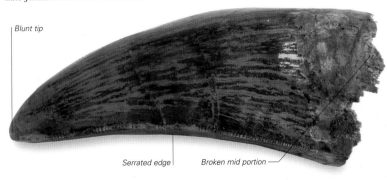

Blunt tip

Serrated edge | Broken mid portion

Dromaeosaurus

Name: *Dromaeosaurus*.
Meaning: Running or swift lizard.
Grouping: Reptile, Dinosaur, Theropod, Dromaeosaur.
Informal ID: Raptor.
Fossil size: Claw length 5cm/2in.
Reconstructed size: Nose–tail length 2m/6½ft.
Habitat: Varied land habitats, mainly woodlands.
Time span: Middle to Late Cretaceous, 110–65 million years ago, for the group.
Main fossil sites: Mostly Northern Hemisphere.
Occurrence: ◆

The dromaeosaurs were medium-sized carnivorous dinosaurs of the Mid to Late Cretaceous Period, which were often called raptors, meaning 'hunters' or 'thieves'. They include the well-known *Velociraptor*, the human-sized *Deinonychus*, whose name means 'terrible claw', and the larger, 5m-/16½ft-long *Utahraptor*, which had large eyes and long, grasping hands. Most of these dinosaurs ran quickly on their back legs and had strong arms with long-clawed fingers. *Dromaeosaurus* was the first of the group to be named, but it is among the least well known from the fossil record. Remains of several *Deinonychus* have been found together, indicating that it may have hunted in packs, as the dinosaur version of today's wolves. The exciting feature of the group is the 'terrible claw' or 'killing claw' on each foot.

Below: The extremely effective 'killing claw' was found on the second toe of each rear foot. Usually it was held up, clear of the ground. But it could be swung down fast in an arc at the toe joints as the dinosaur kicked out, to slash wounds in its prey or to inflict terrible injuries on its enemies.

Slim, knife-like proportions

Toe joint

Curved end, sliced like a machete

OSTRICH- AND BIRD-FOOT DINOSAURS

The ostrich-dinosaurs, or ornithomimosaurs, appeared late in the Age of Dinosaurs. Their skeletons are remarkably similar in size and overall proportions to the largest living bird, the ostrich, so they were probably the fastest runners of their time. Bird-foot dinosaurs, or ornithopods, were plant-eaters and included one of the best-known and most-studied dinosaurs of all, Iguanodon.

Ornithomimus

Built for speed, *Ornithomimus* walked and ran on its two long, slim, yet powerful rear legs. The legs' proportions were typical of a sprinter, with a shortish, well-muscled thigh, longer shin and lengthy foot bones, but comparatively short, stubby, clawed toes. Like other ostrich-dinosaurs, *Ornithomimus* had a long, low head, large eyes and a beak-like mouth that lacked any teeth. The dinosaur probably snapped up any kind of food available, from plant seeds and leaves to small creatures, such as insects, worms and little lizards. The ostrich-dinosaurs evolved fairly late in the Age of Dinosaurs.

Below: This bone is the terminal phalanx at the toe-tip. The claw is flattened and narrow, rather than sharp, probably to provide good traction as the dinosaur sped along, rather than for slashing at prey. Only the three middle toes of each rear foot touched the ground, with the central toe being longest.

Name: *Ornithomimus*.
Meaning: Bird mimic.
Grouping: Reptile, Dinosaur, Theropod, Ornithomimosaur.
Informal ID: Ostrich-dinosaur.
Fossil size: Bone and claw 11cm/4¼in long.
Reconstructed size: Nose–tail length 4m/13ft.
Habitat: Open scrub.
Time span: Ostrich-dinosaurs: Mainly Late Cretaceous, 85–65 million years ago.
Main fossil sites: Chiefly Asia, North America.
Occurrence: ◆ ◆

Terminal joint

Flattened base of claw

Joint with medial phalanx

Thescelosaurus

A medium-sized plant-eater, *Thescelosaurus* was probably a cousin of *Iguanodon*, although it has also been linked to another family of ornithopods called the hypsilophodonts. It had a smallish head, strong body, shorter front legs, powerful rear legs and a long, tapering tail. The beak-like front of its mouth indicates that it snipped and cropped vegetation, which it chewed with long rows of crushing cheek teeth. One celebrated specimen of *Thescelosaurus*, nicknamed 'Willo', supposedly has its heart preserved as a fossil, an incredibly rare instance of soft-tissue preservation. However, the 'heart' may be a natural mineral formation.

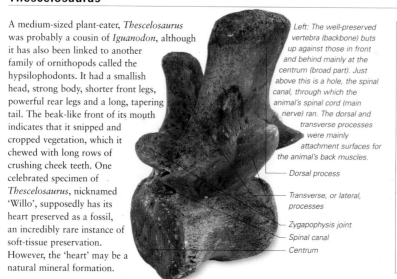

Left: The well-preserved vertebra (backbone) buts up against those in front and behind mainly at the centrum (broad part). Just above this is a hole, the spinal canal, through which the animal's spinal cord (main nerve) ran. The dorsal and transverse processes were mainly attachment surfaces for the animal's back muscles.

Dorsal process

Transverse, or lateral, processes

Zygapophysis joint

Spinal canal

Centrum

Name: *Thescelosaurus*.
Meaning: Marvellous or wonderful reptile.
Grouping: Reptile, Dinosaur, Ornithopod.
Informal ID: Thescelosaurus.
Fossil size: Vertebra 6cm/2¼in long.
Reconstructed size: Nose–tail length 3–4m/ 10–13ft.
Habitat: Woods, scrub.
Time span: Late Cretaceous, 70–65 million years ago.
Main fossil sites: North America.
Occurrence: ◆

Iguanodon

A large dinosaur fossil at an Early–Mid Cretaceous site in Western Europe may well be from *Iguanodon*. Large herds of these big plant-eaters roamed the region for tens of millions of years. *Iguanodon* probably stooped down to carry its 5-tonne bulk on all fours, but it might also have reared up to run mainly on its larger back legs. One of its special distinguishing features was a bony spike on the thumb (the first digit of the front paw), which it might have used as a weapon with which to jab at enemies. As part of *Iguanodon*'s original reconstruction in the nineteenth century, this spike was erroneously placed on the animal's nose!

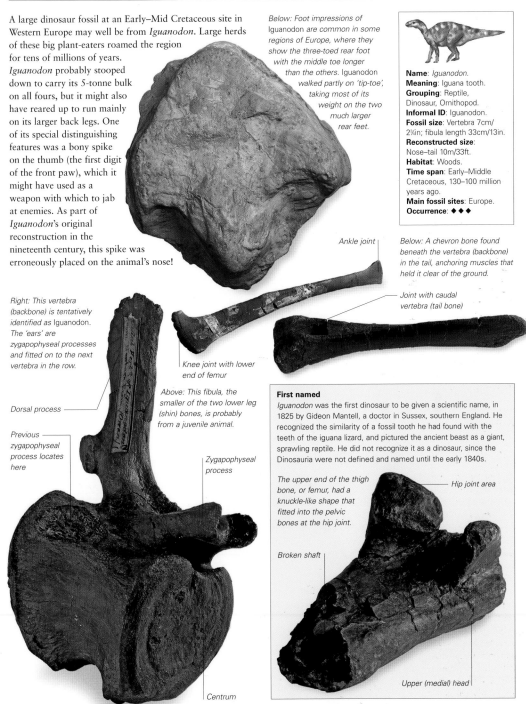

Below: Foot impressions of Iguanodon *are common in some regions of Europe, where they show the three-toed rear foot with the middle toe longer than the others. Iguanodon walked partly on 'tip-toe', taking most of its weight on the two much larger rear feet.*

Name: *Iguanodon*.
Meaning: Iguana tooth.
Grouping: Reptile, Dinosaur, Ornithopod.
Informal ID: Iguanodon.
Fossil size: Vertebra 7cm/2¾in; fibula length 33cm/13in.
Reconstructed size: Nose–tail 10m/33ft.
Habitat: Woods.
Time span: Early–Middle Cretaceous, 130–100 million years ago.
Main fossil sites: Europe.
Occurrence: ◆ ◆ ◆

Ankle joint

Below: A chevron bone found beneath the vertebra (backbone) in the tail, anchoring muscles that held it clear of the ground.

Joint with caudal vertebra (tail bone)

Right: This vertebra (backbone) is tentatively identified as Iguanodon. The 'ears' are zygapophyseal processes and fitted on to the next vertebra in the row.

Knee joint with lower end of femur

Above: This fibula, the smaller of the two lower leg (shin) bones, is probably from a juvenile animal.

Dorsal process

Previous zygapophyseal process locates here

Zygapophyseal process

First named

Iguanodon was the first dinosaur to be given a scientific name, in 1825 by Gideon Mantell, a doctor in Sussex, southern England. He recognized the similarity of a fossil tooth he had found with the teeth of the iguana lizard, and pictured the ancient beast as a giant, sprawling reptile. He did not recognize it as a dinosaur, since the Dinosauria were not defined and named until the early 1840s.

The upper end of the thigh bone, or femur, had a knuckle-like shape that fitted into the pelvic bones at the hip joint.

Hip joint area

Broken shaft

Centrum

Upper (medial) head

DINOSAURS AND PTEROSAURS

Among the last of the dinosaurs were the hadrosaurs, or duck-bills, which probably evolved from Iguanodon-like ancestors. They flourished especially in North American and Asia. Pterosaurs have been traditionally regarded as flying 'reptiles', but most had furry bodies and were probably warm-blooded. Different types lived all through the Mesozoic Era and perished at its end.

Edmontosaurus

Joint with wrist

Joint with finger

Ossified tendon

Below: Viewed from inside the mouth, the lower part of this jaw bone is missing, which reveals new teeth growing and pushing up to replace the old ones. This renewal was a continuous process. The teeth have sharp ridges for grinding plant matter.

Above: This metacarpal (hand bone) has ossified tendons – the tough, stringy ends of muscles that have, themselves, became mineralized like bone during life.

One of the largest of the hadrosaurs, *Edmontosaurus* had the typical flattened, toothless, duck-beak-like front to its mouth, with batteries of hundreds of huge cheek teeth behind for grinding its tough plant food. Fossils of many individuals of this dinosaur have been found together, and remains come from widely spaced locations in North America, from Colorado to as far north as Alaska, leading to suggestions that it migrated in huge herds.

Name: *Edmontosaurus.*
Meaning: Edmonton lizard.
Grouping: Reptile, Dinosaur, Hadrosaur.
Informal ID: Duck-billed dinosaur.
Fossil size: Tooth battery 6cm/2¼in long; metacarpal 31cm/12¼in.
Reconstructed size: Nose–tail length 13m/42½ft.
Habitat: Wooded areas, valleys, hillsides.
Time span: Late Cretaceous, 75–65 million years ago.
Main fossil sites: North America.
Occurrence: ◆ ◆

Triceratops

The largest of the ceratopsians, or horned dinosaurs, was the well-known *Triceratops*. It was probably hunted by the meat-eating *Tyrannosaurus*, and both were among the very last of the dinosaurs, surviving right up to the mass-extinction event of 65 million years ago. Its 50 or so cheek teeth were very small compared with the size of the dinosaur's massive head, with its nasal horn, twin eyebrow horns and wide, ruff-like neck frill. The front of *Triceratops'* mouth was a toothless beak used for snipping and plucking vegetation. The remains of several *Triceratops* of different sizes found in the same location indicate that this dinosaur herded in mixed-age groups.

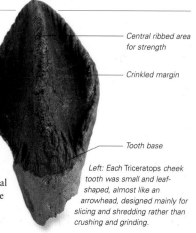

Central ribbed area for strength

Crinkled margin

Tooth base

Left: Each Triceratops *cheek tooth was small and leaf-shaped, almost like an arrowhead, designed mainly for slicing and shredding rather than crushing and grinding.*

Name: *Triceratops.*
Meaning: Three-horned face.
Grouping: Reptile, Dinosaur, Ceratopsian.
Informal ID: Triceratops.
Fossil size: Tooth 3.5cm/1½in tall.
Reconstructed size: Nose–tail length 9m/30ft.
Habitat: Woodlands, forests.
Time span: Late Cretaceous Period, 70–65 million years ago.
Main fossil sites: North America.
Occurrence: ◆ ◆

Pterosaur (unidentified)

Pterosaurs were the principal flying creatures during much of the Mesozoic Era (birds evolved at the end of the Jurassic Period but did not appear to attain great size or numbers until the Tertiary Period). Some pterosaurs, such as *Pteranodon* and *Quetzalcoatlus*, were the largest flying creatures ever, with wingspans in excess of 12m/40ft. The front limbs had evolved into wings, held out mainly by the very elongated bones of the fourth finger. A pterosaur's body had many weight-saving features, including tube-like, hollow bones – as in modern birds – so their remains are very rare and fragmentary.

Right: This may be a fragment from the ulna, one of the lower arm bones, which would have been found at the front of the wing. The arm, wrist and hand bones held out the inner section of wing, but the outer half of each wing's span was supported by the creature's fourth finger. The fossil, from near Oxford, England, dates from the Bathonian age of the Mid Jurassic Period, about 160 million years ago.

Name: Pterosaur.
Meaning: Wing lizard.
Grouping: Pterosaur.
Informal ID: Pterodactyl.
Fossil size: Fragment length 4cm/1½in.
Reconstructed size: Wingspan 2m/6½ft.
Habitat: Coastal regions.
Time span: Group spanned almost the whole Mesozoic Era, 225–65 million years ago.
Main fossil sites: Worldwide.
Occurrence: ◆

Fragmented end ———

——— Shaft

Rhamphorhynchus

Name: *Rhamphorhynchus.*
Meaning: Beak snout.
Grouping: Pterosaur, Rhamphorhynchoid.
Informal ID: Pterodactyl.
Fossil size: Total width approx 1m/3¼ft.
Reconstructed size: Wingspan up to 1.8m/6ft.
Habitat: Coasts, lagoons.
Time span: Late Jurassic, 170–140 million years ago.
Main fossil sites: Throughout Europe.
Occurrence: ◆ ◆

Right: This is not an original fossil, but a cast of a Solnhofen specimen. The contrast between the bones, sharp teeth (which suggest a fish diet), and the preserved soft tissues of the wing and tail membranes has been etched using special paint.

This genus has given its name to the first of the two main groups of pterosaurs, the Rhamphorhynchoidea. They were the earliest types, living mainly to the end of the Jurassic Period, and each had a long trailing tail, unlike the tail-less Pterodactyloidea that succeeded them in the Cretaceous. Various species of *Rhamphorhynchus* are known, varying in wingspan from 40cm/16in to almost 2m/6½ft. Many of their fossil remains occur in the fine-grained limestone of Solnhofen, southern Germany, along with those of the earliest bird, *Archaeopteryx*, and the small dinosaur *Compsognathus*. Long ago, this area was a shallow lagoon into which various creatures fell, died, sank and then quickly became buried for detailed fossilization.

Wing membrane | Fourth finger bone | Skull | Tail membrane | Tail bones

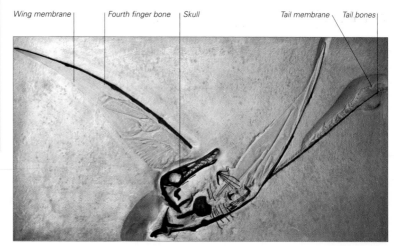

BIRDS

The first known bird is Archaeopteryx, *dating to the Late Jurassic Period, 155–150 million years ago. During the Cretaceous, several further groups of birds appeared, but most died out with the dinosaurs. The only survivor is the Neornithes, to which all present-day 9,000-plus species belong. Bird fossils are rare because their bones were lightweight, fragile and hollow, and were soon scavenged or weathered.*

Archaeopteryx

Among the world's most prized and precious fossils are those of *Archaeopteryx*, the earliest bird so far discovered. Its remains come only from the Solnhofen region of Bavaria, Germany. There, the very fine-grained Lithographic Sandstone (so named because it was formerly quarried for printing) has preserved amazing details, including the patterns of feathers, which are very similar to those of modern flying birds. *Archaeopteryx* had dinosaurian features, such as teeth in its jaws and bones in its tail, but also bird features, including proper flight feathers (rather than fuzzy or downy ones). It probably evolved from small, meat-eating dinosaurs called maniraptorans or 'raptors', but it was perhaps a side-branch of evolution and left no descendants.

Right: One of only seven known fossils of Archaeopteryx, this is termed the 'London specimen'. The neck is arched over its back, a common death pose for reptiles, birds and mammals. The wings show their spread feathers and the legs were strong, with three weight-bearing toes on each foot.

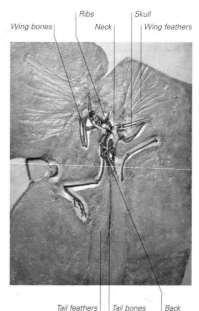

Ribs | Skull
Wing bones | Neck | Wing feathers

Tail feathers | Tail bones | Back

Name: *Archaeopteryx*.
Meaning: Ancient wing.
Grouping: Aves, Archaeornithes.
Informal ID: First bird, early bird.
Fossil size: Total width 30cm/12in.
Reconstructed size: Nose–tail length 50–60cm/20–24in.
Habitat: Wooded areas, tropical islands.
Time span: Late Jurassic Period, 155–150 million years ago.
Main fossil sites: Throughout Europe.
Occurrence: ◆

Phalacrocorax

Freshwater and marine birds, such as ducks, geese, gulls, cormorants and waders, are more likely to be preserved than many woodland species, whose remains were quickly scavenged or rotted. A bird that falls into water may quickly be covered by current-borne sand, mud or other sediments. This keeps away oxygen so that aerobic decomposition cannot occur, but fossilization can. Many bird fossils are of species that live around lakes or along seashores. This specimen belongs to the same genus, *Phalacrocorax*, as modern cormorants and shags.

Humerus
Wrist
Hand
Keel | Fingers
Radius and ulna

Left: The cormorant has powerful wing bones for swimming underwater after its food, and a long, hook-tipped, sharp-edged beak for grabbing slippery prey, such as fish.

Name: *Phalacrocorax*.
Meaning: Finger raven.
Grouping: Aves, Pelicaniform.
Informal ID: Cormorant.
Fossil size: Beak–tailbone length 60cm/24in.
Reconstructed size: Beak–tail length 80cm/32in.
Habitat: Seashores, inland waters.
Time span: Tertiary Period, Pliocene Epoch, about 2 million years ago.
Main fossil sites: Worldwide.
Occurrence: ◆

Feathers, nests, eggs and prints

Bird nests and footprints have been preserved as trace fossils in many parts of the world. This process may also occur with the nesting ground burrows of birds such as penguins, and tree holes, such as those made by woodpeckers. The nest shown below is still complete with its eggshells. It may have been raided by a predator or suffered some other sudden catastrophe just after the young hatched, since the empty shells are usually removed by the parent bird or trampled into fragments by the hatchlings. Moulted feathers are another relatively common fossil find for birds, having been shed, fallen into shallow freshwater or marine sediments and then quickly been buried to prevent decomposition. This tends to happen in habitats with still or slow-flowing water, where the sediments are less disturbed.

Below: This nest is a pseudofossil – a relatively recent specimen that has become infiltrated and mineralized (petrified, or 'turned to rock') due to the action of water, probably from a splashing spring in a limestone area. It is the typical cup-shaped nest of a small songbird, perhaps in the tit family. The nest diameter is 8cm/3⅛in.

Above: These trace fossils of bird footprints, from Utah, USA, date to the Eocene Epoch, 53–33 million years ago. They were probably made by a presbyoniform, an early type of bird related to the duck and goose group, Anseriformes. Presbyornis itself looked like a combination of duck and flamingo, and stood 1m/3¼ft tall.

Left: Moulted feathers were often preserved in the fine sediments of lake mud, such as this one from the Oligocene Epoch, about 25 million years ago. Feather width 2cm/⅘in.

— *Separated barbs*

Below: Preserved examples of bird eggs include those from waterfowl, such as ducks and geese. Their nests are usually built low, just above the water's surface, and the eggs may fall in and down into the soft, muddy bottom intact if a predator upsets the nest. This may be a duck's egg from the Oligocene Epoch about 25 million years ago. Length 4.5cm/1¾in.

SYNAPSIDS (MAMMAL-LIKE REPTILES)

*Synapsids get their name from the pattern of openings, or windows, on the sides of the skull bone, with
one main opening in each upper side. Synapsids appeared about 300 million years ago and during the
Permian Period they were the main large land animals, with both plant- and meat-eaters. They faded as
the dinosaurs came to domination in the Triassic, but some of their kind gave rise to the first mammals.*

Dimetrodon

One of the early carnivorous synapsids, *Dimetrodon* is well
known from the many remains that have been discovered in
the Red Beds of Texas, USA. All parts of its skeleton have
been found and they show a long, low predator with big,
powerful jaws and sprawling legs, which could probably run
rapidly. The most unusual feature of *Dimetrodon* was a tall
flap of thin skin on its back, held up by a series of spine-like
bony rods from the backbone. This has led to its common
names, such as 'sail-back' or 'fin-back'. The skin may have
helped to absorb heat from the sun,
allowing *Dimetrodon* to warm up
faster and move more quickly
than its contemporaries. A similar
reptile from the same time and
place, plant-eating Edaphosaurus,
also had the feature. During the
later Mesozoic Era, various
dinosaurs, such as the huge
predator *Spinosaurus*, had a
similar feature.

*Below: This piece of jaw from
Oklahoma, USA, shows the
powerful teeth with which
Dimetrodon seized its prey. In life,
sharp, full-grown teeth would be
more than 10cm/4in long.*

Name: *Dimetrodon*.
Meaning: Two forms of teeth.
Grouping: Reptile,
Synapsid, Pelycosaur.
Informal ID: Sail-back
(sometimes incorrectly
called 'dinosaur').
Fossil size: Jaw piece
length 5cm/2in.
Reconstructed size:
Nose–tail length 3m/10ft.
Habitat: Mosaic of scrub
and swamp.
Time span: Early
Permian, 295–270 million
years ago.
Main fossil sites: North
America, Europe.
Occurrence: ◆ ◆

Diictodon

Dicynodonts were squat, barrel-shaped, strong-legged
creatures, in body bulk not unlike modern pigs. But they
had typical reptile scales and the front of the mouth was
extended into a hooked 'beak', probably used for plucking
bits of plant food from trees and bushes. Their name means
'two tusk teeth' and most types had two large, tusk-like
teeth in the upper jaw. These were probably for defence
against enemies, or display at breeding time. *Diictodon* is
one of the first vertebrates for which fossils suggest sexual
dimorphism: that is, males and
females have different features.
In this case, of about a
hundred skeletons found,
only some had tusks –
presumably the males.

*Below: This fossil skull from
Beaufort West District, Cape
Province, South Africa, clearly
shows large tusks and was
presumably a male. Some
Diictodon remains have been
found in burrows, perhaps
excavated by the beak-like mouth
or blunt-clawed feet. The tusks
lack digging-type wear marks.*

Name: *Diictodon*.
Meaning: Two tusk
teeth.
Grouping: Reptile,
Synapsid, Dicynodont.
Informal ID: Mammal-
like reptile.
Fossil size: Skull length
14cm/5½in.
Reconstructed size:
Nose–tail length 1m/3¼in.
Habitat: Mixed.
Time span: Late
Permian, 260–250 million
years ago.
Main fossil sites: Regions of
southern Africa.
Occurrence: ◆

Beak-like front of mouth

Tusk

Eye socket

Lystrosaurus

Name: *Lystrosaurus*.
Meaning: Shovel or spoon reptile.
Grouping: Reptile, Synapsid, Therapsid, Dicynodont.
Informal ID: Lystrosaurus, herbivorous mammal-like reptile.
Fossil size: Skull length 22cm/8¾in.
Reconstructed size: Total length 70cm–1m/28in–3¼ft.
Habitat: Riverbanks, lakesides, shallow fresh water.
Time span: Early Triassic, 245–230 million years ago.
Main fossil sites: Southern Africa, Asia, Antarctica.
Occurrence: ◆ ◆ ◆

Lystrosaurus was a type of small 'reptilian hippopotamus'. With its barrel-like body, short tail and stubby limbs it probably spent its time wading in shallow water. Here it would feed on lush aquatic foliage that it pulled up and gathered into its mouth using the paired tusks projecting from the upper jaw. The distribution of fossils from this dicynodont (see also *Diictodon*, opposite) encompasses southern Africa, Russia, China and India, and in 1960 its remains were also discovered in Antarctica. This is compelling evidence for the union of most of these landmasses at the time – the Permian Period – into the one great southern supercontinent known as Gondwana.

Left: Lystrosaurus fossils, such as this skull, are very common in certain parts of southern Africa, including the Karoo, where they give their name to the Lystrosaurus Zone of the Early or Lower Triassic. The rocks are sandstones and shales, which indicate a moist climate with regular flooding. The local distribution of Lystrosaurus remains suggests that these herbivores lived in groups, or herds, and grazed in freshwater shallows. In this skull, the eye socket is upper left and the snout is to the right.

Cynognathus

Name: *Cynognathus*.
Meaning: Dog jaw.
Grouping: Reptile, Synapsid, Therapsid, Theriodont.
Informal ID: Cynognathus, mammal-like reptile.
Fossil size: Skull length 40cm/16in.
Reconstructed size: Nose–tail length 1.8m/6ft.
Habitat: Open country, scrub, semi-desert.
Time span: Early Triassic, 240–230 million years ago.
Main fossil sites: Southern Africa, South America.
Occurrence: ◆

One of the Early Triassic synapsids, and also one of the largest in its group called the cynodonts, *Cynognathus* was a strongly built carnivore with powerful jaws. It had three types of teeth, similar to those of the Carnivora mammals that would evolve some 200 million years later. These teeth consisted of small front incisors for nipping and nibbling, large pointed canines for ripping and stabbing, and molar teeth with coarse serrations for chewing and shearing. The cynodonts, 'dog teeth', were among the longest-surviving of all the synapsids, living from the Late Permian to the Middle Jurassic. They were mammal-like in many features and deserve their casual name of 'mammal-like reptiles' – especially since some of their kind probably gave rise to the true mammals.

Right: The skull shows a large flange on top where powerful jaw muscles were anchored. The jaws could open wide partly due to an upward-angled component at the rear of the lower jaw, the coronoid process, which formed the jaw joint.

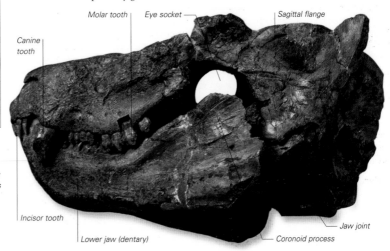

Molar tooth | Eye socket | Sagittal flange

Canine tooth

Incisor tooth

Lower jaw (dentary)

Coronoid process

Jaw joint

EARLY MAMMALS

The first mammals appeared almost alongside the first dinosaurs, more than 200 million years ago.
However, throughout the Mesozoic Era they were mostly small, vaguely shrew-like hunters of insects and
other small creatures. Only a few of them exceeded the size of a modern domestic cat, one being
the recently discovered koala-sized Repenomamus.

Megazostrodon

Below: This fossil specimen shows the main body or trunk region with the pelvis (hip) bone, rear limb and foot bones. In overall appearance, Megazostrodon and other very early mammals probably looked like the tree-shrews or tupaids of today, although they belonged to a very different mammal group, the triconodonts, which all became extinct.

Tibia (shin)

Femur (thigh)

Vertebrae (backbones)

Pelvis

Foot bones

One of the earliest known mammals is *Megazostrodon*, the remains of which are known from Late Triassic rocks in Southern Africa. Around the same time, similar small mammals were appearing on other continents. When studying fossils, the key features of a mammal include the bones that form the lower jaw (dentary/mandible) and the jaw joint, and the alteration of what were formerly jaw bones to the three tiny bones, called auditory ossicles, in each middle ear. Large eye sockets indicate *Megazostrodon* was nocturnal, at a time when the day-active dinosaurs were beginning to dominate life on land. Its teeth show it probably hunted insects in the manner of today's shrews.

Name: *Megazostrodon*.
Meaning: Large girdle tooth.
Grouping: Mammal, Triconodont.
Informal ID: Early mammal, shrew-like mammal.
Fossil size: 3cm/1¼in.
Reconstructed size: Nose–tail length 12cm/4¾in.
Habitat: Wooded areas, scrub regions.
Time span: Late Triassic to Early Jurassic, 210–190 million years ago.
Main fossil sites: Regions of southern Africa.
Occurrence: ◆

Monotreme

Today there are just a handful of monotremes (egg-laying mammals), including the duck-billed platypus, *Ornithorhynchus*, of Australia, and the echidnas (spiny anteaters), *Tachyglossus* and *Zaglossus*, of Australia and New Guinea. Of all living mammal groups, the monotremes have the most ancient fossil record, stretching back 100 million years. The living platypus shows reptilian traits, such as limbs angled almost sideways from the body, and of course egg laying. However, its defining mammal features include its three middle ear bones, warm-bloodedness, a fur-covered body and feeding its young on milk.

Left: A monotreme fossil molar tooth is compared with the teeth in a skull of a modern duck-billed platypus, Ornithorhynchus. There is great similarity in the cusp (point) pattern. The fossil tooth is dated to 63 million years ago, just after the mass extinction at the end of the Cretaceous Period.

Name: *Ornithorhynchus* (living platypus).
Meaning: Bird beak or bill.
Grouping: Mammal, Monotreme.
Informal ID: Platypus.
Fossil size: Fossil tooth 1cm/⅜in across.
Reconstructed size: Unknown, possible head–body length 50cm/20in.
Habitat: Unknown, possibly fresh water.
Time span: Early Tertiary, 65–60 million years ago.
Main fossil sites: (This specimen) South America; Australia.
Occurrence: ◆

Diprotodon

About one-fifteenth of all living mammal species are marsupials, or pouched mammals, with the biggest being the red kangaroo of Australia. Many other species lived during the Tertiary and Quaternary Periods. One of the largest was *Diprotodon*, which became extinct relatively recently, perhaps just 30,000 years ago. It was a plant-eater resembling a wombat, with a large snout and stocky body. But it was huge, almost the size of a hippo. The protruding nasal area may have supported very large nostrils or the muscles for an elongated snout or shortish, mobile trunk, similar to the modern tapir. Various diprotodontids came and went from the Oligocene Epoch onwards, and their living relations include the wombats themselves, as well as kangaroos and koalas.

Below: An important distinguishing feature of the diprotodonts, as seen in this fossilized skull, was a single pair of lower front incisor teeth, which pointed forwards, and from which the name is derived. There was also a long gap, or diastema, as seen in rodents living today, between the front teeth and rear chewing teeth.

Name: *Diprotodon*.
Meaning: Two prominent/ forward teeth.
Grouping: Mammal, Marsupial, Diprotodont.
Informal ID: Giant wombat.
Fossil size: Skull length 50cm/20in.
Reconstructed size: Head–body length 3m/10ft.
Habitat: Woods, forests.
Time span: Pleistocene to 30,000 years ago.
Main fossil sites: Australia.
Occurrence: ◆ ◆

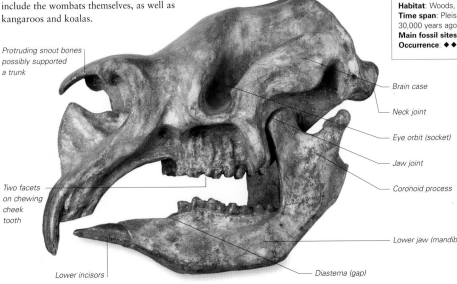

Protruding snout bones possibly supported a trunk

Two facets on chewing cheek tooth

Lower incisors

Brain case

Neck joint

Eye orbit (socket)

Jaw joint

Coronoid process

Lower jaw (mandible)

Diastema (gap)

Protungulatum

The ungulates are the hoofed mammals, which today include horses, rhinos, giraffes, hippos, deer, cattle, sheep and goats. One of the earliest examples was *Protungulatum*. Its fossils have been associated with those of dinosaurs. But the dinosaur fossils from the end of the Mesozoic Era may have been eroded from earlier rocks and then mixed with *Protungulatum*'s remains during the start of the Tertiary Period, about 60 million years ago. *Protungulatum* may have had claw-like 'hooves', but very few remains of this animal are known, other than teeth, which show the trend towards the broad, crushing teeth of later ungulates, designed to masticate tough, fibrous plant foods.

Right: This tiny tooth is from the Hell Creek Formation of Montana, USA. It has two larger pointed cusps, a broken section where other cusps would have stood, and a double-root anchored in the jaw bone. Protungulatum probably fed on fruits and soft plants, but could still chew small creatures, such as insects, as its ancestors had done.

Cusps

Root

Root

Name: *Protungulatum*.
Meaning: Before ungulates.
Grouping: Mammal, Condylarth.
Informal ID: Early hoofed mammal.
Fossil size: 5mm/³⁄₁₆in.
Reconstructed size: Nose–tail length 40cm/16in.
Habitat: Woodland.
Time span: Late Cretaceous, 70–65 million years ago, or early Tertiary (see main text).
Main fossil sites: Regions of North America.
Occurrence: ◆

SMALL MAMMALS

The original mammals hunted insects and similar small creatures. The second-largest group of living mammals are bats, with about one-fifth of all mammalian species. They appeared almost fully evolved early in the Tertiary Period. Other small mammal groups moved from eating insects to mainly plants, including the rats, mice and other rodents, and rabbits and hares, called lagomorphs.

Palaeochiropteryx

Bats have lightweight, fragile bones and fossils are rare. *Palaeochiropteryx* is known from beautiful specimens in the Messel fossil beds of Germany. The front limbs have become wings, with arm and hand bones holding out what would have been a very thin, stretchy flying membrane, or patagium. The first digit (thumb) was a claw for grasping and grooming, the second digit was also clawed (unlike most modern bats) and the very long bones of the third digit extended to the end of the wing. The hip area had become small, with the rear limbs used mainly for hanging.

Below: This specimen has been prepared from the front side (skull lowermost), then that side embedded in resin and the other, rear side likewise cleaned and prepared. It shows the whole skeleton with one wing folded over the backbone. The resin prevents the fragile oil-shale fossil, typical of Messel specimens, from degenerating.

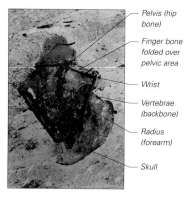

— Pelvis (hip bone)

— Finger bone folded over pelvic area

— Wrist

— Vertebrae (backbone)

— Radius (forearm)

— Skull

Wing shapes
The wings of *Palaeochiropteryx* were relatively short in span, but broad from front to back, indicating a fast, manoeuvrable flier. Bats that soar and swoop more, without sudden directional changes, have longer, narrower wings. The same principles apply to the wings of birds.

Name: *Palaeochiropteryx*.
Meaning: Ancient hand wing.
Grouping: Mammal, Chiropteran.
Informal ID: Bat.
Fossil size: Whole specimen about 7cm/2¾in across.
Reconstructed size: Nose–tail length 5–6cm/2–2¼in; wingspan 20cm/8in.
Habitat: Woods.
Time span: Eocene, 50 million years ago.
Main fossil sites: Throughout Europe.
Occurrence: ◆

Leptictidium

Most insectivores eat not only insects, but also worms, spiders and other small creatures. *Leptictidium* was a ground-dwelling animal that could bound along at speed, something like a modern rat-kangaroo, using its powerful rear legs and extremely elongated hind feet (as in a real kangaroo). Its long, pointed skull suggests an animal with a lengthy, fleshy snout, which was probably equipped with long whiskers for probing, feeling and sniffing food. The large eye sockets indicate a nocturnal lifestyle. The leptictid group began to expand just before the end of the Cretaceous Period and became common and diverse during the early part of the following Tertiary Period.

Left: This well-preserved skeleton of Leptictidium *has been caught in a very lifelike, bounding pose, showing the enormously long tail that in life would have helped it with balance. The tail was flicked to one side to twist the body to the other side when darting about and changing direction at great speed.*

Name: *Leptictidium*.
Meaning: Delicate/Graceful weasel.
Grouping: Mammal, Insectivore.
Informal ID: Shrew.
Fossil size: Nose–tail length 65cm/25½in.
Reconstructed size: As above.
Habitat: Woods, Forests.
Time span: Eocene, 50–40 million years ago.
Main fossil sites: Europe.
Occurrence: ◆

Palaeolagus

Rabbits and hares are often confused with rodents, but they are a separate group, known as the lagomorphs, or 'leaping shapes'. A lagomorph has two pairs of gnawing or nibbling teeth in the upper and lower front of the skull, rather than one main pair, as in rodents. The jaw joint allows chewing by side-to-side motion using the five or more pairs of molar (cheek) teeth (seen in living rabbits), rather than the up-and-down motion of rodents. *Palaeolagus* was one of the earliest rabbits, after the hares and rabbits split off from the other main group of lagomorphs, called the pikas.

Below: This fossilized skull of a young Palaeolagus *clearly shows the sharp gnawing teeth, which were tipped with ridges of enamel, prominently situated at the front of the animal's skull. The femur is the thigh bone and the tibia is the main shin bone, and both of these are greatly elongated in lagomorphs to provide them with the power they require for jumping and running to avoid being taken as prey. This specimen is from South Dakota, USA.*

Name: *Palaeolagus*.
Meaning: Ancient leaper.
Grouping: Mammal, Lagomorph.
Informal ID: Rabbit.
Fossil size: Skull length 4cm/1½in.
Reconstructed size: Nose–tail 25cm/10in (adult).
Habitat: Woods.
Time span: Oligocene, 30–25 million years ago.
Main fossil sites: North America.
Occurrence: ◆

Gnawing incisors

Gap (diastema)

Eye socket

Molars

Ankle joint end of tibia

Tibia

Hip joint end of femur

Knee joint

Lepus

Hares are generally bigger than rabbits, with a body form even further adapted to leaping and jumping. The rear leg bones are long for greater leverage as powerful muscles extend the leg joints to fling the animal forward. The genus *Lepus* includes modern hares such as the common or brown hare and mountain hare. Early lagomorphs probably lived in woodlands, since the grasslands where most rabbits and hares live today did not begin to spread until the Miocene, from about 25 million years ago.

Below: This recent subfossil of a lower leg or shin bone (tibia), shows details such as the foramina (tiny holes) for blood vessels and nerves passing into the bone's interior. It was found in a cave, suggesting that the hare could have been sheltering. If it had been a prey item, the bone may well have been cracked and gnawed.

Enduring design
Lagomorph design has proved to be very successful, with changes in their dental and skeletal development mostly occurring only very gradually. Although worldwide in distribution now, lagomorphs were naturally absent from only a few regions of the world: most of the islands of Southeast Asia, Australia, New Zealand, the island of Madagascar, southern South America and Antarctica.

Name: *Lepus*.
Meaning: Leaper.
Grouping: Mammal, Lagomorph.
Informal ID: Hare.
Fossil size: Length 14cm/5½in.
Reconstructed size: Nose–tail length 45cm/18in.
Habitat: Open scrub, grass, moor.
Time span: Quaternary, within the last 2 million years.
Main fossil sites: Worldwide.
Occurrence: ◆ ◆

Head (knee joint end)

Shaft

Ankle joint end

RODENTS AND CREODONTS

Rodents make up almost half of all the living mammal species and first appeared as squirrel-like creatures some 60 million years ago. Creodonts were the main carnivorous mammals from about 60 to 30 million years ago. Different types looked like bears, wolves, cats and hyaenas. However, they had all but faded away by 10 million years ago, to be succeeded by the Carnivora of today.

Mimomys

The vole genus *Mimomys* had shown rapid evolution during the past five million years, with 40 or more species being recognized based on features including their molar teeth. *Mimomys* was widespread, especially across Europe, but seems to have died out about half a million years ago. The many species of *Mimomys* are described mainly from the shape and features of the teeth, especially the lower first molar and the upper third molar. The molars had roots, which are lacking in the modern equivalent of *Mimomys* – the water vole, *Arvicola*.

Below: This Pleistocene specimen is from Mimomys savini, a type of water vole, from the West Runton Gap, near Cromer, Norfolk, England.

High crown formed of prisms

Lower right first molar

Toothless gap between molars and incisors

Lower jaw (mandible)

Name: *Mimomys*.
Meaning: Mouse-like.
Grouping: Mammal, Rodent, Myomorph.
Informal ID: Vole, Water vole.
Fossil size: Length of jaw fragment 8mm/⅓in.
Reconstructed size: Head–body length 5–8cm/2–3⅛in.
Habitat: Woods, wetlands.
Time span: Quaternary Period, Pleistocene Epoch, within past 1.75 million years.
Main fossil sites: Worldwide.
Occurrence: ◆ ◆

Vole clock
Remains of *Mimomys* are very common in the fossil record and have been accurately dated in some well-studied rocks. Because of this, they can be used as a 'vole clock' to date zones of unfamiliar rocks and the fossils they contain.

Rodent teeth

Long, curved, sharp teeth are the common remains of many rodents, from mice, voles and rats, to larger types, such as beavers. They are the front teeth, or incisors, and in the rodent group they keep growing throughout the animal's life. As the rodent gnaws and nibbles, the teeth are continually worn down. The layer of enamel on the front is hardest and shows least wear, with the dentine just behind being eroded faster. This gives an angle to the tooth tip, which forms a self-sharpening cutting ridge. Rodents have two pairs of incisors – one pair in the top jaw and one pair in the lower – which distinguishes them from similar gnawing-nibbling herbivores, such as rabbits.

Cutting tip

Crown

Root end

Left: This rodent, a coypu (nutria), shows the upper of its two pairs of long, gnawing, continuously growing, self-sharpening front teeth, incisors, which are characteristic of the group. They are used for tackling woody or nutty foods and also for defence – they can inflict severe wounds on breeding rivals or predators.

Left: These Quaternary Period specimens came from a cave in Mid Glamorgan, Wales. Most of the length of the tooth is root, anchored in the maxilla (upper jaw) or mandible (lower jaw).

Name: Incisors.
Meaning: Cutting.
Grouping: Mammal, Rodent.
Informal ID: Rodent front teeth.
Fossil size: Length around curve 1.3cm/½in.
Time span: Palaeocene, 60 million years ago, to today.
Main fossil sites: Worldwide.
Occurrence: ◆ ◆ ◆

Hyaenodon

The creodont genus *Hyaenodon* was long-lasting, widespread and varied. Some species were as small as stoats, others as big as lions – however, they were not closely related to stoats or lions, since these hunters belong to the more modern Carnivora group. The *Hyaenodon* genus originated in the Late Eocene Epoch, probably in Asia or Europe. Some members spread to North America and possibly Africa, and one Asian kind survived into the Pliocene Epoch, some five million years ago – then the whole creodont group went extinct. See the next page for a comparison with the jaws and teeth of a 'true' hyaena.

Below: This specimen of Hyaenodon comes from Oligocene rocks in Wyoming, USA. The skull is long and low, with a large snout indicating a good sense of smell. It is deep-jawed, with immensely strong jaw bones. These would be able to crack and chew tough food, such as gristle and bone, in the manner of the modern scavenging hyaena.

Name: Hyaenodon.
Meaning: Hyaena tooth.
Grouping: Mammal, Creodont, Hyaeonodont.
Informal ID: Ancient 'hyaena'.
Fossil size: Skull length 15cm/6in.
Reconstructed size: Head–body length 1.2m/4ft.
Habitat: Wood, scrub.
Time span: Oligocene, 30–25 million years ago.
Main fossil sites: Europe, Asia, North America.
Occurrence: ◆ ◆

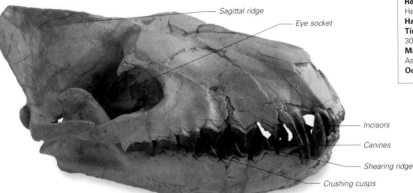

Sagittal ridge

Eye socket

Incisors

Canines

Shearing ridge

Crushing cusps

Hesperocyon

Name: Hesperocyon.
Meaning: Western dog.
Grouping: Mammal, Canid.
Informal ID: Hesperocyon, prehistoric mongoose, 'first dog'.
Fossil size: Skull length 15cm/6in.
Reconstructed size: Total length 80cm/32in.
Habitat: Mixed.
Time span: Oligocene to Early Miocene, 30–20 million years ago.
Main fossil sites: North America.
Occurrence: ◆ ◆

This vaguely mongoose-like animal was an early canid, or member of the mammal family Canidae. This embraces wolves, dogs, foxes, jackals and their kin. Modern canids tend to be long-legged, dogged pursuit specialists. *Hesperocyon* had a long, bendy body and short legs, more like those of a weasel or stoat. Perhaps it lived like a mustelid, squirming through dense undergrowth and following prey animals such as rodents into their burrows. Its fossils come mainly from Nebraska, USA. The species is usually named as *Hesperocyon gregarius*, since the distribution of remains suggests they lived in groups.

Below: Hesperocyon's skull is long and low, with an elongated snout. The shape is suited for movement in restricted places, such as tunnels. Complete specimens show 42 teeth rather than the usual 44, due to a missing molar on each side in the rear of the upper jaw. Details of the inner ear region and the small ear bones (ossicles) mark this as a true canid rather than another type of carnivorous mammal.

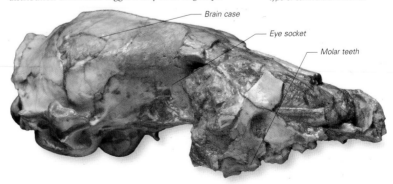

Brain case

Eye socket

Molar teeth

CARNIVORES (MAMMALIAN)

A 'carnivore' is generally any flesh-eating or predatory creature. However, Carnivora is also the official name for the mammal group containing hunters ranging from weasels, stoats and otters, through civets, raccoons, foxes, wild dogs and wolves and hyaenas to cats and bears. Many other prehistoric mammal groups had carnivorous members, especially the credonts, but these were not members of the Carnivora.

Hyaena/Crocuta

The hyaena of the Carnivora should not be confused with *Hyaenodon* of the Creodonta. These two groups differed in several important features, including the bones of the ear and foot, the pattern of teeth, and a relatively smaller brain among the creodonts. In the Carnivora, the rearmost upper premolar tooth moves down against the first lower molar tooth so that their sharp edges work together like shearing blades – this is called the carnassial pattern. Hyaenas have a slight variation on this pattern, with certain teeth missing. The massive jaws, with fairly tall-crowned but blunt teeth, exert 'spots' of pressure – essential for cracking gristle and bone.

Below: This section of mandible (lower jaw) shows the massive crunching cheek teeth, and the sharp shearing surface. The specimen may be from the cave hyaena, Hyaena spelea, or the modern spotted hyaena of Africa, Crocuta crocuta, but at a location from which hyaenas have long since disappeared – Devon, England.

Name: *Hyaena/Crocuta*.
Meaning: Hyaena.
Grouping: Mammal, Carnivore, Hyaenid.
Informal ID: Hyaena.
Fossil size: Jaw section length 10cm/4in.
Reconstructed size: Nose–tail length 1.6m/5¼ft.
Habitat: Scrub, plains.
Time span: Pleistocene, within the past 1.75 million years.
Main fossil sites: Europe, Africa.
Occurrence: ◆

Smilodon

Name: *Smilodon*.
Meaning: Knife tooth.
Grouping: Mammal, Carnivore, Felid.
Informal ID: Sabre-tooth cat, sabre-tooth 'tiger'.
Fossil size: Canine length 17cm plus/6½in plus.
Reconstructed size: Head–body length 2m/6ft.
Habitat: Varied terrain, mainly wooded.
Time span: Pliocene, to 12,000 or even 10,000 years ago.
Main fossil sites: North America.
Occurrence: ◆ ◆ ◆

One of the best-known and most recent sabre-tooth cats, *Smilodon*, was approximately the size of a modern lion. Its massively elongated canines were wide from front to back, but were narrower from side to side, giving them an oval cross-section. The rear edge of each canine was serrated to cut through hide and flesh more easily. *Smilodon*'s jaws opened extremely wide – to an angle of 120° – allowing the animal to use its canines as effective stabbing or slashing weapons. The chest, shoulders and neck were all heavily muscled to give great power for its deadly lunges. The cat may have sliced open the belly of a larger victim, causing it to bleed to death, or it could have severed the spinal cord or windpipe of smaller prey.

Left: The remains of many hundreds of Smilodon have been recovered from Rancho La Brea in Los Angeles, California, USA. In Pleistocene times, surface 'lakes' of oozing tar trapped many creatures, from huge herbivores like mammoths and bison, to predators attracted by their struggles, including dire wolves and sabre-tooth cats.

Shearing (carnassial) cheek teeth

Stabbing canine teeth

Ursus

Name: *Ursus spelaeus*.
Meaning: Cave bear.
Grouping: Mammal,
Carnivore, Ursid.
Informal ID: Cave bear.
Fossil size: Skull length
50cm/20in.
Reconstructed size:
Head–body up to 3m/10ft.
Habitat: Varied.
Time span: Pleistocene,
within the past 2 million years,
to today (for the genus).
Main fossil sites:
Throughout Europe.
Occurrence: ◆ ◆

Eye socket
Sagittal crest
Thick zygomatic arch (cheek bone)
Long, low snout
Canine tooth

Above: The ursid skull is typically robust with large surfaces for muscle anchorage.
Below: A Pleistocene lower jaw from Sophie's Cave in Bavaria, Germany.

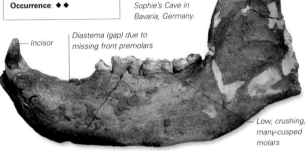

Incisor
Diastema (gap) due to missing front premolars
Low, crushing, many-cusped molars

The ursids, or bears, make up one of the smallest living families within the mammal group Carnivora, but it contains the largest and most powerful members. *Ursus spelaeus* was the fearsome-looking European cave bear that our prehistoric ancestors would have encountered across Central and Southern Europe during the most recent Ice Ages, perhaps less than 10,000 years ago. Its probable diet was chiefly plants rather than meat. Its remains are often found in caves where the bears regularly rested, sheltered, gave birth and hibernated, sometimes in large numbers – but never woke up. The bones and teeth were used in rituals and as decoration by Neanderthal people more than 30,000 years ago.

Canid

Head (articulates with hip bone)
'Knuckle' of knee joint
Neck
Shaft

Name: Canid femur.
Meaning: Dog thigh bone.
Grouping: Mammal,
Carnivore, Canid.
Informal ID: Wolf or fox.
Fossil size: Length 13cm/5in.
Reconstructed size:
Head–body length 50–60cm/
20–24in.
Habitat: Mixed.
Time span: Pleistocene,
within past 1.75 million years,
to today.
Main fossil sites: Worldwide.
Occurrence: ◆ ◆

The canids are the wolf, dog and fox family of the mammal group Carnivora. They are characterized by having long snouts and they are long-legged running specialists, built for 'dogged' pursuit of prey, rather than a stealthy stalk-and-pounce like the cats. Canids first appeared in the fossil record in the Eocene Epoch, some 40 million years ago. Many species have come and gone, including the dire wolf, *Canis dirus*, of North America – many specimens of which have been excellently preserved in La Brea Pits of tar pools in Los Angeles (see opposite). The grey wolf (common wolf), *Canis lupus*, has a long fossil history and is the ancestor of all breeds of modern domestic dog.

Above: This well-preserved Quaternary-dated specimen is from Mid Glamorgan, Wales and is less than 2 million years old. The femur, or thigh bone, shows the typical ball-shaped head that fitted into the bowl-like socket of the animal's hip bone, or pelvis. The ball is carried on an angled neck, as is the case in most mammalian species.

HERBIVOROUS MAMMALS – EQUIDS

Many types of mammals eat plants, from tiny voles and mice to huge elephants. The main group of large herbivores is the ungulates, or hoofed mammals. Their toe-tips evolved from claws into hard hooves, and most have become alert, swift browsers or grazers. The perissodactyls – odd-toed ungulates, with an odd number of toes per foot – include the equids, which are horses and zebras, and also rhinos and tapirs.

Hyracotherium

Name: *Hyracotherium*.
Meaning: Hyrax beast.
Grouping: Mammal, Perissodactyl, Equid.
Informal ID: Early horse.
Fossil size: Length of teeth section 2.5cm/1in.
Reconstructed size: Head–body length 60cm/24in.
Habitat: Woods.
Time span: Early Eocene, 50 million years ago.
Main fossil sites: Northern Hemisphere.
Occurrence: ◆ ◆

The evolution of horses is probably one of the best-studied of all prehistoric mammal sequences. One of the early members was *Hyracotherium* from the Eocene Epoch, some 50–45 million years ago. The reason that this name refers to the hyrax, a smallish African mammal said to be the elephant's closest living relative, rather than a horse, is because of a mistaken identity that occurred back in the nineteenth century. A more apt alternative has occasionally been proposed, *Eohippus*, meaning 'dawn horse'. *Hyracotherium* was a small woodland herbivore, about the size of a fox terrier, widespread across North America, Europe and Asia. It had four toes on each front foot and three on each back foot.

Below: This specimen of three molar teeth embedded in their jaw bone is from the Willwood Formation rocks near Powell, Wyoming, USA. It is dated to the Wasatchian Age (Stage), at the start of the Eocene Epoch. The molars are low-crowned, suited for chewing soft forest leaves.

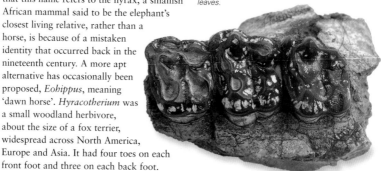

Miohippus

Below: The skull shows the incisors for nipping and nibbling leaves, and the premolars and molars for shredding. The snout has become longer than in previous horse types, with a slightly dished upper surface, and the deep lower jaw was operated by large chewing muscles.

By the Oligocene Epoch some horses were becoming larger, able to run swiftly through open woodlands. There were no wide-open grassy plains as yet – grasses were only just beginning to evolve. *Miohippus* lived from 30 to less than 25 million years ago and had three toes per foot, all of which touched the ground. However, the middle toe was larger than the other two and bore most of the weight.

Miohippus was about the size of today's sheep, standing some 60cm/24in at the shoulder, and had a longer, dished snout, a longer neck and longer legs that earlier horses. It lived at the same time as another horse genus, *Mesohippus*, with probably the former giving rise to further, larger types of horses.

Name: *Miohippus*.
Meaning: Lesser horse.
Grouping: Mammal, Perissodactyl, Equid.
Informal ID: Early horse.
Fossil size: Skull length 18cm/7in.
Reconstructed size: Head–body 1m/3¼ft.
Habitat: Open woodland regions.
Time span: Oligocene, 30–23 million years ago.
Main fossil sites: North America.
Occurrence: ◆ ◆

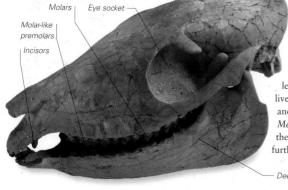

Molars
Eye socket
Molar-like premolars
Incisors

Deep mandible (lower jaw)

Merychippus

Continuing the horse trend in increased body size, *Merychippus* appeared about 17 million years ago in North America, and stood about 1m/3¼ft at the shoulder – the same height as a 10-hand pony of today. This coincided with the appearance of grasses and their spread to form wide-open plains in the drier climate of the Miocene. The neck and legs of *Merychippus* were even longer than its ancestors. There were still three toes on each foot, but the central one (digit 3) was now much larger and the two outer ones hardly touched the ground at all and bore no weight. The position of the eyes allowed better all-round vision to scan the open habitat for enemies.

Right: The overall proportions of the skull of Merychippus *were approaching those of the modern horse, with the jaw deeper and eyes set farther back, partly to accommodate the long roots of the molar teeth. The brain was also larger in relation to body size, allowing it to process information from the keen senses of sight, smell and hearing.*

Higher snout profile

Enlarged nasal chamber

Higher-crowned molar teeth

Name: *Merychippus*.
Meaning: New-look horse.
Grouping: Mammal, Perissodactyl, Equid.
Informal ID: Prehistoric horse.
Fossil size: Skull length 26cm/10¼in.
Reconstructed size: Head–body length 1.5m/5ft.
Habitat: Open grassy plains.
Time span: Miocene, 17–15 million years ago.
Main fossil sites: North America.
Occurrence: ◆ ◆

Equus

By the late Miocene Epoch, the *Merychippus* line had undergone 'explosive radiation' and quickly produced several new branches of the horse evolutionary tree. One of these led to the genus *Equus*, which includes the modern horse, *Equus caballus*, as well as zebras and asses. About 1 million years ago there were more than 20 *Equus* species on almost every continent, except Australia and Antarctica, but many went extinct. The modern horse carries all its weight on only one large hoof-capped toe per foot, with small 'splints' representing the toes to either side.

Name: *Equus caballus*.
Meaning: Work horse.
Grouping: Mammal, Perissodactyl, Equid.
Informal ID: Modern horse.
Fossil size: Tooth length 5cm/2in.
Reconstructed size: Head–body 2m/6½ft.
Habitat: Grasslands, open woodland.
Time span: Quaternary, within past 1.75 million years, to today.
Main fossil sites: North America, then almost worldwide.
Occurrence: ◆ ◆

Left: The horse's cheek or molar teeth are high-crowned, or hypsodont: that is, adapted for lengthy chewing of abrasive, low-nutrition foods. The teeth grow continually, with strong cement-covered crests to resist wear.

Right: This skull of a modern horse, Equus caballus, *is from Pleistocene rocks in Germany. The nipping and nibbling incisors are separated by a long gap from the row of powerful chewing cheek teeth. There are signs of damage to the sinus area, perhaps caused by an injury.*

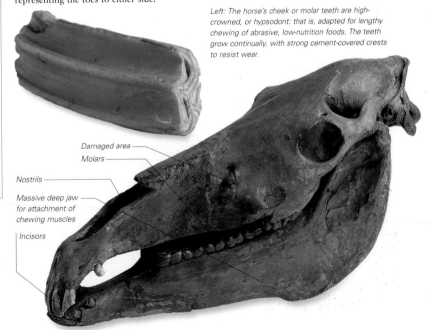

Damaged area

Molars

Nostrils

Massive deep jaw for attachment of chewing muscles

Incisors

HERBIVOROUS MAMMALS (CONTINUED)

Many mammal groups contained herbivores during their prehistory, such as the xenarthrans – sloths, anteaters and armadillos. The elephant group have always been plant-eaters. Like the rhinos, they were once more numerous, diverse and widespread. Over millions of years, scores of species of rhinos and elephants ranged greatly in a wide variety of habitats, compared with the very few living species today.

Glyptodon

South America was cut off as a giant island for much of the Tertiary Period, and many animals evolved there that were found nowhere else. The glyptodonts were big, heavily armoured, armadillo-type creatures that mostly wandered the pampas regions of grassy scrub over the past few million years. They had no teeth at the front of the mouth, and the rear or molar teeth lacked hard enamel covering, but they grew continuously to replace the wear from grinding up tough grasses and other plant food. The most recent glyptodonts survived until a few thousand years ago.

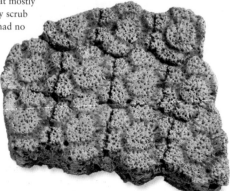

Right: The dome-shaped 'shell' consisted of bony polygonal (many-sided) plates or scales fused together to make a rigid covering, like an upturned bowl.

Name: *Glyptodon*.
Meaning: Sculptured or grooved tooth.
Grouping: Mammal, Xenarthran.
Informal ID: Prehistoric armadillo.
Fossil size: Overall width 15cm/6in.
Reconstructed size: Nose–tail length 2m/6½ft.
Habitat: Plains.
Time span: Pleistocene, within past 2 million years.
Main fossil sites: South America.
Occurrence: ◆ ◆

Mammuthus

Root

There were several types of mammoth, all members of the elephant family. Most widespread was the woolly mammoth, which ranged across all northern continents and grew a very long, thick, hairy coat to keep out the bitter cold of the recent Ice Ages. It may have used its long, curving tusks to sweep away drifted snow to get at plants beneath. This mammoth is known from mummified and/or frozen specimens in the ice of Siberia, Northern Asia. Prehistoric Neanderthal people and those of our own species hunted mammoths and used their tusks and bones for building, tools, utensils and ornaments.

Chewing surface

Above: A milk (deciduous) molar tooth of a young mammoth, shed to make room for the adult tooth.

Right: Adult shoebox-sized tooth, composed of a stack of plates of enamel-covered dentine, bound together by dental cement, which wore into a series of sharp ridges. This is known as the loxodont or 'washboard' form of dentition.

Name: *Mammuthus primigenius*.
Meaning: Earth-burrower (from old Russian 'mammut').
Grouping: Mammal, Proboscid.
Informal ID: Woolly mammoth.
Fossil size: Adult tooth 24cm/9½in long.
Reconstructed size: Head–body length 3.5m/11½ft, excluding trunk and tail; tusks up to 4m/13ft.
Habitat: Tundra, steppe, open plains.
Time span: Pleistocene, within past 2 million years.
Main fossil sites: Throughout Northern Hemisphere.
Occurrence: ◆ ◆

Protitanotherium

Brontotheres, meaning 'thunder beasts' (also called titanotheres), were relatives of horses and rhinos, which made their appeared during the Early Eocene Epoch, about 50 million years ago, but then died away after 20–15 million years. Brontotheres were mainly large to very large browsing mammals, feeding on the soft, juicy leaves and other plant matter of mixed woodlands and forests. *Protitanotherium* probably led something of a rhino-like existence.

Name: *Protitanotherium*.
Meaning: Before titanic beast.
Grouping: Mammal, Perissodactyl, Brontothere.
Informal ID: Brontothere.
Fossil size: Tooth 8cm/3in long.
Reconstructed size: Head–body length 3m/10ft.
Habitat: Mixed woodland and forests.
Time span: Middle Eocene, 45 million years ago.
Main fossil sites: North America, East Asia.
Occurrence: ◆ ◆

Left: Brontothere cheek teeth were relatively low-crowned, designed for cutting soft vegetation such as new forest leaves, rather than grinding tough material. In most cases, the molars have a characteristic W-shaped outer shearing blade, or ectoloph. This molar (cheek) tooth from South Dakota, USA, shows wear along the ridges.

Coelodonta

The woolly rhinoceros, *Coelodonta*, is well known from plentiful fossil, subfossil and mummified remains, as well as from specimens frozen in ice, and also the cave paintings and remnant tools, utensils and ornaments of Neanderthal people and our own prehistoric ancestors. Like the woolly mammoth, its long, coarse hair kept out the snow and wind on the tundra and steppe of the recent Ice Ages. The woolly rhino probably originated in East Asia about 350,000 years ago and spread to Europe, but it never reached North America. It survived in some areas to perhaps just 10,000 years ago. The front horn reached a length of more than 1m/3¼ft in older males.

Name: *Coelodonta*.
Meaning: Hollow tooth.
Grouping: Mammal, Perissodactyl, Rhinocerotid.
Informal ID: Woolly rhino.
Fossil size: Vertebral body width 6cm/2⅜in; tooth length 6cm/2⅜in.
Reconstructed size: Head–body length 3.5m/11½ft.
Habitat: Cold grassland, scrub, steppe, tundra.
Time span: Pleistocene, within the past 1 million years.
Main fossil sites: Europe, Asia.
Occurrence: ◆ ◆

Left: The woolly rhino vertebra (backbone) has a massively strong build with large processes (flanges) for muscle attachment. The deeply dished centrum (main body) fitted into the corresponding bulge of the next vertebra, all along the vertebral column. This meant great strength, but little flexibility, of the back.

Dorsal process

Centrum

Transverse process

Left: This fossil of an upper molar tooth has been positioned with its chewing surface facing down – as it would have been in life – showing the roots.

HERBIVOROUS MAMMALS (CONTINUED)

By far the largest living group of ungulates (hoofed animals) are even-toed ungulates or artiodactyls, with an even number of hoof-tipped toes per foot. They include pigs (suids), hippos, camels and llamas (camelids), deer (cervids), giraffes (giraffids) and the enormous bovid family of antelopes, gazelles, cattle, sheep and goats. All have a rich fossil history across all the continents except Australia and Antarctica.

Merycoidodon

This sheep-like browser from the Oligocene Epoch, some 30 million years ago, had a mixture of features reflecting pigs and camels. It also had a body shape reminiscent of a short-legged early horse, but with four toes on each foot. It has also been known as *Oreodon*, meaning 'mountain tooth', after the region where its early fossils were found. *Merycoidodon* gave its name to a group of similar plant-eaters that were successful in North America during the Oligocene Epoch. They evolved into various forms, including squirrel-like tree-climbers and bulky, hippo-shaped swamp-dwellers. However, the group declined during the Miocene and became extinct soon after.

Below: The teeth of Merycoidodon *were medium- to high-crowned and had a characteristic crescent shape, suited for slicing and chewing soft forest vegetation.*

Name: *Merycoidodon*.
Meaning: Ruminant tooth.
Grouping: Mammal, Artiodactyl, Merycoidodont (Oreodont).
Informal ID: Merycoidodont, sheep-pig.
Fossil size: Four-tooth section length 3cm/1⅛in.
Reconstructed size: Head–body length 1.5m/5ft.
Habitat: Woodlands, forested areas.
Time span: Oligocene, 30–25 million years ago.
Main fossil sites: North America.
Occurrence: ◆ ◆

Sus

The living wild boar or wild pig, *Sus scrofa*, is one of the widest-ranging of all wild hoofed animals, being found in woods and forests across most of Europe and Asia. It is also the ancestor of the domestic pig, which is now one of the world's most common larger mammals, distributed with humans in almost every habitable corner of the globe. Along with deer and similar creatures, wild boar feature in the cave art of prehistoric humans, and their tusks were valued in decorative and ornamental items, such as necklaces and headdresses. The genus *Sus* first appeared in Asia during the Miocene, and several different species arrived in Europe at various times. These included the wild boar itself in the Late Miocene. It also spread to Africa, but disappeared there in Neolithic (New Stone Age) times.

Below: Male boar have nibbling incisors and long, curving tusks, which are the lower canines, used for defending themselves as well as battling with rival males at breeding time. This Pleistocene tusk specimen is from Burwell in Cambridgeshire, England.

Root

Name: *Sus*.
Meaning: Pig, swine.
Grouping: Mammal, Artiodactyl, Suid.
Informal ID: Boar, Pig.
Fossil size: 12cm/4¾in around curve.
Reconstructed size: Head–body length 1.5m/5ft.
Habitat: Forested areas, open woodlands.
Time span: Quaternary, within past 1 million years.
Main fossil sites: Europe, Asia, North Africa.
Occurrence: ◆ ◆

Macraucheria

During South America's time as an isolated island continent through much of the Tertiary Period, many types of ungulates, known as meridiungulates, evolved there. They resembled various hoofed mammals in other regions, such as horses, camels, pigs and wild cattle. The litopterns, or 'simple ankles', were mostly like camels and horses, and showed the same trend to fewer toes per foot, usually the odd number of three or one. However, there were differences from true horses in the foot and ankle bones and, particularly, in the paired lower leg bones, of radius/ulna and tibia/fibula, which did not fuse as in horses. *Macraucheria*'s nostrils were high on its skull, suggesting it may have had a trunk.

Left: This particularly historic specimen was collected in Argentina in 1834 by naturalist Charles Darwin, on his around-the-world voyage on the Beagle. It is the right front foot, showing the three-toed structure for standing on the 'tip toe' hooves.

Original Royal College of Surgeons registration number

Phalanges (toe bones)

Name: *Macraucheria*.
Meaning: Big llama, large neck.
Grouping: Mammal, Meridiungulate, Litoptern.
Informal ID: Prehistoric camel-llama.
Fossil size: Height 30cm/12in.
Reconstructed size: Head–body length 3m/10ft.
Habitat: Open woodlands.
Time span: Pleistocene, within past 2 million years.
Main fossil sites: South America.
Occurrence: ◆ ◆

Bison

There are two living species of the hulking, sharp-horned wild cattle called bison, in America and Eastern Europe. Formerly, other species roamed most northern lands, being hunted by prehistoric people and featuring in their cave art and rituals. Various types of bison have inhabited North America and Europe since Pliocene times. The horns are true horns – that is, they grow continually with a bony core covered by an outer horny material, unlike the antlers of deer, which are shed yearly.

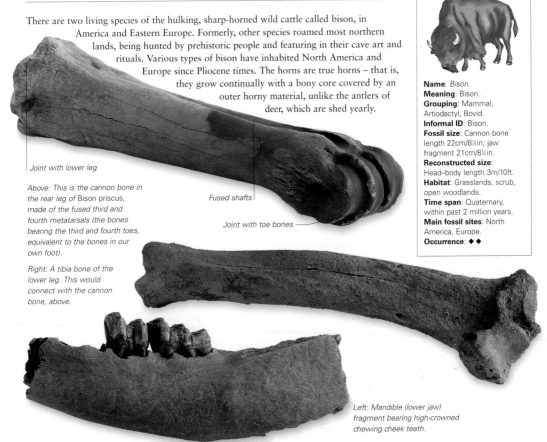

Joint with lower leg

Above: This is the cannon bone in the rear leg of Bison priscus, *made of the fused third and fourth metatarsals (the bones bearing the third and fourth toes, equivalent to the bones in our own foot).*

Right: A tibia bone of the lower leg. This would connect with the cannon bone, above.

Fused shafts

Joint with toe bones

Left: Mandible (lower jaw) fragment bearing high-crowned chewing cheek teeth.

Name: *Bison*.
Meaning: Bison.
Grouping: Mammal, Artiodactyl, Bovid.
Informal ID: Bison.
Fossil size: Cannon bone length 22cm/8½in; jaw fragment 21cm/8¼in.
Reconstructed size: Head–body length 3m/10ft.
Habitat: Grasslands, scrub, open woodlands.
Time span: Quaternary, within past 2 million years.
Main fossil sites: North America, Europe.
Occurrence: ◆ ◆

HERBIVOROUS MAMMALS (CONTINUED), MARINE MAMMALS

The deer family, Cervidae, was one of the last main groups of large herbivorous mammals to evolve, in the Miocene Epoch from some 13 million years ago. In contrast, the whale and dolphin family, Cetacea, was one of the first mammal groups to take to the sea, becoming fully aquatic over 50 million years ago.

Megaloceros

Megaloceros is known as the 'Irish elk' because many of its remains have been discovered in Irish peat bogs, and its spreading antlers resemble those of the modern elk (moose). However, it is, in fact, a closer relative of today's fallow deer and its remains have been found across most of the north of Europe and Asia. The antlers spanned almost 4m/13ft, weighed an incredible 50kg/110lb and – unlike the horns of bovids (cattle, sheep and goats) – they grew anew each year. In most living species, only the male deer possess antlers, and this was the case in *Megaloceros*. It is possible that the 'giant elk' *Megaloceros* survived until less than 10,000 years ago in parts of Central Europe, as it was pictured and hunted by early people.

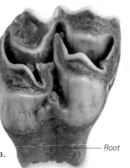

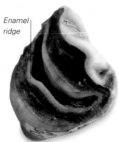

Left and below: The high-crowned molars have complex wavy lines. These are formed by cusps that have become elongated in a front-rear direction and folded, giving extra cutting and grinding ridges of enamel. Deer and some cattle have this tooth pattern, known as the selenodont dentition.

Root

Enamel ridge

Name: *Megaloceros*.
Meaning: Gigantic large horn.
Grouping: Mammal, Artiodactyl, Cervid.
Informal ID: Giant elk, Irish elk.
Fossil size: 2.5cm/1in.
Reconstructed size:
Head–body length 2.5m/8¼ft.
Habitat: Woods, scrub, moor.
Time span: Pleistocene, within the past 400,000 years.
Main fossil sites:
Europe, Asia.
Occurrence: ◆ ◆

Rangifer

Name:
Rangifer.
Meaning:
Single-file walker.
Grouping:
Mammal,
Artiodactyl,
Cervid.
Informal ID:
Reindeer (Europe),
Caribou (North America).
Fossil size: Length 25cm/10in.
Reconstructed size:
Head–body length 2m/6½ft.
Habitat: Tundra, steppe, cold forests.
Time span: Quaternary, within past 1 million years.
Main fossil sites: Throughout the Northern lands.
Occurrence: ◆ ◆

Among the modern deer species living today, only the reindeer, or caribou (*Rangifer*), has antlers in the females as well as in the males. *Rangifer* fossils from almost one million years ago crop up in Alaska, North America, and in Europe the fossils date to nearer half a million years ago. In more recent times, *Rangifer*'s fossils are associated with those of mammoths, woolly rhinos and similar Ice Age mammals around the north of the Northern Hemisphere. Neanderthal people of 40,000 years ago fashioned reindeer antlers, bones and other parts into tools and utensils. They are also depicted in cave art around 15,000 years old. Domestication may have occurred in the Aerhtai Shan (Altai Mountains) region – now in west Mongolia close to the Russian border – some 5,000 years ago.

Main beam

Brow tine

Boss

Left: Deer antlers grow each year, and are mainly used by males for battles during the rutting season, after which they are shed, generally in the spring (younger deer slightly later than this). They are made of bone that grows directly from the skull at the blunt lower end or boss. Elk antlers can grow at more than 2.5cm/1in in a day; deer antlers lengthen at less than half this rate.

Cetacean

After the end-of-Cretaceous mass-extinction event of dinosaurs and many marine reptiles, mammals lost no time in taking to the seas and becoming highly evolved as marine hunters. The first cetaceans – members of the whale, dolphin and porpoise group – date from the Eocene Epoch, more than 50 million years ago. All are meat-eaters, with toothed cetaceans – such as dolphins, porpoises and sperm whales – specialized as hunters of fish and squid, and the great whales as filter-feeders of much smaller prey, such as shrimp-like krill. The fossil record of cetaceans shows loss of rear limbs and extreme adaptation of the skull and jaws, especially in the great or baleen whales. In these animals, all teeth are lost, the skull bones form long curved arches and prey is sieved from the water by fringe-edged, comb-like baleen plates, which rarely survive preservation.

Name: Cetacean, skeleton of Mysticete.
Meaning: Mysticete refers to the upper lip (as in moustache).
Grouping: Mammal, Cetacean, Mysticete.
Informal ID: Baleen whale skeleton.
Fossil size: Length 9m/30ft.
Reconstructed size: As above.
Habitat: Seas, oceans.
Time span: Eocene, 50 million years ago, to today.
Main fossil sites: Worldwide.
Occurrence: ◆ ◆

Tail region | Main vertebrae (backbones) | Rib cage | Cervical vertebrae (neck bones) | Brain case area of skull | Front upper skull (rostrum)

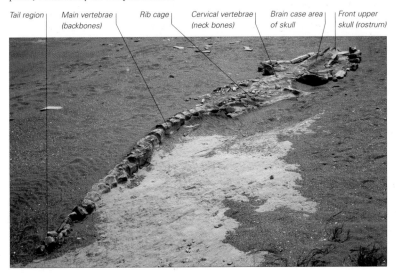

Left: These in situ *remains are from near Sacaco, Peru, and show most of the fossilized skeleton of a 9m/30ft great whale from the Tertiary Period. Once located on the ocean bed, the region is now a desert.*

Cetacean tympanic bone

The ears of most whales are very different from those of other mammals. There is no ear flap, or even an ear canal opening, as we have. In water, sounds travel as ripples of water pressure that vibrate a channel of fat and bone in the whale's lower jaw – this is the route for sounds entering the ear. These vibrations are passed on to tympanic bones (rather than our own flexible skin-like eardrum), which, in turn, pass them to the nerve centre of the inner ear. Whales use many types of sounds for communication, especially when males 'sing' at breeding time. In addition, toothed whales – from dolphins to sperm whales – use them, like bat squeaks and clicks in the air, for the echo-location of prey and obstacles.

Name: Cetacean tympanic bone.
Meaning: Whale 'ear bone'.
Grouping: Mammal, Cetacean.
Informal ID: Whale ear bone.
Fossil size: Length 10cm/4in.
Reconstructed size: Total animal length 20m/65ft.
Habitat: Seas, oceans.
Time span: This specimen Early Pleistocene, almost 2 million years ago.
Main fossil sites: Worldwide.
Occurrence: ◆ ◆

Right: The ear bones of a large whale are stout and bulky, passing vibrations from the lower jaw area to the inner ear. They are made of very hard, dense bone and fossilize well, being common in some marine deposits. Each specimen has a unique shape.

PRIMATES

The two main groups of living primates are prosimians, or strepsirhines – lemurs, lorises, pottos and bushbabies – and the anthropoids, or haplorhines – all monkeys (including marmosets and tamarins), the lesser apes or gibbons, the great apes and humans. Main group features are grasping hands and often feet with nails rather than claws, flexible shoulders, forward-facing eyes, and a large brain for body size.

Plesiadapis

A profusion of *Plesiadapis* fossils in north-east France, in the Cernay region, indicate that this was once a common creature. Remains are also frequent in North America. *Plesiadapis* had the primate hallmarks of gripping fingers and toes suited for a life in the trees, and a long tail for balancing. Its body and hands were primate-like, but its tooth pattern was more similar to that of the rodents, with large incisors and a gap, or diastema, between these and the chewing molars at the rear. With other features, this had led to debate over whether *Plesiadapis* was a true early primate or close relative. It was relatively long-limbed and could probably move fast and with agility, both through the branches and across the ground.

Diastema (gap)

Molar tooth

Above: This is a section of the left side of the mandible (lower jaw). The large incisor teeth were procumbent – angled forwards, as in some rodents – and were perhaps used for digging into bark to get at wood-boring grubs or release the sticky sap.

Name: *Plesiadapis*.
Meaning: Near Adapis (another primate genus).
Grouping: Mammal, Primate, Prosimian, Plesiadapid.
Informal ID: Lemur.
Fossil size: Length 3.3cm/1¼in.
Reconstructed size: Head–tail length 80cm/32in.
Habitat: Woods, forests.
Time span: Late Palaeocene and Eocene, 55 to 45 million years ago.
Main fossil sites: North America, Europe.
Occurrence: ◆ ◆

Megaladapis

Below: This Quaternary specimen comes from near Ampoza, in southwest Madagascar. It has prominent incisors with a small gap (or diastema) to the rear of these. The mandible (lower jaw) is deep and strong with large areas for anchoring powerful chewing muscles. The probable diet of this creature was leaves and fruits.

Large nostril area possibly supported a small trunk

Prominent canine

The giant lemur *Megaladapis* was the size of living great apes today, such as the orang-utan or perhaps even the gorilla. Like all living lemurs, its fossils come from the island of Madagascar, off the east coast of Africa, although earlier members of the lemur group occurred across North America and Europe as well as Africa. *Megaladapis* had a stout body and relatively short limbs. Weighing in at more than 50kg/110lb, it probably clambered between large branches in the slow, deliberate manner of a sloth. There is evidence that this huge lemur survived to at least 2,000 years ago, and possibly less than 1,000 years ago – perhaps humans played a part in its eventual extinction.

Cranium (braincase)

Orbit not completely enclosed by bone

Deep mandible

Name: *Megaladapis*.
Meaning: Big Adapis (another primate genus).
Grouping: Mammal, Primate, Prosimian, Lemurid.
Informal ID: Giant lemur.
Fossil size: Skull length 29cm/11¼in.
Reconstructed size: Head–body length 1.5m/5ft.
Habitat: Woods, forests.
Time span: Pleistocene, less than 2 million years ago.
Main fossil sites: Madagascar.
Occurrence: ◆

Australopithecus afarensis

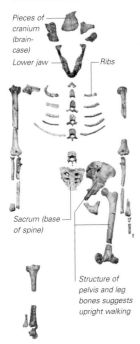

Pieces of cranium (brain-case)

Lower jaw

Ribs

Sacrum (base of spine)

Structure of pelvis and leg bones suggests upright walking

A combination of fossil, DNA and biochemical evidence suggests that human lines of evolution separated from those of our closest living relatives, the chimpanzees, sometime around 6–8 million years ago. Early fossil finds of the human family, hominids, indicate that through time they gradually became taller, more upright and larger-brained. However, newer finds show that a mix of species came and went over the past few million years. Some of the earlier ones were more like recent humans, while some of the later ones 'reverted' to being more ape-like. The australopithecines included several species that lived from about 4 to 1 million years ago, all in Africa. They ranged from large and powerful, or robust, types with massive jaws and teeth, to smaller and more slender, or gracile, types, such as the famous 'Lucy'. She was once thought to be female, but in fact 'she' could have been a male, 'Lucifer'. *Australopithecus afarensis* is a relatively well-known species of early human, with fossilized remains coming from more than 300 different individuals. The remains exhibit many ape-like features and an average brain size equivalent to that of a modern chimpanzee.

Left: The 'Lucy' specimen (AL 288-1) of Australopithecus afarensis *was found near Hadar, Ethiopia, in 1974 by Donald Johanson and his team. At a remarkable two-fifths complete, it is dated to just over 3 million years old.*

Name: *Australopithecus afarensis*.
Meaning: Southern ape from Afar.
Grouping: Mammal, Primate, Hominid.
Informal ID: 'Lucy', ape-man, ape-woman.
Fossil size: Femur (thigh bone) length 25cm/10in.
Reconstructed size: Overall height 1m/3¼ft.
Habitat: Open woodland, open savannah.
Time span: Pliocene, from 4 to 3 million years ago.
Main fossil sites: Africa.
Occurrence: ◆

Homo neanderthalensis

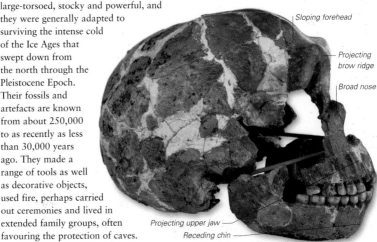

Name: *Homo neanderthalensis*.
Meaning: Human from Neander (in Germany).
Grouping: Mammal, Primate, Hominid.
Informal ID: Neanderthal, 'cave-man'.
Fossil size: Skull front–rear 20cm/12.5in.
Reconstructed size: Adult height 1.5–1.65m/5–5½ft.
Habitat: Open woodland, steppe, tundra, rocky areas.
Time span: Pleistocene, from 250,000 years ago, to perhaps 30,000 years ago.
Main fossil sites: Europe, West Asia.
Occurrence: ◆ ◆

Following the australopithecines such as 'Lucy', and in some cases overlapping them in time and place, came members of our own genus, *Homo*. *Homo neanderthalensis* was named after the Neander Valley region of Germany, the discovery site in 1856 of the first fossil skeleton to be scientifically described. Neanderthal people were strong-limbed, large-torsoed, stocky and powerful, and they were generally adapted to surviving the intense cold of the Ice Ages that swept down from the north through the Pleistocene Epoch. Their fossils and artefacts are known from about 250,000 to as recently as less than 30,000 years ago. They made a range of tools as well as decorative objects, used fire, perhaps carried out ceremonies and lived in extended family groups, often favouring the protection of caves.

Below: This skull of a female Homo neanderthalensis *is from Tabun at Mount Carmel, Israel. It is dated to around 100,000 years ago. Most Neanderthals had brains as large as, or even bigger than, our own.*

Sloping forehead

Projecting brow ridge

Broad nose

Projecting upper jaw

Receding chin

GLOSSARY

Terms for mineral optical effects are glossed on p52; lustre on p53; crystal system on p48; and crystal form and habit on p49.

abdomen rear or lower part of the body in many animals.

absolute dating estimating the age of rocks and fossils by physical means, such as the radioactive minerals they contain.

accessory any mineral not essential to the rock's character.

accreted terrane belt of rock welded to the edge of a continent by subduction.

accretion various meanings, including gradual growth of a body such as a grain by addition of new material to the surface.

acidic rock rock rich in silica.

aggregate mass of rock or mineral particles.

agnatha literally 'without jaws', usually applied to fish that lack jointed jaws, including the first fish and modern lampreys and hagfish.

allochromatic owing its colour to impurities.

alloy a manmade combination of two metals.

alluvial deposited by rivers.

alpine cleft mineral-rich fissure typical of the Alps.

amber the hardened fossilized resin or exudate from certain plants, mainly coniferous trees.

ambulacra curved plates associated that make up the shell or test of an echinoid (urchin).

anion negatively charged ion.

aphanitic of grains too fine to see with the naked eye.

aptychi in fossils, plate like structures that were possibly the 'jaws' or 'teeth' of molluscs such as ammonoids.

arenite sedimentary rock with sand-size grains.

articulated in fossils, as in life, when parts are found still attached to each other at joints.

assemblage zone strata dated by a group of fossils

association minerals that commonly occur together.

asthenosphere hot, partially molten layer of the Earth directly below the lithosphere.

astrobleme eroded remains of a meteorite impact crater.

banded iron formation narrow iron-rich layers of rock.

batholith large mass of usually granite plutons at least 100 sq km (39 sq miles) in area.

bedding plane the boundary between layers of sedimentary rock formed at different times.

bedrock solid rock that lies beneath loose deposits of soil and other matter.

Benioff-Wadati zone sloping zone of earthquake centres in a subduction zone.

binomial nomenclature the traditional system of two names, genus and species, for identifying organisms, both living and extinct.

biogenic formed by living things.

biostratigraphy dating layers of rocks with fossils.

byssal threads strong threads by which various invertebrates, such as mussels, attach themselves to a firm surface.

calcareous consisting predominantly of calcium-rich minerals such as calcium carbonate, as in limestones and chalks.

carapace the hard outer covering over the head and body, especially in crustaceans.

carat unit of weight for gems, equivalent to 200 milligrams.

Carnivora the mammal group that includes dogs, bears, otters, mongooses, civets, cats and similar hunting members.

cartilage a tough, springy substance that forms the skeleton of sharks, and parts of the skeleton in other vertebrates.

cation positively charged ion.

caudal to do with the tail or rear end of the body, as in the caudal or tail fin of fish.

cementation the stage in lithification when cement glues the sediment particles together.

chalcophile of elements such as copper and zinc that have an affinity for sulphur.

cladistical analysis determining the relationships of living things and grouping them according to the possession of unique characters derived from a common ancestor.

cladogram a branching diagram showing relationships between living things determined according to cladistical analysis.

clast fragment of broken rock.

clastic sediment sedimentary rock made mainly of broken rock fragments.

cleavage the way a mineral breaks along certain planes.

compaction stage in lithification when water and air is squeezed out of buried sediments by the weight of overlying deposits.

comparative anatomy studying the structure or anatomy of living things and comparing their parts, organs and tissues to suggest origins, groupings and evolutionary relationships.

concretion distinct nodules of materials in sedimentary rocks.

contact metamorphism when rocks are metamorphosed by contact with hot magma.

contact twin twinned crystals in which each crystal is distinct.

continental drift the slow movement of the continents.

convergent evolution when two dissimilar organisms come to resemble each other superficially, due to adaptation to similar conditions and lifestyles.

coprolites fossilized droppings or dung.

core the centre of the Earth.

country rock the rock that surrounds a mineral deposit or igneous intrusion.

craton ancient part of a continent unaltered for at least one billion years.

crust the top layer of the earth, attached to the upper mantle.

cryptocrystalline with crystals so small that they cannot be seen even under an ordinary microscope.

crystal form the way in which the different faces of the crystals are arranged.

crystal habit the typical shape in which a crystal or cluster of crystals grows.

crystal system one of the different groups into which crystals can be placed according to how they are symmetrical.

cuesta low ridge with one steep side and one gently sloping.

cusp a mound-like projection or point, as might appear on a tooth.

D" or D double prime the transition zone between the Earth's mantle and core.

Dana number number assigned to each mineral according to the classification system devised by James Dwight Dana.

detrital made of rock fragments.

diagenesis all the processes that affect sediments after deposition including compaction and lithification.

diaphaneity the degree to which a mineral is transparent.

disseminated deposit mineral deposit created by infilling pores and cracks in igneous rock.

DNA De-oxyribonucleic acid, the genetic substance that contains the instructions for life, passed from generation to generation.

dorsal to do with the upper side or back.

drift geology the geology of loose surface deposits.

dyke sheetlike igneous intrusion, either near vertical, or cutting across existing structures.

Ediacaran the geological period lasting from about 600 to 542 million years ago. This was added to the system in 2004.

effusive volcano volcano that erupts easily flowing lava.

element the simplest most basic substances, such as gold, each with its own unique atom.

Era vast portion of geological time lasting hundreds of millions of years.

erosion wearing away of rocks and other materials by the forces of weather such as rain, sun, wind, snow and ice.

evaporite a natural salt or mineral left behind after the water it is in has dried up.

exoskeleton a body casing or framework on the outside, as in insects and shelled animals.

exposure where a rock outcrop is exposed at the surface.

extrusive igneous rock type of rock that forms when volcanic lava cools and solidifies.

facies assemblage of mineral, rock or fossil features reflecting the conditions they formed in.

fault a long fracture in rock along which rock masses move.

feldspathic rock containing feldspar.

felsic rock rich in feldspar and silica, typically light in colour.

fissure volcano volcano which erupts through a long crack.

flood basalt plateau formed from huge eruption of basalt lava from fissure.

foid abbreviation of feldspathoid.

foliation flat layers of minerals in metamorphic rock formed as minerals recrystallize under pressure.

foramen a gap, hole, opening or window in a surface or object.

forearc basin region on the trench side of an island arc in a subduction zone.

fossil correlation cross-checking of fossils between separate rock outcrops, used in rock dating.

fractionation the way in which the composition of a magma changes as crystals separate out when it melts and refreezes.

fracture the way in which a mineral breaks when it does not break along planes of cleavage.

frond the leaf of a fern.

gastroliths 'Stomach stones', swallowed by animals such as sauropod dinosaurs into the digestive tract, to help grind tough plant food.

geode hollow globe of minerals that can develop in limestone or lava, often lined with quartz.

geological column diagram showing the successive layers of strata that have formed over geological time, with the oldest at the bottom, youngest at top.

glacial a period in an Ice Age when ice sheets spread, or anything related to glaciers.

graded bedding layers of rocks which show a decrease of grain size, from coarse at the bottom to fine at the top.

granular texture of rock with visible, similar-sized grains.

groundwater water existing below ground. See also *phreatic water*.

guard the pointed, often bullet-shaped structure in the rear of a belemnoid's body.

heterocercal in fish, when the tail (caudal fin) has two unequal lobes.

homocercal in fish, when the tail (caudal fin) has two equal lobes.

hydrothermal related to water heated by magma.

hydrothermal deposit mineral deposit formed from mineral-rich hydrothermal fluids.

hypabyssal igneous rocks in small intrusions such as dykes.

ichnogenus a genus known only from trace fossils or signs it has left, such as footprints or droppings, rather than fossils of bodily matter.

idiochromatic mineral getting its colour from its main ingredients.

igneous rock rock that has solidified from molten magma.

impact crater crater formed by the impact of a meteorite.

index fossil key fossil used for correlating strata.

index mineral mineral that typifies a metamorphic facies.

intraclast any sedimentary rock fragment that originated within the area of the rock's formation.

intrusion emplacement of magma into existing rock.

ion atom given an electrical charge by gaining or losing electrons.

island arc curved chain of volcanic islands in a subduction zone.

isostasy the natural buoyancy of the Earth's crust, making it rise and sink as its weight changes.

isotope variety of atom of an element that has a different number of neutrons (uncharged particles in its nucleus).

joint crack in rock created without any appreciable movement on either side, often at right angles to the bedding.

karst typical limestone scenery characterized by caverns, gorges and potholes.

kimberlite igneous rocks rich in volatiles, normally forming pipes.

labial to do with lips or the front of the mouth.

Large Igneous Province vast outflow of lava, typically on the ocean bed.

laterite weathered, soil-like material in the tropics rich in iron and aluminium oxides.

lava erupted magma.

leaflet segment of a compound leaf or frond.

leucocratic of light-coloured igneous rocks.

lithification change of loose sediments into solid rock.

lithosphere the rigid outer shell of the Earth containing the crust and upper mantle, broken into tectonic plates.

lode cluster of disseminated deposits.

lutite sedimentary rock made mainly of clay-sized grains.

macro/microconch molluscs in which one sex (usually male) is small, with a reduced shell, while the other sex (usually female) is larger.

mafic of rocks rich in magnesium and ferric (iron) compounds, equivalent to basic.

magma molten rock.

magma chamber underground reservoir of magma beneath a volcano.

mantle the zone of the Earth's interior between the crust and the core, made of hot, partially molten rock.

mantle convection circulation of material in the magma driven by the Earth's interior heat.

mantle plume long-lasting column of rising magma in the mantle.

mass extinction when a large proportion of organisms die out at a particular time, usually due to some form of rapid global change.

massive of rocks with even texture, or a body of a mineral without any distinct crystals.

matrix In palaeontology, the rock and other material in which a fossil is embedded.

melt a mass of liquid rock.

metamorphism the process in which minerals in rocks are transformed by heat and pressure to create a new rock.

metasomatism metamorphic process in which minerals are transformed by hot solutions penetrating rock.

meteoric water water that comes from the air, typically in reference to groundwater.

meteorite a chunk of rock from space that reaches the Earth's surface.

Mohs scale the scale of relative hardness of minerals devised by Friedrich Mohs.

mummification when preservation of dead organisms is aided by severe drying out.

MVT Mississippi Valley Type mineral deposits, formed as limestone is altered by hot solutions.

native element element that occurs naturally uncombined with any other element.

neural to do with nerves, or the upper parts of bones or plates.

nodule rounded concretion.

nuée ardente fast-moving clod of scorching ash and gases created by a volcanic eruption.

ocean spreading the process in which oceans widen as new rock is brought up at the mid-ocean ridge.

oolith small, usually calcareous accretions in a rock.

oolitic made mostly of ooliths.

operculum the door-like cover to the shell opening certain molluscs, such as snails.

ophiolite grouping of mafic and ultramafic igneous rocks, including pillow lavas, that were once part of the sea floor.

oral to do with the mouth or feeding area.

ore natural material from which useful metals can be extracted.

orogeny mountain-building.

outcrop area of rocks occurring at the surface.

oxidation zone upper layer of mineral deposit where minerals are altered by oxygen and acids in water.

paedomorphosis when the adult retains features or characteristics usually seen only in the young or juvenile phase.

palaeomagnetism the study of the Earth's magnetism, and how it affected and aligned magnetic minerals in rocks as they formed.

pectoral to do with the front limbs or front side area of an animal (in humans, the shoulders).

pegmatite very coarse-grained igneous rock, usually found in veins and pockets around large plutons and rich in rare minerals.

pelagic living in the open sea.

pelvic To do with the rear limbs or rear side area of an animal (in humans, the hips).

penetration twin mineral crystal twin in which the crystals have grown into each other.

permafrost ground which never unfreezes, as found in the far north of Asia today.

Period major portion of geological time lasting tens of millions of years.

permeable of rock that lets fluid or gas to seep through it easily.

permineralization when certain substances, such as the tissues of a once-living organism, are replaced by inorganic minerals, as happens during certain forms of fossilization.

petrification turning to stone or rock.

phaneritic of grains visible with the naked eye.

phenocryst relatively large crystal in igneous rock.

phloem the tiny tubes in a vascular plant that transport sap and similar fluids.

phragmocone the cone-like inner shell in molluscs such as belemnites.

phreatic water groundwater in the saturation zone of rock below the water table.

pillow lava lava formed on the sea bed consisting of pillow-shaped blobs.

pisolith pea-sized, usually calcareous accretions.

pisolitic made mostly of pisoliths.

placer deposit of valuable minerals washed into loose sediments such as river gravels.

planktonic floating with the ocean currents.

pleural to do with the lungs or breathing anatomy of animals; a pleural spine is the body spine of a trilobite.

pluton any large intrusion.

polymorphs different minerals created by different crystal structures of the same chemical compound.

porous full of voids (holes).

porphyritic of an igneous rock containing lots of phenocrysts.

predator an animal that gains its food by active hunting.

primary mineral mineral that forms at the same time as the rock containing it.

proboscis a long, flexible, snout-or trunk-like part on the head of an animal.

protolith the original rock forming a metamorphic rock.

pseudomorph mineral that takes the outer form of another.

punctuated equilibrium a form of evolution where long periods of stability are interrupted by times of rapid change, in contrast to the traditional view of slow, gradual change.

pyroclast fragment of solid magma plug ejected during a volcanic eruption.

radiometric dating the dating of rocks from radioactive isotopes within it.

radula the file-like tongue of certain animals, especially the snail and slug group of molluscs.

regional metamorphism metamorphism over a wide area typical of fold mountain belts.

relative dating estimating the age of fossils from the rocks they are contained within, and from the rocks and fossils in adjacent layers.

rift valley trough-shaped valley bounded by parallel faults.

rudite sedimentary rock consisting mostly of at least gravel-sized grains.

schistosity banding of minerals in schists created by parallel growth of mineral crystals.

scutes protective hard, bony, shield-like plates or scales found in some fish and reptiles.

secondary mineral mineral formed by alteration of primary minerals.

sediment solid grains that have settled out of water.

seismology study of earthquakes.

semelparity when an organism reproduces only once in its lifetime, often dying soon after.

serpentinization alteration of ultramafic rocks to serpentine by hydrothermal fluids.

sexual dimorphism when the male and female of a species exhibit different physical characteristics (other than reproductive parts).

siderophile of elements such as cobalt and nickel that have an affinity for iron.

silicic of igneous rock such as granite rich in silica, making it acidic.

sill sheetlike igneous intrusion, either near horizontal, or following existing structures.

skarn typically mineral deposit created by the alteration of limestones by metasomatism.

soft tissues nerves, blood vessels, muscles and similar soft parts of a body.

sporangia the spore-producing units found on many seedless plants, such as ferns.

streak mark made by a mineral rubbed on unglazed porcelain.

strike direction of fold or fault.

subduction zone boundary between two tectonic plates where one plate descends into the mantle beneath the other.

sutures firm, tight joints, usually marked by lines, as in the shell of an ammonoid or the skull of a vertebrate.

taphonomy the study of how an organism decays over time and produces a fossil.

taxonomy grouping or classifying organisms, living or extinct, along with the reasons and principles underlying their classification.

tectonic plate one of the 20 or so giant slabs into which the Earth's rigid surface is split.

tenacity how a mineral deforms.

tetrapod four-legged, usually referring to vertebrate animals with four limbs.

thoracic to do with the chest, the front or upper body of an animal.

trace fossils fossils which record the signs, marks or parts left by a living thing, rather than being actually part of the living thing.

turbidity current swirling undersea current.

twin paired mineral crystals growing together.

vascular to do with tubes, pipes and vessels, as in the sap and water tube systems in a vascular plant, or the blood vessel network of an animal.

vein thin deposit of minerals formed in cracks.

ventral to do with the underside or belly.

vesicle small cavity formed by gas bubbles in lava flow.

water table level below which rock is saturated by groundwater.

whorl a complete turn of a shell or casing, such as in a snail or ammonite shell.

xylem the tiny tubes in a vascular plant that transport water and dissolved minerals, usually up from the roots.

INDEX

PICTURE ACKNOWLEDGEMENTS

Note: t = top; b = bottom; m = middle; l = left; r = right.

Artworks
Peter Bull drew all colour geological artworks included in the book, except for p63 (Silicate shapes) which was drawn by Anthony Duke. Anthony Duke also supplied the crystal structure diagrams appearing throughout the *Minerals' Directory*. Othr artworks are as follows: Andrey Atuchin 289br (*Giganotosaurus*); Peter Barrett 264bl (Jurassic scene); 282b (freshwater scene); 283 (panel); 283br (dinosaur tracks); 288mr (*Baryonyx*), 206br (*Pteranodon*); 317bl (*Nautilus*); 317 (panel, left, modern coelacanth); Stuart Carter 286bl (*Iguanodon*); Anthony Duke 267 ('recording geological time'), 276b ('how a fossil forms'); 283br (dinosaur footprints map); 319tr (human migration map). All panel reconstructions as credited below; 341br (panel); Denys Ovenden all artworks on pages 316–17 except for Nautilus and modern coelacanth (see Peter Barrett, above); 319 (panel). Reconstructions of fossils appearing in the panels on pages 328–503 were drawn by Anthony Duke, except for the following which were drawn by Samantha J. Elmhurst www.livingart.org.uk: 471t; 473b; 476; 477t; 478; 479t; 480; 484t; 487b; 488b; 489; 490; 491b; 492t; 493; 494–5; 496t; 497b; 498t; 498b; 499b; 500

Photographs
25br, 33tr, 42–3, 125, 131, 137, 219 © British Geological Survey; 91, Peter Frank © 2005 Canadian Museum of Nature, Ottawa, Canada; 231 © China Clay Museum, Cornwall; 262–3, 266 (panel), 268tr, 268bl, 272bl, 275tr, 280tr, 281tl, 284tl, 284tr, 285 (panel), 288bl, 289tr, 290tr, 291tr, 292tr, 299tr, 309tr, 309br, 310tl, 310bl, 311tr, 311mr, 312tr, 312mr, 315bl, 315br, 317tr, 318bl, 318br, 447 © Corbis (www.corbis.com); 2, 33tl, 34m, 65tl, 67, 109b, 119t, 119b, 139b, 157, 169b, 199bl, 199br, 258–9 © Department of Earth Sciences, University of Bristol; 99 © David L Reid, Department of Geological Science, University of Cape Town, South Africa; 197 © Shane Dohnt; 56bl © Getty Images (www.gettyimages.com); 10–11, 60–1, 141 © David Huston, Geoscience Australia; 201, by permission of the Illinois State Museum and the artist, Robert G Larson; 23tr © Marli Miller; 13tl, 29tr, 52, (in Optical Effects panel: chatoyancy, play of colour, pleochroism), 54bl, 55tr, 85, 107t, 115, 167, 195b, 205, 211, 213, 221, 223, 235t, 235b, 239, 260tr, 260bl, 260br, 261b, 265tr, 270tr, 270bl, 278bl, 279tl, 284br, 285tl, 285tr, 286tl, 286tr, 287tl, 287tr, 287 (panel), 288tr, 290bl, 291 (panel), 292tr, 292bl, 293bl, 293 (panel), 294tr, 296bl, 297tl, 297bl, 297 (panel), 298–9 (all photographs except 299tr), 305tr, 305br, 305 (panel), 307br, 307 (both images in panel), 308tr, 310tr, 311 (panel), 312bl, 314tr, 314bl, 315tr, 315mr, 315 (panel), 316bl, 317tl, 318t, 325b, 329t, 332t, 334b, 342, 343b, 352t, 352b, 363t, 363b, 363 (panel), 377t, 377b, 390t, 403t, 403b, 405t, 411 (panel), 442t, 443t, 450b, 459b, 460t, 461t, 462–3, 468–9 (except 468t), 475b, 481b, 482t, 485, 486t, 487t, 491b, 492b, 499t, 502–3 © Natural History Museum Picture Library, London (www.piclib.nhm.ac.uk); 14–15, 38bl, 79, 175, 183 © Nature Picture Library; 121, Earth Observatory, 191, Jet Propulsion Laboratory © NASA; 19br, 320–1, 327, 332 (panel), 333 (panel), 336 (panel), 337 (panel), 338 (panel), 342 (panel), 355 (panel), 361, 392 (bottom-left photograph of *Eryma*), 430 (panel), 458 (panel), 465b (left-hand photograph of *Rutiodon*), 490 (coypu) © NHPA (www.nhpa.co.uk); 57b © Andrew Wygralak, Northern Territory Geological Survey; 287bl (Crystal Palace dinosaurs) © Mary Evans Picture Library (www.maryevans.com); 16bl, 24, 28, 31tr, 33b, 35t, 36bl, 38tl, 44t, 44b, 46t, 55tl, 55br, 119, 133, 149, 179, 185, 189, 207, 241, 245, 255 © Oxford Scientific; 195 © Ted Rieger; 75t, 84t, 84b, 85t, 87b, 91b, 96b, 125b, 126t, 149b, 153t, 156t, 193b, 209b © The School of Earth, Atmospheric and Environmental Sciences, University of Manchester, England; 19tl, 75, 139, 159, 324–5 (all photographs except panel), 466b, 501t © Science Photo Library, www.sciencephoto.com; 14–15, 40tr, 56t, 77, 143, 209, 257 © Still Pictures Ltd; 173 © Els Slots; 101 © Anthony G Taranto Jr, Palisades Interstate Park; 163 © United States Geological Survey; 73, 135, 243 © Ian Coulson, University of Regina, Saskatchewan, Canada;

All other photographs are © Anness Publishing Ltd. The publishers would like to thank the following for loan of featured specimens:
Richard Tayler for the majority of specimens photographed for the Directory of Rocks and Minerals (pages 61–257), with additional material from his friends and associates. The specimens were photographed by Martyn Milner.
The Museum of Wales, Cardiff, Wales: 265bl (all except 265tr), 274tr (claw and tooth), 275tl (bivalves), 275 (panel), 275mr (leaf), 276tr (trilobite), 277 (panel), 278ml (ammonite), 279 (panel), 279b (fish), 281bl (frog skull), 282tr (vertebrae), 295br, 296tr, 297br, 300 (panel), 301tl, 301tr, 303tl, 303 (panel), 304tl, 304tr, 304b (shoal), 306tl (display), 306tr (toe and hoof), 326 (all), 328t, 328b, 329b, 330–1 (all), 333t, 333bl, 334t, 335t, 335b, 336b, 337t, 339t, 340–1 (all), 343t, 344t, 345t, 345br, 346b, 347b, 348–9 (all except 348tr), 350–1 (all except 351tr), 353b, 354–5 (all except panel), 356–7 (all), 358–9 (all except panel), 360 (all), 362t, 365t, 366b (both of *Raphidonema*), 367b, 368b (both of Ediacaran medusoid), 369b, 372t, 373 (all), 374t, 375t, 376t, 376b (all *Archimedes*), 381b, 382t, 383 (all except 383tr), 385t, 386 (panel), 387b, 394t, 394b, 395b, 396b, 403 (panel), 404 (all), 405b, 406 (panel), 408–9 (all), 410t (both *Lopha*), 411 (all *Gryphea* except centre), 412–3 (all), 414b, 415t (*Viviparus*), 416–7 (all except 416t), 418–9 (all), 420t, 421t, 421b, 422–3 (all), 424–5 (all except 424t), 425 (all), 428–9 (all), 430b (*Nautilus*), 431t, 431 (panel), 432b, 432 (panel), 433t, 437t, 437 (panel), 439 (panel), 440t, 441b, 442b, 443b, 444–5 (all), 446m, 446r, 448bl, 444–5 (panel), 450t, 451t, 451b, 452t, 452b, 454t, 454b, 456–7 (all except 457b), 458tr (*Branchiosaurus*), 459t, 461 (both *Nanopus*), 464t, 464 (panel), 465t, 465br, 466–7 (all except 466t), 468t, 469 (all), 472 (all), 474b, 475t, 476–7 (all), 478 (all), 479 (caudal vertebra), 480–1 (all except 481b), 483 (moulted feathers and bird eggs), 487b, 490–1 (all except 491b), 493 (all except 493m), 494–5 (all except 495m), 496 (all photographs), 497t, 498 (all photographs), 499 (top and bottom photographs of bison), 500–1 (all photographs except 501t);
Booth Museum of Natural History, Brighton & Hove City Council, England: 260mr (hyaena jaw), 264trm (trilobite), 274bl (fish), 280ml (fern), 281tb (pond snails), 300tr, 301ml, 301mr, 302 (all except 302tr), 303br, 304ml, 304mr, 304b; 313br, 313 (panel), 338t, 339b, 344bl, 344br, 353t, 353m, 364t, 364 (panel), 360t, 367t, 374b, 383 (top *Calymene*), 388t, 389b, 390b, 391t, 396t, 398–9 (all except 398bl), 400b, 401t, 402t, 406t, 406b, 407 (all), 410b, 411t (largest *Gryphea* in centre), 415 (bottom *Viviparus*), 424t, 426t, 433b, 434b (both *Marsupites*), 435 (panel), 436 (all), 438–9 (all except panel), 440b, 441t, 441 (panel), 446l, 448–9 (all except *Hybodus*, bottom-left), 453 (all), 455 (all), 466t, 470b, 473 (all), 479 (all except caudal vertebra), 483bl (pseudofossil), 493t, 493m, 495m, 497bl, 497br, 499 (middle bison);
Manchester Museum, England: 272tm (sponge), 276tr (chain coral), 317bl (nautiloid), 333 (middle, *Calamites*), 338b (both *Neuropteris*), 345 (bottom, *Stigmaria*), 346t, 365b, 368t, 370–1 (all), 372b, 375 (all *Spirorbis* worm casts), 378–9 (all), 380–1 (all except 381b), 382b, 384–5 (all except 385t), 386t, 386b (both *Agnostus*), 387t, 388 (panel), 389t, 391b, 392 (both *Eryma*), 393 (all), 395t, 397 (all except panel), 398b, 400t, 401t (two *Trichopterans*, bottom), 401b (*Cupedid*), 402b, 414t, 416t, 420b, 426b, 430t (top *Nautilus*), 431b, 432 (middle and bottom in panel), 434t, 435t (both *Eucalyptocrinites*), 435b, 457b, 458t, 474t.

Additional photography credits
Photograph of *Sequoia* fossil 281 (panel) © Rich Paselk, Natural History Museum, Humboldt State University, CA, USA; photograph of Dorothy Hill 289 (panel) © The Physical Sciences and Engineering Library, University of QLD, Australia; Kings Canyon 307bl © Amy-Jane Beer/Origin Natural Science; balloon vine 358 (panel) © Jackie Miles and Max Campbell; amber 397 and 401 (panel, top-left) © John Cooper; fossil operculum 415 (panel) © Dr Bill Bushing, Starthrower Educational Media, www.starthrower.org; brittlestar 437b © R. Shepherd; © iStockphoto.com 9mr, /Tammy Bryngelson 9ml, /Michael Gray 9t, /Jason Reekie 1, /Wolfgang Steiner 8b.